Resource Recovery from Waste

Humans generate millions of tons of waste every day. This waste is rich in water, nutrients, energy, and organic compounds. Yet waste is not being managed in a way that permits us to derive value from its reuse, whilst millions of farmers struggle with depleted soils and lack of water. This book shows how Resource Recovery and Reuse (RRR) could create livelihoods, enhance food security, support green economies, reduce waste and contribute to cost recovery in the sanitation chain.

While many RRR projects fully depend on subsidies and hardly survive their pilot phase, hopeful signs of viable approaches to RRR are emerging around the globe including low- and middle-income countries. These enterprises or projects are tapping into entrepreneurial initiatives and public-private partnerships, leveraging private capital to help realize commercial or social value, shifting the focus from treatment for waste disposal to treatment of waste as a valuable resource for safe reuse.

The book provides a compendium of business options for energy, nutrients and water recovery via 24 innovative business models based on an in-depth analysis of over 60 empirical cases, of which 47 from around the world are described and evaluated in a systematic way. The focus is on organic municipal, agro-industrial and food waste, wastewater and fecal sludge, supporting a diverse range of business models with potential for large-scale out- and up-scaling.

Miriam Otoo is a Research Economist, leading the Research Group on Resource Recovery and Reuse at the International Water Management Institute (IWMI).

Pay Drechsel is Principal Researcher at the International Water Management Institute (IWMI), leading IWMI's Strategic Program on Rural-Urban Linkages and the related Research Flagship of the CGIAR Research Program on Water, Land and Ecosystems (WLE).

RESOURCE RECOVERY FROM WASTE

Business Models for Energy, Nutrient and Water Reuse in Low- and Middle-income Countries

Edited by Miriam Otoo and Pay Drechsel

First published 2018 by Routledge

2 Park Square, Milton Park, Abingdon, Oxfordshire OX14 4RN
605 Third Avenue, New York, NY 10017

First issued in paperback 2021

Routledge is an imprint of the Taylor & Francis Group, an informa business

© 2018 selection and editorial matter, International Water Management Institute; individual chapters, the editors and authors

The right of Miriam Otoo and Pay Drechsel to be identified as the authors of the editorial material, and of the authors for their individual chapters, has been asserted in accordance with sections 77 and 78 of the Copyright, Designs and Patents Act 1988.

All rights reserved. No part of this book may be reprinted or reproduced or utilised in any form or by any electronic, mechanical, or other means, now known or hereafter invented, including photocopying and recording, or in any information storage or retrieval system, without permission in writing from the publishers.

Trademark notice: Product or corporate names may be trademarks or registered trademarks, and are used only for identification and explanation without intent to infringe.

British Library Cataloguing-in-Publication Data
A catalogue record for this book is available from the British Library

Library of Congress Cataloging-in-Publication Data
Names: Otoo, Miriam, editor. | Drechsel, Pay, editor.
Title: Resource recovery from waste : business models for energy, nutrient and water reuse in low- and middle-income countries / edited by Miriam Otoo and Pay Drechsel.
Description: New York, NY : Routledge, 2018. | Includes bibliographical references and index.
Identifiers: LCCN 2017028495| ISBN 9781138016552 (hardback) | ISBN 9781315780863 (ebook)
Subjects: LCSH: Recycling (Waste, etc.)—Developing countries—Econometric models. | Recycling (Waste, etc.)—Developed countries—Econometric models.
Classification: LCC TD794.5 .R4555 2018 | DDC 628.4/4580684—dc23
LC record available at https://lccn.loc.gov/2017028495

Typeset in Helvetica
by Keystroke, Neville Lodge, Tettenhall, Wolverhampton

Cover photos left to right: Shebeko/Depositphotos.com; IWMI; Hamish John Appleby/IWMI

ISBN 13: 978-0-367-77877-4 (pbk)
ISBN 13: 978-1-138-01655-2 (hbk)

Contents

Editors and authors   xi
Acknowledgements   xiii
Foreword by Guy Hutton   xv

SECTION I: BUSINESS MODELS FOR A CIRCULAR ECONOMY: INTRODUCTION   1

1. Business models for a circular economy: Linking waste management and sanitation with agriculture   3
   Pay Drechsel, Miriam Otoo, Krishna C. Rao and Munir A. Hanjra

2. Defining and analyzing RRR business cases and models   16
   Miriam Otoo, Solomie Gebrezgabher, Pay Drechsel, Krishna C. Rao, Sudarshana Fernando, Surendra K. Pradhan, Munir A. Hanjra, Manzoor Qadir and Mirko Winkler

SECTION II: ENERGY RECOVERY FROM ORGANIC WASTE   33
Edited by Krishna C. Rao and Solomie Gebrezgabher

Recovering energy from waste: An overview of presented business cases and models   34

3. Business models for solid fuel production from waste   39
   Introduction   40

   Case – Briquettes from agro-waste (Kampala Jellitone Suppliers, Uganda)   41

   Business model 1: *Briquettes from agro-waste*   51

   Case – Briquettes from municipal solid waste (COOCEN, Kigali, Rwanda)   61

   Case – Briquettes from agro-waste and municipal solid waste (Eco-Fuel Africa, Uganda)   72

   Business Model 2: *Briquettes from municipal solid waste*   82

4. Business models for in-house biogas production for energy savings   91
   Introduction   92

   Case – Biogas from fecal sludge and kitchen waste at prisons   93

   Case – Biogas from fecal sludge at community scale (Sulabh, India)   103

Case – Biogas from fecal sludge at Kibera communities at Nairobi (Umande Trust, Kenya) 114

Business model 3: *Biogas from fecal sludge at community level* 124

Case – Biogas from kitchen waste for internal consumption (Wipro Employees Canteen, India) 133

Business model 4: *Biogas from kitchen waste* 142

5. **Business models for sustainable and renewable power generation** 149
   Introduction 150

Case – Power from manure and agro-waste for rural electrification (Santa Rosillo, Peru) 152

Case – Power from swine manure for industry's internal use (Sadia, Concordia, Brazil) 162

Case – Power from manure and slaughterhouse waste for industry's internal use (SuKarne, Mexico) 172

Business model 5: *Power from manure* 182

Case – Power from agro-waste for the grid (Greenko, Koppal, India) 193

Case – Power from rice husk for rural electrification (Bihar, India) 203

Business model 6: *Power from agro-waste* 215

Case – Power from municipal solid waste at Pune Municipal Corporation (Pune, Maharashtra, India) 222

Business model 7: *Power from municipal solid waste* 232

Case – Combined heat and power from bagasse (Mumias Sugar Company, Mumias District, Kenya) 238

Case – Power from slaughterhouse waste (Nyongara Slaughter House, Dagorretti, Kenya) 248

Case – Combined heat and power and ethanol from sugar industry waste (SSSSK, Maharashtra, India) 257

Case – Combined heat and power from agro-industrial wastewater (TBEC, Bangkok, Thailand) 268

Business model 8: *Combined heat and power from agro-industrial waste for on- and off-site use* 278

6. **Business models on emerging technologies/bio-fuel production from agro-waste** 284
   Introduction 285

Case – Bio-ethanol from cassava waste (ETAVEN, Carabobo, Venezuela) 286

Case – Organic binder from alcohol production (Eco Biosis S.A., Veracruz, Mexico) 296

Business model 9: *Bio-ethanol and chemical products from agro and agro-industrial waste* 307

SECTION III: NUTRIENT AND ORGANIC MATTER RECOVERY 315
Edited by Miriam Otoo

Nutrient and organic matter recovery: An overview of presented business cases and models   316

7. Business models on partially subsidized composting at district level   321
   Introduction   322

   Case – Municipal solid waste composting for cost recovery (Mbale Compost Plant, Uganda)   324

   Case – Public-private partnership-based municipal solid waste composting (Greenfields Crops, Sri Lanka)   333

   Case – Fecal sludge and municipal solid waste composting for cost recovery (Balangoda Compost Plant, Sri Lanka)   341

   Business model 10: *Partially subsidized composting at district level*   351

8. Business models on subsidy-free community-based composting   359
   Introduction   360

   Case – Cooperative model for financially sustainable municipal solid waste composting (NAWACOM, Kenya)   362

   Business model 11: *Subsidy-free community-based composting*   371

9. Business models on large-scale composting for revenue generation   378
   Introduction   379

   Case – Inclusive, public-private partnership-based municipal solid waste composting for profit (A2Z Infrastructure Limited, India)   381

   Case – Municipal solid waste composting with carbon credits for profit (IL&FS, Okhla, India)   391

   Case – Partnership-driven municipal solid waste composting at scale (KCDC, India)   400

   Case – Franchising approach to municipal solid waste composting for profit (Terra Firma, India)   411

   Case – Socially-driven municipal solid waste composting for profit (Waste Concern, Bangladesh)   422

   Business model 12: *Large-scale composting for revenue generation*   434

10. Business models on nutrient recovery from own agro-industrial waste   447
    Introduction   448

    Case – Agricultural waste to high quality compost (DuduTech, Kenya)   450

    Case – Enriched compost production from sugar industry waste (PASIC, India)   459

    Case – Livestock waste for compost production (ProBio/Viohache, Mexico)   468

    Business model 13: *Nutrient recovery from own agro-industrial waste*   478

11. Business models on compost production for sustainable sanitation service delivery — 484
    Introduction — 485

    Case – Fecal sludge to nutrient-rich compost from public toilets (Rwanda Environment Care, Rwanda) — 487

    Business model 14: *Compost production for sustainable sanitation service delivery* — 496

12. Business models for outsourcing fecal sludge treatment to the farm — 504
    Introduction — 505

    Case – Fecal sludge for on-farm use (Bangalore Honey Suckers, India) — 508

    Business model 15: *Outsourcing fecal sludge treatment to the farm* — 516

13. Business models on phosphorus recovery from excreta and wastewater — 523
    Introduction — 524

    Case – Urine and fecal matter collection for reuse (Ouagadougou, Burkina Faso) — 527

    Business Model 16: *Phosphorus recovery from wastewater at scale* — 538

SECTION IV: WASTEWATER FOR AGRICULTURE, FORESTRY AND AQUACULTURE — 547

Edited by Pay Drechsel and Munir A. Hanjra

Wastewater for agriculture, forestry and aquaculture: An overview of presented business cases and models — 548

14. Business models on institutional and regulatory pathways to cost recovery — 553
    Introduction — 554

    Case – Wastewater for fruit and wood production (Egypt) — 556

    Case – Wastewater and biosolids for fruit trees (Tunisia) — 569

    Case – Suburban wastewater treatment designed for reuse and replication (Morocco) — 584

    Business model 17: *Wastewater for greening the desert* — 595

15. Business models beyond cost recovery: Leapfrogging the value chain through aquaculture — 604
    Introduction — 605

    Case – Wastewater for the production of fish feed (Bangladesh) — 606

    Case – A public-private partnership linking wastewater treatment and aquaculture (Ghana) — 617

    Business Model 18: *Leapfrogging the value chain through aquaculture* — 631

16. Business models for cost sharing and risk minimization — 639
   Introduction — 640

   Case – Viability gap funding (As Samra, Jordan) 642

   Business model 19: *Enabling private sector investment in large scale wastewater treatment* 656

17. Business models on rural–urban water trading — 664
   Introduction — 665

   Case – Fixed wastewater-freshwater swap (Mashhad Plain, Iran) 670

   Case – Flexible wastewater-freshwater swap (Llobregat delta, Spain) 679

   Business model 20: *Inter-sectoral water exchange* 691

   Case – Growing opportunities for Mexico City to tap into the Tula aquifer (Mexico) 698

   Case – Revival of Amani Doddakere tank (Bangalore, India) 710

   Business model 21: *Cities as their own downstream user (Towards managed aquifer recharge)* 720

18. Business models for increasing safety in informal wastewater irrigation — 728
   Introduction — 729

   Business model 22: *Corporate Social Responsibility (CSR) as driver of change* 733

   Business model 23: *Wastewater as a commodity driving change* 745

   Business model 24: *Farmers' innovation capacity as driver of change* 760

SECTION V: ENABLING ENVIRONMENT AND FINANCING — 775

19. The enabling environment and finance of resource recovery and reuse — 777
   Luca di Mario, Krishna C. Rao and Pay Drechsel

   Frugal innovations for the circular economy: An epilogue — 801
   Jaideep Prabhu

   Index — *804*

Editors and authors

Editors

Miriam Otoo holds a PhD in Agricultural Economics from Purdue University, and specialized since she joined the International Water Management Institute (IWMI) in 2011 in the economics of waste reuse, business development and entrepreneurship, agricultural markets and productivity in developing countries. Her research focuses on understanding the linkages between agriculture, sanitation and organic waste management to enhance food security via the analysis of business opportunities in the waste reuse sector in Africa, Asia and Latin America. Miriam led the development of a multi-criteria methodological framework which has formed the basis for the development and assessment of waste reuse business models for their feasibility, replicability and scaling-up potential in low- and middle-income countries. In 2008, Miriam received the Norman E. Borlaug LEAP Fellowship Award for her doctoral research.

Pay Drechsel holds a PhD in Environmental Sciences and is a principal researcher and research program leader at the International Water Management Institute (IWMI), based in Colombo, Sri Lanka. Pay worked before for the International Board for Soil Research and Management (IBSRAM) and has 25 years of working experience in the rural-urban interface of developing countries, coordinating projects and programs addressing the safe recovery of water, nutrients and organic matter from domestic waste streams for agriculture, with a special interest in wastewater irrigation, urban and peri-urban agriculture, and the cutting edge of applied inter-disciplinary research. Pay has authored over 300 publications, half in peer-reviewed books and journals. He has worked extensively in West and East Africa, and South and Southeast Asia. In 2015, Pay received the Development Award for Research from the International Water Association.

Authors

Abasi Musisi, Kampala Jellitone Suppliers (KJS) Ltd., Kampala, Uganda

Alexandra Evans, Independent Consultant, London, UK

Andrew Adam-Bradford, Coventry University, UK

Ashley Muspratt, Pivot Ltd./Waste Enterprisers Holding, Kigali, Rwanda

Binu Parthan, Sustainable Energy Associates, Kottayam, India

Charles B. Niwagaba, Makerere University, Kampala, Uganda

Doraiswamy Ramadass Naidu, Advanced Centre for Integrated Water Resources Management, Government of Karnataka, India

George K. Danso, Government of Alberta, Edmonton, Canada

Guy Hutton, United Nations Children's Fund (UNICEF), New York, USA

Hari Natarajan, Independent Consultant, Energy access, New Delhi, India

Heiko Gebauer, EAWAG: Swiss Federal Institute of Aquatic Science and Technology, Dübendorf, Switzerland

Ishara Atukorala, La Trobe University, Melbourne, Australia

Jack Odero, Gamma Systems Limited, Nairobi, Kenya

Jaideep Prabhu, Judge Business School, University of Cambridge, UK

Jasper Buijs, Sustainnovate (http://sustainnovate.nl), Arnhem, The Netherlands

Javier F. Reynoso-Lobo, Independent Consultant, Monterrey, Mexico

Joginder Singh, Punjab Agricultural University, Ludhiana, India

Kamalesh Doshi, Simplify Energy Solutions LLC, Ashburn, Virginia, USA

Katharina Felgenhauer, International Water Management Institute (IWMI), Accra, Ghana

Krishna C. Rao, International Water Management Institute (IWMI), Bangalore, India

Lars Schoebitz, EAWAG: Swiss Federal Institute of Aquatic Science and Technology, Dübendorf, Switzerland

Lesley Hope, Ruhr-Universität, Bochum, Germany

Linda Strande, EAWAG: Swiss Federal Institute of Aquatic Science and Technology, Dübendorf, Switzerland

Linus Dagerskog, Stockholm Environment Institute, Stockholm, Sweden

Louis Lebel, Unit for Social and Environmental Research, Chiang Mai University, Chiang Mai, Thailand

Luca di Mario, Asian Development Bank, Manila, Philippines

Manzoor Qadir, Institute for Water Environment and Health, United Nations University, Hamilton, Canada

Marudhanayagam Nageswaran, Independent Consultant, Pondicherry, India

Miriam Otoo, International Water Management Institute (IWMI), Colombo, Sri Lanka

Mirko Winkler, Swiss Tropical and Public Health Institute, Basel, Switzerland

Munir A. Hanjra, International Water Management Institute (IWMI), Colombo, Sri Lanka

Nancy Karanja, University of Nairobi, Nairobi, Kenya

Patrick Watson, I-Dev International, Lima, Peru; San Francisco, USA; Nairobi, Kenya

Paul Skillicorn, Lyndon Water Limited, Kingston-upon-Thames, UK

Pay Drechsel, International Water Management Institute (IWMI), Colombo, Sri Lanka

Priyanie Amerasinghe, International Water Management Institute (IWMI), Colombo, Sri Lanka

Pushkar S. Vishwanath, Tide Technocrats Private Limited, Bangalore, India

Sampath N. Kumar, Tide Technocrats Private Limited, Bangalore, India

Sena Amewu, International Water Management Institute (IWMI), Accra, Ghana

Solomie Gebrezgabher, International Water Management Institute (IWMI), Accra, Ghana

Sudarshana Fernando, International Water Management Institute (IWMI), Colombo, Sri Lanka

Surendra K. Pradhan, Aalto University, Aalto, Finland

Acknowledgements

We gratefully acknowledge the support of the Swiss Agency for Development and Cooperation (SDC), the International Fund for Agricultural Development (IFAD) and the CGIAR Research Program on Water, Land and Ecosystems (WLE) which resulted in this publication.

The here presented book or 'business model catalogue' is an output from a joint project (2011–2015) between the International Water Management Institute (IWMI), the Department of Sanitation, Water and Solid Waste for Development (SANDEC) of the Swiss Federal Institute of Aquatic Science and Technology (EAWAG), the International Centre for Water Management Services (CEWAS), the Swiss Tropical and Public Health Institute (Swiss TPH) and the World Health Organization (WHO), and follow-up studies (2016-2017) funded by WLE. The editors duly acknowledge the contribution of each partner.

The project was initiated by a landscape analysis implemented by the International Water Management Institute (IWMI) for the Bill & Melinda Gates Foundation (BMGF). The study suggested that a key factor for supporting cost recovery at scale in the sanitation sector would be the introduction and implementation of 'business thinking' along the sanitation value chain and in particular in the domain of Resource Recovery and Reuse (RRR).

Disclaimer: The opinions expressed in this book are the respective authors' own and do not reflect the views of the funding agencies or collaborating partner institutions. Maps and country boundaries are only indicative and not to scale.

Graphics and design: Michael Dougherty
Editorial support: Justin Dupre-Harbord, Robin Leslie

Foreword

Rapid increases in the human population and in consumption per capita threaten to stretch the planet's capacity to sustain growth beyond its limits. The time has come to move away from the 'take, make, dispose' paradigm of production and consumption, which has dominated global society since the Industrial Revolution, towards what has been termed a 'Circular Economy', which incorporates recycling into the production-consumption cycle.

The Circular Economy concept offers multiple benefits, which have gained recognition in recent decades under various guises: including ecological economics, green growth and sustainable development. The United Nations Agenda 2030 and its 17 Sustainable Development Goals (SDGs) acknowledge the environmental limits to growth and human well-being. The environment features prominently in many targets of the SDGs – particularly SDG 2 on food security and sustainable agriculture, SDG 6 on water reuse and water for ecosystems, SDG 12 on waste recycling and reuse and SDG 15 on restoring degraded soils, to name a few.

What the green development has lacked so far, however, thus limiting its success, are workable business models that incentivize economic agents to act on the basis of social and environmental concerns, and consider these as concrete bottom lines in their business decisions. As a result, efforts to mainstream Corporate Social Responsibility (CSR) have relied mainly on the conscience of business leaders, appealing to their sense of responsibility for social and environmental concerns. The proponents of CSR policies have rarely justified them in terms of their most important bottom line, the financial one. The goal of the Circular Economy, on the other hand, is for business leaders to assess business viability not only in the short term but for the future generations who will demand their services. The idea is for businesses to internalize the wider environmental costs and benefits in their production decisions and to make consumers complicit in these decisions.

This publication showcases real examples from around the world, demonstrating how plant nutrients, energy and water can be recovered from what is currently viewed as 'waste' – avoiding their unregulated disposal into the environment and associated costs (e.g. health costs, clean-up costs), while also capturing the financial value associated with reuse of the treated or recycled resource. Like a catalogue, compiled mostly from low- and middle-income countries, the book covers a wide range of value propositions to maximize cost recovery and social or financial benefits. It is impossible to underestimate the importance of recovering resources, particularly from food waste in growing urban centers, for the benefit of the water, energy, nutrient and carbon cycles. If these case studies and the models derived from them can inform broader programs aimed at scaling up good practices, they will contribute importantly to the achievement of many SDG targets, including SDG 11 on more resilient cities.

What these case studies demonstrate is that businesses working towards a Circular Economy can create social and financial value beyond cost recovery. However, the success of these business models relies on the presence of an enabling environment, such as laws and regulations, strong capital markets, consumer advocacy, and so on, to attract private capital and expertise. These findings underline the critical role of governments in making the Circular Economy a reality.

The catalogue fills a significant gap in the literature and should prove useful not only for today's investors and policy makers but also for the curricula of engineering, economics, environmental and business schools. This will help sensitize the next generation of decision makers to the opportunities inherent in the Circular Economy.

On behalf of the editors and authors, we strongly recommend the catalogue to readers working at the interface between waste management, sanitation and other sectors, such as agriculture, and urge them to make good use of this timely and valuable publication.

Guy Hutton
Economist
Senior Adviser, UNICEF

SECTION I – BUSINESS MODELS FOR A CIRCULAR ECONOMY

Introduction

1. BUSINESS MODELS FOR A CIRCULAR ECONOMY: LINKING WASTE MANAGEMENT AND SANITATION WITH AGRICULTURE

Pay Drechsel, Miriam Otoo, Krishna C. Rao and Munir A. Hanjra

Business models for a circular economy: Linking waste management and sanitation with agriculture

Urbanization is the pre-eminent global phenomenon of our time. Currently, urban areas account for 75% of the world's natural resource consumption, while producing over 50% of the globe's waste on just 2–3% of the earth's land surface (UNEP, 2013). Without recycling, cities will continue to constitute vast sinks for food waste including valuable crop nutrients and organic matter, while millions of rural, peri-urban or urban farmers struggle with depleted soils to feed the growing urban population. Yet, it is not only the loss of valuable, and in part, finite resources, but also the costs of poor waste management, i.e. environmental pollution and the production of avoidable greenhouse gases (GHG) which threatens sustainable urban growth. Halving, for example, the current rate of food wastage would greatly support waste management while reducing GHG emissions by 22–28% (WEF, 2016). So far, the environmental costs of poor waste management are usually externalized and the market incentives to reduce waste are minimal.

While global demand projections for water, food and energy predict continuous and significant growth, the declining reserves of the non-renewable phosphorus, copper and zinc resources (Holmgren et al., 2015) reinforce the need for more investments in resource recovery and reuse across the food, waste and sanitation sectors (Ellen MacArthur Foundation, 2017; TBC, 2016).

While Europe continues setting an example with the implementation of a first action plan on the circular economy (EC, 2016), more attention should be given to natural resource loops in low- and middle-income countries, especially in the tropics where soils are poor and nutrient depletion is high and commercial fertilizer is basically unaffordable. Minimizing resource loss and returning resources into the food production process is essential in particular in drier climates where every drop of water counts and organic matter is needed for sustaining soil fertility as natural biomass production is low.

Aside from the reduction of food waste along the food chain, resource recovery allows to capture value even from apparently 'wasted' resources (FAO, 2011). In particular, domestic and agro-industrial waste is rich in water, nutrients, energy and organic compounds. Yet, in most parts of the world, this waste is not being managed in a way that permits us to derive value from its reuse, although resource recovery is nothing new. Closed loop systems linking food waste and food production have been practiced for generations in many rural societies. However, population growth and urbanization in particular have increased distances and polarized food flows towards urban centres where agricultural reuse opportunities for food waste are limited.

But cities are not only 'hungry'; they are also 'thirsty'. Van Rooijen et al. (2005) crafted the term 'Sponge City' to visualize the **urban metabolism** which is absorbing freshwater from its periphery while discharging wastewater which has a high potential to support ecosystem services and food production in water-scarce regions, if wastewater treatment and safe reuse can be achieved. If not, this water will be a threat to food safety and public health.

In fact, due to limited treatment capacities, the various domestic waste streams, solid as well as liquid, form a significant part of the unwanted urban footprint. The resulting pollution constitutes not only the paramount environmental and health challenges that today's exploding cities and their surroundings are facing, but also a significant economic challenge in countries where waste collection and treatment cannot be financed through taxes and fees (Kennedy et al., 2007; Le Courtois, 2012). This mismatch puts into question the sustainability of urban growth where it is the fastest unless alternative business models are put in place (Muradian et al., 2012; Villarroel Walker et al., 2012).

In the context of resource poor countries, it is more than opportune to argue for a circular metabolism, as increasingly promoted in many developed nations, where waste segregation and recycling contribute to overall system resilience (UNEP, 2017) and the values of *green growth*, i.e. an economy without degrading the environment. In this regard, the urban waste challenge – including fecal matter generation – can offer immense and scalable opportunities for entrepreneurs through transforming waste from domestic and agro-industrial sources into low-carbon assets for use in agriculture and other sectors (Figure 1). This is strongly supported by the Sustainable Development Goals (SDG) targeting for example water reuse (SDG 6), renewable energy (SDG 7) and waste recycling and reuse (SDG 12), which can help to restore degraded soils (SDG 15) for sustainable agriculture and food security (SDG 2) and resilient cities (SDG 11). Especially wastewater and the different organic fractions of municipal waste streams offer a significant potential for the support of a 'biocycle economy' (Ellen MacArthur Foundation, 2017).

These opportunities for value creation from resources that would otherwise be irretrievably lost also allow for cost savings and/or cost recovery in the sanitation sector; for example in the case of composting which, depending on scale, reduces municipal solid waste volumes and transport costs with the potential to enhance the lifetime of landfills with less GHG emissions. Furthermore, by moving increasing amounts of biological material through anaerobic digestion or composting back into the soil, a circular economy approach will reduce the need for chemical fertilizers and soil amendments (Box 1).

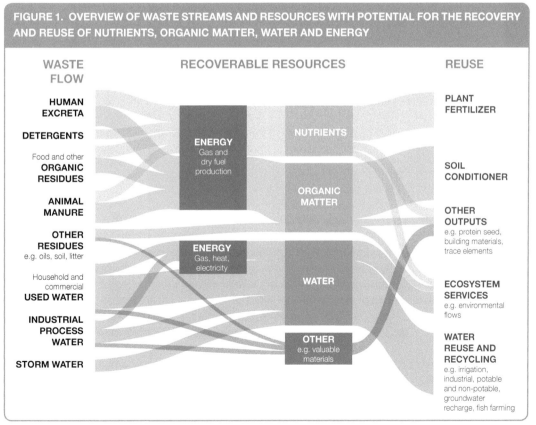

FIGURE 1. OVERVIEW OF WASTE STREAMS AND RESOURCES WITH POTENTIAL FOR THE RECOVERY AND REUSE OF NUTRIENTS, ORGANIC MATTER, WATER AND ENERGY

Source: Andersson et al., 2016, modified.

> **Box 1. The potential of organic waste for the circular economy**
>
> The World Economic Forum has estimated potential global revenues from the biomass value chain (production of agricultural inputs, biomass trading and biorefinery outputs) as high as USD 295 billion by 2020. Cities, as major concentrators of materials and nutrients, and the power of generating over 80% of the global Gross Domestic Product (GDP), will play a major role on the 'biocycle economy'.
>
> **Return of food waste:** If 100% of consumption-related food waste and 50% of other food waste generated today were returned to the soil, it could replenish 5 million tonnes of nitrogen, phosphorus and potassium (N, P, K) reserves, substituting for 4% of current N, P, K consumption.
>
> **Return of animal manure:** If all the nutrients from the current stocks of cattle, chicken, pig and sheep manure were captured, they would yield an astounding 345 million tonnes of N, P, K annually – more than twice the world's current consumption. Using animal manure also improves soil structure and organic content and reduces commercial fertilizer loss.
>
> **Return of human waste:** Human waste also contains significant amounts of N, P, K. If nutrients contained in the waste of the world's population were captured, they would amount to 41 million tonnes, representing 28% of the current N, P, K consumption.
>
> In theory, the organic sources of N, P, K fertilizer recovered from food, animal and human waste streams could on a global scale contribute up to 2.7 times the nutrients contained within the volumes of chemical fertilizer currently used.
>
> Further analysis is needed to assess what share of organic fertilizers could be returned to the soil in a cost-effective way. In OECD countries, for example, an estimated 177 million tonnes of municipal organic waste are produced annually, of which 66 million tonnes are so far valorized in composting or anaerobic digestion. The market value of N, P, K in this fraction is estimated at USD 121 million per year, and adds an estimated 5 million tonnes of stable carbon (and 10 million tonnes of carbon in total) to OECD soils every year in the form of compost/digestate. ISWA (2015) estimate that around 58 million tonnes additionally could feasibly be collected and valorized.
>
> Source: Ellen MacArthur Foundation, 2013, 2017; ISWA 2015.

As farm soils need organic material, especially on highly weathered tropical soils, closed loop processes appear to be a win-win situation (Drechsel and Kunze, 2001). It is estimated that halving the current rate of food wastage could meet over a fifth of caloric needs by 2050, reducing required cropland by 14% (WEF, 2016). The reality is however, that resource recovery and reuse (RRR) has been until now more theory than practice. RRR remains challenged where awareness for 'green' values and opportunities is less developed, public perceptions do not favour reuse or municipal capacities are too constrained to make the required investment. Developing countries spend around USD 46 billion annually on waste management, and it is estimated that they should spend another USD 40 billion to cover the current service delivery gap (Le Courtois, 2012). The total costs are expected to surpass USD 150 billion by 2025. The additional capital investments required for safe fecal waste management in support of the SDGs target 6.2 amount to about USD 49 billion per year (Hutton and Varughese, 2016).

In their daily struggle with the service delivery gap, many municipalities consider RRR a task for the future, once their current challenges are under control. What Onibokun (1999) called 'Managing the

Monster' is in fact often absorbing as much as half of the municipal budget in many low-income countries (Le Courtois, 2012).

Accepting the limitations of the public sector, an opportunity is to leverage private capital based on the value of the recovered resources (Otoo et al., 2012; Le Courtois, 2012). This could also support a conceptual transition from 'treatment for safe disposal' to 'design for reuse' (Murray and Buckley, 2010; Huibers et al., 2010). In such a postmodern sanitation system (Ushijimaa et al., 2015), incentives for financing sanitation could be shared between 'front-end users' and 'back-end users' building on demand for the products of sanitation and waste management to motivate a combined finance model and more robust operation and maintenance of complete sanitation systems (Murray and Ray, 2010). This would require a supportive regulatory and finance environment and well-designed partnerships agreements.

However, the lessons learned so far have also shown that closed loop processes do not manifest themselves through the promotion of composting, water reuse or – for example – *ecological sanitation*. What is often described as an engineering challenge (*'Reinvent the Toilet'*) and in fact is often driven by technology development, like for the removal of unwanted struvite in wastewater treatment plants ('phosphorus recovery'), is increasingly understood as an institutional, social and economic challenge. There is significant need for investments in market research, bankable business models for cost recovery, stakeholder buy-in and innovative partnerships, especially if scalability and sustainability are targeted (Guest et al., 2009; Le Courtois, 2012; Beltramello et al., 2013; Hanjra et al., 2015; Verstraete and Cornel, 2014). Countless failed composting projects began with significant amounts of grant funding but eventually collapsed due to their inability to support their operational costs (World Bank, 2016).

Given the common situation of the waste and sanitation sectors, especially in Africa and Asia, the term 'business models' might appear to be out of place. However, exactly where every step towards cost recovery counts, the thinking has to change (Koné, 2010). While for example wastewater treatment was and is first of all a 'social business model' with a strong economic justification and returns on investments through safeguarding public health and the environment, a second (reuse-based) value proposition can offer incentives for private sector engagement, that leverage private capital to help realize commercial or social value. However, what sounds in theory promising often faces fundamental structural barriers. In fact, 88% of developing country governments have no cost recovery efforts at all for water and sanitation (Muspratt, 2016a).

There are multiple bottlenecks faced by both the public sector and/or the emerging private sector across most low- and middle-income countries. These include financing challenges, unsupportive regulations and slow approval processes, but also missing the capacity to present viable business plans for penetrating the reuse market. In particular, organic waste composting is often more driven by cost savings than revenue generation (Box 2) which can potentially undermine those SDG targets, which will count actual 'reuse'. Thus, private sector participation in waste management is not a panacea for success in promoting Resource Recovery *and Reuse* unless the companies understand how to approach the reuse market (e.g. Rouse et al., 2008) and can count on an enabling environment (see Chapter 19). In particular in Africa, smaller start-ups struggle with bureaucracies and financing (Muspratt, 2016ab), while larger companies, that can accommodate delays, succeed. In India, for example, several firms have emerged that treat today the waste collected by municipalities without any charge, while revenue is generated exclusively by recycling the waste collected (Furniturwala, 2012). In this regard, urbanization is not only posing challenges but also opportunities compared with rural areas, such as market proximity, shorter transport distances, higher purchasing power, export hubs and economies of scale that can attract private capital, if the enabling policy environment is in place and *de facto* functional.

> ### Box 2. Cost savings as a driver for resource recovery (and reuse)
>
> Where land prices go up in urban vicinity, communities do not accept hosting a landfill. This is resulting in increasing transport costs for municipal waste disposal to remote areas where land is still abundant. As transport can be their major cost factor, many waste managers show a strong interest in composting as a means to reduce waste volumes and transport costs (Drechsel et al., 2010). If the compost is eventually 'burned', distributed for free, or becomes a revenue stream, from the waste management perspective this is often of lower relevance, especially where (i) contracts are based on the processed waste volume, but not on the sale or reuse of the recovered resource, or (ii) the gains from volume reduction outweigh any expected returns from compost marketing.
>
> The same applies to those waste-to-energy projects, which are designed (and financed) for absorbing municipal solid waste (MSW) in order to reduce municipal service costs as a whole. Energy production is in these cases often only a secondary revenue stream, while MSW sorting cost and its low calorific value constrain the business. Another example are enterprises engaged in the collection of human excreta from non-sewered sanitation systems, which might engage in composting, primarily to reduce the costs of waste disposal, and not because of expected compost revenues.
>
> In addition, energy or phosphorus recovery within wastewater treatment processes is largely driven by cost reduction. The recovery of phosphorus, for example, prevents damage of pipes and valves through unwanted precipitation. The resulting savings in chemicals otherwise needed to remove the crystals can more or less finance enterprises specialized in P recovery while the generated P-fertilizer is a side product which is often struggling to find more than a niche market (Otoo et al., 2015; see also Business Model 16).

A second prominent RRR bottleneck is the understanding of the impact and related value of planned interventions compared with the counterfactual 'business as usual'. Internalizing any possible externalities especially on human and environmental health is important to attract public subsidy as a well-justified revenue stream. When the environmental and societal benefits of investments in sanitation and waste management are accounted for, most RRR projects will be viable (ADB, 2011; Andersson et al., 2016). However, while benefits can be easily and fully internalized by governments and citizens, they are very difficult for a private company to monetize (Muspratt, 2016a). On the other hand, the private sector is under increasing pressure to accept corporate social responsibility (CSR), account for its own externalities, and engage in mitigation measures.

Corporate social and environmental responsibility

The call for corporate responsibility is echoed in SDG 12.6 (Encourage companies, especially large and transnational companies, to adopt sustainable practices and to integrate sustainability information into their reporting cycle). While CSR has a high potential to support a circular economy, its success at national level will not only depend on the private sector but also how governments, which carry the responsibility for achieving the SDGs, will 'encourage' firms to take part (Fogelberg, 2015). For example, section 135 of India's Companies Act 2013 requires (on a "comply-or-explain" basis) that firms satisfying specific size or profit thresholds spend a minimum of 2% of their average (pre-tax) net profit on CSR; moving a voluntary CSR contribution into a law. The risk is that this transforms CSR more into an offset tax than social or environmental consciousness, as a company can choose to contribute, e.g. to funds of the Central Government or the State Governments for socio-economic development (Grant Thornton India LLP, 2013), independently of the company's own practices and challenges, e.g. in view of responsible resources management.

A closer monitoring is provided by independent CSR assessment agencies, as for instance by the Newsweek Green Rankings. The ranking is based on eight key performance indicators, including waste generation/recovery/reuse, GHG emissions, energy and water demands and so forth. Companies failing to disclose data for the rigorous analysis by Newsweek and partners would receive a score of '0', thus negatively affecting their overall performance and public image. The rise of the social media is in this regard an important factor. When catering to global markets, big companies sell millions of products every day. However, any negative press can set off within the shortest period a series of consequences via social media that may be detrimental to a product or brand's image. These days this puts much higher pressure on companies to maintain their image compared to a decade ago and a number of rating agencies support these efforts (Novethic, 2013).

Corporate social and environmental responsibility can thus directly and indirectly trigger and support RRR. The key words are "responsible and sustainable sourcing of raw materials", including direct commitments to the circular economy (Box 3).

Sustainable sourcing is increasingly receiving attention as consumers and other stakeholders want to know where their food comes from and how it was produced. Supply chain audits can have a far reach and catalyze environmental consciousness at an unexpected pace and far from the company's home. In one of the reported cases in this catalogue (Chapter 18), local private textile suppliers offered their own government to co-finance wastewater treatment plants to be able to comply with the responsible sourcing criteria of their European buyers as otherwise they would no longer be accepted, resulting in financial crisis.

Box 3. Towards a circular economy in the food sector

Based on CSR principles global companies such as Cargill, Nestlé, Starbucks, Unilever etc. support in many low-income countries extension services, traders and farmers, e.g. in view of access to inputs and markets along the companies' value chains. Social and environmental commitments include responsible sourcing of raw materials and a high commitment to personal and product safety, resource recovery and zero waste schemes, or for example the provision of fortified but affordable food. In larger companies, these commitments are part of the corporate value proposition and monitored through audits and certifications by independent accredited bodies issuing sustainability rankings and indices. The same applies to agricultural input suppliers like BASF and its resource use efficiency optimizing 'Verbund' principle. Global food company, Danone, to give another example, has announced in 2016 a new partnership with the global waste-management company, Veolia, to embed circular economy principles inside the company and to promote them widely. Danone aims for systemic change to preserve natural resources and to move to a more circular value chain. Danone was recently awarded the Environment Top Performance prize by the Environmental and Social Governance (ESG) ratings agency Vigeo, among 1,300 companies assessed. The company has circular economy projects like recycling by-products from yoghurt production for animal feeds, fertilizer and energy.

See also: www.mckinsey.com/business-functions/sustainability-and-resource-productivity/our-insights/toward-a-circular-economy-in-food?cid=eml-web (accessed November 7, 2017)

The next step of corporate responsibility is the monetary valuation of the ecosystem services that are positively or negatively affected, and to integrate these financial values into corporate accounting. Negative balances could be offset through carbon or ecosystem credits (NSW, 2007; The Rockefeller Foundation, 2015). Internal carbon pricing, which is of particular interest for RRR, is now becoming a widely used tool helping companies shift to lower-carbon business models. Over 1,200 companies reported to CDP, formerly the Carbon Disclosure Project, in 2016 that they are currently using an internal price on carbon or plan to do so within the next two years (CDP, 2016).

To avoid that offsetting becomes the main investment and a license for 'business as usual', green accounting requires shared definitions, indicators and methodologies for measuring and monitoring impacts to allow public sector investment ideally in the same area of concern, e.g. in wastewater treatment (DeLonge, 2012; Meyers and Waage, 2014).

From business cases and opportunities to business models

With three SDG supporting directly RRR, and an increasing attention to the synergies between CSR and the circular economy, the objectives of this book are:

> To show scalable options for RRR as a value proposition to stimulate business thinking in the interface of sanitation, waste and agriculture.

> To build capacity for a more integrated and inclusive approach to the recovery of water, carbon, nutrients and energy from domestic and agro-industrial waste for reuse.

> To provide opportunities for local business model adaptation across low-income countries, where the public sector struggles to finance closed loop processes through household taxes and fees, and start-ups struggle with an only slowly emerging enabling environment.

Chosen from about 150 public and private RRR projects and enterprises, of which over 60 were analysed in detail, this catalogue presents a selection of 47 empirical business cases (Figure 2), from which 24 business models were extracted. Chapter 2 provides some background into the methodology and definitions used for the selection and analysis of the cases and models. A separate catalogue looking at 18 institutional business models for managing the 'ultimate' food waste, i.e. fecal sludge, including resource recovery and reuse as fertilizer and energy source, has been published separately (Rao et al., 2016).

Our understanding of the term 'business model' follows Osterwalder and Pigneur (2010), i.e. a business model describes the rationale of how a firm or organization creates, delivers and captures value in economic, social, cultural or other contexts. In our case, the common value proposition is the creation of a useful resource from material which otherwise would be wasted. Given the multitude of domestic and industrial RRR options, **in this publication we are looking mostly at those options where either the waste derives from the food chain and/or the recovered resources support the food chain**. In other words, most presented cases and models are limited to the recovery of (i) water, (ii) crop nutrients and carbon (organic matter) and (iii) energy, derived from domestic and agro-industrial waste, including food waste, wastewater and excreta. By limiting the scope to the food chain, other recyclable resources like glass, plastic or metal are not addressed.

In order to increase the probability of replication in low- and middle-income countries we tried to focus mostly on cases and experiences in Asia, Africa and Latin America, operating at community or city scale, i.e. we exclude individual household- or farm-based efforts for resource recovery and reuse. A few cases from high income countries, with potential for replication in other parts of the world,

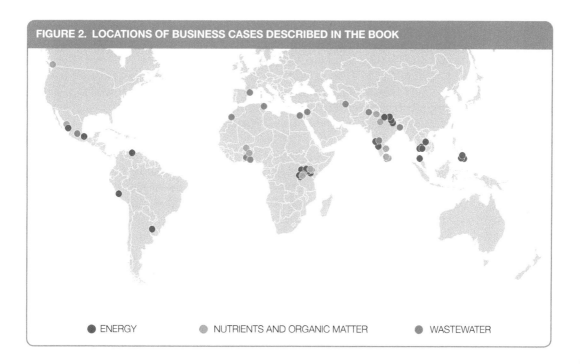

FIGURE 2. LOCATIONS OF BUSINESS CASES DESCRIBED IN THE BOOK

● ENERGY   ● NUTRIENTS AND ORGANIC MATTER   ● WASTEWATER

are also included. The description of cases and models followed defined templates (see Chapter 2) bridging between the needs of students in business schools looking for detailed case studies and those of investors in need of a compact information, which was not easy to combine.

In the literature, the term 'business model' is commonly used for a broad range of informal and formal business processes, structures and purposes resulting in very diverse interpretations and definitions. Similarly, many options exist to name or cluster business cases and models in categories, especially in the young domain of sustainable development and green economy where existing examples are fragmented (George and Bock 2011; Beltramello et al., 2013). It is important in this context to stress that the term '**business**' should not imply that 'business models' have to be profit-oriented or able to achieve through their value proposition full cost recovery. **In sectors, like waste and sanitation, which usually rely on public financing, any scalable efforts towards cost recovery or cost savings are already a paradigm shift and should be seen as a step in the right direction, next to the creation of social and environmental value**. Reduced expectations are in particular required in view of water reuse in agriculture. In many situations, the direct revenues from selling treated wastewater to farmers are small, given that fresh water prices are often subsidized or groundwater freely accessible. However, the situation can change if further value propositions are added, such as the use of the water for fish feed and fish production, energy recovery or treatment for industrial or potable reuse (Rao et al., 2015). In those cases, the full recovery of operational and maintenance costs, or even the recovery of capital costs, can be possible as the examples in the book show. But more common and equally important are those cases where operational cost recovery varies between 10 and 90% and it is critical to analyse what prevents a waste-based venture from moving up the scale.

In cooperation with different business schools, the catalogue adopted the **extended Business Model Canvas** (Osterwalder and Pigneur, 2010) to visualize the different business models, including their externalities. Externalities are very important as the waste and sanitation sectors not only benefit society but also are prone to environmental and human health risks. Hence, an important requirement for any type of waste management scheme, including resource recovery, is the need to safeguard public health. Risk management and mitigation for safe waste handling and reuse are thus essential components of the sustainability and acceptance of any RRR business model, especially where the waste might contain fecal matter or other chemical contaminants. This was emphasized through collaboration with the World Health Organization (WHO) and development of the *Sanitation Safety Planning* (SSP) concept, which supports the operationalization of the safe use of wastewater, excreta and greywater in agriculture and aquaculture (WHO, 2015; Andersson et al., 2016).

This catalogue with its cases and models targets a community more interested in business opportunities than technical solutions. The description of technologies as far as they relate to the value proposition or particular safety measures remains throughout brief except where business models are technology driven. With its focus on low- and middle-income countries, the catalogue does not include those high-tech solutions for RRR, which first have to show their replicability and sustainability in the context of these regions as stressed, e.g. by Wang et al. (2015), Nhapi and Gijzen (2004), Murray and Drechsel (2011) or Libhaber and Orozco-Jaramillo (2013).

Although neither the presented cases nor models cover the whole spectrum of agriculture related RRR value propositions, this catalogue is the most profound analysis and comprehensive compilation made so far to show the business side of RRR in the interface of sanitation, waste management and agriculture in low-income countries. While in some cases it was not possible to obtain from the private or public sector the requested financial information, or only under a non-disclosure agreement, the models should provide enough information to be an excellent starting point for business schools and investors to approach this so far uncharted sector.

An analysis, which is cutting across several of the presented cases and models was presented by Rao et al. (2015) for water reuse, Gebrezgabher et al. (2015) for energy recovery and Otoo et al. (2015) for nutrient and organic matter recovery and reuse.

References and further readings

Andersson, K., Rosemarin, A., Lamizana, B., Kvarnström, E., McConville, J., Seidu, R., Dickin, S. and Trimmer, C. 2016. Sanitation, wastewater management and sustainability: From waste disposal to resource recovery. Nairobi and Stockholm: United Nations Environment Programme and Stockholm Environment Institute.

Asian Development Bank (ADB). 2011. Toward sustainable municipal organic waste management in South Asia: A guidebook for policy makers and practitioners. Mandaluyong City, Philippines: Asian Development Bank.

Beltramello, A., Haie-Fayle, L. and Pilat, D. 2013. Why new business models matter for green growth (OECD Green Growth Papers, 2013-01). Paris: OECD Publishing.

CDP. 2016. Embedding a carbon price into business strategy. www.cdp.net/en/reports/downloads/1132.

DeLonge, M. 2012. Greenhouse gas costs and benefits from land-based textile production. www.fibershed.com/wp-content/uploads/2014/01/Appendix-H.pdf (accessed Nov. 5, 2017).

Drechsel, P. and Kunze, D. (eds.). 2001. Waste composting for urban and peri-urban agriculture: Closing the rural-urban nutrient cycle in sub-Saharan Africa. Wallingford: IWMI/FAO/CABI.

Drechsel, P., Cofie, O. and Danso, G. 2010. Closing the rural-urban food and nutrient loops in West Africa: A reality check. Urban Agriculture Magazine 23: 8–10. www.ruaf.org/closing-rural-urban-food-and-nutrient-loops-west-africa-reality-check (accessed Nov. 5, 2017).

Ellen MacArthur Foundation. 2013. Towards a circular economy (Vol 2: Opportunities for the consumer goods sector). www.ellenmacarthurfoundation.org/assets/downloads/publications/TCE_Report-2013.pdf (accessed Nov. 5, 2017).

Ellen MacArthur Foundation. 2017. Urban Biocycles. www.ellenmacarthurfoundation.org/publications/urban-biocyles (accessed Nov. 5, 2017).

European Commission (EC). 25 October 2016. Juncker Commission presents third annual Work Programme: Delivering a Europe that protects, empowers and defends. Strasbourg. http://europa.eu/rapid/press-release_IP-16-3500_en.htm (accessed Nov. 5, 2017).

Fogelberg, T. 2015. Thinking bigger: Business engagement in the SDGs. http://sd.iisd.org/guest-articles/thinking-bigger-business-engagement-in-the-sdgs (accessed Nov. 5, 2017).

Food and Agriculture Organization of the United Nations (FAO). 2011. Global food losses and food waste: Extent, causes and prevention. Rome. www.fao.org/docrep/014/mb060e/mb060e00.pdf (accessed Nov. 5, 2017).

Furniturwala, I. 2012. Setting the trend in waste processing, a story from India. Private Sector Development 15: 18–21. https://www.proparco.fr/en/waste-challenges-facing-developing-countries (accessed Nov. 5, 2017).

Gebrezgabher, S., Rao, K., Hanjra, M.A. and Hernandez, F. 2015. Business models and economic approaches for recovering energy from wastewater and fecal sludge. In: Drechsel, P., Qadir, M. and Wichelns, D. (eds.). Wastewater: An economic asset in an urbanizing world. Springer: Dordrecht, Heidelberg, New York, London. pp. 217–245.

George, G. and Bock, A.J. 2011. The business model in practice and its implications for entrepreneurship research. Entrepreneurship Theory and Practice 35(1): 83–111.

Grant Thornton India LLP. 2013. Implications of Companies Act, 2013: Corporate social responsibility. http://gtw3.grantthornton.in/assets/Companies_Act-CSR.pdf (accessed Nov. 5, 2017).

Guest, J.S., Skerlos, S.J., Barnard, J.L., Beck, M.B., Daigger, G.T., Hilger, H., Jackson, S.J., Karvazy, K., Kelly, L., Macpherson, L., Mihelcic, J.R., Pramanik, A., Raskin, L., Van Loosdrecht, M.C.M., Yeh, D. and Love, N.G. 2009. A new planning and design paradigm to achieve sustainable resource recovery from wastewater. Environmental Science and Technology 43(16): 6126–6130.

Hanjra, M.A., Drechsel, P., Mateo-Sagasta, J., Otoo, M. and Hernandez, F. 2015. Assessing the finance and economics of resource recovery and reuse solutions across scales. In: Drechsel, P., Qadir, M. and Wichelns, D. (eds.). Wastewater: Economic asset in an urbanizing world. Springer: Dordrecht, Heidelberg, New York, London. pp. 113–136.

Holmgren, K.E., Li, H., Verstraete, W. and Cornel, P. 2015. Sate of the art compendium report on resource recovery from water. London: IWA resource recovery cluster.

Huibers, F., Redwood, M. and Raschid-Sally, L. 2010. Challenging conventional approaches to managing wastewater use in agriculture. In: Drechsel, P., Scott, C.A., Raschid-Sall, L., Redwood, M. and Bahri, A. (eds). Wastewater irrigation and health: Assessing and mitigating risk in low income countries. London: Earthscan. pp. 287–301.

Hutton, G. and Varughese, M. 2016. The costs of meeting the 2030 Sustainable Development Goal targets on drinking water, sanitation, and hygiene. Washington, DC: World Bank/Water and Sanitation Program (WSP).

International Solid Waste Association (ISWA). 2015. Circular economy: Carbon, nutrients and soil (Report 4). www.iswa.org/fileadmin/galleries/Task_Forces/Task_Force_Report_4.pdf (accessed Nov. 5, 2017).

Kennedy, C., Cuddihy, J. and Engel-Yan, J. 2007. The Changing Metabolism of Cities. Journal of Industrial Ecology, 11: 43–59.

Koné, D. 2010. Making urban excreta and wastewater management contribute to cities' economic development: A paradigm shift. Water Policy 12: 602–610.

Le Courtois, A. 2012. Municipal solid waste: Turning a problem into resource. Private Sector Development 15: 2–4. www.proparco.fr/webdav/site/proparco/shared/PORTAILS/Secteur_prive_developpement/PDF/SPD14/revue_SPD15_UK.pdf (accessed 13 March 2017).

Libhaber, M. and Orozco-Jaramillo, A. 2013. Sustainable treatment of municipal wastewater. IWA's Water 21 (October 2013): 25–28.

Meyers, D. and Waage, S. 2014. Environmental profit and loss: The new corporate balancing act. www.greenbiz.com/blog/2014/02/18/environmental-profit-and-loss-new-corporate-balancing-act (or https://goo.gl/9sDYop; accessed Nov. 5, 2017).

Muradian, R., Walter, M. and Martinez-Alier, J. 2012. Hegemonic transitions and global shifts in social metabolism: Implications for resource-rich countries (Introduction to the special section), Global Environmental Change 22(3): 559–567.

Murray, A. and Buckley, C. 2010. Designing reuse-oriented sanitation infrastructure: The design for service planning approach. In: Drechsel, P., Scott, C.A., Raschid-Sally, L., Redwood, M., Bahri, A. (eds). Wastewater irrigation and health: Assessing and mitigation risks in low-income countries. London: Earthscan-IDRC-IWMI. pp. 303–318.

Murray, A. and Drechsel, P. 2011. Why do some wastewater treatment facilities work when the majority fail? Case study from the sanitation sector in Ghana. Waterlines 30(2): 135–149.

Murray, A. and Ray, I. 2010. Wastewater for agriculture: A reuse-oriented planning model and its application in peri-urban China. Water Research 44(5): 1667–1679.

Muspratt, A. 2016a. Make room for the disruptors: while desperate for innovation, the sanitation sector poses unique structural challenges to startup companies. www.linkedin.com/pulse/make-room-disruptors-while-desperate-innovation-sector-muspratt (accessed 31 Jan 2017).

Muspratt, A. 2016b. How do we leverage the speed and innovation of small companies in the inherently slow and bureaucratic sanitation sector? www.linkedin.com/pulse/how-do-we-leverage-speed-innovation-small-companies-slow-muspratt? (accessed 31 Jan 2017).

Nhapi, I. and Gijzen, H.J. 2004. Wastewater management in Zimbabwe in the context of sustainability. Water Policy 6: 501–517.

New South Wales (NSW, Sydney, Australia). 2007. BioBanking: Biodiversity banking and offset scheme (Scheme Overview). Sydney: Department of Environment and Climate Change.

Novethic. September 2013. Overview of ESG rating agencies. www.novethic.com/fileadmin/user_upload/tx_ausynovethicetudes/pdf_complets/2013_overview_ESG_rating_agencies.pdf (or https://goo.gl/RuAoK8; accessed Nov. 5, 2017).

Onibokun, A.G. 1999. Managing the monster: Urban waste and governance in Africa. Ottawa: IDRC.

Osterwalder, A. and Pigneur, Y. 2010. Business model generation. Hoboken: John Wiley & Sons.

Otoo, M., Drechsel, P. and Hanjra, M.A. 2015. Business models and economic approaches for nutrient recovery from wastewater and fecal sludge. In: Drechsel, P., Qadir, M. and Whichelns, D. (eds). Wastewater: Economic asset in an urbanizing world. Springer: Dordrecht, Heidelberg, New York, London. pp. 247–268.

Otoo, M., Ryan, J.E.H. and Drechsel, P. 2012. Where there's muck there's brass: Waste as a resource and business opportunity. Handshake (IFC) 1:52–53.

Rao, K.C., Kvarnström, E., Di Mario, L. and Drechsel, P. 2016. Business models for fecal sludge management (Resource Recovery and Reuse Series 6). Colombo, Sri Lanka: International Water Management Institute (IWMI), CGIAR Research Program on Water, Land and Ecosystems (WLE).

Rao, K., Hanjra, M.A., Drechsel, P. and Danso, G. 2015. Business models and economic approaches supporting water reuse. In: Drechsel, P., Qadir, M. and Whichelns, D. (eds). Wastewater: Economic asset in an urbanizing world. Springer: Dordrecht, Heidelberg, New York, London. pp. 195–216.

Rouse, J., Rothenberger, S. and Zurbruegg, C. 2008. Marketing compost: A guide for compost producers in low- and middle-income countries. Duebendorf: EAWAG/SANDEC.

The Rockefeller Foundation. 2015. Incentive-based instruments for water management: Synthesis review. Pacific Institute, Foundation Center, Rockefeller Foundation. http://pacinst.org/wp-content/uploads/2016/02/issuelab_23697.pdf (accessed Nov. 5, 2017).

Toilet Board Coalition (TBC). 2016. Sanitation in the circular economy: Transformation to a commercially valuable, self-sustaining, biological system. www.toiletboard.org/media/17-Sanitation_in_the_Circular_Economy.pdf (accessed Nov. 5, 2017).

United Nations Environment Programme (UNEP). 2013. City-level decoupling: Urban resource flows and the governance of infrastructure transitions. A report of the working group on cities of the International Resource Panel. Nairobi, Kenya.

United Nations Environment Programme (UNEP). 2017. Resilience and resource efficiency in cities. UNEP: Nairobi, Kenya.

Ushijimaa, K., Funamizu, N., Takako Nabeshimab, Hijikata, N., Ito, R., Sou, M., Maïga, A.H. and Sintawardani, N. 2015. The postmodern sanitation: Agro-sanitation business model as a new policy. Water Policy 17 (2): 283–298.

Van Rooijen, D.J., Turral, H. and Biggs, T.W. 2005. Sponge city: Water balance of mega-city water use and wastewater use in Hyderabad, India. Irrigation and Drainage 54: 81–91.

Verstraete, W. and Cornel, P. 2014. Resource recovery: The processes, the products and the mind-sets. IWA's Water 21 (February issue): 50–51.

Villarroel Walker, R., Beck, M.B. and Hall, J.W. 2012. Water – and nutrient and energy – systems in urbanizing watersheds. Frontiers of Environmental Science and Engineering in China 6(5): 596–611.

Wang, X., McCarty, P.L., Liu, J., Ren, N.-Q., Lee, D.-J., Yu, H.-Q., Qian, Y. and Qu, J. 2015. Probabilistic evaluation of integrating resource recovery into wastewater treatment to improve environmental sustainability. Proceedings of the National Academy of Sciences of the United States of America 112(5): 1630–1635.

World Bank. 2016. Sustainable financing and policy models for municipal composting (Urban development series knowledge papers 24). Washington, DC: World Bank.

World Economic Forum (WEF). 2016. The Global Risks Report 2016 (11th Edition). Geneva: World Economic Forum.

World Health Organization (WHO). 2015. Sanitation safety planning manual for safe use and disposal of wastewater, greywater and excreta. Geneva: World Health Organization.

2. DEFINING AND ANALYZING RRR BUSINESS CASES AND MODELS

Miriam Otoo, Solomie Gebrezgabher, Pay Drechsel and Krishna C. Rao with support from Sudarshana Fernando, Surendra K. Pradhan, Munir A. Hanjra, Manzoor Qadir and Mirko Winkler

Defining and analyzing RRR business cases and models

The objective of this second chapter is to explain how the cases were selected and analyzed and how the authors derived the business models. The starting point was the identification of 'promising' empirical resource recovery and re-use (RRR) enterprises and governmental projects. In other words, the presented models are essentially not theoretical but have been tried – in most cases – in the context of low- or middle-income countries. 'Promising' in this context means that the cases, which informed the models, moved beyond a fully-subsidized pilot stage or were never designed as such, and aim at cost recovery or profit with potential for replication and scaling up. It does not mean that the selected cases are flawless, and there are many lessons to learn from their challenges. With some exceptions, every model presented in the catalogue derived its information from several empirical cases, which allowed extracting and flagging their strengths and opportunities as well as possible weaknesses and threats.

For the purposes of this catalogue, we define RRR **business cases** as:

> Business cases are entities, like enterprises, governmental projects or public-private partnerships (PPPs), that are engaged in the productive and safe recovery of water, nutrients, organic matter and energy from domestic and agro-industrial waste streams (including wastewater) by utilizing the recovery and/or re-use value of waste to generate revenue or recover costs in support of waste management and/or a healthy or more productive environment.

With the objective of showing scalable options, the presented cases are usually operating at community or city scale, i.e. household- or farm-based efforts in RRR have not been included.

Guided by Osterwalder and Pigneur (2010), a **business model** is defined in this catalogue as follows:

> A business model describes how a business creates, delivers and captures value; essentially the entire solution comprising the core aspects of the business – business process (e.g. technology), target customers, produce, infrastructure, organizational structures, trading practices, operational processes and policies, and the strategies it implements to achieve its objectives (be they for cost recovery, profit maximization, social impact, etc.).

Serving different target groups of this book, the presentation of empirical RRR business cases and models was challenging. While business schools might prefer detailed case studies, practitioners or decision makers will prefer a compact overview. The analysis of the cases and development of related business models does not come with the well-established base of literature and guidance that we are accustomed to from more conventional business sectors (George and Bock, 2011). Moreover, the assessment of both formal and informal RRR business cases requires significant groundwork to understand the factors that drive their success and likely sustainability, replicability and scalability barriers, particularities and opportunities. The analysis thus required the development of a suitable methodology, taking into consideration different types of readers, as well as both the micro- and macro-environment that cases operate in, while being flexible to cope with possible data gaps.

Assessment of RRR business cases

The business model concept

It is imperative that the concept of business modelling is clearly defined and more so in the context of resource recovery and re-use of waste. In the past two decades, the business model concept has become an increasingly pertinent concept in management theory and practice and has received substantial attention from academics and business practitioners (Magretta, 2002; Hedman and Kalling, 2003; Osterwalder et al., 2005; Shafer et al., 2005; Zott et al., 2011). Numerous definitions of the concept have been proposed although no particular terminology has so far been accepted in the domain of RRR (Bocken et al., 2014). In general, a business model describes how a business creates, delivers and captures value. In the RRR or eco-innovation context, the generic value proposition is the recovery of a useful resource from material which would otherwise be wasted. The related direct or indirect benefits can be savings, cost recovery, profits, welfare benefits, or an improved reputation (Beltramello et al., 2013; Hanjra et al., 2015).

In order to understand and operationalize the business model concept, Osterwalder and Pigneur (2010) described a business model as consisting of four core elements which can be disaggregated into nine building blocks that, taken together, create and deliver value. These four core elements describe a firm's:

1) *Value proposition* which distinguishes it from other competitors through the products and services it offers to meet its customers' needs;
2) *Customer segment(s)* the firm is targeting, the channels a firm uses to deliver its value proposition and the customer relationship strategy;
3) *Infrastructure* which contains the key activities, resources and the partnership network that are necessary to create value for the customer; and
4) *Financial aspects* (costs and revenues) which ultimately determine a firm's ability to capture value from its activities and break even or earn profit.

Based on these core elements, Osterwalder and Pigneur (2010) describe a business model through a canvas of nine components. There are different possibilities to extend or modify the canvas.[1] In this catalogue, we use an extended canvas by the same authors, which considers, with two additional components, possible positive and negative externalities (Figure 3). This extension is particularly important for the waste and sanitation sectors given related risks for human and environmental health, but also significant social benefits.

The business model canvas also provides many of the details needed to understand if a particular model could be viable in a different context than where it was used so far. However, the canvas does not provide information of the external business environment, like competition, regulations and the enabling business environment in general (see Chapter 19) which can be captured through RRR feasibility studies (Otoo et al., 2016).

Nomenclature and classification of RRR business models

Bocken et al. (2014) provide a structural approach towards business model categories in the domain of sustainability. The models described in this book fall in general under the archetype 'create value from waste' where we also find the concepts of 'closed loop' and 'circular economy'. However, while we could argue that water, energy and nutrients are indeed materials which are continually recycled through the production system, an alternative term could be 're-materialization', i.e. the innovative sourcing of materials from waste, creating entirely new products such as high-quality fertilizer or energy (Clinton and Whisnant, 2014). Business models within any of these structures could be categorized based on the type of waste, type of recovered resource, type of value proposition, partnership or ownership, or modes or scale of revenue generation (Evans et al., 2013).

These models can be very dynamic as with increasing environmental awareness and technical options, waste management approaches are continuously redesigned to optimize their value proposition. This includes their ability to capture so far missed RRR opportunities and values such as through carbon trading or biodiversity offset programs (Bocken at el al., 2013).

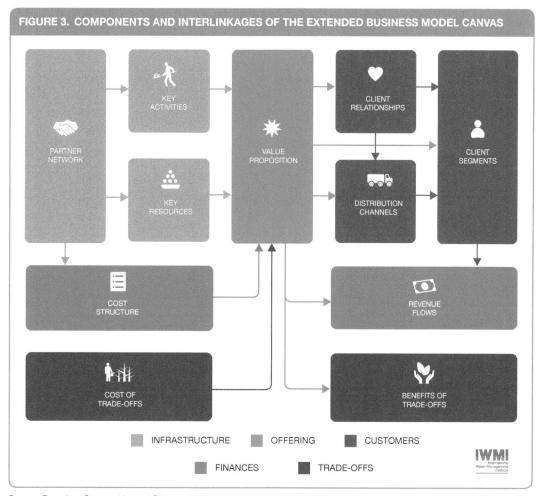

Source: Based on Osterwalder and Pigneur, 2010.

Given the paucity of a common terminology, the business model names and structure used in the following chapters were in large based on pragmatic reasoning and consent, not any particular academic discourse, and can be further developed and adapted as needed. Wastewater models might for example be distinguished by the agricultural end-product, energy projects by the business approach they use, nutrient cases by the way of waste valorization, while factors like the type of financing or PPP might allow other categories. The ideal categorization will thus vary between different readers of this catalogue and their objectives.

One possible classification of the models presented in this catalogue is to start with the main value-added product for reuse, means a) energy recovery, b) nutrient and organic matter recovery and c) water reuse. As any business model is driven by its objective, the next step considered in the decision tree could be the overall *business objective*, followed by the *business model* itself (Table 1).

CHAPTER 2. DEFINING BUSINESS CASES AND MODELS

TABLE 1. A POSSIBLE CATEGORIZATION OF THE PRESENTED RRR BUSINESS MODELS

VALUE-ADDED PRODUCT	SECTOR	OBJECTIVE	BUSINESS MODEL
Water Reuse	Public; Public/private	Cost recovery	Wastewater for greening the desert
			Enabling private sector investments in large-scale wastewater treatment
	Public/private	Welfare/profit maximization	Leapfrogging the value chain through aquaculture
	Public/ Informal Public/ private	Welfare maximization	Cities as their own downstream users
			Inter-sectoral water exchange
			Corporate social responsibility as driver of change
			Wastewater as a commodity driving change
			Farmers' innovation capacity as driver of change
Nutrient and organic matter Recovery	Public / private sector	Cost recovery	Subsidy-free community based composting
			Partially subsidized composting at district level
	Public and/ or Private Sector	Welfare/profit maximization	Large-scale composting for revenue generation
			Compost production for sustainable sanitation service delivery
		Cost savings	Nutrient recovery from own agro-industrial waste
			Phosphorus recovery from wastewater at scale
	Private and/or Informal sector	Cost savings	Outsourcing fecal sludge treatment to the farm
Energy Recovery	Public Sector	Cost recovery	Power from municipal solid waste
	Private Sector	Profit maximization	Briquettes from agro-waste or municipal solid waste
			Bio-ethanol and chemical products from agro- and agro-industrial waste
		Profit maximization/ Cost Savings	Combined heat and power from agro-industrial waste for on- and off-site use
		Profit and Welfare maximization	Power from agro waste
			Combined heat and power from agro-industrial waste for on- and off-site use
		Cost savings/Welfare maximization	Biogas from fecal sludge and kitchen waste
			Power from manure

Criteria and process for the selection of analyzed RRR business cases

The cases presented in this catalogue were selected in different steps each using different criteria. The main objective of the exercise was to understand drivers of success and sustainability strategies; and based on the analysis of different related cases, to extract/construct generic business models, which summarize innovative and promising components of these businesses with potential for scaling up and out in (other) low- and middle-income settings including emerging economies. Following an initial screening of about 150 cases suggested by the literature, media and experts, over 60 empirical re-use cases were analyzed in detail of which 47 are presented here. As some operate in different locations, the actual number of cases is larger. These selected cases allowed for the development of 24 generic business models, which are also presented.

For the first selection round, the cases had to provide evidence, as much as possible, of the following:

i. Operation in Africa, Asia or Latin America, with special consideration for wastewater re-use cases in the Middle East and Northern Africa (MENA) regions;
ii. Conversion of waste into one or more of the following outputs: nutrients, biomass, energy or water for agriculture (i.e. waste becomes an asset and compensates for resources in short supply);
iii. Generation of revenues from RRR or supporting, at least, cost savings;
iv. Transactions (will) support cost recovery and ideally also parts of the sanitation chain financially;
v. Replicability in low- and/or middle-income countries at scale, i.e. not only at the level of one household or farm;
vi. Distinct creation of social and/or environmental benefits; and
vii. Likelihood of data accessibility.

The empirical investigation of the preselected 60+ RRR businesses was based on a template (Box 4) with questions tailored to the different waste streams and recovered resources. Information was obtained, wherever possible, through local data collection by project staff or consultants, i.e. in direct interaction with the businesses, or remotely via email, explaining the purpose and background of the study and incentives[2] for collaboration. Depending on the sensitivity of the case/business entity, and/or its responsiveness, in-depth literature surveys combined with expert consultations were also employed.

Box 4. Business case assessment template

1) **Context and background:** Describes the wider perspective on the history and development of the business. It also describes the geographical location and the government policy on re-use activities within which the business is operating. Most of the information contained in this section is gathered from business entities or secondary literature.
2) **Market environment:** Describes the needs in the market that drive the existence and development of the business, i.e. it describes what the business does and how it serves market needs. The assessment of the market environment was also supported by a literature review.
3) **Macro-economic environment:** Discusses briefly the global or national market conditions or economic infrastructures that enable or represent a supportive factor or a constraint to the business. Relevant information on the macro-economic environment was gathered from country policy reviews and other relevant literature.
4) **Business model description:** Describes the RRR business case by applying the business model canvas as illustrated in Figure 3. This section discusses the linkages between the elements of the business model and focuses on answering: why the business model works, the core element for its functioning and the essence of the business model. Most of the information was gathered from business entities.

Where identified value propositions are analysed separately, associated descriptions use the same background colour within the canvas. In the case where characteristics relate to several or all value propositions, a color coding, different from those of the value propositions is used.

5) **Value chain and position in the chain:** Describes the value chain in which the enterprise positions itself. This section applies Porter's five forces methodology (Porter, 1985) to describe the critical relationships with suppliers, partners, customers and other value chain actors.
6) **Institutional environment:** Describes institutional responsibilities and any legal or regulatory factors in the respective country that support or represent a constraint to the business.
7) **Technology and processes:** Describe the technology or process used by the business. The status of the technology as to whether it has been commercially proven, its local appropriateness and risks associated with the technology are also examined.
8) **Funding and financial outlook:** Describes the source of financing for the enterprise. Where data are available, the key capital and operational cost, revenue streams and cash flow statements are presented.
9) **Socio-economic, health and environmental impact:** Discusses not only the socio-economic impact of the business in terms of, for example, number of jobs created, health and environmental benefits, but also experienced or possible negative externalities.
10) **Scalability and replicability potential:** Discusses the potential for scaling up/out the business in other geographical locations or settings.
11) **SWOT analysis:** This section summarizes the model looking at its strengths, weaknesses, opportunities and threats.
12) **Contributors, links and references:** Acknowledge local and international experts and business staff who assisted in data gathering, web links and literature used to compile the case.

The collected data were analyzed using a combination of the multicriteria approach, business model canvas and strengths, weaknesses, opportunities and threats (SWOT) analysis. Depending on data availability and time, the amount of gathered data/information varied. In several cases, financial data were, for example, only available under the condition of non-disclosure, or insufficient for any financial analysis or representative presentation.

Development of RRR business models

The key objective for the assessment of existing RRR business cases was to understand their success, drivers, challenges and sustainability strategies and, based on these cases, construct generic business models with the potential for scaling up and out in other settings. Thus, instead of building theoretical RRR business models, the presented models are based on existing cases, or in other words, each model comes with several application examples. Only a few models were derived from just one case and only one was formulated on promising developments without a particular empirical case. This concerns the potential of corporate social responsibility for addressing unsafe wastewater use in the informal irrigation sector where the priority value proposition would be risk reduction.

The Business Model Canvas (BMC) was the main tool used for the development of RRR business models, based on the 11 fundamental building blocks (see Fig 3). The strength of the BMC lies in its simplicity and ability to provide a holistic overview of the essential components of the business model that the firm leverages. The BMC is best used as a pre-business planning activity to map out the various options a business has for adopting a particular business strategy. In addition, the

BMC allows for stepping away from the details of technological innovations and focusing on the best-fit business organizational form that will support successful implementation and adoption of the technological innovation. The BMC can be used to map existing models (such as the presented cases in this catalogue) and develop new models as adaptations to existing ones or entirely different ones. However, as mentioned above, the canvas only addresses parts of a business case, and requires additional information.

The presented business models draw strongly on the analyzed business cases, supported by additional information from related cases in the literature and interviews. Each model represents an optimized generic business model building on the success factors of its supporting cases, with different degrees of innovation, while incorporating strategies that address identified or likely shortcomings. These relate in particular to the analysis of possible health risks (IFC, 2009) to identify likely hot spots for risk monitoring (WHO, 2015).

The business model description follows, like the case description, a standard template (Box 5) with exception of some wastewater models, which are based on only one case, and follow a hybrid of both templates. Compared with the business cases, some additional components of the model presentation require further explanation. This concerns, in particular, the assessment of potential risks and risk mitigation measures and the summary assessment based on selected criteria.

Box 5. Business model description template

Business value chain: Describes the basic concept behind the business, explaining the different partners and their roles, the organizational structure (public, private etc.), the overall business process flow and value chain, the technology and financial arrangements.

Business model description: Describes the linkages between the elements of the business model canvas (Figure 3) and focuses on answering: why the business model works, the core element for its functioning and the essence of the business model, including information on partners and financial aspects to the extent available. Where identified value propositions are analysed separately, associated descriptions use the same background colour within the canvas. In the case where characteristics relate to several or all value propositions, a color coding, different from those of the value propositions is used.

Alternative model scenarios: Describe the option for alternate models derived from the parent model.

Potential risks and mitigation measures: Describe the potential risks associated with the business model and related mitigation measures. The risks considered include market, competition, technology performance, political and regulatory risks, social equity, and environmental and health risks.

Business performance: Summarizes the potential for scaling up/out or for replicating the business in other geographical locations or settings. It also describes in general how the business model has been appraised based on five performance criteria (cost recovery/profitability, scalability, replicability, social impact and environmental impact). It provides an overview of the conditions under which the business model should be undertaken and which factors, such as those related to land, investment and finance, should be given particular consideration.

Business risks and risk mitigation

An optimized business model will seek to minimize business risks. These can include but are not limited to: a) market risks, b) competition risk in both input and output markets, c) technology performance risk, d) political and regulatory risks and e) the risk of undermining social equity. Thus, the business models presented here tried to capture possible risks based on the analysis of their supporting cases. As business-related risks are context-specific, the risk section can only touch on the possible complexity. For market risks, the key factors considered were, e.g. changes in supply and demand, as well as likely sources of competition and ease of entry into the market, which depends again on location-specific market structures. Technological performance risks are related to whether the technology is commercially proven and if there are anticipated challenges with repair and maintenance from a developing country perspective. As fledgling businesses and their sustainability are largely influenced by their enabling environment, political, regulatory and financial instruments to rectify, for example, market failures (e.g. price subsidies), are briefly addressed. However, given its crucial role, Chapter 19 provides more details and examples on how regulatory mechanisms and finance instruments can shape an enabling environment for RRR. Finally, social equity related risks were assessed in view of poverty alleviation (employment) and gender inclusiveness.

To illustrate the qualitative assessment steps and criteria used, further details for the (i) health and environmental risks and (ii) social equity risks are provided in the following:

(i) Health and environmental risk assessment

Given that RRR businesses deal with potentially harmful source materials, special attention was given to environmental and health risks. Although the 'models' imply, per definition, full compliance with safety measures, it is important to flag critical control points and common mitigation measures. Given the generic nature of the models for possible application in different countries, the risk assessment had to remain generic. In the instance of a model being implemented, a concrete and site-specific risk assessment will be needed, taking into consideration the actual technology, scale of the enterprise and possible risk factors in the environment, such as groundwater proximity (Otoo et al., 2016; Winkler et al., 2017).

The risk assessment drew from the studied cases although it was not applied to the same extent to the cases themselves, which generally followed local safety standards and regulations. Reported or observed deviations were analyzed if they represented generic shortcomings to be captured for the related models. Some of the presented business models have submodels in which, for example, an alternative institutional set up was suggested. In such cases the assessment was conducted for the generic model. However, if submodels implied, for instance, a change in technology or inputs and outputs possible implications were marked. Following the structure of the catalogue each of the main categories – (1) energy, (2) nutrient/organic matter and (3) wastewater – were analyzed for key exposure groups and risk pathways. Models on water and nutrient recovery, for example, usually have farmers as users of the generated product, while the possible risk groups continue along the food chain. The situation is obviously different for energy models with biogas, electricity or briquettes as the final product. Based on this analysis, a generic risk assessment template was developed following the source-pathway-receptor model.

The four key exposure groups are shown in Table 2.

TABLE 2. THE FOUR EXPOSURE GROUPS

RISK TYPE	EXPOSURE GROUPS
1. Occupational risk on site	Workers, employees
2. Occupational risk off site	Farmers/users of RRR products
3. Consumption risk	End users
4. Social environment	Communities near treatment facilities

Table 3 shows typical pathways linking exposure groups with potential risks. In some countries, natural resources themselves are considered as receptors (e.g. water resources in the United Kingdom). In this analysis, air, water and soil were mainly considered as pathways rather than receptors. Table 2 also presents common mitigation measures that can be put in place to prevent likely risks.

TABLE 3. EXPOSURE PATHWAYS AND MITIGATION MEASURES

EXPOSURE PATHWAY	DESCRIPTION	TYPICAL MITIGATION MEASURES
Direct contact	Handling, sorting, mixing, collecting, transportation	Protective wear – boots, gloves, coats and overalls, and good hygiene
Insects	Breeding sites for carriers and vectors	Insect spraying, cleaning, netting
Air	Aerosols, particulates and gases	Protective wear – goggles and masks, ear plugs, wind barriers (e.g. tree belts), coverage of waste piles
Water and soil	Effluent, leachate and leakages	Avoid untreated discharge, support e.g. phytoremediation
Food	Insufficiently treated waste products used in farming	On-farm risk (contact) reduction, crop restrictions, produce washing and/or boiling

The level of risk was categorized as low, medium or high considering: nature of exposure (direct, indirect, external, internal, etc.), intensity of exposure (severity and probability), and required effort of mitigation (**simple** like via safety gear; **advanced**, e.g. via emission reduction; **substantial,** e.g. via addition treatment). Emphasis is placed on likely hazards, not all theoretically possible hazards:

(a) Direct contact

Low risk	Contact with hand and foot during operations possible (or use of less hazardous waste). Contact can be easily avoided by employing simple risk mitigation measures.
Medium risk	Contact with skin during operations likely. This can be easily avoided by employing more advanced mitigation measures.
High risk	Contact with skin during operations is difficult to avoid, unless by applying substantial mitigation measures.

(b) Insects (flies, mosquitoes, etc.)

Low	Process creates unfavourable conditions for breeding and waste materials have low pathogen levels. Risks can be avoided by employing simple mitigation measures.
Medium	Process creates favourable conditions for breeding or involves materials (feces) with high pathogen loads, but risks can be avoided by employing advanced mitigation measures.
High	Process creates favourable conditions for breeding and/or deals with high pathogen loads which are difficult to avoid unless by employing substantial mitigation measures.

(c) Air (aerosols, dust, particulates, gases, machinery sound, etc.)

Low	Low emission and noise which can be avoided by employing simple mitigation measures.
Medium	Significant emission and/or noise which can be avoided by employing advanced mitigation measures.
High	Significant emissions and/or noise which are difficult to avoid unless by employing substantial mitigation measures.

(d) Water and soil (leachate, leakages, etc.)

Low	Low leachate production or only partially treated effluent potentially released to the environment which can be avoided by employing simple mitigation measures.
Medium	High leachate production or partially treated effluent potentially released to the environment. This can only be avoided by employing advanced mitigation measures.
High	High leachate production or untreated effluent potentially released to the environment and it can only be avoided by employing substantial mitigation measures.

(e) Food chain

Low	Low risk of microbiological contamination which can be avoided by employing simple mitigation measures such as produce washing, smoking or boiling.
Medium	Microbiological contamination which can be avoided by employing mitigation measures that require more efforts such as investments in drip kits for irrigation and compliance monitoring.
High	Chemical contamination (e.g. heavy metals) which is possible but difficult to mitigate, unless via substantial mitigation measures, such as further waste sorting or additional treatment steps.

For more details on exposure pathways, risk evidence and mitigation, please see Stenström et al. (2011) and WHO (2015), and the application example to RRR business models by Winkler et al. (2017).

The overall risk assessment for each model used the following scale and risk mitigation symbols:

(ii) Social equity related risks

Equal employment opportunities and other gender-specific benefits or burdens were analyzed as far as possible for each business model. The assessment of equality considered in particular how far either men or women might be (dis)advantaged in engaging in the waste valorization process, as an entrepreneur or worker, or as a direct beneficiary of the resulting products. The assessment was qualitative and considered positive implications for (a) common gender roles, like time spent for water or fuel collection; and (b) comfort at home/workspace through the provision of improved services or clean energy (clean air, studying after sunset/girl literacy). The assessment also considered gender-specific disadvantages related to (i) the recommended technology, (ii) business-related job opportunities as well as (iii) gender-specific occupational health risks. Each analyzed model displayed between 0–3 factors which were given equal weightage. The most common factors providing advantages for women relate to energy production for the benefit of households, allowing women to save time for collecting external fuel, as well as a healthier (fire- and smoke-free) working environment. The most common factor to advantage men was related to gender-specific labor roles, like construction work or truck driving. Particular advantages for one group do, however, not imply a direct risk or disadvantage for the other. The judgement, which remains without local context tentative and preliminary, has been summarized in a pictorial balance beam reflecting possible gender specific dis/advantages.

WOMEN ADVANTAGE — MEN ADVANTAGE

Performance potential

For the last part of the business model template, the suggested models were evaluated for their performance potential, expecting a triple bottom line based on the following indicators/criteria: a) profitability/cost recovery, b) social impact, c) environmental impact, d) scalability and replicability and e) innovation. Each criterion was evaluated on a three-level scale based on the average score of a three-level ranking of the constituent parameters (Table 4). The ranking of the parameters and the resulting ranking of the indicators was based on a combination of quantitative and qualitative data sourced from empirical cases and application of the Delphi method[3], respectively.

TABLE 4. GUIDELINES FOR RANKING OF BUSINESS MODELS

INDICATORS	GUIDING QUESTIONS	PARAMETERS	SCORE
Profitability/ cost recovery	What is the level of operational profits/ cost recovery achieved by the business model on an annual basis?	Loss making	1
		Break-even	2
		Profit	3
	How many revenue streams does the business model depend on and how strong are these revenue line items?	One strong revenue source	1
		Two or more revenue sources with one strong revenue line	2
		Two or more revenue sources with two strong revenue lines	3

TABLE 4. CONTINUED

	How many of these factors represent a risk of increased costs to the business model? Factors are: 1) high worker and managerial skill requirements, 2) diverse customer base, 3) diverse products, 4) need for R&D and 5) self-distribution of product to end customer	More than 3 factors applicable	1
		2–3 factors applicable	2
		0–1 factor applicable	3
Social impact	How many jobs are created/provided by the business model compared with the range of all the business cases within the same section (energy or nutrients or water)?	Low	1
		Medium	2
		High	3
	Number of people with increased positive health impact from the business model compared with the range of all the business cases within the same section (energy or nutrients or water).	Low	1
		Medium	2
		High	3
	How many of these factors does the business model have an improved/increased positive impact on? Factors are: 1) water security, 2) food security, 3) energy security, 4) improved living standards, 5) reduced governmental costs for waste management services (sanitation), health services and 6) gender	Meets 0–2 factors	1
		Meets 2–4 factors	2
		Meets more than 4 factors	3
Environmental impact	What quantity of waste is being processed/re-used compared with the range of all the business cases within the same section (energy or nutrients or water)?	Low	1
		Medium	2
		High	3
	How many of these factors does the business model have an improved/increased positive impact on? Factors are: 1) health of waterbodies, 2) reduced GHG emissions, 3) soil fertility, 4) renewable source/raw material and 5) reduced deforestation	Meets 0–1 factor	1
		Meets 2–3 factors	2
		Meets more than 3 factors	3
Scalability and replicability	How many of these factors limit the replication potential of the business model elsewhere? Factors are: 1) new technology, 2) policies and regulations, 3) strong institutional capacity, 4) specific waste availability 5) market demand and 6) ambiguity of product acceptance	Meets more than 4 factors	1
		Meets 3–4 factors	2
		Meets 0–2 factors	3
	What is the ease of scaling the business model vertically and horizontally?	Low potential for vertical AND horizontal scaling	1
		High potential for either vertical OR horizontal scaling	2
		High potential for BOTH vertical and horizontal scaling	3
	How easy is it to finance the business model elsewhere?	Investment is HIGH and financing is UNIQUE	1
		Investment is HIGH and financing is COMMON	2
		Investment is LOW and financing is UNIQUE	2
		Investment is LOW and financing is COMMON	3

Innovation	How innovative is the technology or process?	Known technology or process	1
		Relatively new to developing countries (technology transfer)	2
		New to the world	3
	How innovative are the partnership arrangements?	No partnerships required	1
		Partnerships within the same sector	2
		Partnerships cross-cutting different sectors (PPP, R&D, finance)	3
	How innovative is the product or value proposition?	Standard product and value proposition	1
		Relatively new product or value proposition	2
		New to the world	3

The overall appraisal of the indicators for each business model is represented in a radar diagram (Figure 4). It is important to note that this is an overview assessment and any actual implementation of any RRR business model will require a context-specific and more detailed ex ante feasibility assessment (Otoo et al., 2016).

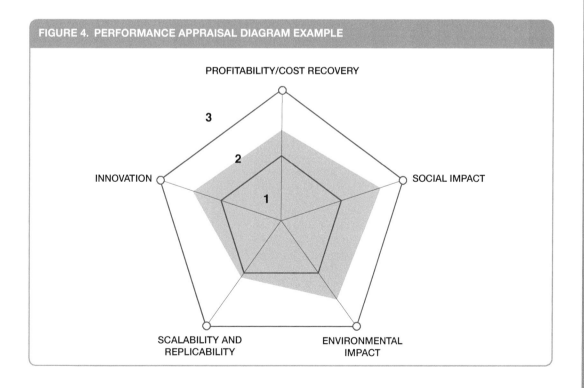

FIGURE 4. PERFORMANCE APPRAISAL DIAGRAM EXAMPLE

Limitations

The information provided in the case studies refers to the **time of their individual assessment** between 2012 and 2017. The authors regret any possible error or missed update. Case descriptions are detailed to serve students as case studies, probably too detailed for practitioners and investors, while there are still many other criteria to assess business cases and describe business models, which we were not able to capture. This concerns for example the history and timeline of the cases, the personal engagement, contacts and investments of the entrepreneurs, their experiences with seeking an appropriate business partner and lessons learned *vis-à-vis* their local regulatory, financial and administrative challenges, or the difference between the official and *de facto* enabling environment. Other limitations faced, concern the availability or accessibility of (in particular financial) data and common lack of quantitative impact assessments. Data access from private enterprises was challenging. In several cases, they were unwilling to provide financials or information on the technology. In such cases, the authors had to rely on secondary sources with their limitations. While the private sector had its reasons to withhold data, accessing data from the public sector came with its own challenges related to their availability, like in many parts of Africa. Often only older data were available, if any. As business operations are dynamic, data of the presented cases will change over time. Finally, the investment ranges stated for the business models are largely based on the analysed case studies, and could be larger.

References and further readings

Beltramello, A., Haie-Fayle, L. and Pilat, D. 2013. Why new business models matter for green growth (OECD Green Growth Papers, 2013-01). Paris: OECD Publishing.

Bocken, N.M.P., Short, S.W., Rana, P. and Evans, S. 2013. A value-mapping tool for sustainable business modelling. *Corporate Governance* 13(5): 482–497.

Bocken, N.M.P., Short, S.W., Rana, P. and Evans, S. 2014. Review: A literature and practice review to develop sustainable business model archetypes. *Journal of Cleaner Production* 65: 42–56.

Clinton, L. and Whisnant, R. 2014. Model behavior: 20 business model innovations for sustainability. http://10458-presscdn-0-33.pagely.netdna-cdn.com/wp-content/uploads/2016/07/model_behavior_20_business_model_innovations_for_sustainability.pdf (accessed Nov. 5, 2017).

Evans, A., Otoo, M., Drechsel, P. and Danso, G. 2013. Developing typologies for resource recovery businesses. *Urban Agriculture Magazine* 26: 24–30.

George, G. and Bock, A.J. 2011. The business model in practice and its implications for entrepreneurship research. *Entrepreneurship Theory and Practice* 35(1): 83–111.

Hanjra, M.A., Drechsel, P., Mateo-Sagasta, J., Otoo, M. and Hernandez-Sancho, F. 2015. Assessing the finance and economics of resource recovery and reuse solutions across scales. In Drechsel, P., Qadir, M., Wichelns, D. (Eds.). Wastewater: Economic asset in an urbanizing world. Dordrecht, Netherlands: Springer. pp.113-136.

Hedman, J. and Kalling, T. 2003. The business model concept: Theoretical underpinnings and empirical illustrations. *European Journal of Information Systems* 12: 49–59.

International Finance Corporation (IFC). 2009. Introduction of health impact assessment. Washington DC; International Finance Corporation.

Magretta, J. 2002. Why business models matter. *Harvard Business Review* 80(5): 86–92.

Osterwalder, A. and Pigneur, Y. 2010. *Business model generation*. Hoboken: John Wiley & Sons.

Osterwalder A., Pigneur, Y. and Tucci, C.L. 2005. Clarifying business models: Origins, present, and future of the concept. *Communications of the Association for Information Systems (AIS)* 16(1): 1–25.

Otoo, M., Drechsel, P., Danso, G., Gebrezgabher, S., Rao, K. and Madurangi, G. 2016. *Testing the implementation potential of resource recovery and reuse business models: from baseline surveys to feasibility studies and business plans* (Resource Recovery and Reuse Series 10). Colombo, Sri Lanka: International Water Management Institute (IWMI); CGIAR Research Program on Water, Land and Ecosystems (WLE).

Porter, M.E. 1985. *Competitive advantage*. New York: Free Press.

Shafer, S., Smith, H. and Linder, J. 2005. The power of business models. *Business Horizons* 48(3): 199.

Stenström, T.A., Seidu, R., Ekane, N. and Zurbrügg, C. 2011. *Microbial exposure and health assessments in sanitation technologies and systems* (EcoSanRes Series, 2011-1). www.ecosanres.org/pdf_files/Microbial_Exposure_&_Health_Assessments_in_Sanitation_Technologies_&_Systems.pdf (accessed Nov. 5, 2017).

Winkler, M.S., Fuhrimann, S., Pham-Duc, P., Cissé, G., Utzinger, J. and Nguyen-Viet, H. 2017. Assessing potential health impacts of waste recovery and reuse business models in Hanoi, Vietnam. *International Journal of Public Health* 62(Suppl 1): 7–16.

World Health Organization (WHO). 2015. *Sanitation safety planning manual for safe use and disposal of wastewater, greywater and excreta.* Geneva: WHO.

Zott, C., Amit, R. and Massa, L. 2011. The business model: Recent developments and future research. *Journal of Management 37(4)*: 1019–1042.

Notes

1 See for example: www.ppplab.org/wordpress/wp-content/uploads/2016/04/PPPCanvas_Example.pdf (accessed November 5, 2017).
2 Exposure of success to the donor community, inclusion in this catalogue, participation in follow-up conferences, feasibility studies for their models in different continents.
3 https://en.wikipedia.org/wiki/Delphi_method (accessed November 5, 2017).

SECTION II – ENERGY RECOVERY FROM ORGANIC WASTE

Edited by Krishna C. Rao and Solomie Gebrezgabher

Recovering energy from waste: An overview of presented business cases and models

Access to affordable and sustainable energy is key to economic prosperity and sustainable development in developing countries. Energy plays a critical role not only in ensuring quality of life at individual or household level but also as one of the factors of production whose cost affects other goods and services (Amigun et al., 2008). Access to energy or the lack of it affects all facets of development: social, economic and environmental aspects. It is the key to sustaining the livelihood of the poor as well as ensuring industrial development of a country. Energy is crucial for achieving almost all of the Sustainable Development Goals (SDGs), from eradication of poverty through advancements in health, education, water supply and industrialization, to combating climate change (UN, 2016). SDG 7 is dedicated to the access to affordable, reliable, sustainable and modern energy for all, with target 7.2 calling for a substantial increase of the share of renewable energy including power derived from solid and liquid biofuels, biogas and waste.

With the aim of achieving a more sustainable natural environment while providing reliable and affordable energy to different sectors of the economy, interest in alternative sources of energy as a means of reducing dependence on fossil fuels has grown. Studies have shown that energy demand will increase during this century by a factor of two or three while about 88% of this demand is met by fossil fuels (IEA, 2006). The negative effects of the conventional energy sources coupled with the limited capacity of current energy infrastructure and the increase in energy demand have spurred interest in alternative sources of energy which are environment friendly and renewable.

Around 3 billion people cook and heat their homes using solid fuels (i.e. wood, charcoal, coal, dung, crop wastes) on open fires or traditional stoves. Such inefficient cooking and heating practices produce high levels of household (indoor) air pollution which includes a range of health-damaging pollutants such as fine particles and carbon monoxide. About 4.3 million people a year die from the exposure to household air pollution (WHO, 2016).

Under increasing deforestation, the global waste to energy market was valued at USD 24 billion in 2014 and it is expected to reach USD 36 billion by 2020 – a growth rate of 7.5% (Figure 5). Waste-to-energy is a waste treatment process to generate energy in the form of electricity, heat or fuel from both organic and inorganic waste sources. In this book, the focus is only on cases and models targeting energy generation from biomass (organic waste). While recovering energy from organic waste streams is essential to ensure energy security and sustainable development, waste-to-energy solutions still face numerous barriers including high investment cost, inadequate policy support and insufficient revenue generation due to limited experience with business or cost recovery models. This section addresses this last void, while opportunities and barriers in the enabling environment are discussed in Chapter 19.

In this section of the catalogue, waste-to-energy conversion process in all the business cases and models can be broadly presented as in Figure 6. The energy recovery models and cases use one of the waste streams (agro-waste, agro-industrial waste and effluent, livestock waste, fecal sludge and organic fraction of municipal solid waste) to produce energy products in solid (briquette), liquid (bio-fuel/ethanol) and gaseous (producer gas and biogas) forms. These energy products are used to generate heat, electricity or fuel for transport.

The energy recovery chapter describes in total 9 business models derived from 19 business cases, and these 9 business models can be broadly classified into 4 categories:

- Production of Solid Fuels from Waste.
- Sustainable and Renewable Power Generation.
- Institutional (In-house) Biogas for Energy Savings.
- Emerging Technologies for Bio-fuel Production from Agro-waste.

Energy products made from waste can be in one of the three physical forms and a relatively straightforward process is to convert waste into solid fuel by transforming organic fraction of municipal solid waste, market waste and agricultural residues into briquette fuel. This is an emerging scalable model in Sub-Saharan Africa particularly in East Africa and there are similar observations in Asia (**Business models 1 and 2: Briquettes from agro-waste** and **Briquettes from municipal solid waste**). Briquettes are a form of solid fuel produced by compacting loose biomass residues into solid blocks that can be burned for heat energy and can substitute traditional biomass based energy sources such as charcoal and firewood for domestic or institutional cooking as well as for industrial heating processes. The business cases highlight different strategies and processes such as simple technology for ease of maintenance, research and development (R&D) for right combination of different agro-waste to produce high calorific value briquettes as is the case in Kampala Jellitone Suppliers; franchise models to scale operations as is the case in Eco-Fuel Africa in Uganda; and implementation of a public-private partnership (PPP) to get contracts of waste collection as is the case in COOCEN, which is a women cooperative in Rwanda.

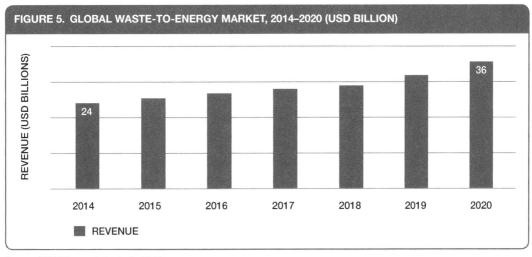

FIGURE 5. GLOBAL WASTE-TO-ENERGY MARKET, 2014–2020 (USD BILLION)

Source: ZION Research Analysis, 2016.

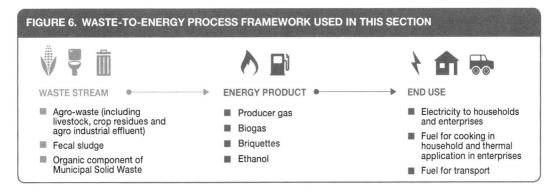

FIGURE 6. WASTE-TO-ENERGY PROCESS FRAMEWORK USED IN THIS SECTION

One of the most common waste-to-energy solutions that is widely implemented in developing countries is production of biogas from organic waste. Biogas can be produced from nearly all kind of biological feedstock – various organic waste streams including human waste (Holm-Nielsen et al., 2009). **Business models 3 and 4: Biogas from fecal sludge at community level** and **Biogas from kitchen waste** present institutional biogas models for energy savings. The business case examples are from India, Nepal, the Philippines, Rwanda and Kenya which highlight successful partnership with local authorities, non-governmental organizations and communities for successful implementation.

In this section of the catalogue, biogas production is demonstrated at different scales with the lowest scale of biogas production at the institutional level and large-scale production at industrial level. As the target stakeholder is industries in the later, the scale of waste generated is higher resulting in higher gas production and thus enabling to generate electricity from biogas. This is the case for livestock industry which generates biogas from manure for onsite use (**Business Model 5: Power from manure**). The case examples presented demonstrate rural electrification models from livestock waste along with innovative financing mechanisms of using carbon credits to invest in the technology. For example, Sadia, a company from Brazil, processes meat, and in order to mitigate the social and environmental impacts associated with livestock production systems, it has installed bio-digesters on the farms within its supply chain on a Build, Operate and Transfer (BOT) basis. Sadia uses carbon credit method to finance biogas systems on the farms that supply meat to the processing factory while taking the responsibility of registration of the project as a CDM and the management of the carbon credit revenues.

In addition to business models that highlight power generated from manure, there are also other business models that use agro-waste or municipal solid waste (MSW) to generate electricity (**Business Models 6–8: Power from agro-waste; Power from municipal solid waste (MSW); and Combined heat and power from agro-industrial waste for on- and off-site use).** Agro-processing industries, such as sugar and palm oil factories, and slaughterhouses in low-income countries, are diversifying into creating by-product value addition through co-generation and bioethanol production. The energy production technologies are either owned and operated by the factory or are installed by an external private entity on a Build, Own, Operate, Transfer (BOOT) model. These business models allow agro-industries to be self-sufficient in energy while securing additional revenue streams by exporting excess electricity to the national grid and trading carbon credits. The cases here also highlight social enterprise models for rural electrification.

In this section of the Resource Recovery and Reuse (RRR) catalogue, while the focus is on innovative energy recovery business models with relatively simple technology, there are also few business models and cases which use more sophisticated and high investment cost energy solutions. There is limited focus on advanced technologies to produce biogas, syngas and liquid fuels except in the case of **Business model 9: Bio-ethanol and chemical products from agro and agro-industrial waste** which highlights production of biofuel from cellulosic sources such as agro-waste produced from mills processing cassava, rice, wheat, coffee and so on. The model also covers processing of vinasse waste generated during ethanol production. Vinasse can be used to produce an organic binder (lignosulfonates) which has numerous applications across many industries.

Further business cases and models where energy generation plays a role are presented in the section **on wastewater treatment for reuse**.

Waste-to-energy business cases and models described in this section demonstrate improved economic viability from RRR to provide not only environmentally beneficial solutions along with increased energy access to governments, donors, entrepreneurs and non-government organizations in developing countries but also offer larger socio-economic benefits from safe waste management. By adopting these solutions, they not only help meet the ever-increasing demand for energy but also pull out millions of underserved communities from extreme poverty in an environmentally responsible manner. For increased energy security and to meet SDG 7 indicators, there is a need to triple investments in sustainable energy infrastructure per year from USD 400 billion to USD 1.25 trillion by 2030 (UN, 2016) and waste-to-energy RRR business models and cases provide a means to achieve not only SDG 7 indicators, but also, for example, SDG 12.5 to substantially reduce waste generation through prevention, reduction, recycling and reuse.

References and further readings

Amigun, B., Sigamony, R. and Von Blottnitz, H. 2008. Commercialization of biofuel industry in Africa: A review. Renewable and Sustainable Energy Reviews, 12: 690–711.

Holm-Nielsen, J.B., Al Seadi, T., Oleskowicz-Popiel, P. 2009. The future of anaerobic digestion and biogas utilization. Bioresource Technology, 100: 5478–5484.

United Nations (UN). 2016. Affordable and clean energy: Why it matters. http://www.un.org/sustainable development/wp-content/uploads/2016/08/7_Why-it-Matters_Goal-7_CleanEnergy_2p.pdf (accessed 6 Nov. 2017).

United Nations (UN). 2016. Progress towards the sustainable development goals. Economic and Social Council: Report of the Secretary General (E/2016/75).

World Health Organization (WHO). 2016. Burning opportunity: Clean household energy for health, sustainable development, and wellbeing of women and children. Geneva: WHO.

Zion. 2016. Waste to Energy (Thermal and Biological Technology) Market: Global Industry Perspective, Comprehensive Analysis, Size, Share, Growth, Segment, Trends and Forecast, 2014–2020. http://www.marketresearchstore.com/report/waste-to-energy-market-z47278 (accessed 20 Feb. 2017).

3. BUSINESS MODELS FOR SOLID FUEL PRODUCTION FROM WASTE

Introduction

Urban and rural populations in developing countries predominantly depend on traditional biomass fuels such as charcoal and firewood for cooking due to lack of affordability and access to modern fuels. Despite more than a decade of work to reduce domestic air pollution sources, progress toward universal access to clean cooking fuels remains far too slow. According to the World Health Organization (WHO, 2016), almost 3.1 billion people still rely on polluting, inefficient energy systems, such as biomass, coal or kerosene, to meet their daily cooking needs – a number virtually unchanged over the past decade. The same applies to heating and lighting. For instance, almost half of all African households across the 25 countries surveyed by WHO rely primarily upon highly-polluting kerosene lamps, compared to about 30% of households surveyed in South-East Asia. Women and girls bear the largest health burden not only from domestic pollution sources, but often also from related fuel-gathering tasks. For instance, available survey data from 13 countries showed that girls in sub-Saharan African homes with polluting cook stoves spend about 18 hours weekly collecting fuel or water, while boys spend 15 hours. In homes mainly using cleaner stoves and fuels, girls spend only 5 hours weekly collecting fuel or water, and boys just 2 hours (WHO, 2016). There are also environmental impacts, such as deforestation and climate change, associated with the consumption of charcoal and firewood due to the unsustainable nature of their production and use.

Overdependence on firewood has resulted in reduced availability and consequently necessitates the efficient utilization of agricultural residues and municipal solid waste as a source of heating and cooking fuel by transforming them into alternative fuel products called briquettes. The briquette business model aims to tap into this potentially vast market by providing urban and rural populations with affordable and environmentally friendly alternative efficient fuel products. In developing countries, the briquette industry is gaining momentum in certain regions such as in East Africa and Asia. The empirical business cases, which led to this business model, are primarily from East Africa as there is more experience in briquette business in this region.

The business models (**Business Models 1 and 2: Briquettes from agro-waste** and **Briquettes from MSW**) highlight production of briquettes from different waste streams, carbonized or non-carbonized, and distinguish between end users such as households, and commercial and institutional users which have different needs and requirements. Competition from alternative fuels, firewood and charcoal, is a major threat to the success of this business model. Thus, to compete with alternative fuels, different strategies are used such as targeting of segmented market, designing an efficient and effective value chain using local technology to reduce production cost and providing products with consistent quality.

References and further readings

World Health Organization (WHO). 2016. Burning opportunity: Clean household energy for health, sustainable development, and wellbeing of women and children. Geneva: WHO.

CASE
Briquettes from agro-waste (Kampala Jellitone Suppliers, Uganda)

Solomie Gebrezgabher and Abasi Musisi

Supporting case for Business Model 1	
Location:	Kampala, Uganda
Waste input type:	Agricultural farm waste/residues (saw dust, millet husks, ground nut shells, wheat bran, maize combs, coffee husks)
Value offer:	Briquettes (Clean cooking fuel), briquette burning stoves
Organization type:	Private
Status of organization:	Operational since 2001 (briquette business)
Scale of businesses:	Medium
Major partners:	Fuel from Waste Research Centre, Danish International Development Agency (DANIDA), United States Africa Development Foundation (USADF), Deutsche Gesellschaft für Internationale Zusammenarbeit (GIZ)

Executive summary

Kampala Jellitone Suppliers (KJS) is a limited company located in Kampala, Uganda that produces non-carbonized briquette from agricultural residues. KJS has been operational since 1981 and at the time of the assessment employed over 100 people, 70% being women. The company started with roasting coffee using diesel burners, followed by a bakery that used firewood ovens. The baking and roasting propelled the need to look for an alternative fuel source and gave rise to the production of briquettes made from agricultural waste. This has led to KJS becoming the first large scale non-carbonized briquette producer in Uganda and wining the ASHDEN Global Green Awards in June 2009. Its clients now include institutional and commercial users who previously used wood fuel and charcoal for cooking and heating. KJS provides them with briquettes which have high heating value and consistent properties and burn longer than alternative cooking fuel, as well as selling efficient briquette-burning stoves. The company has also set up the Fuel from Wastes Research Centre (FWRC), an NGO which conducts innovative research and development in suitability of agricultural wastes for briquetting, briquette making, and designing and manufacturing of briquette burning stoves.

KEY PERFORMANCE INDICATORS (AS OF 2012)	
Land use:	2.4 ha
Capital investment:	USD 698,964
Labor:	100 full-time workers and 400 external laborers along the value chain
Operation and Maintenance (O &M) cost:	0.240–0.260 USD/kg of briquette
Output:	1,680 tons of briquettes per year based on one shift operation
Potential social and /or environmental impact:	Savings to users of 0.08–0.32 USD/kg compared to charcoal, CO_2 emission savings of approx. 6.1 ton CO_2/ton of briquettes, additional income to farmers – USD 3 to USD 14 per ton of input
Financial viability indicators:	Payback period: 14.5 years Post-tax IRR: 7% Gross margin: 10%

Context and background

Kampala Jellitone Suppliers, Kampala, Uganda was founded in 1976 to produce cosmetic products from petroleum jelly. KJS diversified into coffee processing and baking, using liquefied petroleum gas (LPG) as the fuel. In 1992, KJS started to look for cheaper alternative fuels. The production of briquettes was initially started to meet internal energy needs for coffee roasting and bakery, but KJS soon recognized the potential and became a large-scale producer of non-carbonized briquettes. As well as manufacturing briquettes which provides a cleaner, cheap and easy to handle cooking fuel, it also supplies efficient briquette-burning stoves. The initial business set up was supported by the Danish Embassy through Danish International Development Agency (DANIDA), which funded a feasibility study on biomass briquetting and assisted KJS to buy the first briquetting machine, and carry out research in briquetting technology. The company is now selling briquettes to 35 institutions including schools, hospitals and factories. It is financed by its founder and own income, as well as grants from DANIDA (USD 100,000) and the United States African Development Foundation (USADF) (USD 85,000) for developing business plans and staff training.

Market environment

Biomass is still the most important source of energy for the majority of the Ugandan population. About 90% of the total primary energy consumption is generated through biomass, which can be separated in firewood (78.6%), charcoal (5.6%) and crop residues (4.7%). Firewood was most commonly used by rural households (86%) while charcoal is commonly used in urban areas (70%). In Kampala, 76% of the population use 205,852 tonnes per year of charcoal as their main source of fuel for cooking. The urban household use accounted for about 70% of that demand while commercial establishments, such as hotels, accounted for 25%. The charcoal use is estimated to increase at 6% per year, which matches the rate of urbanization. High demand for wood fuels used inefficiently results in overuse and depletion of forests. About 90,000 hectares (equals 900 km²) of forest cover are lost annually, which leads to fuel wood scarcity in rural areas and increasing price levels of charcoal and fuel wood. The production of charcoal is carried out under primitive conditions with an extremely low efficiency at 10–12% on weight-out to weigh-in basis and an efficiency rate on calorific value basis at 22%. At the same time, households use biomass in a very inefficient way as the three-stone fire is still widely used.

Non-carbonized briquettes serve as a replacement to natural firewood and raw biomass fuel. They offer greater energy per unit weight than wood or raw biomass but release as much smoke. Consequently, these are more appropriate for industrial/commercial processes or institutions where emissions can be controlled. Customers like the convenience of buying, handling and storing briquettes. The cooks like the reduced smoke, heat and charcoal dust, and faster cooking. Table 5 shows the prices of briquettes and other competing fuels in Kampala. The financial savings are significant where charcoal has been used in the past. One primary school now spends USD 24 (51,000 USh) per day on briquettes, instead of about USD 32 (69,000 USh) per day on charcoal.

TABLE 5. PRICES OF BRIQUETTES AND ALTERNATIVE FUELS (DEC 2011)

FUEL TYPE	PRICE (USD/KG)
Eco-Fuel Africa briquettes	0.17
Firewood	0.24
Kampala Jellitone Suppliers Ltd. briquettes	0.28
Informal producers briquettes	0.40
Charcoal	0.60

Source: Ferguson, 2012; Personal communication with Eco-Fuel Africa; Personal communication with KJS

In Uganda, there are 180,000 schools and a wide range of agricultural and food processing businesses that could use briquettes. Institutional stoves cost around USD 740 (1.6 million USh). About 65% of customers pay KJS for the stove in installments, others pay the full cost at the time of installation. KJS recently dropped the domestic users due to lack of briquette stoves on the market to match the briquettes whereas for the other segments, the briquettes can be used without modifications in the existing stove. Hence, there is a considerable opportunity and scope to expand production and supply the existing client base. Recent increases in charcoal prices, as shown in Figure 7, have created an opportunity for briquette businesses to serve these users.

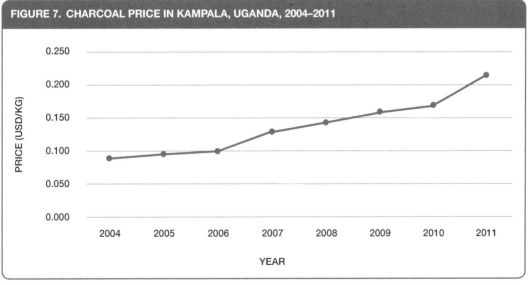

FIGURE 7. CHARCOAL PRICE IN KAMPALA, UGANDA, 2004–2011

Source: Uganda Bureau of Statistics, 2010 and 2012.

Macro-economic environment

The biomass has historically been a cheap and accessible source of fuel for Uganda's population but this is unlikely to continue. The FAO reported that between 1990 and 2005 Uganda lost 26% of its forests (78% in areas around Kampala), and the National Environment Management Authority (NEMA)

State of the Environment Uganda 2008 report predicts that this deficit will lead to complete depletion of the nation's forests by 2050. The unsustainable levels of the charcoal production operations are increasingly a source of environmental concern, especially considering that slow-growing, hard-wood tree species are targeted without plans for replacement planting.

The contribution of firewood and charcoal to Uganda's GDP is estimated at USD 48 million and USD 26.8 million respectively (UNIDO, 2015). The fact that the biomass wood industry represents a significant economic activity implies that wood fuel will continue to be the dominant source of energy in Uganda for the foreseeable future. This has implications for briquette business as the success of briquette business depends on its price competitiveness to the wood fuel/charcoal. In terms of employment, biomass production creates nearly 20,000 jobs for Ugandans.

Business model

KJS implements a value-driven business model. The establishment and partnership with the Fuel from Wastes Research Centre has enabled KJS to be innovative in its use of varieties of agricultural waste, in making consistent quality briquettes and in designing efficient stoves (Figure 8). Briquettes are sold via distributors while briquette stoves are customized and installed at the user's site. The company provides its briquette and stove customers pre-sales and post-sales support by giving a training/demonstration on how to use the products. It also conducts sensitization and training workshops for farmers on the best ways possible to preserve the agricultural wastes by milling it before delivery to allow the transport of larger quantities as well as for end users on how to use the briquettes and stoves effectively and efficiently to get value for their money. Thanks to these practices, KJS has been making profits for the last five years and has plans to scale up its operations by targeting industries which rely on biomass for industrial energy supply, such as cement factories, bricks, tile production, etc.

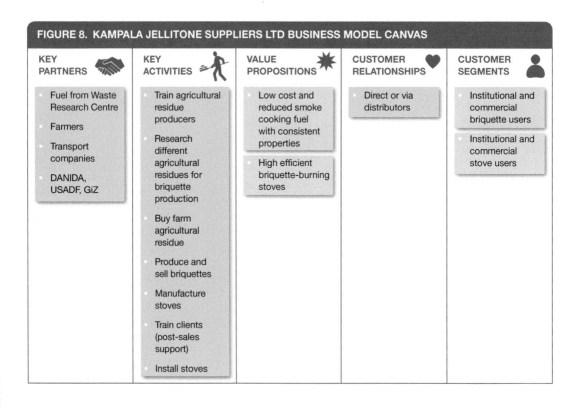

FIGURE 8. KAMPALA JELLITONE SUPPLIERS LTD BUSINESS MODEL CANVAS

KEY PARTNERS	KEY ACTIVITIES	VALUE PROPOSITIONS	CUSTOMER RELATIONSHIPS	CUSTOMER SEGMENTS
Fuel from Waste Research Centre Farmers Transport companies DANIDA, USADF, GiZ	Train agricultural residue producers Research different agricultural residues for briquette production Buy farm agricultural residue Produce and sell briquettes Manufacture stoves Train clients (post-sales support) Install stoves	Low cost and reduced smoke cooking fuel with consistent properties High efficient briquette-burning stoves	Direct or via distributors	Institutional and commercial briquette users Institutional and commercial stove users

CASE: BRIQUETTES FROM AGRO-WASTE

KEY RESOURCES
- Financial resources
- Laboratory
- Research expertise
- Land, building, equipment, labor
- Agricultural residue
- Training
- Name, brand, ASHDEN, reputation

CHANNELS
- Direct or via Distributors
- Pre-sales awareness and after-sales support to users

COST STRUCTURE
- Investment cost (Land, building, briquetting machines)
- Environmental impact assessment cost
- Operational cost, marketing and packaging cost
- Tax = 18% VAT and 6% withholding for all governmental institutions KJS supplies
- R&D
- Stove installation services

REVENUE STREAMS
- Sales of briquettes
- Sales of stoves

SOCIAL & ENVIRONMENTAL COSTS
- Laborers' health risk due to handling of waste and/or inorganic/foreign particles such as glass and plastic
- Loss of jobs (livelihood) for charcoal and wood fuel traders

SOCIAL & ENVIRONMENTAL BENEFITS
- Contribute to reduction of deforestation
- Reduction of environmental pollution
- Reduction of open burning of agricultural residues
- Energy saving
- Creation of jobs/additional income for farmers
- Improved household/users' health

Value chain and position

KJS is overlooking all the activities across the value chain from research, supply of inputs to final sales of briquettes and stoves (Figure 9). KJS conducts its own research in briquette making and stove manufacturing through the Fuel from Wastes Research Centre, a research NGO set up by the company.

KJS's customer segments include institutional and commercial users which previously used firewood for cooking and heating. Although prices of firewood are high which gives briquettes a competitive advantage, buyers can easily shift back to firewood as briquettes are used without modifications in the existing stoves. Threat from existing briquette businesses or new entrants is low. KJS is the first large-scale non-carbonized briquette producer in Uganda. The majorities of briquette producers in Uganda are small-scale and are targeting household customer segment. Furthermore, a high investment cost is required to start up a large-scale briquetting business.

Input suppliers (farmers) are key partners as KJS depends on their reliable supply of agricultural waste. The processing of commercial crops generates large volumes of biomass residues including rice husks, coffee pulp and maize stalks. These, along with sawdust from sawmills and furniture factories, often go to waste. Residues are usually simply dumped in large heaps which are then burned to dispose of them. Data provided by the government in the Uganda Renewable Energy Policy 2007 suggests that 1.2 million tons of agricultural residues are available each year.

KJS briquetting business has created employment opportunities and has generated additional incomes to its agricultural residue input suppliers. KJS employs about 100 staff at the factory, and also uses contractor to collect the residues from the agricultural processors and sawmills and other

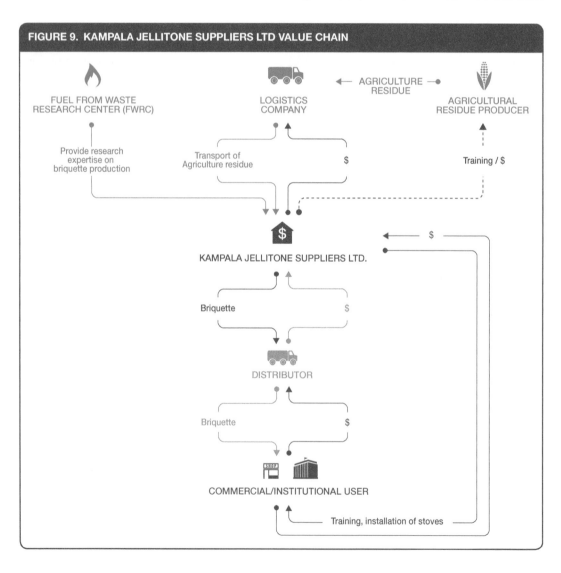

FIGURE 9. KAMPALA JELLITONE SUPPLIERS LTD VALUE CHAIN

haulage companies to deliver briquettes to customers. The residue producers are paid between USD 3 and USD 14 (6,000 and 30,000 USh) per ton of residue and earn extra income from something that was once regarded as waste. KJS pays a higher price for processed feedstock (already milled) and are seeking to supply farmers with milling machines in an attempt to improve transport efficiency.

Institutional environment

In order to support alternative clean energy initiatives, government strategy on the demand side is dissemination of more energy efficient technologies (Renewable Energy Policy, 2007). Furthermore, with support from the UNDP, the government is implementing key interventions in charcoal production which includes increasing the charge that the National Forestry Authority levies on charcoal burners. This provides an opportunity for alternative fuels to compete further with the cost of charcoal.

Several initiatives to conserve biomass resources have been undertaken by government and the private sector, including NGOs. These include the promotion of improved stoves and afforestation. However, the impact of these efforts is still limited.

Technology and processes

A study conducted by KJS funded by DANIDA in 2002 identified 16 possible agricultural farm waste/residues, such as coffee husk, rice husk, sawdust, wheat, groundnut husks, etc., that could be used for making briquettes. Before production takes place, the agricultural waste undergoes intensive tests to ascertain different characteristics including burning characteristics, ash content and the calorific value (Figure 10). At the factory, the residues are sieved (to remove large pieces, glasses and stones), pulverized using a hammer mill and dried to a moisture content of 13% using a flash drier in addition to sun-drying. Each agricultural residue is then blended by pouring it into a separate hopper which feeds it into a mixing machine to get a homogeneous mixture of different materials with the required

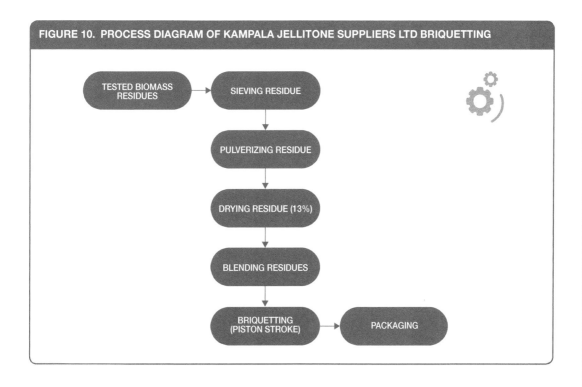

FIGURE 10. PROCESS DIAGRAM OF KAMPALA JELLITONE SUPPLIERS LTD BRIQUETTING

proportions. The mixed biomass is fed into the briquetting machine which compresses it using a piston stroke. KJS operates two imported electrically-powered piston machines with a combined capacity of 1.25 tonnes per hour (3,500 tonnes per year) as well as an industrial drier for drying feedstock. However, these machines do not operate at full capacity, limited by the throughput of the feedstock drying process. Under pressure, the natural lignin in the agricultural residues binds the particles together to form a solid block and thus the use of binders is not necessary in this process. Finally, the agricultural wastes are compressed into a solid particle with a heat value of about 14.5 MJ/kg and packed in sacks (40 kg) ready for delivery. The sacks are held in a dry store until delivery to the customers. KJS has also designed an efficient briquette-burning stove, for institutions such as schools and colleges and for food processing industries. The stove is made from fired bricks with a grate and combustion chamber and a chimney to remove the smoke and is constructed on site by KJS staff.

Funding and financial outlook

The total investment cost is estimated to be USD 698,964 (Table 6). The owner invested own cash towards 85% of the total investment and the remaining was obtained from donors. Operational cost including cost of input, labor, utilities, operating and maintenance is estimated to be approx. 238 USD/ton. Marketing and packaging costs are estimated to be approx. 16.3 USD/ton. To meet growing demand, the enterprise plans to expand production. For this it needs to procure 5 briquetting machines with production capacity of 750 kg/hr, trucks to deliver farm residues, agricultural milling machines and other equipment. The whole project requires about USD 2 million. The United States African Development Foundation (USADF) promised to finance about 12.5% (USD 250,000) of the total capital needs.

KJS produced and sold about 1,530 tons of briquettes at a price of 282.8 USD/ton and installed 1,309 institutional stoves for USD 740 in 2009. KJS's sales are estimated to be 1,680 ton of briquettes at a price of 282.8 USD/ton (Table 7).

KJS have registered their venture as a CDM project in Uganda and with support from the Belgian Embassy are aiming to develop an appropriate methodology for carbon financing.

TABLE 6. KJS INVESTMENT AND OPERATIONAL COST OF THE BRIQUETTE UNIT

ITEM	AMOUNT (USD)
Investment cost:	
Land	232,200
Buildings	227,272
Machinery / equipment	234,492
Environmental impact assessment	5,000
Total investment cost	**698,964**
Operational costs:	USD/ton
Input cost	129.2
Labor	23.52
Operating and maintenance	41.92
Utilities	42.16
Marketing	12.16
Packaging	4.16
Vehicle maintenance	1.00
Depreciation	8.00
Total operational costs	**262.12**

TABLE 7. FINANCIAL SUMMARY OF KJS BRIQUETTE BUSINESS (YEAR)[1]

ITEM	AMOUNT (USD/YEAR)
Total revenue from briquette sales (1680 ton @ 282.8 USD/ton)	475,104
Total production cost	440,362
Net income	34,742
Net cash flow	48,182
Payback period (Year)	14.5
Internal rate of return (IRR) (%)	7%

Socio-economic, health and environmental impact

In agriculture-based countries like Uganda, there is a vast natural supply of biomass found in the form of agro and forest residues. Often these residues are simply burned in the fields. This is not only an unfortunate waste of an energy source, but it is also a cause for increased pollution in local regions. In combination with an energy-efficient stove, briquette use contributes to reduction of deforestation, helps fight climate change and enables the end user to save money. This business reduces the amount of biomass waste that is discarded, decreasing the incidence of fires and its associated risk and avoids the release of methane due to its decomposition. The briquettes manufactured from the agricultural waste are much cleaner to burn than coal. This business further provides communities with economical and safer sources of energy for cooking. The sale of agricultural wastes by farmers to the factories creates additional source of income thereby improving the incomes of the farmers.

A study by the University of Makerere estimated that 1 ton of briquettes replace 1.2 tons of firewood and 0.3 tons of charcoal. KJS's annual production of 1,680 tons would replace about 2,016 tons of firewood and 504 tons of charcoal. Assuming CO_2 emissions of 1.55 tons and 14.02 tons per ton of firewood and charcoal respectively, this is equivalent to saving emissions of 3,125 tons of CO_2 from wood equivalents and 7,066 tons of CO_2 from charcoal equivalents.

Scalability and replicability considerations

The key drivers for the success of this business are:
- Regulations against cutting down trees.
- Increased charge that the National Forestry Authority levies on charcoal burners.
- Rising prices of charcoal and fuelwood.

KJS is a promising business case with significant potential for scaling-up and replication in Uganda and in other low-income countries where there are regulatory frameworks on use of firewood/charcoal. This business could potentially be up-scaled and replicated in urban centres where access to both raw material and high potential markets for the briquettes exist and charcoal prices are high. KJS's existing clients have a consumption estimated at 1,200 tons per month. KJS's production is just 140 tons per month. There is considerable scope to expand production to 1,060 tons per month from a new briquetting factory and supply the existing client base. KJS has more demand than it can address, mainly because of limited drying capacity. The company is in the process of moving to a new and larger factory to increase production. The project is labor intensive involving the collection of agricultural residues that were formerly burnt as more and more farmers are taking benefit of added income.

Summary assessment – SWOT analysis

The key strength of KJS is application of strategic practices such as conducting its own research in briquette making and stove manufacturing, which enables it to make briquettes with high energy value and consistent properties (Figure 11). KJS maintains good partnership with its input suppliers and good customer relationship. In addition to that, the fact that it won the ASHDEN award will boost its image. The weaknesses of KJS are its challenge to meet market demand due to its inability to maintain consistent supply of briquettes. Opportunities arise from the fact that there is increasing government support for renewable energy and increasing prices of substitute products which result in significant potential demand for briquettes in the future. KJS aims to reduce deforestation and GHG emissions and this presents opportunities for KJS to earn carbon credit sales by registering the business as a CDM project. Competition from alternative fuel providers and availability of and competition for needed raw materials are the largest external threat.

FIGURE 11. SWOT ANALYSIS FOR KJS

	HELPFUL TO ACHIEVING THE OBJECTIVES	**HARMFUL** TO ACHIEVING THE OBJECTIVES
INTERNAL ORIGIN ATTRIBUTES OF THE ENTERPRISE	**STRENGTHS** • Research expertise and innovation through blending of different residues • Product diversification by selling complementary stoves • Strong partnership with suppliers of input • Good customer relationship through training and installing briquette stoves onsite • Simple substitute for wood without stove modifications • Good image due to winning of the ASHDEN award • Briquettes less expensive than wood and charcoal	**WEAKNESSES** • Loss of household customer segment due to lack of briquette stoves • Failure to maintain consistent production, performance quality and supply of briquettes • High transportation cost of agricultural residues from rural areas • Initial start-up cost • Dusting and high noise levels in production areas • Lack of finance required for expansion
EXTERNAL ORIGIN ATTRIBUTES OF THE ENVIRONMENT	**OPPORTUNITIES** • Significant potential demand for briquettes and briquette stoves • Increasing price, diminishing supply and high demand of substitute products – charcoal and fuelwood • Unused agricultural waste • Carbon credit – registering the business as a CDM project • Government support for renewable energy • Cooperation with rural groups and support from local councils	**THREATS** • Competition from suppliers of raw materials and other dry fuel suppliers, especially from price-driven enterprises • Lack of financing • Customers' behavior – Habitual excess fuel loading • Unstable grid power • A lack of appropriate regulatory, framework and policy • A lack of standards and quality assurance • Domestic markets remain difficult to penetrate due to the lack of awareness (and acceptance) among household consumers and difficult distribution in rural areas

Contributors

Ronald Ssebaale Lukoda, Kampala Jellitone Suppliers (KJS) Ltd., Uganda
Johannes Heeb, CEWAS, Switzerland
Jasper Buijs, Sustainnovate; Formerly IWMI
Josiane Nikiema, IWMI, Ghana
Kamalesh Doshi, Simplify Energy Solutions LLC, Ashburn, Virginia, USA

References and further readings

Ashden awards case study/Kampala Jellitone Suppliers Ltd., Uganda / Summary. https://www.ashden.org/winners/kampala-jellitone-suppliers (accessed 6 Nov. 2017).

Ferguson, H. 2012. Briquette business in Uganda: The potential for briquette enterprises to address the sustainability of the Ugandan biomass fuel market. London: GVEP International.

Kampala Jellitone Suppliers Ltd. Business plan.

Kampala Jellitone Suppliers Ltd. 2012. www.jellitone.com (accessed Oct. 29, 2012).

Musisi, A. 2014. Interview with A. Musisi (KJS) by Solomie Gebrezgabher via online questionnaire. Kampala, Uganda. March 19, 2014.

Sanga, M. 2012. Interview with M. Sanga (Eco-Fuel Africa) by Charles Niwagaba via Skype. September 5, 2012.

Sebitt, A. 2005. Project Idea Note (PIN) for Kampala Jellitone Suppliers Ltd. Makerere University Kampala.

Uganda Bureau of Statistics. 2012. Statistical abstract. www.ubos.org/onlinefiles/uploads/ubos/pdf%20documents/2012StatisticalAbstract.pdf (accessed Nov. 5, 2012).

Uganda Bureau of Statistics. 2010. Statistical abstract. www.ubos.org/onlinefiles/uploads/ubos/pdf%20documents/2010StatAbstract.pdf (accessed Nov. 5, 2012).

United Nations Industrial Development Organization (UNIDO). 2015. Baseline report of clean cooking fuels in the East African community: Draft report. open.unido.org/api/documents/4677132/download/Baseline%20Report%20of%20Clean%20Cooking%20Fuels%20in%20the%20East%20African%20Community (accessed Nov. 6, 2017).

Case descriptions are based on primary and secondary data provided by case operators, insiders, or other stakeholders, and reflect our best knowledge at the time of the assessments 2015–2016. As business operations are dynamic, data can be subject to change.

Note

1. KJS has recently introduced additional briquetting line which produces carbonized briquettes to support clients who prefer carbonized briquettes as a replacement to charcoal. However, data was not available to incorporate cash flows into the business case.

BUSINESS MODEL 1
Briquettes from agro-waste

Krishna C. Rao and Solomie Gebrezgabher

A. Key characteristics

Model name	Briquettes from agro-waste
Waste stream	Agricultural farm waste/residues (saw dust, millet husks, ground nut shells, wheat bran, maize combs, coffee husks, etc.)
Value-added waste product	Briquettes (clean cooking fuel)
Geography	Region with ease of availability of crop residue and lack of ease in availability of fuel wood
Scale of production	Medium scale; 1,000–2,000 tons per year of briquettes
Supporting cases in this book	Kampala, Uganda
Objective of entity	Cost-recovery [ ]; for profit [X]; social enterprise [ ]
Investment cost range	Approx. USD 200,000 to 450,000
Organization type	Private
Socio-economic impact	Reduction in deforestation and environmental pollution, reduced indoor air pollution resulting in improved health for household and employment generation
Gender equity	Beneficial to women and children using fuel with less indoor air pollution than firewood; time savings in fuel collection for women

B. Business value chain

The business model is initiated by either a standalone private enterprise or agro-industries such as coffee processing units or rice mills that generate large quantities of crop residues as waste. The business processes crop residues such as wheat stalk, rice husk, maize stalk, groundnut shells, coffee husks, saw dust etc. and converts them into non-carbonized briquettes as fuel. Non-carbonized briquettes serve as a replacement to natural firewood and raw biomass fuel. They can also be offered as a replacement fuel among rural populations where firewood is still dominant. Further commercial processes such as drying of crop, drying of tea, curing of tobacco and firing of ceramics/brick can also make use of briquettes. The key actors in the business value chain are the suppliers of crop residue such as farmers and agro-industries, product distributors and end users of the product: households and energy intensive industries (Figure 12).

The characteristics of the agricultural waste including burning characteristics, ash content and the caloric value are first ascertained before making briquettes. The process of briquetting involves sieving of agricultural waste to remove large content such as glasses and stones, pulverizing, drying, mixing of different materials with the required proportions, briquetting using high pressure compression such

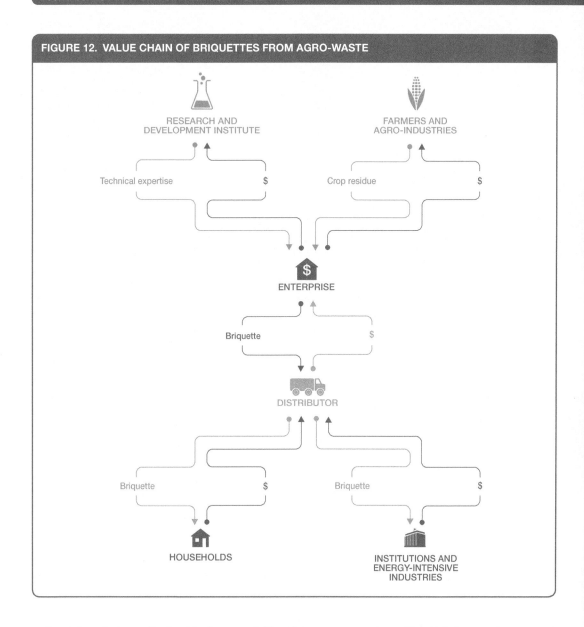

FIGURE 12. VALUE CHAIN OF BRIQUETTES FROM AGRO-WASTE

as by piston stroke and using binding agent. The high pressure and resulting high temperature causes the lignin (the natural woody material in plants) to flow and bind the material together. The action of the piston pushes the material through a dye, to make a continuous rod about 50 mm in diameter. The rod cools in the air and breaks into 'sticks' or briquettes about 400 mm long. As multiple crop residues with differing calorific value are the raw material input, it is ideal for the enterprise to collaborate with a research institution to find a suitable combination of crop residue to produce briquettes with higher calorific value and consistent quality.

There are two technologies for making briquettes, reciprocating ram/piston press and screw press technology. The screw pressed briquettes are generally found to be superior to the ram pressed solid briquettes in terms of their storability and combustibility. While the briquettes produced by a piston press are completely solid, screw press briquettes on the other hand have a concentric hole which gives better combustion characteristics due to a larger specific area. The screw press briquettes are also homogeneous and do not disintegrate easily.

Another option is to produce carbonized briquettes or charcoal from crop residues by burning them in low-oxygen atmosphere. The resulting charred material is compressed into carbonized briquettes. Carbonized briquettes can act as a replacement for charcoal for domestic and institutional cooking and heating, where they are favoured for their near-smokeless use. Moreover, briquettes can be used as fuel for gasifiers and generators to generate electricity or powering boilers to generate steam. The business model described in this section is focused on using briquettes as fuel for thermal applications only.

C. Business model

The primary value proposition of the business model depends on the entity initiating the business model. For an agro-industry generating large quantities of crop residue, the value proposition is to dispose the crop residue to mitigate risks from negative externalities of social and environmental impact and, in the process, incur savings from reduced energy costs. However, for a standalone private enterprise the value proposition is to use crop residue to provide briquettes to households, institutions, such as schools and prisons, and small and medium enterprises that need fuel for heating (Figure 13). The business model described in this chapter presumes the operation for a standalone private enterprise.

The briquettes are delivered to the customers either through direct sales, network of distributors or micro-franchising[1]. The direct sales requires large human resource of sales and marketing team and thus has related challenges associated with managing large staff base. The business requires developing strategic partnerships with farmers and agro industries to ensure reliable supply of crop residues at an agreed price. The key activities of the business model are procurement and processing of crop residue, briquette production and sales. To improve the production efficiency and product quality, training of farmers can be a useful activity so that farmers provide crop residue with lower moisture content and store crop residue in an appropriate manner to reduce moisture content. Research and development (R&D) would be a useful activity to streamline a process that delivers higher calorific value product. However, the cost-benefit of R&D should be assessed and ideally partnership with a research institution would mitigate the risk of need for high-skilled labor.

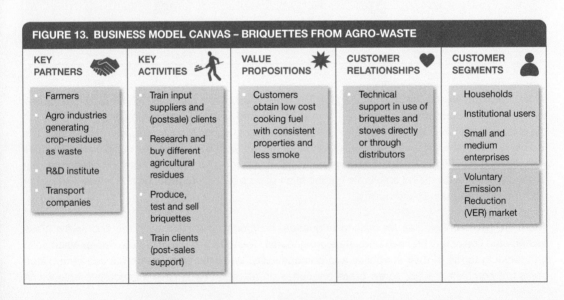

FIGURE 13. BUSINESS MODEL CANVAS – BRIQUETTES FROM AGRO-WASTE

KEY PARTNERS	KEY ACTIVITIES	VALUE PROPOSITIONS	CUSTOMER RELATIONSHIPS	CUSTOMER SEGMENTS
• Farmers • Agro industries generating crop-residues as waste • R&D institute • Transport companies	• Train input suppliers and (postsale) clients • Research and buy different agricultural residues • Produce, test and sell briquettes • Train clients (post-sales support)	• Customers obtain low cost cooking fuel with consistent properties and less smoke	• Technical support in use of briquettes and stoves directly or through distributors	• Households • Institutional users • Small and medium enterprises • Voluntary Emission Reduction (VER) market

BUSINESS MODEL 1: BRIQUETTES FROM AGRO-WASTE

KEY RESOURCES	CHANNELS
- Financial resources - Laboratory - Research, engineering, training and extension expertise - Crop residue - Network of distributors - Marketing and sales force - Contracts with institutional users and small-medium enterprises	- Direct interaction with users as well as agricultural waste suppliers/farmers - Pre-sales awareness and after-sales support to users - Distributors - Micro-franchisees

COST STRUCTURE	REVENUE STREAMS
- Investment cost – Land, building, and machinery - Operational cost - Transportation, labor, utilities, maintenance, marketing and packaging, training of farmers and distributors/micro-franchisees and voluntary emissions reductions costs	- Briquette sales - VER sales

SOCIAL & ENVIRONMENTAL COSTS	SOCIAL & ENVIRONMENTAL BENEFITS
- Laborers' health risk due to handling of waste and/or inorganic/foreign particles such as glass and plastic - Loss of jobs (livelihood) for charcoal and wood fuel traders	- Contribute to reduction of deforestation - Reduction of environmental pollution - Reduction of open burning of agricultural residues - Energy saving - Creation of jobs/additional income for farmers - Improved household/users' health - Saves time in the case of time spent in collecting firewood - Contribute to improving the educational opportunities among girls who previously missed school to fetch firewood

The business enterprise's key capital costs are building and machinery and primary operational costs are transportation, labor, utilities and marketing. Briquette sales is the only revenue source unless the enterprise is able to tap into the carbon market. A briquette enterprise is potentially eligible for carbon offset depending on the type of fuel replaced and the baseline used to calculate benefits from reduced greenhouse gas emission. In comparison to fossil fuels, briquettes produce net lower greenhouse gas as the raw material inputs are already part of the carbon cycle. Even for regions with high deforestation where wood is used as fuel, briquettes from crop residue will make a strong case for carbon benefits. However, briquette enterprises are unlikely to be individually able to apply for Clean Development Mechanism (CDM) projects due to associated transaction costs, and therefore the preferred route would be to apply via producer associations or for carbon offset on Voluntary Emission Reductions (VERs).

D. Alternative Scenarios

The business model can incorporate two additional value propositions in addition to briquette production from crop residues: a) produce low cost compost, a by-product from briquette production and b) vertical integration of business by manufacturing and selling improved cook stoves and ovens (Figure 14).

Scenario I: Compost production

Production of briquettes results in generation of crop residual waste, which can be used to produce compost. The compost can be either sold or given away to the farmers on good will basis and strengthen their relationship with farmers for reliable supply of crop residue. The additional key activity required for this value proposition is production of compost and related costs incurred. The sales and distribution process will be similar to sales of briquettes.

Scenario II: Manufacturing of improved cook stoves

The business model offers scope for vertical integration as the briquette enterprise could potentially manufacture improved cook stoves and ovens that use the briquettes produced by the enterprise. The improved cook stoves have high social benefits for households especially for women and children through reduced indoor air pollution. In addition, with improved cooking efficiency and reduced fuel consumption, household would earn savings. The business model does not require significant alteration to its distribution process. The additional key activity required is for the manufacturing of improved cook stoves, which has related capital and operational costs. Similar to briquette production, R&D is a required activity to design the cook stoves and oven that meet the customer's requirements. The product also requires specific marketing and awareness campaign.

E. Potential risks and mitigation

Market risks: Briquettes are targeted for households that do not have access to fossil fuels and that are dependent on firewood for cooking. This customer segment has low market risks in the urban areas due to scarcity of firewood. However, in the rural areas in developing countries the market risks for households as customers is high due to free availability of firewood if picked up from forest/plantation on community land. The business should target diverse customer base to mitigate these risks. It is preferred for the business to have both household and institutional customers. The business could also get into bulk contractual arrangement with institutional customers and have assured sales.

Competition risks: Briquettes have strong competition risks from competing alternative products such as charcoal, wood, kerosene and liquefied petroleum gas (LPG). Fuel choice typically depends on availability, consumer preference, price, convenience and at times social status associated in using certain types of fuels like LPG. Ideally briquette should be targeted to customer segment that uses firewood and charcoal because briquettes can be more competitive, convenient and efficient.

Risks associated with stoves are similar to briquette and there are multiple suppliers of different types of stoves in the market. In the case of compost as mentioned above the enterprise could give it away for free as goodwill measure to the farmers in exchange for assured reliable supply of crop residues which can be procured either directly from the farm gate or have the farmers deliver the agro-waste for a fee. A key risk in procuring crop residue from farmers is that with time they are likely to demand higher price. To mitigate this risk in addition to giving low cost compost for free, the enterprise should target different types of farmers cultivating different crops so as to negate the rising input cost or have a longer-term agreement with the farmers.

Technology performance risks: The technology used is mechanical compressing with a binding agent or pyrolysis. The technology has been widely used commercially and is proven. It doesn't require high skills for operating it and doesn't have complications towards repair and maintenance.

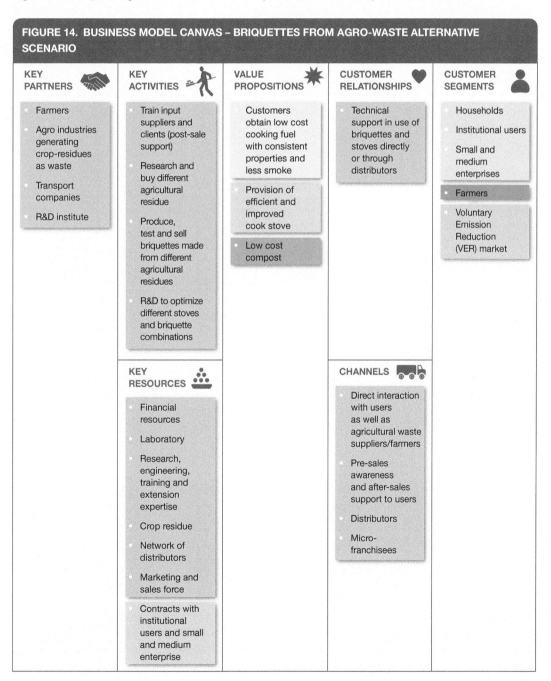

FIGURE 14. BUSINESS MODEL CANVAS – BRIQUETTES FROM AGRO-WASTE ALTERNATIVE SCENARIO

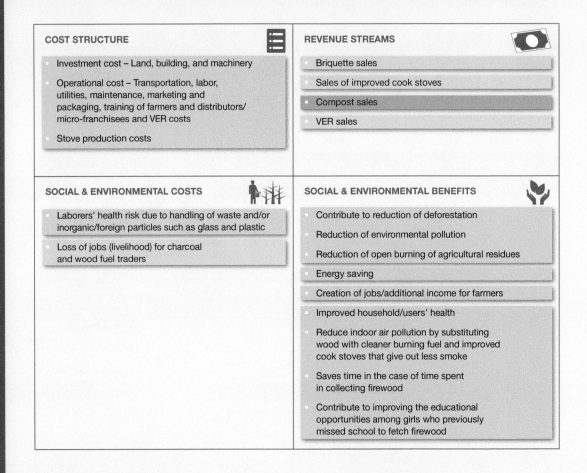

Political and regulatory risks: In most developing countries, cooking is a social issue and governments in developing countries provide subsidy for fuels such as kerosene and LPG. These competing products are priced lower than briquettes and hence can pose significant risks to the business. Diversifying customer base and including energy intensive small and medium enterprise as primary customers can considerably negate this risk. Increasing government support through financial incentives and policies that promote renewable energy reduce this risk considerably in the long term.

Social-equity-related risks: The model is considered to have more advantages to women as culturally in developing countries women collect fuel wood and do the cooking at household. The model provides employment opportunities and additional revenue for farmers to sell their crop residues. The users of briquettes are low-income households who are using other unhealthy and inefficient fuels, or more costly ones.

Safety, environmental and health risks: Organic waste when left in open begins to decay and releases methane, which is more damaging to the environment than carbon dioxide. The waste-processing technologies are not without problems and pose a number of environmental and health risks if appropriate measures are not taken. The safety and health risks to workers are present and thus standard protection measures should be put in place (Table 8). There is a potential risk for those households where less harmful cooking fuels such as LPG, kerosene or electricity are replaced by biomass briquettes especially without introduction of safer cooking stoves.[2]

BUSINESS MODEL 1: BRIQUETTES FROM AGRO-WASTE

TABLE 8. POTENTIAL HEALTH AND ENVIRONMENTAL RISK AND SUGGESTED MITIGATION MEASURES FOR BUSINESS MODEL 1

RISK GROUP	EXPOSURE ROUTE					REMARKS
	DIRECT CONTACT	AIR	INSECTS	WATER/ SOIL	FOOD	
Worker	LOW	LOW				Health risk for households might increase or decrease depending on the quality of the used fuel. Possible exposure to air and noise pollution
Farmer/User						
Community		LOW				
Consumer		MEDIUM				
Mitigation measures	[boots icon]	[mask, headphones, respirator icons]				

Key: ☐ NOT APPLICABLE ▨ LOW RISK ▨ MEDIUM RISK ■ HIGH RISK

F. Business Performance

This business model is rated high on profitability followed by environmental impact (Figure 15). The business model has a strong revenue source and diverse customer base. It has potential for additional revenue source from sale of stoves and VERs. The environmental impact is specifically high for regions with deforestation.

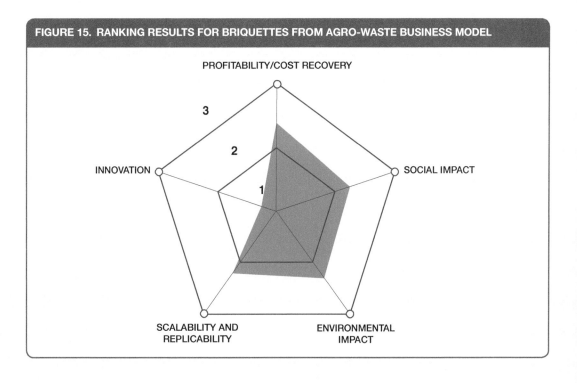

FIGURE 15. RANKING RESULTS FOR BRIQUETTES FROM AGRO-WASTE BUSINESS MODEL

The business model has high potential for replication in developing countries as there are no limiting factors such as new technology, special policies and regulations, institutional capacity, waste availability that can limit replication of the business model. It can be scaled horizontally and has potential for vertical scaling by expanding into the business of selling pressing machines for briquettes through a franchising model and getting into manufacturing of improved stoves. It also has a potential to be implemented in agriculture intensive regions and which have high usage of firewood and charcoal for cooking. The model is straightforward with no sophisticated or innovative financing and technology requirements and hence scores low on innovation.

Notes

1. Micro-franchising borrows the traditional franchising concept with scaled-down business concepts found in successful franchise organizations. It operates as a micro-enterprise following proven marketing and operational concepts with systematic replication. The concept is predominant in delivering services to the poor along the lines of microfinance and microcredit. Micro-franchise entrepreneur has similarities to an agriculture extension worker and typically such an entrepreneur sells multiple product like seeds, fertilizers, water filters, Fast Moving Consumer Goods (FMCG) etc.
2. Winkler, M.S., Fuhrimann, S., Pham-Duc, P., Cissé, G., Utzinger, J., Nguyen-Viet H., 2017. Assessing potential health impacts of waste recovery and reuse business models in Hanoi, Vietnam. Int J Public Health 62 (Suppl 1): 7–16.

CASE
Briquettes from municipal solid waste (COOCEN, Kigali, Rwanda)

Andrew Adam-Bradford and Solomie Gebrezgabher

Supporting business case for Business Model 2	
Location:	Kigali, Rwanda
Waste input type:	Municipal solid waste (MSW)
Value offer:	Briquettes (Clean cooking fuel) and compost
Organization type:	Cooperative/Public-private partnership (PPP)
Status of organization:	Operational since 2002
Scale of businesses:	Medium
Major partners:	Kigali City Council, UNDP, city residents

Executive summary

Coopérative Pour La Conservation De L'Environement (COOCEN), established in 2002, is a women's cooperative that delivers waste collection and briquette production service in the low-income Nyamirambo District of Kigali, Rwanda through the implementation of a public-private partnership (PPP) with the Kigali City Council. The PPP is based on the delivery of waste collection services by COOCEN, and as a component of the partnership, the Kigali City Council provided 7 ha of land in Nyamirambo District for COOCEN from where the primary waste sorting and briquette production takes place. At the time of the assessment, the cooperative collected waste from more than 4,000 households for a fee, while till now demand for briquettes constantly exceeds production. The reason is that COOCEN is the sole supplier of fuel briquettes to 16 prisons in Rwanda, which has become a sustained market segment. The cooperative provides solutions to various issues related to the environment and to living conditions of communities. COOCEN contributes to cleaning of the city and provides sanitation services to local communities, which benefit from improved health and sanitary conditions. It contributes to reduction of CO_2 emissions and to reduction of deforestation by avoiding the burning of firewood. In addition to these, the cooperative generates employment, mostly to women.

KEY PERFORMANCE INDICATORS (DATA AS OF 2012)					
Land use	7 ha				
Capital investment:	USD 162,075				
Labor:	110 workers (90% women)				
O&M cost:	USD 94,875				
Output:	1,500 ton/year (retailed at 0.122 USD/kg)				
Potential social and/ or environmental impact:	Job creation and income generation (women members earn 50 USD/month), households benefit from improved sanitary and health conditions, avoided burning of firewood of 1,200 tons/year, CO_2 emission saving of 297 tons/year				
Financial viability indicators	Payback period:	3	Post-tax IRR:	N.A.	Gross margin: 42%

Context and background

COOCEN initially focused on waste collection in a congested urban area that previously had no waste collection facilities or services. The cooperative expanded its operations and constructed a briquette production plant in the low-income Kimisagara Sector of Nyamirambo District. It collects waste from 4,000 households for a fee, sorts waste, extracts organic fragment and produces briquettes from organic components through the implementation of a strategic PPP with the Kigali City Council. As the component of the partnership, the Kigali City Council provided 7 ha of land to COOCEN. The project obtained further financial assistance from the Global Environmental Facility's Small Grants Programme (GEF-SGP), implemented by the United Nations Development Programme (UNDP). Environmental conservation is a key component of the COOCEN strategy and was an instrumental aspect to securing project grants from the European Union (EU) and UNDP. At the assessment time, the cooperative produced and sold around 1,500 tonnes of briquettes per year to schools, prisons and factories.

Market environment

In the capital of Rwanda, Kigali, wood and charcoal are the primary sources of fuel used for cooking and heating, causing major environmental problems such as deforestation and pollution. Charcoal is the preferred fuel for urban households, serving 51% of households, and demand is pushing up the price. As Rwanda also faces a serious wood fuel deficit, there is a need for alternative sources of fuel. Between 2007 and 2012, the amount of municipal solid waste (MSW) grew almost fourfold. COOCEN's environmentally friendly briquette made from MSW is retailed at 0.122 USD/kg while the price of charcoal in 2014 was 0.20–0.22 USD/kg. However, about 1.6 kg of MSW briquettes will be required to replace 1 kg of charcoal. So far, demand from its customer segments constantly exceeds production particularly as COOCEN is the sole supplier of fuel briquettes to the Rwandan prison service, which has become a long-term customer. COOCEN anticipates an increase in briquette demand as a result of the rising price of charcoal coupled with the government policy to protect the environment and promote alternative sources of energy. COOCEN is aiming to increase production however, there are constraints due to limited production capacity and availability of capital. COOCEN also acknowledges that overtime competition in briquette production will increase and thus it aims to improve the manufacturing process and the quality of the final product.

Macro-economic environment

The primary energy supply in Rwanda is dominated by wood, which accounts for about 80% of the supply, of which 57% is direct supply and 23% for charcoal (Ndegwa et al., 2011). There is a combined per capita demand of wood (both for fuelwood and charcoal) of 1.93 kg/person/day, which creates an unsustainable situation because it largely surpasses the production capacity of 0.46 kg/capita/day. Rwanda lost 37% of its forest cover (around 117,000 ha) between 1990 and 2010. Firewood is associated with environmental, social and health problems, stemming from deforestation and the emissions from wood and charcoal burning respectively. Furthermore, population growth is intensifying deforestation and causing more environmental degradation.

Most of the charcoal is consumed in Kigali, and the main supply areas are the rural areas of Southern and Western Provinces, where charcoal is produced using the traditional earth mound kilns with an efficiency of merely 12%. The chief actors in the supply chain are also poor and unable to invest in the expensive and more efficient biomass conversion technologies – a factor resulting in massive wastage of the wood fuel resource. There are significant health and social benefits of transitioning to charcoal, but it is likely to increase the pressure on the limited wood supplies. The country is taking a 'green economy' approach to economic transformation as a priority. Although fuel wood consumption is expected to increase in the short-term, the long-term strategy of the Government of Rwanda is to reduce fuel wood consumption to 50%.

Unlike in many African countries, the demand for wood fuel is met through forest plantations, mostly of eucalyptus, which are owned by the state or districts and by private entities. About 450,000 hectares or 17% of the country is covered by forests, with 46% being natural forests and the rest public and private plantations. Sixty-five percent of the plantations are state and district owned, while institutions and private citizens own 9% and 25% respectively. Thirty percent of the state forests is left for soil protection, which reduces the amount of plantations that can be harvested to 194,000 hectares.

Vision 2010, the Rwanda development strategy has identified a target of increasing the production of wood for fuel and other uses through the expansion of forest and tree cover to 30% of the national land area by 2020. The wood fuel sector is a major economic activity in Rwanda employing about 20,000 people, which in turn support about 300,000 people (Ndegwa et al., 2011). However, Rwanda still faces a serious wood fuel deficit, which directly impacts the availability and affordability of biomass energy including charcoal production. This gives an opportunity for briquette businesses to fill the charcoal supply and demand gap.

Business model

Figure 16 shows the business model for COOCEN. The cooperative collects waste, extracts organic fragment and produces briquettes from organic components and efficient briquettes cook stoves through the implementation of a PPP with the Kigali City Council. As a component of the partnership, the Kigali City Council provides a site (7 ha of land) for COOCEN from where the primary waste sorting and briquette production takes place. Thus, COOCEN's principle business idea is providing a waste collection service to the local community and then converting the organic waste into fuel briquettes, which are sold to prisons, schools, brick factories and in some cases to households. Initially, the cooperative had difficulties motivating the residents to pay for waste collection services. However, through its awareness campaigns about waste, sanitation and the environment, the cooperative was able to change peoples' attitude. Therefore, waste collection fee and sales of briquettes are the two major revenue streams while selling compost and improved cooking stoves are minor revenue streams. The compost is supplied to the Kigali City Council and is used for city greening and urban amenities including flowerbeds, parks and green walls on the steep urban roadsides.

FIGURE 16. COOCEN BUSINESS MODEL CANVAS

KEY PARTNERS
- Community
- Kigali City Council
- Rwanda Environment Management Authority (REMA)
- Other organic waste producers (such as peat)
- UNDP

KEY ACTIVITIES
- Collection of MSW and other organic waste (such as peat)
- Organic fraction separation
- Production and sales of briquettes and compost
- Promotional campaigns

KEY RESOURCES
- Consumables (MSW and peat)
- Human resource
- Capital
- Land as provided through partnership
- Equipment
- REMA certification

VALUE PROPOSITIONS
- Waste collection service
- Environment friendly briquettes that are cheaper than charcoal and wood (price leadership)
- Organic fertilizer (compost)
- Purposely built briquette stoves

CUSTOMER RELATIONSHIPS
- Direct with households for collection of waste
- Short and long-term contract for sale of briquettes with Institutional customers

CHANNELS
- Direct personal help at point of source of MSW with households
- Selling of briquettes to households directly
- Supply of briquettes to institutional customers directly

CUSTOMER SEGMENTS
- Community
- Households
- Kigali City Council
- Prisons, schools, brick factories, households

COST STRUCTURE
- Investment cost
- Building = 67% of total investment cost
- Machinery = 33% of total investment cost
- Operational cost = (input transportation – MSW and peat, labor, disposal of inorganic waste in landfill, Vehicle rental cost, utilities, maintenance, marketing and awareness campaigns)

REVENUE STREAMS
- Waste collection fee (major revenue stream)
- Briquette sales (major revenue stream)
- Compost sales (minor revenue stream)
- Stoves sales (minor revenue stream)

SOCIAL & ENVIRONMENTAL COSTS
- Possible human health risk when treating MSW

SOCIAL & ENVIRONMENTAL BENEFITS
- Contributes to MSW management
- Generates income and employment
- Creates environmental sanitation awareness
- Saves time and energy for users
- Reduces deforestation
- Reduce GHG emissions

Value chain and position

COOCEN is vertically integrated i.e. it owns the waste collection and briquetting business (Figure 17). As per the recent estimates, about 1.8–2 kg of waste containing 59–65% food waste, is generated per person per day in Kigali (Bazimenyera et al., 2012). Kigali City produces about 100 tons of waste on a daily basis and volume of waste is expected to increase as the population of the city increases.

Currently, the demand for briquettes exceeds supply. COOCEN has a long-term offtake contract with 16 prisons in Rwanda since 2007. The substitute products for briquettes are wood and charcoal. The prices of these substitute products are higher than briquettes, and nowadays, wood is increasingly difficult to get in Rwanda due to government regulations against cutting down trees. With more stringent regulations on cutting down trees and with government policy that promote renewable energy sources, the demand for briquettes from institutions and factories will increase in the future and hence substitute power is low. However, the Rwanda Vision 2010 targets to increase production of wood for fuel through the expansion of forest and tree cover. This may result in more wood available, possibly at a lower price and consequently may dampen briquette market strength.

Moreover, new briquette businesses with more efficient technologies and better product qualities pose a threat to COOCEN due to the fact that its briquette operation is not efficient as it uses mechanical process and heavily relies on uncertain weather conditions to dry its inputs and briquettes. There is

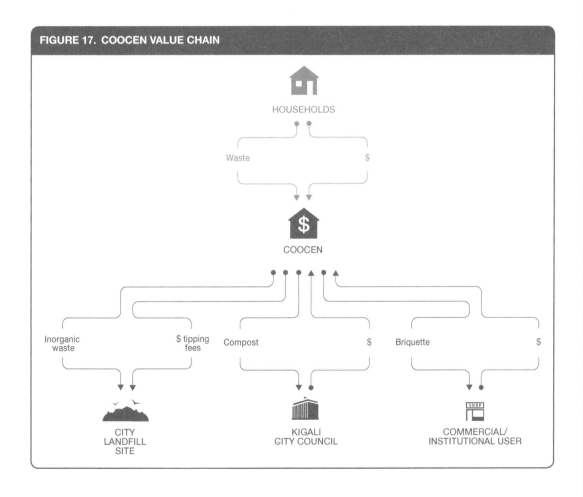

FIGURE 17. COOCEN VALUE CHAIN

also a possibility of installation of bio-digesters at institutions (like prisons) to self-supply biogas for cooking and heating applications. However, it is anticipated that the market for briquettes will grow, which will drive the revenues of briquette business to rise. COOCEN is also looking at the possibility of recycling plastics as an additional income generating activity.

Institutional environment

The main policy objective of the government of Rwanda for the biomass sub-sector is to improve the sustainability of biomass by improving efficiency of use of wood, improving charcoal production methods, facilitate fuel switching from traditional biomass energy carriers toward modern biomass energy technologies, including modern carriers, and cleaner fuel alternatives. The proposal is to decentralize implementation of biomass programmes to the local government levels to improve the impact on the end users, streamline implementation and speed up dissemination. The government has put in place very strict tree harvesting regulations and only licensed persons with tree harvesting permits are allowed to cut trees, including those from private lands.

The Rwandan government initiated an Improved Cook Stove (ICS) programme in the late 1980s or 1990s to combat deforestation. Various programmes have been implemented since, which has led to a penetration rate of "improved'" stoves of over 60% in 2012. However, the World Health Organization (WHO) suggests that some "improved" cook stoves still have emissions 20 times above safe air quality levels and there is a need to provide standards for further improvements. Given that around 85% of all energy in the country is in the form of biomass used for cooking, such an intervention on improving cook stove standards could be one of the most significant interventions in the energy sector.

The Ministry of Infrastructure (MININFRA) is the lead Ministry responsible for developing energy policy and strategy, monitoring and evaluation of projects and programmes implementation. The Department of Energy within MININFRA governs energy policy in Rwanda. The government is targeting to ensure that 80% of households have access to improved cook stoves by 2017 and 100% of households by 2020. The government supports sensitization workshops and training seminars on the economic use of improved cook stoves. This will boost demand for modern and improved cooking technologies, increasing private sector motivation to invest in this business and reduce the use of inefficient and traditional three-stone wood stoves.

COOCEN has received institutional support in the form of two grants and the provision of land from the Kigali City Council. COOCEN has also been licensed to carry out waste collection services and the project has been certified by the Rwanda Environment Management Authority (REMA) although no specific laws, regulations or policies are in place for briquette production. Rwanda Environment Management Authority (REMA) has the mandate to coordinate, oversee and implement environmental policy. Kigali City has partnered with UNDP for support in areas of technical, financial and maintenance techniques on waste management.

In pursuant of Law no. 39/2001 of 13 September 2001, Rwanda Utilities Regulatory Agency (RURA) was established with a mandate to regulate sanitation services. RURA principal mandate is to ensure consumer protections from uncompetitive practices while ensuring that such utilities operate in an efficient, sustainable and reliable manner. RURA gives consent to any city or town, company, or sector cell, public/private, to acquire and operate a dump site. It is responsible for improvement in the delivery of sanitation services including waste disposal and management. The Rwanda Development Board (RDB) also plays the lead role in investment mobilization and promotion for the energy sector, acting as a gateway and facilitator. RDB is developing briquette standards for minimum performance and energy requirements.

Technology and processes

COOCEN collects waste from households and brings it to the local COOCEN station where sorting teams separate the organic and inorganic fractions. The organic fraction is solar-dried and then mechanically ground into smaller particles which are then pressed into cylinder compact briquettes (Figure 18). The mechanical technologies that are used for shedding and briquette pressing are based on locally manufactured electricity-driven machines that are easy to operate, maintain and repair. COOCEN also investigated methods of improving the energy efficiency of the briquette through blending of different organic inputs. Consequently, peat is now added as it increases the conformity of the crude materials and also improves the briquette energy efficiency. Peat is a heterogeneous mixture of more or less decomposed plant (humus) material that has accumulated in a water-saturated

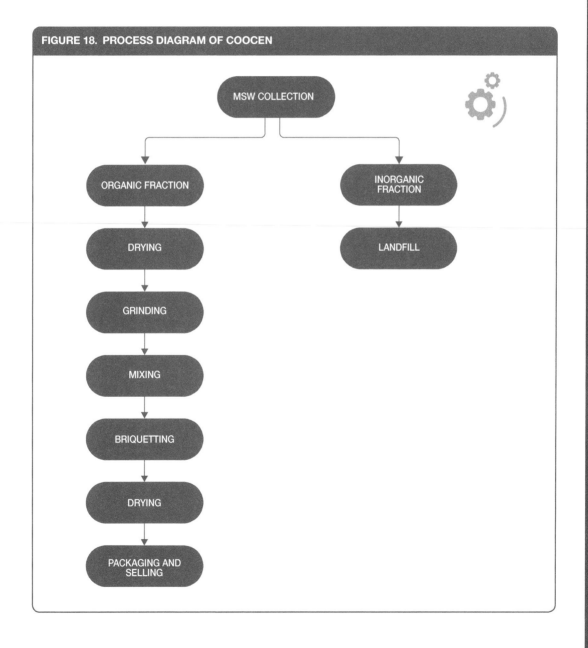

FIGURE 18. PROCESS DIAGRAM OF COOCEN

environment and in the absence of oxygen. Peat is sedentarily accumulated material consisting of at least 30% (dry mass) of dead organic material. Peat is also less compactable than organic waste and thus it provides density to the briquette. However, peat increases production costs due to the extraction and transportation costs.

COOCEN is regularly facing technical constraints due to seasonal changes in the weather pattern and due to the limited processing capacity of the briquette pressing machines. In the rainy season, it takes longer to dry the organic matter which can take up to one week to dry before the organic waste is ready for shredding and mixing with other organic fractions. With respect to the processing capacity, COOCEN is equipped with two manually-operated mechanical briquette-pressing machines which have the capacity to produce 10 tons of briquettes per day, but with automated machines this could be increased to 30 tons per day. However, funds to invest in this technology are hard to get.

Funding and financial outlook

The total capital investment of COOCEN is USD 162,075 comprising of building which accounts for 67% of the total investment and machinery accounting for 33% of the total investment (Table 9). The project secured funding of USD 162,075 from an EU grant during its establishment phase in 2002. In 2007, 7 ha of land was provided by the Kigali City Council for the briquette production plant in Nyamirambo District. In the same year, COOCEN received further financial assistance with a grant of USD 43,760 from the UNDP GEF Small Grant Programme and a bank loan to the value of USD 24,311 was also secured.

TABLE 9. COOCEN INVESTMENT, OPERATIONAL AND MARKETING COST

ITEM	AMOUNT (USD)
Investment cost (USD):	
Land	Free
Buildings	108,590
Machinery / equipment	53,485
Total investment cost (USD)	**162,075**
Operational costs (USD/year):	
Waste transportation and collection	42,788
Electricity	49
Water	175
Wages and salaries	38,898
Repairs and maintenance	9,724
Marketing	3,241
Depreciation	10,805
Total operational costs	**105,680**
Revenue (USD/year)	
Sales of briquettes (1,500 ton at 122 USD/ton)	183,000
Sales of compost (50 ton at 5.67 USD/ton)	284
Total revenue from briquette and compost sales	183,284
Revenue from waste collection service	144,000
Profit before tax (PBT) – briquette business	**77,604**
Profit before tax (PBT) – waste collection and briquette business	**221,604**

Socio-economic, health and environmental impact

COOCEN provides waste collection service to communities, contributes to cleaning of the city and provides sanitation services to local communities (more than 4,000 households) which benefit from improved health and sanitary conditions. Emissions from wood fuel stoves without proper ventilation contain poisonous fumes that can cause respiratory and other human health impacts on women and children, who are traditionally charged with the duty of cooking in Africa. Many more suffer respiratory illnesses resulting in reduced productivity, quality of life and exert an additional burden to the community. The improper waste management can result into bad odor, methane gas explosions, risks of garbage landslides and groundwater pollution. However, from an air quality perspective, also dry fuel can result in net negative health impacts if households do not use safer cooking stoves, or switch from gas to briquettes (Winkler et al., 2017).

COOCEN improves both the efficiency of cook stoves in order to close the gap between supply and demand of fuelwood and charcoal. Harvesting of trees for fuel wood and making charcoal contribute to pressures on forests. Briquettes are more efficient and burn more cleanly, preventing release of excess greenhouse gases that are contributing to climate change. COOCEN's briquetting project contributes to reduction of deforestation by avoiding the burning of 1,800 tons of firewood per year or the cutting of at least 9,000 trees per year, which represents around 9 ha of forest plantation (GEF, 2012). The project has also contributed to reduction of 297 tons of CO_2 emissions per year.

The project not only prevents pollution by implementing better waste management, it also recycles materials that would otherwise go to waste. In addition to these, the cooperative employs 110 persons, mostly women, who earn at least 50 USD per month. In Rwanda, nearly 60% of the population lives below the poverty line, with almost 40% living in extreme poverty on less than USD 0.90 per day (http://www.feedthefuture.gov/country/rwanda). In terms of employees' safety and health, employees are equipped with gloves, protective masks and boots to protect them from injuries and respiratory diseases while manipulating garbage.

Scalability and replicability considerations

The key drivers for the success of this business are:
- Strong partnership with city municipality.
- Regulations against cutting down trees.
- Government policy that promote renewable energy sources.
- Rising prices of fuel wood and charcoal.

The briquette making project by COOCEN has shown the importance of empowering community based organizations as key actors in environmental protection. It also demonstrated that socio-economic benefits are key for project sustainability. Kigali City of Rwanda is making progress towards solid waste management partly because of the house-to-house collection system and a franchise system which involved collection, treatment, recycling and disposal of residues. This system in Kigali City is worthy of emulation by cities in other developing countries. High charcoal price is a pre-requisite for the business to be up-scaled and replicated in other regions. There are already on-going projects which demonstrate the replicability of COOCEN in Kigali. For example, a larger-scale project supported by UNIDO, where the biggest garbage collection company in Kigali started to make and promote use of briquettes at the start of 2011 is evidence that this business can be scaled-out and replicated in other cities. Since the enterprise requires procuring municipal solid waste, developing strong partnership ties with city municipalities is important for reliable supply of input.

Summary assessment – SWOT analysis

The strength of the cooperative business emanates from the fact that it is vertically integrated coupled with a strong marketing strategy and securing of offtake contracts with its customers (Figure 19). Government support for alternative sources of energy and rising prices of wood and charcoal are seen as key opportunities for the business. However, the cooperative is facing technical constraints due to limited drying capacity particularly during rainy season and processing capacity of the briquette pressing machines, limited human and institutional capacity, and limited availability of capital which hinder expansion of the business. It is also anticipated that overtime competition in briquette production will increase. COOCEN has a strategy to improve its manufacturing process and the quality of the final product. The major threats to the business are power shortages, lack of a well-coordinated institutional framework to manage existing and prospective investments, lack of clear technology standards and regulations, as well as unclear processes for approving investments.

FIGURE 19. SWOT ANALYSIS FOR COOCEN

	HELPFUL TO ACHIEVING THE OBJECTIVES	HARMFUL TO ACHIEVING THE OBJECTIVES
INTERNAL ORIGIN ATTRIBUTES OF THE ENTERPRISE	**STRENGTHS** • Vertically integrated cooperative with abundant availability of organic waste and long term contract with customers • Diversified revenue streams from waste collection and briquette sales • Strong marketing strategy effectively created market	**WEAKNESSES** • Limited drying capacity particularly during rainy season • Limited processing capacity of the mechanical briquette machines • Part of the collected MSW is still dumped to landfill sites • Lack of finance to invest in more automated machines and to expand business • Limited human and institutional capacity
EXTERNAL ORIGIN ATTRIBUTES OF THE ENVIRONMENT	**OPPORTUNITIES** • Government support for alternative source of energy • Rising prices of wood and charcoal • Briquette market growth • Possibility of recycling plastics as an additional income generating activity	**THREATS** • Competition from other briquette manufacturing businesses • Seasonal changes in the weather pattern affects production • 2010 Rwanda development strategy includes developing the wood fuel production, is a counter-force to briquette market growth • Power shortages • Lack of a well-coordinated institutional framework to manage existing and prospective investments • Lack of clear technology standards and regulations

Contributors

Jasper Buijs, Sustainnovate; Formerly IWMI
Johannes Heeb, CEWAS, Switzerland
Josiane Nikiema, IWMI, Ghana
Kamalesh Doshi, Simplify Energy Solutions LLC, Ashburn, Virginia, USA

References and further readings

Feed the future. www.feedthefuture.gov/country/rwanda (accessed June 24, 2013).

Global Environmental Facility (GEF). 2012. Briquette as an alternative fuel to fuel wood and prevent deforestation, Kigali, Rwanda: GEF, UNDP. http://goldcoastoptometrist.com/Images/img/112002313.pdf (accessed Nov. 6, 2017).

Ndegwa, G., Breuer, T. and Hamhaber, J. 2011. Wood fuels in Kenya and Rwanda: Powering and driving the economy of the rural areas. *Rural 21*, 02/2011, p. 26–30.

United Nations Development Programme (UNDP). 2012. Comparative experience: Examples of inclusive green economy approaches in UNDP's support to countries. UNDP. https://goo.gl/vh2szj (accessed Nov. 5, 2017).

Winkler, M.S., Fuhrimann, S., Pham-Duc, P. et al. 2017. Assessing potential health impacts of waste recovery and reuse business models in Hanoi, Vietnam. *International Journal of Public Health* 62 (Suppl 1): 7–16.

Case descriptions are based on primary and secondary data provided by case operators, insiders, or other stakeholders, and reflect our best knowledge at the time of the assessments 2012/13. As business operations are dynamic data can be subject to change.

CASE

Briquettes from agro-waste and municipal solid waste (Eco-Fuel Africa, Uganda)

Solomie Gebrezgabher and Charles B. Niwagaba

Supporting case for Business Model 2	
Location:	Lugazi Town, Buikwe District, Uganda
Waste input type:	Agro-waste, municipal solid waste
Value offer:	Briquettes (Clean cooking fuel), biochar
Organization type:	Private
Status of organization:	Operational (since 2010)
Scale of businesses:	Small
Major partners:	National Bureau of Standards, Calvert Foundation (equity), Global Catalyst Initiative (grant)

Executive summary

Eco-Fuel Africa (EFA), located in Lugazi Town, Buikwe District, Uganda, converts farm and municipal waste into briquettes and biochar fertilizer. With good understanding of local fuel usage conditions, EFA ingeniously developed simple, low-cost, easy-to-use technologies – *kilns* for carbonization of waste and *eco-fuel press machine* – to convert it into briquettes, which are cheaper than charcoal and other

KEY PERFORMANCE INDICATORS (AS OF 2012)						
Land use	0.4 ha and 0.8 ha in 2 sites					
Capital investment:	USD 10,500 owner's investment and USD 60,000 from donors; in 2013, USD 372,892 capital required to expand the business					
Labor:	19 full-time and 3 part-time workers					
Total cost of operation in 2012:	USD 98,259 per year					
Output:	200 tons of briquette per year sold for 170 USD/ton					
Potential social and/or environmental impact:	Household savings 200 USD/year, women retailers earn 1,825 USD/year, 1,500 farmers earn 360 USD/year, 43 micro-franchisees earn 1,728 USD/year, job creation, avoidance of GHG emissions and improvement of educational opportunities for women					
Financial viability indicators:	Payback period:	N.A.	Post-tax IRR:	N.A.	Net profit	USD 3,000

briquettes. EFA implements a micro-franchising system whereby it trains its important chain actors (i.e. rural farmers, micro-franchisees and women retailers) to produce and distribute its briquettes to its final customers (i.e. poor households). The project, in addition to combating deforestation and climate change, generates jobs, creates entrepreneurs through its micro-franchising scheme and boosts rural incomes. In addition to the positive effect from the business, a portion of the business' income is donated to tree-planting initiatives to restore destroyed forests.

Context and background

In Uganda, over 90% of the household energy is derived from biomass, mainly firewood and charcoal. The continuous dependence on firewood and charcoal contributes to deforestation. As forests disappear, gathering of firewood, which is mainly done by women and children, becomes difficult. Inspired by the problem of collection of firewood, and by the problems girl children were going through in missing school to fetch firewood, as well as the rate at which Africa was losing forest cover, EFA set out to find a solution. The enterprise invented a simple technology, which can be used by poor communities to convert farm and municipal waste into briquettes and biochar fertilizers. The briquette made, known as 'green charcoal' is a carbon neutral cooking fuel that is made from renewable biomass waste such as sugarcane waste, coffee husks and rice husks. In Uganda, the institutional setting in waste management and recycling supports innovations in renewable energy. However, at the assessment time, no statutory guidelines were available for carbonization and charring.

Market environment

The enterprise's target customer segment is households living in villages, who rely on firewood and charcoal for fuel. Uganda has faced rising charcoal prices due to, among other factors, increased levies on charcoal burners by the government of Uganda in recent years. Between 2009 and 2011, the price of charcoal increased from 0.25 USD/kg to 0.60 USD kg (an increase of 140%) (Ferguson, 2012). With soaring charcoal prices and increased awareness about the problems related to charcoal use, there is increased demand for *cheap* and *clean fuel* for cooking. Briquettes can serve as a direct replacement for firewood and charcoal. This gives EFA the opportunity to tap into the growing market where charcoal prices are rising. It is also planning to tap into other market segments such as small enterprises (restaurants) and institutions (schools) in the near future. Market competition is relatively moderate. Although there are a number of other producers producing briquettes such as Kampala Jellitone Suppliers Ltd. (KJS) and other small informal producers of briquettes, EFA's briquettes are cheaper as it uses mechanical methods with very little electricity input which keeps costs lower than those of competitors (Table 10). EFA business has a great potential as it has a low investment cost while at the same time, the product has a high market demand.

TABLE 10. PRICES OF BRIQUETTES AND ALTERNATIVE FUELS (DEC 2011)

FUEL TYPE	PRICE (USD/KG)
Eco-Fuel Africa briquettes	0.17
Firewood	0.24
Kampala Jellitone Suppliers Ltd. briquettes	0.28
Informal producers briquettes	0.40
Charcoal	0.60

Source: Ferguson 2012; Personal communication with Eco-Fuel Africa; Personal communication with KJS.

Macro-economic environment

In Uganda, wood is by far the most important source of energy, even though the importance of petroleum and hydroelectric power is growing. The contribution of firewood and charcoal to Uganda's GDP is estimated at USD 48 million and USD 26.8 million respectively (UNDP, 2011). In terms of employment, biomass production creates nearly 20,000 jobs for Ugandans. The fact that the biomass wood industry represents significant economic activity implies that wood fuel will continue to be the dominant source of energy in Uganda for the foreseeable future. This has implications for briquette business as the success of briquette business depends on its competitiveness to the wood fuel/charcoal.

In September 2002, the Government of Uganda adopted a new energy policy. The main policy goal is to meet energy needs of the Ugandan population for social and economic development in an environmentally sustainable manner by substantially using modern renewable energy. The overall policy goal is "to increase the use of modern renewable energy, from 4% to 61% of the total energy consumption by the year 2017." There is still limited use of efficient wood fuel, charcoal stoves and biogas in households, institutions and industries. To support alternative clean energy initiatives, government strategy on the demand side is dissemination of more energy efficient technologies (Renewable Energy Policy, 2007).

Furthermore, with support from the UNDP, the government is implementing key interventions in charcoal production which includes increasing the charge that the National Forestry Authority levies on charcoal burners. This provides an opportunity for alternative fuels to compete further with charcoal.

Business model

Briquettes are sold to households via women retailers (Figure 20). The business invented two low-cost and energy-efficient technologies, namely low-cost kiln, which carbonize agricultural waste, and briquetting machine, also called eco-fuel press machine. EFA has invented simple tailor-made briquetting technology which does not need electricity to operate and which can be easily used and maintained by people with limited skills. EFA leases the kilns to farmers and provides the farmers training on how to convert their agricultural waste into charcoal powder using the kilns. The eco-fuel press machine is used by the micro-franchisee to convert charcoal powder brought from farmers to clean burning briquettes. The micro-franchisees sell all the briquettes to EFA, which packages and sells them to its network of women retailers.

Through micro-franchising, EFA have created a decentralized network of village based micro-factories using their already tested technology and business model to convert locally sourced biomass waste into briquettes (green charcoal) and making it easily accessible through women retailers to local people. This eliminates the need to transport biomass waste and green charcoal over very long distances, keeps the cost of green charcoal down which makes it affordable and creates local sustainable jobs.

Value chain and position

The briquette value chain involves three important actors, namely farmers, micro-franchisee and women retailers (Figure 21). EFA is the focal point in the value chain. It is involved in technology transfer and in training each of the chain actors. It provides training to the farmers to convert their agricultural waste into charcoal powder using kilns invented by EFA. The kilns are made out of old oil drums and provided to farmers on a lease-to-own basis. The farmers sell the powder directly to EFA or to the local micro-franchisee. The charcoal powder is then converted into briquettes using the eco-fuel press machine. The press machine is designed to ensure that it can be operated and maintained by local people with no or little formal education. EFA recently invented a low-cost briquetting machine

CASE: BRIQUETTES FROM AGRO AND MUNICIPAL WASTE

FIGURE 20. EFA BUSINESS MODEL CANVAS

KEY PARTNERS
- National Bureau of Standards (UBOS)
- Calvert Foundation
- Global Catalyst Initiative

KEY ACTIVITIES
- Producing and selling briquettes
- Training of farmers on how to carbonize
- Production and sale/lease of production hardware (kilns, pressing machines)
- Research and development
- Maintenance of franchised machines
- Network (franchisees, farmers) management

VALUE PROPOSITIONS
- Less expensive and clean cooking fuel
- Simple, steady franchise income generation by producing briquettes
- Income generation from selling charcoal powder
- Low cost organic fertilizer

CUSTOMER RELATIONSHIPS
- Network of distributors
- Staff trainers from EFA
- Staff trainers from EFA

CUSTOMER SEGMENTS
- Households
- Micro-franchisee (briquette pressing)
- Farmers (carbonize agro waste)

KEY RESOURCES
- Technical competency
- Kilns
- Eco-fuel press machine
- Agricultural residues
- Capital
- EFA brand
- Networks

CHANNELS
- Women retailers
- Direct sales at EFA
- Direct lease

COST STRUCTURE
- Investment cost (land, kilns, eco-fuel press machine)
- Research on technology
- Buy-back of briquettes made by franchisees
- O&M cost (training to farmers and micro-franchisee, maintenance of machines, labor cost)

REVENUE STREAMS
- Sales of briquettes
- Franchising fees
- Lease of kilns

SOCIAL & ENVIRONMENTAL COSTS
- Potential loss of income for firewood and charcoal traders
- Possible workers' health risk while manually handling waste
- Air quality decline where households used gas before

SOCIAL & ENVIRONMENTAL BENEFITS
- Contribute to slowing the rate of deforestation
- Reduce indoor air pollution
- Reduce GHG emissions
- Contribute to better MSW management
- Creation of entrepreneurs and jobs
- Contribute to improving the educational opportunities among girls and women

called eco-fuel press which compresses charcoal powder bought from farmers into clean burning fuel briquettes without using electricity.

Each micro-franchisee can make enough fuel briquettes to meet energy needs of at least 250 local households. EFA mainly makes money from micro-franchising through leasing the technology. Micro-franchisees also pay EFA for training and business support. The micro-franchisees sell all the briquettes to EFA which are packaged and sold to its network of women retailers. Most of these women are illiterate. EFA trains these women thoroughly in areas such as basic book keeping, marketing and customer service. EFA builds a kiosk for each of the selected women after 3 days training, which they use as a retail shop to sell EFA's briquettes to final users. EFA's women retailers sell other items like fruits and vegetables in addition to EFA's briquettes at the kiosks.

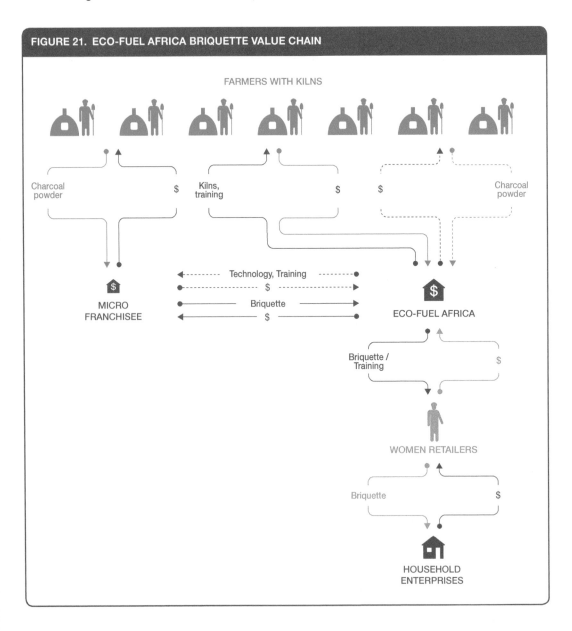

FIGURE 21. ECO-FUEL AFRICA BRIQUETTE VALUE CHAIN

Each community-based briquetting micro-factory needs 10 farmers with kilns who supply the char needed by the factory to make fuel briquettes, five local retailers to sell briquettes to the final consumers in the local community and five employees to run the machines, handle the packaging and distribution of the fuel briquettes. Each of these farmers can earn up to 30 USD/month in extra income. These farmers are also able to use these kilns to make organic fertilizers called biochar which helps them to increase their farm yields by over 50%. Each of these micro-franchises will earn at least 1,728 USD/year. Each of these micro-retailers can earn up to 152 USD/month in extra income. These people are from local community with limited skills.

EFA is making profits and has plans to scale up its business and to serve other customer segment such as small enterprises. EFA has a challenge of attracting funding, which has slowed down their expansion plans. They are growing the business slowly, utilizing internally generated funds. As a long-term strategy, EFA intends to construct a training centre in Lugazi, Buikwe District, Uganda. Investment is already made on two acres of land in this area valued at USD 13,000 where the training centre will be constructed. This centre will enable the enterprise to adequately train its micro-franchisees, farmers with kilns, women retailers and other stakeholders.

Looking at the supply side of the value chain, EFA sources its input from various farmers. It relies on their farm productivity and the resulting farm residue to produce the briquettes. Supplier power is weak as the reuse of farm residue and MSW is very limited in Uganda and thus the farm residues have low market value. But in the future, with the emergence of more briquettes, compost and other reuse businesses, supplier power is expected to be higher. Furthermore, new businesses with more automated and efficient technology and a resultant low-priced briquette pose a threat to EFA whose operations are mechanical and less efficient. On the demand side, EFA targets households who previously relied on firewood for cooking. Experience has shown that, even where cleaner fuels are available, households often continue to use simple biomass fuel as they are more familiar with it. EFA must maintain a price that is lower than firewood/charcoal as households will easily shift to firewood. Buyer power thus plays an important role. There is also the threat of substitutes which exists when the demand for the product is affected by a change in price of a substitute product. Market competition from existing briquette businesses is low. There are few briquette businesses (less than 10) which are operating at the same scale of operation as EFA and only one business operating at a larger scale (about 2,000 tons/year). Most of the briquette businesses are small scale and informal. So far, EFA has a competitive advantage over other producers since it is retailing the briquettes at a lower price and demand is constantly outstripping supply. Briquetting industry is in its infancy in Uganda and even with the emergence of more businesses, the increase in market growth (expanding market) would result in increased revenues. The market for briquettes can grow based on households, institutions and industrial sectors shifting to briquettes for their fuel demand.

Institutional environment

The body charged with the duty to oversee and regulate activities in waste management is the National Environment Management Authority (NEMA). It is responsible for ensuring that waste management activities, e.g. recycling, is carried out in a sustainable manner and do not pollute the environment. Others institutional agencies include the Ministry of Energy and Mineral Development (MEMD). The MEMD produced an energy policy for Uganda in 2002 and a renewable energy policy for Uganda in 2007. This was reinforced in the first National Development Plan (NDP) 2010, and in the current NDP II, 2015–2020.

The Renewable Energy Policy 2007 called for innovations and research in waste management and recycling. To promote the conversion of municipal and industrial waste to energy, the government will provide incentives for the conversion of wastes to energy and put in place fiscal measures that will discourage open burning or disposal of wastes without extracting their energy content. This will cover the conversion of waste to energy through direct combustion, gasification or biological conversion to biogas and therefore wastes will become part of the energy resource base. To foster this development, MEMD will work with municipal authorities and industries that generate lots of waste in developing this potential. Appropriate incentives shall be put in place to promote the conversion of waste to energy. This could be through the Credit Support Facility (CSF), tax waivers and other incentives.

However, no statutory guidelines are available for carbonization and charring. The government is implementing key interventions in charcoal production and is increasing levies on charcoal burners. The Uganda National Bureau of Standards (UNBS) is a key institution, charged with ensuring that products on the market including packaged charcoal meet certain quality standards. However, all charcoal in the market in Uganda is produced and sold by the informal sector and is therefore not certified.

Technology and processes

One of the most common variables of the biomass briquette production process is the way the biomass is dried out. Manufacturers can use torrefaction or carbonization, based on increasing degrees (temperatures, oxygen) of pyrolysis. Researchers concluded that torrefaction and carbonization are the most efficient forms of drying out biomass, but the use of the briquette determines which method should be used but all of them involve heating biomass with little or no oxygen to drive off volatile gasses, leaving carbon behind. The EFA invented a low-cost kiln made out of old oil drums. The kiln carbonizes agricultural waste to produce charcoal powder through pyrolysis. The charcoal powder is sieved and converted into briquettes and the remaining coarse material is mixed with compost and used as organic fertilizer or as biochar (Figure 22).

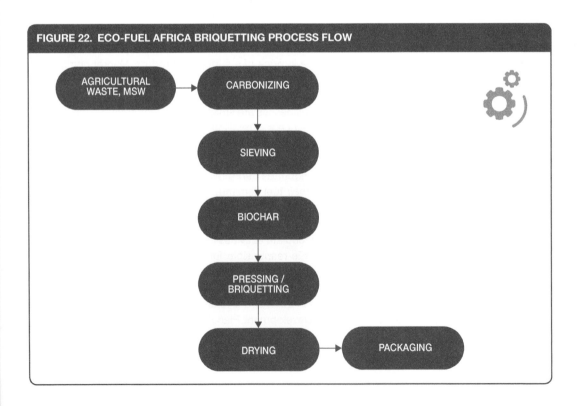

FIGURE 22. ECO-FUEL AFRICA BRIQUETTING PROCESS FLOW

Compaction is another factor affecting production. Some materials burn more efficiently if compacted at low pressures, such as corn stover grind. Other materials such as wheat and barley straw require high amounts of pressure to produce heat. There are also different press technologies that can be used. A piston press is used to create solid briquettes for a wide array of purposes. EFA has also invented a low-cost briquetting machine called eco-fuel press, which compresses charcoal powder bought from farmers into clean burning fuel briquettes. The eco-fuel press requires no electricity and is easy to use. Previously, the machines used by the enterprise were powered by electric motors which required constant monitoring and expensive repairs. With the prevailing unreliable electricity grid, production stoppage was a major problem. The new machine makes much denser briquettes which are more resistant to transport than briquettes produced using the old machine. There are no binders involved in this process. The natural lignin in the wood binds the particles of wood together to form a solid briquette.

The finished briquettes are dried through sun drying which can take up to three to four days. The briquettes are finally packaged in clear plastic bags printed with the enterprise's logo. The technologies invented and used by EFA are simple and low-cost, require no specialized skills and are suitable for the local conditions. EFA provides the workers with hand gloves.

Funding and financial outlook

EFA started with a capital of USD 500 from personal equity. It received a grant of USD 10,000 from the Ugandan government. In 2011, EFA received a grant of USD 20,000 from Calvert Foundation and USD 40,000 from Global Catalyst Foundation. Part of the revenues generated by the business and the grants received are invested to expand the business. With support from the Unreasonable Institute (https://unreasonablegroup.com), EFA was able to raise more than USD 3 million in funding and to be profitable, earning USD 1.2 million in revenue (https://vimeo.com/146802104).

Socio-economic, health and environmental impact

Use of EFA's cooking briquettes reduces the rate of deforestation, avoids GHG emissions, reduces indoor air pollution and improves educational opportunities among girls and women by eliminating the need for collecting wood.

In 2015 and after receiving attention by different investors, EFA was able to claim the following impact:
- Brought clean cooking fuel to over 105,000 households served daily. These households are now able to save up to half of the money they previously spent on charcoal from wood, and with these cost savings, they are able to improve their household living conditions like cooking more consistent meals. Over 57,500 marginalized girls enabled to enrol, stay and study in school. Some of these girls could not previously attend school because they had to walk arduous distances to gather wood for their households.
- Increased incomes and food harvests of 3,500 farmers, about 40% of which are women, who use EFA technology to convert farm waste into organic fertilizers (biochar). Farmers earn on average 360 USD/year in extra income as a result of EFA's project.
- Turned 2,300 local women into micro-retailers of clean cooking fuel. All these women had no jobs before they started retailing for EFA. These women now earn about 1,825 USD/year from clean energy retail businesses.
- 500,000 acres of forests saved in averted deforestation.
- About 127,650 tons of CO_2 mitigated every year.

Scalability and replicability considerations

The key drivers for the success of this business are:
- Regulations against cutting down trees.
- Increased charge that the National Forestry Authority levies on charcoal burners.
- Rising prices of charcoal.
- Government policy that promote renewable energy sources.
- Access to both sufficiently dense community networks and rural markets without electricity.
- Charismatic leader with a business plan gaining international attention.

EFA's business model is based on low-cost and simple technologies that can easily be used by local communities. Within Uganda, EFA was planning to expand to all regions of Uganda by 2015 and to up-scale its operations by building a bigger factory near industrial sources of sugarcane waste to meet growing demand. This business model is highly replicable in other low-income countries where firewood is predominantly used, where wood is scarce and where agricultural waste or municipal solid waste is abundant. With raising more capital to improve its technology and with the franchise model, the business could be out-scaled to other regions in sub-Saharan Africa, latest by 2010. For this business to be out-scaled to or replicated in other regions, high charcoal prices and presence of regulatory frameworks on use of firewood are required.

FIGURE 23. SWOT ANALYSIS FOR EFA

	HELPFUL TO ACHIEVING THE OBJECTIVES	HARMFUL TO ACHIEVING THE OBJECTIVES
INTERNAL ORIGIN ATTRIBUTES OF THE ENTERPRISE	**STRENGTHS** • Low-cost technology • Dynamic and skilled entrepreneur • Well distributed production and micro-franchising system • Access to rural markets with no electricity • Good relationship with chain actors and investors	**WEAKNESSES** • Poor logistics in transporting briquettes to retailers • Lack of local technical and institutional capacity and finance to improve technology • Lack of standardization of the briquettes • The low-margin, high-volume nature of the business with insufficient profit margins for green charcoal producers
EXTERNAL ORIGIN ATTRIBUTES OF THE ENVIRONMENT	**OPPORTUNITIES** • Possible patenting of technology (IP) • Good opportunity for up-scaling through franchising • Good image through its tree-planting initiatives • Increasing prices of substitute products (charcoal) • Increasing demand and market growth of briquettes • Supportive local community	**THREATS** • Competition for input may raise prices of inputs • Low farm productivity (harvest fail) may lead to shortage of supply of farm waste • Lack of finance may slow down expansion and limit research efforts • Competition from other similar products and technologies • Inadequate policy, regulatory and institutional framework and lack of product quality and standards

Summary assessment – SWOT analysis

The key strengths of the business are its application of low-cost technology coupled with a well distributed production and franchising system which contributed to the competitive advantage that EFA has over its competitors (Figure 23). The franchise scheme presents EFA a good opportunity to expand its business. However, lack of finance may slow down expansion and limit research efforts to improve technology.

Contributors

Sanga Moses, Eco-Fuel Africa
Johannes Heeb, CEWAS
Jasper Buijs, Sustainnovate; Formerly IWMI
Josiane Nikiema, IWMI

References and further readings

Eco-Fuel Africa (EFA), 2012. https://www.facebook.com/ecofuelafrica1 (accessed Nov. 6, 2017).

Ferguson, H. 2012. Briquette business in Uganda: The potential for briquette enterprises to address the sustainability of the Ugandan biomass fuel market. London: GVEP International.

Ministry of Energy and Mineral Development (MEMD). 2002. The energy policy for Uganda. Amber House Kampala, Uganda: MEMD.

Ministry of Energy and Mineral Development (MEMD). 2007. Renewable energy policy for Uganda. Amber House Kampala, Uganda: MEMD.

Musoba, E. 2012. Fuel efficiency of wood cooking stoves in communities adjacent to Kibale Forest National Park, Uganda. Unpublished MSc. dissertation. College of Agricultural and Environmental Sciences, Makerere University.

Rawsthorn, A. July 2012. Cleaning up the African kitchen. The New York Times. www.nytimes.com/2012/07/30/arts/design/cleaning-up-the-african-kitchen.html (accessed March 13, 2012).

World Food Programme (WFP). 2009. Safe access to firewood and alternative energy in Uganda: An appraisal report.

Case descriptions are based on primary and secondary data provided by case operators, insiders, or other stakeholders, and reflect our best knowledge at the time of the assessments 2012–2015. As business operations are dynamic, data can be subject to change.

BUSINESS MODEL 2
Briquettes from municipal solid waste

Krishna C. Rao and Solomie Gebrezgabher

A. Key characteristics

Model name	Briquettes from municipal solid waste (MSW)
Waste stream	Organic waste – Organic component of MSW and agro-waste (crop residues)
Value-added waste product	Briquettes used as clean cooking/heating fuel
Geography	Region with lack of ease in availability of fuel wood
Scale of production	Small scale (<300 tons per year) and medium scale (300–1,500 tons)
Supporting cases in the book	Kigali, Rwanda; Kampala, Uganda
Objective of entity	Cost-recovery [ ]; for profit [X]; social enterprise [ ]
Investment cost range	USD 30,000 to USD 450,000 for medium scale
Organization type	Private or cooperative public-private partnership
Socio-economic impact	Reduction in deforestation and environmental pollution, reduced indoor air pollution resulting in improved health for household and employment generation, improved educational opportunities for girls
Gender equity	Beneficial to women and children using fuel with less indoor air pollution than firewood; time savings for girls in fuel collection which can be used for education.

B. Business value chain

The business model is initiated by either a standalone private enterprise or a cooperative under public-private partnership (PPP) where a private entity partners with the municipality to manage the solid waste generated by the city. The business processes organic component of municipal solid waste (MSW) and convert it into briquettes that can be used as clean fuel. The key stakeholders in the business value chain are the waste suppliers – either household or the municipality, product distributors and end-users of the briquettes (household and businesses) (Figure 24).

The process of briquetting involves reducing moisture content in the organic waste, which is shredded and the biomass is compressed at high temperature and using a binding agent. The organic component in MSW consists of multiple substances with different calorific values. Collaboration with a research institution or in-house research laboratory will help in developing suitable process to produce briquettes with higher calorific value. Another option is to produce charcoal from organic waste by carbonizing/ burning it in low-oxygen atmosphere. The resulting charred material is compressed into briquettes.

BUSINESS MODEL 2: BRIQUETTES FROM MUNICIPAL WASTE

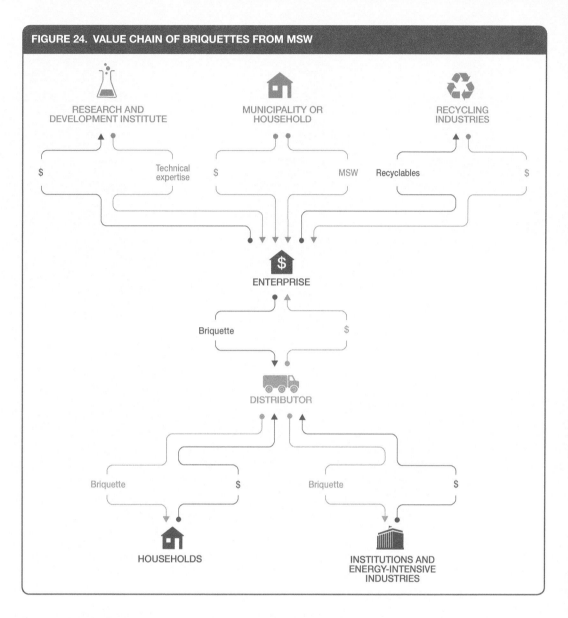

FIGURE 24. VALUE CHAIN OF BRIQUETTES FROM MSW

The briquette is used as fuel for cooking and/or thermal energy in small and medium industries and households. Briquettes can also be used as fuel for gasifiers to generate electricity or powering boilers to generate steam. The business model described is focused on using briquette as fuel for cooking and thermal applications only. The process of making briquettes from MSW requires segregation of organic component, which results in recyclables such as plastics, paper and glass that have good resale value.

C. Business model

The primary value proposition of the business model depends on the entity initiating the business model, which is either a standalone private entity or cooperative. For both the entities, producing high-quality briquettes for households and institutions such as schools and prison and small and medium enterprises who need fuel for cooking and heating is a common value proposition. However, for a PPP, providing waste collection and waste management service is the primary value proposition (Figure 25).

FIGURE 25. BUSINESS MODEL CANVAS – BRIQUETTES FROM MSW

KEY PARTNERS
- Municipality
- Community
- R&D institute
- Other organic waste producers

KEY ACTIVITIES
- Collection of MSW and other organic waste
- Organic fraction separation
- Production and sales of briquettes
- Awareness raising and promotional campaigns
- R&D for briquette combinations
- VER management and administration processes

VALUE PROPOSITIONS
- Environment-friendly briquettes from MSW that are cheaper than charcoal and fuelwood (price leadership)
- Waste collection and management service
- Sale of recyclables recovered from segregation of MSW

CUSTOMER RELATIONSHIPS
- Direct with households for collection of waste
- Short and long-term contract for sale of briquettes with institutional customers direct or with distributors network

CUSTOMER SEGMENTS
- Household and institutional users
- Small and medium enterprises
- Municipality
- Recycling industry
- Voluntary Emissions Reduction (VER) market

KEY RESOURCES
- Consumables (MSW and organic waste)
- Human resource
- Capital, land (via partnership) and equipment
- Certification
- Network of distributors
- Contracts with institutional users
- Research expertise
- Marketing and sales force

CHANNELS
- Direct personal help at point of source of MSW with households
- Supply of briquettes to institutional customers directly
- Distributors
- Micro-franchisees

BUSINESS MODEL 2: BRIQUETTES FROM MUNICIPAL WASTE

COST STRUCTURE

- Investment cost – Land, building, and machinery
- Operational cost – Transportation – MSW and organic waste, labor, disposal of inorganic waste to landfill, utilities, maintenance, marketing and packaging, training of distributors/micro-franchisees and VER costs

REVENUE STREAMS

- Briquette sales
- Waste collection and management fees
- Sale of recyclables
- VER sales

SOCIAL & ENVIRONMENTAL COSTS

- Potential loss of income for firewood and charcoal traders
- Potential health risks for workers at production facility
- Potential environmental risk if the waste is not treated and disposed properly
- Increased health risks if households switch from gas to briquettes

SOCIAL & ENVIRONMENTAL BENEFITS

- Reduces deforestation and GHG emissions
- Saves time in the case of time spent in collecting firewood
- Contribute to improving the educational opportunities among girls who previously missed school to fetch firewood
- Contributes of improved MSW management
- Generates income and employment
- Creates environmental sanitation awareness

The briquettes are delivered to the customers either through direct sales, network of distributors or micro-franchising[1]. The direct sales involves managing a large human resource base for sales and marketing staff. The business requires developing strategic partnerships with municipality to procure MSW and it would likely require contractual arrangements with the municipality. The business will have to collect MSW from the municipal landfill site or have the municipality garbage trucks deliver MSW to the plant. The business can also organize collection of MSW directly from households at a collection fees.

The key activities of the business model are MSW collection and processing, briquette production and sales. Since MSW consists of both organic and inorganic material, the business enterprise must undertake segregation of waste to separate out organic material, which is the key raw material for briquette production. Research and development (R&D) would be a useful activity to ensure high quality and calorific value of the product. However, the cost-benefit of R&D should be assessed and ideally partnership with a research institution would mitigate the risk of need for high-skilled labor.

Key capital costs are building and machinery and primary operational costs are transportation, labor, utilities, marketing and packaging. Briquette sales and waste collection and management fees are the key revenue source. A briquette enterprise is potentially eligible for carbon offset depending on the type of fuel replaced and the baseline used to calculate benefits from reduced greenhouse gas emission and hence there is potential for increasing revenue from sale of carbon. Depending on the scale of MSW processed and managed, the briquette enterprise could apply for Clean Development Mechanism (CDM). However, due to associated transaction costs, a preferred route would be to apply for carbon offset on Voluntary Emission Reductions (VERs). MSW consists of inorganic waste such as plastics, paper and glass that has high resale value and sale of recyclables is another revenue source for the business model.

D. Alternative scenarios

The business model can incorporate two additional value propositions in addition to briquette production from MSW: a) produce low cost compost, a by-product from briquette production and b) vertical integration of business by manufacturing and selling improved cook stoves and ovens (Figure 26).

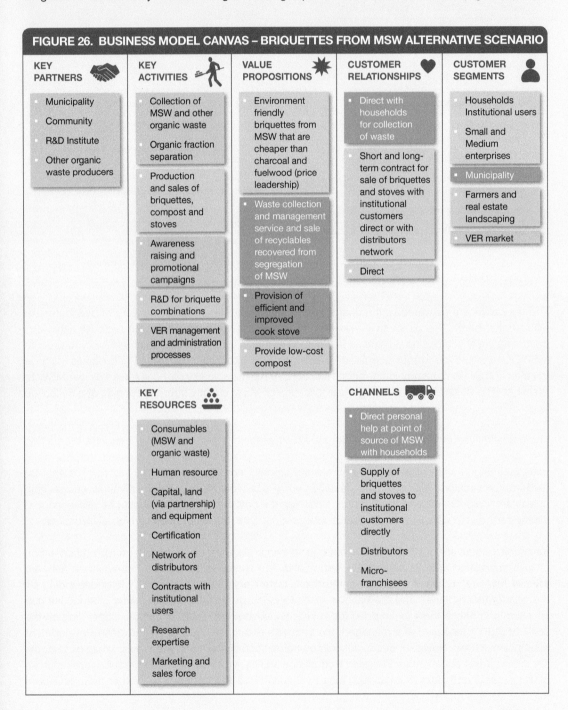

FIGURE 26. BUSINESS MODEL CANVAS – BRIQUETTES FROM MSW ALTERNATIVE SCENARIO

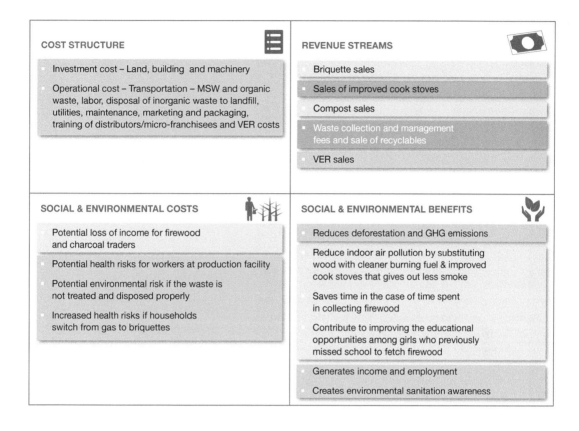

Scenario I: Compost production

Production of briquettes results in generation of organic residual waste which can be used to produce compost. The compost can be either sold to the farmers or landscapers. The additional key activity required for this value proposition is production of compost and related costs incurred. The sales and distribution process will be similar to sales of briquettes.

Scenario II: Manufacturing of improved cook stoves

The business model offers scope for vertical integration since the enterprise into briquette production could potentially manufacture improved cook stoves and ovens that use the briquettes made by the enterprise as fuel for cooking or heating. The sales of stoves could potentially stabilize sale of briquettes as it entices users to not switch to a competing/substituting product. The improved cook stoves have high social benefits for households especially to women and children through reduced indoor air pollution. In addition, with improved cooking efficiency and reduced fuel consumption household incurs savings. The business model does not require significant alteration to its distribution process. The additional key activity required is for the manufacturing of improved cook stoves which has related capital and operational costs incurred by the enterprise. Similar to briquette production, R&D is a required activity to design the cook stoves and oven that meet the customer's requirements. The product also requires specific marketing, certification from independent organization and awareness campaign.

E. Potential risks and mitigation

Market risks: The key customer segment for briquettes are households that do not have access to fossil fuels and are dependent upon firewood and charcoal for cooking. Market risks are high as the willingness to pay is significantly lower among households using firewood for cooking. The business should target diverse customer base to mitigate these risks. It is preferred for the business to target household, institutions and small and medium enterprise as customers. The business could get into long-term bulk contractual arrangement with institutional customers and have assured sales.

Competition risks: The briquette product has strong competition risks from competing products like charcoal, wood, kerosene and LPG. Fuels choice typically depends on consumer preference, price, convenience and at times social status associated in using certain types of fuels like Liquefied Petroleum Gas (LPG). Ideally, briquette should be targeted to customer segment that uses firewood and charcoal because briquettes can be more competitive and efficient. This customer segment has lower competition risks in the urban areas due to scarcity of firewood. However, in the rural areas in developing countries the competition risks for households as customers is high due to free availability of firewood from nearby plantations and forest.

Improved cook stoves and compost have competition risks as there are multiple suppliers of different types of stoves and compost in the market. Stove sales can potentially stabilize briquette market as it lowers chance of customers switching to competing products.

Technology performance risks: The technology used is either mechanical compressing with or without the binding agent or pyrolysis and mechanical compressing. The technology has been widely used commercially and is proven. It doesn't require high skills for operating it and doesn't have complications towards repair and maintenance of equipment.

Political and regulatory risks: In most developing countries, fuel for cooking for household is a social issue and the governments provide subsidy for fuels such as kerosene and LPG. Such fuels are also more reliable and convenient to use. If these competing products are priced lower or even slightly higher than briquettes, it can pose significant risks to the business. Diversifying customer base and including energy intensive small and medium enterprise as primary customers can considerably negate this risk. Increasing government support through financial incentives and policies that promote renewable energy reduces this risk considerably in the long term.

Social-equity-related risks: The model is considered to have more advantages to women as culturally in developing countries women collect fuel wood and do the cooking at household. The model provides employment opportunities in the enterprise producing briquettes. The users of briquettes are low-income households who are using other unhealthy and inefficient or more costly fuels.

Safety, environmental and health risks: The safety and health risks to human arises when processing any type of waste. The risks are even higher when processing MSW. Labor in such enterprises should be provided with appropriate gloves, masks and other appropriate tools to handle the waste to ensure their safety. The risk of environment pollution is high if leachate from MSW is untreated and seeps into groundwater or other natural water bodies. Organic waste when left in open begins to decay and releases methane, which is more damaging to the environment than carbon dioxide. The waste processing technologies are not without problems and pose a number of environmental and health risks if appropriate measures are not taken (Table 11). There is a potential risk for those households where less harmful cooking fuels such as LPG, kerosene or electricity are replaced by biomass briquettes. The risk is lower where also safer cooking stoves will be introduced (Winkler et al., 2017).

TABLE 11. POTENTIAL HEALTH AND ENVIRONMENTAL RISK AND SUGGESTED MITIGATION MEASURES FOR BUSINESS MODEL 2

Risk group	Exposure route					Remarks
	Direct contact	Air	Insects	Water/Soil	Food	
Worker	HIGH	LOW				Health risk for households might increase or decrease depending on the quality of the used fuel. Exposure to sharp objects in MSW, air and noise pollution possible. Fly control measures for MSW and leakage control for composting are required.
Farmer/User						
Community		LOW		LOW		
Consumer		MEDIUM				
Mitigation measures	(gloves/boots)	(mask, ear protection, dust)	(insect control)	(leakage control)		

Key: ☐ NOT APPLICABLE   ▨ LOW RISK   ▩ MEDIUM RISK   ■ HIGH RISK

F. Business performance

This business model is rated high on social impact followed by profitability (Figure 27). The business model provides high number of jobs especially when it is involved in the collection of waste from households. The business model has strong revenue sources from sale of briquette and waste collection and management fees, building on a diverse customer base. It has potential for additional revenue sources from sale of stoves, compost and VERs. The environmental impact is specifically high for regions with deforestation and proper treatment of MSW improves the local health of the environment.

The business model has high potential for replication in developing countries as there are not any strong factors such as new technology, special policies and regulations, institutional capacity, waste availability and so on that can limit replication potential of the business model. The business model has can be scaled horizontally and has potential for vertical scaling by expanding into the business of manufacturing of improved stoves. The business model is straightforward with no sophisticated or innovative financing and technology requirements; however, it requires special partnership arrangement with the municipality for waste collection and management.

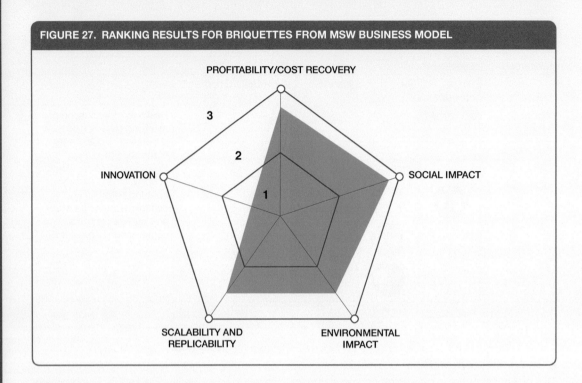

FIGURE 27. RANKING RESULTS FOR BRIQUETTES FROM MSW BUSINESS MODEL

References and further readings

Winkler, M.S., Fuhrimann, S., Pham-Duc, P. et al. 2017. Assessing potential health impacts of waste recovery and reuse business models in Hanoi, Vietnam. International Journal of Public Health 62 (Suppl 1): 7–16.

Note

1 Micro-franchising borrows the traditional franchising concept with scaled-down business concepts found in successful franchise organizations. It operates as a micro-enterprise following proven marketing and operational concepts with systematic replication. The concept is predominant in delivering services to the poor along the lines of microfinance and microcredit. Micro-franchise entrepreneur has similarities to an agriculture extension worker and typically such an entrepreneur sells multiple product like seeds, fertilizers, water filters, Fast Moving Consumer Goods (FMCG) etc.

4. BUSINESS MODELS FOR IN-HOUSE BIOGAS PRODUCTION FOR ENERGY SAVINGS

Introduction

Energy recovery from fecal sludge or kitchen waste through the installation of biogas systems provides opportunities for domestic, institutional and industrial sectors to save on energy costs by using biogas produced onsite for cooking, power generation and lighting. While household biogas installations are very common, experience in institutional biogas systems is limited and is gradually gaining traction in developing countries in Asia and Africa. The consensus is that the larger onsite biogas units that are run by institutions such as schools, hospitals and prisons to manage their waste have proved to have higher viability than the small-scale household bio-digesters.

There are a number of examples where energy recovery from fecal sludge and kitchen waste through the installation of biogas systems has been a success in institutions such as schools, hospitals, prisons and other institutions consisting of large number of residents (**Business model 3: Biogas from fecal sludge at community level** and **Business model 4: Biogas from kitchen waste**).

The business cases presented under these business models are from India, Nepal, the Philippines, Rwanda and Kenya. These businesses were selected as they present a unique example of successful partnership of local authorities, non-governmental organizations and communities.

CASE
Biogas from fecal sludge and kitchen waste at prisons

Krishna C. Rao and Kamalesh Doshi

Supporting case for Business Model 3	
Location:	Nepal, the Philippines and Rwanda
Waste input type:	Fecal sludge, wastewater, kitchen waste
Value offer:	Biogas, bio-fertilizer
Organization type:	Public entity
Status of organization:	First biogas plant operational since 2000
Scale of businesses:	Small, medium and large
Major partners:	International Committee of Red Cross (ICRC); Local technology partners in Nepal, Philippines, Rwanda

Executive summary

The International Committee of Red Cross (ICRC) under its water and habitat unit has implemented numerous institutional biogas sanitation systems across prisons in Rwanda, Nepal and the Philippines in partnership with local organizations for the last 10 years. Biogas sanitation systems are seen as a promising technology for institutional settings in developing countries as they combine effective treatment of human excreta and kitchen waste in a safe and environmentally-friendly manner, while at the same time generating a renewable fuel source for cooking while reducing indoor air pollution and a nutrient-rich fertilizer. The projects reduce the prison costs and contribute to reduction of deforestation from reduced use of fuelwood.

ICRC's prison biogas plants use human waste and in some cases kitchen waste to generate biogas, which is used as fuel for cooking in the prison. The biogas systems consist of fixed dome digesters of varying sizes according to the number of detainees in each prison. The digesters used are of size 10 m^3, 20 m^3, 35 m^3 and 100 m^3 with one to two digesters in each prison. The prison biogas projects resulted in improved sanitation of prisons thereby reducing health risks of inmates. The Nepal prisons are much smaller and the number of detainees ranges from 106 to 270. The prisons in Rwanda are large and typically house around 5,000 detainees.

KEY PERFORMANCE INDICATORS (AS OF 2009)						
Capital investments:	USD 12,960 for all three prisons in Nepal; USD 27,700 for Cagayan de Oro City prison, Philippines; and USD 74,000 for 500 m³ plant in Rwanda					
O&M cost:	2% of capital investment					
Output:	25–62 L/person/day of biogas (higher when kitchen waste is used)					
Potential social and/or environmental impact:	Improved health of detainees, improved sanitation of prisons, reduced air pollution in the kitchen/reduced GHG emissions, reduced deforestation, renewable source of energy for cooking and better landscaping (by use of bio-fertilizers)					
Financial viability indicators:	Payback period:	1.5–5.4 (Nepal prisons)	Post-tax IRR:	N.A.	Gross margin:	N.A.

Context and background

Prisoners are among the world's most discriminated groups often suffering from detrimental sanitary conditions. The main objectives of the systems that are implemented by the ICRC are to improve the sanitary conditions, reduce the health risks and provide a renewable and smoke-free source of cooking fuel. From 2002 to 2009, the ICRC helped build 13 biogas systems in 11 prisons of Nepal (Kaski, Chitwan, Kanchanpur), the Philippines (Cagayan de Oro, Davao, Sultan Kuradat, Manila, Cradle) and Rwanda (Muhanga, Gikongoro, Cyangugu), two of which (Manila and Cradle) were not functioning during surveys between 2009 and 2011.

In 2007, an agreement between ICRC and the local expert partner Biogas Sector Partnership Nepal (BSP-N) was signed to implement five biogas sanitation systems in three district jails in Nepal. The Philippines' prisons, managed by the Bureau of Jail Management and Penology (BJMP), are overcrowded and underfinanced due to a legal system that is unable to keep up with the influx of new suspects. The ICRC implemented biogas systems in several jails including the Cagayan de Oro City jail, which houses more than 1,000 prisoners. In 2009, the BJMP banned the use of firewood in its prisons, due to the deforestation issues it was creating. In Rwanda, the Kigali Institute of Science Technology and Management (KIST) in partnership with ICRC installed large-scale biogas systems including the construction of the system, providing on-the-job training to both civilian technicians and prisoners. Half of the construction cost was paid by ICRC and overseen by ICRC and National Prison Services. The first prison biogas plant (Cyangugu) started its operation in 2001 and has since run with no problems. Since then, KIST has installed biogas digesters in almost half of the 30 prisons in the country. With even the national newspaper reporting on it, it is hoped that more NGOs and government agencies will see the value in small-scale projects like this that not only address sanitation, but also financial, social, and other environmental issues too.

Market environment

According to a study done by the International Centre for Prison Studies (ICPS) in 2008, more than 10 million people are held in penal institutions throughout the world. For every 100,000 population, western Africa and southern African countries have 47 and 219 prisoners respectively. In Asia, South Asian countries has a median rate of 42, while in eastern Asian countries, it is 155. Prison population are growing across the world.

Prisons mostly use fossil fuels or firewood for cooking and incur significant costs. Biogas is a reliable alternative to reduce cost incurred from consumption of these fuels. The need for the biogas systems also arose mainly due to lack of proper sanitation systems in prisons and the associated health risks, risk of groundwater pollution from outdated sceptic tanks, the high costs of obtaining firewood for fuel, as well as the increase in deforestation and GHG emissions. The biogas systems reuse fecal sludge and wastewater in a safe manner to produce cleaner energy thereby solving most of the problems faced by prisons.

Macro-economic environment

In developing countries including Nepal, the Philippines and Rwanda, consumption of firewood far exceeds annual production, causing deforestation. In order to curb this problem, governments are launching different programs and strategies. The Government of Nepal has a long tradition, dating back 1975, for promoting biogas in Nepal through the provision of low-interest-rate loans and subsidies for biogas systems to promote the technology.

In the Philippines, prisons managed by BJMP are posing environmental and fiscal problems due to high prison populations which are consuming much of the BJMP's budget in their need for food. In 2009, the BJMP banned the use of firewood in its prisons and started installing biogas digesters in prisons.

Under the Economic Development and Poverty Reduction Strategy (EDPRS), one of the long-term strategies of the Government of Rwanda is to reduce fuelwood consumption from 94% to 50% through the installations of biogas digesters in both residential homes as well as institutions with a large number of residents such as schools, hospitals and prisons.

Business model

The key value proposition of the biogas plant is to provide improved sanitation service to prison inmates using biogas digesters to process and treat the human waste generated at the prison facility (Figure 28). In the process, the system provides two additional value proposition: a) biogas as a cooking fuel and b) bio-slurry from biogas plant that can be converted into organic compost. Biogas replaces fossil fuels and firewood while the bio-slurry turned into organic compost is used onsite for growing crops and trees. Both these value propositions result in savings for the running of prison operations in terms of money spent on fuelwood or fossil fuel and fees for emptying sceptic tanks. In the Philippines, the prison has organized its inmates to undertake baking activities to create new livelihood opportunities for inmates. Potentially, this could be a source of revenue where baked goods cooked using biogas from human waste is sold in nearby towns.

Value chain and position

The prisons in Nepal, the Philippines and Rwanda rely on the ICRC and local partner organizations to provide the infrastructure, equipment, maintenance toolkit and training for the implementation and operation of a biogas system. The maintenance and operation of the system falls in the hand of prison staff or detainees (Figure 29). ICRC partners with local organizations to provide technical support for the implementation of biogas. By means of user trainings/workshops, the detainees are informed about the biogas system and its operations and maintenance (O&M) requirements of cleaning and flushing toilets, benefits of adding kitchen waste and countermeasures in case of blockages. The vast majority of detainees perceived the biogas systems positively, mainly because it provides a smoke-free source of cooking fuel that contributes to money saving and because it improved the hygienic conditions in and around the prison.

In 2007, with the help of ICRC, biogas sanitation systems were installed in three Nepalese District jails. All five fixed-dome digesters (sizes of 10 m^3, 20 m^3 and 35 m^3) revealed gastight domes and showed high process stability with no accumulation of inhibitory substances. Kitchen waste is added to three out of five digesters to enhance gas production. In the case of the other two digesters, the organic waste is sold to pig farmers.

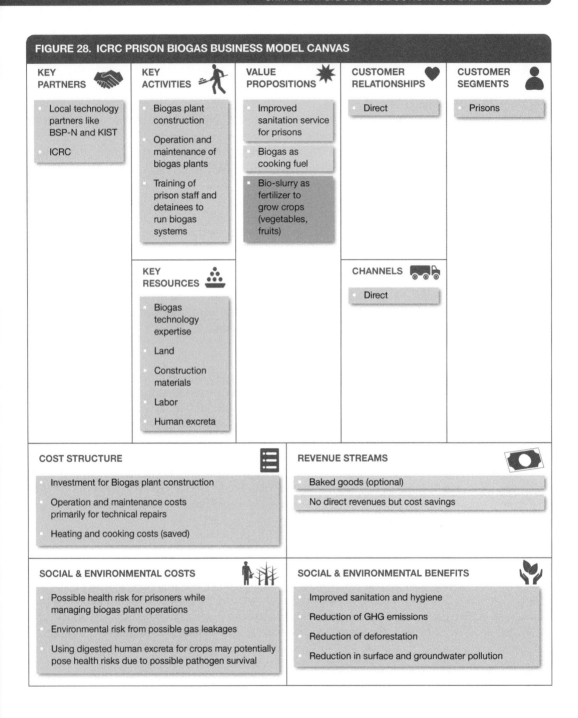

FIGURE 28. ICRC PRISON BIOGAS BUSINESS MODEL CANVAS

Rwanda has 13 prisons with around a total of 54,000–58,000 prisoners in 2007–2015, who previously consumed around 10 tonnes of firewood a day, costing of around 1 billion Rwandan francs (USD 1.7 million). The Rwanda Correctional Services (RCS) started building large bio-digesters in all 13 prisons, financed by the Ministry of Internal Security, and Penal Reform International, with contribution from KIST. Biogas is used for more than 60% of all cooking fuel. KIST staff manage the construction of the system and provide on-the-job training to both civilian technicians and prisoners. They have

CASE: BIOGAS FROM FECAL SLUDGE AND KITCHEN WASTE

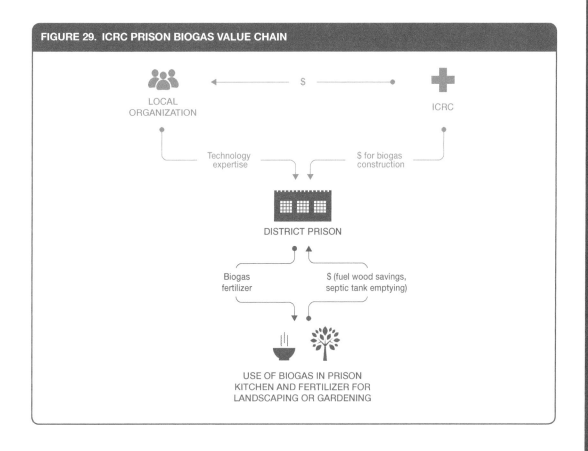

FIGURE 29. ICRC PRISON BIOGAS VALUE CHAIN

constructed a drainage line to take effluent from a prison biogas plant to a test site where experiments are conducted into growing water hyacinth in the effluent treatment ponds. The water hyacinth will be harvested and added to the biogas plants to increase gas production above the rate possible using sewage alone. Rainwater harvesting will be used to dilute effluent as it enters some of the treatment ponds, allowing fish to be farmed there as well. To date the project has trained a total of 150 artisans and technicians, out of which three private businesses have so far been established. It is planned that by 2013, no prison will be using firewood. After at least one year of operation, 11 systems out of the 13 implemented systems were in operation with satisfactory process parameters with daily biogas production ranging between 26 L/person and 62 L/person (obtained in prisons where kitchen waste was added to the digester). The first prison biogas plant started its operation in 2001, and has run with no problems since then.

Institutional environment

Biogas Sector Partnership Nepal (BSP-Nepal) is a professional organization involved in developing and promoting appropriate rural and renewable energy technologies, particularly biogas, effective in improving livelihood of the rural people. It was established in 2003 under District Administration Office, Kathmandu under the Social Organisation Act of the Government of Nepal. The Biogas Support Programme (BSP) was established by the SNV Netherlands Development Organisation (SNV) with the objective of promoting a wide-scale use of biogas as a substitute for fuelwood, agricultural residues, animal dung and kerosene that are generally used for cooking and lighting in most rural households in Nepal. BSP-Nepal is the implementing agency of BSP. BSP-Nepal was established as an NGO in 2003 to take over the implementation responsibility of BSP, which formerly was managed directly by the SNV.

BJMP is an attached agency of the Department of the Interior and Local Government mandated to direct, supervise and control the administration and operation of all district, city and municipal jails in the Philippines with pronged tasks of safekeeping and development of its inmates.

The College of Science and Technology of the **University of Rwanda**, the former KIST in Kigali, Rwanda is the first technology-focused institution of higher education to be created by the Rwanda government in November 1997. Within KIST, the Centre for Innovations and Technology Transfer (CITT) is mandated to transfer technical innovations, managerial and entrepreneurship skills into community applications. In 2005, KIST won an international environmental award in recognition of its innovative work at Cyangugu Prison, the ASHDEN Award for Sustainable Energy. Since winning the ASHDEN Award, KIST has worked in a variety of areas towards goals. They have begun work on a research project to investigate the use of porous volcanic rock inside the digesters. It is hoped that the rock will increase the surface area available for the anaerobic bacteria. After realizing the success of biogas in prisons, the Government of Rwanda introduced the National Domestic Biogas Program (NDBP) to develop a commercial and sustainable domestic biogas sector under the Ministry of Infrastructure with financial and technical support provided by SNV, Deutsche Gesellschaft für Internationale Zusammenarbeit (GIZ) and the Dutch Ministry of Foreign Affairs (DGIS).

Technology and processes

The biogas system uses a number of individual fixed dome-type digesters, each ranging 10–100 m^3 in volume and built in an excavated underground pit. Toilet waste is flushed into the digesters through closed channels, which minimize smell and contamination. The digester is shaped like a beehive, and built up on a circular, concrete base using bricks made from clay or sand-cement. Biogas is stored on the upper part of the digester. The gas storage chamber is plastered inside with waterproof cement to make it gastight. On the outside, the entire surface is well plastered and backfilled with soil, then landscaped. From the manhole cover, the gas is piped underground towards the kitchen where it is used for cooking purposes. The continuous input of waste and the gas pressure push digested effluent out of the bio-digester to a stabilizing tank and from there, to a solid/liquid separation unit. The stabilizing tank allows additional gas production. The solids are composted for three months and then used as fertilizer in the prison gardens. The fertilizer is only used for crops that stand above ground, such as papaya, maize, bananas, tree tomato and similar crops (Figure 30).

The biogas reactor is an anaerobic, sealed chamber that serves as a primary settling tank, with relatively fast passage of the liquid effluent through the chamber and digestion of much of the settled sludge by anaerobic bacteria. In this way, it is much like a septic tank, except that its sealed nature allows all of the biogas (a mixture of methane and carbon dioxide that is released from anaerobic digestion) to be captured and used. Since most of the organic matter is converted to biogas, sludge production is relatively low. The settled sludge usually remains in the unit for several years and, when removed, is relatively pathogen-free, requiring only some post-composting to ensure sterility.

Satisfactory operation of a biogas system can be achieved if adequate attention is given to the site selection and dimensioning of the system. For this, it is crucial to understand the local climatic and geotechnical conditions, sanitary habits, waste flows and power relations in the prisons. Kitchen waste addition can boost (even double) the biogas production, but its use might be in conflict with potential competitors (e.g. local farmers who use it as animal feed). To deal with high fluctuation in detainee numbers, it is advised to install digesters in series instead of a single large one. It is absolutely essential to install condensation traps at the lowest points of the gas pipe. Vapour, a natural component of biogas, condenses in the pipe and eventually leads to blockage of the pipeline so that the gas does not reach the kitchen anymore. Regular emptying of these water traps is crucial.

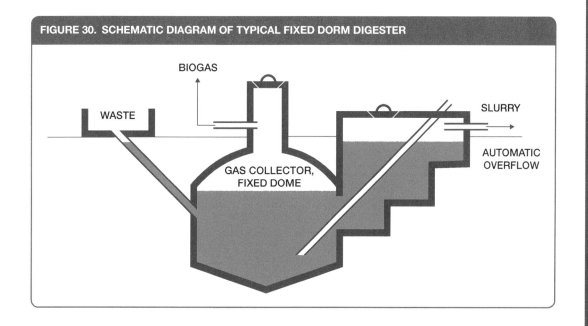

FIGURE 30. SCHEMATIC DIAGRAM OF TYPICAL FIXED DORM DIGESTER

Regarding digester volume, 100 L of digester volume is required per person, (e.g. a prison with 200 detainees needs a 20 m³ digester). This is based on the estimation that 3.3 L/pers. of diluted substrate (feces, urine and flush water) is added and a Hydraulic Retention Time (HRT) of 30 days is envisaged.

Funding and financial outlook

Funding for the implementation of the prison biogas systems was provided by the ICRC in partnership with local institutions. The average cost of a biogas system is 250 USD/m³ in Nepal, 230 USD/m³ in the Philippines and 300 USD/m³ in Rwanda. The total cost of all five digesters in Nepal amounts to USD 12,960, for the Philippines USD 27,700 and for a 500 m³ plant in Rwanda, it is about USD 74,000 which is paid for by the Ministry of Internal Security. The operational and maintenance costs are 2% of the total investment cost.

Implementation of biogas systems is an advantage for the prison management in terms of savings from substitution of fuel. The firewood savings are 22.5 tons/year in Nepal, approx. 40 tons/year in the Philippines and 6.35–7.35 tons/year in Rwanda. For the project in Rwanda, all 75,500 L of biogas are used for cooking. For an average 6,000-person prison population, where the prison once paid RWF 1,000,000 per month for firewood, the cost has been reduced to RWF 800,000 per month.[1] The savings in one year are RWF 200,000 x 12 = 2,400,000. Given the total investment of RWF 19,000,000 for the system, the payback period is roughly eight years. When evaluated on the basis of kerosene replacement, the payback period would be seven years. However, further savings are realized with the use of the improved manure in place of the imported mineral fertilizer. On the other hand, the post-treatment rids the wastes of the health hazards that may otherwise result in costly medical care and the loss of productive labor. The treatment of wastes in this way generates biogas, which can offset firewood consumption by 80%, thus mitigating the rate of deforestation and conserving the environment. On the bases of the demonstrated benefits, resources should be mobilized to utilize anaerobic treatment technology in a wider application for greater impact.

Socio-economic, health and environmental impact

The biogas systems are favourably perceived by the vast majority of detainees as energy systems rather than as sanitary treatment systems due to general improved living conditions of detention by:

1) Improving sanitary and hygiene conditions in toilets with cleaner surrounding areas resulting in less outbreaks of disease and fewer complaints from neighbours about odor and overflowing feces.
2) Reducing local deforestation by eliminating the need for firewood.
3) Reducing the emission of greenhouse gases by using the carbon-neutral biogas and surface and groundwater pollution.
4) Reducing the surface and groundwater pollution and removing sight and smell of the sewage, and their related health risks not only for prisoners but also for neighbouring areas, caused by the input of essentially untreated wastewater into the local environment.
5) Providing a renewable and smoke-free source of cooking fuel saving fuelwood as well as time for cooking and pot cleaning and removing unpleasantly hot and smoke-filled environment for the chefs. The biogas supply was assured unlike shortages of fuelwood.
6) Reducing costs to the prison by reducing the need for the purchase of cooking fuel.
7) Empowering the lives of the prisoners by engaging them in a new income-generation activities like inmate-run bakery that is fuelled in part by the biogas, keeping them busy and learning useful skills that they can apply when released.
8) Providing sludge as bio-fertilizer to benefit crop/vegetables production and fuel plantations by restricted irrigation, reducing costs further by eliminating the need for fertilizer purchases and also promotes sustainability by using the waste for local food production.
9) Development of private biogas companies by providing technical and business training to the civilian graduates, technicians, as well as prisoners. To date, over 30 civilians and 250 prisoners have received training, and three private biogas businesses have been started in Rwanda.

Scalability and replicability considerations

The key drivers for the success of this business are:
- Government support for renewable energy and improvements in living conditions in prisons.
- Existing funding resource to finance the investment for construction of biogas plant.
- Strong partnership with local partners, technology provider and prison authorities.
- High cost of fuel wood for cooking.
- Lack of proper sanitation systems in prisons and associated health and environmental problems.

With regard to operational aspects, the total organic waste management including issues of strong ownership and responsibilities for maintenance work needs to be examined carefully before dimensioning a biogas system and detainees should be convinced of the benefits of human waste feeding into the digester. The organization of anticipated kitchen waste feedstock has to be elaborated with the responsible persons. In terms of slurry use, there is a certain risk of public perception in the use of human excreta based product on food. Studies have shown that restricted irrigation is possible, and instead of promoting the use of slurry on vegetables, the irrigation of banana trees seems to be promising. If not properly operated and maintained, the adverse effects such as methane emissions or health risks of leaking gas pipes in the kitchen can clearly exceed the benefits.

Given above mentioned constraints that could be fairly resolved, this business case has high potential for widespread replication and scaling across institutions providing residency such as prisons, hostels, hospitals, hotels and so on. CITT has already undertaken smaller installations in three residential schools in Rwanda. To ensure the replication of this success, the United Nations Development Programme (UNDP) is supporting a KIST-implemented biogas project for Kigoma Prison, with funding from the Netherlands Embassy. In addition, the Ministry of Internal Security, in partnership with the Red Cross, has plans to provide three prisons per year with biogas systems.

Summary assessment – SWOT analysis

The key strength of the business case is its strong local partnership with the prison authorities and local technology supplier to install the system and train prison inmates and officials to operate the plant (Figure 31). The weakness is in reliance on donor money to install the infrastructure. The business has threats of health risk of cultivating crops from bio-slurry. If the human waste is not treated appropriately killing pathogens, there is health risk for operators and inmates who come in direct contact with the bio-slurry. The business has high potential for replication and opportunities to expand its revenue source to include carbon offset.

FIGURE 31. ICRC PRISON BIOGAS SWOT ANALYSIS

	HELPFUL TO ACHIEVING THE OBJECTIVES	HARMFUL TO ACHIEVING THE OBJECTIVES
INTERNAL ORIGIN ATTRIBUTES OF THE ENTERPRISE	**STRENGTHS** - Continuous supply of input - Strong partnership with prison authorities and local technology suppliers - Reduced public expenditure to run prisons from reduced fuel cost - An inclusive business model that improves sanitation for prisoners – a discriminated group - Short payback period for an investment that improves sanitation	**WEAKNESSES** - No financially viable plan set up for financing repair and maintenance of the plant - Heavy dependence on donor money to finance construction of biogas plant - Lack of strong management arrangement at the operational level to take the business to scale - Irregular feeding, and lack of operation and maintenance - Relatively high requirements of strong ownership and responsibilities for maintenance work
EXTERNAL ORIGIN ATTRIBUTES OF THE ENVIRONMENT	**OPPORTUNITIES** - Potential to increase revenue by bundling the biogas projects at different prisons and applying for carbon offset - Potential to expand the business model to prepare cooked food and provide it to local market - High potential for replication at other institutions providing food and residential services like hospitals, schools, hotels, etc.	**THREATS** - Potential health risk from direct human contact with bio-slurry used for crop production and associated public perception risk - Leakage of gas causing environment damage - Resistance to change and perceptions of people on use of sludge from fecal waste

Contributor
Radheeka Jirasinha, Consultant, Colombo, Sri Lanka

References and further readings

Butare, A. and Kimaro, A. 2002. Anaerobic technology for toilet wastes management: The case study of the Cyangugu pilot project. World Transactions on Engineering and Technology Education 1(1): 147–151.

Gauthier, M., Oppliger, A., Lohri, C. and Zurbrugg, C. 2012. Ensuring appropriateness of biogas sanitation systems for prisons: Analysis from Rwanda, Nepal and Philippines. Geneva, Switzerland: International Committee of the Red Cross (ICRC).

ICRC. April 2012. Water, sanitation, hygiene and habitat in prisons: Supplementary guidance. http://www.icrc.org/eng/assets/files/publications/icrc-002-4083.pdf (accessed Nov. 6, 2017).

Kigali Institute of Science and Technology (KIST). 2005. Biogas plant providing sanitation and cooking fuel in Rwanda. Rwanda Technical Report: ASHDEN Awards. https://www.ashden.org/winners/kist (accessed Nov. 6, 2017).

Lohri, C., Vogeli, Y., Oppliger, A., Mardini, R., Giusti, A. and Zurbrugg, C. 2010. Evaluation of biogas sanitation systems in Nepalese prisons. IWA-DEWATS Conference, Indonesia, 2010.

United Nations Environment Programme (UNEP). 2011. Biogas for the Cagayan de Oro City jail: An ICRC-funded environmental and livelihood project in the Philippines. https://goo.gl/WaAJnD (accessed Nov. 6, 2017).

Winkler, M.S., Fuhrimann, S., Pham-Duc, P. et al. 2017. Assessing potential health impacts of waste recovery and reuse business models in Hanoi, Vietnam. Int J Public Health 62 (Suppl 1): 7–16.

Walmsley, R. 2015. World prison population list (Eleventh edition). Institute for Criminal Policy Research. http://www.icpr.org.uk/media/41356/world_prison_population_list_11th_edition.pdf (accessed Nov. 6, 2017).

Case descriptions are based on primary and secondary data provided by case operators, insiders, or other stakeholders, and reflect our best knowledge at the time of the assessments (2015/2016). As business operations are dynamic, data that can be subject to change.

Note
1   RwF 800,000 was depending on the year about USD 1,100–1,400 in the assessment period.

CASE
Biogas from fecal sludge at community scale (Sulabh, India)

Solomie Gebrezgabher and Hari Natarajan

Supporting case for Business Model 3	
Location:	New Delhi, India
Waste input type:	Fecal sludge
Value offer:	Hygienic and affordable sanitation services; biogas and compost
Organization type:	NGO
Status of organization:	Operational (since 1970)
Scale of businesses:	Large
Major partners:	National government, local government bodies

Executive summary

The Sulabh International Social Service Organization (Sulabh), an Indian NGO, was founded in 1970 to develop a low-cost, easy-to-implement, environmentally-friendly and socio-culturally-acceptable toilet solution at the household level. Sulabh has also proved through its pay-and-use public toilet model that low-income people are willing to pay for use of toilet facilities that are clean and hygienic. The key technological solutions include the Sulabh Flush Compost toilet for households, the Public Toilet Complex and the Public Toilet Complex with a biogas system. Sulabh is noted for achieving success in the field of cost-effective sanitation, liberation of scavengers, social transformation of society, prevention of environmental pollution and development of non-conventional sources of energy.

The NGO implements a build, operate and transfer (BOT) model for public toilets. For the construction of the public toilets, Sulabh is approached by the municipality or other local government agencies and private sponsors to build a public toilet in a specific location. The agency is responsible for capital expenditures while Sulabh takes care of the operational and maintenance expenditure for 30 years. Sulabh charges a consultation fee of 20% of the project cost, which is the primary source of income that covers the overheads and administrative costs and sustains the operations of the overall organization. Sulabh has thus far installed over 7,500 public pay-and-use toilet complexes and 200 public toilets with biogas systems in 26 states of India. Owing to this technology, Sulabh has been able to liberate over 60,000 scavengers, offering programs to reintegrate them into society and has, through its public toilet complexes contributed in the field of community health and hygiene and environmental sanitation.

KEY PERFORMANCE INDICATORS (AS OF 2012)					
Land	1.75 m²/toilet seat and 8.28 m²/twin pits				
Water requirements:	1.5–2 L/flush at household toilet and 3–4 L/flush at public toilets				
Capital investment:	No data available (charges 20% of project cost to cover operational costs)				
Labor:	2–3 full-time per public toilet complex				
O&M cost:	Public toilet complex 10,320 USD/year				
Output:	1.2 million household toilets; more than 7,500 public toilet complexes; 200 public toilet complexes with biogas plant				
Potential social and/or environmental impact:	1.2 million household obtained basic sanitation facility; 60,000 scavengers freed; over 50,000 jobs; 19,000 masons trained; Improved community health, hygiene and environmental sanitation, conserve water, use of biogas for cooking and to generate electricity for toilet complex lighting				
Financial viability indicators:	Payback period:	5–6	Post-tax IRR:	N.A.	Gross margin:	N.A.

Context and background

The local government bodies in cities and urban areas in India are entrusted with the responsibility of providing basic civic amenities including sanitation facilities. Lack of service coverage of a large proportion of the population, poor quality of service delivery and limited revenue generation are the universal problems faced by the local government bodies in view of the rapid urbanization and fast-growing slum and low-income population. The Sulabh is an Indian-based NGO noted for achieving success in the field of cost-effective sanitation, liberation of predominantly women scavengers, social transformation of society, prevention of environmental pollution and generation of biogas. Biogas is utilized for cooking, lighting through mantle lamps, electricity generation and being converted into energy to be used for lighting streetlights and other uses. The sludge at the bottom of the digester can be used as fertilizer. Recycling and use of human excreta for biogas generation is an important way to get rid of health hazards from human excreta without any manual handing of excreta at any stage. Under the system, only human excreta with flush water is allowed to flow into biogas plant for anaerobic digestion.

The key technological solutions included the Sulabh Flush Compost toilet (FCT) for households, the public toilet complex (PTC) and the public toilet complex with a biogas system (PTC-biogas). The social NGO has now become the international pioneer in pay-and-use toilets. Sulabh has thus far constructed over 1.2 million FCT, over 7,500 PTC, and 200 PTC-biogas of capacity 35–60 m³ per day in different parts of India. This solution has been universally accepted by the state and central governments in India and the cost of the same is covered to a large extent by subsidies/grants. Sulabh takes 30 years maintenance guarantee for the toilet complexes constructed by it by collecting a fee of pay per use. There are 60,000 volunteers working with Sulabh that include technocrats, managers, scientists, engineers, social scientists, doctors, architects, planners and other non-revenue staff. This solution has also gained recognition from several multilateral development agencies such as the World Health Organization (WHO), United Nations Children's Fund (UNICEF) and United Nations Development Programme (UNDP) and has been taken up for adoption in other developing countries in southern Asia and Africa.

Market environment

The 2011 census indicates that nearly 50% of the households (18.6% urban households and 69.3% rural households) in India still do not have basic sanitation facilities. The problem lies not only in provision of appropriate toilets but also in inducing a behavioural change among the target beneficiaries. This was further compounded by the fact that there were neither affordable solutions available in the market nor were there solution providers that could cater to the differing needs across different geographies. Sulabh plays the role of a catalyst and a partner between the official agencies and the users for the construction, operation and maintenance of public sanitation facilities. Sulabh has proven that poor slum communities are willing to pay for improved water and sanitation services and that such operations can be financially viable. Sulabh has constructed 1.2 million flush compost toilets, while 120 million households lack basic sanitation facilities. This indicates that there is still an opportunity for Sulabh to further scale up its operations and reach unserved population.

Cooking is the most convenient use of biogas. Biogas burners are available in a wide-ranging capacity from 0.2–2.8 m^3 biogas consumption per hour. It burns with a blue flame and without soot and odor. The biogas mantle lamp consumes 0.05–0.08 m^3/hour having illumination capacity equivalent to 40 W electric bulbs at 220 V. Motive power can be generated by using biogas in dual fuel internal combustion (IC) engine using 20% diesel and 80% biogas. Recently, Sulabh has modified the generator, which does not require diesel and runs on 100% biogas. About 30 m^3 of biogas is equivalent to 17 m^3 of natural gas, about 30 litres of butane (LPG), 24 litres of gasoline or 21 litres of diesel oil. For the safe reuse of human waste from public toilets, housing colonies, high-rise buildings, hostels and hospitals, a technology is developed for complete recycling and reuse of excreta through biogas generation and on-site treatment of effluent through a simple and convenient technology for its safe reuse without health or environmental risk. The treated effluent is colorless, odorless and pathogen-free and is safe for discharge into any water body without causing pollution. It can also be used for cleaning of floors of public toilets in water-scarce areas.

Sulabh in collaboration with UN-HABITAT, Nairobi has trained professionals from 14 African countries for their capacity development towards achieving the initial Millennium Development Goal (MDG) for sustainable development in water and sanitation, which predated the current Sustainable Development Goals (SDG), and trained more than 50,000 people to work in the construction and maintenance of community toilets in India.

Macro-economic environment

One of the challenges in the successful dissemination of basic sanitation facilities to communities is convincing the poor to use a toilet instead of the outdoors and convince them to pay for use of public toilet. This is because the hygiene practices of communities are deeply rooted in cultural and religious values. Therefore, the success of a business involved in sanitation service depends not only on installing the appropriate toilet models but also on the interaction between a complex and diverse range of institutions, processes and actors (both public and private).

An estimated 50% of all Indians, or close to 600 million people, still do not have access to any kind of toilet. Among those people who live in urban slums and rural environments are affected the most. Goal 7 of the MDG called on countries to halve by 2015 the proportion of people without improved sanitation facilities (from 1990 levels); while India had its even more ambitious goal of providing "Sanitation for All" by 2012, established under its Total Sanitation Campaign.

The restructured program moves away from the principle of state-wise allocation of funds, primarily based on poverty criteria, to a demand-driven approach. The successful state program moved from a high-subsidy to a low-subsidy regime, with investment of funds in building awareness and increasing sanitation coverage through public-private partnerships with non-profit organizations such as Sulabh.

Business model
Sulabh's target customer segment is the poorer section of the society, particularly urban slums with no access to basic sanitation facilities and users of public toilet complexes (Figure 32). Sulabh implements a build-and-transfer model for household toilets and a build, operate and transfer (BOT) model for public toilets. In the case of public toilet complexes, Sulabh was the first to introduce a pay-and-use system to cover the costs of maintenance of the toilet complex. For the construction of the public toilets, Sulabh is approached by the municipality or other local government agencies and private sponsors to build a public toilet in a specific location. The agency is responsible for capital expenditures while Sulabh takes care of the operational and maintenance expenditure for 30 years and trains toilet complex operators on how to run the public toilet. Sulabh charges the project sponsors a consultation and implementation fee of 20% of the project cost. In addition to its creating technologically and socially-efficient solutions, one of the strongholds of the organization is its partnership with the local governments, local authorities, international organizations and local communities. This partnership coupled with community participation has made a substantial impact in improving the sanitation services to the poor.

Value chain and position
The value chains for Sulabh's public toilet and household toilet are depicted in Figure 33. In the case of public toilet complexes, Sulabh is typically approached by a local government body or private entity for establishing a public toilet in a specific location. Based on a survey, Sulabh determines the appropriate capacity of the toilet complex and designs and constructs the same and operates and maintains it for 30 years based on fees collected on pay-per-use basis. For the construction, operation and maintenance of the toilet complexes, the organization plays the role of a catalytic agent between the government, local authorities and the users of toilet complexes.

Institutional environment
Recently, the Government of India has significantly increased the financial support for family-size biogas plant and also launched two schemes, mainly Biogas Fertilizer Plant (BGFP) and Biogas Power generation. In the case of household toilets, a large part of the costs is covered through central and state government subsidies and incentives. Under the scheme, community toilet complexes are to be established only when there are space constraints in the community that prevent the installation of household toilets. In the case of public toilet complexes, Sulabh is invited by a local government agency or private sponsor to construct and operate a toilet complex with or without biogas plant in a specified location. Land, as well as the funds for construction of the toilet, is provided by the sponsoring agency. In such cases, the cost (USD 4,000) is borne by central government, state government and the community in the ratio of 60:30:10. In the case of public toilets with biogas, 75% of the additional capital costs are subsidized by the government.

Technology and processes
In the case of public toilet complex with biogas digester and associated treatment plant for the effluent, a floating dome type was first tried. Abundant quantity of gas was produced for cooking and lighting but there was foul smell because human excreta, after decomposition, used to float. Moreover, using the floating dome type resulted in lower biogas production during winters. Finally, Sulabh switched over to the fixed dome biogas digester, with some change in the design. The digester is built underground into which excreta from public toilets flows under gravity. Inside the digester biogas is produced due to anaerobic fermentation by the help of methanogenic bacteria. The biogas, thus produced, is stored in inbuilt liquid displacement chamber.

CASE: BIOGAS FROM FECAL SLUDGE AT COMMUNITY SCALE

FIGURE 32. SULABH BUSINESS MODEL CANVAS

KEY PARTNERS
- Municipality/local authorities
- Local community
- Private sponsors

KEY ACTIVITIES
- R&D on toilet technologies
- Build and transfer household toilets
- Build, operate, maintain and transfer public toilets
- Consultation

KEY RESOURCES
- Technology/expertise
- Construction materials
- Human resource
- Users (human excreta)

VALUE PROPOSITIONS
- Households get access to basic sanitation facilities
- Municipality get better management of human waste and communities get access to toilet facilities for a nominal user fee
- Private sponsors get an opportunity to contribute to fighting scavenging and promote good sanitation practices

CUSTOMER RELATIONSHIPS
- Subsidies and incentive programs to finance individual toilets
- Invitation

CHANNELS
- Direct
- Toilet operators (complexes)

CUSTOMER SEGMENTS
- Central and state governments, benefiting individual households
- Local governments or private sponsors benefiting communities

COST STRUCTURE
- Construction cost:
- Household toilet
- Public toilet complex; public toilet with biogas
- Utilities (water, power)
- Operation and maintenance (staff cost, chemicals)
- Toilet technology development costs

REVENUE STREAMS
- Implementation/installation fee (20% of project cost)
- Consultation, implementation and maintenance fee (20% of project cost)

SOCIAL & ENVIRONMENTAL COSTS
- Using digested human excreta for crops may potentially pose health risks due to possible pathogen survival
- Potential leakage of methane gas from biogas systems

SOCIAL & ENVIRONMENTAL BENEFITS
- Improve community health, hygiene and environmental sanitation
- Elimination of practice of scavenging
- Saving for municipality from better management of human waste
- Safe processing and disposal of human excreta
- Conserve water from reduced usage for flushing toilets
- Change attitudes of people towards sanitation
- Create job opportunities
- Diffusion of innovation in toilet technologies

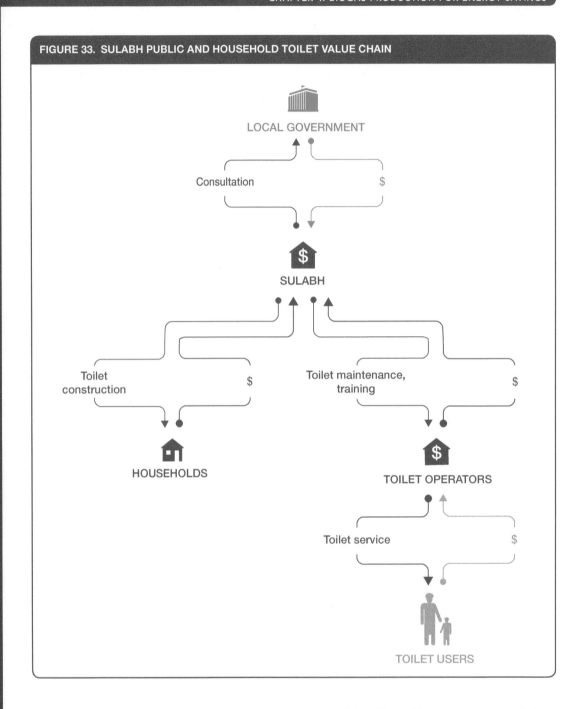

FIGURE 33. SULABH PUBLIC AND HOUSEHOLD TOILET VALUE CHAIN

The design developed by Sulabh does not require manual handling of human excreta and there is complete recycling and resource recovery from the wastes. During biogas generation, due to anaerobic condition inside digester, most of the pathogens are eliminated from the digested effluent making it suitable for using it as manure. Sulabh has also carried out a series of experiments on biogas generation from water hyacinth (an aquatic weed) after harvesting, drying and pulverizing it; vegetables, fruit and household kitchen wastes with or without mixing with human wastes. Better results were obtained when human waste and vegetable waste were fed in combination.

CASE: BIOGAS FROM FECAL SLUDGE AT COMMUNITY SCALE

One cubic foot of biogas is produced from human excreta per person per day. Human excreta based biogas contains 65–66% methane, 32–34% carbon dioxide and rest the hydrogen sulphide and other gases in traces. To convert biogas into energy, earlier the engine was run on diesel and biogas with a ratio of 20:80 and have now shifted to the battery system, where the engine is run 100% on biogas. A public toilet used by about 2,000 persons per day would produce approximately 60 m^3 of biogas which can run a 10 kilovolt-ampere (KVA) gen set for 8 hrs a day, producing 65 kilowatt hours (kWh) of power.

After a series of experiments a simple and convenient technology named Sulabh Effluent Treatment (SET) are invented to further treat effluent of biogas plant and turn it into a colorless, odorless and pathogen-free manure. The technology is based on sedimentation and filtration of effluent through sand, aeration tank and activated charcoal followed by exposure to ultraviolet rays. The effluent treatment plant consists of a series of filtration steps through sand and activated charcoal, followed by UV treatment, which eliminates not only the bad odor but also the bacterial content. The Biochemical Oxygen Demand (BOD) and Chemical Oxygen Demand (COD) of the wastewater is reduced significantly, with BOD being less than 10 mg/L post treatment, which is safe for agriculture, aquaculture, discharge into water bodies – practically safe for all purposes except drinking. The residue water from the plant can be used, too, as bio-fertilizer because it contains phosphorus, nitrogen and potassium. A detailed diagrammatic representation of Sulabh model is given in Figure 34.

The institute has successfully demonstrated its use as hydroponics, i.e. soil-less culture of plants. The effluent is first dried in earthen pots kept in sunlight where, owing to the evaporation of the liquid, the concentration of nutrients increases. It is filtered with a thin plastic mesh. Some trace elements are added in the filtered effluent. Such effluent is completely odorless. Various plants have been grown

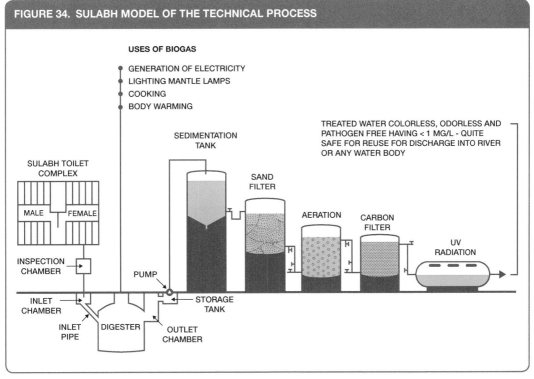

FIGURE 34. SULABH MODEL OF THE TECHNICAL PROCESS

Source: Pathak, 2015.

exclusively on such an effluent when mixed (5–10% by volume) with tap water. Plants can be grown in glass bottles or any other jars and kept inside or outside the room. Such technology is useful for the culture of rare plants like cactus and other ornamental plants.

Funding and financial outlook

While Sulabh, as a not-for-profit, can access grants and donations. It does not depend on external agencies for finances and meets all the financial obligations through internal resources. Sulabh charges the project sponsors (local government or private sponsors) a consultation and implementation fee of 20% of the project cost and also takes on the maintenance responsibility for the toilet complex for a period of 30 years from user's charges. In the case of public toilet facilities, land and cost of construction is met by local body while maintenance is met from user's charges. Estimated project cost for public toilet facility is USD 4,000, financed by central government, state government and the community in the ratio of 60:30:10. In the case of public toilet with biogas plant, 75% of the additional cost of the biogas plant (USD 4,000) is financed by the government.

For a typical toilet complex that caters to approximately 2,000 users per day, annual revenues (assuming 50% of users are paying customers) are USD 10,800 whereas the operating costs are USD 10,320, thereby leaving very little surplus to cover capital costs. Within Sulabh's portfolio of 7,500 toilets, around 50% are generating enough revenues to be self-sustaining and profitable. The maintenance of the other toilet complexes is cross subsidized from the income generated from toilet complexes in busy and developed areas. In the case of public toilet with biogas, the gas is used for heating/electricity requirements of the toilet complex and thus results in cost recovery through reduced requirement for LPG/kerosene. Estimated payback period for the toilets with biogas plant is 5–6 years.

The complex with biogas plant recovers about USD 7,000 per year in terms of savings in diesel for power generation when an average of 1,000 people use the toilets, generating 30 m^3 biogas, equivalent to 21 L of diesel costing 0.9 USD/L. In addition, it reduces the GHG emissions by capturing methane and converting it to CO_2 during combustion in internal combustion engines.

Socio-economic, health and environmental impact

Sulabh is one of the pioneers in improving the sanitation levels in the country by shifting people from the practice of open defecation to use of toilets. Sulabh has developed toilet designs that cater to varying income levels and locations. It has installed more than 1.2 million household toilets and maintains more than 7,500 public pay-and-use facilities in different states of India. A Sulabh public toilet complex employs two to six persons. It has provided training to 19,000 masons to build low-cost twin-pit toilets using locally available material. Owing to this technology, Sulabh has been able to liberate over 60,000 scavengers, offering programs to reintegrate them into society.

Human excreta contains full spectrum of pathogens. Unsafe disposal of human excreta facilitates the transmission of oral-fecal diseases, including diarrhoea and a range of intestinal worm infections such as hookworm and roundworm. In this technology, most of these pathogens are eliminated in anaerobic condition inside the digester. Cost of collection of sewage and operation and maintenance of the system are very low. Provision of toilets connected to biogas digesters has helped communities gain access to sanitation and an inexpensive energy source. No manual handling of human excreta is required. It is aesthetically and socially accepted. The toilet requires only 1.5 to 2 L of water for flushing and thus conserves water. In addition to conserving and reusing water the system has additional inbuilt advantage of reducing greenhouse gas effect arising out of carbon dioxide and methane production due to degradation of human waste.

Due to design of leach pit (Sulabh Toilet) produced carbon dioxide is diffused in soil through honey combs. It does not escape in atmosphere as in other cases. During anaerobic digestion of human wastes during biogas production, methane is produced that is used for different purposes. Methane as such is not left to escape in the atmosphere. Thus, both these technologies are helping in reducing greenhouse effect and thus improve environment. Besides using biogas for different purposes, the plant effluent can also be used as manure or discharged safely into any river or water body without causing pollution. Treated effluent is safe to reuse for agriculture, gardening or discharge into any water body. In drought-prone areas, treated effluent can be used for cleaning floors of public toilets. If discharged into the sewer, pollution load on a sewage treatment plant (STP) will be much lower. Thus, biogas technology from human wastes has multiple benefits, i.e. sanitation, bioenergy and manure.

At the household level, manure obtained from a family of five members in a year is approximately 200 kg (40 kg/person/year). Assuming that manure obtained is utilized for agriculture purposes, the family saves 19 USD/year from using the manure (assuming a cost of 0.09 USD/kg of manure).

Scalability and replicability considerations

Key drivers for the success of this business are:
- Partnership with local governments, local authorities, international organizations/donors and local communities.
- Central and state government support and incentives.
- Low-cost and locally available technology.
- Movement toward low subsidy regime.
- User payment per use to fund O&M of the complex.

Sulabh's low-cost, environmentally-friendly and socio-culturally-acceptable toilet technologies are suited for up-scaling and replication in other developing countries. The Sulabh movement originated in one town of India but has now spread to 26 states in India. Such facilities should be provided on a pay-and use basis at all places of congregation where 'people throng in large numbers for worship and meditation' there is a need of a decentralized system based on biogas generation technology that is not only cost effective but also easy in operation and maintenance. The hygiene practices of communities are deeply embedded in cultural and religious values and therefore convincing the poor to use a toilet instead of the outdoors and to pay for the construction of a toilet, are great challenges. Moreover, the Sulabh toilet model, while being suitable for dry areas is unsuitable for those with a high water table such as coastal zones or those receiving high degree of rainfall, because of water logging of the pits.

Sulabh technology has been recognized not only by the Government in India but also by governments in other countries and by several international development agencies. In collaboration with UN-HABITAT Nairobi, Sulabh has imparted training to engineers, planners, administrators and entrepreneurs from 14 African countries which include Ethiopia, Mozambique, Uganda, Cameroon and Burkina Faso, Kenya, Tanzania, Cote d'Ivoire, Mali, Ghana, Rwanda, Senegal and Zambia. They have also been trained as part of achieving the Millennium Development Goals set for the sustainable development in water, sanitation, health and hygiene sectors. Sulabh technical team had gone to Ethiopia and Bangladesh for giving training on Sulabh Technologies. The Sulabh model has also been adopted by a number of countries, including China, Bhutan, Bangladesh, Afghanistan, Burkina Faso, Ghana, Kenya, Mali, Nigeria, Senegal, Tanzania and Zambia for expansion and promotion of sanitation facilities. Hence, it can be asserted that Sulabh's technologies have long since passed the test of replicability and scalability. However, for the Sulabh model to be successfully replicated in other countries, close coordination and partnership between the government, local authorities and NGOs backed by community participation is very important.

Summary assessment – SWOT analysis

In addition to its creating technologically and socially efficient solutions, one of the strongholds of the organization is its partnership with the local governments, international organizations and the local communities (Figure 35). Dependence on invitation from government and availability of public funds restricts ability to scale. However, the fact that this NGO is highly recognized by other governments and international NGOs presents opportunity for it to be expanded to other locations.

FIGURE 35. SWOT ANALYSIS FOR SULABH

	HELPFUL TO ACHIEVING THE OBJECTIVES	HARMFUL TO ACHIEVING THE OBJECTIVES
INTERNAL ORIGIN ATTRIBUTES OF THE ENTERPRISE	**STRENGTHS** • Strong partnership with governments and NGOs • Low-cost and socio-culturally acceptable technology • Strong capabilities of organizing communities and social mobilization campaigns • Highly recognized by other governments and international NGO	**WEAKNESSES** • Sulabh largely identified with one individual, the founder • Dependence on internal funds limits ability to scale • Centralized decision making • Technology suitable only in dry regions limits ability to scale • Limitations of NGO for raising capital funds from financial institutions • Health risks from residual accumulation of toxic materials and pathogens
EXTERNAL ORIGIN ATTRIBUTES OF THE ENVIRONMENT	**OPPORTUNITIES** • Out-scaling & up-scaling combination opportunity in other locations through franchise model • Raising more funds from other sources (grants from other NGOs) • Multiple benefits of biogas plants (sanitation, energy reduction in GHG emissions and water supply in dry areas) • Political will and recognition of benefits	**THREATS** • Dependence on invitation from government and availability of public funds restricts ability to scale • Too low revenue collection by pay per use for cost recovery and financial sustainability • Uncertainty on willingness to pay by poor users • Social constraints and psychological prejudice to use of human waste materials

Contributors
Johannes Heeb and Leonellha Barreto-Dillon, CEWAS, Switzerland
Jasper Buijs, Sustainnovate, The Netherlands; Formerly IWMI
Josiane Nikiema, IWMI, Ghana
Kamalesh Doshi, Simplify Energy Solutions LLC, Ashburn, Virginia, USA

References and further readings
Chary, V.S., Narender, A. and Rao, K.R. March 2003. Serving the poor with sanitation: The Sulabh approach. Third World Water Forum, PPCPP Session, Osaka.

Hansen, S. and Bhatia, B. 2004. Water and poverty in a macro-economic context. Norwegian Ministry of the Environment. https://goo.gl/4kHdVb (accessed Nov. 6, 2017).

Heierli, U., Hartmann, A., Munger, F. and Walther, P. 2004. Sanitation is a business: Approaches for demand-oriented policies. Swiss Agency for Development and Cooperation, Bern https://www.ircwash.org/resources/sanitation-business-approaches-demand-oriented-policies (accessed Nov. 7, 2017).

Ministry of Drinking Water and Sanitation, Government of India. 2012. Guidelines: Nirmal Bharat. http://www.mdws.gov.in/documents/guidelines (accessed Nov. 7, 2017).

Pathak, B. 2011. Sulabh sanitation and social reform movement. International NGO Journal 6:014–029.

Pathak, B. 2015. Innovation in Water & Sanitation Technology. https://www.slideshare.net/indiawaterportal/sanitation-water-technologies-developedsulabh-internationalindovation-201523-january-2015 (accessed Nov. 7, 2017).

Ramachandran, K. 2009. Satisfying solution for a compelling need makes Sulabh an entrepreneurial success. Vikalpa, 34(1): 109–111.

Ramani, S.V., SadreGhazi, S. and Duysters G. 2012. On the diffusion of toilets as bottom of the pyramid innovation: Lessons from sanitation entrepreneurs. Technological Forecasting and Social Change 79: 676–687.

Sulabh International Social Service Organization. www.sulabhinternational.org.

Case descriptions are based on primary and secondary data provided by case operators, insiders, or other stakeholders, and reflect our best knowledge at the time of the assessments (2015/2016). As business operations are dynamic, data can be subject to change.

CASE

Biogas from fecal sludge at Kibera communities at Nairobi (Umande Trust, Kenya)

Solomie Gebrezgabher, Jack Odero and Nancy Karanja

Supporting case for Business Model 3	
Location:	Nairobi, Kenya
Waste input type:	Fecal sludge
Value offer:	Sanitation service, energy for cooking and compost
Organization type:	Civil Society Organization (CSO) registered as a trust
Status of organization:	Operational since 2004
Scale of businesses:	Small
Major partners:	Athi Water Service Board (AWSB), Nairobi Water and Sewerage Company, Umande Trust, Local community, Water service providers

Executive summary

Umande Trust–TOSHA 1 is one of the bio-centres that Umande Trust has successfully implemented by working with residents organized in formal groups within the informal settlements to improve access to safe, adequate and affordable bio-sanitation and to provide income generating opportunities for the community. Umande Trust is a civil society organization (CSO) which works closely with community groups, public sector agencies, local government and peer civil society organizations to design, plan and construct bio-centres. The bio-centre is a multi-purpose facility consisting of toilet facilities, a rental space, a meeting hall and a bio-digester. It provides a range of services to the community, i.e. toilet service to the community, biogas cooking facility to women street food vendors, bio-slurry to farmers and a rental space to private businesses. Umande Trust offers technical support and builds capacity of the members of TOSHA 1 to run the bio-centre successfully. Using a pay-for-use revenue model, the bio-centre makes an average net income of nearly USD 1,100 per month. In order for the business to be successfully undertaken, the local communities are important stakeholders in the whole project, thus making community-led strategy the key success factor for the bio-centre. TOSHA 1 is used by an average of 1,000 people per day, making it one of Nairobi's busiest toilets and a producer of biogas. It is a good example of an environmentally-friendly approach to providing sanitation services through safe processing and disposal of human excreta while creating livelihood and jobs to the members of the community-based organization (CBO). A community-led approach to sanitation contributes to capacity building of the community and also to changing people's attitude towards use of human waste as a source of energy and business.

CASE: BIOGAS FROM FECAL SLUDGE AT KIBERA, NAIROBI

KEY PERFORMANCE INDICATORS (AS OF 2012)	
Land:	0.01 ha
Capital investment:	USD 22,500 for construction of each bio-centre; USD 10,000 for advertisement/campaign
Labor:	Skilled and unskilled labor for construction and running the bio-centre
O&M cost:	3,720 USD/year
Output:	Toilet facility 1,000 users/day; Biogas capacity 54 m^3
Potential social and/or environmental impact:	Improved community health, hygiene and environmental sanitation, improved livelihood and capacity building of community, job creation, reduced environmental pollution
Financial viability indicators:	Payback period: 3 years IRR: 33% Profit margin: 77%

Context and background

Kenya with urbanization rate of 5% per annum has more than 1,800 low-income informal settlements with a total estimated population of 12.5 million. The informal settlements of Nairobi cover about 5% of the total residential land area but they are inhabited by over 50% of the city's total population. The characteristics of an informal settlement (slum) are: lack of basic water at affordable prices, sanitation by public or private toilets and other infrastructural services; unplanned, underserved, high density, poor neighbourhood without legal recognition or rights. Kibera is the largest slum in Nairobi and the largest urban slum in Africa. The 2009 Kenya Population and Housing Census reports Kibera's population as 170,070. About 85% of households buy water from privately or communally-owned kiosks at prices four to five times higher per litre than tariffs charged by the Nairobi Water and Sewerage Services Company. TOSHA 1 is a bio-centre within the informal settlements of Kibera managed by TOSHA[1], a CBO that is supported by Umande Trust in Kenya. The bio-centres are bio-sanitation units that provide secure and adequate access to sanitation and income generation by converting human waste into biogas and liquid fertilizer. Once the bio-centre is constructed, Umande Trust provides technical support and trains the CBOs to run and operate the bio-centre.

Market environment

Despite a number of efforts by a range of actors to improve sanitation, the majority of households in urban areas lack access to healthy and affordable sanitation facilities. Over 60% of the Nairobi population lives in informal settlements with only 25% of the households having access to a private toilet facility while 68% of informal settlement dwellers rely on shared toilet facilities and 6% have no access to toilets and have to use open areas or flying toilets. In addition to household sanitation problems, there is also the challenge of household energy for the urban poor. The high and rising cost of fuel (e.g. kerosene, charcoal, firewood) has been a challenge for the urban poor. The little resources allocated to sanitation were basically used for awareness creation and hygiene education, leaving the development of sanitation infrastructure lagging behind. Residents therein are often unhealthier than their rural counterparts because they are deprived of basic public social services, such as health care, water supply, sanitation and garbage disposal. Slum dwellers, exhibit relatively high mortality rates because they are less likely to access preventative and curative medical care despite their proximity to the best hospitals and clinics located in cities.

One of the support projects of the Athi Water Service Board (AWSB) is the construction of bio-centres in selected areas of informal settlements. Bio-centres are not only initiated in informal settlements but also in community facilities, such as schools and churches, making it easy and quick to reach a large number of people in the communities.

Macro-economic environment

For the purposes of achieving the initial Millennium Development Goals (MDGs) on water and sanitation, the Kenya government, in its Vision 2030, proposed improving waste management accessible to all through the design and application of economic incentives and the commissioning of public-private partnerships (PPPs) for improved efficiency in water and sanitation delivery. The government had also initiated a program to replace the slum with a residential district consisting of high-rise apartments, and to relocate the residents to these new buildings upon completion. The apartments are being built in phases in line with the government's budgetary allocations, and a few apartments in phase 1 of the project have been occupied.

Business model

TOSHA runs TOSHA 1, one of the profitable bio-centres set up by Umande Trust. It is used by an average of 1,000 people per day, making it one of Nairobi's busiest toilets and a producer of biogas (Figure 36). Using a pay-for-use revenue model, the bio-centre makes profit of about USD 1,100 per month. Umande Trust offers technical support and builds capacity of the members of TOSHA 1 to run the bio-centre successfully. In order to gain acceptance for the innovative sanitation approach, public awareness campaigns and trainings were done by Umande Trust. The demand for biogas and bio-slurry as an alternative source of energy and fertilizer is slowly gaining popularity. However, due to cultural and social beliefs and preferences a lot of public awareness campaigns have had to be

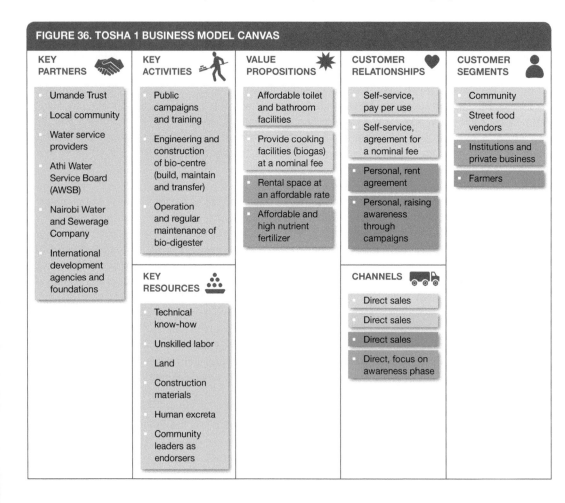

FIGURE 36. TOSHA 1 BUSINESS MODEL CANVAS

KEY PARTNERS	KEY ACTIVITIES	VALUE PROPOSITIONS	CUSTOMER RELATIONSHIPS	CUSTOMER SEGMENTS
• Umande Trust • Local community • Water service providers • Athi Water Service Board (AWSB) • Nairobi Water and Sewerage Company • International development agencies and foundations	• Public campaigns and training • Engineering and construction of bio-centre (build, maintain and transfer) • Operation and regular maintenance of bio-digester	• Affordable toilet and bathroom facilities • Provide cooking facilities (biogas) at a nominal fee • Rental space at an affordable rate • Affordable and high nutrient fertilizer	• Self-service, pay per use • Self-service, agreement for a nominal fee • Personal, rent agreement • Personal, raising awareness through campaigns	• Community • Street food vendors • Institutions and private business • Farmers
	KEY RESOURCES		CHANNELS	
	• Technical know-how • Unskilled labor • Land • Construction materials • Human excreta • Community leaders as endorsers		• Direct sales • Direct sales • Direct sales • Direct, focus on awareness phase	

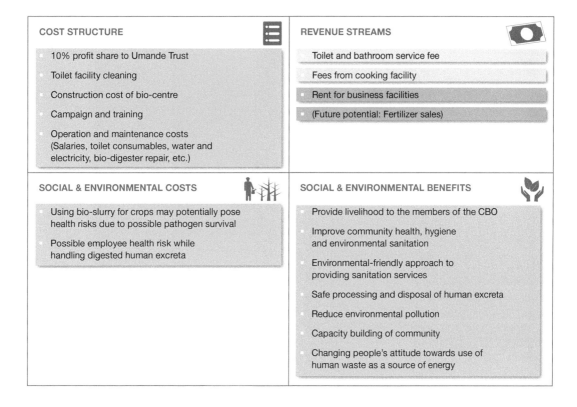

undertaken to break the stigma attached to using human waste by-products. The main customers using the facilities are the nearby residents, food vendors (mostly women), institutions and business people working at the market where the bio-centre is situated. Income generated from the bio-centre is shared amongst the members and 10% of its profit is given to Umande Trust.

Value chain and position

The value chain for TOHSHA 1 is depicted in Figure 37. Umande Trust is responsible for the plan, design and construction of the bio-centre and the success of the bio-centre by offering advisory support and also spearheading campaigns to promote the business activities. Before setting up a bio-centre, Umande Trust identifies, evaluates and selects existing organized groups or CBOs to run the bio-centre. Stakeholder workshops are undertaken to identify the site and the community is involved in selection of the best site. The user surveys, GIS mapping and participatory urban appraisal (PUA) processes ensure that individuals and community groups generate data on existing sanitation conditions and demands.

Once the project is completed, the CBO is trained on how to operate and manage the bio-centre. The TOSHA ensures that customers who come to use the facilities are served. The bio-sanitation centres charge 3 KSh per visit. The water kiosks within the bio-centres charge a flat rate of 5 KSh per 20 L jerry cans. Residents use facilities on a pay-for-use basis once a sponsoring agency provides investment funds for construction. It has also implemented two payment innovations: Beba pay (which exited the Kenyan market in 2015), a cashless system and Kopokopo, which reduce the use of cash handing and promote accountability.

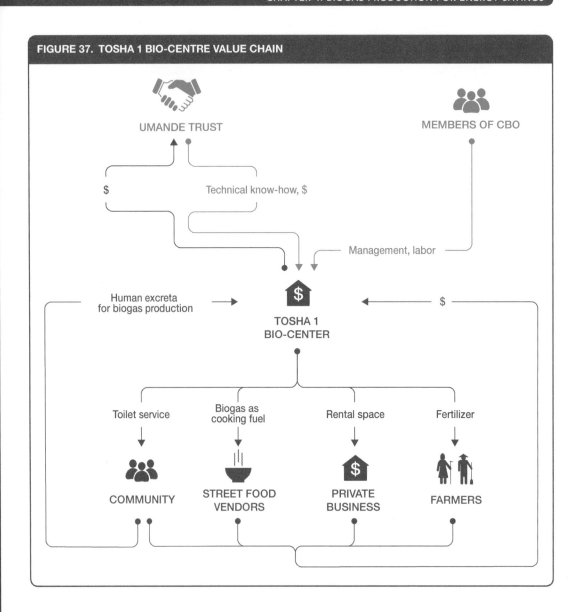

FIGURE 37. TOSHA 1 BIO-CENTRE VALUE CHAIN

Institutional environment

The newly adopted Kenya Constitution has spelled out sanitation to be a human right, and the Ministry of Water and Irrigation (MWI) has drafted a new water policy and bill to align itself to the new constitution. Many sanitation stakeholders have embarked on efforts to improve sanitation situation in Kenya. These stakeholders include government ministries, national corporations and non-governmental Organizations (NGOs). The National Environment Management Authority (NEMA) is a government agency responsible for the management of the environment and the environmental policy of Kenya. The Ministry of Health (MOH) aims at an open defecation-free country by 2017 and has developed a roadmap to achieve this. The installation of bio-centres is part of the intervention targets under Nairobi Informal Settlements Water and Sanitation Improvement Programme (NISWASIP). Athi Water Service Board (AWSB) is one of the eight water boards under the Ministry of Environment, Water and Natural Resources created to bring about efficiency, economy and sustainability in the provision of water and sewerage services in Kenya. The AWSB focuses on water and sanitation services to informal settlements by constructing bio-centres within its area of jurisdiction. Nairobi City Water and Sewerage Company is tasked to connect water to the bio-centres after completion. The plan was also aligned with the specific Millennium Development Goals (MDGs) on water and sanitation i.e. halving the proportion of people without sustainable access to safe drinking water and basic sanitation by 2015.

Before the start of operation, to set-up a bio-centre, an environmental impact assessment (EIA) is required. There are two types of quality standards applicable to the bio-slurry: the agricultural standard and the NEMA standard. These quality standards require that the bio-slurry has to meet acceptable standards in order for it to be re-used in the farm and also to be safely disposed to the environment. The bio-slurry should be within recommended environmental guidelines.

The Water Services Trust Fund (WSTF) has developed a national sanitation concept for up-scaling public sanitation. Up-Scaling Basic Sanitation for the Urban Poor in Kenya-UBSUP-Kenya is a five-year program which is implemented by WSTF with support from the Deutsche Gesellschaft für Internationale Zusammenarbeit (GIZ). The program is financially supported by KfW, a German financial cooperation, and the Bill and Melinda Gates Foundation (BMGF) and in kind by GIZ. The on-site sanitation systems will have a strong focus on sustainable fecal sludge management.

Technology and processes

TOSHA bio-centre consists of toilets, showers, operator's office, meeting hall, restaurant and a bio-digester. The technology used at TOSHA bio-centre is an adapted replication of the sanitation systems by Sulabh International Social Service Organisation in India. The bio-digester unit is an anaerobic treatment technology that produces biogas and a digested slurry that can be used as liquid fertilizer. The bio-digester is fed with the fecal sludge from the sanitation facilities equipped with flush toilets constructed within the bio-centre serving nearly 600–800 people per day. Each of the three states of matter are rendered useful: gas for the production of energy, liquid i.e. the treated water recycled and the treated solid waste used as fertilizer. The success of this technology depends on the proper construction of the bio-digesters. This means skilled labor is required during the setting-up phase. Umande Trust prefers this type of bio-digester due to the simplicity of construction, operation and maintenance.

The bio-digester is a fixed dome 54 m^3 reactor comprising of brick-constructed dome chamber that has been built below ground. In principle the hydraulic retention time (HRT) should be a minimum of 15 days in hot climates and 25 days in temperate climates as per the design criteria. Average HRT of the bio-digester is 20 days at an ambient average temperature of 25 °C. During operations, water is necessary for proper decomposition of the waste. Hence the operators ensure that despite water shortage, water is made available for flushing the toilets. However, due to insufficient water supply from the water service providers, water is flushed manually using tins from water drums that have been placed within the toilets. In order to avoid foreign objects (non-biodegradable) entering the bio-digester, a sieve is placed at the entrance to trap them. Once waste products enter the digestion chamber, gases are formed through fermentation. The gas forms in the sludge but collects at the top of the reactor, mixing the slurry as it rises. As gas is generated, it also exerts a pressure and displaces the slurry into an expansion chamber. When the gas is removed, the slurry will flow back down into the digestion chamber. The pressure generated can be used to transport the biogas through pipes. The slurry that is produced is rich in organics and nutrients and is almost odorless. No further treatment is done on the slurry, as tests done by Umande Trust, confirm that it is safe to be applied directly by farmers on their farms.

Funding and financial outlook

The general approach used by Umande Trust to construct the bio-centres for CBOs is that they obtain funds to construct the facility while the community would provide all unskilled labor required for construction. The bio-centre requires nearly 100 m^2 of land to construct it. The bio-centre was constructed on private land and hence there was no cost of land acquisition. The cost incurred to construct the bio-centre was nearly USD 22,500 (in 2006) and an additional USD 10,000 for campaigns and training to sensitize the community on the new technology to ensure successful implementation of the project (Table 12).

The main expenses incurred during the operation of the bio-centre include: cleaning the sanitation facilities, operation and maintenance of the facilities and employees to manage the facilities. As part of handing over the facilities to the CBO, Umande Trust ensures that the operators are well-trained to effectively operate and maintain the bio-centre. Campaigns to eradicate the stigma of using biogas and the bio-slurry are also undertaken by Umande Trust. This is done by involving leaders and respected men in the society to endorse the innovation during commissioning of the project.

TABLE 12. TOSHA 1 INVESTMENT, OPERATIONAL AND MARKETING COST

ITEM	AMOUNT (USD)
Investment cost:	
Construction cost	22,500
Campaign	10,000
Total	32,500
Annual operating and maintenance:	
Salary	1,800
Toilet paper	1,200
Water and electricity	240
Maintenance costs	100
Exhaustion services	380
Total cost	3,720

TABLE 13. FINANCIAL SUMMARY OF TOSHA 1 BIO-CENTRE

ITEM	AMOUNT (USD/YEAR)
Toilet services	13,920
Biogas	720
Rent (cyber)	1,200
Total revenue	15,840
Total cost	3,720
Operating profit	12,120
Payback period (years)	3 years
Net present value (NPV) (USD)	50,000
Internal rate of return (IRR) (%)	33%

The major income stream is from toilet services accounting for 88% of the total revenue (Table 13). TOSHA 1 is hardly getting any revenue from the bio-slurry due to personal distaste over using it as fertilizer. Assuming that profits and cash flows will continue to be the same in the future and assuming that useful life of the project is 20 years, the payback period is 3 years, the NPV is USD 50,000 and the internal rate of return is 33%. A future plan of the bio-centre is to containerize the biogas and sell it to individuals, institutions and hotels as an alternative source of energy.

Socio-economic, health and environmental impact

TOSHA 1 is a good example of an environmental-friendly and sustainable approach to providing sanitation services through safe processing and disposal of human excreta. About 1,000 men, women, youth and children from poor slum households have benefited from the bio-centre. The availability of business premises to individual business owners at an affordable rate has contributed to creating entrepreneurs in the community. The provision of clean toilets and bathrooms as well as a cooking area at a very affordable rate have improved the lives of the community residing near the bio-centres. In addition to these, a community-led approach to sanitation contributes to capacity building of the community and also to changing people's attitude towards use of human waste as a source of energy and business.

Amongst the key benefits would be improved health status of the general population with resultant significant savings from medical bills and improved social relations among neighbours who would now be sharing the same facilities. The project promotes renewable energy, helping the shift from the usual wood, charcoal, kerosene and gas to biogas for cooking. This helps in improving energy efficiency, reducing carbon dioxide emissions and alleviating pressure on forests. It is also cheaper and relatively affordable. The consistent participation of communities as shareholders (not stakeholders) is designed to promote ownership, the sharing of responsibilities and profits accruing from community-managed water and sanitation initiatives.

Scalability and replicability considerations

The key drivers for the success of this business are:
- Lack of access to healthy and affordable sanitation facilities for households in urban areas.
- Government strategy, vision 2030 to improve waste management through application of economic incentives and appropriate PPPs.
- Community-led strategy to promote use of biogas and bio-slurry as alternative energy sources and fertilizer respectively.
- Sponsorship from various multilateral, bilateral and national entities.

Umande Trust is replicating this initiative on varying scales in Kenya using funds from sponsors who were encouraged by the success of the existing bio-centre projects. Umande Trust is planning to scale up the sizes of the bio-centres by increasing the size of the bio-digester and the number of latrines as well as commercial facilities. It plans to construct one large bio-centre where the other bio-centres can be emptied into so that they can generate more gas with a possibility of generating electricity. They also plan to construct a bio-centre that use solar and wind energy. With availability of financial resources the goal is also to containerize biogas or pipe it to nearby hotels to increase their sales. In order for this business to be successfully up-scaled and replicated in other locations, community-led strategy with sponsorship from various multilateral, bilateral and national entities is a key to the successful implementation of the project. Moreover, continued campaigns to promote use of biogas and bio-slurry as alternative energy sources and fertilizer respectively are needed.

Umande Trust and the communities have also formed a Sanitation Development Fund (SANDEF), a self-sustaining fund which loans out the funds needed to undertake a sanitation project. Government and NGOs can make a donation to SANDEF. It is after a project is completed that the loan will be repaid to SANDEF and those funds can be loaned out again for another project, hence multiplies the impact of donations.

Summary assessment – SWOT analysis

Key strengths of this business are implementing of community-led strategy and multiple revenue streams, which reduce the risk of failure (Figure 38). However, the business is highly subsidized and high dependence on donor grants pose sustainability issues.

FIGURE 38. SWOT ANALYSIS FOR TOSHA 1

	HELPFUL TO ACHIEVING THE OBJECTIVES	HARMFUL TO ACHIEVING THE OBJECTIVES
INTERNAL ORIGIN ATTRIBUTES OF THE ENTERPRISE	**STRENGTHS** • Community-led strategy • Different revenue streams reduces risk of failure • Strong partnership with local government and NGOs • Low cost and environment-friendly proven technology	**WEAKNESSES** • No immediate market for bio-slurry • Highly subsidized (may pose difficulty in scaling-up) • Limited technical and business skills of the community • Inadequate space for sanitation development in low income urban areas
EXTERNAL ORIGIN ATTRIBUTES OF THE ENVIRONMENT	**OPPORTUNITIES** • Up-scaling potential • Raising more funds from other sources (NGO or bilateral and multilateral entities) • Development of a point-sales biogas market • Unmet opportunities of ensuring access to basic sanitation for low income urban areas	**THREATS** • Stigma against using human by-products and cultural barriers may lead to costly campaigns • High dependence on donor grants pose sustainability issues • Initial reluctance and lack of interest by community • Too many fragmented actors in sanitation sector • Most resources allocated on awareness creation rather than development of infrastructure

Contributors

Johannes Heeb, CEWAS
Jasper Buijs, Sustainnovate; Formerly IWMI
Josiane Nikiema, IWMI
Kamalesh Doshi, Simplify Energy Solutions LLC, Ashburn, Virginia, USA

References and further readings

Agence Française de Développement (AFD) and Athi Water Services Board (AWSB). 2010. Evaluation report: Assessment of impacts of ablution blocks project in informal settlements of Nairobi.

Herbling, D. 2012. How Nairobi bio-centres reap from human effluence. Business Daily Nov. 18, 2012 (https://goo.gl/BueanG; accessed Nov. 6, 2017).

Kenya Vision 2030. 2007. Republic of Kenya. www.theredddesk.org/sites/default/files/vision_2030_brochure__july_2007.pdf (accessed June 26, 2013).

Munala, G.K., Mugwima, B.N., Omotto, J. and Rosana, E. 2015. Managing Human Waste in Informal Settlements: Bio-centres in Kibera Informal Settlement, Kenya. http://www.huussi.net/wp-content/uploads/2015/06/Munala_Full-paper_DT2015.pdf (accessed Nov. 7, 2017).

Water Services Trust Fund. 2012. Up-scaling basic sanitation for the urban poor in Kenya (UBSUP-Kenya).

Case descriptions are based on primary and secondary data provided by case operators, insiders, or other stakeholders, and reflect our best knowledge at the time of the assessments (2015–2016). As business operations are dynamic, data can be subject to change.

Note

1 TOSHA stands for Total Sanitation and Hygiene Access and was founded in 2004 as a civil society network by the Umande Trust (in Swahili Tosha means 'adequate'). Over time many CBOs left TOSHA to establish their own bio-centres, resulting in a decentralized approach with wide outreach.

BUSINESS MODEL 3
Biogas from fecal sludge at community level

Krishna C. Rao and Solomie Gebrezgabher

A. Key characteristics

Model name	Biogas from fecal sludge at community level (while providing sanitation services)
Waste stream	Fecal sludge/night soil and/or kitchen waste from public toilets and residential institutions (like prisons)
Value-added waste product	Biogas for cooking and lighting; bio-fertilizer
Geography	Applicable to residential institutions and public toilets that provide toilet facilities to underserved communities
Scale of production	Small to medium scale; as small as 10 m^3 up to 200 m^3 of biogas per day
Supporting cases in this book	Nepal, the Philippines and Rwanda, India, Kenya
Objective of entity	Cost-recovery [X]; For profit [X]; Social enterprise [X]
Investment cost range	About USD 10,000 to USD 85,000
Organization type	Private and public-private partnership
Socio-economic impact	Environment-friendly cooking fuel, reduced deforestation, air pollution and greenhouse gas emissions, improved fecal sludge management results in improved sanitation and health and reduced pollution of local water bodies, employment generation and better landscaping (by use of bio-fertilizers)
Gender equity	Clean indoor air and working environment; creating new and sustaining existing public toilets reducing personal risks especially for women and girls

B. Business value chain

The business model is initiated by either enterprises/NGOs, providing sanitation services such as public toilets or by residential institutions such as hostels, hospitals and prisons that produce large quantity of human waste. The business concept is to construct new toilets with collection and transfer of waste to bio-digester, process and treat human waste and/or kitchen waste in a bio-digester to generate biogas. Biogas can be used for internal use for lighting and cooking or sold to nearby households and businesses. The bio-fertilizers can be used for landscaping or vegetables gardens within the complex or nearby. Value chain of the enterprise varies depending on the entity initiating the business model.

Toilet complex business model

The key stakeholders in this business value chain are the toilet users, technology supplier to install the biogas plant and its maintenance, local bodies and agencies/donors for funding the capital cost, end

BUSINESS MODEL 3: BIOGAS AT COMMUNITY LEVEL

users of the additional service (rental space) and value added product (biogas and potentially fertilizer) from the toilet complex (Figure 39).

The ownership and operation of toilet complex can be either by an entrepreneur, a community-based organization (CBO) or municipality. One of the roles of the municipality is to provide land for the toilet complex. Human waste from the toilet complex is fed directly to bio-digester and the biogas generated can be used within the toilet complex for lighting and heating water. The enterprise can also sell biogas to nearby households and businesses as fuel for cooking. Depending on the land space availability, the toilet complex can rent out a space within the complex to a private business such as newspaper/book stand, small neighbourhood retail store and so on. The business could potentially make fertilizer from the bio-slurry output from the bio-digester, which can be used either for landscaping purpose

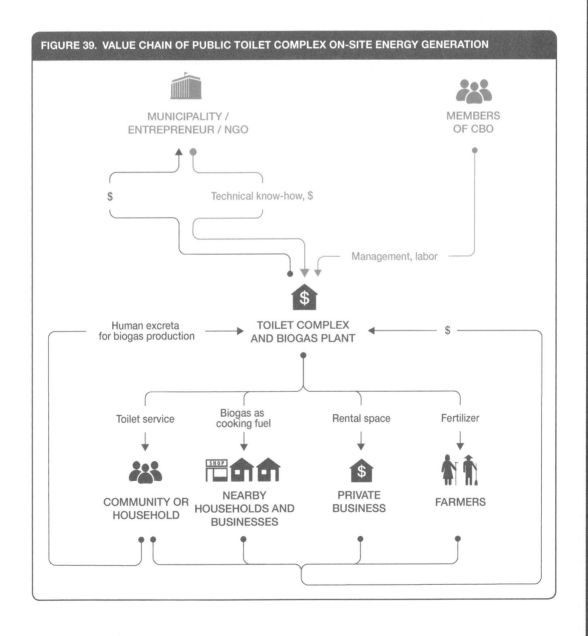

FIGURE 39. VALUE CHAIN OF PUBLIC TOILET COMPLEX ON-SITE ENERGY GENERATION

around the toilet complex or sold to farmers. However, since the product is made from human waste, the enterprise needs to take specific care to ensure the product is free from pathogens.

Residential institution business model

The key stakeholders in this business value chain are the management of the residential institution (e.g. prisons), residents (e.g. inmates of the institution), biogas operator and the kitchen management at the residential institution (Figure 40).

The concept in this business is to utilize the concentrated source of human waste generated by residents of the institution to generate biogas, which can be used within the institution premises as fuel for cooking. The process consists of sending human waste from toilets to biogas digester and the biogas produced is used in the kitchen for cooking. As an additional source of income generation,

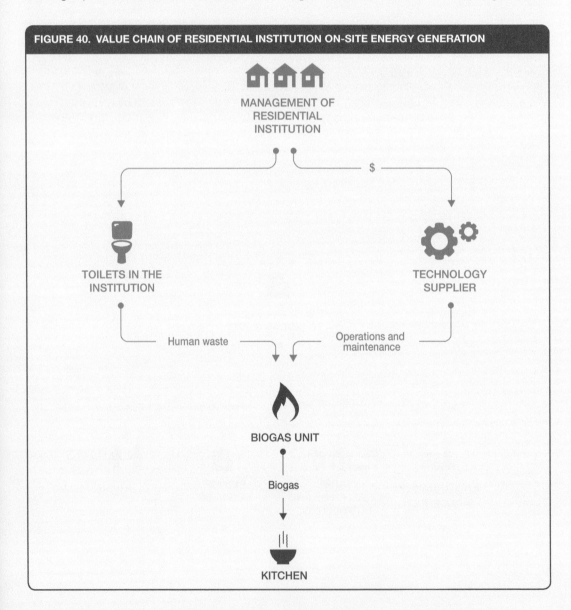

FIGURE 40. VALUE CHAIN OF RESIDENTIAL INSTITUTION ON-SITE ENERGY GENERATION

potentially the residential institution's inmates can undertake business activity of baking or making processed food that can be sold in nearby town/city. The biogas can also be fed with other organic waste such as kitchen waste and biomass (leaf litter) generated within the institution premises. The bio-slurry from the bio-digester can be used towards landscaping or as fertilizer for growing vegetables within the institution premises under very specific and strict safety protocols.

Both the business models are eligible for sale of carbon as the biogas is generated from human waste. In both these business models, there is scope for private technology enterprise that could get into Build, Own and Operate (BOO) or Build, Own, Operate, Transfer (BOOT) arrangements. The private entity could bring all investment to set up the biogas technology while the institution provides land and sends human waste to the biogas plant. The private entity designs, constructs and maintains the biogas unit until BOOT period is expired after which it assists the institute to operate the unit.

C. Business model

The primary value proposition of the business model for both public toilet complex and a residential institution is to provide improved safe sanitation service in an environmentally responsible manner and secondary value proposition is to generate biogas from human waste. The value proposition remains same irrespective of the entity driving the business initiative which is either the public toilet complex, residential institution or private biogas technology supplier that uses BOO or BOOT approach to provide improved human waste management service to the institution along with provision of environment friendly fuel for cooking. The business model also offers value proposition of providing organic compost from bio-slurry, the output from the biogas plant.

The business model canvas is significantly different for public toilet complex and residential institution (Figure 41). In addition to the value propositions mentioned above, the public toilet complex with biogas plant can offer rental space for a business with access to a uniform group of customer segment.

Public toilet complex enterprise has multiple revenue sources. The primary revenue is from the fees collected for usage of toilet. Other revenue sources are from sale of biogas, compost and rental income. The model requires partnership with municipality or local community for access to land to build the toilet complex.

In the residential institution business model (Figure 42), the biogas is used primarily for internal consumption thereby the institution incurs substantial savings from avoided fuel purchase for cooking. The business model offers scope to sell either entire or excess biogas to nearby households and businesses. In the case of compost, the residential institution can use it internally for landscaping or growing vegetables for internal consumption. The inmates of residential institutions could be organized to undertake business activity of making processed food such as snacks and bakery product and in the process have additional revenue source from sale of these products.

Residential institution model requires developing partnership with biogas technology supplier whose assistance is critical in the initial stages towards operation and maintenance until a local labor is trained. The key activities include the production of biogas and the key resources are land, equipment, biogas technology and access to human waste.

Both these business models are eligible for carbon offset, however the biogas plant size is small to be viable to apply for Clean Development Mechanism (CDM) projects due to associated transaction costs and preferred route would be to apply for carbon offset on Voluntary Emission Reductions (VERs) market.

CHAPTER 4. BIOGAS PRODUCTION FOR ENERGY SAVINGS

FIGURE 41. BUSINESS MODEL CANVAS FOR PUBLIC TOILET COMPLEX

KEY PARTNERS
- Equipment suppliers
- Municipality or community
- Carbon trading partners
- Water service providers
- International development agencies and foundations

KEY ACTIVITIES
- Sanitation service
- Processing human waste
- Biogas generation
- Distribution and sale of biogas
- Compost production
- Public campaigns and training
- Managing Voluntary Emission Reduction (VER) process

VALUE PROPOSITIONS
- Provision of improved, safe and affordable sanitation service
- Environment-friendly fuel for cooking/heating and lighting
- Organic compost made from bio-slurry by-product from biogas digester
- Provide a convenient space with direct access to a large customer segment of similar outlook

CUSTOMER RELATIONSHIPS
- Direct interaction

CUSTOMER SEGMENTS
- Households in the local community
- Households and street vendors
- Farmers and household with garden
- Business
- Carbon market

KEY RESOURCES
- Organic waste
- Equipment and technical know-how
- Land
- Labor
- Community leaders as endorsers

CHANNELS
- Direct sales at the enterprise
- Carbon market agents

COST STRUCTURE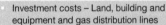
- Investment costs – Land, building and equipment and gas distribution lines
- O&M costs – toilet facility cleaning, toilet papers and consumables, training, utilities, labor (can be intensive and skilled labor)
- Costs incurred for VER registration and carbon sale

REVENUE STREAMS
- Toilet usage fees (pay-per-use)
- Sale of Biogas
- Sale of carbon credit
- Sale of compost
- Rental space income

SOCIAL & ENVIRONMENTAL COSTS
- Potential leakage of gas
- Potential health risks for workers from direct contact with human waste
- Potential environmental risk if the human waste is not treated and disposed properly
- Using bio-slurry for crops may potentially pose health risks due to possible pathogen survival

SOCIAL & ENVIRONMENTAL BENEFITS
- Improved human waste management and treatment at the source
- Reduced local pollution from improved human waste management and treatment
- Reduced GHG emission
- Create job opportunities
- Capacity building of community
- Changing people's attitude towards use of human waste as source of energy

BUSINESS MODEL 3: BIOGAS AT COMMUNITY LEVEL

FIGURE 42. BUSINESS MODEL CANVAS FOR RESIDENTIAL INSTITUTION

KEY PARTNERS	KEY ACTIVITIES	VALUE PROPOSITIONS	CUSTOMER RELATIONSHIPS	CUSTOMER SEGMENTS
• Equipment suppliers • Carbon trading partners • Local, central and state government agencies for partial funding	• Sanitation service • Processing human waste • Biogas generation • Distribution and sale of biogas • Compost production • Training of staff • Managing VER process	• Provision of improved safe sanitation service • Environment friendly fuel for cooking/heating and lighting • Organic compost made from bio-slurry by-product from biogas digester	• Direct interaction	• Inmates of residential Institution • Nearby households and businesses • Carbon Market
	KEY RESOURCES • Organic waste • Equipment and technical know-how • Land • Skilled labor		**CHANNELS** • Direct sales at the enterprise • Carbon market agents	

COST STRUCTURE	REVENUE STREAMS
• Investment costs – land, building and equipment and gas distribution lines • O&M costs – training, utilities, labor (can be intensive and skilled labor) • Costs incurred for VER registration and carbon sale • Savings from cooking fuel and compost	• Sale of biogas • Sale of cooked food • Sale of carbon credit

SOCIAL & ENVIRONMENTAL COSTS	SOCIAL & ENVIRONMENTAL BENEFITS
• Potential leakage of gas • Potential health risks for workers from direct contact with human waste • Potential environmental risk if the human waste is not treated and disposed properly • Using bio-slurry for crops may potentially pose health risks due to possible pathogen survival	• Improved human waste management and treatment at the source • Reduced local pollution from improved human waste management and treatment • Reduced GHG emission • Reduction in deforestation • Create job opportunities

D. Potential risks and mitigation

Market risks: The market risk is different for both toilet complex and residential institution business model. Market risks hardly exist if the business is initiated by the residential institution; however, in the case of toilet complex, it has risks of community or household willing to use the toilet facility. In both the cases, there is a potential social implication of willing to use biogas and bio-fertilizers generated from human waste for cooking and landscaping purpose.

Competition risks: Biogas competes with LPG, kerosene or other traditional cooking fuels such as fuel wood and charcoal. In most developing countries, kerosene and LPG are subsidized for domestic consumption and thus biogas should be produced at a lower cost than these competing products to get buy-in from end users. For end users who use charcoal and fuelwood, there is a need for additional investment in cooking stoves for them to shift to biogas use. However, for the residential institutions with biogas plant, expense incurred for purchasing cooking fuel is reduced significantly. On a long-term basis and before the life cycle of the biogas plant, investment cost of the plant and its operation cost is completely recovered by the residential institution.

Technology performance risks: The technology used is anaerobic digestion, which is well established and mature. However, the type of digester required could potentially be sophisticated and might not be available in developing countries, and in addition the technology requires skilled labor. It is ideal for the business to transfer the technology from a market where it is widely implemented and have their staff trained in repair and maintenance of the technology. The extra care will have to be taken by operators to make sure that digested slurry is free from pathogens before using it as fertilizer.

Political and regulatory risks: In most developing countries, price of cooking fuels such as kerosene and Liquefied Petroleum Gas (LPG) are subsidized for domestic consumption. Such government policies can diminish the economic advantage offered by the biogas supplied to households and in unlikely case, if the policy is extended to commercial entities, the business model is unviable. Lately, governments are encouraging green initiatives by providing incentives such as financial assistance, concessional loans and depreciation benefits. Policies supporting green initiatives make this business model highly attractive.

Social-equity-related risks: The public toilet complex model offers greater benefits to women from increased privacy rather than defecating in the open. The biogas generated from the toilets, if used in household for cooking, would again benefit women due to use of cleaner fuel. The biogas used internally for energy savings and for residential institutions is gender neutral. Both the models mostly offer energy savings and the benefit is accrued by the institution or toilet complex. Employment opportunities, while limited, benefit the marginalized. Improved sanitation in the case of public toilet complex model benefits the underserved.

Safety, environmental and health risks: Safety and health risks to human arise when processing any type of waste but the risks are further increased when dealing with human waste. Labor in such enterprises should be provided with appropriate gloves, masks and other appropriate tools to handle the waste to ensure their safety. Ideally, the enterprise should have strict safety policies as the potential for direct human contact with human waste is very high. The risk of environment pollution is high if human waste is not treated properly and is disposed of openly leading to groundwater or surface-water pollution. The environmental risks associated with the anaerobic digestion units include possible leakage of gas and these emissions should be controlled. Compost from bio-slurry has high risks of pathogens, if not treated properly. However, if proper operation procedures are followed, the risks are reduced significantly (Table 14).

BUSINESS MODEL 3: BIOGAS AT COMMUNITY LEVEL

TABLE 14. POTENTIAL HEALTH AND ENVIRONMENTAL RISK AND SUGGESTED MITIGATION MEASURES FOR BUSINESS MODEL 3

RISK GROUP	EXPOSURE ROUTES					REMARKS
	Direct contact	Air	Insects	Water/Soil	Food	
Worker	HIGH RISK					Direct contact risks exist for workers if fecal sludge is wrongly handled, and if also compost is produced. Food produced with bio-slurry compost could have pathogen exposure and should be cooked for safe consumption.
Farmer/User	LOW RISK					
Community			LOW RISK	MEDIUM RISK		
Consumer					MEDIUM RISK	
Mitigation Measures	🧤	😷	🧴	🚱	🍲	

Key: ☐ NOT APPLICABLE   ▨ LOW RISK   ▨ MEDIUM RISK   ■ HIGH RISK

E. Business performance

This business model scores high on scalability and replicability followed by environmental impact (Figure 43). The business model has high potential for replication in developing countries and, except for the social acceptance of the product from human waste, there are no factors that limit its potential for replication. The business model offers horizontal and vertical scaling by expanding business to other sectors, such as compost and selling cooked food; however, expansion to these other businesses is a theoretical possibility. The environmental impact scores high because of high replication potential that

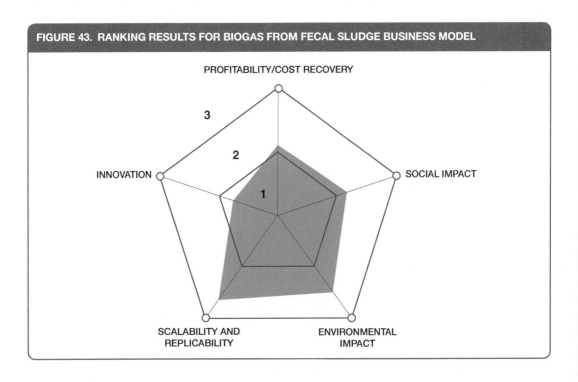

FIGURE 43. RANKING RESULTS FOR BIOGAS FROM FECAL SLUDGE BUSINESS MODEL

the business model offers that could result in safe management of human waste and in the process reduced pollution of groundwater and surface water resulting in lesser damage of ecosystem and its services.

The business model scores reasonably well on social impact largely from the treatment and safe management of human waste. Depending on the type of entity initiating the business, the revenue source can vary. However, even in the toilet complex business model, which offers multiple revenue source, these will remain modest, while building on cost savings based on reduced fuel expenses. The business model scores low on innovation as the technology and financing required is fairly straightforward.

CASE

Biogas from kitchen waste for internal consumption (Wipro Employees Canteen, India)

Kamalesh Doshi, Krishna C. Rao and Binu Parthan

Supporting case for Business Model 4	
Location:	Bangalore, India
Waste input type:	Food waste, kitchen waste and sewage sludge
Value offer:	Energy for cooking using biogas from kitchen waste
Organization type:	Private
Status of organization:	Operational since 2008
Scale of businesses:	Medium
Major partners:	Mailhem Engineers Pvt. Ltd (for biogas technology, plant and O&M services)

Executive summary

Established in 1945, Wipro Ltd is a large business international conglomerate, with a revenue of over USD 7.3 billion and more than 75,000 employees in India. Wipro Ltd provides comprehensive IT solutions and services, including systems integration; information systems outsourcing; IT-enabled services; package implementation; software application, development and maintenance and research and development services to corporations globally and also produces lighting, engineering, personal and medical products. Wipro operates a large canteen catering to 5,000 to 5,500 employees in their Bangalore headquarters and generates about 1,500 kg of canteen and kitchen waste per day. As a part of its corporate social responsibility, Wipro supported the initiative to convert kitchen and food waste from the employee's canteen to biogas for cooking in the canteen.

Wipro partnered with Mailhem Engineers Pvt Ltd, a waste management technology firm, to install, operate and maintain the biogas plant capable of treating three tons per day (1,095 tons/year) of canteen waste. Mailhem has indigenously developed bio-methanation technologies with modified upward anaerobic sludge blanket technology that treat all types of solid and liquid waste having large percentage of suspended solids. About 69,300 to 74,250 m^3 of biogas is produced annually. The biogas has replaced Liquefied Petroleum Gas (LPG) as the cooking fuel, saving four 19-kg LPG cylinders per day, leading to annual fuel cost saving of USD 24,480 at price of USD 17 per 19-kg cylinder for commercial applications and an increase in brand equity along with generating employment for four people. Around 108 tonnes of bio-sludge is generated annually, which is used as manure in the gardens on Wipro's campus and 3 m^3/day of overflow water is fed to sewage treatment plant in the premises.

KEY PERFORMANCE INDICATORS (AS OF 2014)	
Land use:	300 m² (20 m x 15 m)
Water requirements:	1,500–1,800 L per day
Power consumption:	25 kWh/day
Capital investment:	USD 100,000
Labor:	3 full-time employees and 1 part-time employee
O&M cost:	USD 10,320/year
Output capacity:	210–225 m³/day of biogas; 2 tonnes/day of sludge 1,500 kg/day of canteen waste, vegetable and fruit peels + 12 m³ of organic sludge from the existing sewage treatment plant (STP)
Potential social and/or environmental impact:	Jobs for 4 people created, waste reused without being discharged in municipal waste; carbon emissions offset from avoided municipal waste landfill and also replacement of LPG which otherwise would have been used for cooking; carbon emissions saved 37.26 tons CO_2/year from waste recycling and 306.77 tons CO_2/year from LPG saved
Financial viability indicators:	Payback period: 3.5 years Post-tax IRR: > 51% Gross margin: 25%

Context and background

The biogas plant was initiated by Wipro in 2008 as part of its corporate social responsibility (CSR) initiative. The immediate goal was to manage the kitchen waste in the environmentally acceptable manner with long-term goal of use of waste management initiatives to reduce the company's environmental footprint and to achieve corporate sustainability. It was financed by Wipro through its internal revenues and constructed and installed by Mailhem. Collection and cooking duties are performed by Wipro, while segregation, digestion and production services are performed by Mailhem. The plant is located, adjacent to the kitchen at Wipro's headquarters in Bangalore. The input for the plant is organic canteen and food waste with 80% moisture that is generated in the employees' canteen. Before the biogas plant, it was a tedious task for staff to pack the huge amount of waste in polyethylene bags and hand them over to the civic body almost daily. It is now easy to dump this waste into the biogas plant after some segregation.

Market environment

As Wipro provides the inputs and uses the product within campus, there are no external dependency for the project. Since the project uses canteen and food waste, which otherwise needs to be disposed of, there is no cost of inputs to the project. The product, biogas, is consumed within the campus saving the cost of LPG for cooking. The price of LPG for commercial users is around USD 17 per 19 kg cylinder. LPG for corporations are not subsidized in India, and that is an additional factor motivating Wipro to invest in the technology and generate its own cooking fuel.

Macro-economic environment

There is an emerging focus on green technology in India. CSR also called corporate conscience, corporate citizenship or responsible business is a form of corporate self-regulation integrated into a business model. CSR policy functions as a self-regulatory mechanism whereby a business monitors and ensures its active compliance with the spirit of the law, ethical standards and national or international norms. With some models, a firm's implementation of CSR goes beyond compliance and engages in "actions that appear to further some social good, beyond the interests of the firm and that which is required by law." The aim is to increase long-term profits through positive public relations, high ethical standards to reduce business and legal risk, and shareholder trust by taking responsibility for corporate actions. CSR strategies encourage the company to make a positive impact on the environment and stakeholders including consumers, employees, investors, communities and others. Wipro's primary motivation for this venture is to showcase the company's corporate social responsibility efforts, better manage its kitchen and food waste, produce biogas and use it on site to reduce its LPG consumption for cooking.

Business model

The value proposition for Wipro (Figure 44) is to minimize cost and to better manage its waste and be environmentally responsible. Wipro serves as both the customer and the producer since it is the supplier of waste and consumer of end products. The key activities include the production of biogas from food waste for use in its kitchen and produce organic compost to be used in landscaping within its campus and the key resources are land, equipment, biogas technology and sourcing of the waste. The produced biogas has resulted in substantial savings from avoided LPG purchase, has created jobs for four individuals and contributed towards reducing pollution of water bodies and natural habitats. In the event of surplus biogas and compost produced, Wipro could sell the energy to neighbouring households and compost to urban households to use them in their garden or to urban farmers. Wipro also would qualify to sell the carbon offset from this investment.

Value chain and position

The value chain consists of collection and segregation of kitchen and food waste, digestion of waste in biogas digester, production of biogas and finally the use of biogas in the kitchen for cooking and use

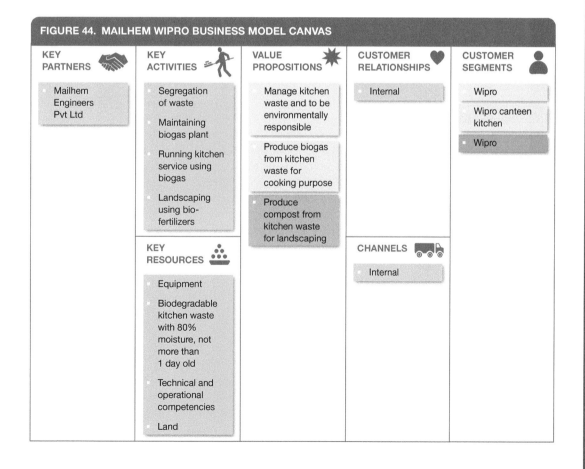

FIGURE 44. MAILHEM WIPRO BUSINESS MODEL CANVAS

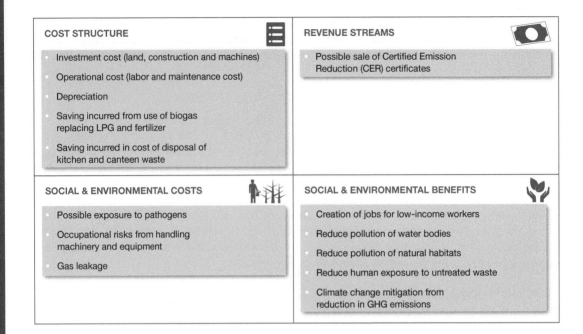

of digested slurry as organic fertilizer for landscaping in the campus. A critical relationship pertaining to the investment is with the kitchen manager and staff who make available kitchen waste with the right moisture and biodegradable content specifications and who again use the biogas for cooking food. Another important relationship is with the Mailhem engineers who supplied the technology and usually also operate and maintain the system. Wipro pays for the kitchen operation and the operation and maintenance of the system (Figure 45).

Institutional environment

In India, the concept of CSR is governed by clause 135 of the Companies Act 2013. The CSR provisions within the Act is applicable to companies with an annual turnover of USD 180 million and more, or a net worth of USD 9 million and more or a net profit of USD 0.9 million and more. The Act encourages companies to spend at least 2% of their average net profit in the previous three years on CSR activities. The government's suggested CSR activities include measures to eradicate hunger; promote education, environmental sustainability, protection of national heritage and rural sports and make contributions to prime minister's relief fund. The new rules, which will be applicable from the fiscal year 2014–2015 onwards, also require companies to set up a CSR committee consisting of their board members, including at least one independent director. The new Act requires that the board of the company shall, after taking into account the recommendations made by the CSR committee, approve the CSR policy for the company and disclose its contents in their report and also publish the details on the company's official website, if any, in such manner as may be prescribed. If the company fails to spend the prescribed amount, the board, in its report, shall specify the reasons.

While the CSR spending by the top 100 Indian companies is estimated at USD 0.86 billion per annum, the Indian Institute of Corporate Affairs anticipates that about 6,000 Indian companies will be required to undertake CSR projects in order to comply with the new guidelines, with many companies undertaking these initiatives for the first time. Some estimates indicate that the CSR spends in India could triple to USD 2.6 billion a year. This combination of regulatory as well as societal pressure has meant that companies have to pursue their CSR activities more professionally. A large number of

CASE: BIOGAS FROM OWN KITCHEN WASTE

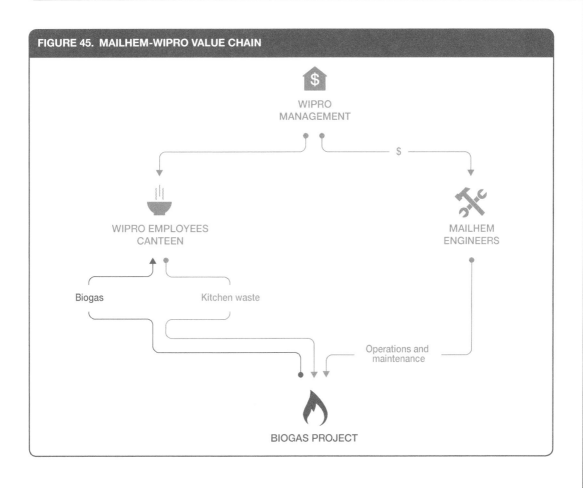

FIGURE 45. MAILHEM-WIPRO VALUE CHAIN

companies are reporting the activities they are undertaking in this space in their official websites, annual reports, sustainability reports and even publishing CSR reports.

The Government of India provides economic incentives in the form of accelerated depreciation benefits that can be claimed to offset tax obligations of the firm that invests in renewable energy technology. The Ministry of New and Renewable Energy Sources (MNRE) takes an active part in biogas projects and researches and provides subsidy incentives towards capital costs. The laws in the state of Karnataka mandate the management of organic waste by local bodies. Municipal solid waste management is carried out by the local bodies and not the establishments which generate the waste.

Technology and processes

Fresh (less than 1 day old) biodegradable kitchen waste with 80% moisture is used as the primary input. The facility used high-rate modified up-flow anaerobic sludge blanket (UASB) bio-methanation technology, in which the biomass is retained as a blanket and kept in suspension in the lower part of the digester. The UASB is a high-rate suspended growth in which a pre-treated raw influent is introduced into the reactor from the bottom and distributed evenly. "Flocs" of anaerobic bacteria will tend to settle against moderate flow velocities. The effluent passes upward through, and helps to suspend, a blanket of anaerobic sludge. A particular matter is trapped as it passes upward through the sludge blanket, where it is retained and digested. Digestion of the particular matter retained in the sludge blanket and breakdown of soluble organic materials generate gas and relatively small amounts of new sludge. The rising gas bubbles help to mix the substrate with the anaerobic biomass. The biogas, the liquid fraction and the sludge are separated in the gas/solid/liquids phase separator, consisting of the gas collector dome and a separate quiescent settling zone. The settling zone is relatively free of mixing effect of the gas, allowing the solid particles to fall back into the reactor; the clarified effluent is collected in gutters at the top of the reactor and removed. Maintenance of the sludge blanket is an important factor in the efficient operation of these digesters. The plant can treat all types of solid and liquid waste having large percentage of suspended solids. The end products are biogas, organic manure and treated water for gardening. The technology is indigenously developed by Mailhem and is locally available and any component that needs to be replaced can be sourced or fabricated locally.

The design and performance of anaerobic digestion processes are affected by many factors. Some of them are related to feedstock characteristics, reactor design and operation conditions. The prerequisites for production of biogas are a lack of oxygen, a pH value from 6.5 to 7.5 and a constant temperature of 35–45 °C (mesophilic) or 45–55 °C (thermophilic). The digestion period or retention period is typically between 10 and 30 days depending upon the type of digestion employed. The anaerobic digestion systems of today operate largely within the mesophilic temperature range. The operation and management of the project is handled by Mailhem and the technicians employed for O&M have been trained and supervised by Mailhem (Figure 46).

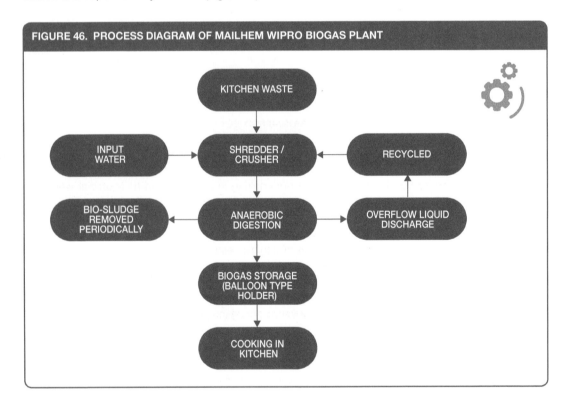

FIGURE 46. PROCESS DIAGRAM OF MAILHEM WIPRO BIOGAS PLANT

Funding and financial outlook

The facility was set up at the Wipro campus, using equity from the company's internal finances. The plant and machinery cost USD 100,000 to set up. The primary input, kitchen waste, is sourced through Wipro's internal kitchen and is free of charge. Operation and maintenance costs amount to USD 10, 320 per year (Table 15).

TABLE 15. MAILHEM-WIPRO FINANCIALS

Key capital costs	Land building	No additional charge (uses own land)
	Plant, machinery and civil construction	100,000 USD
Operating costs	Kitchen waste and sewerage sludge costs	No additional charge (from own kitchen and sewerage treatment plant)
	Operation and maintenance	10,320 USD/year
Financing options	Equity from Wipro	100,000 USD
Outputs	Biogas and digested slurry	Supplied to Wipro free of charge and internally consumed to save cost of LPG and organic fertilizers for landscaping

The major savings is from avoided cost of LPG. The investment has resulted in onsite management of waste generated in the employees' canteen and displaces 27.36 tons of LPG, which the company would have purchased otherwise at the cost of USD 24,480 in 2014 prices per year. There are minor savings in the form of avoided purchase of manure for the garden, but these are relatively small. The internal rate of return for the investment is 27.34%. Thus, the investment of USD 100,000 is recovered within less than four years.

Socio-economic, health and environmental impact

In most of cities and places, kitchen waste is disposed in landfill or discarded which causes public health hazards and diseases like malaria, cholera, typhoid. Inadequate management of wastes like uncontrolled dumping bears several adverse consequences. It not only leads to polluting surface and groundwater through leachate and further promotes the breeding of flies, mosquitoes, rats and other disease-bearing vectors. It also emits unpleasant odor and methane which is a major greenhouse gas contributing to global warming.

As the organic fraction accounts for the larger part of the municipal solid waste, anaerobic digestion thereof at its source of generation, a decentralized level, would be an appropriate solution to reduce the amount of waste dumped and/or landfilled as it minimizes transport costs and provides renewable energy and organic fertilizer. Carbon dioxide produced after the burning of methane contributes lesser to climate change than methane. The carbon emissions saved by Wipro's biogas plant is 37.26 tons CO_2/year from waste recycling and 306.77 tons CO_2/year from LPG saved.

The efficient utilization of organic waste in biogas plants creates a cycle of economic sustainability: continuously generated by-products that can be profitably employed to produce electricity, and/or heat. This reduces the accumulation of waste which production plants would otherwise have to dispose of, often at great cost. The benefit is two-fold: The impact on the environment is reduced and the value-added chain is optimized. On top of that, the campuses can use digested slurry/residues as valuable fertilizer and displace harmful chemical fertilizers.

This and other similar initiatives have resulted in a green image for Wipro's electronic city campus, and the employees consider this initiative as a positive aspect of the company's sustainability efforts. The investment has provided full-time employment to three people and part-time employment to one person.

Scalability and replicability considerations
The key drivers for the success of this business are:
- Ease of available bio-degradable waste within the campus.
- Willingness of chefs to use biogas for cooking.
- Availability of land near the canteen within the campus.
- Strong partnership with Mailhem to install, operate and maintain the plant.
- Strong financials with shorter payback period.
- Business is incentivized by green opportunities from market and showcase its corporate social responsibility.

At the present location, the generated kitchen waste is fully utilized by the biogas plant. The digester has the capacity to process three tonnes of waste, and around two tonnes is processed almost daily. Wipro would like to replicate this model to other campuses in the long term. 'Greening' of businesses is also rapidly becoming an important consideration in corporate India, and corporate headquarters with common kitchen facilities could be motivated to uptake such efforts. The corporations and small and medium enterprises (SMEs) can consider such endeavours in the name of CSR. Modest tax breaks or accelerated depreciation offsets offered for such projects will encourage replication, thus boosting the businesses of firms that specialize in renewable energy by providing stimulus for further development of technology. However, businesses would require land for putting up such biogas plants.

Summary assessment – SWOT analysis
The key strength of this case is the sufficient resources available to support the investment and Wipro's mandate to undertake initiatives like this under its CSR program (Figure 47). The weakness is the lack of in-house technical capability to manage biogas plant and land required for the biogas plant. Wipro's biogas plant occupies 300 m^2 of land. The investment does not have any significant threats unless in unlikely situation of heavy price subsidies on LPG for commercial use. The opportunity is huge as this can be easily replicated in other Wipro campuses and also applicable to campuses of other large business corporations and institutions.

References and further readings
Confederation of Indian Industry. 2011. Case Study on the Wipro Biogas Plant. Sohrabji Godrej Green Business Centre. www.greenbusinesscentre.com/msg/renewable-e4.html (accessed Nov. 7, 2017).

Newsweek 2012. Newsweek's green rating of 500 global companies. http://www.newsweek.com/2012/10/22/newsweek-green-rankings-2012-global-500-list.html (accessed Nov. 7, 2017).

FIGURE 47. MAILHEM WIPRO SWOT ANALYSIS

	HELPFUL TO ACHIEVING THE OBJECTIVES	**HARMFUL** TO ACHIEVING THE OBJECTIVES
INTERNAL ORIGIN ATTRIBUTES OF THE ENTERPRISE	**STRENGTHS** • Saving in cost of transportation of waste to landfill sites • Easier to maintain hygienic conditions in the premises and elimination of malodors • Well-known technology • Sufficient internal resources to support the waste to energy project • The business model is financially attractive on the basis of avoided cost of LPG purchases • Existence of a corporate sustainability strategy encouraging environment and waste management initiatives • Corporate policy of having own facilities for providing food to employees responsible for central waste generation, collection, segregation and energy conversion	**WEAKNESSES** • Waste management is not part of core business strategy and considerations at Wipro • Lack of in-house technical capability at Wipro for managing and operating the waste to energy plant • Land required for biogas plant is significant and it could be difficult during expansion of the plant if land is not easily available; replication of similar plants in other campuses are dependent upon land available • Source dependent composition of waste • Every biogas plant is different • Negative pressure in biogas system can cause explosion • High upfront cost for potential assessments and feasibility studies
EXTERNAL ORIGIN ATTRIBUTES OF THE ENVIRONMENT	**OPPORTUNITIES** • Favourable policy and regulatory environment for industrial and commercial waste in India; • Renewable energy policy of India • Availability of technical and management expertise and support partners for waste management in India • Opportunities to replicate the kitchen-linked biogas plant in other Wipro campuses as well as other corporate campuses and institutions on a business case, rather than as part of corporate sustainability efforts • Mandated requirements as per Companies Act 013, as per recent amendments	**THREATS** • Changes in the price for LPG directly affect the financial attractiveness of the business; however, LPG prices are unlikely to fall • Lack of awareness of biogas opportunities • Direct animal feeding is an equal or favoured solution to reduce the amount of organic waste (in semi-urban or rural areas)

Case descriptions are based on primary and secondary data provided by case operators, insiders, or other stakeholders, and reflects our best knowledge at the time of the assessments (2015/2016). As business operations are dynamic, data can be subject to change.

BUSINESS MODEL 4
Biogas from kitchen waste

Krishna C. Rao and Solomie Gebrezgabher

A. Key characteristics

Model name	Biogas from kitchen waste
Waste stream	Kitchen and food waste
Value-added waste product	Biogas as fuel for cooking through anaerobic digestion of kitchen and food waste
Geography	Institutions with large kitchen facility to cook for large number of people
Scale of production	Medium scale; About 100–300 m³ of biogas per day used to cook food for around 3,000 to 7,000 employees and about 1 to 4 tons/day of sludge as compost
Supporting cases in this book	Bangalore, India
Objective of entity	Cost-recovery [X]; For profit [X]; Social enterprise [ ]
Investment cost range	About USD 75,000 to 125,000
Organization type	Private
Socio-economic impact	Environment-friendly cooking fuel, reduced greenhouse gas emissions; carbon emissions offset from avoided municipal waste landfill and also replacement of LPG/coal/liquid fuel which otherwise would have been used for cooking, improved organic waste management results in reduced pollution of local water bodies and employment generation
Gender equity	Mostly gender neutral; but with benefits through improved indoor air and working environment for women operators/chefs

B. Business value chain

The business model could be initiated by institutions such as industries, hostels, hospitals, prisons and schools with large cafeteria and kitchen facility that generate large quantities of kitchen waste and food waste. Alternatively, it can be initiated by a technology supplier who provides waste management solution to the institutions. The business concept is to process organic waste from kitchen and cafeteria to generate biogas. Biogas can be used for internal use to cook food in the cafeteria's kitchen or can be sold to nearby households and businesses. Biogas can also be used to generate electricity. The digested slurry (compost) can be used within the institution for landscaping or sold to local farmers (Figure 48).

The key stakeholders in the business value chain are the waste suppliers, institution, technology supplier and end users of the product – the institution itself or household and businesses. The biogas plant can also be fed with other organic waste such as biomass (leaf litter) and sewage sludge generated within the institution premises. The business is eligible for sale of carbon as the thermal energy for cooking is generated from sustainable biomass source instead of using LPG/coal or other liquid fuels

BUSINESS MODEL 4: BIOGAS FROM KITCHEN WASTE

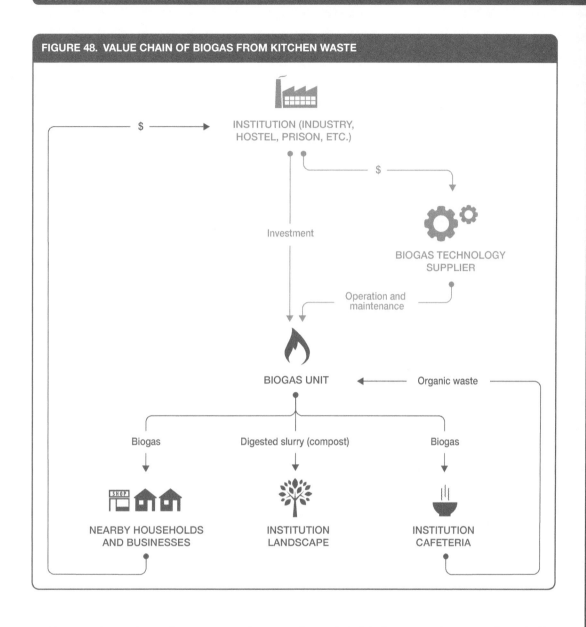

FIGURE 48. VALUE CHAIN OF BIOGAS FROM KITCHEN WASTE

and improved organic waste management as organic waste left in the open releases methane to the atmosphere. Alternatively, this business model can use the biogas to generate electricity especially if the institution is not connected to grid.

In the business model, there is scope for a private technology enterprise, an energy service company, who could get into Build, Own, Operate, Transfer (BOOT) arrangements with the institution. The private entity could bring all investment to set up the biogas technology while the institution provides land and kitchen waste inputs. The private entity designs, constructs and maintains the biogas unit, and sells biogas and digested slurry to the institution. This is done until the BOOT period is expired, after which it transfers ownership of the plant to the institution and assists it to operate the unit on a chargeable basis. The BOOT period can range from three to five years until the investment made by the private technology enterprise is recovered.

C. Business model

The value proposition of the business model varies to some extent with the ownership and the entity driving the business initiative. In the case where the institution is the owner and is driving the initiative, the emphasis is on the management and cost reduction, while it is more on the service provision in the BOOT case. In both cases, the value proposition is to better manage organic waste generated by the institution in an environmentally responsible manner, as well as to generate biogas from food waste for kitchen use. As mentioned above, depending on the connectivity to the grid, the model can offer either electricity or biogas as cooking fuel (Figure 49). Finally, the model also offers the provision of organic compost using the bio-slurry output from the biogas plant that can be used for internally, e.g. for landscaping or sold to local farmers.

If the institution is owner and operator of biogas unit and uses it internally, it incurs substantial savings from avoided fuel purchases for cooking. The business model has scope to sell either entire or excess biogas generated to nearby households and businesses. The model requires the development of a partnership with a biogas technology supplier whose assistance is critical in the initial stages of operation and maintenance until local labor is trained. The key activities include the production of biogas and the key resources are land, equipment, biogas technology and sourcing of the waste.

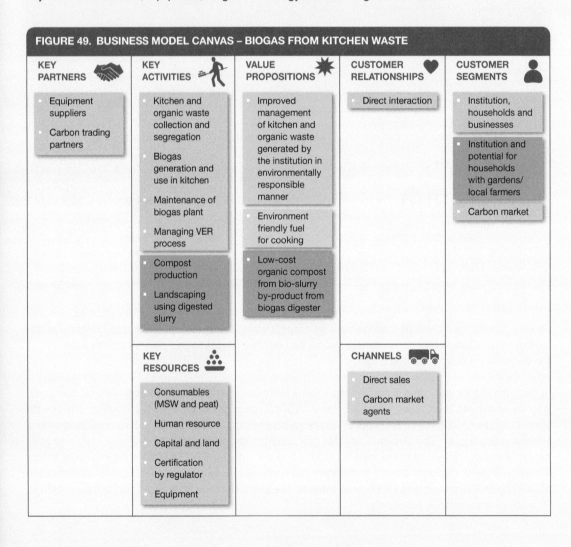

FIGURE 49. BUSINESS MODEL CANVAS – BIOGAS FROM KITCHEN WASTE

BUSINESS MODEL 4: BIOGAS FROM KITCHEN WASTE

COST STRUCTURE	REVENUE STREAMS
• Investment costs – land, building and Equipment and gas distribution lines • O&M costs – training, utilities, labor (can be intensive and skilled labor) • Costs incurred for VER registration and carbon sale • Savings – cooking fuel, compost costs and cost of disposal of kitchen and cafeteria waste	• Sale of surplus biogas • Sale of carbon credit • Sale of surplus digested slurry (compost)

SOCIAL & ENVIRONMENTAL COSTS	SOCIAL & ENVIRONMENTAL BENEFITS
• Potential leakage of gas • Potential occupational health risks from handling machinery and equipment for workers at production facility • Potential environmental risk if the organic waste is not treated and disposed properly, particularly possible exposure to pathogens	• Creation of jobs for low-income workers • Reduction of pollution of water bodies and natural habitats • Reduction of human exposure to untreated waste • Climate change mitigation and reduction in GHG emissions • Improved waste management and treatment at the source contributes to reduced MSW management for the government

The business model primary revenue is from sale of biogas to nearby household and business; however, as mentioned above, when biogas is internally used, it offers operational costs savings for the institutions from avoided fuel purchase for cooking. In addition to energy, a key output from biogas plant is digested slurry (compost), which is rich in nutrients and can be processed to make organic compost that can be used for landscaping within the institution premises. Thus, there is additional savings for the institution from avoided purchase of fertilizer. The biogas model is eligible for carbon offsets. If the biogas plant size is too small to be viable to apply for Clean Development Mechanism (CDM) projects due to associated transaction costs, another preferred route would be to apply for carbon offset on the Voluntary Emission Reductions (VERs) market, or bundle with other similar projects for combined registration as CDM project.

D. Potential risks and mitigation

Market risks: The market risk does not exists if the business is initiated by the institution and the biogas and compost are used internally within the institution. However, in the case of private biogas supplier initiating the business, there is potential risk of the institutions' willingness to participate for a BOOT arrangement. Based on the economics the institution is likely to incur savings and there are no high risks associated except if the biogas plant is not treating the organic waste properly and is causing environmental pollution. If the business has high dependence on sale of carbon credit for its viability, the volatility of carbon credit market puts the sustainability of this reuse business under risk. In such scenarios, the business has to diversify its revenue streams by using biogas and compost productively so as not to entirely depend on the sales of carbon credits.

Competition risks: The business risk for the output is present if the competing fuel source provides higher economic benefits. However, in this business model, there is the cost incurred by the institution to purchase cooking fuel is reduced significantly with biogas plant installed. With short payback periods of three to five years, before the life cycle of the biogas plant, investment cost of the plant and its operation cost is completely recovered.

Technology performance risks: The technology process used is anaerobic digestion, which is well established and mature. However, the type of digester required could potentially be sophisticated and might not be available in developing countries, and in addition the technology requires skilled labor. It is ideal for the business to transfer the technology from a market where it is widely implemented and have their staff trained in repair and maintenance of the technology to mitigate the performance risks.

Political and regulatory risks: In most developing countries, price of cooking fuels such as kerosene and Liquefied Petroleum Gas (LPG) are subsidized for domestic consumption. If the government has similar policies for commercial entities and institutions, it can diminish the economic advantage offered by the biogas plant and hence making this business model less viable. Lately, governments are encouraging green initiatives by providing incentives such as concessional loans and accelerated depreciation benefits. Policies supporting green initiatives make this business model highly attractive.

Social-equity-related risks: The model does not have social equity risks. The model is mostly gender neutral and the benefits are accrued by the institution generating waste with limited or no employment creation.

Safety, environmental and health risks: Safety and health risks to humans arise when processing any type of waste. Laborers in such enterprises should be provided with appropriate gloves, masks and other appropriate tools to handle the waste to ensure their safety. The risk of environment pollution is high if leachate from kitchen waste seeps into groundwater or other natural water bodies. The waste processing technologies are not without problems and pose a number of environmental and health risks if appropriate measures are not taken. The environmental risks associated with the anaerobic digestion units include possible leakage of gas and these emissions should be controlled. Organic waste when left in open begins to decay and releases methane, which is more damaging to the environment than carbon dioxide. There is a very limited chance that the compost made from digested slurry potentially could have risks of pathogens (Table 16).

BUSINESS MODEL 4: BIOGAS FROM KITCHEN WASTE

TABLE 16. POTENTIAL HEALTH AND ENVIRONMENTAL RISK AND SUGGESTED MITIGATION MEASURES FOR BUSINESS MODEL 4

Risk group	Exposure					Remarks
	Direct contact	Air	Insects	Water/Soil	Food	
Worker	LOW	LOW				Risk to workers through direct contact with waste and compost can be mitigated using protective equipment and gear.
Farmer/user	LOW					
Community						
Consumer						
Mitigation Measures	🧤	😷	🚫🪰			

Key: ☐ NOT APPLICABLE ▨ LOW RISK ▧ MEDIUM RISK ■ HIGH RISK

E. Business Performance

This business model is rated high on cost-recovery and positive environment impact followed by replicability (Figure 50). The business model doesn't have a strong revenue source. However, it is based on cost recovery through savings incurred from avoided fuel expense. In addition, due to it environment friendly aspects, it offers scope for revenue from sale of carbon. The business model has a high potential for replication in developing countries with no limiting factors except for the technology. However, on scalability it scores low.

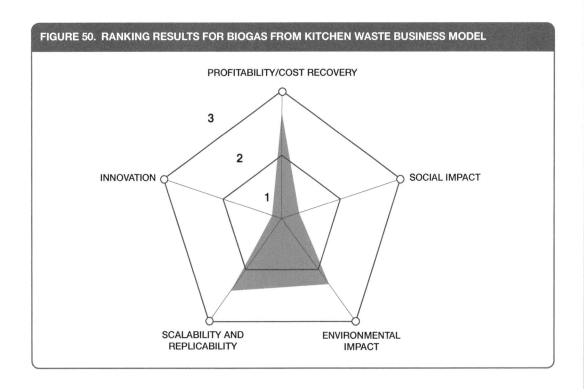

FIGURE 50. RANKING RESULTS FOR BIOGAS FROM KITCHEN WASTE BUSINESS MODEL

The environmental impact scores high because of high replication potential that the business model offers that could result in safe management of organic waste and reduced burden on the government machinery to manage solid waste. In addition, it offers reduced greenhouse gas emissions. The business model scores low on innovation and social impact. The technology and financing required is fairly straightforward. The social impact scores low due to low job creation and on a comparative basis with other business models managing solid waste, net impact on social development indicators is low.

5. BUSINESS MODELS FOR SUSTAINABLE AND RENEWABLE POWER GENERATION

Introduction

Over 1.2 billion people do not have access to electricity, a majority of them living in countries in Africa and Asia. SDG 7 thus calls to ensure access to affordable, reliable, sustainable and modern energy for all, and to increase substantially the share of renewable energy in the global energy mix by 2030, including energy derived from waste, biogas and biofuels (SDG 7.2). In developing countries, governments are promoting power generation from various agro-industrial waste and municipal solid waste (MSW) streams to improve access to energy and ensure long-term security of power supply. The recent high energy prices, coupled with environmental and financial incentives such as carbon financing and modern biomass energy options such as biomass-based energy generation, are becoming economically attractive in low-income countries. Depending on the waste source and end use of the power generated along with the ownership structure of the entity generating power, the business model can take various forms:

- **Business model 5: Power from manure** – Livestock management and the related industry are important components to the growth of the economy and an important source of livelihood in developing countries. Livestock industry results in large quantities of livestock manure, which if not managed properly pollutes waterways and generates greenhouse gas emissions. However, it also presents an opportunity to harness energy in the form of biogas or electricity on a commercial scale. Business Model 5 demonstrates that through sustainable market mechanisms, successful commercial biogas systems could be implemented.
- **Business models 6 and 7: Power from agro-waste and power from MSW** – In this business model, a social enterprise or private entity which is not a public utility generates power and sells electricity either to utilities or directly to households or businesses. The models are typically focused in regions with communities that do not have access to reliable energy and is one of the key modes to achieve SDG 7 indicators. The business cases discussed take different ownership structure and are initiated by a standalone private enterprise or are set up as social enterprise or as public-private partnerships. While there is increasing need for waste-to-energy plants, high variations in the calorific value of unsorted wet MSW make the business highly dependent on subsidies, which are justified given the large waste volume reduction. However, in this energy section, the focus is on organic waste valorization and not waste in general, thus we did not include waste incineration cases in our selection.
- **Business model 8: Combined heat and power from agro-industrial waste for on- and off-site use** – Majority of large-scale agro-industries such as sugar processing factories, cassava, palm oil and slaughterhouse industrial units in developing countries are diversifying into usage of agro-industrial waste produced during the process into a value-added by-product through co-generation. Energy generation from own agro-industrial waste also referred to as on-site energy generation model is driven by the need for agro-processing units to reduce their energy costs. In addition, these units explore new revenue streams from selling excess energy. The power generation technologies are either designed, constructed, owned and operated by the agro-industrial processing factory or are installed by an external private entity on a Build, Own, Operate, Transfer (BOOT) model. The business model offers a multi-value proposition as it not only allows agro-industries to be self-sufficient in energy while disposing of their waste sustainably, but also secures additional revenue streams by exporting excess renewable electricity to the national grid along with trading of carbon credits.

These business models have been successfully implemented in Latin American, African and Asian countries with cases presented from Brazil, Peru, Mexico, India, Kenya and Thailand. Policies, regulations and institutions play crucial roles in the successful implementation of these business models through appropriate national policies, programs and fiscal incentives. For example, a number of policy reforms in the Kenyan power sector have liberalized the energy-generation sector thereby paving the way for independent power producers (IPPs) such as Mumias Sugar Company (MSC) to participate in power generation. A number of domestic and international programs to support bagasse-based cogeneration in India were launched which promoted the advancement of co-generation plants in India. These support programs include extension of loans for cogeneration by the Asian Development Bank (ADB) through the Indian Renewable Energy Development Agency (IREDA), capital and interest subsidies, research and development support, accelerated depreciation of equipment, a five-year income tax holiday and excise and sales tax exemptions by the Ministry of Non-Conventional Energy Sources (MNES).

The cases on livestock industry will for instance, demonstrate the role of industry such as the meat or dairy industry in promoting sustainable development in the livestock sector through the implementation of innovative financing schemes to set up biogas systems in the livestock farms to be energy self-sufficient while earning additional revenue through carbon credit market (Sadia case in Brazil). The cases will also highlight effective partnership amongst a range of stakeholders and community-led strategies coupled with market mechanisms to lead to the successful implementation of a rural electrification program.

Thus, depending on local conditions in the respective countries on renewable energy policy, institutional set-up and power purchase agreements, the cases provide a broad range of innovative partnership structure, value chain, market and pricing mechanisms.

CASE

Power from manure and agro-waste for rural electrification (Santa Rosillo, Peru)

Patrick Watson and Krishna C. Rao

Supporting case for Business Model 5	
Location:	Santa Rosillo, San Martin, Peru
Waste input type:	Livestock waste and other agro-waste
Value offer:	Biogas, manure and slurry, electricity efficient waste and water management, climate change mitigation and adaptation, Economic development through renewable Energy Increase of local food production, and added income streams
Organization type:	Public and non-government organization
Status of organization:	Owned by Municipality of Huimbayoc, Operational since 2010; plant managed by the community including its O&M
Scale of businesses:	Small electricity generation plant of 16 kW supplying power to 42 families
Major partners:	Comercial Industrial Delta SA (CIDELSA), SNV, Regional Government of San Martin, Cordaid, Fact, Practical Action

Executive summary

Santa Rosillo, a rural community in the deep jungle of the Peruvian Amazon in northern Peru, is more than 16 to 21 hours away from the nearest city, Tarapoto, and is only accessible by boat and on foot. Santa Rosillo consists of 42 households (220 people) who have an average monthly income ranging between USD 23 and USD 47. Due to the extreme remoteness of the village, prior to this project, most of the community did not have access to electricity and relied on candles, batteries and lighters for domestic lighting. Approximately 12% of the population had access to electricity through private diesel generators.

In 2010, SNV Netherlands Development Organisation (SNV), a non-profit international development organization, in partnership with the regional government initiated a rural electrification project to install two bio-digesters in the village linked to a power generator and mini-grid to provide electricity to the community. The community's primary economic activity is livestock and agriculture (cocoa), and all organic waste is fed into the two bio-digesters. The biogas generated is fed into the electricity generator and electricity is distributed to each house. The installed electrical capacity is 16 kW which provides electricity to 42 houses, the local doctor's office, the local college and public lighting for approximately

5.3 hours per day. Approximately 60% of the slurry by-product produced by the bio-digesters is then used as fertilizer to improve the soil quality of the communal grazing area, while the remaining 40% is sold to local farmers. Comercial Industrial Delta SA (CIDELSA), a Peruvian engineering company, supplied the two lagoon bio-digesters for the project.

KEY PERFORMANCE INDICATORS (AS OF 2013)						
Land use:	3,000 m^2 (including community grazing area for animals)					
Water	50,000 L/year					
Capital investment:	USD 130,519					
Labor:	1 x system operator / administrator (full-time)					
O&M cost:	USD 0.57 per kWh (total levelized cost of electricity over a life of 20 years)					
Output:	16 kW for 5.3 hours/day, supplying 85 kWh/day of electricity, Biol (solid fertilizer) and Biosol (liquid fertilizer)					
Potential social and/ or environmental impact:	42 households now have access to electricity; it has reduced the environmental pollution from manure and improved livelihood of remote community					
Financial viability indicators:	Payback period:	N.A.	Post-tax IRR:	N.A.	Gross margin:	N.A.

Context and background

The Santa Rosillo community is located in the district of Huimbayoc, 190 km from the city of Tarapoto in northern Peru. The community's main activities are agriculture, livestock and forestry, all of which generate organic waste, which was not being utilized prior to this project. Because of its remote location, the community is not connected to the national energy grid and had very limited access to gas or electricity, leaving its 42 families reliant on diesel generators or candles for power and lighting. In 2010, SNV in alliance with Practical Action and local partners commissioned the installation of two bio-digesters by CIDELSA, a company with over 10 years of experience building and installing bio-digesters. The project in Santa Rosillo was the pilot installation for SNV's rural electrification program, "BIOSINERGÍA: Access to energy with biofuels in the Peruvian Amazon." The National Public Investment System, a government investment initiative, funded the grid connecting the power generators to the village whilst the foundations CORDAID and FACT funded the installation and equipment costs for the power generators.

Market environment

Prior to the installation of the bio-digesters, only 12% of the population had access to electricity, generated through private generators and solar panels. Average usage was 2.5 hrs/day and the monthly cost ranged from USD 9.6 to over USD 465 per household/family. The families with higher costs were those that had small businesses such as a restaurant or furniture shop. The remaining 88% used battery-powered lights and candles for lighting. This project provides electricity to 100% of the community of Santa Rosillo. One of the principal advantages of this project is the anticipated low cost of electricity compared to other available forms of rural electrification. Each family is undergoing a grace period until January 2013, at which point a flat rate of USD 6 (average cost prior to project) will be charged until an exact wattage consumption has been determined. The final cost will be determined by demand, operational and maintenance costs, in addition to the population's ability to pay (approximately USD 6/month).

In Santa Rosillo, the cattle alone produce approximately 300 kg of manure daily that is now used as fuel rather than contaminating the local waterways. If this project proves successful it is intended that it will be rolled out to other rural communities throughout Peru.

Business model

The project has two key value propositions: providing electricity service to houses and businesses and providing fertilizer to farmers in Santa Rosillo (Figure 51). The municipality, donor agency and local organization played a key role in mobilizing the community and financial resource to establish electricity service provision. Since the project results in carbon offset, there is potential for generating revenue from sales of carbon.

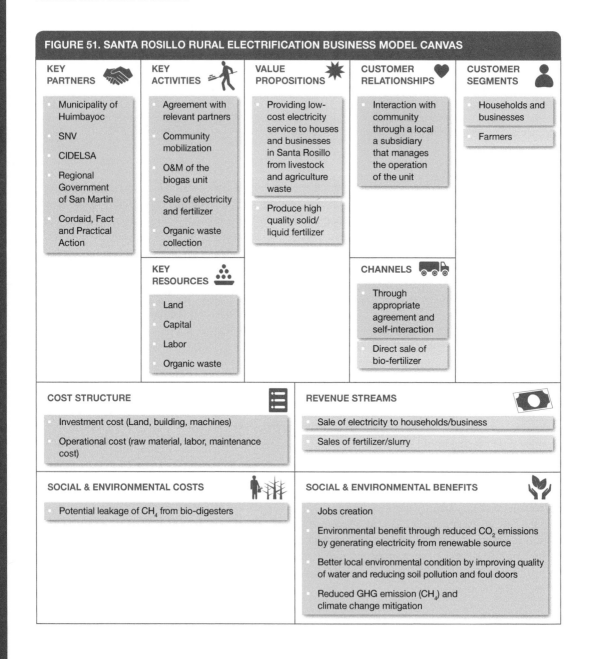

FIGURE 51. SANTA ROSILLO RURAL ELECTRIFICATION BUSINESS MODEL CANVAS

Value chain and position

Santa Rosillo's two bio-digesters are fuelled by the waste produced by the community's cattle (approximately 60), which are kept in a partial barn. Cattle spend 12 hrs/day in a communal pen, and two members of the community are responsible for collecting excrement once daily. Each cow produces approximately 5 kg of manure per day, yielding a total of 300 kg, which is loaded into the bio-digesters on a bi-weekly basis.

The project is designed as a cooperative structure to provide energy service at a cost lower than the possible alternatives. Figure 52 displays the overall management structure. The following describes the role of key actors in the process:
- Communal Energy Services Unit (USEC): The USEC team, made up of two people from the community, is responsible for operating and maintaining the system on a daily basis in direct and constant contact with both users and community authorities.

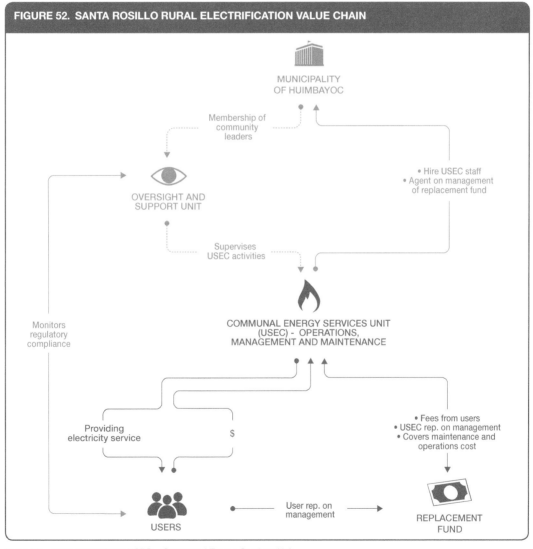

FIGURE 52. SANTA ROSILLO RURAL ELECTRIFICATION VALUE CHAIN

Note: RF = replacement fund; USEC = Communal Energy Services Unit.

- Municipality of Huimbayoc: The municipality owns the power generation and distribution system and hires USEC to maintain and operate it. The municipality voluntarily undertakes equipment inspection and maintenance checks alongside USEC and subsidizes the maintenance costs that cannot be covered by revenues from the system.
- Oversight and Support Unit: Comprised primarily of community leaders; this unit is responsible for monitoring USEC and user compliance.
- Users: The community members of Santa Rosillo, who will pay a monthly fee for electricity used, once established. Each user will have electricity meter installed in their homes and sign an electricity supply contract.

Income generated from service fees will be used to create a revolving fund designed to cover operational, management and maintenance costs. The revolving fund will be managed by three people from the USEC, a user representative and a municipal agent.

Institutional environment

Electricity access in rural Peru is challenging because of the mountainous terrain and scattered settlements. Low energy consumption and limited purchasing power per household add to the challenge. Investors are therefore not attracted to these projects unless the state provides the right financial incentives and other necessary requirements. Renewable energy is a largely unexploited market in Peru and small-scale renewable energy (biogas, biofuels, small-hydro and solar energy) provides less than 1% of national energy supply.

In 2008, the Peruvian government passed a legislative decree to promote inclusion of renewable energy, which includes biogas. This has helped renewable energy production growth exponentially. The government of Peru projects renewable energy to provide 7% of the national energy supply by 2017. The Ministry of Energy and Mines and its General Directorate of Rural Electrification (MINEM-DGER) have 437 rural electrification projects clustered into 35 groups. The total investment is estimated at USD 418 million and will benefit 1.2 million people (Mitigation Momentum, 2015). Additionally, DGER is implementing 16 other special projects which will benefit 150,000 people; approximately USD 140 million will be invested.

The government of Peru has identified rural development, environmental protection and energy security as national priorities. They have developed a legal and regulatory framework that promotes competition and investment in the sector and, more recently, have successfully developed mechanisms to promote the use of our vast renewable energy resources. The 2013–2022 National Rural Electrification Plan produced by MINEM, in concordance with an Energy Universal Access Plan, establishes a policy for the sector with the aim is to raise the rural electrification rate from 87% to 95% by 2016. The national electrification rate has increased from 55% in 1993 to 87.2% in 2012. The National Rural Electrification Plan 2013–2022 provides strategic direction to provide access to electricity to 6.2 million people in the next 10 years. Peru is undertaking efforts to increase access to energy via auctions for solar photovoltaic systems, grid extension, mini-grids with hydro, solar and wind. The Law for the Promotion of Investment in Renewable Energy Generation[3] grants competitive advantages to projects for renewables.

The following policies have been established to address these issues:
- Use of renewable energy in electricity generation (2008), amended 2011: The government is promoting the use of renewable energy resources by providing tax concessions to qualifying projects, e.g. biomass, wind, solar, geothermal, tidal and hydropower.

- Non-conventional renewable energy resources in rural areas (2005): This law provides additional tax concessions to qualifying projects that promote the use of non-conventional renewable energy in rural communities.
- Investment in electricity generation using water and other renewable resources (2008): In order to incentivize the investment in a renewable energy infrastructure, all renewable energy projects benefit from accelerated depreciation for tax purposes.

Technology and processes

The two bio-digesters, each with a 75 m^3 capacity, produce biogas for electricity (16 kW), and bio-fertilizer. Figures 53 and 54 depict bio-digesters and the power generation system, which in turn is connected to a micro power grid, extending electricity to each family.

The amount of biogas generated will depend on the quality of cow manure collected, as a rough estimate 1 kg of cow manure will generate about 40 L of biogas.

Despite popular belief, the amount of waste going in the digester is almost equal to the amount coming out. However, the quality of the waste is altered for the better (less odor, better fertilizer, organic load reduced, less polluting). Waste coming out of the digester can be separated (solid/liquid): the solid part can be composted (Biol) and the liquid part can be used as liquid fertilizer (Biosol) or can be treated further and disposed.

Funding and financial outlook

The total project cost of USD 130,519 was funded as grant through a public-private partnership between the Regional Government of San Martin (GRSM) (30%), FACT and CORDAID (68%), in addition to each beneficiary family contributing USD 59 (2%) (Table 17).

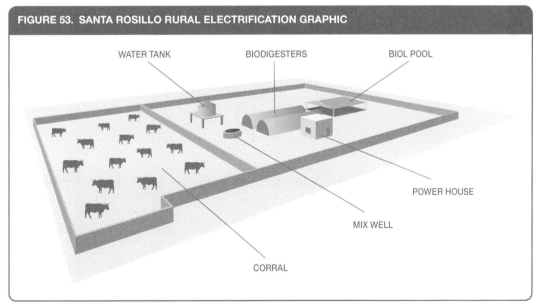

FIGURE 53. SANTA ROSILLO RURAL ELECTRIFICATION GRAPHIC

Source: Veen, 2014; modified.

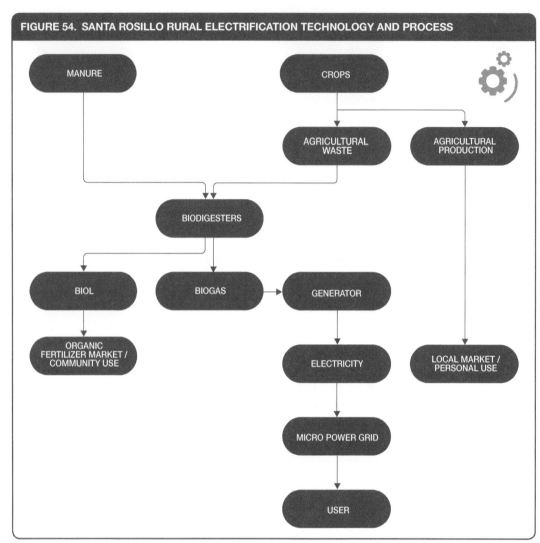

FIGURE 54. SANTA ROSILLO RURAL ELECTRIFICATION TECHNOLOGY AND PROCESS

Source: Veen, 2014; modified.

TABLE 17. INVESTMENT SOURCE AND AMOUNT

ITEM	AMOUNT (USD)
Regional Government of San Martin	39,285
FACT & CORDAID	88,305
Beneficiary (USD 59/household)	2,930

Each family was given a grace period until January 2013, at which point a flat rate of USD 6 (average cost prior to project) will be charged until exact wattage consumption has been determined. The final cost will be determined by demand, operational and maintenance costs, in addition to the population's ability to pay (approximately USD 6/month). In this case, the electricity generator is created to meet a need in the community, which pays for the service. In turn, these revenues allow the proper operation and maintenance of the company, so the service becomes sustainable over time. The sale of slurry will provide an additional income stream.

The following assumptions were built into the financial projections (Table 18):
- Total generating capacity of 619,040 kWh over a 20-year life, slowly ramping up to full capacity over the first 10 years. The generating capacity of the bio-digesters is 16 kW whereas the demand of the community is 13 kW.
- Demand for electricity will grow at a continual pace of 2.5% annually, primarily driven by population growth. Demand is expected to be approximately 17 kW by 2022, slightly higher than the capacity of the two bio-digesters (16 kW).
- Service fees are based on estimated operational and maintenance costs for equipment repairs, replacement parts and servicing. Total estimated costs are USD 3,516 in Year 1, reaching USD 5,621 by Year 20. This assumes a growth rate of 2.5% in line with demand.
- The bio-digesters can produce roughly 1,041 L of slurry per day which, with a potential sales value of USD 0.05/L, will generate a monthly income of approximately USD 1,500. It is estimated that only 40% of the generated slurry will be sold, helping to cover O&M costs that are not covered by the monthly fee income.

TABLE 18. SANTA ROSILLO FINANCIALS

PROJECT FINANCIAL SUMMARY				
USD	YR0	YR1	YR2	YR3
Investment				
GRSM	(39,284.56)			
FACT & CORDAID	(88,304.84)			
Community investment (~USD 59 per family)	(2,929.69)			
(+) Income		11,198.21	11,286.10	11,376.18
Annual usage income (50 families)		3,515.63	3,603.52	3,693.60
Growth rate			2.5%	2.5%
Sale of Slurry (41% total production)		7,682.58	7,682.58	7,682.58
(−) Costs		(11,263.17)	(11,263.17)	(11,263.17)
Operational / Maintenance		(11,263.17)	(11,263.17)	(11,263.17)
Cashflow	(130,519.09)	(64.96)	22.93	113.01

Source: Authors.

Socio-economic, health and environmental impact

The project benefits the Santa Rosillo community, its approximately 50 families. More than 220 people now have access to electricity, allowing them to improve their living conditions. Children now have more hours of light to do homework, enhancing the learning process. There are improved teaching conditions in schools, enabling the use of computers. The slurry (effluent from bio-digesters) can be used to enhance crop yields, further improving family incomes. There will also be a reduced likelihood of domestic accidents from the use of candles and better illumination in the house, as well as improved conditions in the community health centre, including the ability to now refrigerate drugs and vaccines.

By extracting methane out of waste and using it to produce heat and/or electricity, we ensure that the waste will not degrade in an open environment, therefore reducing direct methane atmospheric emissions. By managing and reusing the livestock excrement, the project has substantially reduced pollution of the local rivers and lakes. Moreover, the energy provided by the biogas is likely to displace fossil fuel which is the main contributor to GHG emissions. By installing a digester the farmer can profit from the biogas by reducing doors and enhancing the fertilizing value of the manure. The project requires less area than aerobic compost, reduces the volume and weight of landfills, produces a sanitized compost and nutrient rich liquid fertilizers, maximizes the benefits of recycling and in the process improves the air quality through improved odor and reducing groundwater contamination.

Scalability and replicability considerations
The key drivers for the success of this business are:
- Strong financial support from municipality, donors and local agencies.
- Strong community participation.
- Simple low-cost model.
- Ease of available animal and agro-waste.

The community of Santa Rosillo is the representative of a large number of rural Amazonian villages within Peru and in other countries with similar conditions that are not connected to the electric grid or other natural resources (e.g. sun or strong water current for solar or hydro power) required for alternative micro-power solutions, but that have high volumes of unused organic waste (e.g. from agriculture or livestock). According to the Peruvian national Census (2007), there are 2.2 million households in rural areas, 36% of which do not have access to the national grid equating to approximately 800,000 households. Many such communities are heavily reliant on cattle and other livestock that produce significant waste, which traditionally causes on-going water pollution and land degradation.

Summary assessment – SWOT analysis
The key strength of the project is its low-cost alternative electricity solution for a remote region and a proven technology that can readily use abundant available waste source (Figure 55). The weakness of the project is its inconsistency in provision of reliable electricity and in addition if the community size grows and when the demand for electricity increases, the electricity generated will not be sufficient to meet new higher demand.

The key threat to the project is from the remoteness of the site. In the event of system failure, due to lack of local technical know-how and available skill, time taken to repair the unit will be longer. This can result in the community losing faith in the overall project. However, based on the success of Santa Rosillo's electrification project, it has very high potential for replication and it can help the Peruvian government define policies to ease the replication of the business with minimum obstacles.

Contributors
Carlos Fernandez, I-DEV International
Cinthya Pajares, I-DEV International
Kamalesh Doshi, Simplify Energy Solutions LLC, Ashburn, Virginia, USA

References and further readings
Fiji Renewable Energy Power Project (FREPP). September 2014. Report on recommendations from technology research on waste-to-energy in Fiji. http://prdrse4all.spc.int/sites/default/files/141005final_report_on_recommendation_from_w2e_technology_research.pdf (accessed Nov. 7, 2017).

Mitigation Momentum. 2015. Sustainable energy production from biomass waste in Peru: NAMA proposal November 2015. Mitigation Momentum. https://goo.gl/jp9mHz (accessed Nov. 7, 2017).

FIGURE 55. SANTA ROSILLO RURAL ELECTRIFICATION SWOT ANALYSIS

	HELPFUL TO ACHIEVING THE OBJECTIVES	**HARMFUL** TO ACHIEVING THE OBJECTIVES
INTERNAL ORIGIN ATTRIBUTES OF THE ENTERPRISE	**STRENGTHS** • Simple and cost-effective example of public-private partnership that can be applied in numerous geographies and communities • Provides a relatively low-cost alternative for electricity to communities that cannot gain access to the grid because of their remote location • Uses an abundant waste source to generate off-grid power • Utilizes very simple technology, easy to operate without significant prior technical knowledge • Limited/low operational and maintenance requirements • The production of fertilizer and gas is not heavily dependent on the type of excrement/waste used	**WEAKNESSES** • Low cost to users is highly dependent on sale of slurry, which is anticipated to provide a subsidy • Power from bio-digesters is inconsistent. The continuous supply of service requires complex infrastructure, unavailable waste volumes and costs that would exceed income • Total energy capacity of 16 kWh may not be adequate if the community grows at a higher-than-anticipated rate • High upfront cost may be a barrier to entry for some smaller communities/governments • Payback period highly dependent on ability to sell fertilizer
EXTERNAL ORIGIN ATTRIBUTES OF THE ENVIRONMENT	**OPPORTUNITIES** • Highly replicable model in Peru, and in other countries and communities with similar dynamics (estimated 800,000 applicable households in Peru alone) • Peruvian national policy promotes and provides substantial tax incentives to communities and governments seeking to partner on similar rural electrification strategies. • Potential involvement of microfinance organizations to aid funding of the bio-digester roll-out • Government of Peru promotes renewable energy • High value bio-fertilizer for additional revenue • High quality renewable fuel, biogas has several proven end-use applications • Positive environmental impact • Climate change mitigation and adaptation	**THREATS** • Remote location of project combined with lack of local technical knowhow could lead to under maintenance of and hence failure of the system. • Potential competition from large bio-fertilizer companies • Possible risk from leakage of gas thus having negative perception of health risk to employees may force O&M costs higher

Niccolai, H. 2012. Cost comparison of decentralized electrification systems 16 kW micro-grid systems based on Santa Rosillo project.

SNV. 2012. Project documents on the budget of Project Santa Rosillo.

Soluciones Practicas. www.solucionespracticas.org.pe.

Veen, M. 2014. Rural electrification with biogas in Santa Rosillo, Peru. www.snv.org/public/cms/sites/default/files/explore/download/150520_annual_report_2014_-_appendices_-_re_peru1.pdf (accessed Nov. 7, 2017).

Case descriptions are based on primary and secondary data provided by case operators, insiders, or other stakeholders, and reflects our best knowledge at the time of the assessments 2015/16. As business operations are dynamic data can be subject to change.

CASE
Power from swine manure for industry's internal use (Sadia, Concordia, Brazil)

Heiko Gebauer and Solomie Gebrezgabher

Supporting case for Business Model 5	
Location:	Concórdia, Brazil
Waste input type:	Swine manure
Value offer:	Energy (Biogas to electricity and thermal energy) carbon credit and bio-fertilizer
Organization type:	Private
Status of organization:	Operational since 2003
Scale of businesses:	Large
Major partners:	Swine farmers, Brazilian Development Bank (Amazon Fund), United Nations (Carbon market), Bio-digester vendors, Espírito Santo University (for measurements of biogas)

Executive summary

Sadia is one of the world's leading producers of chilled and frozen foods with approximately 10,000 integrated poultry and pork farms, which supply raw material to its industrial plants. In order to abate the environmental impacts associated with its swine production farms and to institute sustainability into the pork meat supply chain, Sadia designed and implemented the Program for Sustainable Swine Production (3S Program) in 2003. The 3S Program provides swine producers with bio-digesters and is designed to reduce GHG emissions from the more than 3,500 swine producers in Sadia's supply chain and to qualify the emission reductions as a Clean Development Mechanism (CDM) project. Sadia installs the bio-digesters on its swine producers on a B&T (Build and Transfer) basis. The program seeks to bring sustainability to the company's supply chain by providing additional revenue from carbon credits and better working conditions for swine producers, while reducing the environmental impact associated with swine production. The biogas generated at the swine farms is used on-site and thus significantly saving operational costs for the swine farms. The program contributes to improving the local environmental condition by improving quality of water and reducing soil pollution and foul odors. Moreover, the 3S Program is expected to help disseminate environmental education among swine producers and the surrounding community. Through the design and implementation of the 3S Program, Sadia has incorporated environmental sustainability into its revenue design.

CASE: POWER FROM SWINE MANURE FOR INTERNAL USE

KEY PERFORMANCE INDICATORS (AS OF 2012)						
Land:	The bio-digesters installed at the individual swine farms					
Capital investments:	For the whole 3S Program USD 28 million					
Labor:	Provided by the individual swine farms					
O&M cost:	Provided by the individual swine farms					
Output:	Biogas for onsite use; In 2006 290,000 tons of CO_2-eq carbon credit sold and 2.5 million tons CO_2-eq under agreement					
Potential social and/or environmental impact:	CO_2 offset, improved working conditions of swine farms, improvement in local environmental condition, improvements in water quality and reduction of soil pollution and foul doors.					
Financial viability indicators	Payback period:	5 to 10 years*	Post-tax IRR:	N.A.	Gross margin:	N.A.

*Depending on the value of Certified Emissions Reductions Certificates (CERCs)

Context and background

Sadia, established in 1944, is one of the world's leading producers of chilled and frozen foods in Brazil. It is one of the country's main exporters of meat-based products. As of 2008, Sadia had about 20 industrial plants that together produced over 2.3 million tons of food, including chicken, turkey, pork and beef, pasta, margarine, desserts and other products. Recognizing the increasing influence of social and environmental issues associated with swine production systems in its supply chain, Sadia designed and implemented the 3S Program in 2003. Developed and managed by the Sadia Sustainability Institute, the 3S Program seeks to institute sustainability into the pork meat supply chain by improving animal waste management while providing additional revenue to individual farmers from carbon credits.

The 3S Program provides more than 3,500 swine producers with bio-digesters and is designed to reduce GHG emissions from the swine producers and to qualify the emission reductions as a CDM project. With the program, at least three million litres of swine excrement will be processed daily, a volume equal to 5% of the total swine waste volume produced in Brazil. The voluntary program began at three of Sadia's own swine farms, functioning as prototypes to be extended to its outsourced producers.

Market environment

A significant number of swine facilities in Sadia's chain did not have an environmental permit. Environmental licenses were expensive and not all farmers were aware of their importance or how to obtain them. Most of the manure produced was disposed in groundwater, streams and rivers without adequate treatment, and as a result, nearby communities were affected by water and soil pollution, as well as by the unpleasant odor. Through the implementation of bio-digesters under the 3S Program, swine farmers are able to manage and treat swine manure and reduce GHG emissions. Swine farmers have the opportunity to diversify their income generating activities and increase their farm profits through revenues from selling carbon credits and from reduced energy costs as gases captured from the bio-digester are used by the farms. By reducing costs and creating the possibility for diversifying income sources, Sadia hopes to encourage the small producers to stay in the business.

Sadia expected that about 50% of the producers would want to participate in the 3S Program. By early 2007, 96% had signed a contract indicating willingness to participate in the program. The other 4% were large swine farms that are already prepared or were preparing to individually operate in the carbon credit market. The program is implemented in 30% of facilities. In May 2006, Sadia and the Sadia Institute sold the first carbon credits generated by the 3S Program. Sadia sold 290,000 tons from its own farms of which 50,000 tons were sold at 11 €/ton, and the rest were based on the European Allowance Market Index. The European Carbon Fund also bought approximately 2.5 million tons of carbon from the institute to be sequestrated by the swine farms.

Macro-economic environment

Brazil is one of the major producers and exporters of pork accounting for 10% of the world pork production and exports with annual sales of over USD 1 billion. However, the intensive swine production industry resulted in significant environmental impact as a result of the higher amounts of

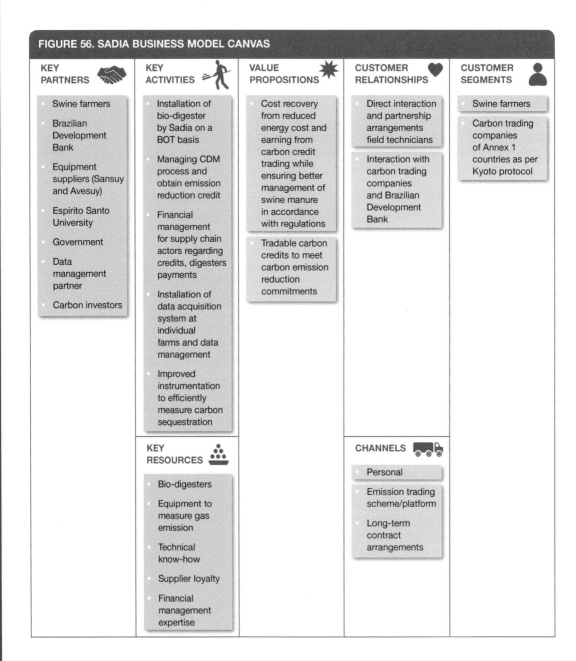

FIGURE 56. SADIA BUSINESS MODEL CANVAS

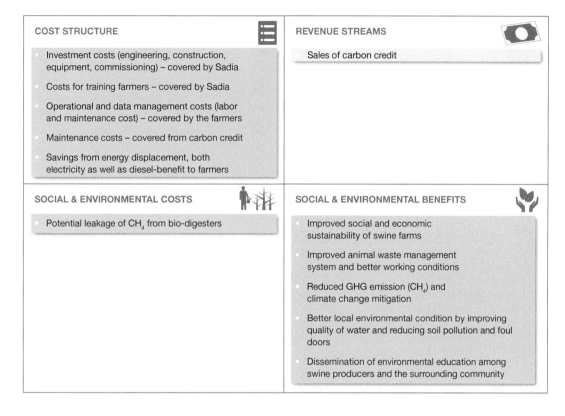

waste generated in these operations. Swine manure has the potential to impact soil, air and water resources requiring proper management, treatment and disposal.

In a mechanism of the Kyoto Protocol, projects in developing countries that mitigate GHG emissions can apply for certificates of emission reduction, most commonly known as carbon credits. These are certificates emitted by an internationally recognized institution, e.g. the UNFCCC, which attests that a certain amount of GHG (usually measured as a ton of CO_2) has been mitigated. Once obtained, these certificates can be traded on the market and exchanged for money. The sale of carbon credits could be enough, according to Sadia's studies, to cover the cost of the bio-digesters. However, applying for the certificates was a rather difficult and expensive process. Brazil has a target to reduce its overall GHG emissions by 36.1% to 38.9% below 1990 levels in 2020. Brazil has been leading in terms of the numbers of CDM project activities after China and India.

Business model

Sadia installs the bio-digesters at its swine producers on a Build and Transfer basis (Figure 56). The participation of swine producers in the program is voluntary. One of the success factors of implementing the 3S Program is the fact that Sadia Institute was able to obtain funds from the Brazilian Development Bank, thus enabling small and medium swine producers to take part in the program. This is a good example of innovative financing mechanism in the waste reuse business. The financing arrangement is that Sadia Institute owns all the equipment installed in the farmers' facilities for the purpose of the 3S Program and is responsible for managing the CDM benefits. The institute trades carbon credits on the carbon trading market. The amount obtained is then shared with farmers, according to each potential emission reduction, after deduction of the investment made in the bio-digesters and in the program implementation and operation costs. In approximately five years, when the farmers finish paying the Institute for it, the bio-digesters and all related equipment will change hands and be owned by the farmers. The program benefits both parties. Sadia is able to increase supplier loyalty and secure supply in the light of environmental regulation. Farmers benefit from improved management of swine manure. Moreover, in addition to creating revenues from carbon credit trading, farmers are able to benefit from cost recovery due to reduced operational costs from using energy produced from the bio-digester and also the by-product from the fermentation process can be used as crop fertilizer or as food for fish breeding.

Value chain and position

Sadia created a non-profit entity called the Sadia Sustainability Institute, an independent non-profit organization in December 2004, to manage the 3S Program and to negotiate the carbon credits (Figure 57). The institute is responsible for managing the 3S Program including unifying the swine producers and building enough carbon credits to create a CDM project. The Sadia Institute (SI) borrowed R$ 65.5 million (USD 36.11 million) from the Brazilian Developmental Bank (BNDES) for starting the implementation of the program. Sadia was the guarantor of the SI's loan for implementing the 3S Program.

The institute first identified the swine producers that could be potential participants. The role of the Sadia Institute is to provide the swine producers with information to procure the bio-digesters, identify the infrastructure needed at each facility and overall administration of the program. Two suppliers (Sansuy and Avesuy) were selected to provide the bio-digesters. Sadia also partnered with Espírito Santo University to develop new measuring equipment for measuring the gas emissions i.e. quantity of methane sequestrated and amount of CO_2 produced. Once the bio-digesters are installed, the famers are responsible for the operation of the bio-digester in their respective farms. The emission reductions qualify for the Kyoto Protocol's CDM program, under which Sadia Institute sells the carbon credits. The resulting surplus would be used to improve the social and environmental conditions of the participating farmers. Farmers pay back for the bio-digester from the carbon credit benefits on an installment basis.

After validation of the biogas equipment by Sadia's engineers using the United Nations Framework Convention on Climate Change (UNFCCC) standards, the farmers will be able to use the biogas. However, if biogas is utilized, no CERs will be claimed for potentially displacing fossil fuels or grid electricity. The treated wastewater and mineralized sludge of the open-air lagoon is used for irrigation of surrounding crops.

Institutional environment

There are no national, state or local requirements providing for GHG emissions of agro-industrial operations (swine production) in Brazil. The state legislation on swine waste in Brazilian states determines that animal waste must have 120 days of retention in a non-permeable open-air lagoon for reduction of the organic load.

Since the 3S Program is registered as a CDM project, both the UNFCCC Kyoto protocol requirements and host country requirements apply. Along with the program implementation, auditing and verification of the program is expected. Such auditing is to be performed every semester by a designated operational entity, an independent auditor accredited by the CDM Executive Board, as determined by the UNFCCC.

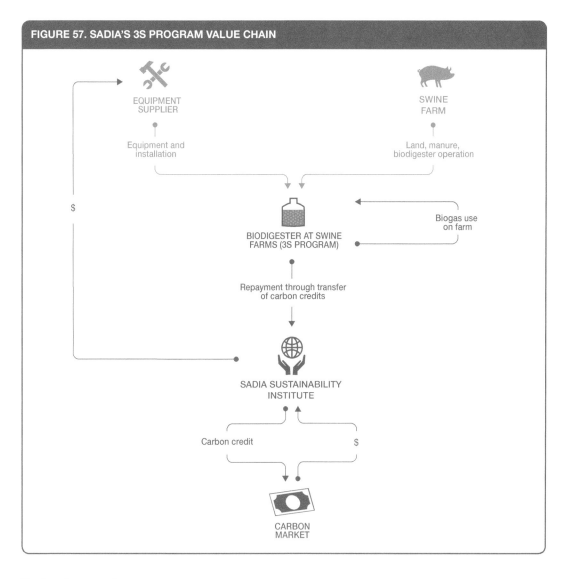

FIGURE 57. SADIA'S 3S PROGRAM VALUE CHAIN

Technology and processes

For all farms included in the 3S Program, the installed equipment follows identical standards. The technology comprises a bio-digester, a combustion system and an open-air lagoon in which to store the treated manure (Figure 58). A process for identifying each farm and data acquisition system is also installed at individual farm. A data system was developed by a software company to ensure that the information about each farm cannot be altered, manipulated or double-counted. This system works with a device called PLC (Programmable Logical Controller) that is installed in each enclosed flare system. It is responsible for the data sources (pressure, temperature, flow, farmer, maintenance and other variables) of the project and where the information is processed. This program operates the system automatically and provides all needed data for each farm. Several technologies are available for manure management in swine farms. However, the selection of a feasible technology should take into account not only technical and economic challenges but also particular farm characteristics. These include the number of housed animals, the available agricultural land for manure application and the opportunities for energy and organic fertilizer production for trading or local consumption.

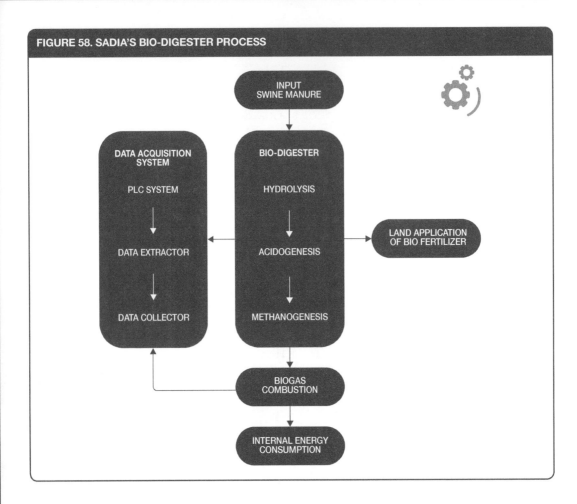

FIGURE 58. SADIA'S BIO-DIGESTER PROCESS

Funding and financial outlook

The Amazon Fund, created in 2008, fosters a low-carbon economy by reducing GHG emissions, and that contributes decisively to improving not only the standard of living and preservation, but also the recovery and the rational use of its natural resources. The Amazon Fund is administered by the Brazilian Development Bank (BNDES) and aims to provide financing services for projects, which aim at the reduction of GHG emissions. This incentivizes companies, such as Sadia, to pursue CDM project activities.

The Sadia Institute (SI) borrowed money from a financial institution (R$ 60 million) from BNDES (UNDP, 2007), approximately USD 33 million) for purchasing and installing the bio-digesters and the combustion system in the outsourced farms. SI owns all the equipment installed in the farmers' facilities for the purpose of the 3S Program. In approximately five years, when the producers finish paying the institute for it, the bio-digesters and all related equipment will change hands and be owned by the swine farmers. The institute negotiates carbon credits with the operational entity under the CDM. The institute takes a percentage of the revenue from the offsets to cover operational expenses, and the remainder is allocated to the producer. Before seeing any income from the credits, the producer pays for the bio-digester; the payment is estimated to take five years of installments.

The process of certification of the carbon credits was rather complex involving several steps and actors. The volume of gas burned as well as the temperature, pressure and other measurements are registered in a computer, which would be constantly monitored by Sadia's field work technicians.

Socio-economic, health and environmental impact

The 3S Program initiated by Sadia provides additional revenue and improved working conditions for Sadia's swine producers, while reducing the environmental impact associated with swine production. The biogas generated can be used as energy in the farm, thus significantly saving operational costs. In addition, new business opportunities are created for the farmers, who can use the by-product from the fermentation process as food for fish breeding or as crop fertilizer and thus improve soil quality. The program contributes to improving the local environmental condition by improving quality of water and reducing soil pollution and foul odors. In addition, the 3S Program is expected to help disseminate environmental education among swine producers and the surrounding community.

Scalability and replicability considerations

The 3S Program was designed and implemented to reduce GHG emissions from swine producers in Sadia's supply chain and to qualify the emission reductions as a CDM project. This program can be used as a demonstration for other supply chains to replicate the program by incorporating environmental sustainability in their revenue design. In order for this program to be replicated in other developing countries with intensive livestock production, government support for projects, which aim at reducing GHG emissions coupled with innovative financing mechanism, is required.

The key drivers for the success of this business are:
- Innovative financing mechanism.
- Availability of financing organizations.
- Partnership among the different actors within the value chain.
- Foreseeing and eliminating regulatory problems relating to production permission in Sadia's swine supply chain.

SI plans to extend the program within its supply chain, including those suppliers that are not swine producers (for example, poultry and beef). SI plans to develop a "Sustainable Site Platform" in which it will give training for new agricultural commodities to be produced by its suppliers to diversify and increase their income. The platform aims to educate the producers on financial and management issues and thus creating entrepreneurs that are better prepared for the market.

Summary assessment – SWOT analysis

The key strength of Sadia is the application of innovative financing mechanism to implement the 3S Program and thus fostering strong partnership with its swine farmers (Figure 59). Availability of financing organizations such as the Amazon Fund which provide financing services for projects which aim at the reduction of GHG emissions and government support for CDM are the key opportunities for the business to further expand its 3S Program. However, sustainability of the 3S Program highly depends on carbon credit sales, the prices of which are volatile.

FIGURE 59. SWOT ANALYSIS FOR SADIA

HELPFUL TO ACHIEVING THE OBJECTIVES

HARMFUL TO ACHIEVING THE OBJECTIVES

INTERNAL ORIGIN — ATTRIBUTES OF THE ENTERPRISE

STRENGTHS
- Innovative financing mechanism
- Strong partnership with swine farmers
- Ensures following environmental regulation in supply chain
- Intensive swine farming (abundance of swine manure)
- Diversified income for farmers
- Natural process of dry fermentation
- The technicians' credibility and respect

WEAKNESSES
- Highly dependent on carbon credit sales
- Farmer's lack of technical know-how to operate bio-digesters
- High technology cost
- High operation and maintenance costs
- A challenge to include the small producers to find a gauge to measure the gas emissions
- Consolidate a large number of carbon sequestrating facilities (bio-digesters)

EXTERNAL ORIGIN — ATTRIBUTES OF THE ENVIRONMENT

OPPORTUNITIES
- Availability of financing organizations
- Government support for CDM
- Carbon credit market opportunities
- High value bio-fertilizer for additional revenue
- High quality renewable fuel, biogas has several proven end-use applications
- Country's participation in the CDM projects

THREATS
- Volatile carbon credit prices
- Possible leakage of CH4 may significantly reduce the carbon credit earnings
- Delay in administrative proceedings
- Environmental laws are not enforced
- Financial challenge the sale realized after the plant is commissioned
- Consolidate a large number of carbon sequestrating facilities (bio-digesters)

Contributors

Johannes Heeb, CEWAS, Switzerland
Jasper Buijs, Sustainnovate; Formerly IWMI
Josiane Nikiema, IWMI, Ghana
Kamalesh Doshi, Simplify Energy Solutions, LLC, Ashburn, Virginia USA

References and further readings

Amazon Fund. 2008. Project document. Amazon Fund. https://goo.gl/kEhZk2 (accessed Nov. 7, 2017).

Hojnacki, A., Li, L., Kim, N., Markgraf, C. and Pierson, D. 2011. Bio-digester global case studies. D-Lab Waste. http://web.mit.edu/colab/pdf/papers/D_Lab_Waste_Biodigester_Case_Studies_Report.pdf (accessed Jan. 16, 2013).

Miranda, B., Sá, C. and Saes, S. 2009. Sadia Case Study: The Challenges of a Diversified Production Chain. Paper presented at the VII International Pensa Conference, São Paulo.

Pereira-Querol, M. 2011. Learning challenges in biogas production for sustainability: An activity theoretical study of a network from a swine industry chain. University of Helsinki Institute of Behavioural Sciences.

Pereira-Querol, M. and Seppanen, L. 2008. Learning as changes in the activity systems: The emergence of on-farm biogas production for carbon credits. In 8th European IFSA Symposium, Clermont-Ferrand, France, July 6–10, 2008.

United Nations Development Programme (UNDP). 2007. Sadia Program for Sustainable Swine Production (3S Program): Bringing sustainability to the supply chain. Value Chains. www.value-chains.org/dyn/bds/docs/740/Brazil_Sadia%20FINAL.pdf (accessed Nov. 7, 2017).

United Nations Development Programme (UNDP). 2008. Creating value for all: Strategies for doing business with the poor. UNDP. https://goo.gl/bTLgXm (accessed Nov. 7, 2017).

United Nations Environment Programme (UNEP). 2009. Towards triple impact: Toolbox for analysing sustainable ventures in developing countries.

Case descriptions are based on primary and secondary data provided by case operators, insiders, or other stakeholders, and reflect our best knowledge at the time of the assessments (2013/2014). As business operations are dynamic, data can be subject to change.

CASE

Power from manure and slaughterhouse waste for industry's internal use (SuKarne, Mexico)

Javier Reynoso-Lobo, Krishna C. Rao, Lars Schoebitz and Linda Strande

Supporting case for Business Model 5	
Location:	Culiacan, State of Sinaloa, Mexico
Waste input type:	Animal and slaughterhouse waste
Value offer:	Biogas to electricity and thermal energy, bio-diesel, compressed Bio-gas, carbon credits and organic bio-fertilizer
Organization type:	Private
Status of organization:	Commercial-scale project under construction
Scale of businesses:	Large
Major partners:	Alberta Innovates–Technology Futures (Technology), Pro Bio (Fertilizer distributor, the group company), National Electricity Commission (Interconnection), United Nations (Carbon market), National Science and Technology Council (Research funding), IGSA (Co-investor[1]), German Biogas Company (Project design and management[2])

Executive summary

Grupo Viz (GV), a family business in the commercial cattle industry, was established in 1969 in Mexico. SuKarne, one of the five business entities of GV group, with annual sales of over USD 2 billion is the third largest feedlot grain-fed company in the world and fifth largest beef provider in North America. SuKarne's business chain produces both animal and slaughter waste, and it sells some of the waste to its affiliated companies (also owned by GV). In 2012, SuKarne began construction of a biogas pilot plant, a first for the cattle industry that uses a mixture of animal waste and residual biomass, with the lagoon's water, to produce biogas for electricity and thermal energy. At full capacity, it is expected to generate approximately 3.2 MW of electricity for self-consumption and 3 MJ to displace boiler diesel with the heat generated. It will also be possible to further treat the biogas to produce liquid fuel or compressed gas to feed the trucks used throughout the whole operation. The plant will also generate organic bio-fertilizer to be sold to an affiliated company. The project was at the time of assessment under commercial-scale development and expected for construction and commissioning in 2015, with possibilities to replicate the model in its other four facilities in Mexico. The plan is that each plant will be self-sufficient in both electricity and thermal energy. The project's feasibility study is registered in the Clean Development Mechanism (CDM) and subject for final approval for a Certified Emission Reductions (CERs) agreement.

KEY PERFORMANCE INDICATORS FOR THE CULIACAN FACILITY (AS OF 2014)					
Water:	67,000 m^3 of slaughterhouse wastewater reutilized per annum				
Capital investment:	USD 12.5 million				
Labor:	9 employees (O&M supervisor, mechanic/electric engineers, 5 operators, 1 technician)				
O&M cost:	Approx. USD 90,000 per annum				
Output:	110,000 tons of animal waste processed per annum to generate 13 million m^3 biogas per annum, or 3.2 MW as electricity and 3 MJ as heat using a combined heat and power (CHP) unit, and approx. 35,000 tons of vermicomposting per annum				
Potential social and/or environmental impact:	Yearly savings of approx. USD 1.8 million in diesel and electricity costs, reduction of approx. 132,000 tons of equivalent CO_2 emissions per annum (US-EPA calculator), generation of a renewable source of energy for the company; Reduced soil, water and air pollution				
Financial viability indicators:	Payback period:	2.9 years	Post-tax IRR:	14.3 %	Gross margin: USD 4.28 million

Context and background

GV is a family business established in 1969 at Culiacan, Sinaloa. Over the years, GV expanded its operation to other parts of the cattle production business value chain and now owns five subsidiary companies operating independently. The five subsidiaries of GV are: a) SuKarne Agro-industrial, a beef, poultry and pork producer, b) Rendimientos Protéicos (Renpro), specializes in processing of tallow, meat and blood meals for livestock and animal feed production, c) Productos Bioorgánicos (ProBio), specializes in the production of organic compost and vermicomposting from animal waste, d) SuKuero, specializes in leather commercialization and e) Agrovizion, an agribusiness dedicated to the promotion and commercialization of agricultural products such as corn, wheat, oats and roughage. This case study is on SuKarne, the largest producer and supplier of beef in Mexico, third largest feedlot grain-fed company in the world and fifth largest beef provider in North America. SuKarne owned at the time of assessment five production facilities around the country, located in the states of Nuevo Leon, Baja California, Michoacan, Durango and Sinaloa. These five facilities maintain a daily average of 425,000 animals confined in open feedlots through the year, and approximately 1,100,000 animals are processed per annum.

Manure is removed from the feedlots twice a year using a scraping system and disposed in piles over lands located near the operation for further degradation through composting processes. Mexican and local state legislation prohibit the unlicensed displacement and/or uncontrolled burning of animal waste, which leaves a huge amount that is left to decay on the ground, thus contributing to the greenhouse gas (GHG) emissions to the atmosphere.

A business opportunity was identified by SuKarne to develop a methane recovery project from the animal waste generated in their five facilities and generate thermal and electric energy in the form of biogas along with organic material for compost, while significantly reducing GHG emissions. In late 2012, SuKarne constructed a dry fermentation pilot plant and throughout 2013 conducted a series of trials which validated the feasibility of the technology and its waste streams for biogas production.

SuKarne is developing the commercial-scale project with Canadian biogas plant providers to construct a large biogas facility near its operation in Culiacan. Prior to this initiative, SuKarne sold its organic waste (feedlot manure) to Pro Bio to produce organic compost and vermicomposting. This case study focuses on the business model for the biogas production facility under development. The project will use animal waste and slaughterhouse waste as feedstock to produce biogas (mainly containing methane). The biogas will be used to generate electric and thermal energy. The expected amount of GHG emissions reduction with the project is on average 132,000 tons of CO_2-eq per annum.

SuKarne's biogas operations and its end users are within its premises or with affiliated companies, and hence there is no competition for procuring waste or sale of energy. The fertilizer produced from the bio-methanation process will be exclusively sold to Pro Bio, an affiliated company in the business of producing organic compost.

Market environment

According to the Mexican Secretariat of Agriculture, Livestock, Rural Development, Fisheries and Food (SAGARPA), 58% of Mexico's land, a total of 113.8 million hectares, is used for beef production. There is a total of 31 million cattle livestock in Mexico owned by 1.13 million breeders: 2 million dairy cattle and 29 million beef cattle. According to the Mexican Ministry of Environment and Natural Resources (SEMARNAT), livestock production has shown an accelerated growth in the past two decades, increasing by 62% in comparison with the 1990s. As a result of this progressive increase in livestock production, 83% of Mexico's emissions from agricultural sector were attributed to livestock production in 2002, equivalent to 8% of the total emissions in Mexico (Table 19). However, the consumption of fossil fuels accounts for 63% of the country's carbon emissions, and a major part of the carbon emissions are from agro-industrial operations such as meat production. There has been no significant action in terms of emission reduction from this part of the livestock sector. Other sectors such as swine farms have developed projects with the support of government programs such as Methane to Markets (M2M) and the CDM, though most account only as far as for biogas burning. So far, there are no other business models in the Mexican livestock industry that transform waste into self-supplied renewable energy at commercial-scale.

TABLE 19. MEXICO EMISSIONS IN CO_2 EQUIVALENT (GG), ADAPTED FROM "METHANE TO MARKETS." SOURCE: SEMARNAT, 2008

EMISSION CATEGORY	2002	PERCENTAGE
1) Energy	389,496.70	70.39%
1A) Consumption of fossil fuel	350,414.30	89.97%
1B) Fugitive methane emissions	39,082.30	10.03%
2) Industrial processes	52,102.20	9.41%
3) Agriculture	46,290.80	8.36%
3A) Livestock	38,527.47	83.23%
3B) Crops	7,763.26	16.77%
4) Waste	65,584.40	11.84%
Total	553,329.40	100%

Macro-economic environment

The livestock operations are prone to serious environmental impacts, such as GHG emissions, odor and water and land contamination, all a result from storage and disposal of animal waste. Confined Animal Feeding Operations (CAFOs) use similar Animal Waste Management System (AWMS) options to store animal residues. These systems emit both methane (CH_4) and nitrous oxide (N_2O) resulting from aerobic and anaerobic decomposition processes (Clean Development Mechanism, 2007). Since approval of the Kyoto Protocol, immense interest in methane recovery has been generated amongst large-scale farms and livestock producers in Mexico, many of who have registered CDM projects.

In addition, Mexico created a strong climate change and renewable energy law in 2012 that targets 30% lower emissions compared to business as usual by 2020 and 50% by 2050.

Business model

SuKarne's methane recovery project has three key value propositions (Figure 60) – production of biogas to generate electricity and heat for self-consumption (displacing electricity purchase from the national grid and diesel costs for boilers and trucks), production of solid/liquid fertilizer from the effluent of the bio-methanation process and sale to Pro Bio and sale of carbon offset both from methane recovered and fossil fuel displacement.

It is key for SuKarne to ensure that all waste generated from its business value chain is brought to the biogas facility. SuKarne partnered with a Canadian technology research centre to help develop the biogas technology to process multiple feedstocks in the biogas digester. The most critical relationships for SuKarne in its methane recovery project is its partnership with co-investor IGSA, biogas plant providers for the design and project management of the facility construction, the national grid for interconnection and electricity supply contracts, affiliates such as Pro Bio for sale of solid/liquid fertilizer and carbon companies buying CER certificates.

Value chain and position

SuKarne manages livestock procurement and production, meat processing, distribution and commercialization (Figure 61). SuKarne captures organic waste generated from several parts of its chain and transforms it to higher-value products such as leather, animal feed, soaps, organic compost and vermicomposting, and with the implementation of this project into biogas as well. The biogas model will merge into the existing model by reutilizing the feedlot manure from the pens, corn stover from the feed mill and paunch content and wastewater from the slaughterhouse to generate renewable energy which will replace fuel and electricity supply used in one of its operations. The composting process in the value chain will shift to take place after the biogas production process.

Institutional environment

Regulatory settings in Mexico require businesses to prepare an environmental impact assessment of the proposed energy generation plants in order to demonstrate that it will not have a negative impact on the environment. This is currently regulated by SEMARNAT. Additionally, every business that intends to generate and/or sell energy must be regulated by the Electricity and Hydrocarbons Regulator (CRE). They grant permits for private self-supply generation, independent power production and co-generation. The CRE has designed several instruments which regulate the relationship between suppliers (Federal Energy Commission and Light and Power Company) and private generators.

Mexico has a progressive policy on climate change and renewable energy commitment to meet its optimistic targets. However, the government will likely have to provide more fiscal incentives. The General Law for Climate Change adopted in May 2012 sets the goal of 35% of energy generated in the country should come from renewable sources by 2024. The law creates a fund for the transition to clean and renewable energy and technologies. These legal instruments are expected to create a better framework to support renewable energy in general and also a future green economy in Mexico. Additionally, implementation of renewable energy reinforcement laws and incentives schemes as well

FIGURE 60. SUKARNE METHANE RECOVERY BUSINESS MODEL CANVAS

KEY PARTNERS
- Local breeders
- Grupo Viz subsidiaries (waste suppliers)
- Federal electricity commission
- Carbon investors
- Biogas technology developer
- Biogas plant provider

KEY ACTIVITIES
- Collection of animal and agro waste
- Production of biogas
- Electricity cogeneration and heat usage
- Production of bio-diesel or compressed biogas
- Production of organic compost
- O&M for production

KEY RESOURCES
- Slaughterhouse waste
- Dry fermentation technology
- Contracts with electricity commission
- Capital investment

VALUE PROPOSITIONS
- Recovery of methane from animal and slaughterhouse waste to create carbon offsets
- Production of biogas for heat and electricity
- Production of solid/liquid fertilizer

CUSTOMER RELATIONSHIPS
- Direct interaction carbon trading companies
- Internal and direct interaction with electricity commission
- Direct interaction and partnership

CHANNELS
- Long term contract agreements

CUSTOMER SEGMENTS
- Carbon trading company
- Grupo Viz subsidiaries (for internal consumption of electricity as well as thermal energy)
- National grid
- Fertilizer affiliate company

COST STRUCTURE
- Investment cost (engineering, construction, equipment, commissioning, cogeneration, transmission)
- Operational cost (labor, maintenance cost, energy, fuel)
- Depreciation
- Savings from energy displacement, both electricity as well as diesel

REVENUE STREAMS
- Sales of carbon offset
- Sale of surplus electricity
- Sales of solid/liquid fertilizer

SOCIAL & ENVIRONMENTAL COSTS
- Potential leakage of methane

SOCIAL & ENVIRONMENTAL BENEFITS
- Job creation
- Reduced water, soil and air pollution
- Reduced carbon emissions
- Displacement of fossil fuel consumption (both for electricity and diesel for boilers and trucks)

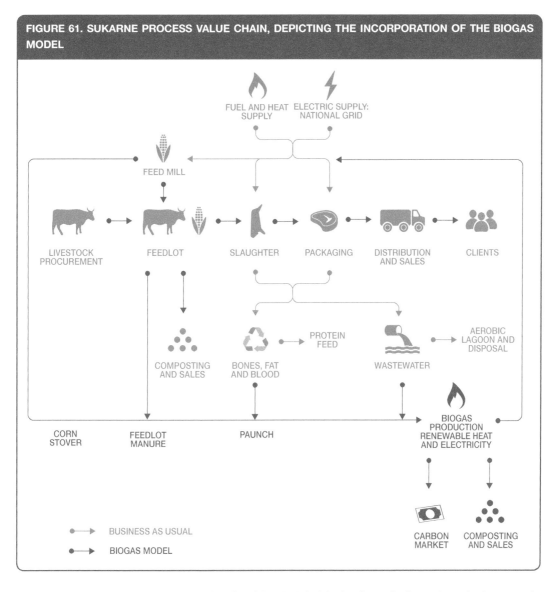

FIGURE 61. SUKARNE PROCESS VALUE CHAIN, DEPICTING THE INCORPORATION OF THE BIOGAS MODEL

as the removal of subsidy on industrial diesel in 2013 in Mexico have further driven the large-scale industries such as SuKarne to adapt and develop new strategies.

Technology and processes

Slaughterhouse effluent has high Chemical Oxygen Demand (COD), high Biological Oxygen Demand (BOD) and high moisture content, which makes it well-suited to anaerobic digestion process. Slaughterhouse wastewater also contains high concentrations of suspended organic solids including pieces of fat, grease, hair, feathers, manure, grit and undigested feed which will contribute to the slowing of the process of biodegrading organic matter. The biogas potential of slaughterhouse waste is higher than animal manure and reported to be in the range of 120–160 m^3 biogas per ton of wastes.

The technology used in SuKarne's methane recovery project is a dry fermentation system including bio-digesters and a percolation tank coupled to a biogas cleaning unit and combined heat and power

(CHP) units to generate electric and thermal energy. The dry fermentation process is an anaerobic digestion technology for solid, stackable biomass and organic waste, which cannot be pumped. It is mainly based on a batch wise operation with a high dry matter content ranging from 20–50% at mesophilic temperatures. It is especially suited for application in semi-arid climates as the water consumption in the process is very low compared to conventional anaerobic digestion systems.

The biogas generated will replace the electricity bought from the national network, as well as the diesel used in boilers and trucks. The facility will consist of 900 m^3 concrete air-gas-tight chambers to manage approximately 110,000 tons of waste per year. The plants will be designed as modular solutions that are scalable according to the amount of waste that is available or energy demand. This will be the first facility of its kind in Mexico and unique worldwide in terms of feedstock characteristics, its source being a commercial cattle feedlot.

Animal waste (manure) is collected at least once every six months from the pens. Internal transport of the waste from the pens to the project site will be done in trucks carrying containers within a distance no longer than 5 km. The collected manure will be transported to the plant site to be shredded and mixed with effluents from the slaughter plant such as paunch content and wastewater to prepare an appropriate organic loading rate. Substrates such as corn stover and wood chips will be incorporated to improve material structure. Prepared substrates will be loaded into the fermentation units and digested to generate biogas. A CHP unit will be used to produce electrical and thermal energy. A biogas-cleaning unit will be incorporated before the CHP unit if necessary. Wood chips will be recovered and reutilized after the composting process.

Equipment and infrastructure required for the project:
- Dry fermentation units and components.
- Substrate mixing equipment and/or machinery.
- Biogas storage and cleaning equipment.
- Combined heat and power unit for cogeneration.
- Complementary equipment and facilities for the modular units.

Funding and financial outlook

SuKarne applied for research funding from the Mexican National Council for Science and Technology (CONACYT) and, in 2012, obtained a USD 320,000 grant for the construction and validation of a dry digestion biogas pilot plant for its Culiacan site. The investment required for the design, construction and commissioning of the large-scale biogas facility is estimated at USD 12.5 million, which will be shared between SuKarne and co-investor partner IGSA. About USD 500,000 was required to develop the mass and energy balances, feasibility study, technologies assessment and selection, pilot plant design and specialized support for the design and implementation of the chosen technology.

SuKarne's methane recovery project requires approximately USD 90,681 for operation and maintenance per annum. The revenue structure consists of savings from electricity and boiler diesel replacement from energy generation, carbon offset and sales from fertilizer. The estimated revenue potential of the plant is approximately USD 1.8 million per annum with an amortization period of 10 years. This investment is projected to result in an IRR of 14.3% with 7 years return at a discount rate of 8% (Table 20).

Note: Unit value refers to the unitary cost of every expense and income; quantity refers to the amount of supply that will be required or sold product as the technology is implemented; value is the total amount per item in US dollars; difference from business as usual (BAU) refers to the additional costs

CASE: POWER FROM MANURE AND SLAUGHTERHOUSE WASTE FOR INTERNAL USE

TABLE 20. SUKARNE METHANE RECOVERY PROJECT FINANCIALS

	UNIT VALUE (USD)	QUANTITY	VALUE (USD)	DIFFERENCE BAU (USD)
Cost				
Fuel	0.54 /L	1,083,022 L/yr	584,832	–
Corn stover	49.70 /t	5,839 t/yr	290,198	48,371
CHP maintenance	0.021 /kWh	23,040,000 kWh/yr	483,840	483,840
O&M / Labor			90,681	90,681
Total				622,892
Saving				
Diesel	0.77 /L	701,736 L/yr	540,337	1,439,893
Electricity	0.087 /kWh	15,986,471.26 kWh/yr	–	1,390,823
Soil replacement	8.41 /t DM	23,342 t/yr	196,306	131,200
Wastewater disposal	0.25 /t	18,247 t/yr	–	4,562
Water	0.54 /t	776,783 t/yr	419,463	99,307
Total				3,065,785
Revenue				
Electricity	0.087 /kWh	7,053,529 kWh/yr	613,657	613,657
Compost	53.50 /t	65,702 t/yr	3,515,057	(98,058)
GHG mitigation (CERs)	10.00 /tCO$_2$e	132,049 t/yr	1,320,490	1,320,490
Gross margin				4,278,982
Payback				2.9 years
Accounting rate of return				14.3%
Capital cost			12,580,057	
Amortization period	10 years			
Interest rate	8%			

incurred by the new business model as well as savings generated, mainly from diesel, electricity and carbon credits.

Socio-economic, health and environmental impact

SuKarne's methane recovery project has high environmental benefits from reduced greenhouse gas emissions. The project ensures proper disposal of animal waste, requires less area than aerobic compost, reduces the volume and weight of landfills, produces a sanitized compost and nutrient rich liquid fertilizers, maximizes the benefits of recycling and in the process improves the air quality through improved odor and reducing groundwater contamination. It also displaces the diesel consumption in the feed mill, trucks and slaughterhouse. Additionally, the project improves the electricity burden on the regional electricity board by reducing its purchase from grid. The economic benefit from improved electricity in the region goes beyond the enterprise. In addition, SuKarne provides additional employment – nine employees for the management of biogas plant and electricity generation operations.

Scalability and replicability considerations

The key drivers for the success of this business are:
- Capital investment from administration board and co-investors.
- No barriers in accessing available in-house animal and agro-waste.

- Supportive environment for environment-friendly initiatives, with many existing livestock projects registered for CDM in Mexico; SuKarne benefits from streamlined process.
- Diesel subsidy removal.
- Favourable policies and incentives by the Government of Mexico.

SuKarne has five livestock production operations across Mexico. Based on the operational viability and profitability of its first large-scale biogas plant, SuKarne plans to separate its biogas operation into a new company and develop similar projects in all its operations across Mexico. This project has replication potential in other large livestock farms but needs to counter the challenges of adequate skilled human resources, investors that understand business risks, carbon credit markets and stable, supportive, government energy policies and financial incentives.

Summary assessment – SWOT analysis

The key strength of the project is its shorter payback period for high investment cost and strong partnership with affiliated companies (Figure 62). The weakness stems from high cost of technology that can deter its promoters from making the investment.

FIGURE 62. SUKARNE METHANE RECOVERY PROJECT SWOT ANALYSIS

	HELPFUL TO ACHIEVING THE OBJECTIVES	HARMFUL TO ACHIEVING THE OBJECTIVES
INTERNAL ORIGIN ATTRIBUTES OF THE ENTERPRISE	**STRENGTHS** • Assured supply of waste • Low O&M and high revenue • Strong partnership with affiliated companies • Natural process of dry fermentation • Food security • Efficient waste and water management • Climate change mitigation and adaptation	**WEAKNESSES** • High technology cost • Requirement of high skilled labor for this technology and research and development • Too much heterogeneity among the livestock production units in relation to their size and use of technology
EXTERNAL ORIGIN ATTRIBUTES OF THE ENVIRONMENT	**OPPORTUNITIES** • Environment stress reduction offers carbon credit market opportunities • Output from biogas plant is high value fertilizer which can be harnessed for additional revenue • Electricity demand is growing and need for renewable energy-based electricity generation increasing in Mexico • Recent changes in legislation on renewable energy control • Classification and separation of waste, translate into increased opportunities for generation • High-quality renewable fuel; biogas has several proven end-use applications • Country's participation in the methane-to-market alliance	**THREATS** • Possible human health risk may lead to investment needs • Possible risk from leakage of gas may force O&M costs higher • Delay in administrative proceedings • Environmental laws are not enforced • Weak national capabilities to design and manage projects to reduce methane emissions • Lack of comprehensive schemes to address the issue of livestock waste • Lack of co-generation equipment for all types of farm sizes and variable methane production

The business offers many opportunities for replication due to the high demand for electricity and heat sources as well as the opportunity to become a sustainable leader in meat industry by significantly reducing fossil fuel consumption, revoking the misconceptions of the livestock industry's contribution to GHG emissions and ultimately helping the future generations live in a cleaner and better world.

Contributors
Kamalesh Doshi, Simplify Energy Solutions LLC, Ashburn, Virginia, USA

References and further readings

Cepeda, J. Interview. November 14, 2012.

Clean Development Mechanism. 2011. Methane recovery and utilization from operations in beef cattle pens from SuKarne. Project Design Document 2.

López, Á. 2012. Manejo sustentable del inventario y extracción de ganado en México; Abasto de carne para el mercado nacional. 2nd Congreso Ganadero, Guadalajara, Jalisco, 2012.

Ministry of Environment and Natural Resources (SEMARNAT). 2008. Mexico profile: Animal waste management methane emissions. Methane to Markets.

Secretariat of Agriculture, Livestock, Rural Development, Fisheries and Food (SAGARPA). 2007. National Livestock Program 2007–2012.

Vizcarra, D. Interview. February 1, 2013.

Case descriptions are based on primary and secondary data provided by case operators, insiders, or other stakeholders, and reflects our best knowledge at the time of the assessments (2015–2016). As business operations are dynamic, data can be subject to change.

Notes
1 IGSA: http://www.igsa.com.mx
2 German partner: http://www.bekon.eu

BUSINESS MODEL 5
Power from manure

Solomie Gebrezgabher and Krishna C. Rao

A. Key Characteristics

Model name	Power from manure
Waste stream	Livestock manure, agro-waste as additional input
Value-added waste product	Biogas, energy (electricity as well as thermal energy), carbon credit, slurry/liquid and solid bio-fertilizer
Geography	Rural regions with livestock farming and large livestock industry
Scale of production	Small, medium to large scale 16 kW up to 5 MW of electricity 22,000 to 700,000 ton CO_2-eq/year
Supporting cases in this book	Santa Rosillo, Peru; Concordia, Brazil; Culiacan, Mexico
Objective of entity	Cost-recovery [X]; For profit [X]; Social enterprise [X] Community development [X]
Waste removal capacity	Manure from small (less than 600 animals), medium (600 to 1,000 animals) and large (more than 1,000 animals) livestock farms
Investment cost range	500–5000 USD/kW for capacities ranging between 1 MW to 3 MW (based on International Renewable Energy Agency or IRENA)
Organization type	Private, public-private partnership (PPP), public and non-profit organization
Socio-economic impact	GHG emission reduction (up to 700,000 ton CO_2-eq/year), improve water quality and reduce air and soil pollution, access to electricity (50 households from 16 kW plant), improved livelihood of remote communities, improved working conditions of slaughterhouse and animal farms
Gender equity	Neutral

B. Business value chain

This business model can be initiated either by livestock processing factories such as meat or dairy processing factories with the objective of ensuring that their products have been produced in an environmentally sustainable way or by small, medium and commercial-sized livestock farms in remote communities to utilize livestock waste to produce off-grid power for rural electrification with the support of regional government and NGOs. Depending on the size of the project, the business can also be registered as a Clean Development Mechanism (CDM) project to earn additional revenue from carbon credit sales.

While the power from manure model can be implemented in different scenarios, the following sections describe power from manure model for: a) carbon credit and sustainable value chain and b) rural electrification.

BUSINESS MODEL 5: POWER FROM MANURE

Power from manure for carbon credit and sustainable value chain

To mitigate the social and environmental impacts associated with livestock production systems, the processing factory (e.g. meat processing factory) installs bio-digesters on the animal farms within its supply chain on a Build and Transfer (B&T) basis. The factory oversees the installation of the bio-digesters on the farms, provides finance for initial capital cost, registers the project as a CDM project and manages the carbon credit revenues. The animal farm operates and maintains the bio-digesters. The factory assists the farmer in loan repayment by trading the carbon credit on behalf of the farmer. After deducting a portion of the receipts for loan repayment, the amount obtained is shared with farmers according to their potential emission reduction. The factory owns all the equipment until such time that the farmer pays back in full. The energy produced from livestock waste is used within the farms resulting in reduced farm operational costs and improved animal waste management, and the bio-fertilizer can be used on farms' own land. This business model seeks to bring sustainability to the entire supply chain by improving animal waste management while providing additional revenue to livestock farmers. In addition to cost recovery from utilizing the energy at the farm, the business model results in additional revenue for the farmer from sales of carbon credits (Figure 63).

Power from manure model for rural electrification

The business model can be commissioned by regional government in villages where there is no access to the national electricity grid or where there is very limited access to gas or electricity and where the

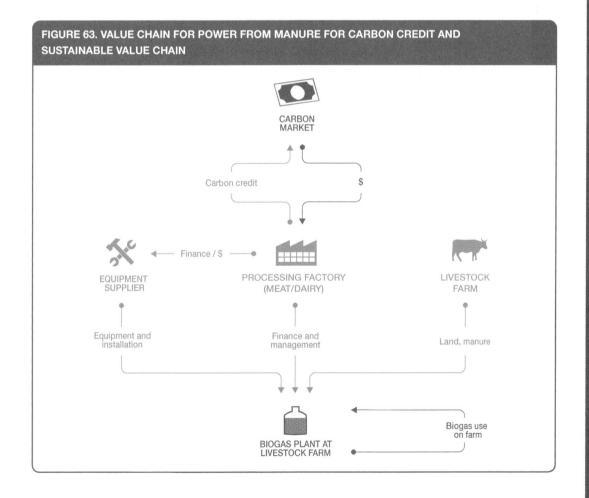

FIGURE 63. VALUE CHAIN FOR POWER FROM MANURE FOR CARBON CREDIT AND SUSTAINABLE VALUE CHAIN

community's primary economic activity is livestock farming. This is done by installing bio-digesters, which are fed with all the livestock and other organic waste from the community to generate biogas, which in turn is fed into the electricity generator and channelled to each house through a newly-installed electricity grid. The by-product from the bio-digesters is used as fertilizer by individual farms within the community to improve the soil quality or can be sold to other local farmers. The project can be financed through a public-private partnership between the regional government and the community with the major part of the funding coming from the regional government. Although the business is financed primarily by government subsidy, the investment can be supplemented with market-based approaches. The project will sustain itself through income streams primarily from monthly electricity usage fees charged to families and secondarily from the sale of slurry. This business has also the potential to earn additional revenue from carbon credit revenues by registering the business as a CDM project (Figure 64).

C. Business model

Business model – Power from manure for carbon credit and sustainable value chain

The processing factory installs bio-digesters at its livestock farmers within its value chain on a Build and Transfer basis in order to reduce GHG emissions from the livestock producers and to qualify the emission reductions as a CDM project. The processing factory could either obtain funds from banks or use own funds to finance the small and medium animal farmers to take part in the program.

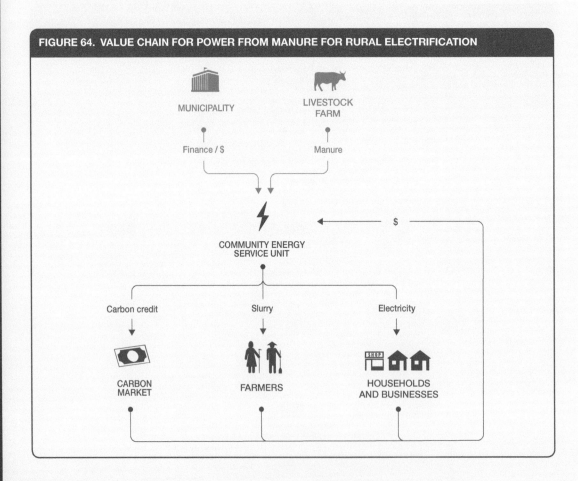

FIGURE 64. VALUE CHAIN FOR POWER FROM MANURE FOR RURAL ELECTRIFICATION

BUSINESS MODEL 5: POWER FROM MANURE

The program benefits both parties. The processing factory is able to increase supplier loyalty and secure supply in the light of environmental regulation and farmers benefit from improved management of animal manure. Moreover, in addition to creating revenues from carbon credit trading, farmers are able to benefit from cost recovery due to reduced operational costs from using energy produced from the bio-digester (Figure 65). The by-product from the fermentation process can also be used as crop fertilizer or as food for fish breeding. The processing factory owns all the equipment installed in the farmers' facilities and is responsible for managing the CDM benefits. The amount obtained from carbon trading is shared with farmers, according to each potential emission reduction and after deduction of the investment made in the bio-digesters including the program implementation and operation costs. The bio-digesters and related equipment will change ownership to the farmer ones the farmer completes payment for the investment cost on installment basis.

Business model – Power from manure for rural electrification

This business model has two key value propositions (Figure 66) – providing electricity service to houses and businesses and providing fertilizer to community farmers. The municipality, donor agency

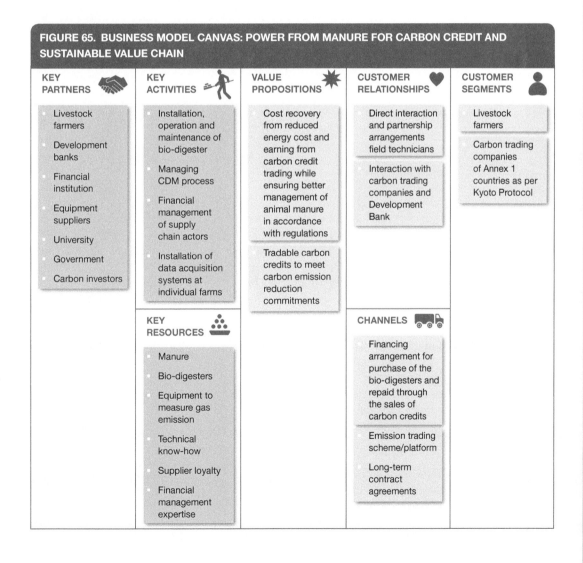

FIGURE 65. BUSINESS MODEL CANVAS: POWER FROM MANURE FOR CARBON CREDIT AND SUSTAINABLE VALUE CHAIN

KEY PARTNERS	KEY ACTIVITIES	VALUE PROPOSITIONS	CUSTOMER RELATIONSHIPS	CUSTOMER SEGMENTS
• Livestock farmers • Development banks • Financial institution • Equipment suppliers • University • Government • Carbon investors	• Installation, operation and maintenance of bio-digester • Managing CDM process • Financial management of supply chain actors • Installation of data acquisition systems at individual farms	• Cost recovery from reduced energy cost and earning from carbon credit trading while ensuring better management of animal manure in accordance with regulations • Tradable carbon credits to meet carbon emission reduction commitments	• Direct interaction and partnership arrangements field technicians • Interaction with carbon trading companies and Development Bank	• Livestock farmers • Carbon trading companies of Annex 1 countries as per Kyoto Protocol
	KEY RESOURCES • Manure • Bio-digesters • Equipment to measure gas emission • Technical know-how • Supplier loyalty • Financial management expertise		**CHANNELS** • Financing arrangement for purchase of the bio-digesters and repaid through the sales of carbon credits • Emission trading scheme/platform • Long-term contract agreements	

SECTION II: ENERGY RECOVERY FROM ORGANIC WASTE

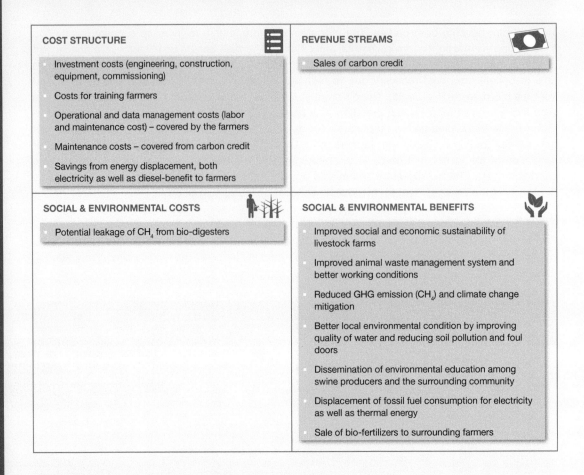

and local organization play key roles in mobilizing the community and securing financial resource to establish electricity service provision. Since the project results in carbon offset, there is potential for generating revenue from sales of carbon.

Alternative business model – Centralized biogas systems for carbon credit and sustainable value chain

An alternative business model is setting up centralized biogas systems owned and operated by farm cooperatives, the members being the participating manure suppliers (Figure 67). Thus, instead of installing bio-digesters at individual farms, manure from several farms within a region is supplied to a central bio-digester. Apart from the manure, the plant can receive various sorts of organic waste to increase energy yield of the system. This centralized system can be implemented by the processing factory on a Build, Operate and Transfer (BOT) model. Centralized system would benefit farmers that cannot individually construct and operate a bio-digester on their own due to capital expense or just don't have the required land, other infrastructures, sufficient number of animals and skilled labor to operate a bio-digester successfully or cost effectively.

The centralized system results in improved economic and organizational framework. It has an economy of scale advantages, and the fact that the centralized system results in a significant supply of energy is an advantage in negotiating contracts for sale of electricity to the state utility. Furthermore, the electricity produced can be supplied to the processing factory and partly used at the farms supplying

BUSINESS MODEL 5: POWER FROM MANURE

FIGURE 66. BUSINESS MODEL CANVAS: POWER FROM MANURE FOR RURAL ELECTRIFICATION

KEY PARTNERS
- Municipality
- Non-profit organization
- Community

KEY ACTIVITIES
- Agreement with relevant partners
- Community mobilization
- Organic waste collection
- O&M of the biogas unit
- Sale of electricity and fertilizer

KEY RESOURCES
- Land
- Capital
- Labor
- Manure and other organic waste

VALUE PROPOSITIONS
- Providing low-cost electricity service to houses and businesses in the community
- High quality solid/liquid organic fertilizer for self-use and sale to farmers

CUSTOMER RELATIONSHIPS
- Interaction with community through local subsidiary that manages the operation of the unit

CHANNELS
- Through appropriate agreement and self-interaction
- Direct

CUSTOMER SEGMENTS
- Household and businesses
- Farmers

COST STRUCTURE
- Investment costs (land, building and machines)
- Operational costs (raw material, labor, maintenance costs, marketing costs and R&D)

REVENUE STREAMS
- Sales of electricity
- Sales of organic fertilizer

SOCIAL & ENVIRONMENTAL COSTS
- Potential leakage of methane (CH_4) from bio-digesters

SOCIAL & ENVIRONMENTAL BENEFITS
- Jobs creation
- Environmental benefit through reduced CO_2 emissions by generating electricity from renewable source
- Improved animal waste management system and better working conditions with savings in cost of handling waste
- Better local environmental condition by improving quality of water and reducing soil pollution and foul doors
- Reduced GHG emission (CH_4) and climate change mitigation

the manure. Thus, the entire supply chain, starting from the farm to processing factory becomes energy self-sufficient. The project can be registered as a CDM project and thus earning additional revenue from carbon credit sales. Other industries or supply chains that are willing to pay premium prices for energy produced in a sustainable way can also be targeted. The bio-fertilizer produced can be used by all the participating farms as bedding for their animals or the excess sold as fertilizer and soil amendments to other farms. The drawback of centralized plants is the costly process of transporting livestock manure, and hence, a well-structured logistics is critical for the success of the business.

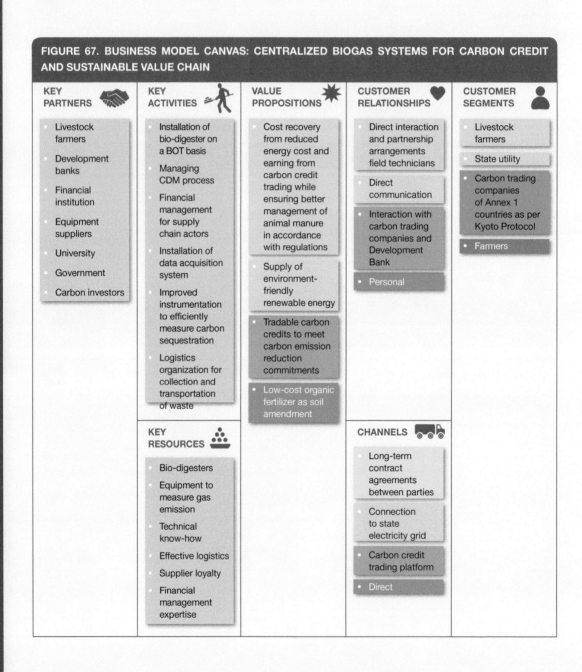

FIGURE 67. BUSINESS MODEL CANVAS: CENTRALIZED BIOGAS SYSTEMS FOR CARBON CREDIT AND SUSTAINABLE VALUE CHAIN

BUSINESS MODEL 5: POWER FROM MANURE

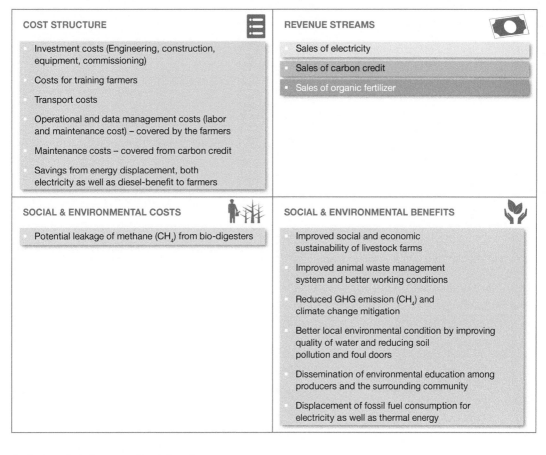

D. Potential risks and mitigation

This section describes the potential risks and mitigation options for power from manure for carbon credit and sustainable value chain.

Market risks: The outputs from this business model are carbon credits sold in the international market, energy used by livestock farms and surrounding communities and bio-fertilizer used on farmers' own land. Market risks exist for the carbon credits as the carbon credit market is volatile which puts the sustainability of the whole reuse business under risk. Thus, the business has to diversify its revenue streams to sale of power, thermal energy and bio-fertilizers to mitigate market risks. For instance, instead of putting bio-digesters in every farm, the farmers can form a cooperative and build a centralized biogas system (alternative scenario), which collects all the manure from member farmers, processes the manure, produces and sells electricity to the national grid. The energy produced can also be supplied to the processing factories and distributed to member farmers. This ensures market for the electricity produced and also the entire supply chain, starting from the farm to processing factory becomes energy self-sufficient. Moreover, it will allow safety monitoring as well as quality control to be centralized if and where required.

Competition risks: In implementing this business model, the processing factory is incorporating environmental sustainability into its revenue design. The risk associated with output market is low. The carbon credits are sold in the international market. In the scenario where centralized biogas systems produce electricity at a large scale, competition risk can be reduced by entering into a long-term power purchase agreement with the state utility, hence ensuring a ready buyer. The electricity can also have a ready buyer when it is supplied to the processing factory and used within the farms. Moreover, other industries or supply chains that are willing to pay premium prices for energy produced in a sustainable way can be targeted.

Technology performance risks: The technologies applied for processing livestock waste are well-established and mature technologies. However, the technologies require skilled manpower to operate and maintain them. Maintaining the performance of the technologies at the standard level is very critical for the economic and environmental viability of the business as the business heavily depends on earnings from carbon credit sales. Farmer's lack of technical know-how to operate bio-digesters may result in leakages of CH_4 which will significantly reduce the carbon credit earnings. The centralized large scale plant will be easier to operate and maintain by skilled labor, which may be difficult at individual farms.

Political and regulatory risks: With the projected electricity demand set to grow, governments are encouraging green power initiatives by putting in place various incentive mechanisms such as concessional loans, feed-in tariff mechanisms and through long-term power purchase agreements. However, it is not advisable to entirely depend on government incentive mechanisms to ensure sustainability of the business. In order to ensure economic viability, the business should diversify its customer base. This can be done by supplying part of the electricity produced to the processing factory and other industries that are willing to pay premium prices for energy produced in a sustainable manner. However, this will also depend on the electricity regulation of the region where the business is operating.

Social-equity-related risks: The beneficiary of the model may change depending on the end use of the energy generated from manure. In the case of rural electrification, underserved communities are the beneficiary while in the case of livestock industry, power is generated for own use. The model offers employment opportunities which could be provided to informal laborer to mitigate any social equity risks the business model may create.

Safety, environmental and health risks: The environmental risks associated with the bio-digesters include possible leakage of CH_4. The safety and health risks to human arise when processing livestock waste.

Proper protection measures should be put in place to protect laborers (Table 21).

BUSINESS MODEL 5: POWER FROM MANURE

TABLE 21 POTENTIAL HEALTH AND ENVIRONMENTAL RISK AND SUGGESTED MITIGATION MEASURES FOR BUSINESS MODEL 5.

RISK GROUP	EXPOSURE					REMARKS
	DIRECT CONTACT	AIR	INSECTS	WATER/SOIL	FOOD	
Worker	HIGH	LOW				Direct contact risk relates to pathogens in livestock manure.
Farmer/user	LOW	LOW				
Community				LOW		
Consumer					LOW	
Mitigation Measures	🧤⚡	😷🎧	🦟	🌊🚱	🚿	

Key: ☐ NOT APPLICABLE ▨ LOW RISK ▨ MEDIUM RISK ■ HIGH RISK

E. Business performance

This business model is rated as high on profitability followed by environmental benefit (Figure 68). The business model is expected to result in a significant reduction of GHG emissions, which consequently is translated into carbon credit sales. Moreover, it is expected to result in promoting sustainable livestock production, generating environmental and social benefits.

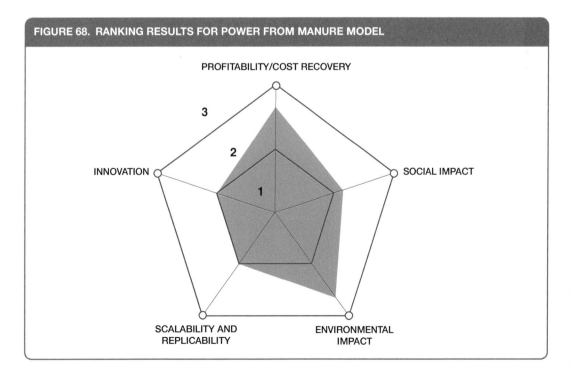

FIGURE 68. RANKING RESULTS FOR POWER FROM MANURE MODEL

The business model has a potential to be implemented in regions where there are intensive livestock farms and where there is government support for CDM projects. Designing of innovative financing mechanisms and having access to finance are essential for the successful implementation of and ensuring sustainability of this business model.

CASE
Power from agro-waste for the grid (Greenko, Koppal, India)

Krishna C. Rao, Binu Parthan and Kamalesh Doshi

Supporting case for Business Model 6	
Location:	Marlanhalli, Koppal, Karnataka, India
Waste input type:	Agro-waste
Value offer:	Power
Organization type:	Private
Status of organization:	Greenko incorporated in 2006 and Ravikiran project operational since 2005 but was acquired by Greenko in 2007
Scale of businesses:	Medium
Major partners:	Investors, like e.g. Blackrock Investment Management, Aloe Private Equity, Impax Asset Management

Executive summary

Greenko Group, a environmentally-driven private Indian company, is an independent power producer utilizing new approaches to clean power, using proven, technologically-advanced systems and processes. Greenko Group has built a portfolio of clean energy projects with risk management strategies through technologically and geographically-balanced portfolio, cluster approach, leveraging the carbon market, raising finances from capital markets, financial institutions and sovereign wealth funds and balancing greenfield and selective acquisitions to generate and sell electricity to the state-owned energy utilities as well as private clients. In the financial year 2006–2007, the business acquired two biomass plants and a 50% interest in a third, including Ravikiran Power project.

The Greenko's 7.5 MW Ravikiran Power project in Marlanhalli, Karnataka state, India was commissioned in June 2005 and buys low-cost agro-waste from local farming villages using large number of biomass supply intermediaries to generate and sale electricity at pre-announced tariffs to regional electricity grid. The Greenko Group is also a part of the CDM process and generates and sells Certified Emission Reductions (CERs), Voluntary Emission Reductions (VERs) and Renewable Energy Certificates (RECs). The project has a significant positive impact on both the local community and environment from carbon offsets and increasing the incomes of local farmers. The company maintains a continuous involvement in localized projects and community programs which centres on education, health and wellbeing, environmental stewardship and improving rural infrastructure.

KEY PERFORMANCE INDICATORS (AS OF 2014)	
Land use:	NA
Water requirement:	50 m³/hr
Capital investment:	USD 6 million
Labor:	9 full-time employees
O&M cost:	USD 2 million/year (including fuel costs)
Output:	46 Gigawatt hours (GWh) net electricity generation at 7.5 Megawatt (MW) output level
Potential social and/or environmental impact:	Nine jobs, 37,468 tCO$_2$eq/year carbon mitigation by avoiding of waste build-up and anaerobic conditions for agro-residues. Jobs created also for biomass collection and transport, an income generation for the local population by sale of agro-residues, significant indirect investment in the region by way of roads schools and civic amenities
Financial viability indicators:	Payback period: 4.7 years Post-tax IRR: 16% Gross margin: 28%

Context and background

Greenko Group was formally incorporated in 2006, founded by Anil Chalamalasetty and Mahesh Kolli, listed on the LSE and raised with a start-up capital through an initial public offering. The main operations of the group are based in India, predominantly in the central and southern states of Andhra Pradesh, Karnataka and, more recently, Chhattisgarh. The group was formed as a vehicle to take advantage of the opportunities for consolidation of the Indian renewable energy market and operate in the two markets of renewable energy supply and CER units provision. Seven of Greenko's projects generate electricity from agro-residues. The company has 289 MW of clean energy capacity from hydro, wind, gas and biomass energy. The company also has a number of projects under development and acquisition and plans to reach 1000 MW capacity in 2015 and 3 GW by 2018. In 2013, Greenko had 309 MW of power generation capacity with 51 MW being commissioned, 446 MW under construction and 1,529 MW of projects under active development.

Ravikiran Power is a 7.5 MW biomass project located at Devinagar Camp, Kampli Road, Gangavathi Taluk of Koppal District, Karnataka. The Ravikiran Power project's location was selected after surveys which indicated adequate availability of the agro-residue, primarily rice husk used by the project as well as proximity to an electrical sub-station for selling the energy generated. Koppal, Raichur and Bellary, which is also called rice bowl of Karnataka along the river Tungabhadra. The project buys 157 tons of agro-residues from local farming villages through a large number of biomass supply intermediaries and uses to assure regular supply at competitive process, providing an income-generating opportunity and a waste management solution. The project also provides employment to local villagers. The project uses rice husk, groundnut shell and bagasse as the major fuel and has a travelling grate, multi-fuel fired boiler. The electricity is sold to Gulburga Electricity Supply Company (GESCOM), which is a state-run regional electricity distribution company.

Ravikiran Power Projects Ltd, the subsidiary of Greenko Group, had entered into a PPA with Gulbarga Electricity Supply Co Ltd (GESCOM) for a period of 20 years, but the PPA was mutually terminated for the year ended March 31, 2013. The company is in discussions with various industrial and commercial customers in the state of Karnataka for the offtake of power generated by Ravikiran Power. No sales of power were made in relation to Ravikiran Power for the year ended March 31, 2014.

Market environment

The main offtaker of the electricity is the state utility. However, it is possible to sell the power directly to other 1 MW electricity consumers using the state's grid network as per the Electricity Act 2003. Such a third-party sale is a financially more attractive proposition due to the economies being driven by avoided cost of electricity supply, albeit it is slightly riskier in realization.

As a result of economic growth, the energy consumption in the country and state are increasing and the market for electricity is growing. The share of the market for the project at the regional level, the market share, is 0.02%. The total potential of biomass power in the state of Karnataka is 1,500 MW for cogeneration using sugarcane bagasse in addition to 1,000 MW from agro-residues. The state of Karnataka is linked to the southern regional electricity grid, which has power and energy deficits and the energy distribution utilities have to resort to cyclical load shedding.

CERs are a type of emissions unit (or carbon credits) issued by the CDM Executive Board for emission reductions achieved by CDM projects and verified by a Designated Operational Entity (DOE) under the rules of the Kyoto Protocol. Methodologies are required to establish a project's emissions baseline, or expected emissions without the project, and to monitor the actual on-going emissions once a project is implemented. The difference between the baseline and actual emissions determines what a project is eligible to earn in the form of credits. One CER equates to an emission reduction of one ton of CO_2 equivalent. Holders of CERs are entitled to use them to offset their own carbon emissions as one way of achieving their Kyoto or European Union emission reduction target. In August 2008, prices for CERS were USD 20 a ton. By September 2012, prices for CERS had collapsed to below USD 5.

The emergence of a secondary market for VERs outside the Kyoto Protocol is driven by corporations and individuals looking to reduce voluntarily their carbon footprint. VERs arise from projects awaiting CDM clearance, special situations (e.g. carbon capture and storage) or smaller projects.

Macro-economic environment

Electricity demand in India is likely to reach 155 GW by 2016–17 and 217 GW by 2021–22 whereas peak demand will reach 202 GW and 295 GW over the same period, respectively. At the national level, India faces an energy shortage of 10.6% and a peak power shortage of 15 GW.

Renewable energy in India comes under the purview of the Ministry of New and Renewable Energy. In order to address the lack of adequate electricity availability to all the people in the country by the platinum jubilee (2022) year of India's independence, the Government of India has launched a scheme called 'Power for All'. This scheme will ensure that there is 24/7 continuous electricity supply provided to all households, industries and commercial establishments by creating and improving necessary infrastructure. A tenfold increase in solar installation rates to 100 GW by 2022, trebling to 60 GW of new wind farms, 10 GW of biomass and 5 GW of small-scale run-of-river hydro has been targeted. In addition, India's private sector has set targets to increase its use of clean energy as a part of global efforts to reduce greenhouse gas emissions in a bid to give a push to clean energy projects in the country.

The country and the state have policies, fiscal and financial incentives to encourage independent power producers, especially those that are using renewable fuels such as agro-residues. The Government of India has provided tax incentives to renewable energy generators in the form of 10-year tax holidays from the usual rate of corporation tax, accelerated tax depreciation of assets and other fiscal incentives. In addition, the government has decided to waive transmission charges for electricity generated from renewable sources. This has been encouraged at state level by the implementation of a tariff structure, which provides base income under long-term PPAs, typically with terms of between 15–20 years. Generation SEBs and state-owned schemes are still the dominant suppliers of energy but independent power producers and captive plants (owned by the end user) have grown significantly since liberalization of the industry began.

Business model

Ravikiran Power has two key value propositions – generation of environment-friendly electricity from crop residues and sales of carbon offset generated by the project (Figure 69). Ravikiran Power project sells the power directly to a state-run electricity distribution company. The project has the potential to have another revenue source through sales of fly ash to brick industry.

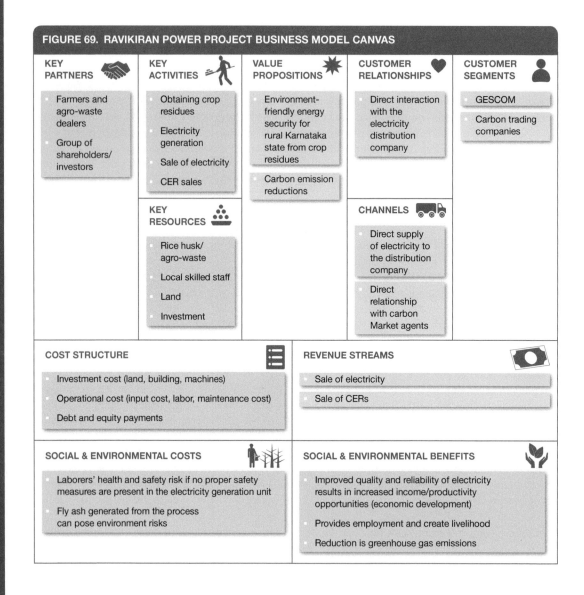

FIGURE 69. RAVIKIRAN POWER PROJECT BUSINESS MODEL CANVAS

Value chain and position

The value chain of the business consists of agricultural production, collection segregation, transport and sales of agro-residues, conversion of agro-waste to energy, transmission of energy, and distribution and sales of energy (Figure 70). The critical relationships the enterprise has are with the agro-residue suppliers who ensure availability of adequate quantity of biomass with good quality at prices that make the business viable. The Ravikiran Power project has partnered with multiple biomass dealers who procure agro-residues from local farmers. The other significant and important relationship the enterprise has is with investors who finance the capital and operating costs and who expect a healthy return on their investments. The project's relationship with a local distribution company that buys the electricity at a contract price plays a significant role in keeping the operation costs of the company lower. However, with only a single buyer of the electricity and also with other energy sources available in the market, there could be a significant threat from substitutes, but with the electricity shortage, it is not applicable. In addition, the project will generate 33,000 CERs per year. A flow-chart illustrating these relationships are given below.

The project is able to offer a good price for agro-residues which otherwise has a low economic value and is able to contribute to meeting the energy and power deficit in the electrical grid, providing additional developmental benefits. The catchment area of the project has enough agro-residues – primarily rice husk and also groundnut shell, arecanut husk and plywood waste. Greenko has secured a proportion of its feedstock supply locally to reduce transport costs through a combination of building relationships with local suppliers and developing its own fast-growing stocks on surplus or adjoining land at its biomass plants. The project processes about 157 tons/day of dry waste with less than 10% moisture content with specific consumption of about 1.23 kgs of waste/kWh. The cost of feedstock used to run a biomass plant represents approximately 50% of power revenue generated. Over time, the company intends to grow up to approximately 30% of its feedstock requirements. This should reduce Greenko's dependency on external providers of feedstock and reduce fuel supply risks.

The fly ash is a lightweight particle captured in exhaust gas by electrostatic precipitators installed before flue gas chimney at the plant. Fly ash is very fine with cement-like properties and has long been used as an additive in cement, concrete and grout, as filler material and as ingredient for bricks. Fly ash bricks now account for about one-sixth of India's annual brick production, saving energy, soil and carbon emissions and putting a toxic waste product to beneficial use.

Institutional environment

The legal and regulatory framework, with implementation of Electricity Act 2003, National Electricity Policy, the National Tariff Policy and the Accelerated Power Development and Reform Program, is conducive for waste-to-electricity projects in Karnataka and generally in India. Electricity generation projects do not require to obtain license to set up, have guaranteed open access to the grid and offered an attractive electricity purchase price and also open access to consumers over 1 MW. There are no regulatory issues relating to the sourcing of the agro-residues. The policy of the Karnataka state environmental agency – Karnataka State Pollution Control Board (KSPCB) covers hazardous waste, battery waste, e-waste and plastics and does not cover agro-residues that are the waste input to the power plant. However, KSPCB environmental policy does cover fly ash, which the plant generates.

The Ministry of New and Renewable Energy has implemented a wide range of programs for the development and deployment of biomass-based power generation. To encourage investment in the sector, fiscal and financial incentives have been provided that include capital/interest subsidy, accelerated depreciation, concessional duties and relief from taxes, apart from preferential tariff for grid power being given in most potential states. To facilitate the development of renewable energy sources

FIGURE 70. RAVIKIRAN POWER PROJECT VALUE CHAIN

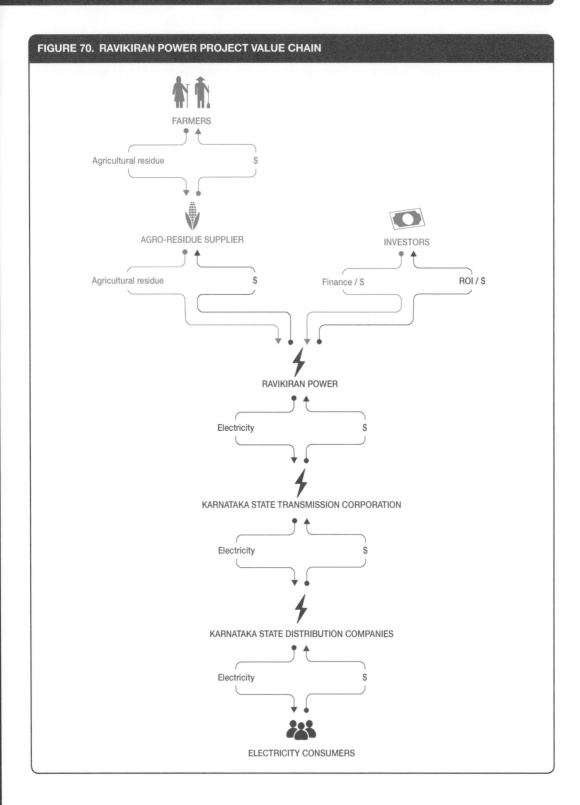

in the state, the Government of Karnataka established Karnataka Renewable Energy Development Ltd (KREDL) on March 8, 1996. KREDL will be responsible for laying down the procedure for inviting of proposals from Independent Power Producers (IPPs) and DPR and evaluation of project proposals, project approvals, project implementation, operation and monitoring. Single-window clearance will be provided. Gulbarga Electricity Supply Company Ltd (GESCOM) has the responsibility for the distribution of electricity in Bidar, Gulbarga, Yadgir, Raichur, Koppal, Bellary districts and purchases power from the project.

Technology and processes

The technology used for waste-to-electricity conversion is the steam cycle with biomass combustion which is proven and used in a large number of thermal (coal, nuclear and biomass) power plants around the world. Biomass combustion is generally suitable for large capacity (few MW and above) and grid-connected applications, whereas biomass gasifier and bio-methanation-based power plants are more suitable for sub-megawatt level and decentralized applications. The agro-residues are combusted in a multi-fuel, travelling grate-water tube boiler to produce steam. The steam passes through a condensing, impulse turbo generator, which generates electricity which is exported at 110 kV to the state grid network. The risks associated with the technology have primarily to do with management of high-temperature steam in the boiler and in the turbo generator. The technology and equipment was locally sourced and spare parts, technicians, operators and technical expertise to service the technology is locally available.

The biomass power projects are normally designed to handle multi-fuel (biomass) along with conventional fossil fuels like coal. Due to poor bulk density, the volume of biomass to be stored and handled poses challenges. The lighter particles of ash flying in furnace get carried away through flue gases and get collected in hoppers below economizer, air pre-heater, duct and electrostatic precipitator. From these hoppers the ash is either collected manually on tractor/lorries or dry ash conveying is carried out. See Figure 71 for Ravikiran Power project's process flow.

Funding and financial outlook

Greenko was incorporated in 2006 and raised its equity through initial public offering by getting listed on the London Stock Exchange. Greenko Group has raised an equity of USD 1.8 million and debt of about USD 4.18 million and plans to raise further its equity in the near future. Ravikiran Power project is one of Greenko's electricity generation projects under its portfolio financed from the debt and equity raised. See Table 22 on financials. In addition, the Ravikiran Power project also benefits from an interest subsidy of 2% on debt from public financial institutions. The total investment cost of the project is about USD 6 million, and plant and machinery is the major cost. The annual total cost of production incurred by the project is about USD 2 million. The fuel and agro-residue cost forms 50% of its operations, followed by debt payment and labor and maintenance. The Ravikiran Power project pays USD 15/ton of agro-residue.

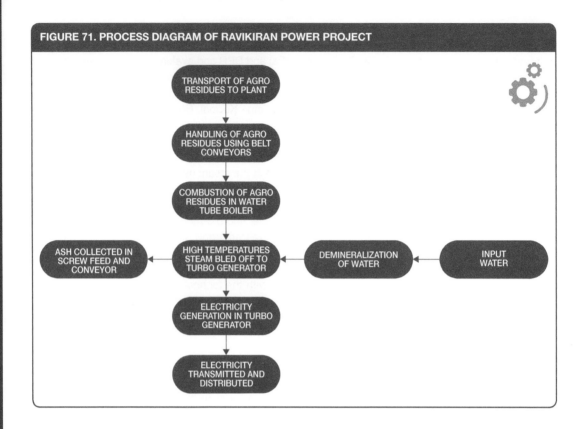

FIGURE 71. PROCESS DIAGRAM OF RAVIKIRAN POWER PROJECT

TABLE 22. GREENKO FINANCIALS (IN MILLION USD)

Key capital costs (2006)	Land building and civil costs	0.56
	Plant and machinery	4.5
	Miscellaneous fixed assets	0.16
	Preliminary and pre-operative expenses	0.6
Operating costs	Fuel/agro-residue costs	1.0
	Operation and maintenance	0.24
	Interest on debt	0.49
	Depreciation	0.29
Financing options	Equity from equity investors	1.8
	Debt from the state bank of India	2.0 at 12%/year interest
	Debt from HUDCO	2.18 at 12%/year interest

Ravikiran Power project has two revenue sources – energy sales and CER sales. Revenue generated from annual CER sales is about USD 0.24 million for 37,468 tCO_2eq/year carbon mitigation. Its annual revenue from sales of energy is an upwards of USD 2.53 million. Taking assumption of net revenue increase of USD 0.05 million every year, Greenko has a payback period of 4.71 years on its investment made in Ravikiran Power project. The investment has an internal rate of return of 16% and gross margin of 28%.

Socio-economic, health and environmental impact

The Ravikiran Power project has a significant economic development impact to the region both from improving the local availability of electricity and by infusing money into the local economy especially to farmers by offering a value of about 15 USD/ton of agro-residue and contributing USD 1 million to the local economy. The project has also created direct employment for nine people and resulted in local job creation in biomass residue collection and transport, in shops and restaurants and in the area. The electricity generated will primarily be used in the local area which can power local agro-industries and commercial establishments. The project directly reduces greenhouse gases which otherwise would have been emitted in the absence of the project. Applying the UNFCCC-approved methodologies for baseline setting, monitoring and verification of emission reductions, it was found that the project resulted in emission reductions of 37,468 tCO_2eq/year. The project has a role in regulating ecosystem services by mitigation of climate change through emission reductions.

Scalability and replicability considerations

The key drivers for the success of this business are:
- Management team with extensive project development, financial and corporate management skills.
- Location of plant near low-cost crop residue and sub-station to sell electricity.
- Support from the Indian government for independent power producers.
- Incentives for renewable energy.
- Electricity shortage and unserved areas.

Greenko has already replicated similar biomass energy projects in six other locations with plants ranging in capacity from 6–8 MW, totalling 34 MW. There are opportunities for scaling up the business but the opportunity would be constrained by the availability of agro-residues within the catchment area of the business. However, the business has considerable scope for replicating the projects in other parts of India.

Summary assessment – SWOT analysis

The key strength of Ravikiran Power project is its resource expertise in managing the power plant (Figure 72). The business has limited weaknesses and it should continue to reduce its reliance on a specific crop residue, e.g. its reliance on rice husk dominates over other crop residues. The business threat is from biomass availability and price fluctuation of biomass residues. The business has an opportunity to replicate the projects in other regions, and in addition, the government of India provides a favourable policy and regulatory environment for independent power producers.

FIGURE 72. RAVIKIRAN POWER PROJECT SWOT ANALYSIS

	HELPFUL TO ACHIEVING THE OBJECTIVES	**HARMFUL** TO ACHIEVING THE OBJECTIVES
INTERNAL ORIGIN ATTRIBUTES OF THE ENTERPRISE	**STRENGTHS** • Cluster approach to project development to optimize use of resources and expertise • Combination of greenfield and selective acquisitions to grow the portfolio of projects • Effective resource mobilization strategy from the capital markets, institutional investors and sovereign wealth funds • Plant location closer to both input and customer. • Avoidance of GHG CO_2 emissions • Positive view of the project by local community • Additional income for rural farmers • Nucleus for other economic activities	**WEAKNESSES** • Plant breakdowns and loss of operating time and revenues • Complex material handling due to different types of agro-residues • Changes in feedstock prices and its seasonal availability could have a significant impact on profitability • Shortage of skilled manpower • Weak biomass availability assessments
EXTERNAL ORIGIN ATTRIBUTES OF THE ENVIRONMENT	**OPPORTUNITIES** • Favourable policy and regulatory environment and incentives for independent power producers in India • Attractive power purchase rates for electricity generated from renewable energy sources • Availability of agricultural waste in the catchment area of the power plant • Availability of financing organizations • Direct third-party sale • Continuing high demand for energy	**THREATS** • Low demand and very low prices for emission reductions from late 2012 onwards • Fluctuations in availability and prices of biomass residues for the power plant • Depreciation of the Indian currency against Euro • Uncertainty in tariff policies • Poor grid stability with prolonged forced or planned outages • Long delays in issue of permits/approvals by authorities • Currency non-convertibility or instability • Channeling of agro-waste for other uses

References and further readings

Greenko Group. Welcome to Greenko Group. www.greenkogroup.com/about#company-profile (accessed August 18, 2017).

Greenko Group. Clean fuel plants 05: Ravikiran Power. www.greenkogroup.com/farms/clean-fuel-plants?page=2 (accessed August 18, 2017).

Greenko Group. Bondholder information: Financial statements. www.greenkogroup.com/investor-relations (accessed August 18, 2017).

Case descriptions are based on primary and secondary data provided by case operators, insiders, or other stakeholders, and reflects our best knowledge at the time of the assessments (2015–2016). As business operations are dynamic, data can be subject to change.

CASE

Power from rice husk for rural electrification (Bihar, India)

Krishna C. Rao, Hari Natarajan and Kamalesh Doshi

Supporting case for Business Model 6	
Location:	Bihar, India
Waste input type:	Primarily rice husk; Currently testing other biomass waste
Value offer:	Electricity to households and small businesses and biochar
Organization type:	Private
Status of organization:	Operational since 2007
Scale of businesses:	Medium
Major partners:	Shell Foundation, Acumen Fund, Bamboo Finance, International Finance Corporation, LGT Philanthropy, CISCO, Ministry of New and Renewable Energy, Government of India, Farmers

Executive summary

Founded in 2008, Husk Power Systems Inc. (HPS) is promoted by first generation entrepreneurs Gyanesh Pandey, Ratnesh Yadav, Manoj Sinha (natives of Bihar, India) and Charles Ransler (USA). It has won several business plan competitions and secured foundation grants in the United States. As a rural empowerment enterprise, HPS has a mission to provide renewable and affordable electricity to rural people in a financially sustainable way. Most of rural Bihar suffers from poor access to modern energy with majority of households relying on either kerosene for lighting or other low-quality energy source, such as candle or batteries. HPS owns, installs, operates and manages decentralized rice husk/biomass gasifier-based 25–100 kW generation and distribution systems to deliver lighting and electrification services to 200–600 households on a "fee for service" basis to households and 5–10 irrigation pump sets and small businesses in rural Bihar. HPS procures rice husk/feedstock at negotiated rates. The consumers prepay a fixed monthly fee, ranging from USD 2–3 to light two fluorescent lamps and one mobile charging station which is at least 30% cheaper than the cost of kerosene and diesel and enables savings of up to USD 50 for each household every year. HPS uses a franchisee-based business model and uses three distinct approaches to deliver electricity services: a) Build-Own-Operate (BOO) b) Build-Own-Maintain (BOM) – operation is managed by a local partner or entrepreneur and c) Build-Maintain (BM) – a local partner/entrepreneur owns and operates the plant. At the time of this assessment, the company had more than 84 plants, enough to provide electricity to over 250,000 people across 300 villages and hamlets and employing 350 people across the state of Bihar.[1]

KEY PERFORMANCE INDICATORS (AS OF 2014)	
Capital investment:	USD 1,300/kW
Labor:	Full-time: 3; Part-time: 5–10
O & M cost:	Estimated to be less than USD 0.15 /kWh
Output:	25–100 kW of electricity
Potential social and/or environmental impact:	Each unit serves about 200–600 households, 5–10 irrigation pump sets and small businesses; improved energy access and cleaner local environment, reduction in GHG emissions, employment generation
Financial viability indicators:	Payback period: 6–8 years; Post-tax IRR: N.A.; Gross margin: 45%

Context and background

Despite significant efforts and resources deployed by the Government of India towards rural electrification, about 480 million citizens residing in about 125,000 villages in India (45% of the total population) do not have access to reliable power. Of those who do, almost all find electricity supply intermittent and unreliable. When grid rationing takes place, villages often receive power only after midnight when "priority" demand from cities and industry is low. This is of little use to rural households and businesses. The Indian government has designated several thousand villages as "economically impossible" to reach via conventional grid. Without electricity, these villagers are forced to live at the whim of natural forces and lack basic communication, education and healthcare infrastructure. Common energy supply options, such as kerosene lanterns or diesel generators, are uneconomical, inefficient and environmentally unfriendly.

The state of Bihar is third largest with 82.9 million population and 12th largest with 94,163 km² geographical size in India. Only 52.8% of villages and 6% of households of the state are electrified, leaving about 85% of the population with no access to electricity. Even the villages connected to grid have frequently interrupted poor quality of power supply. However, Bihar is blessed with fertile soil and good rainfall. It has several geographic and climatic advantages to harness renewable energy. The decentralized electricity generation is the possible solution to reduce transmission losses and to provide electricity to densely populated villages with scattered but large number of small-scale commercial activities. The decentralized power generation can make use of readily-available biomass, which is typically transported out of state. Bihar is a part of the rice belt of India, producing about 4.7 million metric tons of rice per year, generating about 1 million tons of rice husk which is underutilized and is a good source for fuel. Each 32 kW gasifier installed by HPS requires approximately 60 kg of rice husk per hour or 15,000 kg rice husk per month assuming eight hours of operation per day. It was in this context that HPS initiated its operations in 2007, using rice husk as fuel to generate electricity to provide safer, better and cleaner lighting solution at an affordable cost to rural households in Bihar.

Market environment

HPS has identified 25,000 villages as feasible sites within India's rice producing area (Bihar and neighbouring states) for its projects. Promotion of the plants is largely by word-of-mouth and also through local press and media, and their benefits are now well known in Bihar. HPS receives several hundred enquiries about installations each year.

While the minimum services offered by HPS to a household is two light connections and one mobile charging point, a small percentage of households in each mini-grid request and obtain additional supply to power household appliances, such as a fan, television, radio, etc. The cost of services offered by HPS is significantly higher than that of the state utility, but the grid is practically non-existent across Bihar. Even if the grid were to penetrate these areas in future, HPS plants can feed its energy to the grid with minimal additional investment. Also the state of Bihar has numerous private diesel-generator-run electricity providers whose service provision is similar to services offered by HPS. With rising diesel prices, in the long run, it would be difficult for them to compete with HPS.

The residue from gasification is a carbon-rich ash, or biochar, is rich in alkaline components (Ca, Mg and K) high in silica, and this may contribute to the neutralization of soil acidity and to a decrease in the solubility of the phytotoxic metals such as aluminium in the soils. Biochar can bind and release nutrients (N, P, K and Ca) and could reduce nutrient leaching to the subsoil. It also retains water in soils with low plant-available water and helps draining flood-prone areas. It can be used to improve the fertility of the soil for growing rice or vegetables. *Biochar also has appreciable carbon sequestration value. These properties are measurable and verifiable in a characterization scheme, or in a carbon emission offset protocol.*

Macro-economic environment

India has been promoting biomass gasifier technologies in its rural areas to utilize surplus biomass resources such as rice husk, crop stalks, small wood chips and other agro-residues. The goal was to produce electricity for villages with power plants of up to 2 MW capacities. During 2011, India installed 25 rice husk-based gasifier systems for distributed power generation in 70 remote villages of Bihar.

The Electricity Act 2003 de-licenses power generation completely and the techno-economic clearance from the Central Electricity Authority (CEA) has been done away with for any power plant, except for hydroelectric power stations above a certain amount of capital investment. The Independent Power Producers (IPP) can sell electricity to any licensees or where allowed by the state regulatory commissions to consumers directly. However, the act provides for imposition of a surcharge by the regulatory body to compensate for some loss in cross-subsidy revenue to the SEBs due to this direct sale of electricity by generators to the consumers. As per the Act, 10% of the power provided by suppliers and distributors to the consumers has to be generated using renewable and non-conventional sources of energy so that the energy is reliable.

The Government of India launched the Rajiv Gandhi Grameen Vidyutikaran Yojana (RGGVY) – Programme for creation of rural electricity infrastructure and household electrification in April 2005 for providing access to electricity to rural households. As on 30.04.2012, against the targeted coverage of 1.10 lakh[2] un-electrified or de-electrified village and release of free electricity connections to 2.30 crore[3] Below Poverty Line (BPL) households, electrification works in 1.05 lakh un-electrified or de-electrified villages have been completed and 1.95 crore free electricity connections to BPL households have been released under RGGVY. Under RGGVY, electrification of un-electrified BPL households is provided free electricity service connection. Infrastructures created under RGGVY can be used for providing connections to Above Poverty Line (APL) by respective distribution utilities by prescribed connection charges, and no subsidy is available for this purpose. On one side, the program improves access to energy, while on the other side, it creates further problems for India's electricity sector. In addition to RGGVY, there are subsidies provided by the Ministry of New and Renewable Energy for renewable energy projects such as biomass gasification.

Business model

The business offers multiple value propositions, and the primary value proposition is to provide high-quality electricity service to household and businesses in rural areas that have either no access to electricity or it is unreliable (Figure 73). The enterprise uses rice husk from rice farmers and rice mills to generate electricity using biomass gasification technology. The enterprise partners with local community and the government. HPS uses a franchisee-based business model through three distinct approaches to deliver electricity services: a) Build-Own-Operate-Maintain (BOOM), b) Build-Own-Maintain (BOM) – operation is managed by a local partner or entrepreneur and c) Build-Maintain (BM) – a local partner/entrepreneur owns and operates the plant.

Value chain and position

The rice husk (or hull) is the outermost layer of the paddy grain that is separated from the rice grains during the milling process. Around 20% of paddy weight is husk which is largely considered a waste product with no commercial value and is often burned or dumped in the rivers or on landfills. As per estimates, about 1.8 billion kg of rice husk are produced every year in Bihar. The franchise partners procures rice husk from local rice farmers. HPS is dependent upon farmers and rice mills for rice husks and to mitigate any potential shortfall, HPS reaches out to more farmers/rice mills (Figure 74).

The cost of rice husk is approximately USD 0.02–0.025/kg. HPS faced significant challenge in procuring rice husk for the gasifier for a suitable price. At one point suppliers – rice mill operators and farmers – started demanding higher price. HPS countered this by establishing its own rice mills, where it offered milling services at no cost in return for the rice husk. This forced the suppliers to enter long-term contracts at a fixed price. HPS is exploring other input feed stocks such as wheat husk, mustard stems, corn cobs, wood chips, etc. HPS business suffers from substitutes and new entrants.

HPS's value proposition lies in making the plants so simple to operate and maintain that high-school-educated people from the village can be trained to manage and run them. Tars and other particulates in the producer gas can damage equipment, in particular engines, so a key factor for successful operation is the rigorous HPS maintenance program. HPS also requires high safety standards and detailed monitoring. It is through this attention to maintenance and monitoring that HPS plants achieve over 93% availability.

Electricity fees start at USD 2.2 per month for a basic connection of two lights and one mobile phone charging. "Pay for use" service approach is being followed by HPS for raising revenue and supplying electricity at a low cost. Low-cost prepaid meters have been installed that can efficiently regulate the flow of low-watt electricity and reduce electricity theft to less than 5%. One month's deposit is required when a customer signs the supply contract with HPS. The local HPS collector goes from house to house to collect the fee each month in advance and checks that everything is working well. All complaints are logged and followed up. Under the terms of the contract, HPS agrees to provide service for at least 27 days every month and pro-rates the fees if this level is not met. However, average provision is now over 28 days per month (93% availability). All customers are trained in safe use of electricity. The biochar, the residual waste from the plant, is used in making incense sticks, rubber and manure. About 1,200 women have been employed in incense-stick manufacturing.

HPS has set up a first-of-its-kind 'Husk Power University'. The university will serve as a training facility where new recruits and existing staff will be trained in large engine repair and maintenance, facility management and continuous improvement processes. It will help in job creation for Bihar youths, particularly those living in rural areas, and also in enhancing health and safety conditions at the existing operational sites located in rural areas. HPS has developed significant public support in the local community since it not only provides access to electricity but also creates local employment opportunities either through direct employment in the plant or in the making of incense sticks.

CASE: POWER FROM RICE HUSK FOR RURAL ELECTRIFICATION

FIGURE 73. HUSK POWER SYSTEMS BUSINESS MODEL CANVAS

KEY PARTNERS
- Local community and businesses
- Entrepreneur
- Government of India, donors and private investors to fund the enterprise
- Rice farmers and rice mills in Bihar

KEY ACTIVITIES
- Obtaining the rice husk
- Biomass gasification for electricity generation
- Electricity distribution
- Bill collection
- Community mobilization
- Maintenance of existing plants
- Franchise management
- Selling of biochar

VALUE PROPOSITIONS
- Residents and businesses of rural Bihar get higher quality electricity services for a fair price
- Opportunity to start a business based on proven technology in partnership with a successful company
- Additional income source for households making incense stick from biochar
- Carbon offset

CUSTOMER RELATIONSHIPS
- Direct interaction with the users
- Direct relationship with franchisee, technology transfer, maintenance services, and continuous support
- Direct

CUSTOMER SEGMENTS
- Rural residents and businesses in Bihar
- Local entrepreneurs
- Incense stick makers
- Carbon trading companies

KEY RESOURCES
- Developed technology
- Rice husk/agro-waste
- Partnerships
- Local staff
- Mini-grid

CHANNELS
- Mini-grid maintained by HPS
- Direct contact to reach potential franchisee
- Biochar from gasification process of rice husk is given away by HPS to businesses making incense stick

COST STRUCTURE
- Investment cost (land, building, machines)
- Operational cost (input cost, labor, maintenance cost)
- Debt repay and equity

REVENUE STREAMS
- Sale of electricity
- Subsidies
- Build – Maintenance fee from franchise
- Sale of biochar for incense stick making
- Sale of CO2 offset

SOCIAL & ENVIRONMENTAL COSTS
- Potential laborers' health and safety risk if no proper safety measures are present in the electricity generation unit
- Potential safety issues on handling of biochar

SOCIAL & ENVIRONMENTAL BENEFITS
- Improved quality of electricity results in increased income/productivity opportunities (economic development) and healthier environment through reduction on kerosene usage for lighting
- Provides employment and create livelihood
- Reduction is greenhouse gas emissions

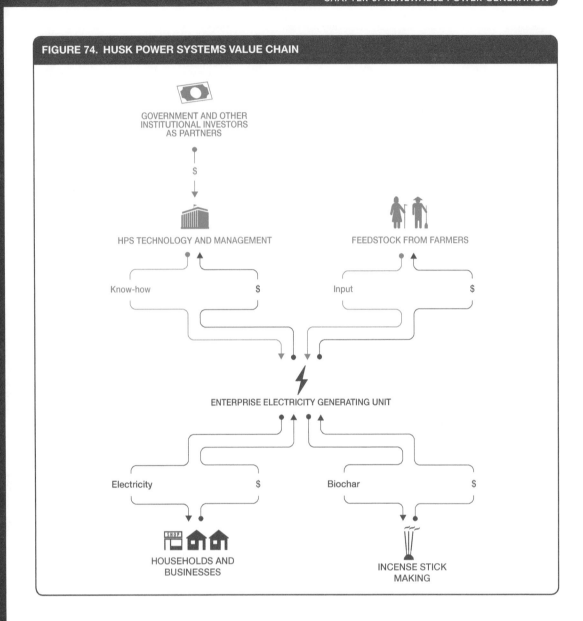

FIGURE 74. HUSK POWER SYSTEMS VALUE CHAIN

Institutional environment

The Ministry of New and Renewable Energy (MNRE) is promoting multi-faceted biomass-gasifier-based power plants for producing electricity using locally-available biomass resources such as wood chips, rice husk, cotton stalks and other agro-residues in rural areas. The biomass-gasifier programs of MNRE supports distributed or off-grid power for rural areas, captive power generation applications in rice mills and other industries as well as tail-end grid-connected power projects up to 2 MW capacities. The program envisages implementation of such projects with involvement of Independent Power Producers (IPPs), Energy Service Companies (ESCOs), industries, cooperatives, Panchayats, SHGs, NGOs, manufacturers or entrepreneurs, industries, promoters, developers, etc. Bihar Renewable Energy Development Agency (BREDA) promotes all renewable energy projects and programs in the state. The ministry is implementing a program for providing financial support for electrification of those remote un-electrified Census villages and un-electrified hamlets of electrified Census villages where grid extension is either not feasible or not cost-effective and not covered under RGGVY.

About 150 MW equivalent biomass gasifier systems have been set up for grid and off-grid projects. More than 300 rice mills and other industries are using gasifier systems for meeting their captive power and thermal applications. In addition, about 70 biomass gasifier systems are providing electricity to more than 230 villages in the country. A system of cross-subsidization is practiced based on the principle of 'the consumer's ability to pay'. In general, the industrial and commercial consumers subsidize the domestic and agricultural consumers. Furthermore, government giveaways such as free electricity for farmers, partly to curry political favour, have depleted the cash reserves of state-run electricity-distribution system. This has financially crippled the distribution network and its purchasing power to meet the demand in the absence of subsidy reimbursement from state governments.https://en.wikipedia.org/wiki/Electricity_sector_in_India - cite_note-140 This situation has been worsened by state government departments that do not pay their electricity bills.

Technology and processes

Biomass gasification is thermochemical conversion of biomass into a combustible gas mixture (producer gas) through a partial combustion route with air supply restricted to less than that theoretically required for full combustion. The HPS solution consists of a gasifier, filters and a gas engine connected to a generator (Figure 75). The gasifier is a down-draft type, where the sack loads of rice husk is loaded from the top into the hopper every 30–45 minutes through to the combustion chamber. Air is drawn through the top, and partial combustion occurs under a restricted supply of oxygen to give energy-rich producer gas, which comprises of hydrogen, carbon monoxide and methane. The residual char drops to the bottom of the chamber and is subsequently removed. The gas that is generated is water-cooled and cleaned through a series of filters made of char or rice husk and finally a cloth filter to eliminate particulate matter. The clean combustible gas is available for power generation in

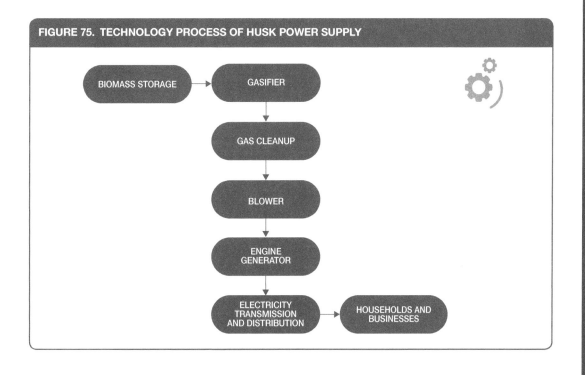

FIGURE 75. TECHNOLOGY PROCESS OF HUSK POWER SUPPLY

diesel-gen-set or 100% producer gas engines, which generates electricity at 240 V, single phase. Thus, electricity generated is distributed at the same voltage level through a single-phase-insulated cable system mounted on bamboo poles (for reduced costs). The distribution network is extended to a maximum of 2 km to keep the losses and voltage drops to acceptable levels. HPS has also developed low-cost prepaid meters (less than USD 8) that allow for better control and reduced theft.

The basic connection provides a household with two 15-W compact fluorescent lights and mobile-phone charging throughout the period each day the plant runs (up to eight hours in the evening). Sometimes, poorer households share a basic connection and get one light each. If a household or business wants to pay more for a higher-power connection, then this can be provided. A fuse blows if the customer attempts to use more than their agreed power.

The key advantages of HPS solution is that the various components of the system have been locally manufactured/adapted, rugged and durable and are simple to operate and maintain. HPS is still conducting research on the technology front to deal with the undesirable tar content of rice husk to explore other potential feedstock and alternative applications and uses of the resultant biochar. HPS has also recently implemented low-cost remote monitoring of its plants for better control and management of the same and shifted to solar - biomass hybrid mini-grids for 24/7 power supply (this information was not available when analyzing the case).

Funding and financial outlook

The capital cost (inclusive of installation) of each plant is less than USD 1,300/kW, and the operational cost is estimated to be less than USD 0.15/kWh. The gross margin at the plant level is expected to be around 45%, but sale of carbon off-sets and sale of biochar towards incense making is expected to each add 10% to the total revenues of a plant. Social enterprises, such as HPS, which step in to address the electricity gap in rural areas of India, are typically funded by a combination of grant and equity, with some support from the government by way of subsidy. HPS received significant grant support to the tune of approximately USD 2 million over four separate tranches from its strategic partner, Shell Foundation which contributed towards the early R&D costs, subsidized a portion of the costs of its high-profile management team, helped ramp up the rate of deployment and attract additional financing. In addition, HPS also raised funding (equity investment) of USD 1.65 million in 2009–2010 from Acumen Fund, Bamboo Finance, LGT Venture Philanthropy, Draper Fisher Jurvetson, CISCO and the International Finance Corporation. HPS also receives a government subsidy of approximately USD 7,100 for each plant from MNRE and the Government of India. Alstom Foundation recently announced a EUR 90,000 grant to upgrade 65 existing power plants by retro-fitting gasifiers using dry cleaning and cooling systems at the plants. The immediate positive impact of implementing the system would be dramatic reduction in water usage – by almost 80% – and also reduce operational cost considerably.

Socio-economic, health and environmental impact

Husk Power Systems has made a tremendous impact in the lives of rural people by supplying affordable electricity. The good quality lighting enables children to study properly and families to relax in the evening, as well as reducing snake and dog bites and petty crimes. Shops and businesses have lower costs and can work more easily even after dark without the need for diesel generators, and some new businesses have started. In one village, mobile-phone ownership increased from 10% to 80% of households after the HPS supply was installed, because previously, people had to go out of the village to have their phones charged. HPS is also using its plants as a channel for promoting and marketing other relevant products from different companies and foundations.

Furthermore, HPS is delivering economic and environmental benefits, as switching from traditional sources of energy by reduction in kerosene use by 6–7 L/month, saving about USD 4.40 per month or twice the cost of a basic connection. The overall portfolio of plants provides direct employment to over 350 people, with additional temporary work created during plant construction. HPS is starting businesses that use the char left over from rice-husk gasification, including that from the manufacture of incense sticks. As of now, more than 1,200 women have been trained (at two plant sites) for manufacturing incense sticks. This enables household to earn up to USD 16 per month and save USD 2.3 on kerosene costs while paying only USD 1.2 for electricity.

The HPS plant makes use of rice husk that is abundantly available but, until recently, was considered as waste. Rice mills are paid about USD 25 per ton of rice husk, so they earn an extra USD 3,000 per year by supplying an HPS plant as well as solving a disposal problem. The burning of rice residues in fields causes severe air pollution in some regions. The alternative, residue incorporation into the soil, in turn causes methane emissions from rice fields, contributing to climate change. Each megawatt of power generated from rice-husk plants has resulted in reduction of CO_2 emissions by about 25,800 every year. These reductions in emissions can be attained with the implementation of 32–33 rice-husk plants. Each plant serves around 400 households, saving approximately 42,000 L kerosene and 18,000 L diesel per year, significantly reducing indoor-air pollution and improving health conditions in rural areas. HPS has also offset a total of 2.2 million units of CO_2 by 2013. Further saving CO_2 from reduced use of diesel.

HPS is developing a program of activities for CDM to gain carbon credits. Moreover, processed wastewater and tar tank water is collected in a settling tank and recycled, which ensures that there is no water pollution. Rice husk char and tar and used filter media are mixed and stored on the ground. HPS is also working to reduce the water consumption in its char removal systems.

HPS makes sure that customers understand how to use electricity safely and that every member of the household agrees to abide by safety rules. HPS has facilitated the education of children of local communities by paying school-fee of USD 0.75 per month. Figure 76 summarizes the social impact of the project.

Scalability and replicability considerations
The key drivers for the success of this business are:
- High demand for electricity.
- Strong partnership with the Government of India.
- Central and state policy that promotes renewable energy and provides good incentives.
- Strong financial support from multiple institutional investors.
- Good availability of uniform fuel input (rice husk).

Given that HPS offers a decentralized, low-cost solution, which leads to lower transmission and distribution losses and makes use of a resource available locally which was earlier considered as waste. It has an immense potential not just across the state of Bihar but also across the entire country and the developing world, where over 1.6 billion people still do not have access to electricity/lighting. HPS has identified 25,000 villages as feasible sites within India's rice-producing belt (Bihar and neighbouring states) for establishing rice-husk-based power plants. Rice husk is a plentiful resource in India and many other countrie., Bihar alone produces 3 million tons/year of paddy, which could provide sufficient husk to supply electricity to 3 million households. HPS technology could therefore be used in many other rice-producing areas, as well as places with other biomass residues. HPS is exploring other avenues to increase its revenue other than fees collected for electricity service. Monetizing carbon offsets from biomass gasifier plant (125–150 tons CO_2/year per gasifier).

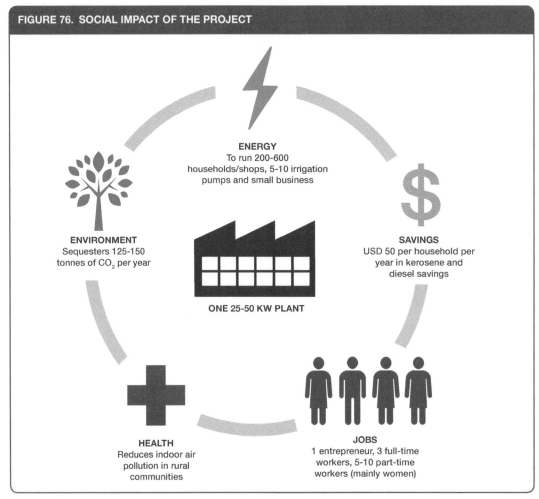

FIGURE 76. SOCIAL IMPACT OF THE PROJECT

Source: Husk Power Systems; modified.

HPS is planning to build a training centre and also provide some training by distance learning. New ideas under development and testing include programmable prepayment meters, char removal systems that cut water use and automated plant monitoring. Other ways of adding value to char are also under investigation.

The conditions across most parts of India, South Asia and sub-Saharan Africa, with regards to access to electricity in rural areas though not as severe, are quite similar, thereby offering significant potential for replication of the solution and business model offered by HPS. HPS started looking for funding to the tune of USD 6–8 million, to help achieve its ambitious target of establishing 3,000 plants to address the electricity needs of 10 million people across 10,000 villages.

Summary assessment – SWOT analysis

The key strength of HPS is its strong partnership with government, institutional investors along with buy-in from households and farmers (Figure 77). The weakness of HPS is its heavy dependence on rice husk as feedstock and subsidies from government. The business has a significant threat from government electrification programs and, in the event of flood or drought, it might not have access

FIGURE 77. SWOT OF HUSK POWER SUPPLY

	HELPFUL TO ACHIEVING THE OBJECTIVES	**HARMFUL** TO ACHIEVING THE OBJECTIVES
INTERNAL ORIGIN ATTRIBUTES OF THE ENTERPRISE	**STRENGTHS** • Strong government partnership to finance infrastructure investment for rural electrification • Strong partnerships, also with donors and social venture finance to fund company's soft costs • An inclusive business model • More than one revenue source (sale of electricity, biochar, franchisee fees and carbon credits) • Local buy-in • Robust technology	**WEAKNESSES** • Too much dependence on rice husk for input feedstock • Technological adaptation required for different biomass inputs • Strong dependence on one revenue source while HPS is diversifying to incense stick production and CO_2 credits sales • Heavy dependence on government subsidy for capital cost • Lack of trained work force at the local level
EXTERNAL ORIGIN ATTRIBUTES OF THE ENVIRONMENT	**OPPORTUNITIES** • Low penetration of rural electrification • Carbon credit market opportunities • Huge potential for scaling and replication in India and other developing countries • Rising crude oil prices weaken competitors • The Government of India incentives to electrify un-electrified areas • Locally available biomass (rice husk) • A severe energy crunch in India • Poor rule of law economic growth	**THREATS** • Threat of entrants and substitutes, especially solar units and cheaper electricity by utilities under rural electrification program • Business operations effected by poor rule of law • Unavailability of rice husk in case of drought or floods • Rice husk popularity could potentially drive up prices of inputs, undermining the business proposition

to rick husk. The business has opportunity to scale up and scale out and is already scaling using franchising model. It could improve its business stability by increasing its revenue source from sale of carbon credits and sale of biochar.

Contributors
Johannes Heeb, CEWAS
Jasper Buijs, Sustainnovate; formerly IWMI

References and further readings

Ashden. Husk Power Systems: 21st century living arrives in Bihar. Ashden. https://www.ashden.org/winners/husk-power-systems (accessed November 7, 2017).

Boyle, G. and Krishnamurthy, A. October 2011. Taking charge: Case studies of decentralized renewable energy projects in India in 2010. Greenpeace India Society. www.greenpeace.org/india/Global/india/report/2011/Taking%20Charge.pdf (accessed August 18, 2017).

Development Alternatives. Case studies on decentralized renewable energy: A case study of rice husk power plant, Patna, Bihar. Development Alternatives. www.devalt.org/knowledgebase/case_studies.aspx (accessed 18 January 2018).

Gurtoo, A. and Lahiri, D. 2011. Empowering Bihar: Policy pathway for energy access. Greenpeace. www.greenpeace.org/india/Global/india/report/Empowering-Bihar-Policy-pathway-for-energy-access.pdf (accessed August 18, 2017).

Hartnell, C. June 2011. Case study: Husk Power Systems featuring an interview with Gyanesh Pandey. Alliance Magazine. www.alliancemagazine.org/feature/case-study-husk-power-systems-featuring-an-interview-with-gyanesh-pandey (accessed August 18, 2017).

Husk Power System. www.huskpowersystems.com (accessed August 18, 2017).

Pokar, K., Chaurasia, S. and Maheshwari, R. March 2012. Innovation at bottom of pyramid – Husk Power System: Electrifying rural India (A case study). SSRN. http://papers.ssrn.com/sol3/papers.cfm?abstract_id=2029639 (accessed November 7, 2017).

Singh, S. November 2010. Empowering Villages: A comparative analysis of DESI Power and Husk Power Systems. IFMR Lead. http://ifmrlead.org/empowering-villages-a-comparative-analysis-of-desi-power-and-husk-power-systems/ (accessed August 18, 2017).

Case descriptions are based on primary and secondary data provided by case operators, insiders, or other stakeholders, and reflects our best knowledge at the time of the assessments (2013–2014). As business operations are dynamic, data can be subject to change.

Notes

1 Towards 2017, HPS increased its promotion of solar - biomass hybrid mini-grids for 24/7 power supply (this information was not available and considered when writing the case).
2 One lakh = 100,000.
3 On crore = 10,000,000.

BUSINESS MODEL 6
Power from agro-waste

Krishna C. Rao and Solomie Gebrezgabher

A. Key characteristics

Model name	Power from agro-waste
Waste stream	Agro-waste (from farmers and agro-industries)
Value-added product	Power (through biomass gasification or combustion)
Geography	Rural areas with large acres of crop cultivation for ease of procurement of crop residues
Scale of production	Small to medium scale; 25–100 kW (gasification) and up to 8 MW (combustion)
Supporting cases in this book	Koppal, India; Bihar, India
Objective of entity	Cost-recovery [X]; For profit [X]; Social enterprise [X]
Investment cost range	Approx. USD 1,000 to USD 1,400 per kW
Organization type	Private
Socio-economic impact	Improved energy access resulting in increased local income and productivity, cleaner local environment, reduced greenhouse gas emissions and employment generation
Gender equity	Clean air working environment supports in particular women where it is replacing kerosene lamps

B. Business value chain

The business model is initiated by a standalone private enterprise, social enterprise or agro-industries such as coffee processing units or rice mills that generate large quantities of crop residues as waste (Figure 78). The business concept is to process crop residues like wheat stalk, rice husk, maize stalk, groundnut shells, coffee husks, sawdust, etc. which has no commercial value and is often burned or dumped in the rivers or on landfills to generate electricity. The electricity can be consumed internally or sold to households, business or local electricity authority or combinations thereof.

The key stakeholders in the business value chain are the suppliers of crop residue: farmers and agro-industries, government as a regulator and/or investor, technology supplier and end users of the product – household and businesses directly or through the local electricity authority. Generating electricity from agro-waste or crop residue can be from one of the following processes: anaerobic digestion through biogas, gasification through producer gas and combustion/incineration through steam. Biomass combustion is generally suitable for large capacity and grid-connected applications, whereas biomass gasifier and bio-methanation-based power plants are more suitable for sub-megawatt level and decentralized applications. The business is eligible for sale of carbon as the electricity is generated from sustainable biomass source. In this business model description, the process used is gasification where crop residue is used as a feedstock for making syngas or producer gas, which contains carbon

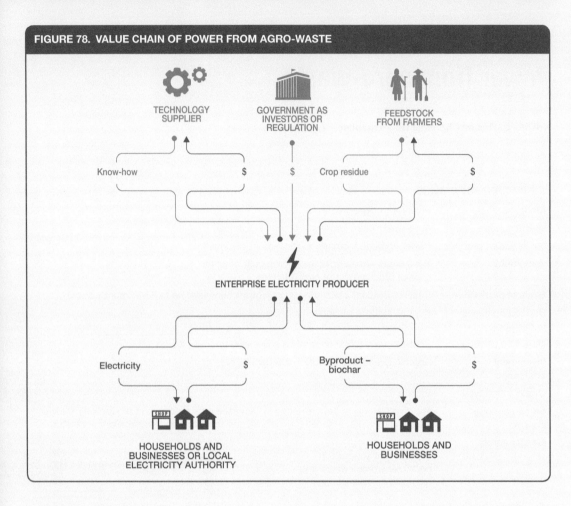

FIGURE 78. VALUE CHAIN OF POWER FROM AGRO-WASTE

monoxide and hydrogen and can be used to generate electricity using modified diesel or gasoline engine generators.

The ownership and operation of the enterprise generating electricity can take different forms. The plant can be designed, constructed, owned and operated by a standalone private enterprise such as agro-industrial processing factory, community-based organization, social enterprise and individual entrepreneur, either on a Build, Own and Operate basis or on a Build, Own, Operate, Transfer (BOOT) basis. In the latter scenario, the private entity brings investment to set up the energy production technology, while the concessionaries i.e. the agro-industrial factories or community provides land and inputs. The private entity designs, constructs and maintains the energy production unit until BOOT period is expired after which it assists the host company to operate the unit. The business model can use BOOT approach and franchise its model to scale up its business operations.

C. Business model

The primary value proposition of the business model (Figure 79) is to be a reliable provider of electricity from sustainable biomass source (agro-waste/crop residue). Depending on the ownership structure, the primary motive of the enterprise varies. This would in turn result in significant differences in the operations and management of the business.

BUSINESS MODEL 6: POWER FROM AGRO-WASTE

FIGURE 79. BUSINESS MODEL CANVAS – POWER FROM AGRO-WASTE

KEY PARTNERS	KEY ACTIVITIES	VALUE PROPOSITIONS	CUSTOMER RELATIONSHIPS	CUSTOMER SEGMENTS
• Government • Equipment suppliers • Farmers and agro-industries • Carbon trading partners • Investors/Donors	• Collection of crop residue • Contractual agreements • Electricity generation and distribution • Community mobilization • Meter, billing and collection • Managing Clean Development Mechanism (CDM) process	• Reliable provision of environment-friendly electricity at a fair price from a sustainable biomass source • Environment-friendly, low-cost biochar • Carbon emission reductions	• Direct interaction with the community and electricity authority • Direct interaction • Direct interaction with carbon market agents	• Households and businesses • Electricity authority • Industries household • Carbon market
	KEY RESOURCES • Crop residue • Technical know-how • Land and skilled local labor		**CHANNELS** • Emission trading platform • Direct electricity connection to community • Direct sales or network of distributors	

COST STRUCTURE	REVENUE STREAMS
• Land costs • Investment costs – building and equipment, transmission and distribution lines • O&M costs – training, input cost (crop residue), utilities, labor (can be intensive and skilled labor) • Debt and equity payments • Costs incurred for CDM/Voluntary Emission Reductions registration and carbon sale	• Sale of electricity • Sale of activated carbon or charcoal (by-product) • Sales of carbon credit

SOCIAL & ENVIRONMENTAL COSTS	SOCIAL & ENVIRONMENTAL BENEFITS
• Potential leakage of gas • Laborers' health and safety risk if no proper safety measures are present in the electricity generation unit • Potential safety issues on handling of biochar	• Improved social and economic sustainability through improved access to electricity • Reduced GHG emission • Create job opportunities

Social enterprise or community business model

A community-based organization or a social enterprise would run such an enterprise with the primary motive of providing reliable electricity to remote and underserved households and small businesses as a service while trying to achieve operational sustainability. This type of business model requires the enterprise to mobilize the community, procure agro-waste from local sources, develop appropriate and agreed pricing and establish transmission and distribution lines to reach out to every customer. The key activities for this business model are labor intensive, as it requires regular maintenance of transmission and distribution lines along with monthly meter, billing and collection activity for each customer. The business on a per capita basis is higher on capital and operation cost. From a cost recovery perspective, the business model is dependent on subsidies from government and donors to cover at the least its capital cost, but it has high potential to be financially viable and recover its operational cost including making marginal profits. Typical electricity generated under this model are in the range of 25–100 kW. The electricity generated is too small in size to be viable to apply for CDM projects unless the business does franchising of its model and bundles these transactions. However, it can access carbon offset on VERs market.

For-profit private business model

A private enterprise with profit maximization motives would get into a power purchase agreement with a local electricity distribution company. The electricity generated is directly fed to a local grid, and the local electricity distribution company pays an agreed price per unit to the enterprise as per the long-term power purchase agreement (PPA). The burden of transmission and distribution of electricity is transferred to the local electricity distribution company. The enterprise is not as labor-intensive as the social enterprise business model on per capita of electricity generated. In addition, this business model installs larger electricity generation plants of up to 8 MW. This is large enough and viable to apply for CDM.

In both the business models described above, it is important for the enterprise to have a strong partnership with farmers and agro-industries to ensure reliable supply of crop residues at an agreed price. The common key activities are procurement and processing of crop residue, electricity generation and sales. To improve the production efficiency, training of farmers can be a useful activity so that farmers provide high-quality crop residue and store-crop residue in appropriate manner to reduce moisture content. Sales of electricity is the primary revenue source with some additional revenue if the enterprise is able to tap into the carbon market.

Gasification process results in a by-product called biochar, which is rich in carbon. Biochar has multiple applications, as it can be sold to household or businesses as fuel to industries to produce activated carbon, and it is also an excellent fertilizer. The business model could potentially increase its revenue through sales of this by-product. The combustion process has fly ash as its waste product, which is used in brick manufacturing.

D. Potential risks and mitigation

Market risks: The electricity generated from processing crop residue is mainly sold to local electricity grid on a long-term power purchase agreement or to household and businesses through a social enterprise or community-based model. In the latter, community mobilization and product pricing are key activities before the enterprise is established, and hence this risk is addressed significantly. In the business model where electricity is sold to local electricity grid, since the demand for electricity is continuing to grow in developing countries and local electricity distribution companies are trying to manage to bridge the gap between demand and supply, the risks are lower. However, in environments where the electricity sector is regulated and the state utility is the sole buyer, the bargaining power of the business producing and selling electricity will be low. If the business has high dependence on sale of carbon credit for its viability, the volatility of carbon credit market puts the sustainability of the business under risk. In such scenarios, the business has to diversify its revenue streams so as not to entirely depend on the sales of carbon credits.

Competition risks: The business risk for the output (electricity) is relatively low. The social enterprise business model has risks from competitive products like solar home lighting system while the for-profit business model selling electricity to local grids has to compete with businesses generating electricity from cheaper fuel source such as coal and hydropower. The business has higher risk in procuring inputs (crop residue) at a price suitable for the business' financial viability. With time, as they realize the revenue potential from crop residue, the farmers are likely to demand higher price. To mitigate this risk the enterprise should target different types of farmers cultivating different crops or have longer-term agreements with farmers. The enterprise can also create its own plantation or agro-processing unit (rice mills) to secure its supply of agro-waste.

Technology performance risks: The technology used is gasification, which is well-established and mature for decentralized applications. The technology has been widely used commercially and is proven. However, the technology requires skilled labor.

Political and regulatory risks: In most developing countries the demand for electricity is projected to grow and governments are encouraging green initiatives by providing incentives such as concessional loans, feed-in tariff mechanisms and through long-term power purchase agreements. However, in regions where electricity is dominated by public sector and regulations do not allow sale of electricity, the business model cannot be established.

Social-equity-related risks: The model is considered to have relatively more advantages to women especially in underserved communities having clean working environment from clean indoor air by replacing kerosene used for lighting with modern energy. The social enterprise business model is geared to ensure no social equity risk arises in the community. However, the same cannot be said about for profit-private business model. The power generated is fed to the grid and this additional power might be used to improve energy reliability in existing regions rather than providing energy to underserved areas. Both social enterprise and for-profit models provide employment opportunities and additional revenue for farmers from sale of crop residues.

Safety, environmental and health risks: The waste-processing technologies are not without problems and pose a number of environmental and health risks if appropriate measures are not taken. The environmental risks associated with the gasification units include possible leakage of gas, and with the combustion unit's emission of flue gas and fly ash. These emissions should be controlled within acceptable limits by putting in place suitable equipment. Organic waste when left in open begins to decay and releases methane, which is more damaging to the environment than carbon dioxide. The safety and health risks to workers are present, and thus standard protection measures should be put in place (Table 23).

TABLE 23. POTENTIAL HEALTH AND ENVIRONMENTAL RISK AND SUGGESTED MITIGATION MEASURES FOR BUSINESS MODEL 6

RISK GROUP	EXPOSURE ROUTE					REMARKS
	DIRECT CONTACT	AIR	INSECTS	WATER/ SOIL	FOOD	
Worker	LOW	LOW				Risk from crop residues can be more physical (sharp edges) than of other nature.
Farmer/user						
Community		MEDIUM				
Consumer						
Mitigation measures						

Key: ☐ NOT APPLICABLE ▨ LOW RISK ▨ MEDIUM RISK ■ HIGH RISK

E. Business performance

This business model is rated high on environmental impact followed by profitability (Figure 80). The environmental impact scores high from the large-scale impact potential that the business model offers along with reduced greenhouse gas emission. The business model has a strong revenue source and

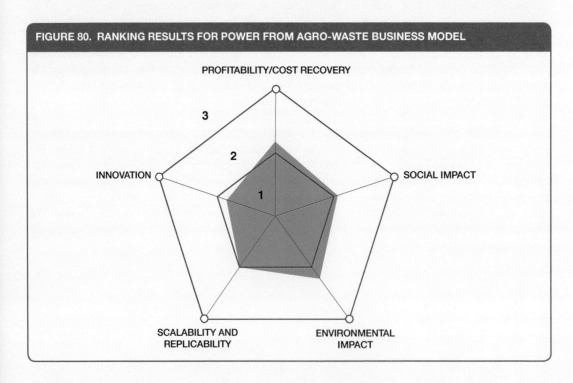

FIGURE 80. RANKING RESULTS FOR POWER FROM AGRO-WASTE BUSINESS MODEL

offers potential for additional revenue source from sale of carbon (VERs) and biochar. The social impact of the business model is dependent upon the customer segment served for provision of electricity. If the electricity is provided to underserved communities, the impact will be higher.

The business model has a high potential for replication in developing countries with availability of waste, technology and institutional capabilities. It can be scaled horizontally and has a potential for vertical scaling by expanding into the business of adding value to biochar for domestic and industrial use. It has a strong potential to be implemented in agriculture intensive regions. The business model scores relatively low on innovation as it is fairly straightforward with no sophisticated technology requirements. However, it may require innovative partnership and financing arrangements.

CASE

Power from municipal solid waste at Pune Municipal Corporation (Pune, Maharashtra, India)

Krishna C. Rao, Binu Parthan and Kamalesh Doshi

Supporting case for Business Model 7	
Location:	Pune, Maharashtra, India
Waste input type:	Municipal solid waste (MSW)
Value offer:	Biogas to electricity
Organization type:	Public
Status of organization:	Biogas plant operational since 2009
Scale of businesses:	Medium
Major partners:	Mailhem Engineers Pvt Ltd, Solid Waste Collection and Handling (SWaCH), Janwani, Cummins India, MITCON, Kirloskar and Maharashtra Plastic Manufacturers Association (MPMA)

Executive summary

The case demonstrates power generation from organic fraction of MSW in Pune Municipal Corporation (PMC) through generation of biogas. With population of more than 31 million, area of 243 km^2 and 48 zones, Pune is the seventh largest metropolitan area in India and the second largest in the state of Maharashtra. The biogas from MSW initiative in Katraj Gaon region in Pune is part of a larger Zero Waste Electoral Ward Initiative. The biogas plant provides street lighting services for about 4 km long Katraj–Kondhwa road in Pune. The bio-sludge is used as a manure in the 112 municipal parks and gardens maintained by PMC. The project is with a partnership between PMC; the Solid Waste Collection and Handling (SWaCH), an NGO, for door-to-door collection of waste and Mailhem Engineers Pvt Ltd, a waste management technology firm to process the MSW collected to produce biogas, electricity and bio-sludge. SWaCH has employed waste pickers to collect segregated waste from households and ensuring it reaches the secondary collection system, while PMC is providing support in various ways.

KEY PERFORMANCE INDICATORS (AS OF 2014)	
Land use:	0.03 ha
Water requirement:	1.25 m³/day
Capital investment:	180,000 USD
Labor:	4 persons, 3 persons at full-time employment and 1 person at half-time employment
O&M cost:	18,000 USD/year
Output:	300–325 m³/day processing 5 tons/day and electricity generation of 144 MWh/year, 180 tons/year of bio-sludge used manure
Potential social and/or environmental impact:	Created 4 jobs, processing a significant share of municipal solid waste of Pune and providing municipal lighting services; Project also reduces 76.1 tCO_2eq of GHG emissions by reducing electricity consumption.
Financial viability indicators:	Payback period: Approx. 6 years	Post-tax IRR: 16%	Gross margin: 62%

Context and background

The average annual MSW generated in India is 120 kg/capita/year. PMC generates 2,550 tons/day of MSW, with 40% to 60% organic matter, and hence is useful for generating energy. Most of the cities in India collect only 60% to 70% of waste actually produced, have insufficient landfill sites and find it difficult to locate new sites at affordable transportation distances. The composition of MSW varies greatly from municipality to municipality (country to country) and changes significantly with time. There is no single approach that can be applied to the management of all waste streams.

The term "digestible wastes" defines organic waste materials which can be easily decomposed by the anaerobic digestion process. The digestible household waste, such as food and kitchen waste, green waste, and most paper waste, includes not only waste from households, but also from institutions, digestible municipal park and garden trimmings, vegetable residues and discarded food from markets and catering businesses, out-dated food from supermarkets, etc. Not all of this organic waste would be suitable for anaerobic digestion. Wood and other lignin containing waste materials are typical examples of organic wastes that are not suitable for anaerobic digestion. PMC is the civic body responsible for providing waste collection and management service to its residents and has initiated a number of waste management projects. Katraj Gaon, as part of admin ward of Dhankawadi, is among the largest electoral wards in terms of area and has population of 15,377 with the blend of high and low-income and nearly 12,000 commercial establishments. Every day about nine tons of waste is collected by waste pickers organized as SWaCH. Nearly three tons of wet waste segregated by waste pickers is sent to biogas plants. Because of the project, the burning and dumping of waste on open plots and public spaces has also reduced considerably. Dry waste collection has also gone up as a result of the efforts and a lot more dry waste is now being sold for recycling. A substantial amount of waste consisting of dry non-saleable and low-value waste and mixed waste, however, still has to be sent to the landfill.

Market environment

The output of the project is biogas, electricity and sludge slurry. Biogas is a methane-rich gas (45–80% methane content), which can be used as renewable fuel for direct combustion for heating applications in commercial and communal kitchens in the city, co-generation (renewable electricity and/or heat generation) or upgraded to bio-methane (typically >94% CH_4) and injected into the gas grid or used for vehicle fuel. Electricity generated in this way will then be used to power streetlights and water and sewerage pumps through a distributed generation-based model. Katraj biogas plant is able to light only 140 street lights which are limited by waste availability. The liquid sludge rich in plant macro and micro nutrients can be used as soil improver and as fertilizer for plants, provided that it meets the strict quality requirements imposed for such application. Its application to land brings humus and slow-releasing macro and micro nutrients to the soil, contributes to moisture retention and improves soil structure and texture. Using compost made from recycling, such as organic wastes, is considered environmentally sustainable.

If a sustainable zero waste system is successfully put to test in such an area, its replicability would be high. PMC has 144 electoral wards, and in 2012, it had 22 biogas plants in operation. The market share is relatively small, and the waste management to municipal street lighting is only provided in one out of 144 electoral wards in Pune. There is a need for more waste management solutions for Pune including composting. PMC spends a considerable amount of resources on municipal street lighting and is looking for waste to energy solutions like the Katraj project. The attractiveness of the opportunity increases with passing time as the waste generation in the city is on the increase as well as the energy prices. One of the challenges for the waste input is segregation and making sure that only the organic waste is sent to the biogas plant.

There are also opportunities for similar waste-to-energy solutions for other cities in India. Competition for this business model is primarily from the energy utilities as the competing product is electricity. However, following historical trends, it is likely that the electricity tariffs will only increase in the future making biogas to electricity from MSW even more attractive. The impact of waste prevention and resource efficiency initiatives is likely to increase in the future, and food waste per capita may well start to decrease.

Maharashtra has an assessed potential to generate 637 MW from MSW, industrial waste and sewerage (out of the country's 3,400 MW potential in the sector). Of this, Maharashtra has achieved just 22.51 MW, and the new grid-connected renewable energy policy, which was approved by the state cabinet recently, aims at generating another 300 MW from such waste.

Macro-economic environment

As per the rules by Central Pollution Control Board (CPCB), every municipal authority shall be responsible for collection, storage, segregation, transportation, processing and disposal of MSWs. Municipal authorities shall adopt suitable technology or combination of such technologies to make use of wastes so as to minimize burden on landfill. The biodegradable wastes shall be processed by composting, vermicomposting, anaerobic digestion or any other appropriate biological processing for stabilization of wastes. Landfilling shall be restricted to non-biodegradable, inert waste and other waste that are not suitable either for recycling or biological processing. All of the Municipal Corporation, including PMC, spends a substantial amount of their annual budget on waste collection and transportation. With the dual objective of catering to the ever-increasing demand for electricity as well as the promotion of environmentally-friendly renewable energy technologies, the State Government of Maharashtra had issued guidelines to encourage power generation from MSWs into electricity. Apart from the pollution it causes, disposal of waste has also become a major problem with the continued depletion of potential landfill sites. As a result, 52.88 MW proposals in four cities in Maharashtra are under active consideration through private sector participation using municipal solid waste as raw material.

The Government of India launched Jawaharlal Nehru National Urban Renewal Mission (JnNURM), a massive city modernization scheme under which state governments and city municipalities can apply for funds to improve city infrastructure. Pune is eligible under this scheme and could potentially access these funds to install the biogas to electricity from MSW at other 143 electoral wards.

Business model

The business model canvas is from the perspective of the entity managing the biogas plant to generate electricity. PMC biogas plant in Katraj Gaon has several interlinked value propositions (Figure 81): production of biogas to generate electricity to provide street lighting services to Katraj–Kondhwa Road in Pune and organic compost produced from slurry and waste output from the biogas plant for landscaping of electoral wards within the Pune municipality. The biogas plant contributes to carbon offset, and therefore there is potential to realize revenue from sales of carbon. Janwani, an initiative of the Mahratta Chamber of Commerce Industries and Agriculture along with PMC, Cummins India Ltd, SWaCH Cooperative, Lions Club and Maharashtra Plastic Manufacturers Association is working towards a common goal and supports the project.

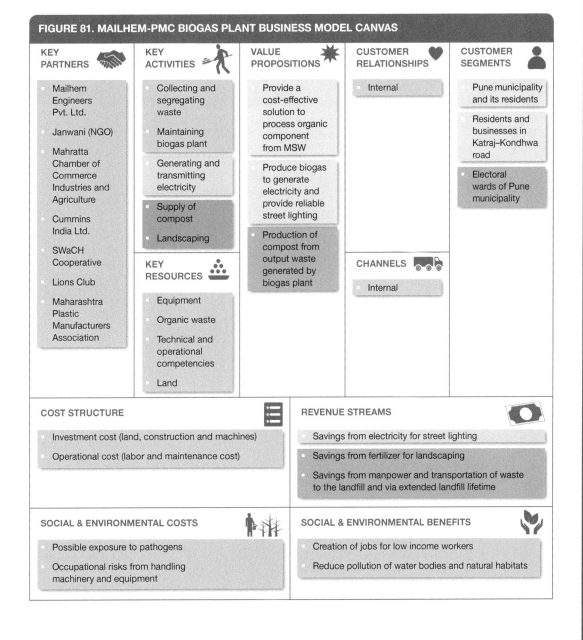

Value chain and position

The value chain consists of waste collection, waste segregation, waste-to-energy conversion and street lighting (Figure 82). The citizens helped by initial segregation of their waste. The waste is then collected and further segregated by SWaCH, which has signed a memorandum of understanding with PMC under which an annual payment is made by PMC for waste management. Each household also pays a fee to SWaCH on a monthly basis for waste collection. Solid waste other than MSW, i.e. garden waste, domestic hazardous waste, e-waste, biomedical waste, hazardous waste, construction and demolition waste, animal carcass, street sweeping, etc. was to be collected by SWaCH for additional user fees. PMC provided equipment, slum subsidy, push-cart maintenance amount, sorting centre and admin desk/office space to SWaCH. Post-segregation of waste, members of SWaCH deliver organic content of the waste collected to the PMC biogas plant, which is linked to electricity generation facility. This electricity is used to provide street lighting to Katraj–Kondhwa Road of the municipal area and also as back-up power for the municipal administration building. PMC does not have required skills or expertise in maintenance of waste-to-energy conversion and municipal street lighting. PMC engaged

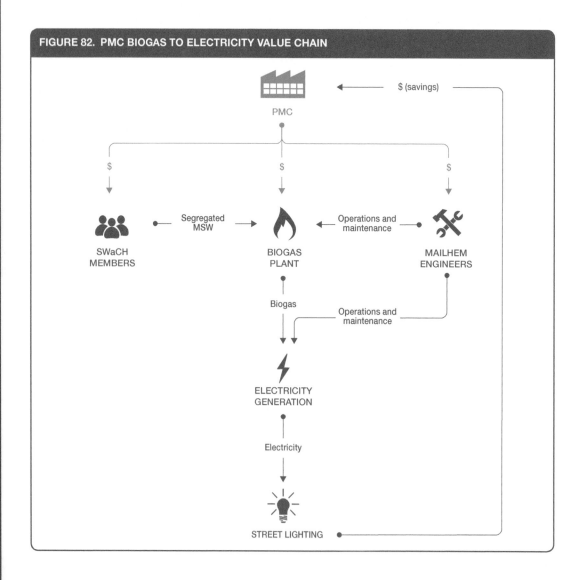

FIGURE 82. PMC BIOGAS TO ELECTRICITY VALUE CHAIN

with Mailhem to install waste-to-energy infrastructure and contracted Mailhem for operation and maintenance of the biogas digester and electricity generator.

The main challenge was to teach citizens to separate biodegradable and non-biodegradable waste, because the two kinds of waste are treated differently, and to collect user fees from citizens/waste generators. Janwani used tools like home visits, announcements from vehicles and street puppet theatre to deliver its message. A second challenge was to create value from the waste by processing all organic or wet waste within the ward and by recycling dry waste. Dry trash like plastic, glass and paper is sold for recycling by the waste-pickers.

Institutional environment

The solid waste management in developing countries has received lesser attention from policymakers and researchers than other environmental problems, such as air and water pollution. However, the legal and regulatory framework in India mandates the treatment of solid waste. According to governmental policies, the organic waste component of MSW has to be bio-digested or composted and the inorganic portion landfilled. There are a number of other relevant waste management policies for controlling hazardous waste, plastics, construction and demolition waste, e-waste, battery waste and MSW. The State Government of Maharashtra has banned the sale and use of plastic bags across the state since 2006 after the Mumbai floods of 2005.

The Government of India has established JnNURM with an aim to encourage reforms and fast-track planned development of identified cities. Focus is to be on efficiency in urban infrastructure and service delivery mechanisms, community participation and accountability of ULBs/parastatal agencies towards citizens. Assistance under JnNURM is additional central assistance, which would be provided as grant (100% central grant) to the implementing agencies. The sectors and projects eligible for JnNURM assistance includes sewerage and solid waste management. The Ministry of New and Renewable Energy (MNRE) promotes power generation from MSW projects by providing a capital subsidy for power generation from MSW of USD 0.3 million per MW, with max of USD 1.55 million per project. Each proposal will be examined and concurred by Integrated Finance Division of the Ministry on a case-to-case basis. The Maharashtra Electricity Regulatory Commission (MERC) has also been very proactive in promoting energy generation from renewable energy sources. MERC has been in the forefront of determining preferential tariffs for renewable energy technologies, with its first tariff order for non-fossil fuel based co-generation projects issued even before the enactment of Electricity Act 2003.

Technology and processes

Anaerobic digestion is a collection of processes by which micro-organisms breakdown biodegradable material in the absence of oxygen. The best practice for bio-degradable waste is separation at source, as they need to be of high quality (i.e. free from physical impurities) in order to ensure stable operation of the anaerobic digestion process. The chemical and biological pollutants, contaminants, toxins, pathogens or other physical impurities must also be strictly monitored and limited to allow safe and beneficial utilization of sludge as fertilizer. Anaerobic digestion can be single stage, multi stage or batch process. Based on the content of total solids (TS) of the substrate to be digested; the anaerobic digestion processes can be low solids (LS), containing less than 10% TS; medium solids (MS), containing about 15–20% TS and high solids (HS), ranging between 22–40% TS (Verma, 2002). The industrial process takes places in a specially designed digester tank, which is part of a biogas plant.

The technology employed for anaerobic digestion is modified up-flow anaerobic sludge blanket. The technology is proven and is used in waste management systems around the world (Figure 83).

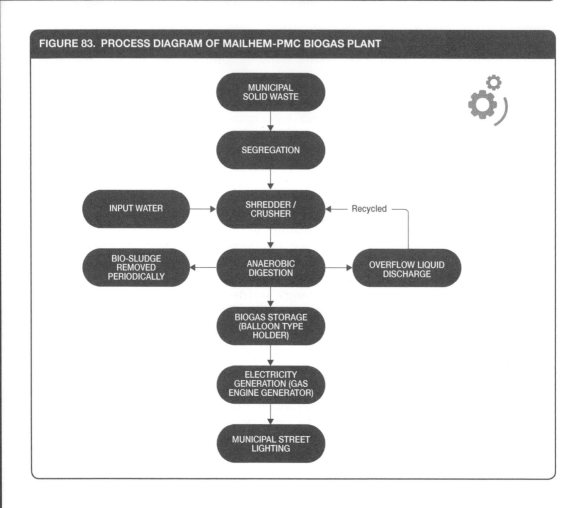

FIGURE 83. PROCESS DIAGRAM OF MAILHEM-PMC BIOGAS PLANT

Maintenance of the sludge blanket is an important factor in the efficient operation of the reactors. The biogas produced is combusted in gas engine coupled to an electrical generator to produce electricity for street lighting. The operation of the technology requires segregated MSW with only the organic portion with 80% moisture. Therefore, segregation of MSW is an important aspect, and the technology may not work properly when the input biomass deviates from the specifications.

The technologies – the biogas digester and electricity generator – are locally available and components that need replacement can be fabricated locally. The operation and maintenance of the plant is managed by Pune-based Mailhem, and the technicians employed have been trained and supervised also by Mailhem. The intellectual property rights for the specific digester construction and commissioning lies with Mailhem.

Funding and financial outlook

The operations of biogas plant at Katraj Gaon electoral ward in Pune started in 2009. The land and building costs were covered by the municipality at an existing facility. Investment was towards plant and machinery cost which was USD 180,000 with PMC financing the entire investment (Table 24). on financials. The annual operation and maintenance cost incurred is about 18,000 USD/year. PMC has a contract with SWaCH to deliver organic waste from MSW, and as a part of providing waste management service to households it pays an agreed amount. There is no additional amount given

TABLE 24. MAILHEM PMC BIOGAS PLANT FINANCIALS

Key capital Costs	Land building	No costs (uses PMCs own land)
	Plant and machinery and civil costs	180,000 USD
Operating costs	MSW costs	No costs
	Operation and maintenance	18,000 USD/year
Financing Options	Equity from PMC	180,000 USD
Revenue	Savings from electricity	About 15,840 USD/year
	Savings on manpower and transportation	About 21,000 USD/year
	Savings on manure	About 10,800 USD/year

to SWaCH for supplying organic waste to the plant. Therefore, no cost is considered for MSW input. In Katraj, Cummins India gave USD 45,000 to Janwani and offered 3,000 employees as volunteers, helping Janwani to create awareness. PMC biogas has indirect revenue sources in the form of savings from electricity and fertilizer. Based on these savings, PMC biogas plant has a payback period of 19 years on its investment with an internal rate of return of 2%. PMC can generate revenue from annual carbon sales. It offsets 76.1 tCO_2eq per year.

Socio-economic, health and environmental impact

Poor solid waste management is a threat to public health and causes a range of external costs. Mixed wastes from municipalities are often landfilled. Landfill deposits pose the risk of uncontrolled air, soil and water pollution. Left to degrade naturally in landfill sites, organic wastes from households and municipalities have very high methane production potential; thus, have a negative impact on the environment. Methane has a very high global-warming potential. Over a period of 100 years, each molecule of methane (CH_4) has a direct global warming potential which is 25 times higher than that of a molecule of carbon dioxide (CO_2). Anaerobic digestion can save up to 1,451 kg CO_2/t of waste treated compared to 1,190 kg CO_2/t in the case of composting. Source separation helps divert organic wastes from landfill, thus reducing the overall emissions of greenhouse gases and the negative environmental and health effects related to these waste disposal methods. In order to decrease the environmental and health effects associated with landfilling, waste management is nowadays moving away from disposal and towards waste prevention, reuse, recycling and energy recovery.

PMC biogas plant has a positive impact on socio-economic, health and environment for the region. It provides full-time employment to three people and part-time employment to one person. The waste collection and street lighting efforts have resulted in residents of Katraj electoral ward getting waste management as well as street lighting services. The waste picking members of SWaCH have increased their daily earnings and has improved their social and economic stature. The plant effectively manages municipal solid waste generated within the Katraj Gaon ward which is one of the biggest divisions in Pune, therefore providing environmental benefits as well as health benefits from proper waste management. The project also displaces electricity for street lighting in Katraj–Kondhwa Road and displaces 144 MWh/year of electricity purchases by PMC otherwise. It also mitigates 76.1 tCO_2eq/year of GHG as a result of the avoided electricity consumption.

Scalability and replicability considerations

The key drivers for the success of this business are:
- Partnerships with SWaCH/NGO to deliver segregated organic waste to the plant.
- Technology partnership for operation and maintenance of the plant.

- Demand for end products – electricity and compost are internal and hence no significant competition risks.
- Government policy toward renewable energy.
- Rising electricity tariffs.

If the Katraj zero-garbage model proves itself, Janwani plans to set up a biogas plant in every ward. The potential for replication is high within the PMC and also in other metropolitan areas in India. The increasing quantities of MSW generation and rising electricity tariffs are likely to make replication opportunities further attractive.

Summary assessment – SWOT analysis

The key strength of the business model is its arrangement to secure reliable supply of organic solid waste for the biogas plant and the operation owned by a large urban body that can easily finance such investments (Figure 84). The business can be easily replicated. With an ever-increasing volume of household waste generated, the business has a high opportunity for replication to other areas of a city and to other cities in India. The weaknesses stem from its low rate of financial returns as the key

FIGURE 84. MAILHEM-PMC BIOGAS PLANT SWOT ANALYSIS

	HELPFUL TO ACHIEVING THE OBJECTIVES	HARMFUL TO ACHIEVING THE OBJECTIVES
INTERNAL ORIGIN ATTRIBUTES OF THE ENTERPRISE	**STRENGTHS** • Established arrangements with reliable partners for waste collection and delivery and operation and maintenance of the waste-to-energy plant • Large urban body with significant internal financial resources for waste management activities, which makes financing of the effort easy • Use of an innovative public-private partnership (PPP) model for waste management	**WEAKNESSES** • Government body as the key partner in the business model and subject to bureaucratic delays • Financial model not attractive based on the avoided cost of electricity and bio-fertilizers purchases • Households only pay a small contribution towards waste management significantly below the cost of waste management • Non-availability of suitable land • Inadequate manpower with the municipal corporation for implementation and compliance verification with MSW rules
EXTERNAL ORIGIN ATTRIBUTES OF THE ENVIRONMENT	**OPPORTUNITIES** • Favourable policy and regulatory environment for waste management and independent power producers in India • Significant opportunities exist for replication of the model to other divisions/ parts of the urban civic body • Availability of an ever-increasing volume of household waste and expected increase in power tariff in other parts of India • Increased yields and beatification of city parks using bio-fertilizers	**THREATS** • Changes in the composition and volume of household waste and the challenge of effective segregation of the waste • Increased public awareness and active commitment and participation of citizens in local collection schemes required

revenue is from electricity savings, which is a relatively small amount in comparison to the amount investment. Therefore, investment does not look financially attractive.

The success of the project requires increased public awareness and active commitment and participation of citizens in local collection schemes. Some of the other constraints are non-availability of suitable land, lack of technical awareness of citizens with respect to waste processing technologies, inadequate waste pickers and manpower with the municipal corporation for implementation and compliance verification with MSW rules. The business model faces significant threat from any changes in composition and volume of input waste.

Contributors
Mailhem Engineers Pvt Ltd

References and further readings

Bhaskar, A. 2011. Zero Waste Electoral Ward Initiative at Katraj Gaon, Pune. Better India. www.thebetterindia.com/3531/zero-waste-electoral-ward-initiative-at-katraj-gaon-pune (accessed November 7, 2017).

Godbole, P. January 2012. Waste no trash. Business Standard. www.business-standard.com/article/beyond-business/waste-no-trash-112011500052_1.html (accessed November 7, 2017).

Mailhem Engineers. www.mailhem.com.

Pune Municipal Corporation. September 2008. Memorandum of understanding between Pune Municipal Corporation and SWaCH. For the 2015 follow-up agreement see http://swachcoop.com/pdf/Agreementforwebsite.pdf (accessed November 7, 2017).

Pune Municipal Corporation. www.punecorporation.org.

Case descriptions are based on primary and secondary data provided by case operators, insiders, or other stakeholders, and reflects our best knowledge at the time of the assessments (2015/2016). As business operations are dynamic, data can be subject to change.

BUSINESS MODEL 7
Power from municipal solid waste

Krishna C. Rao and Solomie Gebrezgabher

A. Key characteristics

Model name	Power from municipal solid waste (MSW)
Waste stream	Organic waste – Organic component of MSW
Value-added waste product	Anaerobic digestion of organic waste to produce biogas to generate electricity and produce organic compost
Geography	Applicable to cities and towns that generate large quantities of organic solid waste
Scale of production	Medium to large scale; About 145 MWh/year to 9 GWh/year of electricity and 180 tons/year to 12,000 tons/year of organic compost
Supporting cases in this book	Pune, India
Objective of entity	Cost-recovery [X]; For profit [X]; Social enterprise [X]
Investment cost range	About USD 180,000 for lower size plant to USD 11 million of large size plant
Organization type	Private and public-private partnership (PPP)
Socio-economic impact	Improved and reliable electricity resulting in increased income and productivity, reduced greenhouse gas emissions, improved MSW management and processing and employment generation
Gender equity	Clean air working environment for women where kerosene lamps are replaced; and waste collection jobs for women

B. Business value chain

The business model is initiated by a standalone private enterprise or a public-private partnership (PPP) where a private entity partners with the municipality to manage the solid waste generated by the city. The business concept is to collect, segregate and process organic component of MSW to generate electricity and compost. The electricity and compost can be sold to households, business or local electricity authority (Figure 85).

The key stakeholders in the business value chain are the waste suppliers, either household or the municipality; regulators-government; investors – municipality or private enterprise; technology supplier and plant operator – private enterprise and end users of the product–household and businesses or municipality. The process of generating electricity from MSW can be done through either incineration to produce heat and steam, gasification or anaerobic digestion. In this business model, the technology process used is anaerobic digestion where the organic component of MSW is segregated and sent to a digester to produce biogas, which is used to generate electricity. The business is eligible for sale of carbon as the electricity is generated from sustainable organic source with improved MSW management as MSW left in the open releases methane to the atmosphere.

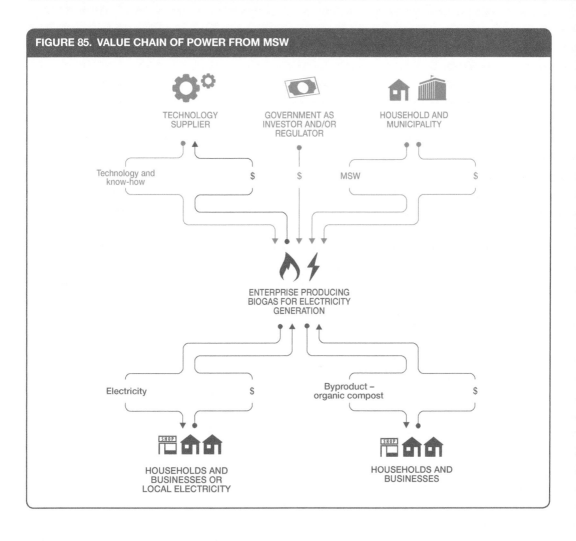

FIGURE 85. VALUE CHAIN OF POWER FROM MSW

The contractual agreement of the PPP and role of private and public entities can take many forms. There are multiple options for the ownership of the plant. The plant can be owned by the municipality or the private enterprise with the concession from municipality to provide land and MSW. If it is owned by private enterprise, it may be on Build, Own, Operate (BOO) basis which is also called energy service company (ESCO) or on Build, Own, Operate, Transfer (BOOT) basis. Under BOOT, the private entity designs, constructs and maintains the energy production unit until BOOT period is expired after which it assists the municipality to own and operate the unit.

C. Business model

The primary value proposition of the business model is to reliably provide electricity from MSW. However, it would vary for a PPP as the primary mandate would be to provide waste collection and waste management service (Figure 86). The model also offers value proposition of providing organic compost which is a by-product of the process.

The business will have to collect MSW from the municipal landfill site or have the municipality garbage trucks deliver MSW to the plant. The business can also organize collection of MSW directly from households, and this would require a larger labor force. The enterprise would sell the electricity either

FIGURE 86. BUSINESS MODEL CANVAS – POWER FROM MSW

KEY PARTNERS
- Government
- Municipality
- Technology and equipment suppliers
- Carbon trading partners

KEY ACTIVITIES
- MSW collection and segregation
- Contractual agreements
- Electricity generation and distribution
- Managing CDM process
- Compost production

KEY RESOURCES
- MSW
- Equipment
- Partnership, technical know-how and operational competencies
- Land and skilled local labor
- Agreement with electricity authority

VALUE PROPOSITIONS
- Reliable provision of environment-friendly electricity from a sustainable biomass source
- Waste collection and management service
- Organic compost from the by-product of anaerobic digestion

CUSTOMER RELATIONSHIPS
- Direct interaction with the community and electricity authority
- Direct interaction with household and municipality
- Direct interaction

CHANNELS
- Direct
- Direct collection from household or municipality
- Direct sales or via distributors
- Via carbon market agents

CUSTOMER SEGMENTS
- Households and businesses
- Electricity authority
- Households and municipality
- Recycling industries
- Farmers and landscapers
- Carbon market

COST STRUCTURE
- Investment costs – Building and equipment, transmission and distribution lines
- O&M costs – Training, utilities, labor (can be intensive and skilled labor)
- Costs incurred for CDM/VER registration and carbon sale

REVENUE STREAMS
- Sale of electricity
- Sales of carbon credit
- Waste collection and management fees
- Sale of recyclables
- Sale of compost

SOCIAL & ENVIRONMENTAL COSTS
- Potential leakage of gas
- Potential health risks for workers at production facility
- Potential environmental risk if the waste is not treated and disposed properly

SOCIAL & ENVIRONMENTAL BENEFITS
- Improved social and economic sustainability through improved electricity access
- Reduced GHG emission and pollution of water bodies and natural habitats
- Contributes to improved MSW management with reduced human exposure to untreated waste
- Create job opportunities

to household or businesses or to a local electricity distribution company. Other key activities for this business model are regular maintenance of transmission and distribution lines along with monthly meter, billing and collection. However, it is preferred by the enterprise to get into a power purchase agreement with the local electricity distribution company and feed the electricity to local grid at an agreed price per unit. The burden of transmission and distribution of electricity is as such transferred to the local electricity distribution company.

Sale of electricity and waste collection and management are the key revenue source. The enterprise is also eligible for carbon offset and since the electricity generated is substantial, it will be viable to apply for Clean Development Mechanism (CDM). MSW consists of inorganic waste such as plastics, paper and glass that have high resale value and sale of recyclables is another revenue source for the business model. In addition to above mentioned revenues, key outputs from the biogas plant are bio-slurry and sludge, which are rich in nutrients and can be processed to make organic compost. The enterprise can sell the compost to farmers and landscapers. Compost can be delivered to the customers either through direct sales, network of distributors or micro-franchising. The direct sales would involve managing a large human resource base for sales and marketing staff.

D. Potential risks and mitigation

Market risks: The electricity generated from processing MSW is primarily sold to local electricity grid on a long-term power purchase agreement. Since the demand for electricity is continuing to grow in developing countries and local electricity distribution companies are trying to bridge the gap between demand and supply, the market risks associated are lower. However, in environments where the electricity sector is regulated, with the tariff decided by the regulatory commission and the state utility is the sole buyer, the bargaining power of the business producing and selling electricity will be low. If the business has high dependence on sale of carbon credit for its viability, the volatility of carbon credit market puts the sustainability of this reuse business under risk. In such scenarios, the business has to diversify its revenue streams so as not to entirely depend on the sales of carbon credits. The business model also sells compost to farmers and landscapers. The market demand for compost can be low in urban areas, but in rural areas there is always high demand from farmers. However, sales in rural areas might significantly increase its transportation cost. The business needs to find the right balance between its urban and rural market.

Competition risks: The business risk for the output (electricity) is relatively low. The business model has to compete with other businesses generating electricity from cheaper fuel source such as coal. The business has higher risk in procuring MSW if it is not able to obtain a contract with municipality. To mitigate this risk the enterprise will have to undertake door-to-door collection of household waste. In the case of compost, competing products are chemical fertilizers, which are subsidized in developing countries; hence, the challenge for compost to be price competitive. The business should diversify its customers across different types of farmers including plantations and agro-forestry that will likely have higher demand for compost. The business could also produce varieties of compost products suitable for different types of crops. However, this in an unlikely scenario if the revenue stream is minor.

Technology performance risks: The technology process used is anaerobic digestion, which is well-established and mature. However, the type of digester required could potentially be sophisticated and might not be available in developing countries, and in addition, the technology requires skilled labor. It is ideal for the business to transfer the technology from a market where it is widely implemented and have their staff trained in operation, repair and maintenance of the technology.

Political and regulatory risks: In most developing countries, the demand for electricity is projected to grow and governments are encouraging green initiatives by providing incentives such as concessional loans, feed-in tariff mechanisms and through long-term power purchase agreements. However, in regions where electricity is dominated by public sector and regulations do not allow sale of electricity, the business model cannot be established. The risks to compost from political and regulatory aspects are similar to electricity. If the regulation does not allow compost from MSW for crop production, it can hamper business viability and restrict the application of such compost to a very specific market and customer segment such as forestry or landscaping.

Social-equity-related risks: The model does not have any social equity risks. The model is considered to have more advantages for women due to clean working environment from clean indoor air by replacing kerosene used for lighting and offering women jobs in waste collection. The employment opportunities are limited for non-skilled workers. The model could potentially improve the working environment of informal workers especially women engaged in waste collection.

Safety, environmental and health risks: Safety and health risks to human arise when processing any type of waste. The risks are even higher when processing MSW. Laborers in enterprises handling such waste should be provided with appropriate gloves, masks and other appropriate tools to handle the waste to ensure their safety. The risk of environment pollution is high if leachate from MSW is untreated and seeps into groundwater or other natural water bodies. The waste processing technologies are not without problems and pose a number of environmental and health risks if appropriate measures are not taken. The environmental risks associated with the anaerobic digestion units include possible leakage of gas (methane) and these emissions should be controlled. Organic waste when left in open begins to decay and releases methane which is more damaging to the environment than carbon dioxide (Table 25).

TABLE 25. POTENTIAL HEALTH AND ENVIRONMENTAL RISKS AND SUGGESTED MITIGATION MEASURES FOR BUSINESS MODEL 7

RISK GROUP	EXPOSURE					REMARKS
	DIRECT CONTACT	AIR	INSECTS	WATER/ SOIL	FOOD	
Worker	MEDIUM	HIGH	LOW			Air exposure if RDF is burnt (incl. organic and non-organic waste) Food based risks if slurry is chemically contaminated, which requires compost monitoring for chemical contaminants
Farmer/User						
Community		HIGH	LOW			
Consumer					HIGH	
Mitigation measures	🧤 ⚡	😷 🎧 🌳	🚫🦟	⚠️	🍲 Pb Hg Cd	

Key: ☐ NOT APPLICABLE ▨ LOW RISK ▨ MEDIUM RISK ■ HIGH RISK

E. Business performance

This business model is scores high on profitability, followed by social impact and environmental impact indicators (Figure 87). The business model has multiple revenue sources: sale of electricity, waste collection and management fees, sales of recyclables and potential for sale of carbon and compost, and the model serves a diverse customer base. Since the business model is involved in MSW management and offers multiple value propositions, it provides direct jobs to about 20 people, and the number is higher if the business is also involved in the collection of waste from households. The environmental impact scores high from the large-scale impact potential that the business model offers from safe MSW management along with reduced greenhouse gas emissions.

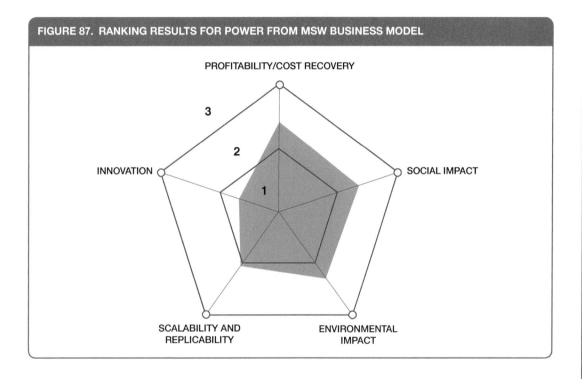

FIGURE 87. RANKING RESULTS FOR POWER FROM MSW BUSINESS MODEL

The business model has high potential for replication in developing countries. It can be scaled horizontally and has potential for vertical scaling by expanding into the compost business. It has a strong potential to be implemented in agriculture intensive regions. The business model scores low on innovation. However, one needs to acknowledge that despite the process is well-known, there is need for a sophisticated technology tailored to specific characteristics of MSW in developing countries and the model requires innovative financing arrangements.

CASE

Combined heat and power from bagasse (Mumias Sugar Company, Mumias District, Kenya)

Solomie Gebrezgabher, Jack Odero and Nancy Karanja

Supporting case for Business Model 8	
Location:	Mumias District, Western Kenya
Waste input type:	Bagasse (Sugarcane waste)
Value offer:	Clean and renewable electricity
Organization type:	Private
Status of organization:	Operational since 1972 (Co-generation unit since 2009)
Scale of businesses:	Large
Major partners:	Out-growers, PROPACO (French development agency)

Executive summary

Mumias Sugar Company Ltd (MSC) generates electricity from its bagasse-based co-generation plant. MSC was primarily established to produce sugar for local production and export. Over the years, the company has created additional revenue from its bagasse-based co-generation plant. The company generates about 34 MW of electricity, 26 MW of which is sold to Kenya Power and Lighting Company (KPLC) based on a long-term power purchase agreement (PPA), while the remaining is used for factory needs and domestic use in the staff quarters. MSC's business model is cost-driven and is based on its strategic access to input from its nucleus estates and out-growers. This guarantees the company reliable and low-cost supply of input. Since the commissioning of the co-generation plant, MSC has experienced stability in electricity supply to run its operations as opposed to the unreliable supply from the national grid. In addition to enabling MSC to be energy self-sufficient, the co-generation plant contributes to mitigating environmental pollution by replacing fossil-based energy production while satisfying the energy demand of the country. The co-generation plant is also beneficial to local populations by contributing to expanding access to electricity supplies in areas otherwise distant from the grid.

CASE: COMBINED HEAT AND POWER FROM BAGASSE

KEY PERFORMANCE INDICATORS (AS OF 2012)					
Land:	0.6 ha (for the co-generation plant)				
Capital investment:	USD 63 million (boiler costs USD 28 million and generator USD 14 million)				
Labor:	20–25 persons full-time and 5 persons on a contract basis (for the co-generation plant)				
O&M costs:	1 million USD/year				
Output:	34 MW of electricity and 100 tons/day of ash				
Potential social and/or environmental impact:	Mitigation of environmental pollution by reducing GHG emissions, reducing fossil fuel dependence, employment generation, expanding access to electricity for rural areas				
Financial viability indicators:	Payback period:	17	Post-tax IRR:	3%	Gross margin:	30%

Context and background

MSC is a registered agro-based publicly-listed company in Kenya that is involved in the growing and processing of sugarcane to produce sugar. In 2014, its ownership is 36% local corporate, 56% local individuals and 8% foreign investors. The company started its operations in 1972 and became the largest sugarcane processor in Kenya both in terms of profitability and scale of operations, crushing (depending on supply) between 1.1 and 2.4 million tons of sugarcane annually and producing 70,000–260,000 tons of sugar. MSC was primarily established to produce sugar for local production and export. Over the years, the company has created additional revenue streams from electricity generation. Construction of the power co-generation plant started in 2007 and was completed in 2009, with an initial finance of USD 35 million provided by the French development agency, PROPARCO. The initial and long-term goals of the co-generation plant were waste management and electricity production, which varied in the last years between 21 and 34 MW of electricity per year.

Market environment

Kenya has an electricity demand of 1,191 MW and installed power capacity of 1,429 MW of electricity. Peak demand load is projected to grow to 2,500 MW by 2015 and 15,000 MW by 2030. The demand for electricity in Kenya outstrips supply despite imports from Uganda. With the projected electricity demand set to grow, the government is encouraging green power initiatives such as co-generation units. Though on-site power production through bagasse co-generation is on the increase, its potential is not fully exploited in the industry.

The Kenya Electricity Generating Company (KENGEN) generates approximately 80% of electricity consumed in the country, while the balance is produced by independent power producers (IPPs), such as MSC. In 2013, MSC sold about 26 MW of electricity to KPLC, which distributes the power through the national grid. KPLC buys power from generators like KENGEN and MSC and is responsible for the transmission, distribution and retail of electricity throughout the country.

MSC with its CDM initiative has also concluded purchase agreements with financial group Japan Carbon Finance Ltd (JCF). Launched in late 2004, the financial group has received committed funds of approximately USD 140 million for its Japan GHG Reduction Fund (JGRF). In addition to JGRF, JCF has established second and third funds to purchase further carbon credit for some of fund providers in JGRF. JCF plans to use the fund to purchase emission reductions credits from CDM and joint implementation (JI) projects in developing and "in transition" economies around the world. JCF has assisted development of various types of CDM/JI projects, such as renewable energy, energy efficiency and waste management, and concluded purchase agreements in more than 30 projects worldwide, including cogeneration plant of MSC.

Macro-economic environment

The Kenyan power sector is characterized by the heavy reliance on hydroelectricity, frequent power outages, low access to modern energy and high dependence on oil imports. With the enactment of the Electric Power Act in 1998, the KPLC was unbundled into three entities: the KPLC that was to carry out transmission and distribution functions to meet demand, the KENGEN to carry out the generation function and the Electricity Regulatory Board (ERB) to regulate the power sector. The Act also allowed IPPs to enter into PPAs with KPLC to add more power into the grid. KPLC has, however, retained the transmission and distribution functions all over the country which hinders the emergence of decentralized IPPs and independent power distributers.

Kenya leads in exploiting renewable energy (RE) sources to meet the challenges of growing demand and addressing the related environment concerns to complement the realization of Vision 2030: "accelerating transformation of the country into a rapidly industrializing middle-income nation by the year 2030". The incentives for RE power include 0% import duties and value-added tax exemption on renewable energy materials, equipment and accessories, feed-in tariffs at a price level that attracts and stimulates new investment in the renewable energy sector. These will have direct impacts on the development of renewable energy in Kenya and on the available energy that KPLC can supply to regional populations. The IPPs were introduced into the sub-sector as a means of redressing the challenge of capacity shortfalls. At least 174 MW of power is supplied by IPPs. MSC generates 34 MW, 26 MW of which is dispatched to the grid. The government has identified the unexploited potential of up to 300 MW from other sugar factories.

Business model

MSC is essentially a sugar factory with a co-generation plant that processes bagasse to produce and sell electricity to KPLC through a long-term PPA (Figure 88). MSC employs a cost-driven business model. Since commissioning of the co-generation plant, MSC has experienced more stability in electricity supply to run its operations, and thus reduced operational costs compared to the unreliable supply from the national grid that it was previously relying on.

Value chain and position

MSC sources its sugarcane from its nucleus estates and its out-growers. Bagasse, produced after sucrose extraction from sugarcane, is used as fuel in the boilers to generate high-pressure steam, which runs generator to produce electricity. Ash, generated by the cogeneration plant, is applied as a soil conditioner in the company's sugarcane plantation (Figure 89).

Although MSC is vertically integrated and owns its nucleus estate, it still heavily depends on out-growers for its input, and thus the supplier power is high. In recent years, MSC has experienced a declining supply of sugarcane from both its nucleus estate and out-growers. This has been to a large extent attributed to the declining soil productivity of the cane fields due to continuous mono-cropping of sugarcane. This situation has led to production that is well below the installed plant capacity and has forced the company to reduce cane crushing and sugar milling to one or two times a week, down from the efficient all-week year-round production. Since MSC waste re-use operations are a direct result of cane crushing, there has been an associated decline in outputs for electricity generation.

MSC has a PPA with KPLC to provide 26 MW of electricity to the national grid. KPLC is responsible for the transmission, distribution and retail of electricity throughout the country, which gives it a high bargaining buyer power. The terms of the PPP are such that if MSC fails to provide the agreed 26 MW, it is penalized by KPLC which deducts a percentage of the revenues accruing to MSC. The company has also been forced to procure sugarcane from far areas and from out-grower farmers and bagasse

CASE: COMBINED HEAT AND POWER FROM BAGASSE

FIGURE 88. MUMIAS SUGAR COMPANY BUSINESS MODEL CANVAS

KEY PARTNERS	KEY ACTIVITIES	VALUE PROPOSITIONS	CUSTOMER RELATIONSHIPS	CUSTOMER SEGMENTS
• Out-growers • PROPARCO (French development agency)	• Operate co-generation plant • Sales of electricity	• Competitively priced and renewable energy based electricity through processing of bagasse in environmental-friendly way • Use of ash as soil enrichment in cane sugar plantation	• Long-term PPA	• KPLC • Nucleus estate (internal use)
	KEY RESOURCES • Equipment • Long-term PPA with KPLC • Sugarcane, bagasse • Technical and operational competencies • Capital • Nucleus estates		**CHANNELS** • Connection to KPLC national grid	

COST STRUCTURE	REVENUE STREAMS
• Capital investment cogeneration – Financed by debt • Operational cost (labor cost, water repairs and maintenance) • Reduced operational costs from using own electricity and thermal energy • Minimized transport costs • Interest on borrowed fund • Tax = 16% VAT	• Electricity sales • Improved soil productivity and yield • Sales of carbon credits

SOCIAL & ENVIRONMENTAL COSTS	SOCIAL & ENVIRONMENTAL BENEFITS
• Possible contamination of water source	• Contribute to mitigating environmental pollution • Contribute to the economy of local community • Generate employment opportunity • Contribute to improved waste management • Expanding access to electricity supplies to local communities

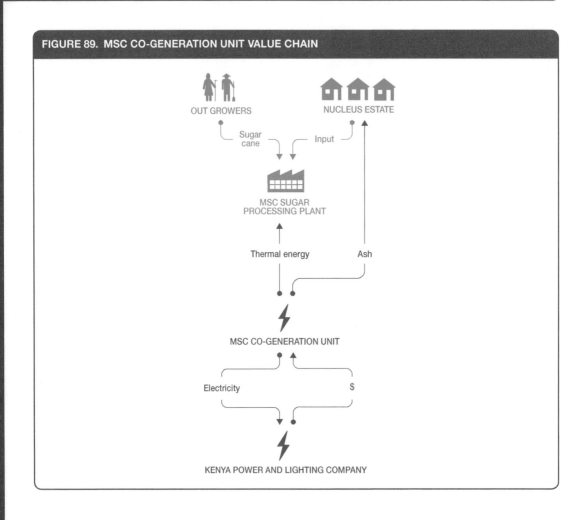

FIGURE 89. MSC CO-GENERATION UNIT VALUE CHAIN

from other sugar companies not producing power/electricity to complement supplies from the nucleus estate and the out-growers. To mitigate this situation, MSC is researching on developing a cane variety that can survive nutrient-depleted soils to ensure sustained supply of sugarcane for crushing. An energy balance initiative has also been commissioned aimed at improving energy consumption to further enhance the export of green energy to the national grid.

The co-generation plant is a CDM project, which qualifies for Carbon Emissions Reduction (CER) certificates. MSC estimates an annual production of 130,000 tons of CERs. Under purchase agreement with JCF as well as the provision of Article 12 of the Kyoto Protocol on CDM, MSC is allowed to sell and JCF to purchase CER certificates. In 2010, MSC was the first Kenyan firm to sell carbon credits, making Ksh 22 million (about USD 270,000).[1]

Institutional environment

In 2008, Kenya launched Vision 2030, a development blueprint aimed at making Kenya a newly industrialized middle-income country providing a high quality of life to all its citizens in a clean and secure environment. Addressing climate change is a top priority of the Government of Kenya. Kenya's Intended Nationally Determined Contribution (INDC) includes an ambitious mitigation contribution of a 30% reduction in GHG emissions by 2030 relative to the business as usual scenario. The plan was developed through a cooperative and consultative process that included stakeholders from governments, private sector, civil society and development agencies.

Kenya's power sector reform was initiated following the enactment of the Electric Power Act 1997 whereby the policy formulation function was retained by the Minister for Energy, while regulatory functions were passed on to an autonomous regulator: Electricity Regulatory Board (ERB) and commercial functions in respect of generation, dispatch, transmission, distribution and supply to various commercial entities. The government amended the Electricity Act to enable the reform and restructuring of the sub-sector in order to prepare it to attract adequate funding, especially from the private sector, for operations and development and to improve financial and technical efficiency of entities involved. With the implementation of reforms, KPLC is now transformed from the de facto vertically integrated structure into a single buyer (purchasing egency) model in which it purchases bulk power from IPPs and the public sector generation company under long-term bilateral PPAs.

The government has been encouraging and supporting green power initiatives such as wind power and co-generation such as the one undertaken by MSC, all with the goal of increasing the installed power capacity of Kenya. The RE department is responsible for leading the planning, development, implementation, promotion and execution of structures for the development and regulation of the RE and energy efficiency through research and planning, development of standards and regulations, compliance and enforcement. RE portal provides easy access to relevant information about entry requirements and procedures for operating a RE power plant, the legal and regulatory framework for such investments (such as tariff regulation) and relevant market information.

Technology and processes

Co-generation is the production of electricity and heat from a single fuel source. For MSC, the fuel source is bagasse. Bagasse produces sufficient heat energy to supply all the needs of a typical sugar mill, with energy to spare. To this end, a secondary use for this waste product is in co-generation, the use of a fuel source to provide both heat energy used in the mill and electricity, which is typically sold on to the consumer electricity grid. Bagasse is the fibrous matter that remains after sugarcane is crushed to extract its juice. The high moisture content of bagasse, typically 40–50%, is detrimental to its use as fuel. Bagasse is an extremely inhomogeneous material, making it particularly problematic for paper manufacture. In general, bagasse is stored prior to further processing. For electricity production, it is stored under moist conditions, and the mild exothermic reaction that results from the degradation of residual sugars dries the bagasse pile slightly.

Figure 90 depicts MSC sugar and co-generation unit process. Bagasse produced after sucrose extraction from sugarcane is used as fuel in the boilers to generate electricity. There are three main steps involved in co-generation power production:

- Steam generation – Bagasse's chemical energy is converted into heat by burning. The heat energy is used in boilers to heat water to produce steam at specified pressures and temperatures.
- Steam turbine operation – Steam from boilers is used to drive the turbines, which convert the heat energy into mechanical energy. This provides the power to turn the equipment for power generation at controlled speeds.
- Power generation – The turbines are used to turn electrical power generators. All the major capital equipment including the boiler and generator were imported.

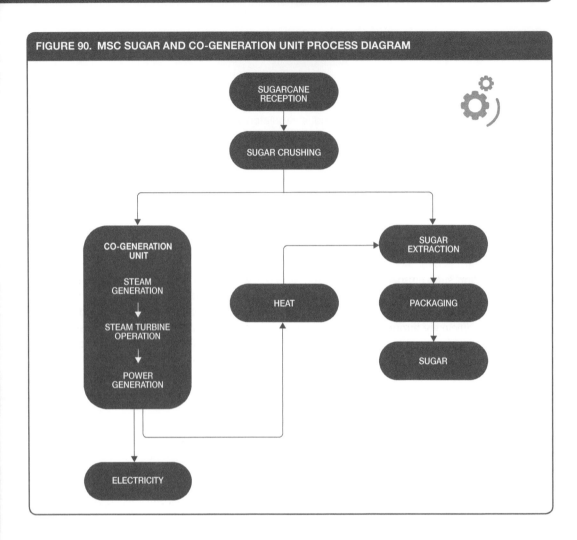

FIGURE 90. MSC SUGAR AND CO-GENERATION UNIT PROCESS DIAGRAM

Funding and financial outlook

The total investment cost of the co-generation unit was USD 63 million, with the boiler and generator taking bulk of the capital costs at USD 28 million and USD 14 million respectively. There was no land acquisition since the power plant was built on a yard that had an old sugar production line. Initial finance of USD 35 million for the co-generation plant was provided by PROPARCO in 2007 as a loan at an annual interest of 6.24%, one-off arrangement fee of 1% and commitment fee of 0.5%. The total amount repayable, loan plus interest and fees, amounts to USD 39.8 million. The loan was repayable over 10 years after a three-year grace period and was used to finance phase one of the project. The second phase of the project was financed by banks in 2009 lending a total of USD 28 million repayable after 10 years following a three-year grace period. The company successfully commissioned the 34 MW co-generation project effective May 11, 2009, leading of sale of 26 MW of power to grid. The commissioning of the power sub-station and the transmission line was done simultaneously. Production, sales and costs of electricity from the co-generation plant are shown in Table 26. The figures for the years 2013 to 2015 were at the time of the case assessment still reported projections.

Taking the steady annual profit of USD 3.68 million and the initial capital of USD 63 million, the plant's payback period is approximately 17 years and assuming useful life of 25 years, Internal Rate of Return

TABLE 26. FINANCIAL DATA OF MSC FOR 2009–2015

YEAR	2009	2010	2011	2012	2013	2014	2015
Electricity exported (kWh)	25,000	25,000	25,000	25,000	25,000	25,000	25,000
Electricity sales (Million USD)	11.93	12.27	12.27	12.27	12.27	12.27	12.27
Cost of electricity (Million USD)	8.36	8.59	8.59	8.59	8.59	8.59	8.59
Profit from electricity sales (Million USD)	3.57	3.68	3.68	3.68	3.68	3.68	3.68
Payback period	17 years						
IRR	3%						

*Assuming useful life of 25 years and discount rate of 10%.

(IRR) is 3%. In addition to the direct income from sale of electricity, MSC further realizes cost savings from using their own generated 8 MW of electricity at a cheaper cost compared to the cost they could have incurred if electricity was supplied from the main grid. The company is also in advanced stages of negotiating with the Japan Carbon Finance Company for sale of carbon credits and projects that it will earn an annual income of USD 800,000 from the sale of credits due to its green initiatives. Besides the power generated, the company also collects the ash from the burnt bagasse that is used as a soil conditioner and applied on the company's nucleus estate. In addition, MSC is planning to introduce new products such as ethanol production, expanding co-generation, new packages for various market segments, capacity expansion and modernization, investment in computer technology and improved supply chain management for overall efficiency in the company.

Socio-economic, health and environmental impact

As sugar mills tend to be located in rural areas (Mumias is a rural town) near sugarcane plantations, co-generation is beneficial to local populations by contributing to expanding access to electricity supplies in areas otherwise distant from the grid. Co-generation, in addition to enabling MSC to be energy self-sufficient, contributes to mitigating environmental pollution by replacing fossil-based energy production while satisfying the energy demand of the country. As it is a locally sourced fuel, bagasse increases the reliability of electricity supply by diversifying sources and reducing fossil fuel dependence. As a biomass fuel, bagasse supplies a raw material for the production of natural, clean and renewable energy, enabling its use to further government targets for renewable energy use. The CO_2 emissions by burning of bagasse are less than the amount of CO_2 that the sugarcane plant absorbed from the atmosphere during its growing phase, which makes the process of cogeneration GHG-neutral.

Furthermore, it boosts employment for neighbouring communities and allows operational personnel to develop skills in operating the equipment and technologies. The co-generation plant employs between 20 and 25 people on regular basis and another five on contract basis. In order to safeguard employees' health and safety, MSC provides personal protection equipment, annual medical check-ups for all staff and safety signs are put at all entrances. Furthermore, to ensure minimal release of pathogens from the burnt bagasse, it has re-engineered the plant.

MSC had experienced difficulties in disposing the bagasse by direct dumping into forests and water bodies before the co-generation project. Not only was the bagasse-making the land derelict, dry bagasse occasionally ignited and caused fire resulting into loss and destruction of property. However, this problem was nipped after commissioning of the power plant since the bagasse produced after crushing is fed directly into the power plant to fire the boilers. Reuse of bagasse has freed up space/land which can now be used productively. The project complies with local environmental and safety standards and aims to be as close as possible to international reference standards.

Scalability and replicability considerations

The key drivers for the success of this business are:
- Reliable supply of inputs as MSC is vertically integrated.
- Strategically situated near the sugarcane source.
- Securing of long-term PPA.
- Government encourages green power initiatives.

The co-generation plant is best suitable where there already exists some infrastructure i.e. a sugar company that is already generating bagasse, with the power co-generation initiated as a plant within the sugar factory. It may not be feasible to set up the co-generation plant as an independent plant that relies on bagasse from external sources. Given the initial high capital costs, such a project requires the involvement of development agencies that can provide finance to offset the initial capital expenditure. This project has the potential to be replicated in countries where there are large sugar manufacturing companies and where there is a government support for RE initiatives. Issuing longer-term licenses and PPAs with good feed in tariffs allow for sufficient time for the investor to pay off project financing debts as well as provides adequate amortization period for the equipment. MSC is planning to expand its sugar production facilities with corresponding co-generation plant by development of a new sugar factory with capacity to crush 6,000 tons of cane a day or acquiring one or more state-owned sugar factories in Kenya, Uganda and Tanzania.

Summary assessment – SWOT analysis

Key strengths of the business are securing of long-term purchase agreement with the state utility while ensuring energy self-sufficiency (Figure 91). Declining supply of sugarcane due to declining soil productivity is a key threat to the business. There is a threat that failure of MSC to deliver agreed electricity to KPLC may result in penalty and loss of income. To mitigate these problems, MSC has the opportunity to register the project as CDM and earn from sales of carbon credits. Furthermore, MSC has the opportunity to explore the development of a cane variety that can survive nutrient-depleted soils.

Contributors

Johannes Heeb, CEWAS, Switzerland
Jasper Buijs, Sustainnovate; The Netherlands, Formerly IWMI
Josiane Nikiema, IWMI, Ghana
Kamalesh Doshi, Simplify Energy Solutions LLC, Ashburn, Virginia, USA

References and further readings

Mumias Sugar Company. Overview of Mumias Sugar Company. Mumias Sugar. www.mumias-sugar.com/index.php?page=Overview (accessed August 18, 2017).

Mumias Sugar Company. Annual Reports 2008–2016. Mumias Sugar. www.mumias-sugar.com/index.php?page=annual-reports (accessed August 18, 2017).

United Nations Economic Commission for Africa. 2007. Making Africa's power sector sustainable. http://repository.uneca.org/handle/10855/15059 (accessed November 7, 2017).

FIGURE 91. SWOT ANALYSIS FOR MSC

	HELPFUL TO ACHIEVING THE OBJECTIVES	**HARMFUL** TO ACHIEVING THE OBJECTIVES
INTERNAL ORIGIN ATTRIBUTES OF THE ENTERPRISE	**STRENGTHS** - Long-term power purchase agreement - Diversified revenue streams - Energy self-sufficient - Clean and renewable energy generation - Nucleus estates provide some guarantee of input - Community support of sugarcane farmers as well as nearby villages	**WEAKNESSES** - Declining sugarcane production due to mono-cropping led to production below installed capacity - Technical dependency on sugar cane - Internal revenue stream dependency on sugar sales and electricity sales - High initial cost with long payback periods
EXTERNAL ORIGIN ATTRIBUTES OF THE ENVIRONMENT	**OPPORTUNITIES** - Earnings from sale of carbon credits - Register as CDM and earn from sales of carbon credits - Development of a cane variety that can survive nutrient depleted soils - Growing electricity demand - Sourcing bagasse from other nearby sugar companies not producing electricity - Potential of decentralize power generation and distribution in remote villages	**THREATS** - Declining supply of sugarcane, and hence bagasse due to declining soil productivity - Failure of MSC to deliver agreed electricity to KPLC may result in penalty and loss of income

Case descriptions are based on primary and secondary data provided by case operators, insiders, or other stakeholders, and reflect our best knowledge at the time of the assessments 2015. As business operations are dynamic, data can be subject to change.

Note

1   https://www.standardmedia.co.ke/business/article/2000208079/government-to-support-nse-introduce-carbon-credits-trading (accessed 18 January 2018).

CASE

Power from slaughterhouse waste (Nyongara Slaughter House, Dagorretti, Kenya)

Jack Odero, Krishna C. Rao and Nancy Karanja

Supporting case for Business Model 8	
Location:	Dagorretti, Kenya
Waste input type:	Slaughterhouse waste (solid and liquid) form
Value offer:	Biogas, power and bio-fertilizer
Organization type:	Private
Status of organization:	Operational since 2011
Scale of businesses:	Small
Major partners:	United Nations Industrial Development Organization (UNIDO), Kenya Industrial Research and Development Institute (KIRIDI) and United Nations Environment Programme (UNEP)

Executive summary

The Nyongara Slaughter House is located in Dagorretti on the outskirts of Nairobi. Dagorretti is an area famous with the presence of slaughterhouses that supply meat to different regions in Nairobi and its environs. The waste generated by the slaughterhouse was polluting Nairobi River and the National Environmental Management Authority (NEMA), an environment regulatory body, was closing slaughter-house units that were not meeting the regulatory norms of treating their waste. This catalysed partnership between Nyongara Slaughter House and UNEP, UNIDO and KIRIDI through the Ministry of Environment to develop and demonstrate a solution to not only treat the waste to produce biogas but also provide monetary benefits to the slaughterhouse units. The biogas operations began in 2011 with biogas used for heating and to generate electricity primarily for refrigeration and lighting purpose. The slurry output from the plant is high in nutrients and is used in cultivation of tomatoes within the slaughterhouse. Based on the success of Nyongara biogas plant, the proprietor of Nyongara Slaughter House wants to set up a business of treating waste from other slaughterhouse units in Dagorretti, Kenya to generate biogas and sell the electricity to slaughterhouse units.

CASE: POWER FROM SLAUGHTER HOUSE WASTE

KEY PERFORMANCE INDICATORS (AS OF 2012)						
Water requirement:	4,000 L/day wastewater input from the slaughterhouse					
Capital investment:	USD 35,000 to 60,000					
Labor:	1 full-time for composting; 2 part-time for biogas plant operations					
O&M cost:	NA					
Output:	25 cubic meter per day of biogas generating 10 kVA electricity and bio-fertilizer					
Potential social and/or environmental impact:	1 full-time and 2 part-time jobs, a cleaner environment through reduced water pollution and CO_2 emissions, reduction of GHG emissions					
Financial viability indicators	Payback period:	3–5 years	Post-tax IRR:	N.A.	Gross margin:	N.A.

Context and background

Dagorretti is a suburb of Nairobi well known for its slaughterhouses, employing over 5,000 people, which were almost shut down in 2009 due to untreated slaughterhouse waste polluting the Nairobi River. The surroundings were stinking, emitting large quantities of methane, and the blood and wash water were seeping into the groundwater. Unreliable grid electricity forced the slaughterhouses to depend on diesel generators, increasing their high-energy bill, accounting for up to 40% of their total cost of production.

Based on the request of the Ministry of Environment, Government of Kenya and as a part of cleaning of the Nairobi River Initiative, UNEP requested UNIDO to develop solutions to manage slaughter-house waste. At the same time, the proprietor of Nyongara Slaughter House was looking out for a solution to manage its waste and comply with NEMA regulations. This led to the collaboration between Nyongara Slaughter House, UNIDO, UNEP, KIRDI under a public-private partnership for this pilot project.

A 15-kW biogas plant was installed at the Nyongara slaughterhouse, with a high-performance temperature-controlled digester (using solar heating), replacing the diesel generator and recovering waste heat to replace wood and charcoal for hot water to clean the slaughterhouse. The generated electricity is consumed for lighting and powering water pumps and compressors for cold storage and processing of hides and skins while mitigating the pollution of Nairobi River. At the time of the interview, the proprietor of Nyongara Slaughter House was planning to initiate conversation with the owners of other slaughter-house units to process their waste and sell electricity to those units and neighbouring households and enterprises.

Market environment

At the time of the assessment, Nyongara biogas plant processed about 300 kg of solid waste per day and generated electricity for four hours in a day. Dagorretti recorded more than 15 slaughter-house units, and on an average, each unit produces about four tons of solid waste and 4,000 L of wastewater per day. Based on the total quantity of waste (60 tons of solid waste + 60,000 L of wastewater) generated from all the slaughter-house units in Dagoretti, it has the potential to meet the electricity demand of all the units and generate surplus electricity. A typical slaughterhouse requires electricity for refrigerating units, water pumping, heating, slaughtering appliances, office equipment and lighting. At the time of the interview, majority of the slaughter-house units were shut down by NEMA, and these slaughter-house units were looking at Nyongara to provide them with a solution to meet the environment regulations and reopen their business. Key competitor for Nyongara biogas plant is from a local electricity authority in Dagoretti. However, the electricity authority is struggling to meet public and private demand. Dagoretti area suffers from severe electricity shortage and majority of the slaughterhouses invest in diesel generators from backup power supply.

Macro-economic environment

Only 25% of the population in Kenya has access to electricity with less than 5% in rural areas. Installed capacity in 2011 was only 1,590 MW, which is very low for a country of 40 million people (40 W per capita – South Africa's figures are roughly 40,000 MW for 50 million people or 800 W per capita). The main problems in the sector are that existing capacity is barely able to meet demand, especially when hydrological conditions dip. There are independent power producers (IPPs) providing around 27% of the generated energy. Small-scale renewables contribute about 3% of installed capacity and is expected to grow to 6% by 2018.

Kenya's Vision 2030 ambition is to be a middle-income country in 18 years' time. This will require system capacity to grow to 15,000 MW by 2030. Rapid growth in capacity is required both to underpin the GDP growth targets and to allow universal access to electricity to be achieved. However, the present situation remains dogged by problems. Severe power shortage and electricity blackouts are putting pressure on the economic growth. While government plans to significantly add new generation capacity, initiatives such as the biogas plant at Nyongara Slaughter House, is a drop in the ocean that is still an urgent need not just from electricity access but also from environment perspective.

Business model

The Nyongara biogas plant exhibits three value propositions (Figure 92): treating waste from slaughter-house units to enable them to meet environmental regulations, generating electricity from the waste and producing bio-fertilizer. The electricity generated is consumed by the unit and surplus is sold to adjacent slaughter-house units and neighbouring households. The outlet slurry from the biogas plant is rich in nutrients and has the potential to be sold as bio-fertilizer.

Value chain and position

Nyongara biogas plant accepts 300 kg of waste per day, generating 9 kWh of electricity at the Nyongara Slaughter House. However, the proprietor plans to treat waste from other slaughter-house units to generate and provide electricity and thermal energy (Hot water for cleaning). The value chain analysis (Figure 93) is based on future plans. The value chain consists of procuring waste from other slaughter-house units and supplying electricity to slaughter-house units and neighbouring households and enterprise. The project will reduce dumping of slaughterhouse waste, and hence lower methane emissions to the atmosphere.

Since other slaughter-house units are looking for a solution to treat their respective waste, the supplier power is low. The slaughter-house units have existing investments in diesel generator, so unless the Nyongara biogas plant can provide electricity at lower costs, buyer power is prominent. In addition, the enterprise has threat from new entrants. The enterprise has potential to add another revenue source through production and sales of nutrient-rich bio-fertilizer from the slurry output of the biogas plant. The slurry is in the meantime used to irrigate the compound, thus providing the hundreds of workers with a cleaner and greener working environment.

CASE: POWER FROM SLAUGHTER HOUSE WASTE

FIGURE 92. NYONGARA BIOGAS PLANT BUSINESS MODEL CANVAS

KEY PARTNERS
- UNIDO, UNEP, KIRDI for providing know-how

KEY ACTIVITIES
- Procuring and treating waste from slaughterhouses
- Biogas to electricity conversion
- Organic fertilizer production from biogas outlet slurry
- Sales

KEY RESOURCES
- Financial resources
- Land, building, equipment, labor
- Agreement with slaughterhouse units
- Slaughterhouse waste

VALUE PROPOSITIONS
- Treating waste produced by slaughterhouse units to enable them to meet local environmental regulations
- Electricity for slaughterhouse operations and cooking for household
- Organic fertilizer for crop cultivation (potential)

CUSTOMER RELATIONSHIPS
- Direct

CHANNELS
- Direct sales

CUSTOMER SEGMENTS
- Slaughterhouse units
- Slaughterhouse units, neighbouring households and enterprise for energy service
- Farmers

COST STRUCTURE
- Investment cost (land, building, briquetting machines)
- Operational cost (labor, maintenance cost)
- Energy savings

REVENUE STREAMS
- Sale of electricity
- Sale of organic fertilizer

SOCIAL & ENVIRONMENTAL COSTS
- Possible human health hazard from direct contact to pathogens that may still exist from the organic fertilizer
- Laborers' health risk due to contact with slaughterhouse waste
- Environmental risks from biogas leakage to the atmosphere

SOCIAL & ENVIRONMENTAL BENEFITS
- Waste from slaughterhouses treated reduces pollution of Nairobi River
- Reduction of human health problems (and related costs) in the locality due to reduction of pollution of water bodies
- Creation of jobs

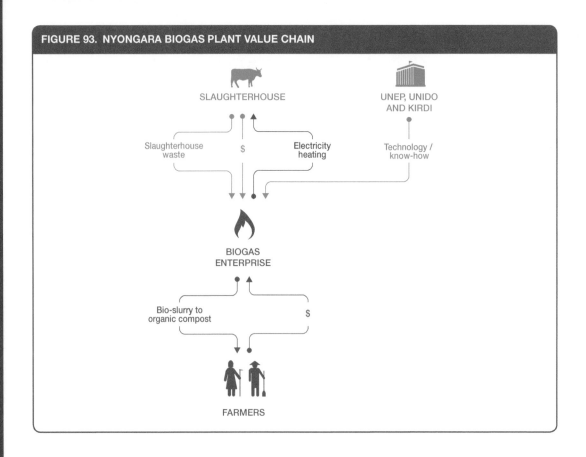

FIGURE 93. NYONGARA BIOGAS PLANT VALUE CHAIN

Institutional environment

The project was developed to address NEMA, a national environmental agency regulation that mandates appropriate treating of the waste generated by slaughterhouses. The project plans an expansion phase to treat waste from other slaughter-house units, and it has buy-in from the Ministry of Energy and Petroleum (MEP), thus showing an acceptance from the government for such operations. The 2006 Energy Act sets up the Energy Regulatory Commission (ERC), an independent regulator meant to formulate licensing procedures, issue permits, make recommendations for further energy regulations, set and adjust tariffs, approve power purchase agreements (PPAs) and prepare national energy plans.

The Energy Act entrusts the MEP to elaborate sustainable renewable energy production, distribution and commercialization frameworks with emphasis on the expansion of local manufacturing sectors and provide specific incentives to existing renewable markets such as bio-digesters, solar systems and hydro-turbines. Renewable energy frameworks will also encourage biomass co-generation – heat and power – and alternative fuel production from sugar mills. The MEP shall also improve levels of international cooperation in the field of technology transfer and financial support. The broad objective of the directorate of renewable energy, one of the four technical directorates of the MEP, is to promote the development and use of renewable energy technologies. Renewable energy support tools included in the Energy Act are income tax holidays for relevant generation and transmission projects; full custom and import duties exemption for exclusive renewable energy equipment. In May 2008, a feed-in tariffs policy on wind, biomass and small hydro-resource-generated electricity was implemented by the MEP.

The feed-in tariff for biomass derived electricity up to 40 MW is USD 0.07 for firm power and USD 0.045 for non-firm power per kWh at interconnection point.

Technology and processes

UNIDO identified, based on the previous implementations at the Bungoma municipal slaughterhouse and the Homa Bay municipal slaughterhouse, the high-performance temperature-controlled (HPTC) biogas digester model marketed by Rottaler as most suitable to process the waste from slaughterhouse (Figure 94). Wastewater and solid waste from Nyongara Slaughter House is transferred daily to the hydrolysis tanks of the biogas plant where complex carbon chains (carbohydrates, proteins and fats) in the organic materials are decomposed into lower organic compounds (amino acid, sugar and fatty acids), which is the main substrate for the methane producing bacteria. These lower compounds are then released into the digester. The digester is an anaerobic (airtight) system in which the methanogens have the best living conditions (37°C) to produce methane out of organic material. Biogas produced in the digester is collected in a balloon (gas holder) from which it is fed either into the generator for

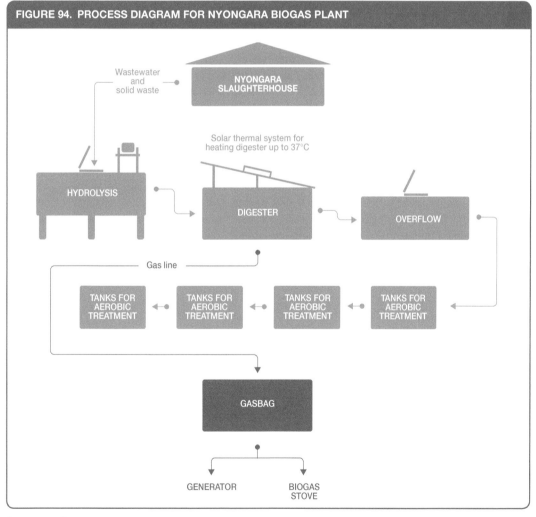

FIGURE 94. PROCESS DIAGRAM FOR NYONGARA BIOGAS PLANT

Source: http://www.kirdi.go.ke

electricity or into the gas compressor to send pressurized gas to the burners. Along with newly-installed solar panels, enough biogas-electricity is being generated to power the slaughterhouse including its cold-storage facilities. The biogas yield of a plant depends not only on the type of feedstock, but also on the plant design, fermentation temperature and retention time. One cubic meter of biogas can be converted only to around 1.7 kWh.

The output slurry from the digester can be further sterilized after the aerobic treatment and used for surface irrigation. Through adding oxygen into the system the BOD and the bacteria can be reduced to an amount which is not effecting the environment and groundwater. Generator operated with 100% biogas supplies the slaughterhouse as well as biogas plant with electricity. Whereas using the gas for direct combustion in household stoves or gas lamps is common, producing electricity from biogas is still relatively rare in most developing countries.

Funding and financial outlook

The capital cost for the Nyongara biogas plant was financed by UNDP, UNIDO, KIRDI and the proprietor of the slaughterhouse. Information on the total investment cost was not provided by the entity. However, it was estimated to be about USD 40,000. Based on limited data provided by the enterprise and from the literature, approximate savings from biogas was calculated. The opportunity cost of using biogas instead of electricity was taken into consideration. In Kenya, 1 kWh of electricity is priced at approximately 16 Kshs (0.2 USD, at an exchange rate of 1 USD = 80 Kshs). Therefore, the 72.6 kWh electricity generated from the biogas will result in daily cost savings of 1,161 Kshs (15 USD) and 34,848 Kshs (436 USD) per month or 5,227 USD per year, with payback period of less than eight years. If the value of heat energy used and revenue by sale of bio-fertilizer is added, the payback period is expected to be in the range of 4–5 years.

Socio-economic, health and environmental impact

The wastes arising from blood and ingesta combined with the large volume of water used to wash off these wastes constituted the greatest proportion of environmental hazards associated with day-to-day slaughterhouse activities. Slaughterhouse wastes pose a serious threat to the environment and the general population at large because of poor waste-management practices which results into adverse impacts on water, land and air (water being the most affected). The adjacent land of most slaughterhouses is often marshy due to improper drainage of wastewater arising from washings in the slaughterhouses. Land pollution occurs when solid wastes such as hides, hooves, horns and ingesta/dung are left unattended on open land. When the rain falls, these wastes are washed into nearby sewerage channels or streams. The project was initiated to assist slaughter-house units to meet the environmental regulations, and thus reduce pollution of Nairobi River. NEMA had closed down all the slaughter-house units in Dagorretti for flaunting environmental regulations. The biogas plant now provides solutions for the closed down units to treat their waste and meet NEMA's requirements.

The economic benefits include reducing the cost of energy from USD 0.20 to USD 0.09 per kW. The biogas produced is used for electricity generation of about 30 KW resulting in reduction in GHG emissions and improving the energy security for the region. The project has cut CO_2 emissions by 108 tons per year. This will not only help to stop deforestation as people look to cleaner, greener sources of fuel but will also assist the country to cut GHG gases and hence their devastating effects.

The biogas plant provides employment in operation and maintenance. Reopening of these closed units will result in restoration of lost jobs. Additionally, if the enterprise gets into production of organic fertilizer from the slurry output, it provides additional benefits to the environment by improving soil quality and carbon offset in comparison to chemical fertilizers.

Scalability and replicability considerations
The key drivers for the success of this business are:
- Strict implementation of environmental regulations from NEMA and closure of slaughter-house units.
- Technology transfer from Germany by UNIDO and financial assistance from UNEP and KIRDI.
- Electricity shortage in Kenya.
- Government promotes renewable energy.

The project could be a blueprint for slaughterhouses across the continent and an important example of how to reduce water pollution from industrial sector. As mentioned earlier, the proprietor of Nyongara biogas plant is planning to scale up the biogas plant and its operations to process waste from all the slaughter-house units. The initial pilot unit is realizing savings from reduced electricity usage and achieving cost recovery. This solution has high replication potential where all slaughter-house units can have an individual retrofit unit within their respective premises. This solution is applicable not only for waste from slaughter-house units but also other organic solid waste. The company is now looking to other products and employment opportunities including poultry feed and pet food as it seeks to be a zero-waste operation. Scientists and engineers from the KIRDI were involved in the implementation from the very beginning of the activity, which enabled UNIDO to transfer the know-how and skills to local technicians, so that the maintenance, replication and up-scaling process would be very smooth.

The two-stage biogas digester technology is widely used for commercial power generation in Europe and the USA. The design was adapted by UNIDO to meet local African requirements (ease of replication, up-scaling and maintenance) and can be implemented in any place where organic waste (food waste, market waste, fish waste, slaughter-house waste, agro-waste, chicken or animal manure) is available. This is the third HPTC biogas project UNIDO is completing in Kenya and, with the support of Government of Kenya through the Ministry of Industrialization, UNIDO would like to implement this model of biogas digesters as a standard in all mid-size slaughterhouses in Kenya. UNIDO will also look into establishing mini tanneries to process raw hides and skin to wet blue, which would further facilitate effective waste treatment as well as energy and employment generation in and around the slaughterhouses. Almost all well-known biogas power plants in developing countries depend on financial support from a third international party. Many new studies come to the conclusion that biogas power plants are not commercially viable without subsidies or guaranteed high prices (approx. 0.20 USD) for the produced outputs.

Summary assessment – SWOT analysis
The key strength Nyongara biogas plant is the reliable technology that can treat the waste to required local environmental norms. In addition, demand for electricity generated is high (Figure 95). The weakness is the investment required is high and financial assistance is needed for small enterprise. The business has the opportunity to both scale up and scale out, and scaling would require increased effort in building sound partnerships. The biggest threat for the business is from competition. The competition is not only from local electricity authority, but also once other slaughterhouse units have access to the technology, they can plan for a similar business approach and eat into Nyongara's market.

Contributors
Johannes Heeb, CEWAS
Jasper Buijs, Sustainnovate; formerly IWMI
Kamalesh Doshi, Simplify Energy Solutions LLC, Ashburn, Virginia, USA

FIGURE 95. NYONGARA BIOGAS PLANT SWOT ANALYSIS

	HELPFUL TO ACHIEVING THE OBJECTIVES	HARMFUL TO ACHIEVING THE OBJECTIVES
INTERNAL ORIGIN ATTRIBUTES OF THE ENTERPRISE	**STRENGTHS** - Availability of waste - Low labor requirement and potentially lower O&M - Able to process waste to meet environmental regulations - Strong technology partnership	**WEAKNESSES** - Dependence on one type of waste and buy in from other slaughterhouse units if they plan to scale up - Significant land required - Potentially financing constraint for additional investment - Import of technology and equipment required - Lack of skilled technicians and engineers - High maintenance cost
EXTERNAL ORIGIN ATTRIBUTES OF THE ENVIRONMENT	**OPPORTUNITIES** - Environment stress reduction offers market opportunities with other slaughterhouse units - High potential for scaling in a cluster set up and can also be easily replicated - Favourable regulatory framework and feed-in tariffs for renewable energy power - Cost-effective internal use of electricity - Electricity shortages and frequent power blackouts	**THREATS** - Possible human and environment health risk from biogas outlet slurry may lead to need for additional investment - Other slaughterhouse units can install their respective biogas plants and sell electricity - Political instability in Kenya - Lack of awareness and local capacity - Lack of finance

References and further readings

Alumasa, P., Fattal, B. and Shuval, H. May 2011. Overview of Conservation Approaches and Technologies – A Programme overview. Canada.

Economic Consulting Associates Ltd. June 2012. Ramboll technical and economic study of small scale grid connected renewable energy in Kenya.

KIRDI. March 2012. Carbon footprint report for Nyongara biogas plant.

Practical Action. 1997. Energy options: An introduction to small scale renewable energy technologies.

Rottaler Modell. www.rottaler-modell.de./rottaler-modell (accessed August 18, 2017). See also https://energypedia.info/wiki/Biogas_Plant_System_-_Rottaler_Modell (accessed November 7, 2017).

Case descriptions are based on primary and secondary data provided by case operators, insiders, or other stakeholders, and reflects our best knowledge at the time of the assessments 2013/2014. As business operations are dynamic, data can be subject to change.

CASE
Combined heat and power and ethanol from sugar industry waste (SSSSK, Maharashtra, India)

Solomie Gebrezgabher and Hari Natarajan

Supporting case for Business Model 8	
Location:	Someshwarnagar, Maharashtra, India
Waste input type:	Bagasse, molasses, spent wash (Distillery effluent)
Value offer:	Electricity/heat, ethanol, pressed mud and bio-fertilizer
Organization type:	Cooperative society
Status of organization:	Operational (since 1962), co-generation unit (since 2010)
Scale of businesses:	Large
Major partners:	Sugarcane farmers (Cooperative society), Government of India, Maharashtra State Government

Executive summary

Shri Someshwar Sahakari Sakhar Karkhana (SSSSK) is a cooperative sugar factory located at Someshwarnagar, taluka Baramati, dist. Pune, Maharashtra that produces sugar from sugarcane grown by its farmer members. In the process producing sugar, it produces bagasse and molasses as waste products. In order to address fluctuations in profits in the sugar unit itself, SSSSK has made additional investments in a distillery unit producing ethanol using molasses; a biogas unit generating biogas using spent wash from the distillery; a cogeneration facility generating combined heat and power using bagasse and biogas and bio-fertilizer using press mud from biogas plant. There is significant demand for the ethanol from companies producing alcoholic beverages as well as pharmaceutical companies. The government has also put in place a requirement of 5% ethanol blending of fuel, which has created a demand for ethanol from petroleum companies. The biogas produced is used internally as input fuel to the boiler while the bio-fertilizer (the discharge from the biogas unit), which is high in organic matter, is distributed at no cost to the farmers in proportion to the cane supplied by them. The electricity generated by the cogeneration unit partially used internally and surplus is sold to the state electricity utility by way of a long-term power purchase agreement at a pre-determined incentive tariff that has been set by the MERC. SSSSK assists the member cane growers to increase cane growth by providing quality seed and tissue culture plantlets, introducing modern techniques of cultivation, integrated nutrient management, irrigation water management and pest management and providing bio-fertilizers to its members. SSSSK's operations have led to significant socio-economic benefits for the local community in terms of the creation of livelihood for member farmers, job creation and improving the quality of basic infrastructure such as roads and access to healthcare and education. Moreover, the whole process results in CO_2 offset due to use of non-fossil fuel for electricity generation.

KEY PERFORMANCE INDICATORS (AS OF 2012)	
Land:	60 ha
Water requirement:	130 m³/hr
Capital investment:	Co-generation unit = USD 20.8 million, Distillery unit = USD 4.4 million, Biogas unit = USD 0.53 million
Labor:	150 full-time and approximately 50 temporary/seasonal persons
O&M cost:	2.5% of capital costs (643,000 USD/year)
Output:	18 MW electricity/year, 5 million L of alcohol/year
Potential social and/or environmental impact:	20,000 farmer members benefit, 200 jobs in the local community, overall socio-economic development in the community – Roads, healthcare and education, CO_2 offset
Financial viability indicators	Payback period: 3–5 years for co-generation distillery. Post-tax IRR: N.A. Gross margin: 10–12%

Context and background

The cooperative ownership structure, prevalent in the sugar industry in Maharashtra, India, enables to undertake activities with a common goal by formation of non-profit economic enterprises for the benefit of their members. SSSSK is a cooperative sugar factory in taluka Baramati, dist. Pune that produces sugar from sugarcane grown by its more than 20,000 farmer members across 40 villages and four talukas. It was established in 1961 with a crushing capacity of 1,016 tons of cane crushed per day (TCD), which has since been increased to 3,600 TCD.

In order to improve the utilization of the waste streams from the process of making sugar, SSSSK has made various investments since its establishment, which include a distillery unit producing ethanol using molasses; a biogas unit generating biogas using spent wash from the distillery and a co-generation facility generating combined heat and power using bagasse and biogas and bio-fertilizer using press mud from biogas plant. The co-generation plant was commissioned on May 21, 2010. In 2011–2012, SSSSK generated 99 million kWh, consumed 29 million kWh and sold 70 million kWh to the grid.

Market environment

Sugar is India's second largest agro-processing industry. There are more than 500 sugar factories with about 5 million hectares of land under sugarcane with an average yield of 70 tonnes per hectare. Biggest problem the sugar industry facing today is surplus production, from 10 lakh (1 million) tonnes in 1950 to over 200 lakh (20 million) tonnes in more recent years. While consumption of sugar is increasing at a steady pace of 4–5% per annum, it does not match the increase in production. As a result, prices of sugar have been steadily sliding over years. With the advancement in the technology for generation and utilization of steam at high temperature and pressure, the sugar factories can also produce significant surplus electricity for sale to the grid using same quantity of bagasse. For example, if steam generation temperature/pressure is raised from 400 °C/33 bar to 485 °C/66 bar, more than 80 KWh of additional electricity can be produced for each ton of cane crushed. The sale of surplus power generated through optimum cogeneration would help a sugar mill to improve its viability, apart from adding to the power generation capacity of the country.

One of the fastest-developing countries in the world today, with economy in transition, India consumes 12.18 quadrillion Btu (Quads) of power, with over 8–10% growth per annum. India's annual per capita energy consumption of 0.7 tonnes of oil-equivalent and its electricity consumption of roughly 835 TWh is less than one-seventh of that of developed countries. It is of utmost importance for business and industry to have adequate, economical, reliable, high-quality power supply. The market for SSSSK's electricity is the state electricity grid. There is significant demand for electricity, given that the state of Maharashtra has been suffering from an energy shortage in excess of 10% for the past decade. In order to address the energy gap, the state electricity regulator (MERC) established an incentive tariff for cogeneration units in 2002 to maximize generation from existing resources. SSSSK's cogenerating unit is capable of generating 18 MW power, 15.8 MW of which is exported to electricity grid based on a long-term power purchase agreement. The incentive tariff set by the regulator is 0.096 USD/kWh. Bagasse-based co-generation of power in India has come to a take-off stage. The lessons learned during the last decade have been extremely useful for achieving accelerated growth in the near future.

There is also a significant demand for alcohol from companies producing alcoholic beverages as well as pharmaceutical and petroleum companies. The government has put in place a requirement of 5% ethanol blending of fuel, which has created a demand for ethanol from petroleum companies. SSSSK is benefiting from government incentives for co-generation units and government regulations in relation to ethanol blending of fuel. The biogas plant has capacity of 14,787 m^3/day which contributes up to 5% input requirements of the boiler, although it was established primarily to address the concerns of the pollution control board with regards to discharge of the spent-wash from the distillery unit.

Macro-economic environment
Bagasse co-generation
The Indian government has set the challenging goal of increasing its electricity capacity six to eight-fold in the next 30 years in the context of significant capacity shortfalls and a financially-ailing electricity sector. The potential from about 575 operating sugar mills spread over nine major states has been identified at 3,500 MW of surplus power by using bagasse as the renewable source of energy. Given that the installed capacity of the total biomass/bagasse-based distributed generation is only 20% of the total estimated resource, the potential benefits of more projects is vast.

The conditions that support the growth of the bagasse-based cogeneration in India started 1992 when a number of domestic and international programs was launched to support the dissemination of bagasse-based cogeneration technology. These support programs include:
- The launching of a national program on promotion of bagasse-based co-generation by the Ministry of Non-conventional Energy Sources (MNES) in 1992 by offering capital and interest subsidies, fiscal incentives, research and development support, accelerated depreciation of equipment, a five-year income tax holiday and concessional import duty, excise and sales tax exemptions.
- Extension of loans for cogeneration by Asian Development Bank (ADB) through the Indian Renewable Energy Development Agency (IREDA).
- International funding for bagasse co-generation from the United States Agency for International Development (USAID).
- The on-going power sector reforms, unbundling of utilities and the enactment of the Electricity Bill 2003, provide further opportunities to sugar mills to emerge as power producers.
- Adoption of Electricity Bill 2003 by states.
- State subsidies and support provided by Maharashtra Energy Development Agency (MEDA).
- Innovative financial mechanisms, including trade of emission reductions from these projects under CDM of the Kyoto Protocol.

- The conducive Central and State Electricity Regulatory Commission's orders on preferential feed-in tariffs.
- Trading of energy in form of renewable energy certificate are allowed since 2010 for which CERC as well as the state ERCs have promulgated various regulations.

There are however barriers to accelerated growth of the bagasse co-generation sector in India such as the non-availability of sustainable policy and regulatory framework regime across different sugar-producing states and no opportunity of exporting electricity outside the state.

Ethanol from molasses

India's transport sector is growing rapidly and accounts for over half of the country's oil consumption whilst the country has to import a large part of its oil needs. In 2002, the government mandated that nine states and four federally-ruled areas will have to sell E-5 (5% ethanol blending) by law from January 1, 2003 which boasted the production of ethanol from sugar factories. The price of ethanol from sugar factories has been set by the government. However, ethanol pricing in India is complicated by differences in excise duty and sales tax across states, and the central government is trying to rationalize ethanol sales tax across the country. This has made ethanol production an uncertain venture and hence hindered the growth of ethanol production.

Business model

The electricity generated by the co-generation unit is purchased by the state electricity utility by way of a long-term power purchase agreement at a pre-determined incentive tariff that has been set by the state electricity regulatory commission (Figure 96). This provides the enterprise a reliable source of revenue for the length of the contract. The incentive mechanisms coupled with the power purchase agreement can make the resulting electricity competitively priced on the open market. The alcohol is sold to producers of alcoholic beverages as well as pharmaceutical and petroleum companies. The biogas produced is used internally as input fuel to the boiler. The discharge from the biogas unit which is high in organic matter is distributed at no cost to the farmers in proportion to the cane supplied by them in order to increase farm productivity, and hence increase reliability and consistency of sugarcane supply to the enterprise.

Value chain and position

SSSSK is vertically integrated and owns its raw materials for the main product (sugarcane) and for its energy-producing units (Figure 97). However, there is a threat that fluctuating sugar prices might force farmer members to shift to other crops. Since area allotted to factory is fixed by the government so as to ensure consistent supply of cane, it becomes all the more important to develop harmonious and good relations with these growers so that they do not switch to alternate cash crops. Hence from Porter's five forces lens, the supplier power is medium.

SSSSK's main buyers are the Maharashtra state utility for its electricity and the petroleum, pharmaceuticals and alcohol companies for its alcohol. The state utility is the only buyer of SSSSK's electricity based on a long-term power purchase agreement. However, the feed-in tariff is decided by the MERC with its terms and conditions, bringing the bargaining power of state utility to medium level. The substitutes for electricity from co-generation unit are electricity from fossil fuel and other renewable energy sources such as biogas, hydropower, wind, solar energy, etc. In the short term, the threat of substitutes is low as SSSSK has a long-term power purchase agreement with the state utility.

The alcohol from the distillery is sold to various industries and buyers with the introduction of the government requirement for 5% ethanol blending of fuel. However, the price of ethanol from sugar

CASE: HEAT, POWER AND ETHANOL FROM SUGAR INDUSTRY

FIGURE 96. SSSSK BUSINESS MODEL CANVAS

KEY PARTNERS	KEY ACTIVITIES	VALUE PROPOSITIONS	CUSTOMER RELATIONSHIPS	CUSTOMER SEGMENTS
• Sugarcane farmers • Maharashtra State Cooperative Bank	• Build, operate, own and maintain co-generation, distillery and biogas unit • Sales of electricity and ethanol • Distribution of the digested solids to the farmers as organic fertilizer	• Electricity produced based on a waste-to-energy approach, contributing to environmental protection and reducing the energy gap in the region • Distillate of various grades of alcohol for production of goods • Organic fertilizer for increasing yield at the sugarcane fields	• Direct • Recurrent purchase based on client satisfaction • Agreement with suppliers of sugar cane that receive organic fertilizer in proportion to the cane supplied by them	• Maharashtra State Electricity Distribution Company (MSEDCL) • Petroleum, alcohol and pharmaceutical companies • Sugarcane farmers
	KEY RESOURCES • Equipment • Bagasse, water • Technical and operational competencies • Capital • Agreement with the Maharashtra State Electricity Distribution Company • Reliable yield at sugarcane fields		**CHANNELS** • Connection to the state electricity grid • Direct sales of distillates • Direct transport of fertilizer to the sugarcane farmers	

COST STRUCTURE	REVENUE STREAMS
• Capital investment (95% financed by debt): Co-generation unit; Distillery unit; Biogas unit • Repayment of debt and dividends on equity • Input cost (bagasse) • Interest on borrowed fund • O&M • Marketing and sales (alcohol)	• Sale of electricity • Sale of alcohol/ethanol

SOCIAL & ENVIRONMENTAL COSTS	SOCIAL & ENVIRONMENTAL BENEFITS
• Flue gas emissions and fly ash from the boiler • Significant water requirement	• Livelihood, advise and support to farmer members • Creation of jobs • Climate change mitigation through use of non-fossil fuels • Contribute to improving quality of basic infrastructure (roads, access to healthcare and education) in rural areas benefitting members

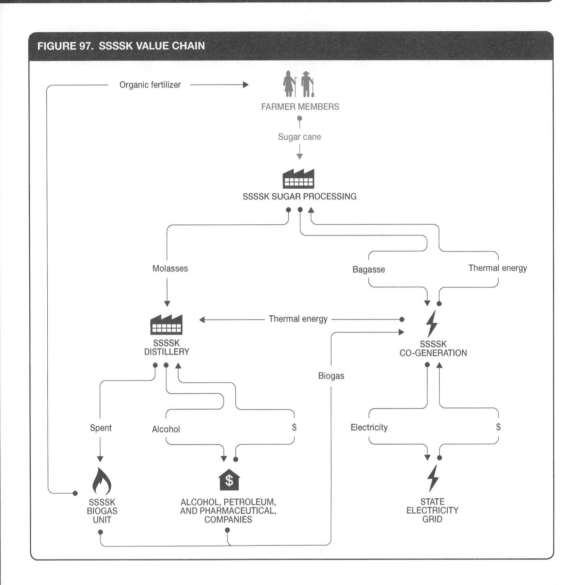

FIGURE 97. SSSSK VALUE CHAIN

factories is set by the government due to which SSSSK has medium bargaining power. The competition among existing companies is low; however, the competition in the sugar commodity market is high due to surplus production. While consumption of sugar is increasing, it does not match the increase in production and this drives sugar price down. The performance of the cogeneration unit is also highly dependent on government subsidy. By-products from the boiler are boiler ash and flue gas for which there are prescribed standards by pollution control board so as to minimize damage to the environment.

Institutional environment

Indian sugar industry comprises of a mix of private and cooperative units and is highly regulated by central and state government bodies across the entire value chain including sugarcane procurement area, pricing of sugarcane and production of alcohol under the Essential Commodities Act 1955 (subsequently amended in 2003). The government has established various support structures, such as the Sugar Development Fund, which provides concessional loans for upgrading and modernization efforts as well as for establishing ethanol plants and power-generation units. Hence, both the co-generation and distillation approaches are well encouraged by the government.

The electricity generated by the co-generation unit is purchased by the state electricity utility by way of a long-term power purchase agreement. The Electricity Bill 2003 enacted subsequently by the Government of India has provided major impetus. The bill has recommended that the states should generate a minimum 10% of energy from renewable sources. It also gave supreme powers to the State Electricity Regulatory Commissions (SERC) for deliberating and deciding tariffs and other terms and conditions for all renewables, including bagasse co-generation. IREDA, a Government of India enterprise and the lending arm of MNES, has provided promotional/development finance for harnessing biomass energy in India over the last 10–12 years.

Industry associations like the Cogeneration Association of India, financial institutions and other stakeholders are pursuing the Central Electricity Regulatory Commission (CERC) to guide SERC to adopt a uniform tariff order for bagasse co-generation in the entire country.

Technology and processes

Sugarcane needs to be crushed within 24 hours of harvesting, else it starts deteriorating resulting in reduced recovery of sugar from the sugarcane. The sugar industry is an energy-intensive industry. Therefore, apart from sugarcane, steam and electricity are essential for running the mill. For this reason, most of the sugar mills have a co-generation unit for supply of steam and electricity. The efficiency of co-generation plant is in the range of 75–90%, as compared to the conventional plant of 35%, because the low-pressure exhaust steam is used for heating purposes in the factory.

The typical recovery of sugar from cane during the process of making sugar is 12%. The balance sugar is available in the molasses, which is a by-product of the process (Figure 98). SSSSK's distillery unit processes the molasses to produce various grades of alcohol such as rectified spirits, extra neutral alcohol, impure alcohol and ethyl alcohol. The distillery unit consists of a multi-stage fermentation process, which is then distilled through separation columns to obtain various grades of alcohol. This unit has a capacity to produce up to 30,000 L of alcohol (95% pure) per day and requires approximately 500,000–600,000 L of water per day. The spent wash from the distillation process, which is high on fructose, is passed on to the biogas plant (18,000 m^3/day equivalent to 40–45 tons of bagasse (input requirements of the boiler for one hour). The biogas contributes up to 5% of the input fuel requirements of the boiler and is generated through a two-stage process comprising of an acid preparation stage followed by methanogenesis. Compost fertilizer is prepared from press mud and spent wash by adding microbial culture with the help of mixing cum aeration machine.

The original setup of the co-generation unit had two low-pressure boilers (16 kg/cm^2 and 21 kg/cm^2) that generate heat to meet the internal process requirements and was capable of generating approximately 2.75 MW power. SSSSK later replaced the low-pressure boilers with a multi-fuel capability, high pressure 100 TPa, 87 kg/cm^2, 570 °C boiler. The input to the boiler includes 42–45 tons/hour of bagasse (supplemented by 18,000 m^3/day of biogas) and 25 m^3/hr of treated water, which is obtained from a reverse osmosis system. The steam generated in the boiler drives a steam turbine, which is connected to a synchronous generator capable of generating 18 MW of power during crushing season. Steam is also extracted at different stages of the turbine to meet the process heat requirements of the sugar and distillery units. Suitable suppliers of equipment were locally available for all the above technologies, with a consultant providing turnkey services for installation, commissioning and preliminary testing.

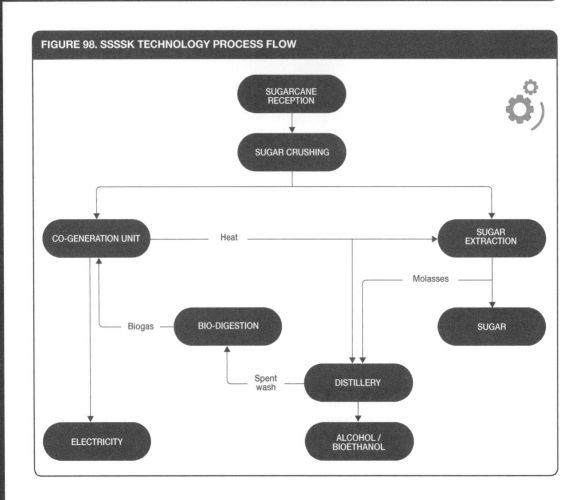

FIGURE 98. SSSSK TECHNOLOGY PROCESS FLOW

Funding and financial outlook

The 18 MW co-generation unit was established at a capital cost of USD 20.8 million. The distillery unit was established at a capital cost of USD 4.4 million. The biogas unit was established at a capital cost USD 0.53 million. The investment for these units was funded predominantly through debt with approximately 5% equity contribution. The loans were obtained under the Sugar Development Fund established by the government, which provided concessional term loans at 4% interest rate and up to 8–10 years duration, and the Maharashtra State Cooperative Bank, which offered standard term loans at market rates. SSSSK has not faced any significant challenges in raising funds to meet its investment needs.

Apart from the revenues from sales of sugar, the revenue streams of SSSSK include sales of electricity from its co-generation unit and alcohol from its distillery unit (Table 27). Out of the 18 MW power generated by the co-generation unit, 15.8 MW is exported to the state electricity grid by way of a

TABLE 27. REVENUE STREAMS OF SSSSK

REVENUE STREAM	QUANTITY	AMOUNT (MILLION USD/YEAR)
Sale of electricity	15.8 MW	7.45
Sale of alcohol	5,000,000 L	4.27

long-term power purchase agreement at a pre-determined incentive tariff of 0.096 USD/kWh. The market price for ethanol has been fixed by the government at 0.5 USD/L, whereas the price for other grades of alcohol could be as high as 0.8 USD/L. In the case of pharmaceutical companies, although the demand is small and not regular, the rates could range from USD 1 to USD 2 per litre.

The operating and maintenance cost for the first year is assumed to be 2.5% of the capital cost. The input cost of bagasse is 36.6 USD/ton. The payback period for the co-generation unit is five years, assuming six months of operation and 100% off-take of the surplus energy generated by the state electricity utility. The payback period for the distillery unit is three years. The CERs from such projects can be sold to international emission reduction buyers generating additional revenue.

Socio-economic, health and environmental impact

In addition to the livelihood provided to over 20,000 farmer members and employment opportunities within the factory to over 150–200 staff, SSSSK has been largely responsible for the socio-economic development in the immediate vicinity. SSSSK provides advice and support on sugarcane cultivation to its farmer members, such as nutrient management, irrigation management, pest control and access to subsidized seeds and fertilizers. SSSSK has also improved the quality of the basic infrastructure such as roads, access to healthcare and education. SSSSK has established six schools, a junior college and a professional science college, which provides preferential admission and reduced fees (50%) to the children of its farmer members.

Moreover, the whole process results in CO_2 offset due to use of non-fossil fuel for electricity generation as well as for transportation. The blending of renewable ethanol in petrol reduces vehicle exhaust emissions and also reduces import burden of the country. The Indian project promoters can sale the CERs internationally and ensure additional financial benefits every year. However, there are also potential environmental costs associated with the operation of the business such as issues related to water usage and flue gas emissions from the boiler. The water requirement for the entire operation, at 130 m^3/hr, is quite high and has posed some challenges on account of insufficient release of water from dams/irrigation canals, especially during poor monsoon seasons, due to competing use for irrigation. Another issue is with respect to the flue gas emissions and fly ash from the boiler. These emissions are controlled within acceptable limits with suitable equipment.

Scalability and replicability considerations

The key drivers for the success of this business are:
- Electricity shortage and concurrent government support mechanisms such as provision of concessional loans and feed-in-tariff mechanisms.
- Government regulations stipulating alcohol blending for fuel.
- Securing of long-term power purchase agreements.
- Consistent supply of input for energy producing units as SSSSK is vertically integrated.

SSSSK is an example of the implementation of well-established and mature technologies in the sugar manufacturing industry. These technologies enable the organization to improve the efficiency of the process, reduce waste and improve the overall economics of the entire operation. This business case also highlights the advantages of a cooperative ownership structure, which leads to significant socio-economic benefits for the local community. SSSSK operates in a highly-controlled and regulated environment, which poses some challenges so far as scaling up prospects and profitability is concerned, but the profitability has been enhanced through investments in cogeneration and distillery units which have been made possible through concessional sources of finance and feed-in tariff schemes available through mechanisms established by the government. This business has the potential to be replicated in other similar sugar factories in the state, within India and other countries where there already exists some infrastructure such as a sugar manufacturing company, and the co-generation, distillery and biogas plant could be initiated as a plant within the sugar factory. In order for this business to be replicated in other countries, government support mechanisms are essential.

Summary assessment – SWOT analysis

Key strengths of the business are its application of well-established technologies, which enabled the business to be energy self-sufficient as well as diversify its revenue streams (Figure 99). However, the processes are water intensive and the profitability of the energy producing units depends on government incentives. The latter is not an immediate threat as government support for renewable sources of energy is likely to increase. Fluctuating sugar prices which may force sugarcane growers to shift to other crops and competition from other sugar producing countries such as China and Brazil threaten SSSSK. The energy producing units might result in reduction of GHG emissions, and this presents opportunities for SSSSK to earn carbon credit sales by registering the business as a CDM project.

FIGURE 99. SWOT ANALYSIS FOR SSSSK

	HELPFUL TO ACHIEVING THE OBJECTIVES	**HARMFUL** TO ACHIEVING THE OBJECTIVES
INTERNAL ORIGIN ATTRIBUTES OF THE ENTERPRISE	**STRENGTHS** • Power purchase agreement • Well-established and mature technologies • Diversified revenue streams • Zero-waste process • Reliable sustainable and self-sufficient energy • Vertically integrated • Business arrangement allows competitive pricing • Implied social benefits	**WEAKNESSES** • Significant water requirement • Dependent on government incentives • Inadequate capacity of various players including sugar mills, financial institutions and regulators • Slow adaptation of modern technologies and modernization of old sugar mills • Weak management of existing facilities
EXTERNAL ORIGIN ATTRIBUTES OF THE ENVIRONMENT	**OPPORTUNITIES** • Demand for electricity is growing and government support for energy from renewable sources is likely to increase • Opportunity for registering the project as a CDM and earn additional revenues from sales of carbon credits • High value products for downstream industries • Huge potential to increase productivity of sugarcane and sugar recovery rate	**THREATS** • Fluctuating sugar prices may force sugarcane growers to shift to other crops and decreasing sugar prices may disrupt business • Competition from other sugar producing countries (China, Brazil) • Competition from fossil-fuel-based energy • Insufficient availability of water from dams due to competing uses for irrigation may pose risk of production stoppage • Ethanol production an uncertain venture due to complex sales tax • Procedural delays of ERCs, SEBs and other agencies • Reduction in yields of sugarcane due to single crop cultivation with overuse of fertilizers and pesticides • Poor financial health of power distribution companies

Contributors
Johannes Heeb, CEWAS, Switzerland
Jasper Buijs, Sustainnovate; Formerly IWMI
Josiane Nikiema, IWMI
Kamalesh Doshi, Simplify Energy Solutions LLC, Ashburn, Virginia, USA

References and further readings

Birla. 2010. Management discussion and analysis. www.birla-sugar.com/Images/Pdf/231.pdf (accessed August 18, 2017).

Chapter 5: Cooperative movement in Maharashtra. http://shodhganga.inflibnet.ac.in/bitstream/10603/2502/11/11_chapter%205.pdf (accessed August 18, 2017).

Kadam, P.P. 2010. Role of co-operative movement in sustaining rural economy in the context of economic reforms: a case study of Ahmednagar district. Chapter 5. PhD thesis, Tilak Maharashtra Vidyapeeth; http://shodhganga.inflibnet.ac.in/handle/10603/2502 (accessed November 7, 2017).

Maruti, K. 2012. The problems of sugar cooperatives in Ahmednagar district. Lokavishkar International E-Journal. www.liirj.org/liirj/apr-may-june2012/24.pdf (accessed November 7, 2017).

SINET. 2007. Indian sugar sector network report: Sector overview and SWOT analysis (SINET). www.scribd.com/document/52580216/sugar1 (accessed August 19, 2017).

The Institute of Chartered Accountants of India. 2010. Technical guide on internal audit of the sugar industry. The Institute of Chartered Accountants of India. https://www.scribd.com/document/90459886/Technical-Guide-on-Internal-Audit-of-Sugar-Industry (accessed November 7, 2017).

Vasantdada Sugar Institute. Sugar statistics: Maharashtra. VSI. www.vsisugar.com/india/statistics/maharashtra_statistics.htm# (accessed August 18, 2017).

Case descriptions are based on primary and secondary data provided by case operators, insiders, or other stakeholders, and reflect our best knowledge at the time of the assessments (2015–2016). As business operations are dynamic, data can be subject to change.

CASE
Combined heat and power from agro-industrial wastewater (TBEC, Bangkok, Thailand)

Louis Lebel and Krishna C. Rao

Supporting case for Business Model 8	
Location:	Bangkok, Thailand
Waste input type:	Wastewater from agricultural industries (starch, palm oil and ethanol)
Value offer:	Build, Own, Operate and Transfer (BOOT), one-stop shop to treat agro-industrial effluent and generate electricity from biogas and CER certificates
Organization type:	Private
Status of organization:	Founded in September 2003
Scale of businesses:	Large – TBEC has processed 6,200,000 m³ of wastewater/year, generated 38,360,000 Nm³ of biogas/year, and 26,500,000 kWh of electricity/year by multiple projects
Major partners:	The Private Energy Market Fund (PEMF), Finland and Al Tayyar Energy (ATE), Morocco (Head office in the United Arab Emirates) (provided investment) and Provincial Electricity Authority (purchased electricity)

Executive summary

Thai Biogas Energy Company (TBEC), founded in 2003, is a one-stop shop for premium biogas Build Own Operate and Transfer (BOOT) projects with strong emphasis on high biogas yield, safety, quality of construction and quality of its human resources. It has implemented a number of biogas projects in Southeast Asia to treat effluents from agro-industrial units, such as palm oil and cassava processing plants. The biogas generated from treating wastewater is used to generate electricity, which is sold to the Thai electricity grid via a provincial electricity authority. Some projects have also received carbon credits for contributing to reductions in GHG emissions. These credits are purchased by companies in Europe. The treated wastewater also has useful mineral and nutrients for plants and is sometimes reused to irrigate rubber trees, or more typically released into public canals. Through its business model, TBEC's investment results in employment of local labor for biogas plant construction. TBEC also shares its revenue, technology and expertise with the host company and provides training to facilitate easy transfer of the project at the end of BOOT period. Since 2016, TBEC has been managed by Asia Biogas Group. It has 8 power plants in Thailand and 1 in Lao PDR. TBEC projects produce 44 million m³ of biogas or 88 GWh equivalent of biogas annually, and reduce greenhouse gas emissions by 320,000 tCO_2e per year.

TBEC is certified ISO 9001, ISO 14001, and follows the guidelines of the International Finance Corporation (IFC) of Thailand on global warming. TBEC is the market leader in the Mekong area for biogas projects for cassava wastewater in Rayong, Kalasin, Saraburi in Thailand, and for the palm oil and rubber industry in Surat Thani. The TBEC Tha Chang Biogas Project won many awards, including Best Biogas Project in Asia Selling Electricity to the Grid at the ASEAN Energy Award in 2010, the Crown Standard from the Thailand Greenhouse Gas Management Organization (TGO) and the designated national authority (DNA) of Thailand and Gold Standard by the World Wide Fund.

KEY PERFORMANCE INDICATORS (AS OF 2013)	
Land:	Land is provided by concessionaries/industry owners
Water requirements:	Most is 'wastewater' output – 25,000 m³ of treated wastewater/day
Capital investment:	Highly project-specific depending on scale, location, labor and benefit sharing arrangements with concessionaires, but as an illustration installing a 1.4 MW biogas power plant involves investment costs of approximately 3.5–3.9 million USD in 2008
Labor:	116 full-time employees (including O&M of multiple plants)
Output:	25,000 m³ of treated wastewater/day; Across several projects, TBEC has processed 6,200,000 m³ of wastewater/year, converted 97,250,000 kg COD/year into around 38,360,000 Nm³ of biogas/year, 26,500,000 kWh of electricity/year and 250,000 tCO_2e/year of CERs
Potential social and/or environmental impact:	Reduced dependence on imported fossil fuels for power generation; CO_2 emission reduction; local jobs in construction of plant; skilled jobs in operation and maintenance; reduced nuisance odors and water pollution
Financial viability indicators:	Payback period: N.A. Post-tax IRR: N.A. Gross margin: N.A.

Context and background

Recognizing the need to reduce GHG emissions to mitigate climate change, TBEC promotes use of cost-effective and environmental-friendly renewable energy such as biogas generated from agro-processing wastewater. TBEC have hired Waste Solutions Ltd, a New Zealand firm of technology developers and consulting engineers, to design the plants. TBEC adopts a BOOT model, bringing in investment to set up biogas plants that treat wastewater from agro-industry factories that provide land and inputs. TBEC finances, designs, constructs, operates and maintains the plant until BOOT term expires. TBEC recovers its costs by producing electricity and selling it to a provincial electricity authority. Training is provided to help the host company after transfer of project. The business has operated projects in Thailand and in Lao PDR and is developing ones in Myanmar, Cambodia and Vietnam. The TBEC has installed and is operating six projects at starch units and three projects at palm oil mills. Examples of plants installed and operated by TBEC include Rayong, starch plant (15,000 m³/day biogas, 1.4 MW of power); Kalasin, starch plant (30,000 m³/day, 2 MW); Saraburi, high fructose syrup from cassava (25,000 m³/day and 1 MW in Lao) and Thachang project at palm oil mill and concentrated latex plant (35,000 m³/day and 2.8 MW). The Thachang project has targeted CO_2 emission reduction of 51,823 tons/year. The construction of the Thachang project started in January 2007 and commissioned in November 2008. Operation started in January 2009, and registration with UNFCCC was in September 2010.

TBEC raised finance from the Private Energy Market Fund (PEMF) in Finland and Al Tayyar Energy (ATE) in Morocco for setting up these plants. PEMF is a private equity fund for alternative energy development and power conservation. It holds about 70% of TBEC. Al Tayyar Energy (ATE) is a clean power development and investment company founded by HRH Prince Moulay Hicham Ben Abdallah Al Alaoui of Morocco. It has head office in the UAE. The company primarily focuses on bio-energies, such as biofuels, biogas and biomass. It also invests in solar, wind and hydroelectric project companies.

Market environment

Thailand is the world's third largest producer of crude palm oil and has one of the largest tapioca processing industry. Most agricultural production processes have significant amounts of organic residue output as a by-product. There are also many underutilized agro-processing waste sources not only in Thailand but also around the region. Due to increasing pressure to reduce GHG emissions, such agro-processing units, the customers of TBEC are looking for ways of treating wastewater from such processing of agricultural products including palm oil or starch from cassava. The waste-to-biogas and power business of TBEC contributes to greater use of renewable energy, allowing the firm to make a profit by selling electricity at preferential prices, as well as carbon credits while improving the environment. The electricity generated is directly sold to the grid of the Provincial Electricity Authority (PEA). Electricity demand is expected to continue to grow over the coming decades despite significant efforts in improving efficiency. Electricity prices are regulated by the government to ensure electricity is priced at a rate which is accessible to both residential and industrial users.

With high quality and safety standards, TBEC is a premium product company with around 10 competitors. For instance, Asia Biogas Company Ltd, Prapob Company and several other newcomers. Most of the new enterprises contract for construction of biogas plants and do not invest and operate the plant. As of 2008, 21 CDM projects in the palm-oil sector were registered with the Thai Greenhouse Gas Management Organization (TGO). The number of approved CDM projects in Thailand is still limited due to the high level of burdensome bureaucratic procedures involved.

Macro environment

The fossil fuels account for 80% of the total energy supply in Thailand. The Government of Thailand targets to increase the share of alternative energy from 6.4% at present to 20.3% of commercial primary energy by 2022, as per the Renewable Energy Development Plan. To achieve the above targets, the Government of Thailand supports the projects by several incentives such as subsidization, soft loan, tax incentive, Board of Investment (BOI), Energy Service Company (ESCO) Fund, CDM, adder cost, etc.

Thailand, with its abundant and varieties of biomass and agricultural wastes, has the great challenge and opportunity for the waste-to-energy projects to supply renewable energy-based electricity. Thailand's Ministry of Energy estimates that the potential of power generation in Thailand from biomass, MSW and biogas is 3,700 MW. Bio-based renewable energy (RE), such as agricultural residues, crops, biogas from biomass and wastes, MSW and biofuels, has shared in a large portion of RE more than 90% of potential RE in Thailand. For example, with 64 palm-oil mills, Thailand had a potential of more than 5 million m^3 of biogas/year from palm oil mill effluent (POME) that can generate more than 50 GWh of electricity/year.

Business model

TBEC develops, designs, finances and operates biogas projects on a Build-Own-Operate-Transfer (BOOT) while the concessionaries provide land and inputs and operates the plant after expiry of BOOT period (Figure 100). The BOOT period is flexible and depends on type and characteristics of individual

CASE: HEAT AND POWER FROM AGRO-INDUSTRIAL WASTEWATER

FIGURE 100. TBEC BUSINESS MODEL CANVAS

KEY PARTNERS	KEY ACTIVITIES	VALUE PROPOSITIONS	CUSTOMER RELATIONSHIPS	CUSTOMER SEGMENTS
• Wastewater producers • PEMF, Finland • Al Tayyar Energy, Morocco • Provincial Electricity Authority	• BOOT biogas projects • Selling the biogas to the boiler for drying process • Selling renewable electricity • Marketing carbon offsets • Producing and selling fertilizer (future option) • Training BOOT customers as part of quality services	• Wastewater treatment for agro-industries through well-serviced BOOT model • Renewable electricity supply to the Thailand state electricity • Carbon offset generation • Organic fertilizer (future option)	• Agreements with agro-industries • Direct PPA with PEA • Direct with carbon trading markets	• Agro-industry that process cassava, palm oil or ethanol • Provincial electricity authority • European institutions purchasing carbon offset • Farmers (future option) needing fertilizer
	KEY RESOURCES • Wastewater • Land • Capital • Skilled labor • Experienced top management team • Technology • Government policies • Longer-term agreements with suppliers • Long-term power purchase agreement with the Electricity Authority		CHANNELS • Direct to biogas plant from the host company • Connect to national grid • Carbon trading markets	

COST STRUCTURE	REVENUE STREAMS
• Infrastructure investment (high) • O&M cost (high) • Training	• Sale biogas to the host company for drying process • Sale of electricity to provincial electricity authority • Sale of CO_2 offset through CDM • Sale of fertilizer (future option)

SOCIAL & ENVIRONMENTAL COSTS	SOCIAL & ENVIRONMENTAL BENEFITS
• In case of leakage of gas, there are consequences to the environment	• Jobs creation • Indirectly increase income to the farmers • Reduce the odor of the wastewater • Environmental benefit through reduced CO_2 emissions by generating electricity from renewable source and reducing pollution (fossil fuels substitution)

projects. It normally takes between 15 to 17 years before the transfer is made to the host company. Thus, key customers are the agro-industrial unit and the entity purchasing electricity which is the Provincial Electricity Authority (PEA) in Thailand. TBEC's unique selling point is quality of product and service, appealing to higher-value markets.

Value chain and position

Figure 101 describes the relationship between some of the key value chain actors in a typical TBEC project. TBEC treats wastewater from agro-processing units (like palm oil) to generate electricity. The major supplier of the plant's raw material is the agro-industry with which TBEC has an agreement to treat waste from the process. Threat to supply of effluent does not exist due to such agreements. The biogas it generates from treating wastewater is used to produce electricity, which is sold to the Thai electricity grid. TBEC has a PPA with the electricity authority, and thus threat of buyer power is low. Electricity as well as thermal energy (heat) could also be sold back to host agro-processing units directly under energy purchase agreement. Carbon credits may be purchased by companies in Europe. The BOOT agreements cover sale of concessions to partners. Thus, the specific role of TBEC in a project can be substantial over time and has certain challenges. Biogas power plants are quite complicated and require careful supervision. Unprofessional management can reduce cost-effectiveness and increase risks. Seasonality of biogas production can cause trouble with production planning.

TBEC develops a project under CDM to obtain CERs and successfully completed United Nations Framework Convention on Climate Change (UNFCCC) registration of all its projects as CDM projects. For example, the actual GHG emission reductions or net anthropogenic GHG removals by sinks achieved during the period of January 1, 2013 to December 31, 2014 were 91,678 tCO_2e against the estimated amount of 51,823 tCO_2e for Thachang project. The examples of the biogas yield from different dry substrates are as follow: 200–400 m^3/ton of cattle manure and dung, 250–450 m^3/ton of pig and chicken dung, 350–700 m^3/ton of energy crops and 700–900 m^3/ton of POME (FNR, 2007 and 2009). A value-added option is to turn the biogas into green gas by removing CO_2 and other gaseous components (H_2S, H_2O) content and increasing the percentage of methane. Compared to the biogas, the green gas and natural gas contain 29% more methane.

Institutional environment

Thailand is one of the first countries in Asia to have a policy to encourage biofuels, cogeneration, distributed generation and the generation of power from renewable energy. Co-generation and the production of power from renewable energy is implemented under the Small Power Producer Program (SPP) of 10–90 MW capacity and Very Small Power Producer Program (VSPP) of less than 10 MW capacity. It became a very effective policy instrument in promoting investment in renewable energy and co-generation. Private power producers sell electricity to the electric utilities under power purchase agreements at a price determined based on avoided cost or users located nearby. The VSPP has a more

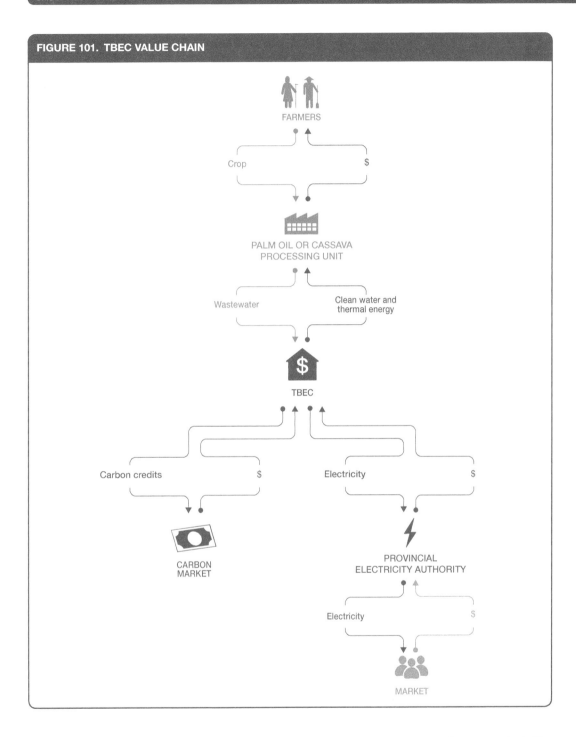

FIGURE 101. TBEC VALUE CHAIN

lenient set of requirements and less complicated power purchase arrangement of 'net metering'. The SPP and VSPP regulations have been amended to be more investor-friendly and practical, including changes to the criteria for qualifying facility, calculation of the avoided cost and interconnection requirements. In addition, the government also launched a program to encourage the renewable energy SPPs by providing an additional tariff for a period of 5–10 years from the Energy Conservation

Fund. The "adder" was determined through a competitive bidding system, which resulted in approval of 14 projects with average "adder" of 0.18 baht per kWh (US¢ 0.56), representing approximately 5% increase from the normal tariff. Financial incentives through soft loans and investment subsidies were expanded in amount and coverage for selected types of renewable energy projects, in particular biogas in pig farms and factories producing tapioca starch, palm oil, rubber sheet, ethanol and other types of agro-industry, municipal wastes and micro hydro. This has given an enormous boost to a number of marginal projects, particularly biogas and municipal waste projects.

The PEA is a government enterprise with prime responsibility concerned with the generation, distribution, sales and provision of electric energy services to the business and industrial sectors, as well as to the general public in provincial areas, with the exception of Bangkok, Nonthaburi and Samut Prakran provinces. The PEA has expanded electricity supply to all areas covering 73 provinces, approximately 510,000 km^2, accounting for 99% of the country's total area.

Technology and processes

TBEC applies a robust, flexible and highly productive Covered Lagoon Bio-Reactor (CLBR) technology suitable for changing volumes and quality of wastewater discharged from industrial factories. Wastewater passes through an anaerobic digestion process through which organic substances such as proteins, carbohydrates and fats are digested by bacteria in a suitable environment and are finally transformed into biogas. The CLBR has uniquely designed mixers, baffles and a thick high-density polyethylene (HDPE) cover with optimized contact with anaerobic bacteria to convert organic matter into biogas. Temperature is a key factor in planning a covered lagoon. Warm climates require smaller lagoons and have less variation in seasonal gas production. Cover materials must be: ultraviolet resistant; hydrophobic; tear and puncture resistant; non-toxic to bacteria and have a bulk density near that of water. The recovered biogas can be used to produce space heat, hot water, cooling or electricity. The biogas is collected in pipes, cleaned and stripped of condensate, dust and hydrogen sulphide and compressed and fed to dedicated biogas engines if used for power generation. The GE Jenbacher engine is designed specifically for gas applications and is characterized by particularly high efficiency, low emissions, durability and high reliability. The engine is designed with a knock control system which increases reliability and availability through control of firing point, output and mixture temperature. The engines gas mixer has been optimized to meet the requirements of modern gas engines and ensure trouble-free operation with biogas. In case of any excess build-up of biogas, the surplus gas will be combusted or flared. The effluent released from the digester is either recycled or sent to a small settling pond where sediment is settled and returned to the digester. The treated waste leaving the treatment system boundary is then pumped to existing water treatment lagoons.

A typical 200 tons-per-day starch factory can produce as much as 25,000 m^3 of methane (4.5 MW) from the cassava wastewater and 16,000 m^3 of methane (2.8 MW) from the cassava pulp. This is equivalent to 40,000 L of heavy fuel oil (HFO) per day or can be used to produce up to 7.3 MW of electricity per hour. Some of the areas of focus for new development are reactor configuration, process control, modelling and optimization for improving biogas yield; use of other feedstock such as solid residual, and energy crops; pre- and post-treatment for digestibility improvement and nutrient recovery; improved biogas clean up processes and upgrading biogas to high value/rich methane gas for fuel cell, vehicle, CNG, etc.

Funding and financial outlook

The investment costs covering project development, design, construction and start-up system depend on the size, location and duration of contract for individual projects. The major investment costs are plant machinery/equipment with minor cost of building and small cost of engineering services and other infrastructure. It should be noted that land and material costs are covered by concession partners. Historically, a key constraint has been reluctance of Thai domestic financial institutions to finance waste-to-energy products. Most financial institutions still define waste-to-energy business as a high-risk business. Unfamiliarity and trust that carbon credits can be saleable to European countries is part of the explanation. For that reason, partners invested their own money in order to initiate the business in 2003. At present, some Thai financial institutions offer refinance since they now realize the business potential.

The main revenue streams are from the sale of biogas and electricity and construction and maintenance under BOT schemes. Carbon credits are still relatively modest. Overall conditions that effect revenue streams include government policies, seasonality and prices. Seasonality is important as unusual seasons or weather conditions have an impact on inputs to the commodity processing factories that, in turn, produce wastes that are turned into energy. Table 28 shows the indicative cost structure of operations expressed in terms of approximate percentage of annual investment cost.

TABLE 28. OPERATIONAL AND MAINTENANCE COSTS OF TBEC

COST ITEM	OPERATIONS COSTS AS A % OF INVESTMENT COST
Equipment (depreciation)	Approx. 65%
Labor	Approx. 10%
Maintenance	Approx. 15%
Electricity	Approx. 5%
Building	Approx. 5%

The financial parameters of the typical project (based on Thachang project) are as follows:

1) Capacity of plant — 2.8 MW
2) Term of BOOT contract — 10 years
3) Investment cost — USD 3.9 million
4) O&M cost — USD 0.2 million
5) Electricity sold to grid per year — 9,644 MWh
6) Average tariff per kWh — USD 0.076
7) Escalation in O&M cost per year — 2%
8) Increase in tariff per year — 2%
9) Average CERs per year — 48,694 tons
10) IRR (without CERs) — 4.44%
11) IRR (with CERs) — 20.60%

Socio-economic, health and environmental impact

The project will create an indigenous renewable electricity resource, replacing power from coal, diesel and natural gas, and will contribute to the development of the region, as well as national economy by reducing Thailand's deficiency of power and need to import fossil fuels. In terms of environmental benefits, the project reduces existing levels of pollutants in wastewater; air pollution; GHG with positive impact on the health of those living around the plant and mitigates global warming by trapping methane. TBEC hires local labor for the construction and operation of biogas plant. The project will directly create more than 10 new jobs, and thus increase stakeholder incomes. It will improve human capacity and diversity of employment opportunity by training project managers, lab technicians and operators.

Scalability and replicability considerations

Key drivers to the success of this business are:
- Strong partnership among different institutions – technology developer, agro-processing businesses and electricity authority and financing institutions.
- Ability to raise finance to set up effective BOOT schemes for various agro-industries.
- Expertise in biogas plant operation.
- The government policy and interest in promoting renewable energy based power.

TBEC already has experience in taking its technology and business model from core operations in Thailand into Lao PDR. TBEC is also in talks with agricultural enterprises in Vietnam and Indonesia to produce biogas from cassava. A bank overseas has already lent EUR 10 million (THB 416 billion) for new projects. They have replicated the model with multiple agro-industries. As the technology can process any organic matter, the business model has potential to reach out to municipalities to process the organic component in the MSW as well.

Summary assessment – SWOT analysis

The key strengths of the business are setting up of effective BOOT schemes, expertise in biogas plant operation and strong partnership with agro-industry (Figure 102). TBEC is branded as a premium

FIGURE 102. SWOT ANALYSIS OF TBEC

	HELPFUL TO ACHIEVING THE OBJECTIVES	HARMFUL TO ACHIEVING THE OBJECTIVES
INTERNAL ORIGIN ATTRIBUTES OF THE ENTERPRISE	**STRENGTHS** - Assured supply of wastewater due to secured rights - Effective BOOT scheme - Strong partnership with agro-industry - Expertise in biogas plant operation - Securing of long-term power purchase agreement - Highly robust technology	**WEAKNESSES** - Complex biological processes - High cost of technology - Requirement of high skilled labor makes recruitment of staff difficult - No immediate market for treated water - Time taken for agreement and partnerships for every new project
EXTERNAL ORIGIN ATTRIBUTES OF THE ENVIRONMENT	**OPPORTUNITIES** - Environmental stress reduction offers environmental credit market opportunities - Treated wastewater has potential for fertilizer due to the process - Expansion potential to other agro-processing plants such as sugar, ethanol and liquor production due to highly robust technology - Electricity demand is growing and need for renewable-energy-based electricity generation increasing in Thailand - Good potential of foreign investment if the incentive policy is retained	**THREATS** - Possible human health risk may lead to investment needs - Possible risk from leakage of gas, thus having negative perception of health risk to employees may force O&M cost up - Seasonality regards biogas production - Volatility of international carbon market

product-service company as it puts more emphasis on quality and safety. However, the technology is high-priced and requires highly-skilled labor. There is no market yet for treated wastewater, but there is an opportunity to use the treated wastewater for agriculture. Growing electricity demand and application of the technology to other agro-processing plants such as sugar, ethanol and liquor production present opportunities for TBEC to expand.

Acknowledgements

Sincere thanks to Pajon Sriboonruang, Chief Operating Officer of the Thai Biogas Energy Co Ltd for agreeing to be interviewed and for providing documents about the company, Chatta Duangsuwan for helping with the interview and site visit and Kalayanee Surapolbhichet for assisting with liaison with TBEC and translation.

Contributors

Nikiema Josiane, IWMI, Ghana
Thai Biogas Energy Company
Johannes Heeb, CEWAS, Switzerland
Jasper Buijs, Sustainnovate; Formerly IWMI
Kamalesh Doshi, Simplify Energy Solutions LLC, Ashburn, Virginia, USA

References and further readings

Chaiprasert, P., Biogas production from agricultural wastes in Thailand. Journal of Sustainable Energy and Environment (2011): 63–65.

FNR. 2007. Fachagentur Nachwachsende Rohstoffe e.V. Renewable resource in industry, 2nd ed. www.fnr.de.

FNR. 2009. Fachagentur Nachwachsende Rohstoffe e.V. Biogas: An introduction, 2nd ed. www.fnr.de.

Foran, T., du Pont, P., Parinya, P. and Phumaraphand, N. Securing energy efficiency as a high priority: Scenarios for common appliance electricity consumption in Thailand. Energy Efficiency 3, no. 4 (2010): 347–364.

Pattanapongchai, A. and Limmeechokchai, B. Least cost energy planning in Thailand: A case of biogas upgrading in palm oil industry. Songklanakarin J Sci Technol 33, no. 6 (2011): 705–715.

Tantrakarnapa, K., Utachkul, U., Aroonsrimorakot, S. and Arunlertaree, C. The potential of greenhouse gas reduction from clean development mechanism implementation in cassava starch and palm oil industries in Thailand. Journal of Public Health (2008): 130–139.

Thai Biogas Energy Co Ltd. About us. www.tbec.co.th/e_about_us.php (accessed August 18, 2017).

Thai Biogas Energy Co Ltd. Our projects. www.tbec.co.th/e_our_project.php (accessed August 18, 2017).

Watcharejyothin, M. and Shrestha, R. Regional energy resource development and energy security under CO_2 emission constraint in the greater Mekong sub-region countries (GMS). Energy Policy 37(2009): 4428–4441.

Watcharejyothin, M. and Shrestha, R. Effects of cross-border power trade between Laos and Thailand: Energy security and environmental implications. Energy Policy 37(2009): 1782–1792.

Case descriptions are based on primary and secondary data provided by case operators, insiders, or other stakeholders, and reflects our best knowledge at the time of the assessments (2015/2016). As business operations are dynamic, data can be subject to change.

BUSINESS MODEL 8
Combined heat and power from agro-industrial waste for on- and off-site use

Solomie Gebrezgabher and Krishna C. Rao

A. Key characteristics

Model name	Combined heat and power from agro-industrial waste for on- and off-site use
Waste stream	Agro-industrial waste – Bagasse from sugar processing factories; Effluent (solid and liquid waste) from agro-industrial units like cassava starch, palm oil and slaughterhouse
Value-added waste product	Electricity, biogas, thermal energy, carbon credit, bio-fertilizer
Geography	Regions with larger agro-industries
Scale of production	Small to large scale 15 KW of power from slaughterhouse waste; 1.4 MW–2.8 MW of electricity annually from effluent from cassava starch and palm oil mills 12 MW–34 MW of electricity from sugar-processing factories
Supporting cases in the book	Mumais district, Kenya; Maharashtra, India; Bangkok, Thailand
Objective of entity	Cost-recovery [X]; For profit [X]; Social enterprise [ ]
Waste removal capacity	About 1.3 million tons of bagasse from crushing 3–4 million tons of sugarcane; 4,000 L/day of wastewater from slaughterhouse; 25,000 m^3/day of wastewater from agro-industrial units
Investment cost range	1.16–1.85 million USD/MW of electricity from sugar-processing factories 2–2.6 million USD/MW of electricity from agro-industrial effluent
Organization type	Private
Socio-economic impact	Reduce environmental pollution by substituting fossil fuel based energy (1.4–34 MW) and by providing better waste management/reducing effluent, reduce fossil fuel dependence, employment generation (5–200 jobs depending on scale)
Gender equity	Access to electricity to local community by replacing kerosene used for lighting resulting in clean working environment for women from clean indoor air

B. Business value chain

This business model can be initiated by industrial factories in order to create additional value and revenue by generating energy from their organic waste by-products. By-products include agro-industrial waste such as bagasse and molasses from sugar-processing factories, and wastewater from cassava, palm oil and slaughter-house industrial factories. The technologies applied and the resulting energy products vary depending on the type of waste processed. These include co-generation units to produce electricity and thermal energy, distillery units to produce ethanol/alcohol and biogas units to produce electricity and thermal energy/heat. Production technologies such as combustion and covered lagoon bio-reactor are suitable for processing bagasse and wastewater to produce biogas. Figure 103 depicts the value chain for on-site energy generation from an agro-industrial waste business model. The electricity produced by the co-generation unit or by the covered lagoon bio-reactor is sold to the state utility on a long-term power purchase agreement. The alcohol/ethanol produced from the distillery unit of sugar-processing factories is sold to petroleum and pharmaceutical companies, while the energy produced by the biogas unit is used on-site as input fuel to the co-generation unit. The discharge from the biogas unit, which is high in organic matter, can be distributed to farmers to be used as bio-fertilizer.

The ownership and operation of the energy-producing units take different forms. The energy-production technologies are either designed, constructed, owned and operated by the factory or are installed by an external private enterprise on a BOOT model. In the latter case, the enterprise brings investment to

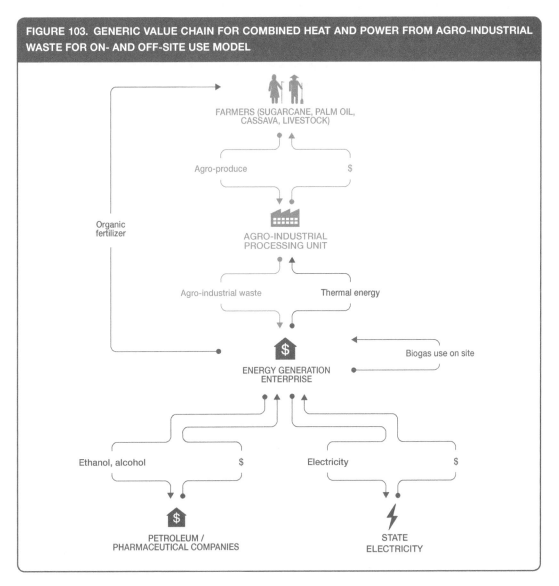

FIGURE 103. GENERIC VALUE CHAIN FOR COMBINED HEAT AND POWER FROM AGRO-INDUSTRIAL WASTE FOR ON- AND OFF-SITE USE MODEL

set up the energy production technology while the concessionaries i.e. the factories provide land and inputs. The enterprise designs, constructs, trains and maintains the energy production unit until the BOOT period expires, after which it assists the host company in operating the unit.

C. Business model

This business model involves processing of waste by-products from an agro-industrial factory in order to generate and sell electricity to the national grid through a long-term power purchase agreement (Figure 104). By-products are heat which can be fed back into the industrial process, resulting in energy savings, and ethanol which can be sold to petroleum and pharmaceutical companies. Additional revenue can be generated from registering the model as a CDM project and earning money from selling certified carbon emission reductions.

D. Potential risks and mitigation

Market risks: The outputs, electricity and alcohol (ethanol), are sold to different markets and hence face different market risks. The electricity is mainly sold to state electricity grid on a long-term power purchase agreement, while ethanol is sold to petroleum or pharmaceutical companies. The growing demand for electricity in developing countries reduces the market risks in terms of ensuring sales.

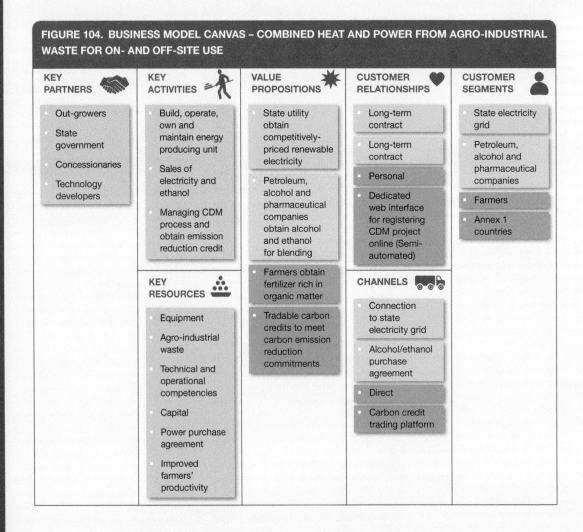

FIGURE 104. BUSINESS MODEL CANVAS – COMBINED HEAT AND POWER FROM AGRO-INDUSTRIAL WASTE FOR ON- AND OFF-SITE USE

BUSINESS MODEL 8: HEAT AND POWER FROM AGRO-INDUSTRIAL WASTE FOR ON- AND OFF-SITE USE

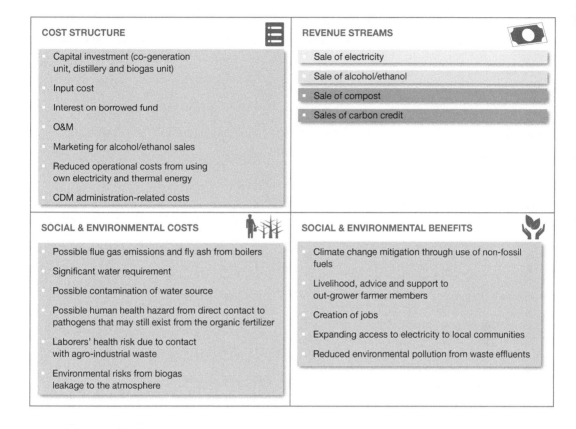

However, in environments where the electricity sector is regulated and the state utility is the sole buyer, the bargaining power of the business producing and selling electricity will be medium. In such situations, the feed-in tariff policy announced by the government will protect the interest of the renewable-energy-based power producers. Ethanol/alcohol is sold to various industries and with the introduction of government requirements for ethanol blending of fuel, ethanol will have various buyers and less market risks. However, in countries where ethanol blending is not mandatory, the business will face competition from other fossil-based substitute products. In sharp contrast to the ensured sales of electricity to state utility, the carbon credit market is considered to be volatile and this puts the sustainability of the whole reuse business under risk if carbon credit sale is the major revenue stream. In such scenarios, the business has to diversify its revenue streams so as not to entirely depend on the sale of carbon credits.

Competition risks: The risk associated with output market is low. The electricity is sold to state utility on a long-term contract, and hence has a ready buyer. With the introduction of government requirement for ethanol blending, ethanol has various buyers and less competition risk. Competition risks exist in the input market. In the case of sugar-processing factory, the cogeneration units are designed to process only a specific kind of input, i.e. the by-product from processing sugarcane and its operations depend heavily on the continuous supply of sugarcane from its suppliers. In scenarios where the inputs are sourced from the sugarcane growers, the competition in the sugar commodity market will affect the decision of sugarcane growers. For instance, fluctuating sugar prices might force farmers to shift to other crops, and this will affect the operations of the cogeneration and distillery units of the sugar-processing factory. This risk can be mitigated by forming a cooperative sugar factory which is vertically integrated and owns the raw materials and agro-waste.

Technology performance risks: The technologies applied for processing agro-industrial waste from sugar-processing factories as well as for processing wastewater are well-established, robust and mature with high flexibility to changing wastewater volumes and quality. However, the technologies require skilled labor for construction as well as O&M of the plant.

Political and regulatory risks: With the projected electricity demand set to grow, governments are encouraging green power initiatives by putting in place various incentive mechanisms such as concessional loans, feed-in tariff mechanisms and through long-term power purchase agreements. Thus, the risk is fairly low.

Social-equity-related risks: The model is considered to have more advantages to women if excess electricity generated by these agro-industries is supplied either for rural electrification or fed to the grid. Since access to electricity to local community will help replace kerosene used for lighting resulting in clean working environment for women from clean indoor air. Modern energy access will also benefit the community from increased productivity. If the energy generated is used for agro-industries internal use, then the model is gender neutral. The social-equity risks from the model are limited; however, the agro-industry could consider under their corporate social responsibility to improve energy reliability in neighbouring community.

Safety, environmental and health risks: The environmental risks associated with co-generation units include possible leakage of gas and emission of flue gas and fly ash. These emissions should be controlled within acceptable limits by putting in place suitable equipment. The safety and health risks to human arise when processing waste from agro-industry, especially meat production. Proper mitigation measures should be put in place to protect laborers, farmers, consumers and surrounding communities (Table 29). Another issue is with respect to the water requirement for the energy-producing units. The water requirement can be high, and thus competes with uses for other purposes such as irrigation. This has important implications in terms of evaluating trade-offs for competing uses.

TABLE 29. POTENTIAL HEALTH AND ENVIRONMENTAL RISKS AND SUGGESTED MITIGATION MEASURES FOR BUSINESS MODEL 8

RISK GROUP	EXPOSURE ROUTES					REMARKS
	DIRECT CONTACT	AIR	INSECTS	WATER/ SOIL	FOOD	
Worker						Risks apply to the use of slaughterhouse waste, and its management, including fly control
Farmer/User						
Community						
Consumer						
Mitigation measures						

Key: ☐ NOT APPLICABLE ▒ LOW RISK ▓ MEDIUM RISK ■ HIGH RISK

E. Business performance

The business model scores high on environmental impact as it avoids environmental pollution from large agro-industrial factories and generates renewable energy on a large scale, substituting fossil-fuel-based energy sources, and thus resulting in reduced GHG emissions (Figure 105). This business model is scalable and replicable in countries where there are large agro-industrial factories and where there is government support such as provision of concessional loans and feed-in tariff mechanisms for renewable energy initiatives and government's directive on blending of ethanol with petrol/gasoline as transportation fuel. The ranking of other factors scores significantly low in comparison to environmental impact.

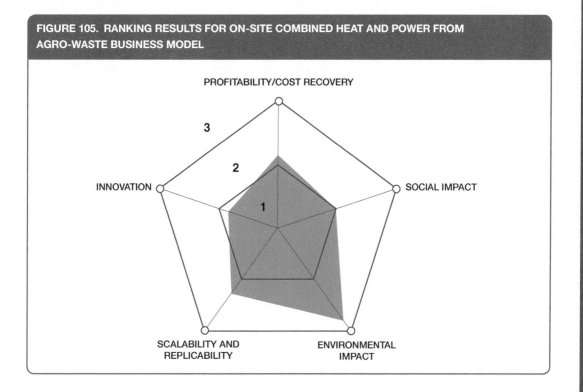

FIGURE 105. RANKING RESULTS FOR ON-SITE COMBINED HEAT AND POWER FROM AGRO-WASTE BUSINESS MODEL

6. BUSINESS MODELS ON EMERGING TECHNOLOGIES/ BIO-FUEL PRODUCTION FROM AGRO-WASTE

Introduction

There has been a significant effort to develop a potentially cost-effective process to produce bio-fuel from cellulosic sources. In our typology of waste covered in this book, important cellulosic sources could be identified for example in agro-industrial waste. Cellulosic bio-fuel sources have the potential to offer greater energy return on investment in comparison to grains (which might better support food security) while they provide environmental benefits by reducing dependency on fossil fuels.

The **Business model 9: Bio-ethanol and chemical products from agro and agro-industrial waste** described in this section highlights production of bio-fuel (bio-ethanol) from agro-industrial waste produced by mills processing cassava, rice, wheat, coffee and so on. The case examples are from Latin America (Venezuela and Mexico). One of the case highlights successful and economically-viable production of ethanol from residual plant waste (yare) associated with cassava flour production. Yare, the juice from the cassava pulp, is poisonous due to its cyanide content and requires proper waste disposal unless there are resource recovery options. The second case study shows a fuel recovery option from a waste generated during alcohol (ethanol) production.

For the presented model to work, since the technology is new and the business cases are in their nascent commercialization stage, it is critical that the patent laws are strong for safeguarding enterprises R&D efforts on the technology and incentives given to competing products such as ethanol from sugarcane or other sources are similar so as to provide a level playing field to all.

CASE

Bio-ethanol from cassava waste (ETAVEN, Carabobo, Venezuela)

Patrick Watson and Krishna C. Rao

Supporting case for Business Model 9	
Location:	ETAVEN, Carabobo, Venezuela
Waste input type:	Cassava waste
Value offer:	Bio-ethanol (as additive to petrol/gasoline as transportation fuel)
Organization type:	Private
Status of organization:	Established in 2007, Business operational since 2012
Scale of businesses:	Medium
Major partners:	University of Carabobo, Ministry of Science and Technology, Libertador Municipality Mayor's Office

Executive summary

ETAVEN C.A. (ETAVEN) is a private Venezuelan company established in 2007 that has patented a process for producing ethanol "Yarethanol" using proprietary rights for the strain of bacteria for fermentation from yare, a by-product of cassava processing. Yarethanol is an ecological, non-fossil, non-poisonous, non-polluting and high demand, renewable fuel. It is produced at 50% of the market price through the patented process with a high yield of 50%.

ETAVEN is situated in the cassava-flour-processing region of Venezuela, which allows it to easily and cost-effectively purchase sub-optimal cassava that cannot be used for other commercial purposes, as well as residual plant waste (yare) associated with cassava flour production. By purchasing and using this waste, ETAVEN has had a significant positive impact on both the local community and environment, reducing the pollution associated with high cyanide run-off from improper disposal of cassava into local rivers and lakes, reducing GHG emissions and increasing the incomes of local cassava farmers by up to 50%. It has a very high social impact due to creation of jobs (>1,500 jobs) fostering agriculture. The company began Yaretanol production in the third quarter of 2012, and through franchise model, it seeks to expand its market beyond Venezuela.

CASE: BIO-ETHANOL FROM CASSAVA WASTE

KEY PERFORMANCE INDICATORS (AS OF 2012)						
Land use:	5 ha					
Water requirement:	8,000 L monthly (water is reused to wash cassava)					
Capital investment:	USD 2.5 million (Site utilized the existing infrastructure of a former sugar cane refinery)					
Labor:	50 plant employees; and 12 university volunteers to analyze and improve the process					
O&M cost:	Approx. USD 375,000 per annum (forecast 2013)					
Output:	30 tons/ day of Yaretanol					
Potential social and/or environmental impact:	Reduced water pollution previously caused by improper yare waste disposal into local rivers, reducing GHG emissions by substituting petrol used for transportation, creation of jobs, Improved incomes of approx. 300 local cassava farmers					
Financial viability indicators:	Payback period:	< 2 years	Post-tax IRR:	> 50%	Gross margin:	99%

Context and background

The cassava is one of the most drought-tolerant crops, capable of growing on marginal soils. The average yield of cassava crops worldwide was 12.5 t/ha in 2010. The cassava plant gives the third highest yield of carbohydrates per cultivated area among the crop plants, after sugarcane and sugar beets. The plant must undergo processing immediately after harvest (within 48 hrs) to remove compounds that generate cyanide. Yare is a regional name for the milky juice arising from pressing bitter cassava that has a high cyanide content. Venezuela produces 60,000 tons of yare per year which traditionally goes unused and improperly discarded.

Until now, the yare is not used in Venezuela as a source of ethyl alcohol (as bio-fuel). ETAVEN undertook research and laboratory experiments for two years and obtained 50% yield of ethanol from yare. Cassava is one of the richest fermentable substances for the production of alcohol. The fresh roots contain about 30% starch and 5% sugars, and the dried roots contain about 80% fermentable substances which are equivalent to rice as a source of alcohol. ETAVEN uses the sub-optimal cassava that cannot be used for other commercial purposes, as well as residual plant waste (yare) associated with cassava flour production. Cassava processing produces annually big quantity of wastes, and if they are not properly managed, they can cause a serious pollution to the environment and human life.

Ethanol has been known to slightly improve gas mileage. It has a high-octane rating of 113 and improves performance while keeping the engine clean. Ethanol also contains 67% more energy than it takes to produce, so it is efficient for your car and for the environment. An important advantage of biofuels is that they can easily be integrated into the existing transport infrastructure, thus avoiding the significant investment costs associated with other renewable options for the transport sector.

In 2008, ETAVEN patented an engineered yeast strain that efficiently produces Yaretanol or ethanol from yare. In early 2012, ETAVEN completed construction of its pilot ethanol plant and began producing ethanol in Q3 of 2012. It then produced approximately 30 tons of ethanol per day, roughly 1% of Venezuela's national consumption of ethanol. ETAVEN ethanol plant is located in Western Venezuela in the Libertador Municipality, which has a robust cassava processing industry, comprising more than 150 producers and over 300 farmers who supply cassava for bread making. However, small cassava roots or diseased plants cannot be used to produce cassava flour for bread, resulting in approximately 40%

of the local cassava going to waste. Unused cassava was traditionally discarded in local waterways (rivers and streams) where it was left to rot and release toxins. Yare is high in cyanide that can leach into the water supply, while plant decomposition releases methane into the atmosphere. ETAVEN procures yare from either farmers or cassava-flour-making units to produce ethanol, which is sold to oil companies in Venezuela to blend it with gasoline.

Market environment

Ethanol production from yare is dictated by both availability of yare and demand for ethanol. Cassava is the third most common source of food in tropical countries after rice and maize with total production reaching approximately 250 million tons in 2011, according to the UN. Within Latin America, Brazil is the dominant player, accounting for approximately 70% of regional production. Considering the high comparative ethanol yield from yare and that it can be used alongside food production, rather than competing with it, there is a significant opportunity in all cassava-producing countries to increase the potential incomes of cassava farmers.

Global consumption of ethanol has surged during the last 10 years, driven by greater environmental awareness, advances in technology that have made ethanol cost-effective and suitable for fuel and growing national interest in energy independence and security. Furthermore, government subsidies and mandates have driven ethanol's growing popularity. A number of additional countries have begun to require a minimum ethanol blend in gasoline. In addition, there has been significant public investment into the ethanol distribution infrastructure to accommodate this increasing production and demand. Global consumption of ethanol increased during 2002–2012 by approximately 500% reaching 1.4 million barrels per day, led by the U.S. and Brazil who accounted for over 85% of total ethanol production and consumption in 2012.

The two key sources of competition for Yaretanol are: 1) other ethanol producers and 2) the oil industry (for petroleum). Approximately 90% of the ethanol used in Venezuela is imported from Brazil at twice the price of Yaretanol, while 1% is produced by ETAVEN and the remaining 9% by other domestic producers. Other sources used to create ethanol are sugarcane and corn.

Though ethanol is a viable substitution of petroleum in combustion engines, it is used only if mandated by the government. In 2006, Petroleos de Venezuela S.A. (PDVSA), the Venezuelan state-owned oil company, announced their "Ethanol Agro-energy Development Project", a USD 1.3-billion initiative. To increase the production of ethanol, PDVSA plans to build 14 ethanol distilleries by 2012 with an output of 20,000 barrels per day of the biofuel. Venezuela imports ethanol to mix in gasoline. The plan's focus has been to double the amount of land used for sugarcane cultivation over the next five years competing with Brazilian sugarcane imports.

Macro-economic environment

ETAVEN is aware of potential obstacles from vested interests including the PDVSA (with their direct interest in sugarcane ethanol), the Government of Venezuela (due to reduction in tax revenue in dollars), importers of ethanol from Brazil and manufacturers of ethanol from Brazil. Hence, ETAVEN is planning to focus on expanding into Latin America (Costa Rica, Panama, Dominican Republic and Peru) and a number of African countries rather than expansion of their market locally. Many countries are striving for energy independence by way of biofuels that do not come from foodstuffs. Significant research has begun to evaluate the use of cassava as the ethanol biofuel feedstock. On December 22, 2007, the largest cassava ethanol fuel production facility was completed in Beihai in China, with an annual output of 200,000 tons, which would need an average of 1.5 million tons of cassava. In November 2008, China-based Hainan Yedao Group reportedly invested USD 51.5 million (£ 31.8 million) in a new biofuel facility that is expected to produce 33 million US gallons (120,000 m³) a year of bioethanol from cassava plants (https://en.wikipedia.org/wiki/Cassava; accessed 18 January 2018).

Business model

ETAVEN's key value propositions (Figure 106) is the production of bio-ethanol from yare for blending with petrol, in the process reducing environmental hazard of pollution of water bodies through leaching of cyanide and reducing in methane emission from natural decomposition. ETAVEN spent initial years in developing technology to process yare to ethanol. Once the technology was ready for commercial production, it formed partnerships to secure procuring of cassava and its by-products from farmers and cassava-processing mills. ETAVEN does not use any middlemen and takes direct responsibility of delivering ethanol to petroleum companies. There is potential for the business model to create additional value making cassava flour and selling cassava shells as animal feed.

Value chain and position

Yaretanol has higher octane rating than petrol as fuel. It is an octane booster and anti-knocking agent, reducing country's dependence on petroleum, source of non-oil revenue for the producing country, and reducing adverse foreign trade balance. ETAVEN has patented its technology and process of producing ethanol from cassava and yare (Figure 107). The company buys cassava directly from approximately 300 local farmers, who sell diseased or small roots unsuitable for use in bread making, or it buys yare produced during pressing for flour production from small-scale cassava flour producers. Both supplier groups provide ETAVEN with waste that cannot be sold or used otherwise; therefore, supplier power is relatively low. Cassava is delivered directly to the plant, while yare is collected by ETAVEN through its fleet of collection tanker trucks. ETAVEN also relies on about 12 university volunteers each month, who evaluate and monitor operations, as well as provide staff training and write key operating manuals.

ETAVEN sells its ethanol to two key clients, Venezuelan petrochemical companies Solven and Inproin. In the context of Venezuela, though the demand for ethanol is high, there is significant buyer power as oil and gas companies can choose whether or not to blend petrol with ethanol. It is yet not mandatory by the Government of Venezuela. If ETAVEN is able to supply consistently necessary amount to oil companies, the buyer power will remain lower as long as they are willing to blend petrol with ethanol. The primary substitute for Yaretanol is ethanol produced from other products, such as sugar cane or corn. It is unlikely that buyers have a propensity to buy ethanol derived from any particular feedstock; therefore, the threat of substitutes is relatively high. Yaretanol has the lowest production cost of USD 0.18 /L, in comparison to USD 0.35 for corn and USD 0.22 for sugarcane. Therefore, it is very competitive on a cost basis.

The ETAVEN can also use starch and ethanol as a base for biopolymers and plastic extract as a base for bio-combustibles. The fermentation by-products with other waste streams, including animal manure and human excrement, can further be anaerobically digested to produce biogas and biological fertilizers. The ethanol can also be used as cooking fuel using specially-designed efficient cook stoves. The project is recognized as CDM to generate carbon credits. The cassava peels can be used with livestock manure as inoculum to generate biogas.

Institutional environment

Though Venezuela is the fifth largest oil exporting country in the world and the industry is a significant source of wealth for the country, the Government of Venezuela has promoted ethanol as a substitute for lead additives in gasoline. The Government of Venezuela did express an interest in 2006 to expand ethanol production (from sugar cane). However, it has not as yet made it mandatory to blend for domestic usage. ETAVEN has patented its proprietary strain of yeast in Panama, since Venezuela exited

FIGURE 106. ETAVEN BUSINESS MODEL CANVAS

KEY PARTNERS
- Farmers and flour-mill processing cassava
- University of Carabobo
- Ministry of Science and Technology
- Libertador municipality mayor's office

KEY ACTIVITIES
- Marketing to petroleum companies
- Farmer training
- Collection of yare and cassava and its by products
- Making ethanol from yare

KEY RESOURCES
- Yare from farmers producing cassava and processing units
- Capital
- Skilled labor
- Patent-protected IP

VALUE PROPOSITIONS
- Production of bio-ethanol from yare for blending with petrol, and in the process reducing environmental hazard

CUSTOMER RELATIONSHIPS
- Direct interaction with petroleum companies

CHANNELS
- Direct supply to petroleum companies

CUSTOMER SEGMENTS
- Petroleum companies

COST STRUCTURE
- Investment cost (Land, building, machines)
- Operational cost (Raw material cost, fuel and products to clean cassava, labor, utilities and maintenance cost)
- Marketing cost
- Depreciation
- IP maintenance
- R&D
- Training

REVENUE STREAMS
- Sale of ethanol

SOCIAL & ENVIRONMENTAL COSTS
- Potential cyanide exposure in local environment

SOCIAL & ENVIRONMENTAL BENEFITS
- Job creation
- Environmental benefit through processing of cassava by-product "yare" that can cause water pollution from improper disposal of cassava into local rivers and lakes
- Environmental benefit through reduced CO_2 emissions by using ethanol from renewable source blended into petrol

CASE: BIO-ETHANOL FROM CASSAVA WASTE

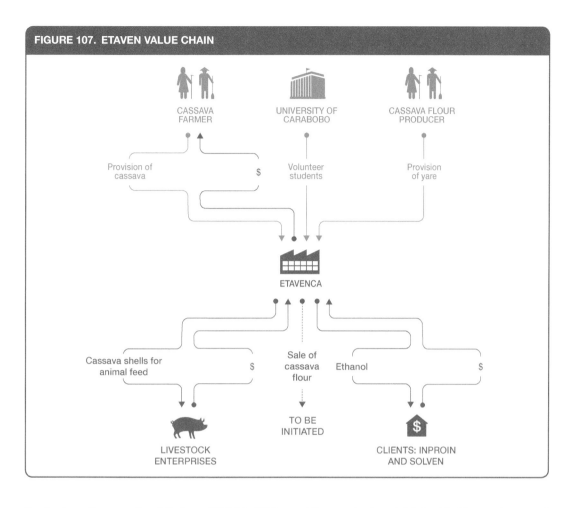

FIGURE 107. ETAVEN VALUE CHAIN

the Andean Community of Nations (CAN) in 2006, and thus, patents established in Venezuela cannot be legally enforced. Blends of E10 (petroleum with 10% ethanol added) or less have been mandated in over 20 countries, spearheaded by the US; however, a required blend has not been implemented yet in Venezuela. The National Council of Scientific and Technological Research (founded in 1967) and the state Ministry of Science and Technology direct and coordinate research activities in Venezuela.

Technology and processes

ETAVEN patented process for producing ethanol from the yare and cassava is as follows (Figure 108): ETAVEN receives two forms of feedstock: 1) yare from the cassava farmers that produce flour and 2) cassava directly from farmers. For cassava farmers that deliver yare, farmers are required to manually press the cassava to extract the liquid and deliver it to nearby collection tanks. ETAVEN owns a fleet of approximately 15 vehicles used to collect yare from each of its collection tanks that are situated near the farms and deliver to a collection center whereby it is aggregated and fed into the plant. In the case of cassava received directly from the farmers, ETAVEN receives the cassava through two different methods; 1) directly from the farmers delivering it to deposits at the plant or 2) utilizing their fleet of vehicles to pick it up directly from farms. For the collected cassava, it is peeled, mashed and then heated up with water to help with the conversion of the starch molecules into sugar, then strained. The cassava skin from peeling process is given away to be used for animal feed. The process of ethanol fuel production involves yeast fermentation of sugars, distillation and dehydration.

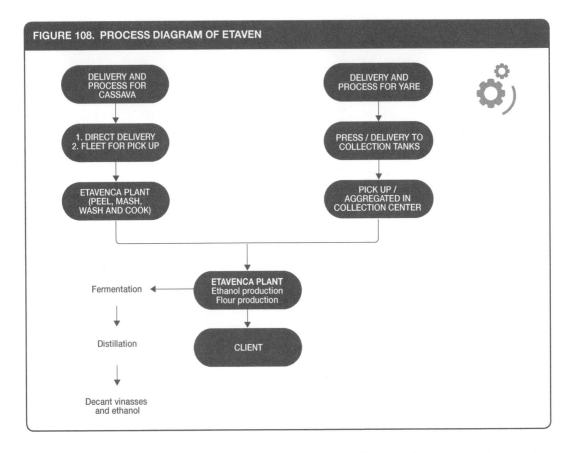

FIGURE 108. PROCESS DIAGRAM OF ETAVEN

The liquids extracted from cassava is fermented with the ETAVENCA strain of yeast in fermentation tanks and undergo distillation process which results in both ethanol and vinasses (residual waste left after distillation) being separated out. The resulting ethanol is then ready to be sold to clients, and the vinasses is treated and disposed. The yield of conversion is about 70–110 L of absolute alcohol per ton of cassava roots depending on the variety and method of manufacture.

There are other optional technologies available for treating cassava-processing waste (from small and large factories). These options include: landfilling, use as animal feed, ensiling of solid residue, fermentation of cassava peel, use of wastewater for irrigation, infiltration of wastewater into the soil, storage in aerobic or anaerobic lagoons and anaerobic digesters. One way is to build anaerobic and aerobic lagoons (ponds) to treat the waste before its disposal. In the condition of anaerobic digestion of cassava waste, cyanide is released in the form of liquor and then liberated by enzymatic and non-enzymatic reactions. This system is very effective and environmental sound but requires a large area of land and large capital investment and therefore is suitable only for the large processing plant. In case cassava processing is of small to medium scale, wastewater can be treated through channelling the waste into shallow seepage areas. The areas, however, should be situated away from natural water sources. Cassava-processing wastewater can also be effectively utilized as a liquid fertilizer, if it is well treated. However, if the waste is not properly treated resulting in its high HCN content that can have a negative effect on plant growth, the use of wastewater for irrigation or as a source of fertilizer should be restricted.

Funding and financial outlook

ETAVEN was established in 2007 by four founding partners who each invested approximately USD 650,000 for a total investment of USD 2.5 million (Table 30) to convert a former sugarcane refinery into the ETAVEN ethanol plant. Total cost was lower than the cost to purchase and restore a brown-field site, in part, because a number of the pre-existing fittings and equipment could be re-used. The most significant capital expenditure items after the land and sugar cane refinery were the four distillation towers, costing approximately USD 75,000 each.

The key production costs for ETAVEN are electricity and raw materials; however, both are low, allowing the company to achieve operating margins in excess of 85%. The feedstock is bought at nominal value (USD 0.02 /L) as it cannot be used for any other purpose, and ETAVEN effectively serves as a waste collection service improving conditions for surrounding farms. The most significant operational cost (approximately 45% in 2012) is for staff wages as the firm employs a substantial workforce of 50 employees. However, only 19 of those employees are officially recorded because the remaining 31 are considered temporary for legal purposes (both are included in the numbers below). From the remaining operational costs, the most significant line items are for fuel (approximately 14% in 2012) and products used to clean the cassava and disinfect equipment (approximately 17%). In addition, ETAVEN runs a significant education budget used to train the *cassaveros* (cassava farmers) in farming best practices and to train staff (approximately 12%).

Production began in June 2012 with monthly sales starting at USD 300,000 and increasing steadily throughout the year to USD 350,000 in December. ETAVEN is forecasting sales to continue growing strongly for the medium term, averaging about 50% growth per annum for the next three years. Due to the low cost of sales, ETAVEN achieved strong operating margins of 99% for the first six months of operation. The net profit margin is forecast to remain stable at about 56% over the forecast period as costs grow in line with revenues.

Socio-economic, health and environmental impact

ETAVEN estimates that their operations have improved the incomes of local cassava farmers and small-scale flour producers by up to 50%. The actual payments to farmers from ETAVEN for cassava and yare are minimal (USD 0.02/L). Therefore, their positive social impact has been primarily through

TABLE 30. ETAVEN FINANCIALS

SUMMARY FORECAST P&L						
USD	2010	2011	2012	2013	2014	2015
Initial Investment	(1,000,000)	(1,000,000)	(500,000)			
Sales			2,287,161	4,190,891	5,573,885	8,249,350
Cost of Sales			(8,922)	(36,786)	(48,925)	(72,409)
Gross Profit			**2,278,239**	**4,154,105**	**5,524,960**	**8,176,941**
Operating and Financing Costs			(230,219)	(450,515)	(601,462)	(847,430)
Human Resources			(107,350)	(216,285)	(216,285)	(216,285)
Profit Before Tax			**1,940,669**	**3,487,306**	**4,707,213**	**7,113,227**
Tax			(659,828)	(1,185,684)	(1,600,452)	(2,418,497)
Net Profit			**1,280,842**	**2,301,622**	**3,106,761**	**4,694,730**
Payback Period from Yr 1 Revenues: <2 years						
IRR: 50%						

indirect channels. By improving the disposal of waste, local farming operations are perceived as more sanitary by larger cassava buyers, who are now willing to purchase cassava from small holder farmers previously viewed as unsuitable suppliers. In addition, through ETAVEN's educational programs for local farmers, small holders have learned to bypass intermediaries to sell their goods directly to buyers who offer higher prices. Finally, the collection and use of waste cassava (approximately 40% of total yield) and yare diverts substantial (approximately 60 tons per day) cyanide-rich waste from local water sources and reduces methane emission that results from the slow rot of cassava in public spaces.

Scalability and replicability considerations

The key drivers for the success of this business are:
- Patented technology and process for making ethanol from cassava and its by-products.
- Lower production cost of ethanol from cassava in comparison to other feedstock such as sugar cane and corn.
- Ease of access to cassava waste and no value or competition for procuring this waste.
- Underdeveloped national ethanol production, inviting price leadership.

ETAVEN's ethanol production process and sourcing model is highly scalable to developing markets where cassava is grown. Yaretanol production technology is relatively simple and cost-effective to implement, requiring only three key steps: extraction, fermentation and distillation before decanting the ethanol and vinasses. Furthermore, cassava is a hardy crop suitable for growth in arid, nutrient-deficient soil, making it popular in sub-Saharan Africa, where there is already a developed cassava industry. ETAVEN intends to franchise its process, targeting Central and South American countries in phase one and further rolling out to Africa and Asia in phase two. The largest potential untapped reserves of cassava are in Nigeria, Thailand, Brazil and Indonesia.

Cassava-to-ethanol plants have already been established in several countries, such as NDZiLO in Mozambique and TMO Renewables in China proving feasibility of the technology. In England, TMO Renewables announced they have advanced to demonstration scale on cassava stalk feedstock with major Chinese fuel and food producers. TMO is now processing an initial shipment of cassava stalk delivered from China, an inexpensive, abundant feedstock underutilized in 2G bio-ethanol. Improved efficiencies at TMO's 0.1-ha demonstration facility are projected to produce ethanol for less than USD 2 per gallon, marking a crucial step toward commercialization. Utilizing cassava stalk, TMO's conversion process will yield 70 to 80 gallons of 2G ethanol per ton of feedstock. The ETAVEN's business is capable of being reproduced at the international scale in tropical climates on the arid soils that are poor in nutrients.

Summary assessment – SWOT analysis

The key strength of ETAVEN is its simple process, low production cost and use of a feedstock that has hardly any other strong alternative uses (Figure 109). The weaknesses of ETAVEN are its unfavourable business environment and limited awareness on ethanol from cassava among policy makers. Due to the limited awareness, there is strong potential for policy makers to give priority for ethanol from sugarcane and corn. Therefore, ETAVEN has significant threat from other established producers of ethanol in its market shed. ETAVEN has strong opportunities to expand its market to other regions in developing countries where cassava is the staple food. In addition, it can further stabilize its business and increase its profitability by vertically scaling its business to capture value from both upstream and downstream parts of its value chain by getting into cassava-flour processing and selling of cassava shells as animal feedstock.

FIGURE 109. ETAVEN SWOT ANALYSIS

	HELPFUL TO ACHIEVING THE OBJECTIVES	**HARMFUL** TO ACHIEVING THE OBJECTIVES
INTERNAL ORIGIN ATTRIBUTES OF THE ENTERPRISE	**STRENGTHS** • Patented technology and process • Relatively simple and cheap process to roll out • Higher ethanol yield from cassava than any other commercially-used feedstock • Uses sub-optimal feedstock thereby not competing with food production • Plant is situated close to a cluster of flour mills and cassava producers helping to reduce operating costs • Have achieved extremely high operating margins on pilot plant, therefore allowing considerable flexibility in potential costs for future expansion	**WEAKNESSES** • Business model reliant on nominal value for purchased cassava; cost feasibility for international expansion would need to be assessed based on market prices • Patented technology virtually invisible; difficult patent enforcement in case of infringement
EXTERNAL ORIGIN ATTRIBUTES OF THE ENVIRONMENT	**OPPORTUNITIES** • Potential of expansion in a number of other cassava-producing countries through franchising • Significant potential for domestic expansion in Venezuela, assuming improved government support for all ethanol production (not just sugarcane) • Potential for vertical scaling and having additional revenue source by getting into cassava-flour making and sales of cassava shells as animal feed • Underdeveloped national ethanol production provides room for maneuver	**THREATS** • Competition or entry into market from other, more established cassava-to-ethanol producers • Domestic pressure from incumbent sugar cane refineries in Venezuela, and government bias toward sugarcane-based ethanol • Unfavorable business environment in Venezuela for ETAVEN • Limited awareness of cassava-to-ethanol opportunity among policy makers and governments who could further benefit from the technology

Contributors
Kamalesh Doshi, Simplify Energy Solutions, LLC, Ashburn, Virginia, USA
Carlos Fernandez, I-DEV International
Cinthya Pajares, I-DEV International

References and further readings
New Tang Dynasty Television. "YARETANOL: Venezuela pioneers alternative fuel from cassava root. YouTube video, 2:31. Posted October 2009. Retrieved from www.youtube.com/watch?v=LAE28nIEwtk (accessed November 7, 2017).

Pernalette Ivaneth, S. Interviewed by Cynthia Pajares via email. March 14, 2014.

Case descriptions are based on primary and secondary data provided by case operators, insiders, or other stakeholders, and reflects our best knowledge at the time of the assessments (2015/2016). As business operations are dynamic, data can be subject to change.

CASE
Organic binder from alcohol production (Eco Biosis S.A., Veracruz, Mexico)

Patrick Watson, Krishna C. Rao and Kamalesh Doshi

Supporting case for Business Model 9	
Location:	Veracruz State, Mexico
Waste input type:	Vinasse waste (from alcohol production)
Value offer:	Clean water and chemical additive (for cement)
Organization type:	Private
Status of organization:	Established in 2011, business operational since March 2013
Scale of businesses:	Pilot plant for Mexican domestic market
Major partners:	Client and supplier of vinasse for plant, Universidad del Medio Ambiente (UMA), Gecco Corp., Industrias ADVIEE, San Jose de Abajo distillery, BiD Network, Green Momentum, New Ventures Mexico.

Executive summary

Eco Biosis S.A (Eco Biosis) is a private Mexican company established in 2011 that has patented an innovative process for producing a chemical additive, an organic binder (BioDisperSis VC®) from the vinasse waste generated in alcohol production. It has launched its pilot factory in March 2013 in Veracruz Ignacio de la Llave in Mexico and is providing BioDisperSis VC® to the construction industry for use as a plasticizer in cement. The Eco Biosis plant is situated alongside an alcohol distillery that provides the vinasse waste remaining after sugarcane alcohol distillation. Eco Biosis receives the vinasse from the distillery free of charge because it offsets the cost of disposal. By utilizing the vinasse waste, Eco Biosis has a significant positive impact on both the local community and environment, reducing the pollution associated with run-off and improper disposal by the refineries into local rivers and lakes. Furthermore, all water extracted from the vinasse during the Eco Biosis process is fed back to the distilleries, thereby reducing overall water usage. This will help the distilleries to earn an environment-friendly enterprise certificate.

CASE: ORGANIC BINDER FROM ALCOHOL PRODUCTION

KEY PERFORMANCE INDICATORS (AS OF 2014)	
Land use:	Pilot Plant: approx. 300 m^2; Expansion plant: approx. 3,000 m^2
Water requirement:	The water used in open circuit cooling system, cleaning system for evaporator and 2 t/hr steam and condensation system is recycled; small quantities of make-up water to recoup losses is used by the plant
Capital investment:	Pilot plant: USD 150,000 for installation of rented equipment and other process equipment and capex (additional USD 700,000 required to buy the plant); Expansion plant (Q2 2015): approx. USD 1,200,000
Labor:	Pilot plant: 12 full-time employees; Expansion plant: 14 full-time employees
O & M cost:	Pilot plant: USD 123,000 per annum; Expansion plant: USD 900,000 per annum
Output:	BioDisperSis VC® production; Pilot plant: 600 tons per annum; Expansion plant: 9,000 tons per annum
Potential social and/or environmental impact:	Employment generation (for 12 employees in pilot plant, 14 in the expansion plant and up to 35 in stage 2 of expansion plant); Reduced water pollution (previously caused by improper vinasse waste disposal into local rivers); Reduced use of water by distillery (as water extracted from vinasse is returned to plants); Substitution of a non-eco-friendly product used today by the cement industry; Simplified logistics result in lower carbon footprint
Financial viability indicators:	Payback period: 1.5 years Post-tax IRR: 34% Gross margin: 22%

Context and background

The alcohol industry in Mexico produces approximately 650 million L of alcohol per year. It also produces about 15–17 L of liquid waste, known as vinasse, for every litre of alcohol produced. Vinasse is the residual effluent left after distillation of the ethanol from fermented wines. It has a low solid content of less than 10% undissolved solids but high content of dissolved solids, organic matter and ashes and has high viscosity, very acidic pH (3.5–6) and very high BOD (17,000–50,000 mg/L). In most cases, it is discharged at very high temperatures (around 90 °C). Vinasse is a potentially highly-polluting effluent that can cause serious health issues, diminish aquatic life, affect productivity of land, contaminate aquifer found lands, and emit methane into our atmosphere if not managed properly.

It is very difficult and costly to treat and dispose of vinasse. Different forms of utilization, treatment and final disposal have been sought for the economical and environmentally-sustainable treatment and disposal to avoid environmentally negative impacts of vinasse. Because of the large quantities of vinasse produced, alternative treatments and uses have been developed, such as recycling of vinasse in fermentation, concentration by evaporation and yeast and energy production. Physical and chemical treatment options of the residue have not been very successful until now, though the high organic content of the residue make it well suitable for biological treatment, especially for anaerobic fermentation. There has been limited success due to the high cost of treating the vinasse before it can be processed. It has unfavourable carbon to nitrogen ratio, lack of important trace elements (like nickel, copper, zinc, etc.) and high content of sulphur reducing the conversion of organic materials into biogas.

The on-site disposal of vinasse by combustion and incineration has also been tried. It generates potassium-rich ash which can be sold commercially. However, it requires considerable amount of energy during pre-evaporation and has difficulties of foaming, salt crystallization and ash fusion. It has been used as organic fertilizer in the cane plantations but can cause salinity problems. Vinasse at lower concentrations may also be used as fodder or as a compost ingredient. In higher concentrations, its chemical properties may affect negatively soils, rivers and lakes if frequently discharged over a longer period of time.

There is no simple, existing way to get rid of vinasse. For this reason, many members of the alcohol industry in low and middle-income countries have chosen to set the problem aside and dispose its waste in an unlawful manner, dumping it into rivers, sewage pipes and land and causing often grave social and/or environmental problems. Thus, green approaches are in demand to address the challenge, building on the hidden resources vinasse offers.

In 2009, Eco Biosis started working on the technology to treat vinasse in collaboration with the Universidad del Medio Ambiente, New Ventures, Fundacion E. and Green Momentum. In 2011, Eco Biosis submitted the patent for a multi-stage dehydration process for treating vinasse and converting it into a commercially valued product, called BioDispersis that is easy to handle and distribute. It acts as a natural dispersing and plasticizing agent that can be used as a substitute for lignosulfonates. The main by-product of the process is clean water, which can be reintegrated in the alcohol production process, helping it achieve greater sustainability standards.

In March 2013, Eco Biosis completed the construction of its pilot plant and began converting vinasse to BioDisperSis VC® (BioDispersis). It expects to operate at 100% capacity producing up to three tons per day for 25 days a month (approx. 600 tons per annum). This is the first of a series of plants the company anticipates building to use its patented technology. Eco Biosis's plant is located in eastern Mexico in the Veracruz state, one of the leading sugarcane-producing and alcohol-refinery states in Mexico. The plant is situated within the site of a sugarcane refinery plant, which produces up to 250 m^3 of vinasse on a daily basis. The refinery provides water, air, electricity, steam and vinasse free of charge to Eco Biosis, because Eco Biosis offers a cost-effective way to dispose of unwanted waste.

Market environment

The organic binder (chemical additive) being produced by Eco Biosis is an ecological substitute for lignosulfonates, water-soluble anionic polyelectrolyte polymers that have a broad range of applications across many industries including cements, agriculture, pesticides, mining, leather tannery, crude industry, livestock, concrete, binding and adhesive and dyes and pigment industries.

The annual global consumption of lignosulfonates in 2013 was approximately 1.24 million tons and grew annually by about 1.5% during the last 13 years. The construction sector leads in demand for lignosulfonates, which are used as a plasticizer for concrete, allowing the concrete to be made with less water while maintaining the ability to flow. In Mexico, the consumption of concrete lignosulfonate is 55,000 tons per annum, which is expected to continue growing by about 2.5–4.5% per annum. Eco Biosis anticipates operations producing 600 tons of lignosulfonates from the pilot plant or approximately 1% of the Mexican market and up to 5,000 tons by 2014 or roughly 10% of the market.

The competitive landscape in Mexico is dominated by Norwegian company Borregaard LignoTech, which has over 60% of the market and produces speciality chemicals for the agro and construction industries. The other key players are Tembec and WestRock (created by merger of Mead Westvaco and RockTenn) which produce lignosulfonates using wood as the primary raw material. The Eco Biosis lignosulfonate substitute, however, has competitive advantages of indigenous supply at fraction of the cost and its green credentials as all of Eco Biosis' competitors use non-sustainable timber as their primary raw material.

Macro-economic environment

The alcohol industry is expected to exceed USD 1 trillion in 2014, with market volume expected to reach approximately 210 billion litres, according to market research firm Market Line. Mexico alone

CASE: ORGANIC BINDER FROM ALCOHOL PRODUCTION

produces over 10 billion litres of vinasse annually, which must be disposed of in accordance with government requirements. Due to the quantity of vinasse produced and high disposal costs, a large amount of run-off ends up in the lakes and rivers causing a significantly negative environmental impact. The regulation around vinasse waste disposal has therefore tightened in recent years, increasing the disposal cost for alcohol distilleries and reducing operating margins. Due to which the domestic alcohol production has fallen in recent years and import of ethanol has increased.

Only a handful of the largest alcohol refineries in Mexico are disposing of vinasse legally, which has led to significant investment into R&D to improve disposal methods. The most common method for disposing of vinasse is through the use of anaerobic reactors and burning the gas or utilizing filtration systems and landfill; however these methods are costly and/or not effective. The Eco Biosis technology provides a profitable solution to disposal methods used by alcohol refineries, in addition to producing a versatile chemical additive that can be used in a number of different industries. Eco Biosis' model provides a solution to vinasse disposal that can be easily replicated on a global scale.

Business model
The key value proposition of Eco Biosis (Figure 110) is to produce lignosulfonates substitutes with superior environmentally-safe moieties from vinasse waste generated during alcohol production, and in the process save water and reduce environment pollution. After spending initial years in developing technology to process vinasse waste to lignosulfonate substitutes, Eco Biosis is running a pilot plant in partnership with an alcohol distillery. Lignosulfonates has multiple applications and Eco Biosis can target multiple customer segments such as concrete, cement, chemical, mining and energy companies. At the time of the interview, Eco Biosis has a contract with a multinational company to supply 100% of BioDisperSis produced during the next six years starting after 2014.

Value chain and position
The key players in the Eco Biosis value chain are alcohol producers as supplier of vinasse, water, utilities and infrastructures, partners for developing technology and business development and clients who will buy BioDisperSis (Figure 111). The San Jose de Abajo distillery provides vinasse for the pilot plant and provided land to Eco Biosis to construct its plant within its factory premises. The distillery's gas, electricity and steam supply are provided free of charge to Eco Biosis operations. Earlier, the distillery was dependent upon water from sugarcane to dilute the waste for its operations, and it had to stop its distillery operations after every harvest. Incorporating the Eco Biosis plant into the distillation process allows the distillery to continue production uninterrupted throughout the year as water recovered from treatment of vinasse by Eco Biosis is sent back to the distillery, therefore positively impacting the profitability of the business. For the expansion plant, Eco Biosis is in negotiations with a number of businesses and hopes to secure a larger and more stable source of vinasse.

Developing the process of treating vinasse to produce lignosulfonate substitute required Eco Biosis to consult with different agencies for technical assistance, product development, use of equipment to start the plant and refine the process. In addition, it also received business development assistance from incubation programs such as BiD Network, New Ventures, Green Momentum and UMA, and in the process gained exposure to investors/funding and overcome legal issues.

Eco Biosis' one key client is committed under contract for the next six years (starting 2014) to buy the entirety of the production of BioDispersis. Eco Biosis is also in advanced conversations with a number of other potential clients who are interested in their product in the long run.

FIGURE 110. ECO BIOSIS BUSINESS MODEL CANVAS

KEY PARTNERS
- Universidad del Medio Ambiente
- Gecco
- Industrias ADVIEE
- San Jose de Abajo distillery
- BiD Network
- Green Momentum
- New Ventures Mexico
- Mexican national Council for Science and Technology (CONACYT)

KEY ACTIVITIES
- BioDisperSis VC° production, storage and sales
- R&D of the product
- Quality control

KEY RESOURCES
- Land
- Capital
- Skilled labor
- Agreement with distillery for supply of vinasse and utilities
- Patent-protected IP
- Green brand image

VALUE PROPOSITIONS
- Producing environment-friendly lignosulfonate substitute from alcohol distillation by-product waste and reduce environment pollution

CUSTOMER RELATIONSHIPS
- Direct interaction with user industries

CHANNELS
- Appropriate agreement and self-interaction

CUSTOMER SEGMENTS
- Construction and chemical companies
- Agricultural companies (in future)
- Livestock feeding companies
- Specialized product-distributing companies
- Mining companies
- Energy companies

COST STRUCTURE
- Investment cost (land, building, machines)
- Operational cost (fuel, labor, utilities, maintenance cost)
- Marketing & sales cost
- R&D
- Patent maintenance costs (substantial: approx. 250,000 over the lifetime of a worldwide patent)

REVENUE STREAMS
- Sale of BioDisperSis (lignosulfonate)

SOCIAL & ENVIRONMENTAL COSTS
- Possible risk of water pollution and environmental hazard if the removal of pathogens and pollutants during the process is not complete and it is discharged into the open

SOCIAL & ENVIRONMENTAL BENEFITS
- Job creation
- Environmental benefit through processing vinasse which can otherwise cause water, land and air pollution
- Environmental benefit from consuming BioDisperSis rather than lignosulfonates made from wood pulp
- Carbon footprint reduction from the local purchase of lignosulfonates instead of importing them

CASE: ORGANIC BINDER FROM ALCOHOL PRODUCTION

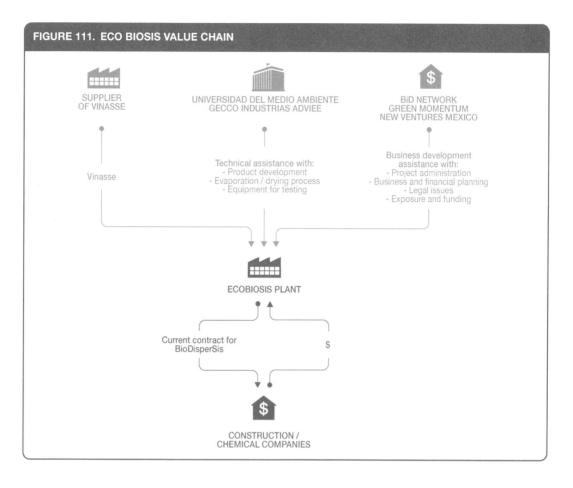

FIGURE 111. ECO BIOSIS VALUE CHAIN

Eco Biosis business has supplier power prominence as the source of vinasse is dependent upon the San Jose de Abajo distillery continuing to supply it to the Eco Biosis plant, in addition to funding the operational costs of the plant. However, supplier prominence is weakened if Eco Biosis plant reduces the operational costs of the distillery. Buyer power prominence and substitutes exist as there is an established market for lignosulfonates. Eco Biosis' hopes to counter it by pricing its product 70% lower than its competitors. The threat of new entrants, using the same process, is limited due to patent protection; however, there are other existing methods of treating vinasse, which could compete with Eco Biosis.

Institutional environment

Prevention and management of waste: Mexico is working on environmental waste reduction to achieve better management of waste through an environmental policy. The president has made policies to reduce global warming a special and personal issue of his administration. In spite of the attention given to the issues, Mexico continues to face serious environmental challenges largely because even when anti-pollution legislation exists, much of it is not being applied and enforced. The Mexican Department for Environmental Affairs implemented a law in 1996 restricting the contamination of national water bodies (NOM-001-ECOL-1996), which proposed a set of contaminant limits for liquids being disposed of. The sample analysis of sugarcane vinasse done by Eco Biosis indicated a total suspended solids content 50 times higher than that specified by the contaminant limits.

Because the technology is untested on an industrial scale, Eco Biosis has encountered certain resistance within the government to build the pilot plant. Eco Biosis has opted not to be classified as a waste treatment service provider, but rather decided to register as a manufacturing company subject to manufacturing sector regulations. Eco Biosis is therefore subject to a different frequency of operating audits than it would otherwise be under the waste treatment classification and has to obtain different more practical permits prior to initiating production.

Technology and processes

Eco Biosis initiated the patenting process for manufacturing BioDisperSis in 2011. Patent approval is pending. Figure 112 presents the process involved in production of BioDisperSis.

Vinasse is received into the plant through a pipeline from the distillery and is passed through a high-temperature clarifier to extract suspended solids (fibres, mud and yeast) before being stored in a tank to start the pre-concentration process. During the pre-concentration process, water is removed and stability agents are added for required physical properties needed for BioDisperSis. The solution then goes through a further filtration process to extract any remaining water. Three products are

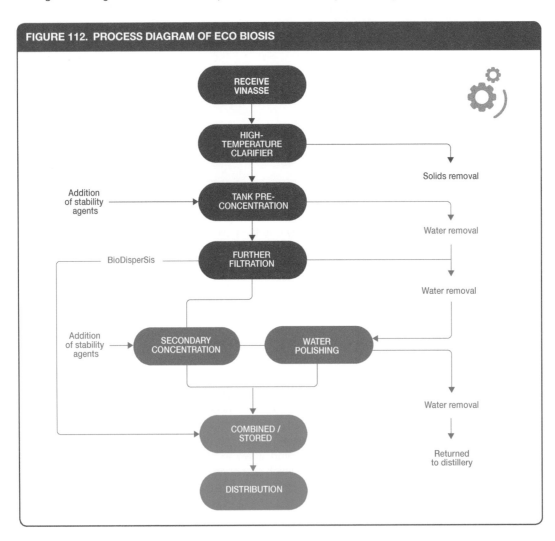

FIGURE 112. PROCESS DIAGRAM OF ECO BIOSIS

derived from this process: water, BioDisperSis and vinasse that needs to go through an additional process of concentration. Water extracted is treated to improve its quality before being returned to the distillery for reuse in the alcohol distillation process. The vinasse passes through an additional low temperature concentration process where other agents are added to the solution before it is ready to be stored and dispatched. The vinasse batch that has gone through the secondary concentration and the BioDisperSis are re-combined and a number of chemical agents are added to the liquid in order to preserve the product quality and durability. The remaining BioDisperSis is then checked for quality control, stored under the appropriate conditions and distributed via the Eco Biosis fleet of distribution trucks.

Funding and financial outlook

The implementation and construction cost for the pilot plant was approximately USD 150,000 and 100% funded by the directors, one other private angel investor and the Mexican National Council for Science and Technology. The total amount spent on the pilot plant does not, however, reflect a standalone build-out of the plant as Eco Biosis utilized second-hand and rented machinery, in addition to renting out the plant, which would have otherwise cost an additional USD 700,000 to acquire (Table 31). The most significant expense items were machinery (USD 80,000), installation (USD 25,000), vehicles (USD 25,000) and electrical costs (USD 10,000). The total investment to date has been approximately USD 400,000; however, the vast majority of this has gone into product R&D. The pilot plant is expected to make a small profit of approximately USD 35,340 on an annualized basis, with revenues of about USD 158,000. The key operational costs for the pilot plant are machinery rental, chemical process and labor, contributing to approximately 79% of the total running costs.

TABLE 31. ECO BIOSIS FINANCIAL SUMMARY

PILOT PLANT FINANCIAL SUMMARY									
USD/MONTH	M1	M2	M3	M4	M5	M6	M7-12	Y1	Y2
Initial Investment	(13,333)	(13,333)	(13,333)					(40,000)	
Revenue (Unit: USD 240)				13,205	13,205	13,205	79,230	118,845	158,460
Costs									
Labor				(1,843)	(1,843)	(1,843)	(11,058)	(16,587)	(22,116)
Chemical Process				(3,850)	(3,850)	(3,850)	(23,100)	(34,650)	(46,200)
Evaporator: Rent				(2,400)	(2,400)	(2,400)	(14,400)	(21,600)	(28,800)
Telephone				(189)	(189)	(189)	(1,134)	(1,701)	(2,268)
Plant Rental and Petty Cash				(1,137)	(1,137)	(1,137)	(6,822)	(10,233)	(13,644)
Distribution				(841)	(841)	(841)	(5,046)	(7,569)	(10,092)
Total Costs				(10,260)	(10,260)	(10,260)	(61,560)	(92,340)	(123,120)
Net Margin				2,945	2,945	2,945	17,670	26,505	35,340
Payback period from pilot plant: 4.5 years									
IRR*: 34%									

*IRR only taken for first 2 years as the pilot plant is not intended to be run on a continual basis but used as a model on which to launch the expansion plant

The pilot plant is being used to prove the quality of BioDisperSis and secure a number of larger-scale contracts in order to start construction of the expansion plant. Eco Biosis is therefore looking to expand (Table 32) from this in two key phases: 1) an initial expansion plant with production capacity of 9,000 tons in 2015–2016 and 2) a full expansion plant coming on-stream in 2017–18 with production capacity of 27,000 tons. Eco Biosis will invest USD 2.6 million in the expansion plant, which will have revenues of USD 1.7 million and breakeven at approximately 45% production capacity. The fully-operational plant will require a further investment of USD 5.4 million and increase potential revenues by up to 300%, with breakeven production of approximately 55%.

TABLE 32. ECO BIOSIS FINANCIALS PROJECTIONS

EXPANSION PHASES FINANCIAL SUMMARY							
USD	2014	2015	2016	2017	2018	2019	2020
Investment	2,587,589		5,368,968				
Revenue		1,668,734	1,768,265	6,332,350	6,710,043	7,110,264	7,534,355
EBITDA		780,143	839,613	3,316,631	3,556,095	3,811,249	4,083,077
Net Profit		368,120	406,656	1,324,082	1,430,042	1,663,888	1,909,755

Eco Biosis has already secured approximately USD 0.8 million in funding for its expansion plant from a number of developmental agencies.

Socio-economic, health and environmental impact

From an environmental perspective, the technology has a significant positive impact as it reduces the contamination of local water bodies through converting unused vinasse into lignosulfonate substitute, and in addition, indirectly improves the livelihood of the local population. Furthermore, Eco Biosis has a negative net water usage as it extracts more water from the vinasse received than it uses in the conversion process, thereby returning water for reuse to the alcohol distilleries and preserving an already scarce supply of potable water. Eco Biosis provides employment to 12 local workers in the pilot plant; however, this will increase to approximately 14 in the expansion plant and up to 35 in the fully-operational plant which is planned to come on-stream in 2018.

Scalability and replicability considerations

The key drivers for the success of this business are:
- Patented technology and process for making BioDisperSis from vinasse.
- Partnerships with alcohol distilleries, allowing extreme low-cost sourcing of inputs.
- Viable lower-cost alternative for vinasse treatment in compliance with regulatory requirement.
- Awareness and market for clean technology solutions.
- Higher-priced substitutes (Lignosulfonates are imported).
- Tightening vinasse disposal regulations.

The Eco Biosis pilot plant feasibility is from multiple factors. Most important is the plant's location within an existing distillery, and in addition receiving services free of charge which would otherwise have had a significant impact on the operational costs. Operational cost savings incurred assist in making a small-scale pilot plant viable. The two key considerations for scaling Eco Biosis are: 1) availability of vinasse as a raw material and 2) demand for lignosulfonate substitute. With continued support of anti-pollution legislation in Mexico, Eco Biosis provides a cost-effective approach to disposing of vinasse legally and can therefore secure significant quantities of the vinasse waste at relatively low cost, enabling it continued domestic expansion.

On a global scale, the alcohol industry continues to grow strongly, expecting to reach USD 1 trillion in 2014, representing almost 210 billion litres. This represents a significant opportunity for the Eco Biosis technology to be utilized in other countries to counter the pollution from vinasse. The demand for lignosulfonate substitutes will continue to grow in the construction industry as it provides an environmental-friendly alternative to wood-pulp-derived lignosulfonates. Furthermore, Eco Biosis can export its product to foreign markets demanding lignosulfonates.

Summary assessment – SWOT analysis

The key strengths of Eco Biosis are the benefits drawn by alcohol distillery and an environmentally-friendly alternative for producing lignosulfonates substitute from vinasse in comparison to mainstream methods of using wood pulp as key input (Figure 113). The weakness of Eco Biosis is high investment required for its expansion. In its future expansion, Eco Biosis might require to alter its process based on the quality of raw material input and could further increase its investment costs. Eco Biosis once has commercially proven and has successfully run its operations for few years. It has strong opportunities to expand both domestic and overseas.

FIGURE 113. ECO BIOSIS SWOT ANALYSIS

	HELPFUL TO ACHIEVING THE OBJECTIVES	**HARMFUL** TO ACHIEVING THE OBJECTIVES
INTERNAL ORIGIN ATTRIBUTES OF THE ENTERPRISE	**STRENGTHS** • A proprietary cost-effective solution to significant global environmental problem • Patent pending that will reduce threat of competition • Beneficial for distilleries as disposing of unwanted waste and getting clean water back • BioDisperSis price competitive with other lignosulfonates • Lignosulfonates are used in a broad range of industries mitigating potential market risk • BioDisperSis is a green alternative to mainstream wood-pulp-derived lignosulfonates • Low water usage throughout process and water extracted from vinasse is returned to distilleries	**WEAKNESSES** • Reliance on operating pilot plant and operating cost subsidies • Have inherent risks with expansion • High level of investment required to launch expansion plant • Have only developed one product with vinasse – BioDisperSis – therefore, significant R&D is still required to open up new markets and process vinasses of different qualities • Capacity of the plant has to be linked to the capacity of the alcohol distillery as transport of vinasse would not be cost effective
EXTERNAL ORIGIN ATTRIBUTES OF THE ENVIRONMENT	**OPPORTUNITIES** • Domestic and overseas expansion as significant supply of vinasse waste that is not being exploited • Source raw material from a number of other vinasse producing industries, e.g. bread making • Expand product line to open up new markets and industries • Growing popularity for sustainable products globally	**THREATS** • Supply of raw material is not secured • Change in quality of raw material will alter the solution properties and slow process • Patent not yet granted

Contributors

Carlos Fernandez, I-DEV International
Cinthya Pajares, I-DEV International
Eco Biosis S.A.

References and further readings

Creixell, M. Interviewed by Patrick Watson via email and telephone. March 25, 2014.

Olguín, E.J., Doelel, H.W. and Mercado, G. 1995. Resource recovery through recycling of sugar processing by-products and residuals. Resources, Conservation and Recycling 15(2): 85–94.

Vargas, M. "Project for the manufacturing of sustainable dispersing bases made from highly polluting wastes." PowerPoint presentation. Retrieved from LinkedIn page of Los Angeles Cleantech Incubator (LACI): www.slideshare.net/LACIncubator/7eco-biosis (accessed November 7, 2017).

Case descriptions are based on primary and secondary data provided by case operators, insiders, or other stakeholders, and reflects our best knowledge at the time of the assessments (2015–2016). As business operations are dynamic, data can be subject to change.

BUSINESS MODEL 9
Bio-ethanol and chemical products from agro and agro-industrial waste

Solomie Gebrezgabher and Krishna C. Rao

A. Key characteristics

Model name	Bio-ethanol and chemical products from agro and agro-industrial waste
Waste stream	Organic waste – Agro-waste from farms and agro-processing factories and vinasse waste generated during ethanol production
Value-added waste product	Bio-ethanol (as additive to petrol/gasoline as transportation fuel) and chemical products (like lignosulfonate substitutes for various industries)
Geography	Regions with large agro-industries
Scale of production	Small to medium scale 20–30 tons of chemical product or ethanol per day from agro or agro-industrial waste
Supporting cases in this book	Carabobo, Venezuela; Veracruz state, Mexico
Objective of entity	Cost-recovery [ ]; For profit [X]; Social enterprise [ ]
Investment cost range	Approx. 150–400 USD/ton of chemical product or ethanol
Organization type	Private
Socio-economic impact	Employment generation (12–50 jobs), improved income of farmers, reduced water, land and air pollution from vinasse and agro-waste, reducing GHG emissions by substituting petrol used for transportation or non-eco-friendly product (like lignosulfonates)
Gender equity	No advantage to a specific gender

B. Business value chain

This business model is owned and operated by either a standalone private entity or agro-industries such as rice mills, coffee, cassava and palm-oil-processing units. The business processes solid or liquid agro-waste or crop residues such as wheat stalk, rice husk, maize stalk, groundnut shells, coffee husks and cassava waste to produce ethanol or chemical products (Figure 114). Specific technology tailored to the quality of available waste needs to be developed for each case, depending on the type of waste. If ethanol is produced, this can be blended with gasoline and used in motor vehicles. This is becoming an increasingly cost-effective renewable solution to transport, as gasoline stations around the world start to provide blended fuel and motor vehicles no longer need any modifications to use this fuel.

The key stakeholders in the business value chain are the suppliers of agro-waste (farmers and agro-industries), technology suppliers and petroleum companies and consumers of ethanol. The process of

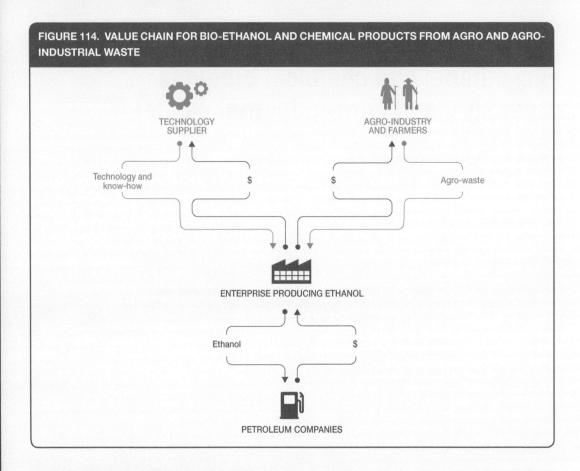

FIGURE 114. VALUE CHAIN FOR BIO-ETHANOL AND CHEMICAL PRODUCTS FROM AGRO AND AGRO-INDUSTRIAL WASTE

generating ethanol uses enzymes to break down cellulose in the agro-waste into fermentable sugars. For the business to be successful, it is important to develop enzymes that break down complex cellulose efficiently and economically. In addition, the business could require developing special strains of yeast or bacteria for improved fermentation processes for better yields of ethanol. These micro-organisms can be engineered to work more efficiently in specific temperatures and acidities, or can be engineered to have new scopes of enzymatic activities or combinations thereof. R&D of such technology is costly and can only be initiated with availability of sufficient R&D capacity, either in-house (for a large company) or with a R&D partner, and with the availability of sufficient funds throughout the technology development stages. The newly-developed materials and/or processes should also be patented to protect the technology to ensure return on investment. However, this represents another substantial cost over the course of life of the patent(s).

Overall, the model contributes to the reduction of environmental and health hazards associated with disposing of waste from agro-processing units, and thus creating a green image for the agro-processing units. The business is eligible for sale of carbon as ethanol is a biofuel and is generated from sustainable biomass source. Furthermore, the production of ethanol results in vinasse, a by-product of waste distillation which can be treated to recover clean water. Finally, there is also the potential to produce value-added animal feed, biogas and bio-fertilizers by further processing of remaining sludge.

C. Business model

The primary value proposition of the business model is production of environment-friendly ethanol from cellulose in agro-waste for blending with petrol/gasoline as transportation fuel. Figure 115 depicts the business model canvas for emerging technologies.

As an additional business line and in consideration of pursuing an emerging technology route, the business should consider developing a product from the vinasse waste generated during ethanol production. One of the business cases described (Eco Biosis) in the energy section of the catalog elaborates on the company's successful efforts to research and produce a low-cost substitute for lignosulfonates from the vinasse. These chemical additives are used extensively in several industries such as construction, agriculture, mining and cosmetics. Benefits in this manner double, as effluent is minimized and cleaned, while another revenue stream is added. Both the Eco Biosis material and the process are protected with patents.

It requires considerable investment in developing the technology and process. The business spends a substantial amount of money on product R&D and subsequent patenting. In order to develop the new product that meets customers' needs at a competitive price and to ensure sales, the business requires collaboration and consultation with different agencies for technical assistance, product development including partnership with an R&D institute, business development and legal assistance including patent protection measures. The business model may require technology validation for which the "launching customer" concept is an ideal strategy. For example, the business starts as a pilot plant in partnership with agro-processing units and gradually increases its scale of operation while at the same time securing off-take contracts with specific buyers. These projects require high-risk money with flexibility of adopting strategies to the business needs for technology and process development.

Once the technology and the process are streamlined for commercial production, it is important for the business to form partnerships with agro-processing units to secure reliable supply of inputs. Hence, this business is either located near or is incorporated into the agro-processing factories as production of ethanol depends, among other things, on the availability of the feedstock. The business receives the feedstock free of charge or pays a nominal value because it offsets the cost of disposal for the agro-processing factory. Incorporating the business into the agro-processing factory is a strategic decision for the agro-processing factory.

An alternative strategy is for the business to buy-in newly-developed technology from a specialized R&D organization that it partners with. This might take longer and have less security but it dramatically reduces the risks associated with the R&D stages. The business may enter a contract R&D arrangement or may invest in participating in a technology development consortium. Still alternatively, the business may adopt an R&D networking strategy in which it vigilantly monitors technology developments within the R&D arena and buys in at a time of interest. It should expect to invest considerable time liaising within the R&D network and the technology transfer channels, with smaller chances of a good match.

A further alternative is for the business to license in strategically-important technology developed and patented by another party. Benchmarks for upfront payments and royalties on sales vary widely and depend on the type of technology, technology maturity and market dynamics. In both two alternatives mentioned, it is important to avail the required critical understanding and capacity in intellectual property rights and legal affairs.

FIGURE 115. BUSINESS MODEL CANVAS – BIO-ETHANOL AND CHEMICAL PRODUCTS FROM AGRO AND AGRO-INDUSTRIAL WASTE

KEY PARTNERS
- Farmers and agro-processing units
- Research institutes
- Funders/investors

KEY ACTIVITIES
- R&D of products
- Production of ethanol/chemical product
- Marketing of product

KEY RESOURCES
- Land
- Capital
- Agreement with agro-processing units and farmers
- Agro-waste
- Technical and operational competencies
- Patent protected IP
- Patent expertise
- Green brand image

VALUE PROPOSITIONS
- Environment friendly ethanol from agro-waste for blending with petroleum
- Producing environment-friendly lignosulfonate substitute from alcohol distillation by-product waste and reduce environment pollution

CUSTOMER RELATIONSHIPS
- Direct interaction with user
- Industries/petroleum companies

CHANNELS
- Direct supply to petroleum companies under contract

CUSTOMER SEGMENTS
- Petroleum companies
- User industries

COST STRUCTURE
- Investment cost (land, building, machines)
- Operational cost (raw material cost, labor, utilities, maintenance)
- Marketing cost
- Depreciation
- R&D
- Patent filing, maintenance and patent attorney costs
- Equity and/or interest on loans

REVENUE STREAMS
- Sale of ethanol
- Sale of chemical products

SOCIAL & ENVIRONMENTAL COSTS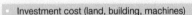
- Possible risk of pollution of water from vinasse

SOCIAL & ENVIRONMENTAL BENEFITS
- Creation of jobs
- Reduced water, land and air pollution through processing of agro-waste and vinasse
- Reduced GHG emissions by using ethanol from renewable source blended into petrol
- Clean water savings
- Environmental benefit from consuming substitutes of lignosulfonates made from wood pulp
- Carbon footprint reduction from the local purchase of lignosulfonates instead of importing them

D. Potential risks and mitigation

Market risks: This business model is offering a new product, which can substitute existing products with an established market. The business faces uncertainty in successfully deploying the new product from R&D and pilot stage to commercialization. Ethanol from agro-waste can be used as a substitute for ethanol from other sources such as sugarcane and corn. This business faces the challenges of developing a viable business and requires extensive marketing and awareness creation among its end users to secure off-take contracts.

Competition risks: The success of ethanol from agro-waste depends on how fast the technology is commercialized and how much it costs compared to established alternative products. The business can avoid competition from existing companies in the market by targeting those buyers which are not served by existing companies or enter a market through strategic positioning by offering the product that is environmentally-friendly and is lower-priced than established alternative products in the market. Ethanol from agro-waste is expected to be less expensive than the alternatives as inputs can be sourced at a low cost, and thus giving this business a competitive advantage over other ethanol producers.

Technology performance risks: Technology is new and was not tested at the assessment time on an industrial scale. Technology development and market introduction are a multi-year, multi-step process, often requiring financial injections at various stages. Capital costs are uncertain when constructing a pilot plant and a commercially-viable demonstration, as the technology is not proven. Hence, there is considerable risk from inability to reach investment coverage at each individual stage. Partnership with an R&D institute is required to move the technology from pilot to commercial scale, and in the process mitigate any risk associated with technology performance.

Political and regulatory risks: There is limited awareness on the technology or process among policy makers. Since the technology is new and not tested on an industrial scale, the business may face challenges from unfavourable business environment, encounter resistance from the government to obtain permits prior to initiating production and go through a lengthy process for obtaining approval for patent. Few governments in developing countries have implemented the policy of mandatory blending of petrol/gasoline with ethanol and such policies will significantly help this business model.

Social-equity-related risks: The model is considered to have no advantage to any specific gender. The benefits of the model are to agro-industries to help manage their waste and the employment opportunities created are for highly-skilled labor. The model could suffer from social-equity risks which can be mitigated from corporate social responsibility initiatives that provide benefits to the community around the plant.

Safety, environmental and health risks: There is possible risk of water pollution and environmental hazard if the production of ethanol from agro-waste does not remove pathogens and pollutants completely and is discharged into the open. Untreated vinasse waste discharged into the open is an environmental hazard and can harm the local ecosystem. However, there are technologies and good practices to prevent this (Table 33).

TABLE 33. POTENTIAL HEALTH AND ENVIRONMENTAL RISKS AND SUGGESTED MITIGATION MEASURES FOR BUSINESS MODEL 9

RISK GROUP	EXPOSURE					REMARKS
	DIRECT CONTACT	AIR	INSECTS	WATER/ SOIL	FOOD	
Worker	LOW	LOW	LOW			Risks to workers from direct contact with the waste can be mitigated using protective gear/equipment.
Farmer/User						
Community		LOW		MEDIUM		
Consumer						
Mitigation measures	👢🧤	😷🌳	🦟	⛰️🏊‍♂️🚫		

Key: ☐ NOT APPLICABLE ▨ LOW RISK ▨ MEDIUM RISK ■ HIGH RISK

E. Business performance

This business model is rated as high in innovation but medium on profitability, social and environmental impact and low on scalability and replicability potential (Figure 116). The business is highly innovative in

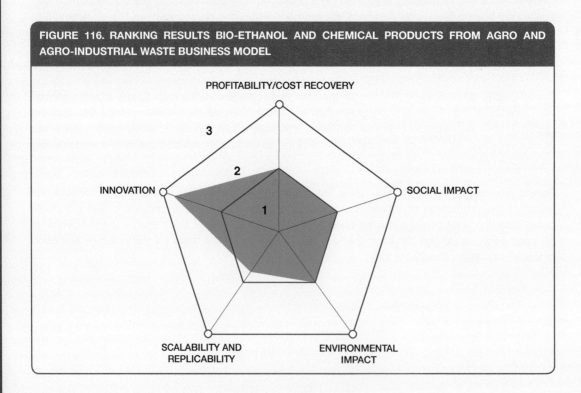

FIGURE 116. RANKING RESULTS BIO-ETHANOL AND CHEMICAL PRODUCTS FROM AGRO AND AGRO-INDUSTRIAL WASTE BUSINESS MODEL

terms of its developing new and patented materials and/or processes. The business is also innovative in creating strategic partnerships with different players in the market, such as input suppliers, technology developers, business development and legal advisors. This business model can result in high returns from its innovative process and strategic partnership. However, the deployment of the new technology into the commercial market requires significant amount of capital, and thus affecting the profitability and future cash flows.

SECTION III – NUTRIENT AND ORGANIC MATTER RECOVERY

Edited by Miriam Otoo

Nutrient and organic matter recovery: An overview of presented business cases and models

Nutrient recovery from organic waste streams such as municipal solid waste, agro-industrial waste, urine and fecal sludge, is high on the development agenda. The increased momentum around nutrient recovery is largely driven by the need to feed the global population with increasingly limited resources under progressing climate change, diminishing global nutrient reserves (peak phosphorus), increasing fertilizer prices and stricter regulations for safeguarding the environment from pollution. In this context, increasing amounts of plant nutrients will be needed to ensure the food security of an expanding global population. However, while a century ago, food waste was locally recycled, urbanization has created a polarizing effect on food flows, thus generating centres of consumption and waste generation. Nutrient recycling is therefore crucial in preventing cities from becoming vast nutrient sinks (Drechsel et al., 2015; Otoo et al., 2012; Otoo et al., 2015). Unfortunately, in most low- and middle-income countries, urban waste management continues to struggle with waste collection and safe disposal making e.g. nutrient recovery only a future target. However, simultaneous efforts are required and possible, also as the waste and sanitation sectors are under pressure to cut costs and show cost recovery. The waste volume reduction through composting and agricultural demand open related opportunities (Drechsel et al., 2015).

Nutrient recovery is additionally of great importance in view of diminishing non-renewable resources, such as phosphorus. As large portions of global phosphate rock deposits cannot be mined efficiently at competitive costs, there is a great debate on when the world will reach a state of 'peak phosphorus' and how far market prices will regulate phosphorus supply (Edixhoven et al., 2013). On the other hand, there is a consensus that the recovery of phosphorus is an increasingly important task, especially given that soils in many tropical developing countries are of very low fertility and fertilizers too expensive. The latter is evident in many African countries and attributed to ineffective policies, and limited and inefficient distribution network. This results in exorbitant market prices, and invariably leading to low fertilizer application rates and decreased agricultural productivity.

Furthermore, nutrient recovery from organic waste streams such as agricultural and agro-industrial waste, the biodegradable fraction of household and market waste, domestic urine and fecal sludge, extends beyond direct economic benefits to health and environmental benefits (ADB, 2011; Hernando-Sanchez et al., 2015; Otoo et al., 2015; Rao et al., 2016). With increasing population growth, nutrients accumulate in consumption centres and contribute to pollution wherever the coverage of waste collection and treatment is insufficient. With progressively limited public funds to support waste management infrastructure and services, particularly in large urban areas in developing countries, nutrient recovery enterprises will be essential for reducing waste quantities and generating revenues from recovered resources to bridge financial gaps (operational and maintenance [O&M] costs) and complement other supportive financing mechanisms for waste management.

There is great potential to close the nutrient recycling loop, support a 'circular economy' and attain cost recovery within the waste sector, and even to create viable businesses. While, many of these efforts have often been limited in size or duration partly because waste is not viewed as a resource and sanitation is a public service rather than a business; there are many interesting and successful examples of cases and business models emerging in developing countries. These cover a wide range of opportunities for waste valorization (Figure 117) and demonstrate significant potential for scalability and sustainability.

OVERVIEW

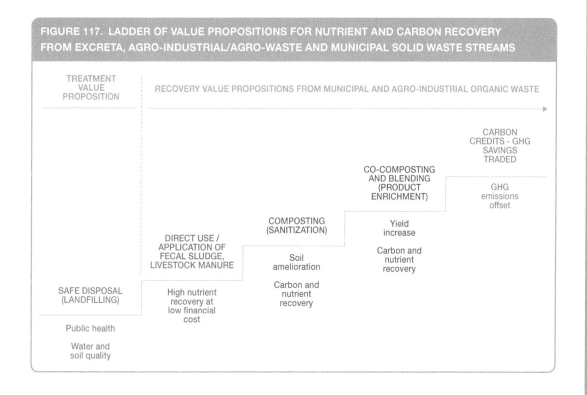

FIGURE 117. LADDER OF VALUE PROPOSITIONS FOR NUTRIENT AND CARBON RECOVERY FROM EXCRETA, AGRO-INDUSTRIAL/AGRO-WASTE AND MUNICIPAL SOLID WASTE STREAMS

Significant investments, mainly public funding, for the set-up and operation of compost facilities is observed throughout the developing world (Kaza et al., 2016). These compost plants are typically large-scale centralized facilities that are able to process huge volumes of waste at a time, but require substantial capital investments, and operational and maintenance costs given the advanced and mechanized equipments used, high-level skill and high energy requirements. Although geared towards full cost-recovery, many of these initiatives are unable to generate sufficient revenues to cover the O&M costs, talk less of recouping capital investments. Municipalities however continue to provide financial support in the form of government grants, subsidies, tax credits, waivers and rebates to bridge the financial gap and ensure sustainability of the compost plants (Kaza et al., 2016; Pandyaswargo and Premakumara, 2014). This is because the net environmental and socio-economic benefits from composting (typically municipal solid waste (MSW) and fecal sludge) outweigh the costs of financial support to the compost plants (**Business model 10: Partially subsidized composting at district level**). In this nutrient recovery section of this Resource Recovery and Reuse (RRR) catalogue, we present three such cases from **Sri Lanka** and **Uganda**, representing different waste streams and options of public-private partnerships.

In view of increasingly shrinking budget allocations for waste management, a notable percentage of compost plants reach the end of their life cycle or in dire need of upgrade and maintenance, especially to improve their production efficiencies and revenues. Decentralized composting enterprises offer some advantages over centralized large-scale systems and are increasingly observed to be financially self-sustaining, particularly for secondary cities and small towns, and even large cities where the local government can allocate land (**Business model 11: Subsidy-free community-based composting**). In instances where technological processes adopted capitalize on abundant local resources (e.g. labor), and models that attribute ownership to communities are encouraged (e.g. cooperatives), high sustainability of the nutrient recovery enterprise has been observed. The presented case study from

Kenya in Chapter 8 shows that subsidy-free community based composting offers a sustainable solution for turning waste into wealth but requires investments in social capital to organize and mobilize the communities.

Looking beyond cost recovery and aiming for profit-making models is imperative if sustainable financial returns on investments are expected (**Business model 12: Large-scale composting for revenue generation**). While the composting concept is applicable across scale, larger composting operations offer greater opportunities for capturing economies of scale benefits, revenue generation and market proliferation. Multiple revenue generation streams beyond compost sales to include sale of energy (electricity) represent additional avenues for nutrient recovery enterprises to become financially viable. The ability for businesses to successfully implement the above value propositions and capture the greatest economic benefits will partly depend on scale and strategic partnerships. The scale element of the model offers access into markets that smaller-scale enterprises are often excluded from such as the energy and carbon credit markets. Although, it is important to note that there are cases where small-scale enterprises form conglomerates to increase accessibility into these markets. The need for strategic partnerships extends beyond those with NGOs for development of waste-based clean development mechanisms (CDM) projects, compost marketers and dealers for increased market share to include municipal authorities for exclusive rights/access to waste streams, research institutes for product and technology innovation, informal workers for increased access to slums and waste segregation efficiency and private sector entities for mitigating fiscal constraints. Mainstreaming private sector participation via public-private partnerships (PPP) can improve production efficiencies and business effectiveness and ensure value for money of public interventions as demonstrated by presented cases from **India** and **Bangladesh** in Chapter 9. Development of high value products (e.g. nutrient-fortified compost tailored for specific crops and soils) based on innovative technologies to enhance competitive advantage in product markets often allow enterprises to mitigate market distortions, for example, in the fertilizer market.

While the first three business models largely centre on food waste and municipal solid waste stream, another set of interesting business models focuses on nutrient recovery from agro-industrial and agro-waste (vegetative and livestock) streams. To ensure business sustainability, largely for compliance with legislative mandates, many agro-processing enterprises are increasingly implementing an additional arm to their main business for converting their waste into organic fertilizers. Conversion of their waste into nutrients is imperative, particularly given that the implicit cost of non-compliance can be significant due to their large operational scale, resulting in potential losses of up to several million dollars in annual revenue (**Business model 13: Nutrient recovery from own agro-industrial waste**). Chapter 10 presents several variants of this model via empirical cases from **Kenya, India** and **Mexico**.

In addition to nutrient recovery from municipal solid waste and agro-industrial waste streams, another set of interesting business models considered in this section focus mainly on fecal sludge and urine reuse for agricultural production. Global mandates to improve access to sanitation (toilets facilities) at the household level in developing countries is notable although some groups such as migratory populations and slum inhabitants still only have marginal access to sanitation products and services. An increasing number of private businesses are setting up public toilet facilities to close this gap, however limited septage collection and treatment can undermine the sustainability of these services. Benefits from nutrient recovery from fecal sludge into value-added products (e.g. urine-enriched compost) for agricultural production are three-fold: a) it significantly reduces the burden for septage collection, treatment and disposal, ensuring a sustainable sanitation service chain; b) it provides sanitation businesses with an additional revenue stream; and c) it provides a sanitized and nutrient-rich compost product for farmers. The latter is an important driver for the business model as farmers have a great demand for the nutrient-rich fecal sludge-based compost (often a substitute for chemical fertilizer) compared to the often low-nutrient MSW-based compost. Chapter 11 describes a case from **Rwanda** where private entities are capturing the commercial value in fecal sludge via nutrient recovery to ensure sustainable delivery of sanitation services (**Business model 14: Compost production for sustainable sanitation service delivery**). It is important to note that while reuse can ensure a sustainable sanitation chain, public toilet fees remain the key driver for financial sustainability of this business model. The case presented here only shows a medium-scale operation and links to the agricultural sector; for a more extensive review on fecal sludge reuse cases and models at different scales and recovered resources, see Rao et al., 2016.

Beyond the formal avenues of septage treatment via nutrient recovery, an interesting model observed in developing countries, is where cesspit truck operators deliver nutrient–rich septage collected from households to farmers' fields instead of designated or unofficial dumping sites – with the latter being more common (**Business model 15: Outsourcing fecal sludge treatment to the farm**). This model is largely driven by farmers' high demand for nutrient-rich septage, therefore bypassing a more formal sanitation process in the form of composting for direct disposal of raw fecal sludge on their farm fields. This practice is increasingly observed in Sub-Saharan Africa and South Asia (Cofie et al., 2009; Drechsel et al., 2011; Evans et al., 2013). The business model presented in Chapter 12, supported by a case from **India**, essentially relegates septage treatment to the farm and importantly reverses the cash flow as farmers pay the cesspit drivers for farm–gate delivery, whereas normally the transporter would have to pay a tipping fee for desludging into a treatment system. Disposal to farmlands outside the city offers a partial waste management solution, however better oversight and occupational and consumer risk reduction measures are critically needed. There are emerging models and cases that aim to increase the safety and usability of fecal sludge via composting, pelletization and blending of fecal sludge-based compost with rock-phosphate, urea/struvite or NPK, among others (see Rao et al., 2016).

Finally, there is also the potential for phosphorus (P) recovery from human excreta (**Business model 16: Phosphorus recovery from wastewater at scale**). The model presented in Chapter 13 demonstrates an opportunity for increased accessibility to phosphorus (in view of diminishing global P resources) for agricultural production and significant prospects for cost recovery if savings in treatment and sludge disposal costs are considered, as until recently phosphate recovery costs still result in prices higher than those of phosphate rock, unless niche markets are targeted. Although different technologies and approaches are possible for P recovery from human excreta, this chapter presents two cases representing the two ends of the opportunity spectrum. One is where urine is collected from unsewered households in **Burkina Faso** and sanitized in storage units for processing into liquid fertilizer (typically occurring at community-scale); and the other is based on phosphorus extraction from sewage treatment using the approach of Ostara in **Canada** as an example. The latter approach is applicable both at a community and large-scale level.

In summary, most of the examples presented in this section demonstrate the potential range of cost recovery to full profitability business models for entities considering nutrient recovery as an avenue for ensuring sustainable delivery of waste management services. Although not exhaustive, the presented cases and models show a tremendous potential for resource recovery and reuse, and private sector participation where the enabling environment is in place. Supportive institutional settings and regulations are important to support the businesses and control the well-known health and environmental risks appropriately, although these may not necessarily be sufficient in guaranteeing the viability of the enterprise (see Chapter 19). Particularly for nutrient recovery enterprises, access to finance, technology and consumers' acceptance will play an important role in facilitating or hindering their sustainability and scalability.

References and further readings

Asian Development Bank (ADB). 2011. Toward sustainable municipal organic waste management in South Asia: A guidebook for policy makers and practitioners. Mandaluyong City, Philippines: Asian Development Bank.

Benson, T., Lubega, P., Bayite-Kasule, S., Mogues, T. and Nyachwo, J. 2012. The supply of inorganic fertilizers to smallholder farmers in Uganda: Evidence for Fertilizer Policy Development. IFPRI Discussion Paper 1228. Washington, D.C.: International Food Policy Research Institute (IFPRI).

Cofie, O.O., Drechsel, P., Agbottah, S. and van Veenhuizen, R. 2009. Resource recovery from urban waste: Options and challenges for community based composting in Sub-Saharan Africa. Desalination 248 (2009): 256–261.

Drechsel, P., Cofie, O.O., Keraita, B., Amoah, P., Evans, A. and Amerasinghe, P. 2011. Recovery and reuse of resources: Enhancing urban resilience in low-income countries. Urban Agriculture Magazine 25 (September 2011): 66–69.

Drechsel, P., Qadir, M. and D. Wichelns (eds). 2015. Wastewater: An economic asset in an urbanizing world. Springer. 282p.

Evans, A., Otoo, M. and Drechsel, P. 2013. Developing business model typologies for resource recovery and reuse-based businesses. Urban Agriculture Magazine 26: 24–30.

Edixhoven, J.D., Gupta, J. and Savenije, H.H.G. 2013. Recent revisions of phosphate rock reserves and resources: Reassuring or misleading? An in-depth literature review of global estimates of phosphate rock reserves and resources. Earth System Dynamics Discussions 4 (2): 1005–1034.

Hernandez-Sancho, F., Lamizana-Diallo, B. and Mateo-Sagasta, J. 2015. Economic valuation of wastewater: The cost of action and the cost of no action. Nairobi, Kenya: United Nations Environment Programme (UNEP). 72p.

Kaza, S., Yao, L. and Stowell, A. 2016. Sustainable financing and policy: Models for municipal composting. Urban development series Knowledge Papers 24. Washington, D.C.: World Bank Group.

Kinobe, J.R., Niwagaba, C.B., Gebresenbet, G., Komakech, A.J. and Vinnerås, B. 2015. Mapping out the solid waste generation and collection models: The case of Kampala City. J. Air Waste Manage. Assoc. 65 (2): 197–205.

Pandyaswargo, A.H. and Premakumara, D.G.J. 2014. Financial sustainability of modern composting: The economically optimal scale for municipal waste composting plant in developing Asia. International Journal of Recycling Organic Waste in Agriculture 2014 (3): 66.

Otoo, M., Drechsel, P. and Hanjra, M.A. 2015. Business models and economic approaches for nutrient recovery from wastewater and fecal sludge. In: Drechsel, P., Qadir, M., Wichelns, D. (eds) Wastewater: Economic asset in an urbanizing world. Springer, Chapter 13, pp. 247–270.

Otoo, M., Ryan J. and Drechsel, P. 2012. Where there is muck, there is money. Handshake – IFC Quarterly journal publication on public private partnerships, World Bank. 1, May 2012.

Rao, K.C., Kvarnström, E., Di Mario, L. and Drechsel, P. 2016. Business models for fecal sludge management. Colombo, Sri Lanka: International Water Management Institute (IWMI). CGIAR Research Program on Water, Land and Ecosystems (WLE). 80p.

7. BUSINESS MODELS ON PARTIALLY SUBSIDIZED COMPOSTING AT DISTRICT LEVEL

Introduction

Many municipalities in large urban areas in developing countries continue to face the challenge of waste management. Limited public funds to support waste management infrastructure and services has resulted in significant environmental pollution as the majority of the generated waste, whether collected or uncollected, is often disposed of untreated in open spaces, waterbodies and/or landfills (Drechsel et al., 2015; Kinobe et al., 2015; ADB, 2011). The long-term effects of these practices include increased human health risks from communities' exposure to untreated waste, and generation of significant quantities of greenhouse gas emissions in the form of methane. This situation is particularly exacerbated for cities characterized by a growing population and rapid urban migration (Sabiiti, 2011).

Policy makers are increasingly challenged to consider other viable options, including market-based approaches that can lead to achieving sustainable solid waste management for current and future generations. Emerging recommendations propose a 'circular economy' which builds on the concept of resource recovery and reuse (RRR), where municipal solid waste (MSW) recycling and reuse offer the opportunity to augment nutrient resources. Nutrient recovery from organic fraction of MSW through composting is increasingly been used as a solution to address the dual challenge of waste management and soil nutrient depletion in large urban areas of many developing countries.

Investments, mainly public funding, for the set-up and operation of compost facilities is growing throughout the developing world. These compost plants are typically publicly-owned, large-scale facilities processing significant quantities of waste at a time. The required operational and maintenance costs can be substantial given the advanced and mechanized equipment used, high-level skill required and high energy requirements. Whilst MSW composting has the potential to generate significant revenues from compost sales and recyclables and most compost plants are geared towards full cost-recovery, the revenues are often insufficient to cover the O&M costs, and less so capital investments. Municipalities are however incentivized to continue providing financial support to ensure the sustainability of the compost plants, as the cost of inaction or alternative existing options such as landfilling and incineration is greater than financial support for operating the compost plants (De Bertoldi et al., 1996; Drechsel et al., 2004; Hutton et al., 2009; EC, 2002). Although justifiable, these governmental instruments may disincentivize compost plants from achieving full cost-recovery given their continued dependence on external support.

On the other hand, although few in number, there are cases of publicly-funded compost plants which started out with the goal of partial cost-recovery but are on a path to financial independence. Key elements of their business strategy are: a) their ability to liaise with urban councils to enact waste tax for institutions that fail to segregate their waste (increased revenue); b) development of different formulations of compost tailored to different customer segments; c) compost product certification and branding; d) sale of carbon credits; e) production of fuel pellets and sale of non-degradable solids (recyclables); and f) improved operational efficiency of technologies. Additional success drivers include the set-up of satellite composting stations at vintage locations close to major customer segments, an avenue for reduction of transport and handling costs. Marketing strategies including free compost samples to first time users on a trial-and-pay basis, and special discounts on bulk purchases can incentivize compost use and upscaling.

While potential opportunities for 'business' in waste reuse are increasingly clear, scaling-up and sustainability of such entities often only emerge as a viable option when public and private actors work together. Case studies across South Asia indicate that while many composting plants hardly survive their pilot phase, successful ones leverage key strategic partnerships with different entities, including community-based organizations and private entities to reduce risk associated with high capital investments, optimize the allocation of resources and activities and increase market access, thus increasing opportunities for profits. Innovative partnerships appear in most cases to have an important role to play where such businesses thrive (ADB, 2011).

In this chapter, we present the **Partially subsidized composting at district level** business model and supporting case examples, and the notable potential it offers for harnessing value from the organic fraction of municipal solid waste. Our examples are not exhaustive and better cases could have been inadvertently omitted due to information and time constraints, but they cover a moderate range of easily accessible cases in selected settings in Sri Lanka and Uganda.

References and further readings

Asian Development Bank (ADB). 2011. Toward sustainable municipal organic waste management in South Asia: A guidebook for policy makers and practitioners. Mandaluyong City, Philippines: Asian Development Bank.

De Bertoldi, M., Sequi, P., Lemmes, B. and Papi, T. (eds). 1996. The science of composting. European Commission, International Symposium. First edition. Springer-Science + Business Media, B.V.

Drechsel P., Cofie, O., Fink, M., Danso, G., Zakari, F. and Vasquez, R. 2004. Closing the rural-urban nutrient cycle: Options for municipal waste composting in Ghana. Final Scientific Report on IDRC project 100376.

Drechsel, P., Qadir, M. and Wichelns, D. (eds). 2015. Wastewater: Economic asset in an urbanizing world. Springer.

EC. 2002. Economic analysis of options for managing biodegradable municipal waste. Final Report to the European Commission by Eunomia Research & Consulting, Scuola Agraria del Parco di Monza, HDRA Consultants, ZREU and LDK ECO on behalf of ECOTEC Research & Consulting, 2002.

Hutton, B., Horan, E. and Norrish, M. 2009. Waste management options to control greenhouse gas emissions – Landfill, compost or incineration? Paper for the ISWA Conference, Portugal, October 2009.

Kinobe, J.R., Niwagaba, C.B., Gebresenbet, G., Komakech, A.J. and Vinnerås, B. 2015. Mapping out the solid waste generation and collection models: The case of Kampala City. J. Air Waste Manage. Assoc. 65 (2): 197–205.

Sabiiti, E.N. 2011. Utilizing agricultural waste to enhance food security and conserve the environment. AJFAND online 11 (6): 1–9.

CASE

Municipal solid waste composting for cost recovery (Mbale Compost Plant, Uganda)

Charles B. Niwagaba, Miriam Otoo and Lesley Hope

Supporting case for Business Model 10	
Location:	Mbale, Uganda
Waste input type:	Municipal solid waste (MSW)
Value offer:	Provision of sustainable waste management services, provision of high quality compost, carbon credits
Organization type:	Government-owned public enterprise
Status of organization:	Operational since 2010
Scale of businesses:	Processes 60 tons of MSW per day
Major partners:	Government of Uganda, National Environment Management Authority (NEMA), Makerere University, National Agricultural Advisory Service, World Bank

Executive summary

Mbale municipal Composting Plant (MCP) is a not-for-profit entity which was started with the primary aim of reducing the quantity of solid waste landfilled and resulting greenhouse gas (GHG) emissions. Additional key drivers have been to: a) reduce open-dumping practices and maintain cleanliness of the city; b) provide an environmentally safe fertilizer alternative for farmers; and c) create jobs for local inhabitants. MCP uses a windrow composting technique and converts approximately 60 tons of waste that it receives daily into a safe organic fertilizer. This initiative is based on a **cost-recovery model** where it seeks to reduce waste management costs faced by the municipality. It mainly generates its revenue from sale of compost and recyclables such as plastics, and plans to engage in carbon trading in the near future as an additional revenue stream. Compost is sold primarily to farmers within Mbale, however MCP's compost product is gradually gaining popularity and is being sold in other regions. Plans are underway to reinvent its marketing strategy by advertising on national television to broaden its market scope. MCP did so far not break even and receives financial support from the government to partially cover its operational and maintenance costs. Additional subsidies are received as operational tax waivers from the National Environment Management Authority (NEMA). MCP has great potential to become financially self-sufficient. It however needs to improve the operational efficiency of its technology to reduce operational costs and invest in product innovation and branding to increase the market demand for its compost product. Additional revenue sources that remain untapped are waste collection fees to be charged to households and businesses. This will however require an instituted mandate by the municipality. Benefits from MCP's activities are substantial and include efficient waste collection systems which have reduced the quantity of openly-dumped waste and consequently improved environmental and human health, and livelihood improvement for workers at the plant and farmers who now have access to affordable and safe fertilizer alternatives.

KEY PERFORMANCE INDICATORS (AS OF 2015)	
Land use	2.8 ha
Capital investment:	USD 350,000
Labor:	30 people
O&M cost:	USD 13,400 per year
Output:	4 tons of compost per day; 95% of incoming solid waste is fully degraded and recycled
Potential social and/or enviornmental impact:	Provision of 21 full time jobs, reduced human exposure to untreated waste, improved environmental health from reduced GHG emissions, enhanced soil fertility and productivity
Financial viability indicators:	Payback period: 10 years Post-tax IRR: N.A. Gross margin: N.A.

Context and background

MCP is a public project administered by the Mbale Municipal Council. Its main goal is to reduce GHG emissions and thus transform municipal organic solid waste into organic compost for agricultural use, thereby improving MSW management in urban areas. The project is part of a national program conducted by NEMA, under the Government of Uganda and World Bank-funded Environmental Management and Capacity Building Project II (EMCBP-II). MCP is located in Doko cell, Namataala industrial region, in the Mbale district. The present location of the site was previously (from 1950s) used as an official government landfill site. The revenue streams of the project are sales from compost and sale of recyclables such as plastics, whereas carbon credit is a planned main revenue, which is anticipated to generate an annual income of USD 25,000–USD 30,000.

Market environment

Most large-scale farming in Mbale is practiced on the slopes of Mt. Elgon, where soil fertility is lost through erosion. Chemical fertilizers and food and agro-waste (not composted) are the primary fertilizers used in restoring the soil nutrients. The nutrients in fresh/un-composted waste are not readily available to the crops. In addition, chemical fertilizers are expensive (approx. USD 1 per kg) and require regular applications throughout the plant growth stages. Mbale composting plant meets the need of the farmers by processing MSW to produce a comparably affordable organic fertilizer and with slow nutrient release into the soil thus requiring fewer fertilization re-applications. The opportunity that MCP exploits lies in the need for affordable and environmentally-friendly fertilizer alternatives by farmers and also sustainable waste management solution to reduce the quantity of landfilled MSW and direct human exposure to untreated waste.

Macro-economic environment

A market condition that could potentially impact MCP's business in Uganda is the price distortions in the fertilizer market from government subsidies for chemical fertilizers. The chemical fertilizer market in Uganda has however never expanded to a significant level due to the ineffective fertilizer policy. The limited use of chemical fertilizers in Uganda is striking and this has also been attributed to the lack of credit to farming households in Uganda. There is neither a large-scale government fertilizer program that provides subsidized fertilizer to farmers nor an active private fertilizer sector that supplies fertilizer at competitive prices (Yamano and Arai, 2010). Additionally, Uganda is landlocked, and with poor transportation system connections to ports, access to the external fertilizer markets is virtually impossible in the country. This represents a great opportunity for waste-based organic fertilizer businesses like MCP who can take advantage of erratic chemical fertilizer prices and the limited number of actors in the respective market.

Business model

Figure 118 presents an overview of MCP's business model. MCP is a socially oriented public entity with the goal to reduce GHG emissions via the conversion of MSW to compost with resulting benefits of a cleaner city and improved agricultural productivity. Initial capital for setting up MCP was received from the central government and the World Bank. It partners with Makerere University and the NEMA for technical support. In implementing its objective, MCP receives and sorts MSW into degradable and non-degradable waste, of which the plastic non-degradable is sold to plastic companies by their workers. Allowing the workers to sell the recyclable waste to recycling companies and earn additional income creates an incentive for the workers to properly and efficiently segregate the waste – reducing MCP's production costs. The compost from processing the degradable waste is sold directly to large scale farmers and sometimes through the National Agricultural Advisory Service's (NAADS) established distribution channels. A major source of revenue for the project is anticipated to be from carbon credit claims, for which it receives support from the World Bank. This anticipated revenue from carbon credits has allowed MCP to sell its compost in the initial phase at a very low price to garner market demand. The current unit price of compost is too low for MCP to break even from sales

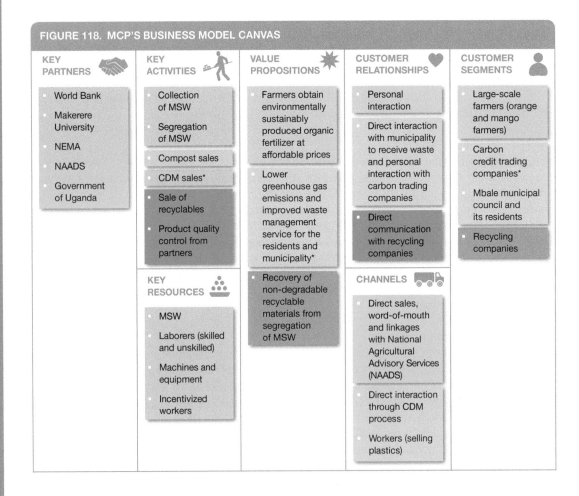

FIGURE 118. MCP'S BUSINESS MODEL CANVAS

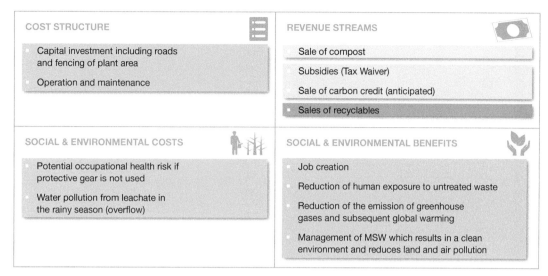

* planned activity

of compost only. Thus eventually MCP will have to increase the product price and ensure revenue receipts from carbon credits in order to fully recover costs. MCP's activities have considerable social and environmental benefits including: a) reduced human exposure to untreated waste; b) reduction in GHG emissions from reduced quantities of landfilled MSW; and c) employment generation to name a few.

Value chain and position

The central government and World Bank provided funds for the set-up of the business and injected money for operations. MCP partners with Makerere University and NEMA for laboratory analysis to ensure product quality as well as technical support. MCP in turn pays for the services rendered by Makerere University. The Municipal Council supplies MCP with raw materials at no cost. MCP has unlimited access to raw materials (MSW) and does not compete with any other company for the waste input (Figure 119). The compost is sold directly to farmers through NAADS at USD 0.04/kg. MCP's compost competes with chemical fertilizers and other organic fertilizers in the market. MCP has only been in existence for a few years however, the compost produced is gradually gaining popularity in the Mbale municipality. An average of 60 tons of compost is sold on a monthly basis. Currently, MCP captures a very small share of both the organic and chemical fertilizer market, but planned product innovation and new marketing strategies can significantly increase this proportion. Plastics and metal scraps obtained during sorting are managed solely by workers and sold to recycling companies to earn them additional income. Carbon credit sales, anticipated to be the main source of income, has not yet been realized and is still under documentation for application.

Institutional environment

According to a 2011 WaterAid report on solid waste management in Uganda, there is no single document of a legally binding nature, either national or regional, that provides comprehensive solid waste management in Uganda. The Public Health Act Cap.281, 2000, Solid Waste Management Strategy (SWMS) December 2002 as revised in 2006, Local Governments Act (1997) revised in 2004, The Constitution of Uganda 1995 (amended 2005) and The National Environment (Waste Management) Regulations, S.I. No 52/1999 provides some coverage for solid waste management in Uganda. Enforcement of regulations have been challenged with weak punitive measures. The ordinance

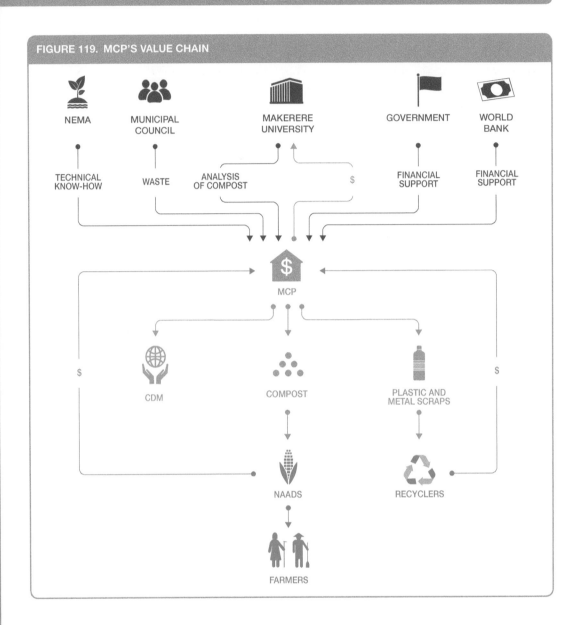

FIGURE 119. MCP'S VALUE CHAIN

proposes a fee for solid waste generator, however it does not provide a mechanism for collection of these fees, and specifically for Kampala city. Additionally, there were so far no laws or legislations that limit the conversion of human excreta into value-added products or its use. However any organic fertilizer product is to be proven safe and must meet certain minimum nutrient levels – as proven from a product certification from the Uganda National Bureau of Standards. While representative of additional costs to waste-based nutrient recovery enterprises such as MCP, certification of their products represents an avenue for product branding and increased market share.

Technology and processes

The production site is composed of an aerobic composting yard made of concrete flooring and a series of sloping double pitched roofs. The dimensions of the yard are 3,405 m² with surrounding drains and it is fully fenced. Municipal solid waste from the urban areas are collected and taken to the composting plant for segregation. At the plant, the waste is sorted according to biodegradable and non-degradable waste. A windrow composting technique is used for the conversion of MSW to compost (Figure 120). The biodegradable waste is aligned in the first windrow (active stage) where decomposition initially takes place after it is moisturized with recycled leachate and water in order to increase its moisture content. A locally manufactured sieving drum is used to manually separate larger particle material from the fine compost. This is a laborious and inefficient component of the production process. The low level of machinery coupled with high volumes of incoming waste make it difficult to completely compost all the biodegradables. The rejects from the sieving process is landfilled instead of being re-composted. Windrows are designed to have a gentle slope which allows leachate to flow by gravity to the leachate tank. The windrow piles are arranged in order of decreasing size from active to maturation stages because the size of compost is expected to decrease with time. Due to constraints in resources, the intended design of transferring compost from one windrow pile to another is not followed, but instead, it is left in one windrow pile from active to the matured stage. The total time for maturation before sieving is eight weeks, but due to characteristics of the waste, like the presence of fibres, it can take as much as twelve weeks for it to be ready for sieving. Sieving of the mature waste is done manually using slanted sieving drums to allow the compost to go through as the rejects (size bigger than the mesh size) go over to a separate area when rotated. The rejects are then landfilled and the compost is sold to farmers.

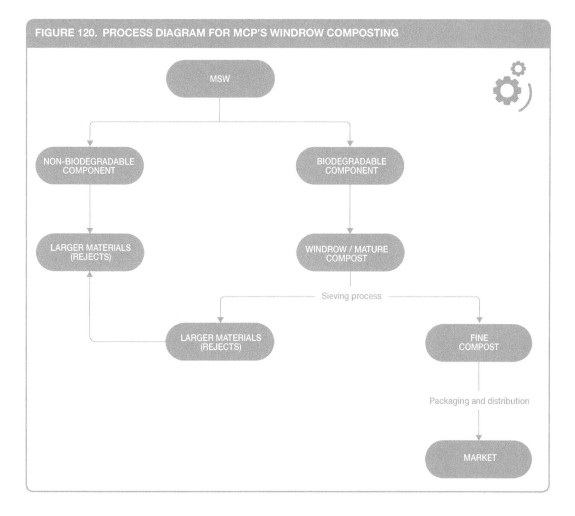

FIGURE 120. PROCESS DIAGRAM FOR MCP'S WINDROW COMPOSTING

Funding and financial outlook

The government of Uganda through the municipal council contributed land (which was formally used as a landfill) to the project. It also contributed USD 40,000, which was used to improve infrastructure (e.g. open up roads) and fence the plant area. These funds were from internally generated local revenue by Mbale Municipal Council. Infrastructure including machinery and equipment was funded by the World Bank with a grant of USD 300,000. The municipal council spends USD 13,400 annually on fuel and machine costs. In June 2011, the World Bank contributed a one-time grant of USD 4,800 to help boost the operational performance of the project. It is anticipated that MCP will be able to recover the investment cost in eight years when carbon credit claims are made in addition to the sales of compost. MCP then will have two main sources of income – sale of carbon credits and sale of compost. A kilogram of compost is sold at USD 0.04 as set by the government of Uganda, averaging related revenue of USD 2,000. Although at the time of the study not breaking even, with annual expected earnings of USD 25,000 to 30,000 from carbon credits, MCP expects to not only recover its costs but make some profits.

Socio-economic, health and environmental impact

MCP's activities have accrued significant benefits to society. Its main activity of converting MSW to compost has reduced the quantity of landfilled waste and will result in the reduction of greenhouse gas emissions. Its activities will also reduce waste management costs associated with land acquisition for landfills and their management. Improved sanitation will result in reduced human exposure to untreated waste and associated costs. Improved soil fertility and agricultural productivity from the use of organic fertilizer has noteworthy implications for smallholder farmer livelihoods and food security. Increased crop yields imply increased incomes for farmers and better livelihoods.

Scalability and replicability considerations

The key drivers for the success of this case are:
- Provision of start-up capital by government;
- Funding support from the World Bank and the government to ensure long term revenue flow from carbon credit;
- Incentives to workers for segregating municipal solid waste by allowing them to sell the recyclable waste to recycling companies and earn additional income;
- Weak national chemical fertilizer market and limited access to external chemical fertilizer markets provide ample opportunity for organic fertilizer production business.

The project currently does not break even and cannot achieve this only from sale of compost without process innovation. The manual nature of the activities, e.g. sorting and sieving, results in a high level of inefficiency and limits scaling up of the enterprise. Whilst there are opportunities for scaling-up and out of the project through mechanization of its production system and exploration of new product markets, continued high dependence on external support may still render the initiative unsustainable. It is also imperative that the suitability of technologies to different contexts and product requirements by different markets be taken into consideration.

Summary assessment – SWOT analysis

The SWOT analysis for MCP is presented in Figure 121 below. The key strength of the business is the initial financial support from the government at the start-up phase and funding from the World Bank to apply for CDM process to ensure a stable revenue source. The plant also has good access roads to the site, making the transportation of waste and compost easier. The key weaknesses of the enterprise are related to the highly labor-intensive operations required for waste segregation and its high dependence on external funding. So far, the enterprise hardly generates any revenue from

FIGURE 121. SWOT ANALYSIS FOR MCP'S BUSINESS

	HELPFUL TO ACHIEVING THE OBJECTIVES	**HARMFUL** TO ACHIEVING THE OBJECTIVES
INTERNAL ORIGIN ATTRIBUTES OF THE ENTERPRISE	**STRENGTHS** - Continuous availability of raw materials (waste) - Financial support from government and World Bank - Low sales price of compost - Low cost of technology - Good access roads to the site - Sales of CDM enabled in near future	**WEAKNESSES** - Manual operations resulting in high level of inefficiency and high cost - Potentially high occupational health risks from waste segregation process if use of protective gear is not enforced – implications for production costs - Poor marketing strategy - Highly dependent on external support - Inadequate composting facilities
EXTERNAL ORIGIN ATTRIBUTES OF THE ENVIRONMENT	**OPPORTUNITIES** - Mechanization of production process and increasing scale of operations - Sorting of waste at source - Waste collection fees - Increase in the scope of product market - Value addition to compost to increase market share and prices - Weak national chemical fertilizer market and limited access to external chemical fertilizer markets provide ample opportunity for organic fertilizer production business	**THREATS** - Subsidized chemical fertilizer - Increasing labor prices may affect production costs

the sale of compost. This, however, offers an opportunity for it to rebrand its compost product and also add value via fortification and pelletization to command higher market prices and increase its sales revenue. Subsequently, the enterprise could mechanize its operations and increase its scale of operations. The primary threat for the business is subsidized chemical fertilizers and increase in labor prices.

Contributors
Krishna Rao, IWMI, Sri Lanka
Jasper Buijs, Sustainnovate, Netherlands; formerly IWMI, Sri Lanka
Josiane Nikiema, IWMI, Ghana

References and further readings
Mbale Municipal Compost plant monthly and annual operational and manual reports.

Personal Comm. with Dr. John Bosco Tumuhairwe, the National Consultant for NEMA on Compost Quality.

Personal Comm. with the Mbale Municipal Compost Plant manager, Mbale Municipality Environment Officer, Data Clerk at MCN and Project Officer, CDM projects at NEMA.

Wateraid. 2011. Solid Waste Management Study in Bwaise II Parish, Kawempe Division, 2011. http://www.wateraid.org/~/media/Publications/Solid-waste-management-study-in-Kawempe-Uganda.pdf (accessed November 8, 2017).

Yamano, T. and Arai, A. 2010. Fertilizer Policies, Price, and Application in East Africa. GRIPS Discussion Papers 10–24. National Graduate Institute for Policy Studies, Tokyo.

Case descriptions are based on primary and secondary data provided by case operators, insiders, or other stakeholders, and reflect our best knowledge at the time of the assessments 2015/16. As business operations are dynamic, data can be subject to change.

CASE

Public-private partnership-based municipal solid waste composting (Greenfield Crops, Sri Lanka)

Miriam Otoo, Lesley Hope and Krishna C. Rao

Supporting case for Business Model 10	
Location:	Matara, Sri Lanka
Waste input type:	Municipal solid waste (MSW)
Value offer:	Provision of waste management services, and a safe and affordable compost
Organization type:	Public-private partnership
Status of organization:	Established and managed by government from 2005 but entered into a public-private partnership with Greenfields (private company) in 2010
Scale of businesses:	Medium; processes between 300 to 400 tons of MSW per month
Major partners:	Municipal council, Tea Research Institute, Coconut Research Institute, USAID

Executive summary

Greenfield Crops (GC) is a public-private partnership-based (PPP) business which was set up to carry out waste management activities in the Matara municipality. GC adopts an open windrow technology to process municipal solid waste (MSW) into compost. It also produces fuel pellets and sells non-degradable material obtained during the sorting of waste. GC has satellite compost stations which are close to local markets and that provide easy access to waste not requiring significant segregation. Compost is sold directly to farmers through a network of dealers. At the time of this study, the company was not making profits but dependent on government funding. The business is still working to improve their management strategies and the quality of the product to increase its marketability. The compost produced is currently perceived as a soil conditioner by the farmers rather than a fertilizer, and thus to increase its market share GC has to invest in product innovation and new marketing strategies. Activities of the business have improved the local environment and prevented contamination of local water bodies (Nilvala River) as hitherto waste was disposed close to a water body.

KEY PERFORMANCE INDICATORS (AS OF 2015)						
Land use:	25 ha including landfill area					
Capital investment:	USD 1,536,688					
Labor:	15 unskilled labor and 3–5 skilled labor/management					
O&M cost:	USD 9,220 per month					
Output:	300–400 tons per month					
Potential social and/or enviornmental impact:	18 jobs created, clean environment at a low cost, production of compost (soil conditioner) and fuel pellets					
Financial viability indicators:	Payback period:	N.A.	Post-tax IRR:	N.A.	Gross margin:	N.A.

Context and background

The Matara composting plant was set up with funds from the "Pillisaru" project, a Government of Sri Lanka initiative under the Central Environment Authority to improve solid waste management in urban centres. It began its operations – handling of MSW in 2005 – but halted operations due to noted sub-optimalities in the management and marketing of the entity. In 2010, GC revived the business through a PPP agreement for a period of seven years, with the first two years being probationary. Under this agreement, the private entity (GC) pays a service fee of USD 1,500 per month to the public entity (municipal council) for using the infrastructure (land, composting facility and machines). The municipality in turn pays GC USD 5 per ton of waste disposed as a tipping fee. Forty tons of waste is collected daily by the municipal council in Matara city and delivered to several different processing sites. GC started satellite compost stations closer to local markets to minimize transportation costs both for waste collection for the municipality and distribution of compost product for the business – thus increasing farmer accessibility to the organic fertilizers. Plans are underway to establish two additional satellite stations in the Eastern Province.

Market environment

Compost sales have been noted on average to be very low in Sri Lanka. This has been attributed to the low nutrient content of the product and inadequate marketing strategies. Standard compost products found on the market penetrate less than 3% of the fertilizer market. This represents an opportunity for initiatives such as GC to penetrate the market by producing high quality compost products. Chemical fertilizers are subsidized in Sri Lanka and this may represent significant competition for GC[1]. The extensive use and over-application of chemical fertilizers have been detrimental to the soils in the Eastern Province of Sri Lanka so although organic fertilizers may be comparably more expensive, there is a growing demand for them. Soil conditioners are needed to bind the soil particles together and GC's compost product can fill this gap. In Matara, 40 tons of waste is generated every day, of which about 60% is organic. Proper and safe disposal has been a challenge and this has caused public protests. The need for sustainable waste management alternatives is unquestionable – thus initiatives such as that of GC will continue to be in demand at least for the few next decades.

Macro-economic environment

As noted, in Sri Lanka, chemical fertilizer is subsidized by the government and has a higher nutrient value – thus representing significant market competition for GC. The subsidized price of a 50kg bag of chemical fertilizer is USD 2.75 and the same quantity of Greenfield compost is sold at USD 3.17, which is comparatively more expensive as farmers will require a greater application with compost quantity than with the former. Another important market condition that affects initiatives such as GC is related to access to funding. Local funding agencies are hesitant to provide loans to waste businesses as they are less cognizant of this business sector and classify it as high risk, and thus this factor represents potential constraints to the development of waste reuse businesses. On the other hand, although international donors are more interested in funding these initiatives, they tend to have a preference for public entities rather than private businesses. New waste reuse businesses will have to take these external market factors into consideration and adopt mitigation measures to ensure their sustainability.

Business model

Figure 122 presents an overview of Greenfield Crops' business model and described from the perspective of the private entity engaged in the public-private partnership. GC is a PPP entity charged with the processing of MSW into organic compost. The organic compost produced is sold in local markets through selected retailers. Plantation farmers such as tea, cinnamon and coconut farmers are the main users of the organic compost produced. Under the PPP, GC is the private entity and the municipal council is the public entity. The composting facility as well as land and other infrastructure were set up by the municipality. GC only manages the business and bears the cost of operations and maintenance. It pays the municipal council for the use of the resources provided, i.e. the composting facility and equipment. The municipal council on the other hand pays GC tipping fees for the disposal and processing of the solid waste. GC also partners with research institutes (Tea Research Institute and Coconut Research Institutes) for product quality analysis and USAID, who provided funds for the establishment of a laboratory. Essential to this model are the satellite compost stations that GC operates. These stations are close to local markets and farmers, resulting in minimizing transportation costs for waste collection for the municipality and distribution of compost product for the business – thus increasing farmer accessibility to the organic fertilizers. GC sells its compost at a flat price exclusive of transportation fee. Traders add on the cost of transportation and sell it at their preferred prices up to a specified limit[2]. A small quantity of recyclables is also sold to recycling units. While this initiative is currently still dependent on government funding, with plans to increase its scale of production via additional satellite stations, full cost-recovery is certainly achievable in the near future. GC's activities have accrued significant benefits to the society including: a) creation of jobs; b) reduced waste management costs; and c) improved environmental health.

Value chain and position

Figure 123 presents an overview of GC's value chain. The initiative receives MSW from the municipal council which pays USD 5/ton as tipping fees for waste disposed and processed. GC, on the other hand, pays the municipal council for use of the composting facility and other infrastructure. Matara municipal council is the sole provider of the MSW and hence has a strong supplier power which would be a major production risk factor for GC. However, given the nature of the PPP agreement, this power cannot be executed by the municipal council and is mandated to deliver the waste to the business. GC partners with the Tea and Coconut Research Institutes for field trials and product quality analysis. Field experiments have shown that there is a tremendous yield increase when GC's compost is used, suggesting a potentially significant demand if farmers do adopt compost use at least as a complementary product. The final compost product is sold directly in the local markets through a network of retailers. GC's key customers are farmers, specifically tea, rubber, coconut, cinnamon and other cash crop farmers. The business entity does not consider the product as an organic compost but rather as a soil conditioner. Since the customers are diverse, buyer power is relatively mitigated. There are no barriers to entry into the composting business, however, the municipal council owns the waste and permission is required and GC currently has the sole agreement with the municipality.

FIGURE 122. GREENFIELD CROPS' BUSINESS MODEL CANVAS

KEY PARTNERS
- Municipal council
- Tea Research institute
- Coconut research institute
- USAID

KEY ACTIVITIES
- Treatment of municipal solid waste
- Production of compost
- Sales of compost and other non-degradable materials
- Management of the satellite compost facilities
- Product quality analysis

KEY RESOURCES
- MSW
- Land
- Composting facilities
- Equipment
- Laborers
- Partnership agreements
- Network of satellite compost stations

VALUE PROPOSITIONS
- Provision of quality soil conditioner and organic fertilizer meeting tea and coconut plantation farmers' requirements
- Provision of improved sanitation and waste management services
- Recovering of non-degradable recyclable materials from segregation of MSW

CUSTOMER RELATIONSHIPS
- Personal relationship with traders (with a price ceiling for negotiation)
- Direct interaction with municipality to provide waste
- Direct relationship

CHANNELS
- Local market through a selected set of retailers
- Compost facility provided by the government, O&M is performed by the private entity
- Direct

CUSTOMER SEGMENTS
- Farmers (cash crop, tea, coconut, cinnamon growers)
- Municipal Council, with the direct beneficiaries being the city dwellers
- Buyers of non-degradable materials

COST STRUCTURE
- Rent of the composting facility
- Operation and maintenance cost (maintenance of machinery, quality control and labor cost)

REVENUE STREAMS
- Sales of compost
- Tipping fee for disposal and processing of solid waste
- Sales of non-degradable materials

SOCIAL & ENVIRONMENTAL COSTS
- None noted based on data provided

SOCIAL & ENVIRONMENTAL BENEFITS
- Reduction of waste generated in the municipality
- Reduces existing waste management costs
- Reduces human exposure to untreated waste
- Creation of jobs

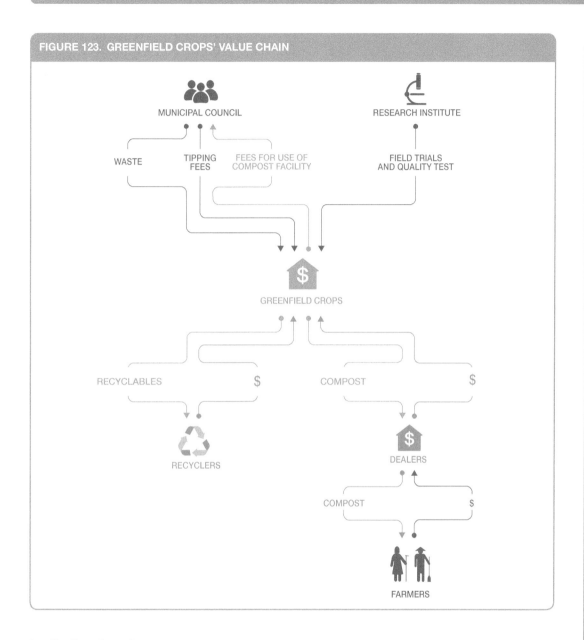

FIGURE 123. GREENFIELD CROPS' VALUE CHAIN

Institutional environment

In 2007, the National Policy on Solid Waste Management was formulated that replaced the 2000 National Strategy for Solid Waste Management which targets waste minimization, reuse of waste, recycling and appropriate final disposal of waste. Under the national policy, the government allocates funds for the capital investment of solid waste management projects. While there are so far no laws that prevent the reuse of treated MSW and fecal sludge, all waste reuse businesses in Sri Lanka require permits, certifications and an approved environmental impact assessment prior to starting operations. The Sri Lanka Standards Institution (SLS) is responsible for the development of national standards for products and services used mainly in the industrial and trade sector. SLS has developed standards for the production and marketing of compost and other organic inputs – SLS 1246:2003, UDC628.477.4 (CEA, 2005). This standard requires quality monitoring of the compost product by certified third party local authority and submission of results to the SLS monitoring committee. Compliance to these standards not only ensures the sustainability of compost businesses but it allows GC to self-brand their product and increase their market share.

Technology and processes

Greenfield Crops uses an open-windrow system for the processing of MSW to compost (Figure 124). The technology is locally manufactured, which reduces the investment cost but also some related maintenance costs as replacement parts can be purchased locally. The MSW is first sorted into degradable and non-degradable fractions. The biodegradable waste is aligned in the windrow where decomposition takes place. Piles are turned once a week to promote aerobic digestion minimizing the odor from decomposition as much as possible. The complete process takes about 45–60 days depending on weather conditions. The duration of each stage also depends on the composition of the waste received. At the end of the composting period, piles are kept for further maturation. The matured waste is then sieved and fibrous materials that degrade slowly are added back to new piles. The sieved material is packaged and sold.

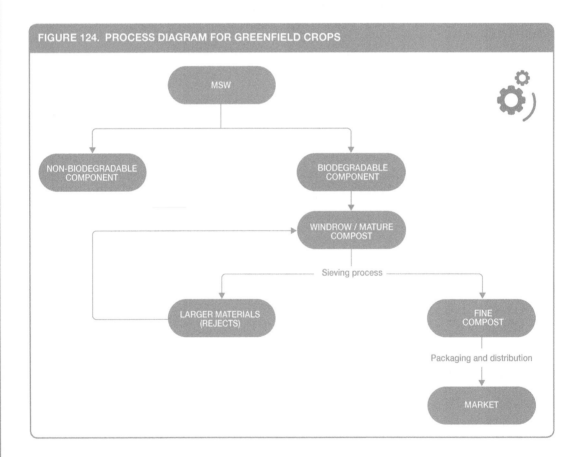

FIGURE 124. PROCESS DIAGRAM FOR GREENFIELD CROPS

Funding and financial outlook

Construction of the composting facility was fully funded by the 'Pillisaru' project of the Central Environmental Authority at a cost of USD 1,540,000. GC spends USD 9,240 per month for its operations. The operational cost includes electricity, fuel, worker wages, repair and maintenance and the service fee. Electricity and fuel cost alone account for 77% of the operation cost. The business generates revenue from sales of compost, non-degradable and tipping fees at a rate of USD 5 per ton of MSW received. The company processes forty tons of waste on a daily basis, amounting to USD 6,000 per month as tipping fees. Monthly sales of compost and non-degradables averages about USD 15,400 and USD 355, respectively. Although representative of 70% of all generated revenue, the enterprise remained unable to sell all of its compost and is working on implementing a new marketing strategy to boost sales and increase its profits.

Socio-economic, health and environmental impact

The PPP has saved the municipal council a significant amount of money which hitherto was used in operating the composting business as it was incurring losses. Additionally, through charges for the use of the composting facility and equipment, it is able to implement a mutual financial sustainability strategy. The activities of GC have rid the municipality of indiscriminate waste disposal while tidying up the city and reducing water pollution. The business has also provided jobs for some low-income earners, but the process of manual sorting, sieving and packaging may present occupational health risks as well if proper mitigation measures are not adhered to.

Scalability and replicability considerations

The key drivers for the success of this business are:
- Given the scale of operations, the PPP arrangement is ideal for this business set-up – public sector constructs the infrastructure and provides the capital cost required for equipment, and private sector brings in sophisticated management and skills to operate the facility.
- Government policy encourages reuse and recycling and sufficient incentives such as tipping fees have been provided to keep the private sector interested in managing the facility.

GC has adopted a system of compost production where compost is produced at vantage points close to local markets. The technology used is simple, and requires limited technological expertise and energy, making it highly replicable. Waste segregation is a primary cost component as well as a major source of inefficiency and thus scaling up may optimise production, as benefits will outweigh costs. A major limitation is the high capital investment requirement for land and especially in localities that are yet to be developed in terms of infrastructure, e.g. roads. This model is highly replicable in large cities with significant waste generation. However, limitations of land availability, competition in the product market and technological adaptations have to be taken into consideration.

Summary assessment – SWOT analysis

The SWOT analysis for GC composting plant is presented in Figure 125. The key strengths of the business are: a) the support from municipal authority, and b) innovative production system of satellite stations which increase its access to the waste source and product markets via reduced transportation costs. A key weakness of the PPP is the high investment requirements for future expansion and the labor-intensiveness of waste segregation. The latter represents a potential risk to the business in the instance where labor wages rise – which would imply the adoption of a new technology or increasing their labor prices to maintain their staff. GC generates a comparably low amount of money from the sales of recyclables. There are opportunities for the business to increase its revenues via value-addition to the plastic materials (via shredding and pelletization) which would command higher prices but also access new markets. Given the success of this public-private partnership, this model could be potentially replicated in other towns and cities in Sri Lanka. Many factors including competition in the fertilizer market, technology adaptation, among others need to be taken into consideration.

FIGURE 125. SWOT ANALYSIS FOR GREENFIELD CROPS

	HELPFUL TO ACHIEVING THE OBJECTIVES	HARMFUL TO ACHIEVING THE OBJECTIVES
INTERNAL ORIGIN ATTRIBUTES OF THE ENTERPRISE	**STRENGTHS** • Soil conditioning characteristics with water and nutrient retention capacity of the product • Support from municipal authority • Compost satellite stations close to waste source and product markets • Adequate and continuous supply of waste • Limited technological expertise needed	**WEAKNESSES** • High investment requirements • Limited access to loan • Inadequate marketing strategies • Nutrient content of compost is fairly low • Technology (sorting) is labor intensive • Technology efficiency dependent on weather conditions
EXTERNAL ORIGIN ATTRIBUTES OF THE ENVIRONMENT	**OPPORTUNITIES** • Production of fuel pellets from non-degradable and increasing sales of recyclables • Replicating similar model for other towns in Sri Lanka • Strong potential for tapping carbon market as an additional revenue source given scale of business • SLS regulation allows for self-branding upon compliance • Government policy allotting funds for solid waste management projects	**THREATS** • High subsidy for chemical fertilizer • Increasing labor wages present potential production risk for business

Contributors

Heiko Gebauer, EAWAG, Switzerland
Jasper Buijs, Sustainnovate, Netherlands; formerly IWMI, Sri Lanka
Josiane Nikiema, IWMI, Ghana

References and further readings

CEA. 2005. Technical guidelines on solid waste management in Sri Lanka. Central Environment Authority, Colombo http://www.cea.lk/web/images/pdf/Guidlines-on-solid-waste-management.pdf or http://librepo.cea.lk/bitstream/handle/1/622/08651.pdf?sequence=2&isAllowed=y (both accessed November 8, 2017).

Personal communication with staff of Greenfield Crops. 2015.

Case descriptions are based on primary and secondary data provided by case operators, insiders or other stakeholders, and reflect our best knowledge at the time of the assessments 2015/16. As business operations are dynamic, data can be subject to change.

Notes

1. Fertilizer subsidy scheme (fixed price for Nitrogen (urea), Phosphorus (TSP), Potassium (MOP) at Rs. 350/50kg) in Sri Lanka was changed in 2016 to a cash payment of Rs. 25,000/ha/year for paddy farmers. (USD 1 = approx. Rs 140).
2. Price information details were not provided.

CASE

Fecal sludge and municipal solid waste composting for cost recovery (Balangoda Compost Plant, Sri Lanka)

Miriam Otoo, Krishna C. Rao, Lesley Hope and Ishara Atukorala

Supporting case for Business Model 10	
Location:	Balangoda, Sri Lanka
Waste input type:	Municipal solid waste (MSW) and fecal sludge
Value offer:	Provision of MSW-based compost ('regular' compost), fecal sludge-based compost ('super' compost) and treated wastewater
Organization type:	Public entity
Status of organization:	Operational since 1999 but was privatized in 2003 and restored to government again in 2005
Scale of businesses:	Small to medium; processes more than 300 tons of MSW/month
Major partners:	Central Environmental Authority, Municipal Council, Universities, LIRNEasia

Executive summary

Balangoda Compost Plant (BCP) is a public entity that converts MSW into compost, and by adding also night soil[1] into super compost, as well as treating of water and trading of recyclables. It was set up to curb environmental and sanitation problems in Balangoda, in particular, indiscriminate disposal of night soil and solid waste accumulation. It uses the open-windrow processing technology to compost municipal solid waste. A simple approach with limited energy requirements is used in treating night soil, where water purifying plants and charcoal filters are used to treat the wastewater in the fecal sludge. Although geared towards cost-recovery and receiving partial financial support from the central government, BCP generates significant income from its multiple revenue streams – sale of compost and recyclables. MSW-based compost and super compost are sold directly to farmers through agro-outlets in local markets. Other government bodies, such as the Urban Development Authority and the Ministry of Agriculture, buy directly in bulk for landscaping. There is however no market for the treated water (leachate product). Resource centres where non-degradable waste is traded have been established in the city centre and at ten schools. BCP purchases segregated non-degradable waste from these resource centres and schools and resells to recycling companies at a higher price. This initiative has significantly reduced direct human contact with untreated waste and provided an improved environment for the community through proper waste management practices in the region. Additionally, it has created jobs and improved infrastructure via the construction of access roads to the project site. It has also caused an attitude change towards waste among the younger generation.

KEY INDICATORS (AS OF 2015)	
Land use:	1 ha
Capital investment:	USD 352,000 including costs of 1 hectare of land
Labor:	17 people (15 unskilled, 2 skilled)
O&M cost:	USD 1,340 per month
Output:	30 tons of compost, 5 tons of super compost and 180,000 litres of treated water, all on a monthly basis
Potential social and/ or enviornmental impact:	17 jobs, production of high quality and affordable compost and super compost, treated water, changed attitude of children to waste, cleaner local environment
Financial viability indicators:	Payback period: N.A. Post-tax IRR: N.A. Gross margin: N.A.

Context and background

The Balangoda Compost Plant (BCP) was set up to process municipal solid waste into compost. The plant started as a project with the mission of providing a solution to the solid waste problem as a community service. However, it gradually evolved into a business while providing community service. Balangoda is situated in Sabaragamuwa province of Sri Lanka, with a population of more than 40,000 and a land area of 16.2 km^2. The plant was started in 1999 and it has undergone several changes in ownership structure; it was set up and managed by the government but privatized after a change in government. The ownership was again transferred to the government when the private entity neglected safe handling of waste and focused completely on profits. Construction cost of the compost plant and the access roads were funded by Central Environmental Authority and Provincial Council. The land was given to the project at no cost by the Land Reform Committee. As a rejuvenated project in 2003, it embarked on cleaning the city in the night including collection of waste which was appreciated by the people and the decision makers resulting in further allocation of funds to improve the plant. By 2005, funding and revenue received was used to purchase the required machines and with the help of the municipality, a resource centre was built to purchase non-degradable waste in the city. The plant procured plastic and polythene pelletizers to add value to the non-degradable waste which reduced related transportation cost from product distribution. In 2008, a fecal sludge treatment plant was established with funds from the "Pillisaru" project of the Central Environmental Authority. Funds were used to construct a receiving tank, 2 sedimentation tanks, a water treatment facility and a drying bed. The majority of the building has been constructed from the funds and subsidies provided by the government. BCP earns revenue from sale of compost, super compost and recyclables. A twelve-year target of making a 'Waste Free City' has been achieved by the plant whereby all generated waste in the city is collected and treated.

Market environment

Waste accumulation in the city caused many problems including unpleasant odor, contamination of water bodies and paddy fields, giving rise to epidemic diseases like Salmonella typhoid and diarrhoea. This has resulted in a great need for the implementation of sustainable waste management solutions. Soils in the Eastern province of Sri Lanka are traditionally very sandy and chemical fertilizers leach out of the soil at a faster rate without the application of soil conditioners. Additionally, the over application of chemical fertilizers has damaged a considerable proportion of the soil structure and has rendered most of the lands unsuitable for agricultural production. Government and farmers in the Eastern province of Sri Lanka realize the importance of organic fertilizer use to mitigate the long-term damage of agricultural lands. This has resulted in an increased demand for organic compost in the Eastern Province. BCP sells an average of five tons of compost per month of which 40% is sold in the locality and 60% in the Eastern province and foresees an increasing trend in demand.

Macro-economic environment

The introduction of cash payment-based subsidies for chemical fertilizers may affect the demand of organic fertilizer in the locality. BCP is largely focusing on the Eastern province fertilizer market where organic fertilizer is in high demand due to the poor structure and declining fertility of the soils. The demand for chemical fertilizers is fairly high country-wide and this has been one of the driving factors for price subsidization by the government. A 50kg bag of chemical fertilizer at the subsidized rate is sold between a range of USD 2.75–3.07 and organic compost produced by Balangoda composting plant is sold for USD 3.14. BCP faces strong competition from both chemical and organic fertilizer businesses (Table 34).

TABLE 34. PRICES OF ORGANIC AND INORGANIC FERTILIZERS IN SRI LANKA 2015

FERTILIZER TYPE	PRICE (USD/50KG)
Organic fertilizer – Balangoda composting plant	3.14
Organic fertilizer – Nawalapitiya	2.74
Chemical fertilizer	2.75 – 3.07

Business model

Figure 126 provides an overview of the Balangoda Composting Plant's (BCP) business model. It is a public entity that processes municipal solid waste (MSW) and fecal sludge (FS) into organic fertilizers, treats leachate (wastewater) and sells recyclables (non-degradable waste). The enterprise was set up with the intent of providing community service and is not profit-oriented. The plant employs a value-driven model, where quality of the product is the main focus. For instance, the nitrogen content of the compost from MSW and FS are 1.68 and 2.9 respectively, compared to an average of 0.5–0.7 found on the market. It also adapts a demand driven approach where compost is sold to farmers that have need of the product, i.e. localities where soils are sandy in nature and thus require soil conditioner to bind soil particles together. Organic compost from both MSW (regular compost) and FS (super compost) are sold to farmers within the locality, as well as to the farmers in the Eastern province of Sri Lanka. As a part of its marketing strategy and to expand its customer base, BCP gave all its first-time customers free compost samples so that they could witness increased yields on their own farms. This has been instrumental in increasing its market share. An additional source of revenue is from the sale of recyclables which are bought from locals and sold directly to recycling companies at higher prices. There is no market for the treated wastewater. BCP has partnered with the Pillisaru Project which contributed funds for the construction of tanks and drying beds required for the production of the super compost. It is important to note that this was a one-time contribution, and whilst a partner to BCP, they are not a key partner in the business model, since it has no continued role in the business. Another key partner has been the local university for laboratory analysis of wastewater and with LIRNEasia[2] for technology development and skills training of the staff. A key success factor of this business has been its ability to liaise with the urban council to enact a waste tax for shops and institutions that fail to segregate waste. This has tremendously reduced the costs associated with waste sorting and sped up the entire production process. Waste resource centres have been implemented in schools. Students have been trained on waste segregation and the benefits of waste reuse which has resulted in an attitudinal change among the young generation.

Value chain and position

Figure 127 provides an overview of BCP's value chain. The Balangoda compost plant is a public entity owned by the municipal council. It receives its major input, i.e. municipal solid waste and fecal sludge, from the council and commercial companies. It partners with universities and LIRNEasia for research

FIGURE 126. BALANGODA COMPOST PLANT'S BUSINESS MODEL CANVAS

KEY PARTNERS
- Central Environmental Authority
- Municipal Council
- LIRNEasia
- Universities

KEY ACTIVITIES
- Collection and separation of waste
- Production and sale of compost and super compost
- Trading of recyclables
- Awareness raising activities at educational institutions

KEY RESOURCES
- Machinery
- Land and labor
- MSW
- License (fertilizer manufacturing and environmental impact assessment)

VALUE PROPOSITIONS
- Provision of improved sanitation and waste management services
- Provision of high nutrient compost and super compost for farming activities and landscaping
- Recycling companies get to buy recyclable wastes at an affordable price

CUSTOMER RELATIONSHIPS
- Personal collection of waste
- Personal help at direct sales (marketing strategy adapted where samples are given to farmers free of charge the first time)
- Personal

CHANNELS
- Direct
- Local market through a well-constructed market chain and direct channel to bulk buyers
- Waste resource centres and school children to collect recyclables and direct sale to recycling companies

CUSTOMER SEGMENTS
- Commercial firms and residents of Balangoda
- Farmers within locality and Eastern province farmers
- Bulk buyers: Government authorities (Urban Development Authority for Landscaping and Ministry of Agriculture)
- Recycling companies

COST STRUCTURE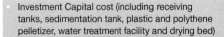
- Investment Capital cost (including receiving tanks, sedimentation tank, plastic and polythene pelletizer, water treatment facility and drying bed)
- Operation and maintenance (maintenance of machinery, quality control and labor cost)
- Purchase of non-degradable waste from resource centres

REVENUE STREAMS
- Waste collection funds from government and waste tax from commercial entities that do not segregate waste
- Waste collection fees collected directly from some commercial entities such as fish market, vegetable market etc.
- Sales of compost and super compost
- Sales from non-degradable waste

SOCIAL & ENVIRONMENTAL COSTS
- Possible pollution of water bodies from leachate

SOCIAL & ENVIRONMENTAL BENEFITS
- Reduction of waste generated in the municipality
- Reduced existing waste management costs
- Possible reduction of pollution of water bodies from reduced indiscriminate disposal of fecal sludge
- Reduced human exposure to untreated waste
- Creation of job opportunities
- Improved food security

into technology development and skill training respectively. The partnership with the university is a win-win situation where students from the universities (Sabaragamuwa and Jayawardena) use the composting site for research activities and the enterprise also benefits from the resulting research outputs. Key products, i.e. compost and non-degradable, are sold directly to locals through the local markets from agro-shops. Government institutions such as Road Development Authority are continuous buyers, but no agreements or partnerships exist between them. Products are supplied on occasional demand. The municipal solid waste used by BCP is collected and managed by the urban council. BCP has the urban council as its primary supplier of waste and hence the supplier power is high. Subsidized chemical fertilizer has a lower price compared to the organic fertilizer and has reduced the demand of organic fertilizer in spite of its nutrient retentive capacity. BCP must thus maintain a price lower than the subsidized chemical fertilizer to penetrate the market. Chemical fertilizer and other organic compost are good substitutes of the organic fertilizer produced by BCP. High price of organic fertilizer attributable to subsidies on chemical fertilizer and high application frequency has created demand for chemical fertilizer over organic fertilizers. The threat of new entrants into municipal solid waste processing is low due to the fact that the urban council owns waste and a permit is required to collect or process waste. In addition, waste recycling businesses are limited by institutional structures.

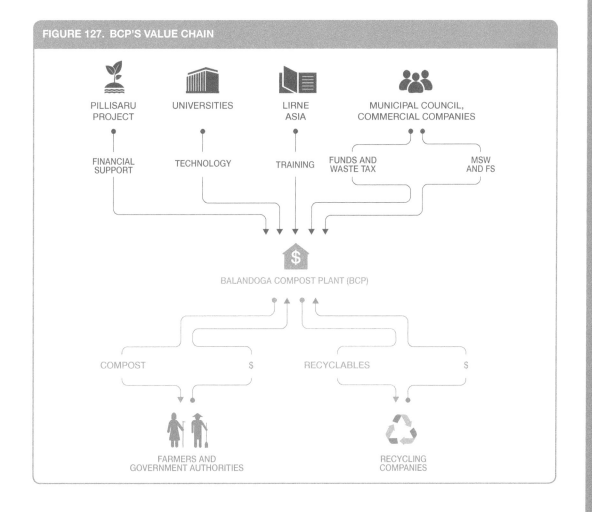

FIGURE 127. BCP'S VALUE CHAIN

Institutional environment

In 2000, the Government of Sri Lanka passed the national strategy for solid waste management that targets waste minimization, reuse of waste, recycling and appropriate final disposal of waste. In 2007, a new policy was formulated and implemented – the National Policy on Solid Waste Management to replace the 2000 National Strategy for Solid Waste Management. Under this new policy, the government annually allots funds for the capital investment of solid waste management projects such as Balangoda. There are currently no laws that limit the reuse of treated MSW or FS. However, all waste reuse businesses in Sri Lanka require permits and certifications prior to starting operations. This is inclusive of an environmental impact assessment to be conducted by a certifiable third party on an annual basis. The Sri Lanka Standards Institution (SLS) is responsible for the development of national standards for products and services used mainly in the industrial and trade sector. The division consists of sections namely agriculture, food, chemicals and cosmetics and textiles. SLS has developed standards for the production and marketing of compost and other organic inputs – SLS 1246:2003, UDC628.477.4 (SLS, 2014). This standard requires quality monitoring of the compost product by certified third party local authority and submission of results to the SLS monitoring committee. Additionally, this standard has set requirements for nutrient levels, biological and microbiological requirements and limits of heavy metals. Compliance to these standards not only ensures the sustainability of compost businesses, but it allows them to self-brand their product and increase their market share.

Technology and processes

Production of MSW-based compost

BCP uses the open-windrow system for the processing of municipal solid waste into compost (Figure 128). The technology has a high rate of recovery for bulky materials, and is thus suitable

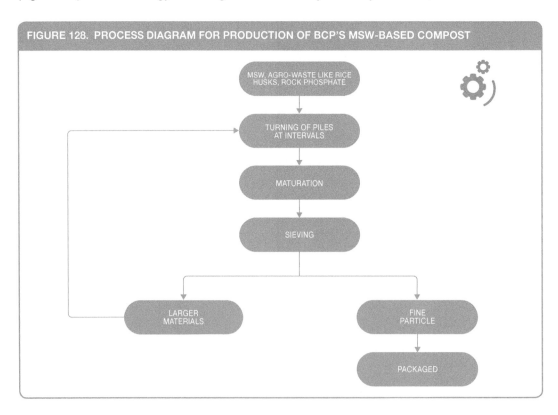

FIGURE 128. PROCESS DIAGRAM FOR PRODUCTION OF BCP'S MSW-BASED COMPOST

for composting large volumes of waste. The equipment is locally manufactured which considerably reduces the investment cost. However, the maintenance cost of equipment is high and it is not space efficient. Sorted MSW is piled up to a size of 5x5x12 feet. Every pile is maintained for six weeks. A temperature of nearly 70°C is maintained inside the pile, which minimizes pathogens including harmful helminths (worms) and fly larvae. Rock phosphate is also added to increase the phosphorus content of the final product. Since the composting site is situated just 25 feet away from the households, efforts are been made to maintain the aerobic conditions thus the piles are mixed at appropriate intervals (at least once a week) maintaining temperature, moisture and amount of air inside the pile. Leachate is collected six hours after the open windrow preparation and this is mixed with water in the ratio of 1:1,000 and sprinkled back on the composting piles for temperature regulation. Once the composting process is over, the piles are left for maturation for one to two weeks where pathogenic fungi such as Aspergillus are eliminated due to the drop of moisture level to around 5%. The compost is then sieved through a 6mm sieve to get fine particles of compost (the stated standard range for particle size is 4mm to 10mm). Before packaging, the moisture level of the compost is increased to 15%.

Production of fecal sludge and municipal solid waste-based co-compost

The treatment of fecal sludge involves a simple approach that does not require any energy except sufficient sunshine (Figure 129). The collected fecal sludge is unloaded into settling tanks and kept there for 45 minutes for the solid material to settle. The liquid portion is then taken into a treatment

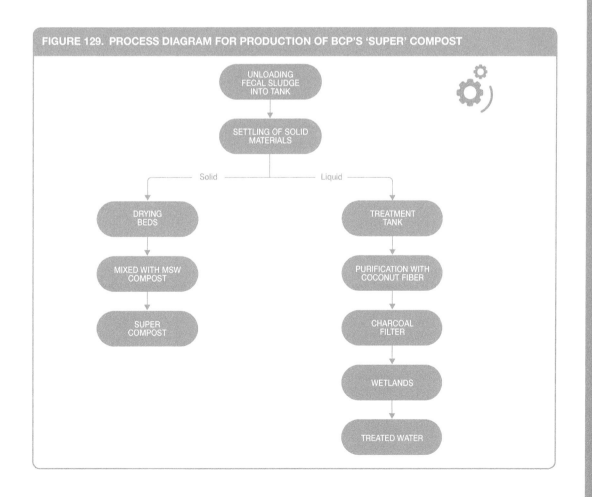

FIGURE 129. PROCESS DIAGRAM FOR PRODUCTION OF BCP'S 'SUPER' COMPOST

tank where coconut fibres are used to create microenvironments rich with micro-organisms that purify the wastewater. This water is passed through a charcoal filter to a constructed wetland for further treatment. The solid matter is sent to two drying beds where they are stored for 28 days. This dried fecal matter is mixed with the MSW-based compost to produce a co-compost (super compost). The product is termed *'super compost'* because the addition of the dried fecal sludge increases the nutrient content and levels of the final product.

Funding and financial outlook

The Central Environmental Authority and the Provincial Council funded the construction of the compost plant and roads at the cost of USD 300,000. The land was provided by the Land Reform Committee at no cost. Operation and maintenance cost is estimated at around USD 1,340 per month. The initial operation costs were catered for by the municipal council until the project began making profit. The Central Environmental Authority together with the urban council funded the construction of the fecal sludge treatment plant at a cost of USD 51,000. Collection centres for non-degradable and associated infrastructure were also established at a cost of USD 3,200. Apart from the recovered resources from organic material, the municipality sells non-degradable materials to recyclers. Acting as middlemen in the business, the urban council doubles the price paid for non-degradable products and earns 100% profit. Collection fees are taken only from several private fish markets, private farms and private meat markets. They are very few in the town. But significant revenue is generated from the waste tax charged to entities that do not segregate their waste. This sums up to USD 3,900 per annum. In 2011, the council made a profit of USD 165 from compost and USD 1806 from sales of non-degradable products. BCP envisions earning additional income from the sale of processed plastics.

Socio-economic, health and environmental impact

The benefits from BCP's activities are multi-fold. This plant has considerably reduced the municipality's waste management cost and also generates additional income beyond cost-recovery. Seventeen workers from the locality are provided with employment. Farmers benefit from the use of high quality and affordable organic fertilizers. The composting plant, in addition to managing municipal solid waste, treats fecal sludge from onsite sanitation systems in Balangoda, thus reducing indiscriminate dumping of fecal sludge. Residents of Balangoda have thus benefited from reduced exposure to untreated waste and improved sanitation which has reduced considerable health risk and surface and groundwater contamination. BCP is an example of an initially fully subsidized compost plant which has been able to transition to a financial self-sufficient business via the implementation of a suitable marketing scheme and strategic partnerships.

Scalability and replicability considerations

The key drivers for the success of this business are:
- Strong funding support from the government and policy that encourages reuse and recycling.
- 2007 government act enabling self-branding.
- Diverse customer base in terms of geographical outreach and strong awareness amongst from farmers in the Eastern province on the need for organic fertilizer.
- Clear awareness among farmers concerning soil degradation and the different effects of organic and chemical fertilizers.

BCP uses a near holistic approach to resource recovery and reuse where almost all waste types, both degradable and non-degradable, are either reused or recycled. The technology adopted is simple, requires limited expertise and energy, making it highly replicable. Waste segregation is a primary cost component while processing waste as well is a major source of inefficiency. BCP mitigates these inefficiencies via the creation of waste resource centres for the segregation of the non-biodegradable waste and thus significantly reducing production costs. A major limitation with implications for replication is the high capital investment requirements for land and in localities that are yet to be developed – infrastructure, e.g. roads. Another challenge to replicating this model is getting support from municipal council to enable a company (private or public) institute a waste tax to reduce the receipt of unsorted waste and essentially minimize costs so as to ensure sustainability.

Summary assessment – SWOT analysis

The SWOT analysis for BCP is presented in Figure 130. The key strengths of the business are: a) the low-cost technology; b) segregated waste delivered to this composting plant; c) funding from government to cover capital and initial operating cost; and d) governmental support to institute waste tax for entities who do not segregate waste. The BCP business however has a couple of weaknesses related to limited land availability for future expansion and its dependency on external entities for waste segregation. In the future if waste resource centres are unable to manage their operation cost, BCP will have to heavily invest in both capital and operational costs for segregating their waste. There are several opportunities in which BCP can tap into: a) compliance to certification standards will not only contribute to the sustainability of the compost business but it will allow them to self-brand their product and increase their market share; b) the enterprise can develop different formulations of

FIGURE 130. SWOT ANALYSIS FOR BALANGODA COMPOST PLANT

	HELPFUL TO ACHIEVING THE OBJECTIVES	HARMFUL TO ACHIEVING THE OBJECTIVES
INTERNAL ORIGIN ATTRIBUTES OF THE ENTERPRISE	**STRENGTHS** • Receives segregated waste thus reducing production cost • Technology compatible with the topography and requires limited energy • Low cost technology • Technology requires low level technical skills or expertise to operate • Strong funding support from government and favourable policy	**WEAKNESSES** • Current land available is sufficient to handle existing quantum of waste, however if they have to expand, availability of land will limit large scale expansion • Enterprise dependent on waste resource centres for segregation
EXTERNAL ORIGIN ATTRIBUTES OF THE ENVIRONMENT	**OPPORTUNITIES** • Developing different formulations of organic compost including pelletization of fecal sludge compost • Municipal tax for non-segregators • 2007 government act, enabling self-branding and obtaining waste management funds • Replication of similar model for other towns in Sri Lanka • Tapping carbon market as an additional revenue source	**THREATS** • Subsidies on inorganic fertilizers represent competition and may impact sales of organic compost • Cultural barriers – farmers unwillingness to use compost mixed with dry fecal matter

compost to meet farmers' requirements; c) with increased scale, BCP can consider tapping into the carbon market as an additional revenue source; and d) increasing government support for solid waste management has created a demand for this model which can be replicated in other towns and cities in Sri Lanka. A significant threat to BCP's business is increasing competition from subsidized chemical fertilizer which may affect the demand for their compost products.

Contributors
Nimal Prematilaka, Balangoda Municipality, Sri Lanka
Jasper Buijs, Sustainnovate, Netherlands; formerly IWMI, Sri Lanka
Josiane Nikiema, IWMI, Ghana

References and further readings
Personal communication with Mr. Nimal Prematilaka (Officer in charge of Balangoda Compost Plant). 2015.

Case descriptions are based on primary and secondary data provided by case operators, insiders or other stakeholders, and reflect our best knowledge at the time of the assessments 2015. As business operations are dynamic, data can be subject to change.

Notes
1. Night soil is a euphemism for human excrement, (formerly) collected at night from households. In our context, it refers to fecal sludge collected from on-site sanitation facilities, like septic tanks and pit latrines.
2. http://lirneasia.net (accessed November 8, 2017).

BUSINESS MODEL 10
Partially subsidized composting at district level

Munir A. Hanjra and Miriam Otoo

A. Key characteristics

Model name	Partially subsidized composting at district level
Waste stream	Municipal solid waste (MSW) and fecal sludge
Value-added waste products	Regular compost, enriched compost, non-degradable recyclables, treated wastewater
Geography	Medium to large urban areas with large quantities of MSW, land availability and access to inexpensive labor
Scale of production	Small to medium, processes about 10–75 tons of MSW/day
Supporting cases in this book	Mbale, Uganda; Balangoda and Matara, Sri Lanka
Objective of entity	Cost-recovery [X]; For profit [ ]; Social enterprise [X]
Investment cost range	On average USD 250,000–370,000 depending on scale
Organization type	Public
Socio-economic impact	Disposal cost savings, new jobs, provision of compost and super compost to plantation farmers, treated water and cleaner environment
Gender equity	Model is fairly gender neutral; where women are engaged in waste segregation, they may earn additional income from sale of recyclables

B. Business value chain

This business model can be initiated by a public entity or through a public-private partnership. The primary goal of the entity is to reduce open-dumping practices (maintain a clean city) and the quantity of waste landfilled, and resulting greenhouse gas emissions through the conversion of MSW and FS into compost. With investments justified based on the net positive environmental and socio-economic benefits, the municipality and/or government authorities often provide the capital investments (land, infrastructure, equipment, others) for the set-up of the compost plants as well as committing to providing continuous support for plant operation and maintenance. The publicly-run waste processing enterprise is often engaged across the entire value chain, i.e. involved in waste collection, segregation, processing, marketing and distribution of the compost. At the input side of the value chain, the public entity–municipality oftentimes owns the city's waste and thus has unlimited access to raw materials (MSW) and does not compete with any other company for the resources input. Collaborations with research institutes are recommended for the adoption of appropriate waste processing and compost production technologies.

This business model has the potential to transition from being subsidy-dependent to full cost-recovery and even profit-making. The efficient allocation of resources and engagement in activities where the business entity has a comparative advantage is critical for sustainability; and innovative partnerships are notable in having an important role to play in this regard. Opportunities for making profits can entice private entities to partner with the public entity and bring win-win outcomes for the stakeholders. In this regard, private sector financing becomes accessible and their strong capacities in product branding and marketing can be tapped into (Kaza et al., 2016). The public entity can sell the compost directly to agricultural producers through a segmented pricing approach to gain more revenue. However, distribution agents and agro-input suppliers/dealers are an efficient channel for accessing the fertilizer market especially if the public entity lacks capacity in marketing and distribution. The option of developing different formulations of compost tailored for specific crops, the sale of non-degradables such as plastics and metals to recycling firms and sale of carbon credits are alternative avenues to generate additional revenue, minimizing subsidy dependency and opportunity to move the model from cost-recovery to profit-maximization (Figure 131).

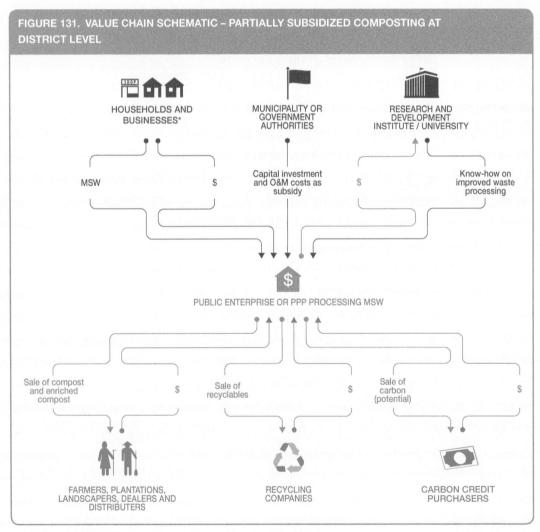

FIGURE 131. VALUE CHAIN SCHEMATIC – PARTIALLY SUBSIDIZED COMPOSTING AT DISTRICT LEVEL

Note: * Under a PPP it is optional if the public or private partner collects the waste.

C. Business model

The business model described here (Figure 132) presumes operation under a public entity with partial subsidies for governmental entities. The model has three value propositions: a) provide improved waste management services to households; b) increase access to environmentally sustainable organic fertilizer to agricultural producers at competitive market prices; and c) provide recycling companies with increased options for purchasing recyclables at competitive prices. Strategic partnership with governmental organizations assure access to capital investments but also recurrent financing for operations and maintenance (Kaza et al., 2016). The provision of waste collection services generates significant revenue for the public entity, received via government payments but also the waste tax they are able to charge to institutions and businesses who fail to segregate their waste. The latter can tremendously reduce segregation costs and speed up the entire compost production process which implies less operational costs and more benefits.

The production of organic fertilizers from MSW and FS imply that farmers have access to fertilizer options. The public entity can sell directly to the end-users and also utilize agricultural extension systems, input suppliers, private dealers or even existing chemical fertilizer distribution channels via partnerships. Implementing a segmented pricing approach, by charging a lower price for bulk sales and market price for retail purchases can increase revenue. By advocating for government incentives similar to those for chemical fertilizers, the compost can be sold to local farmers and farmer organizations at partially subsidized rates through government agencies and agricultural departments to gain a larger share of the fertilizer market. Also, value addition to the compost via fortification and pelletization and branding of the product could be instrumental for greater market penetration and revenue generation. A partnership with a research and development (R&D) institute becomes crucial as the public entity is able to tap into their research capacity to develop competitive compost products for a competitive fertilizer market. As a part of its marketing strategy and to expand its customer base, the public entity can give all its first-time customers free compost samples so that the farmers can see first-hand increased yields on their own farms. This can be instrumental in increasing its market share. An additional source of revenue is from the sale of recyclables which can be purchased from locals and sold directly to recycling companies at higher prices. For efficiency, the public entity can set-up decentralized waste resource centres where informal workers bring and sell the segregated recyclables to them. This value proposition in particular extends the model to be inclusive and provides indirect employment (income) to people that would otherwise be unemployed.

This model, although subsidy-dependent, generates significant environmental and socio-economic benefits that justify governmental support. Reduced open-dumping and burning of waste implies decreased GHG emissions and human exposure to untreated waste. The conversion of MSW and FS to compost is an avenue to improve soil productivity and agricultural yields, but also reduces waste disposal costs, GHG emissions from landfills and chemical fertilizer production. Opportunities to transition the model to financial independence is crucial in view of shrinking municipal budget allocations for waste management.

FIGURE 132. BUSINESS MODEL CANVAS – PARTIALLY SUBSIDIZED COMPOSTING AT DISTRICT LEVEL

KEY PARTNERS	KEY ACTIVITIES	VALUE PROPOSITIONS	CUSTOMER RELATIONSHIPS	CUSTOMER SEGMENTS
• Municipality, city council • Local university/R&D institute • Agro-dealers • Waste resource centre	• Waste collection and segregation • Production and sale of compost • Shredding and pelletization • Purchase and resale of recyclables	• Provision of improved waste management services to households • Farmers have increased access to alternative fertilizer products at a competitive price • Recycling companies have access to recyclables products at an affordable price	• Waste collection through direct contact • Personal help at direct sales • Direct relationship	• Households, commercial entities • Agricultural producers (smallholder farmers, plantations, landscaping companies, local/national agricultural departments) • Recycling companies
	KEY RESOURCES • Machinery • Land and labor • MSW • Permit (composting and environmental impact assessment) • Waste resource centres		CHANNELS • Direct • Direct sales, word-of-mouth • Direct sale	

COST STRUCTURE 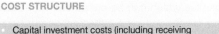	REVENUE STREAMS
• Capital investment costs (including receiving tanks, sedimentation tank, plastic and polythene pelletizer, water treatment facility and drying bed) • Operation and maintenance costs (machinery, infrastructure, labor) • Quality monitoring fee payment • Cost of buying recyclables	• Waste collection funds from government • Waste tax from commercial entities that do not segregate waste • Waste collection fees from commercial entities • Government subsidy (partial) • Sales of compost and other recyclables

SOCIAL AND ENVIRONMENTAL COSTS	SOCIAL AND ENVIRONMENTAL BENEFITS
• Possible pollution of water bodies from leachate • Potential health and occupational safety risks to workers failing to wear protective gear • Potential risk to the public where compost and segregation activities are in close proximity to neighbourhoods	• Reduced existing waste management costs • Creation of job opportunities • Reduction of environmental pollution from untreated waste • Reduced human exposure to untreated waste • Improved yield returns

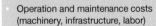

D. Alternate scenarios

In the generic business model described above, a public entity converts MSW and FS to an organic fertilizer for sale to agricultural producers, provides waste collection services to households and resells recyclables at higher prices to commercial firms. This business model can incorporate alternative scenarios to enhance revenue generation and overall sustainability by: i) increasing its scale of operation (large scale) via a public-private partnership, and ii) revenue generation from carbon credits under the CDM.

Scenario I: Large-scale operation as a public-private partnership

Public entities can benefit from economies of scale to further reduce disposal costs and generate significant revenue through composting at a larger scale. This however requires increased capital investments for infrastructure as well as funds to cover operational and maintenance costs. Whilst municipalities are generally able to cover O&M costs, new capital investments can overstretch their budgets. Additionally, publicly-managed compost facilities often show inefficiencies in product innovation and marketing. Many of these shortcomings can be addressed by the business-oriented private sector seeking profits. Tapping into private sector capital and their capacity for management and innovation via public-private partnerships is essential for considerations of scaling-up and transitioning to full-cost recovery models. Public-private partnerships (PPPs) are a well-established means of providing infrastructure and services that public entities have neither the resources nor expertise to supply alone. Under the model described here, a PPP can become a win-win protocol where the public sector gets the opportunity to improve waste management services (waste collection, transportation and proper treatment or disposal) with collaborations from the private sector, while the private sector is given the opportunity to bring a waste business into existence as a profitable endeavour.

For large-scale composting operations, a suitable PPP arrangement could be where: a) the public sector constructs the infrastructure and provides the capital cost for equipment for composting; and b) private sector brings in operational capital and suitable management skills to operate the facility. Under the agreement, the private entity pays a monthly service fee to the public entity for using the already set-up composting infrastructure such as land, machines, composting facility. The public entity in turn collects the waste and pays tipping fees to the private entity for disposal and processing of the municipal solid waste. Under the management contract, the private business entity bears the cost of operation and maintenance. The PPP can establish satellite compost stations to produce compost at vintage points closer to local markets, to minimize transportation costs both for waste collection for the public entity and distribution of compost for the private business entity. This will allow them to sell compost at a flat rate exclusive of transport charges, while traders/retailers can add transportation cost and their own price mark-up to the final sale prices. In addition to compost, recyclables and fuel pellets can be sold to recycling companies and businesses, respectively to increase their revenues and achieve full-cost recovery/profits.

While the potential opportunities of the PPP model are increasingly clear, PPP contracts can be relatively more complicated than conventional procurement contracts. This is because oftentimes all possible contingencies that could arise in long-term contractual relationships are not anticipated beforehand. The sustainability of the model will thus depend on concessions and incentives such as (i) tax assignment and grants for segregation; (ii) advertisement rights for segregation at collection centres; (iii) unit cost payment for collection and transport; (iv) making land available for disposal; (v) buy-back of composting; (vi) tax holidays and other incentives; and (vii) carbon credits, being clearly outlined and agreed upon (ADB, 2011).

Scenario II: Carbon credits

The PPP model fits best where capital and management skills of a private entity can help fill capacity gaps of the public entity. Yet full-cost recovery in the PPP model may still remain elusive at least during the initial years, where economies of scale are not fully realized, and compost prices are still higher than that of subsidized chemical fertilizers. The sale of carbon credits can represent an alternative revenue stream, especially for PPP entities who are still unable to achieve financial break-even and dependent on government financial support. This business model typically requires partnerships/engagements with the local government, national environmental management authorities, private entities and international partners. The application process for carbon credit sale can be lengthy and costly; and in view of volatile market prices, the net return should be taken into account prior to investing in the process.

E. Potential risks and mitigation

The business model presented here was designed and optimized based on the analysis of different case studies (see previous sections). In designing this optimized business model, risks related to safety, local acceptance by the community and business attractiveness for investors were assessed.

Market risks: Risks in the input market are very low as the public entity typically owns the city's waste or is granted exclusive rights by governmental authorities. On the output side, the main risk relates to competition in the 'larger' fertilizer market.

Competition risks: Competition as noted under 'market risks' stem from price distortions in the output market where the compost products compete with often subsidized chemical fertilizers. Product innovation to increase compost nutrient levels, branding via certification, free samples and field trials can help mitigate the negative effects of competition. Satellite composting stations in vantage points and close to its key customers can improve market penetration.

Technology performance risks: The composting technology traditionally used is windrow composting. Depending on the scale, components of the process (e.g. waste segregation) can be mechanized for efficiency. This however implies increased energy requirements which can be costly and if there are energy shortages represent a key challenge for performance. Additionally, the need for advanced-skilled labor represent increased operational costs. On the other hand, if more labor-intensive processes are used, then labor availability (including skills set) and related costs have to be taken into account.

Political and regulatory risks: It is important to note however that policies and regulations differ from country to country and so whilst reuse of fecal sludge may be permissible in Sri Lanka, it may not be allowed elsewhere. Thus, it is important that national and local guidelines and policies are adhered to. Specific to this model, there are low regulatory risks as the public entity will only engage in resource recovery initiatives that are permissible by law as they are financed by public funds. Thus, the plants' practices are very likely to follow the outlined national/local guidelines and policies on waste management activities, and compost product safety.

Social equity related risks: Consideration of the set-up of decentralized waste resource centres for recyclables may offer informal workers the opportunity to sell the segregated recyclables they collect to the plant. This value proposition in particular extends the model to be inclusive and provides indirect employment (income) to people that would otherwise be unemployed. On the other hand, however, improved waste collection, segregation and recycling may limit informal workers access to waste value chain and invariably, income.

Safety, environmental and health risks: The compost product should meet the minimum nutrient level requirements outlined in the respective national/local guidelines via regular quality monitoring. There are potential health risks to different actors along both the sanitation and agricultural value chains, associated with the collection, treatment, processing and use of human excreta (Table 35). In particular, workers that collect the fecal sludge and composted materials and consumers of food products grown with waste-based compost are the groups with the highest level of risk. The provision of protective gear for chamber-emptying operations should be mandatory. From the consumer perspective, microbial testing should be a routine measure for quality assurance of the compost product. Additionally, farmers must be trained on the appropriate application methods for the waste-based fertilizer products. Recommendations of national agriculture agencies must also be implemented in tandem, in association with agricultural extension agents.

TABLE 35. POTENTIAL HEALTH AND ENVIRONMENTAL RISKS AND SUGGESTED MITIGATION MEASURES FOR BUSINESS MODEL 10

RISK GROUP	EXPOSURE					REMARKS
	DIRECT CONTACT	AIR/DUST	INSECTS	WATER/ SOIL	FOOD	
Worker	HIGH					Risk of sharp objects in MSW and fecal contamination. Potential risk of dust, noise and chemical compost contaminants
Farmer/user						
Community				MEDIUM		
Consumer						
Mitigation Measures	(icons)	(icons)	(icons)	(icons)	(icons)	

Key: NOT APPLICABLE | LOW RISK | MEDIUM RISK | HIGH RISK

F. Business performance

The model ranks highest on scalability and replicability as it has a strong potential for implementation in medium and large cities (Figure 133). Depending on the scale of operations, adaptation to the technology and market development may be required. This model is ranked high on environmental and social impact partly due to the large quantities of waste collected and processed which results in reduced indiscriminate waste disposal, reduced human exposure to untreated waste, reduced GHG emissions from landfills and the opportunity for job creation. The inherent dependence on government for financial support makes the model rank very low on profitability. Although generally geared only towards partial cost recovery, the model has potential to transition into full-cost recovery and even profit-making under public-private partnership agreements. The low ranking of the innovation criteria is mainly attributable to the simplicity of the technologies.

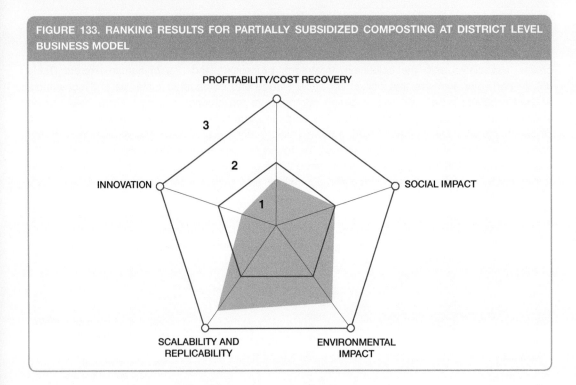

FIGURE 133. RANKING RESULTS FOR PARTIALLY SUBSIDIZED COMPOSTING AT DISTRICT LEVEL BUSINESS MODEL

References and further readings

Asian Development Bank (ADB). 2011. Toward sustainable municipal organic waste management in South Asia: A guidebook for policy makers and practitioners. Manila, Philippines: Asian Development Bank.

Kaza, S., Yao, L. and Stowell, A. 2016. Sustainable financing and policy: Models for municipal composting.Urban development series Knowledge Papers 24. Washington, D.C.: World Bank Group.

8. BUSINESS MODELS ON SUBSIDY-FREE COMMUNITY-BASED COMPOSTING

Introduction

Towns and cities across the developing world continue to face the challenge of managing municipal solid waste (MSW). For smaller towns, the relatively easier availability of land for disposal of MSW and lower costs of transporting the waste to landfills oftentimes represent disincentives for MSW-based composting. However, many of these towns are rapidly transitioning into cities in view of exponential population growth and urbanization; and with limited public funds to support waste management infrastructure and services, there is a dire need to identify and adopt sustainable waste management measures that can handle the significant quantities of waste being generated.

Large-scale centralized composting whilst able to process big quantities of waste at a time tends to be highly mechanized and thus require hefty investments for advanced machineries, significant operation and maintenance costs and a high degree of specialized skills to operate and maintain the plants. Additionally, transportation costs can be substantial as all the waste needs to be transported to disposal facilities often located far from the city. The quality of compost tends to be poor due to the large quantities of unseparated waste with high risks of contamination. Thus, revenue generated from compost sales is often insufficient to cover the capital, operation and maintenance costs. With increasingly shrinking municipal budget for waste management, a large percentage of these compost plants have reached the end of their life cycle or in dire need of upgrade and maintenance. Sustainable funding mechanisms thus become a major factor in the success of national strategies for municipal solid management programs.

Decentralized composting systems offer several advantages over centralized large-scale systems and are increasingly been observed, particularly for secondary cities and small towns, and even large cities where the local government can allocate land. Adopting a labor-intensive, cheap and low technological approach, the business does not require a large capital investment (except for land purchase) or state-of the art machinery, which removes one of the major constraints for business start-ups especially in the developing world context. The decentralised composting approach reduces transportation costs and makes use of low cost technologies based on manual labor and ensures that waste is well-sorted before it is composted. This minimizes many of the problems and difficulties that have led to the failure of large centralized composting plants in the past. There is great potential for the upscaling of this model due to its simplicity. However, poor management and incentives to entities operating the decentralized units often results in poor quality compost (low market demand) and misappropriation of funds, which invariably causes the plant to fail. Studies have shown that whilst most decentralized composting businesses have a non-profit seeking model, these constraints limit cost-recovery and additional public funding is oftentimes required to bridge the financial gap.

Business models with inherently sustainable funding mechanisms (i.e. profit-making model), such as a **subsidy-free community-based** composting initiative, are necessary. As an example, a cooperative model approach to decentralized composting creates a greater incentive for community participation. There is a higher probability of success as benefiting communities are involved in waste collection, separation and composting, plant management and ownership. The sustainability of this model is grounded in strong partnerships and the assured benefits (profit-sharing) accruing to each partner. Voluntary participation via membership fee payments are indicative of the commitment of members and thus ensure success of the enterprise. Municipalities have an incentive to support communities in finding composting sites, developing a proper system for waste collection and disposal of residues, and providing land and funds for construction of composting plants as these initiatives alleviate them of the burden of solid waste management.

INTRODUCTION

In this chapter, we present the business model and a case example that show the concept of subsidy-free community-based composting, and the notable potential it offers by organizing communities into a *cooperative*. The presented case study shows that subsidy-free community-based composting offers a solution for turning waste into wealth, but requires investments in social capital to organize and mobilize the communities.

CASE

Cooperative model for financially sustainable municipal solid waste composting (NAWACOM, Kenya)

Miriam Otoo, Nancy Karanja, Jack Odero and Lesley Hope

Supporting case for Business Model 11	
Location:	Nakuru, Kenya
Waste input type:	Municipal solid waste (including plants and animal waste)
Value offer:	Provision of a safe compost product as a soil conditioner
Organization type:	Cooperative
Status of organization:	Operational since 2002; plant operations had halted at time of last publication review (October 2017)
Scale of businesses:	Processes 28 tons of waste/ day
Major partners:	University of Nairobi, Egerton University, Practical Action, Comic Relief, National Agricultural Advisory Service, World Bank

Executive summary

The Nakuru Waste Collectors and Recyclers Management Cooperative Society (NAWACOM) is a cooperative that has brought together various community-based organizations (CBOs) in the organic waste recovery arena in Nakuru. Their main focus was to take up the waste management challenge in Nakuru town and create an avenue for income generation under the slogan 'turning waste to wealth'. CBOs initially operated as individual entities but transitioned into a cooperative to secure financial support from Comic Relief via Practical Action to scale up their operations. NAWACOM was then formed as the representative umbrella body. The CBOs produce a partially processed compost product from agricultural, household and market waste using a windrow composting technology, which is then sold to NAWACOM. The product is further composted, fortified, packaged and branded under the name Mazingira. The benefits of the decentralization of NAWACOM's activities has ensured that: a) smaller-scale CBOs are still able to financially sustain their businesses by not having to put up significant capital investment for equipment and establishing sound marketing and distribution channels; and b) NAWACOM allocates its resource efficiently – i.e. waste collection and separation is outsourced to communities, reducing high transportation costs. Ninety-five percent of the organic fertilizer is sold directly to farmers through word of mouth and the remaining percentage through agroshops. Revenue streams of the cooperative are mainly from compost sales and member subscription fees. All accrued profits are shared among cooperative members. NAWACOM's activities have helped to significantly reduce the city's waste management costs, reduce human exposure to untreated waste and contribute to the livelihoods of local communities.

KEY PERFORMANCE INDICATORS (AS OF 2015)	
Land use:	0.41 ha
Capital investment:	USD 4,671 excluding land costs
Labor:	6 (2 skilled part-time, 4 unskilled part-time) – excludes employees in the different CBOs
O&M:	USD 9,977 per year
Output:	100–300 tons of compost per season
Potential social and/or enviornmental impact:	Creation of 6 part-time jobs, provison of a nutrient rich organic fertilizer for agricultural production and a clean environment
Financial viability indicators:	Payback period: 5 years Post Tax IRR: N.A. Gross margin: 40%

Context and background

Nakuru town is the fourth largest urban centre in Kenya. It is centred in rich agricultural hinterland with fertile volcanic soils and has an ever developing industrial and tourism industry. Rapid urban growth, which is estimated at 3.4% per annum over the last three years, has resulted in the development of unplanned residential areas and slums; hence garbage heaps are a common sight as the Municipal Council is over-stretched in offering services in solid waste management. To bridge the gap between waste generation and collection, NAWACOM, a cooperative society, in 2002 stepped in with the aim of providing sanitation services and environmental conservation whilst generating revenue. Community-based organizations involved in waste reuse initially operated as individual entities but transitioned into a cooperative to secure financial support from Comic Relief via Practical Action to scale up their operations. In 2006, NAWACOM was registered as a cooperative in accordance with Section 3 of the Cooperative Societies Act (Amended 2004) of the laws of Kenya. Technical support came from Practical Action Kenya, which is an international non-governmental organization while funding was provided by Comic Relief (a UK-based charitable organization). The objective of this partnership was to showcase how community members could contribute towards solid waste management in a sustainable way. The cooperative works by contracting its members (CBOs) to collect and compost organic waste from peri-urban areas of the town (mostly livestock and household waste from famers) and also private waste collectors who sort and compost waste from within the town. At the time of the assessment, membership stood at 94 people, with 55 women and 39 men. Membership recruitment was open to all provided each member shared in the cooperative's vision and was able to pay the annual membership subscription of USD 56.92 (Ksh 5000)[1].

Market environment

The negative effect from chemical fertilizer over-application on soils and water bodies has caused an upsurge in the demand for organic fertilizer use. Farmers have observed declining soil health and decreased crop yields over time, and recognize the need to adopt environmentally sustainable agricultural practices. Additionally, recommended agricultural practices, particularly for the production of exported food products, require the use of organic agricultural inputs. Furthermore, rapid urban population growth in Nakuru city has resulted in the development of unplanned residential areas and slums and subsequently generation of significant amounts of waste. The quantity of generated waste has overstretched the municipal council's budget for waste management. NAWACOM and its community members thus ceased this opportunity to fill in the gap for providing waste management and a safe organic fertilizer for the production of exportable goods.

Macro-economic environment

The Kenyan government highly subsidizes chemical fertilizers. The government's fertilizer subsidy programs began in 2008 with the aim to cushion farmers against seasonal changes in the price of fertilizer. By the end of the 2012/2013 financial year, over 400,000 metric tons of fertilizer, worth Ksh 13.80 billion had been distributed countrywide. The amount of subsidies on chemical fertilizer has grown exponentially in the last few decades and has been mainly attributed to inflation and price fluctuations in the international market. The government has plans to increase its fertilizer subsidy budget allocation to Ksh 15 billion over the next five years. With continued governmental support, chemical fertilizer prices will continue to be more competitive than organic fertilizer prices making it difficult for new businesses to enter the fertilizer market. With a growing need to increase the availability and quality of bio-fertilizers and composts in the country to improve agricultural productivity while maintaining soil health and environmental safety, Kenya will need to set up a scheme to augment the infrastructure for production of quality organic and biological inputs and some level of price subsidy to organic fertilizer producers to make them competitive on the market.

Business model

NAWACOM is a waste processing cooperative that uses household, animal and market waste to produce an organic fertilizer product – Mazingira, which is sold directly to small-scale farmers. As a cooperative, it contracts its members to collect and compost organic waste from peri-urban areas of the town. Essential in its business model is the decentralization of NAWACOM's activities. Members of the various CBOs compost the organic waste resources on their premises and deliver a partially composted product to NAWACOM, who then processes it further to maturation and fortifies it. This has ensured that: a) smaller-scale CBOs are still able to financially sustain their businesses by not having to put up significant capital investment for equipment and establishing sound marketing and distribution channels; and b) NAWACOM allocates its resource efficiently – i.e. waste collection and separation is outsourced to communities, reducing high transportation costs. The price of the partially processed compost, ranging from USD 0.05 to 0.07 per kg, is determined by its nitrogen content and level of pathogens, which are the indicators of quality. This pricing strategy helps NAWACOM maintain a high product quality standard as all members aim to receive the highest purchase price for their compost as possible per the market's willingness-to-pay. The cooperative is value-driven where quality of the product is the main focus. NAWACOM sells the final organic fertilizer product, Mazingira, mainly to small-scale farmers at USD 17.65 per 50kg bag. The cooperative markets their product via word-of-mouth which has proven to be an effective strategy given the high product quality. The cooperative also generates revenue from membership subscription fees at USD 59 per member per annum, which is used to cover operational costs and has ensured continuous operation of the business. The cooperative has nine staffs, of which six form an oversight committee. The remaining three are the executives who are also signatories to the account. NAWACOM instituted an oversight committee to prevent swindling of cooperative funds by the executives. The cooperative partnered with Comic relief and Practical Action Kenya for financial and technical support at the onset of the business. Egerton University and University of Nairobi are the main bodies in charge of the product quality analysis. The compost is fortified with Mijingu Phosphate Rock as a means of increasing the nutrient content and demanding a higher market price. Plans are underway to get a Kenya Bureau Standard Board (KEBS) certification, which will enable NAWACOM to penetrate the large-scale farmers' customer segment. The activities of NAWACOM have contributed to the reduction of cost associated with waste management whilst keeping the city clean. In addition, it has provided cooperative members with an additional income. See Figure 134 opposite for the diagrammatic overview of the business model.

FIGURE 134. NAWACOM'S BUSINESS MODEL CANVAS

KEY PARTNERS
- Agricultural input shops
- Egerton University
- University of Nairobi
- Comic Relief
- Practical Action
- Standard Board (KEBS)
- Municipal Council
- Cooperative members

KEY ACTIVITIES
- Coordinating delivery of partially processed compost by members
- Quality control of incoming compost
- Processing and fortification of compost
- Packaging
- Sale of compost
- Branding, marketing

KEY RESOURCES
- Household, market and animal waste
- Mijingu Phosphate Rock
- Laborers
- Licensing
- Brand name Mazingira

VALUE PROPOSITIONS
- Local farmers obtain a quality-checked, organic fertilizers of high nutrient level at a competitive price

CUSTOMER RELATIONSHIPS
- Personal help at direct sales

CHANNELS
- Direct sales via word-of-mouth (95%)
- Agro-input shops (5%)

CUSTOMER SEGMENTS
- Small-scale farmers

COST STRUCTURE
- Capital investment (land, machinery, licensing)
- Operation and maintenance (labor cost (6 part-time), electricity, land rent, maintenance costs)
- Quality control fees (University of Egerton and Nairobi University laboratories)
- Administrative costs (collection of member fees and selling the compost)
- Transport costs savings

REVENUE STREAMS
- Sales of compost (75%)
- Membership subscription fees (25%)

SOCIAL & ENVIRONMENTAL COSTS
- Manual execution of activities such as sieving and packaging may be a source of occupational health risk

SOCIAL & ENVIRONMENTAL BENEFITS
- Reduced existing waste management costs
- Reduced human exposure to untreated waste
- Creation of jobs (6 part-time jobs)
- Improved food security
- Increase in income for low-income population
- Government savings from reduced importation of inorganic fertilizers
- Community empowerment by means of a cooperative

Value chain and position

NAWACOM is a waste processing cooperative that produces organic fertilizer from MSW. The cooperative's activities are the production, marketing and sale of fortified compost (Figure 135). NAWACOM sources its raw materials (partially processed compost) from its members (CBOs) and is their sole client. Market and household waste are the main waste streams used for the composting activities of the CBOs. Given that NAWACOM as a business entity does not directly source for MSW for processing activities, it faces very low input supply risk. Additionally, MSW is an abundant resource especially in the peri-urban areas, markets and high-density inner city with currently limited alternative use. NAWACOM purchases the partially processed compost from its members at a fee (dependent on nitrogen concentration and pathogen levels) and further processes and fortifies with Mijingu Phosphate Rock. The final product is sold directly to small-scale farmers. Although NAWACOM partners with the University of Egerton and Nairobi University for the fee-based quality analysis of their product, the cooperative's failure to obtain KEBS certification has limited its ability to penetrate new markets. Other organic fertilizers and chemical fertilizers are good substitutes for NAWACOM's organic fertilizer. Additionally, chemical fertilizer is high in demand due its ease of application, high NPK levels and KEBS certification. In 2012, NAWACOM received an order for 500 tons of compost to be supplied over the entire year from a major agricultural input supplier. It was however unable to meet the demand as it is illegal to supply large quantities of compost to agricultural input suppliers with the seal of KEBS. NAWACOM faces fierce competition in the fertilizer market but the acquisition of KEBS certification will increase product demand and ease its penetration into larger customer segments, beyond the about 3,000 farmers it serves per year.

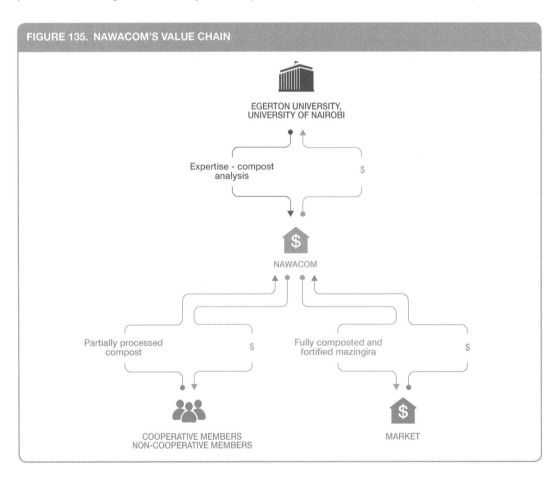

FIGURE 135. NAWACOM'S VALUE CHAIN

Institutional environment

Management of solid waste in Kenya is dealt with under several laws, by-laws, regulations and acts of parliament. Some of them include the Environmental Management and Coordination Act (EMCA) of 1999 and City Council (solid waste management) by-law of 2007 which requires waste reuse businesses assure the safety of all actors involved in the business operations and the quality of the product. To legally engage in composting activities in Kenya, a waste management permit from the City Council (at USD 200 per year) and NEMA (at USD 471 per year) are a requirement and are renewable on a yearly basis. The Kenya Bureau Standard Board (KEBS) is mandated to certify organic products for sales in the country. Compliance to product quality guidelines for compost is largely unregulated in Kenya although KEBS has developed standards and guidelines to meet demand in the country for marketing of organic fertilizer products. Organic fertilizer produced by NAWACOM has yet to meet the standards set by KEBS and this has limited NAWACOM's access into certain market segments.

Technology and processes

NAWACOM works on a contractual basis where suppliers, both members and non-members (although members are given the priority), collect, sort and compost organic waste in their homes for four to six weeks. Windrow composting is the technology used. This technology, although labor-intensive, requires low capital investment and has high rates of resource recovery (Figure 136). The technology however requires significant amounts of space which can be a challenge for small-scale CBOs. It is in this regard, that the CBOs partially compost the organic waste at their own premises and deliver it to NAWACOM's main processing site for maturation and quality check. Once the compost has fully matured, samples are taken to Kenya Agricultural Research Institute (KARI) and Egerton University

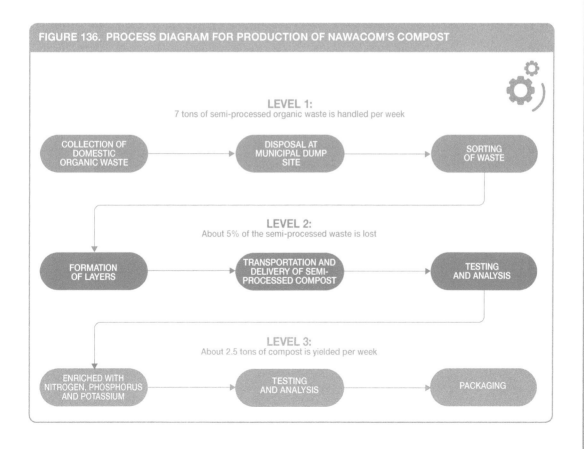

FIGURE 136. PROCESS DIAGRAM FOR PRODUCTION OF NAWACOM'S COMPOST

for quality analysis, mainly to ascertain nitrogen concentration, and pathogen and heavy metal levels. The compost is then transported back to the cooperative's operations site where it is sieved to a finer particle size. The end product (fine compost) is fortified with Minjingu Phosphate Rock and other natural materials to increase its potassium and phosphorus levels to attain an NPK ratio of 2:1.5:1.8. Products are then packaged into 25 and 50kg bags. NAWACOM implements strong internal regulations, ensuring that all persons involved in the compost production process wear protective gear at all times.

Funding and financial outlook

Initial investment for NAWACOM as a community-based organization came from membership subscription, which amounted to USD 3,529 per year. This was barely sufficient to purchase the partially processed compost that NAWACOM further added value to, thus production and operations were low until 2006 when Comic Relief came in to provide financial support. The provision of financial support (USD 47,000) was on condition that the umbrella body – NAWACOM – be registered and operate as a cooperative. The investment provided covered costs of machinery, inputs (partial compost) and licensing. Operation and maintenance cost is estimated at USD 9,976/year and includes costs of labor (six part-time), electricity, land rent and other associated repairs and maintenance costs. The revenue streams of NAWACOM are sales of compost (75%) and membership subscription fees (25%). Compost is a seasonal product and sold in the two agricultural seasons in the year. NAWACOM sells between 3,000–6,000 50kg bags retailing at USD 17.65/bag. This translates into gross revenue of USD 52,000 to USD 105,000 per year. The cooperative has 94 members and the membership fees yield a revenue of USD 5,527 per annum at a rate of USD 58.8 per membership fee per person per year. NAWACOM has been generating profit since the exit of Comic Relief in 2008, indicating that with increased production and demand, the cooperative stands to accrue high profits/benefits to its shareholders.

Socio-economic, health and environmental impact

Economic gains of NAWACOM's activities include environmental, social and human health benefits. Although no absolute figures were provided, environmental benefits can be traced to reduction of pollution due to reduced human exposure to untreated waste and contamination of water bodies from open dumping. NAWACOM has increased the income of considerable number of people through employment and the sales of semi-composted organic waste. The increase in income for these people represents increased purchasing power, which can be translated into improved food security. The cooperative's activities have also had a positive impact on the government budgets as waste collection is done free of charge. An important risk to bear in mind is that related to the manual sieving and packaging of the compost, which may represent a source of occupational health risk if mitigation measures such as wearing of nose mask and gloves are not adhered to.

Scalability and replicability considerations

The key factors driving the success of this business are:
- Farmers have observed declining soil health and decreased crop yields over time and recognize the need to adopt environmentally sustainable agricultural practices.
- Assured high quality product sold at a competitive market price.
- Strong relationships and win-win partnership with its members.
- Innovative pricing strategy for input (partially processed compost) ensuring high quality product.
- Traditional word-of-mouth marketing strategy has proven to be a successful strategy given the assured quality of their product.
- Establishment of an oversight committee has been essential in curbing the misappropriation of cooperative funds.
- Strong commitment of members to the vision of the cooperative.

This model has a high potential of being replicated in developing countries where community involvement in waste management is encouraged. This case is unique in that it is a cooperative that has contracted its members to partially compost household, animal and market waste. The monetary benefits accruing to all parties create an incentive for commitment and success of the business. This model can easily be replicated as the start-up capital is fairly low and the technology is simple and capitalizing on the abundance of labor, requires a lot of land depending on scale. With rapid urbanization, rental and sale prices of land in both urban and peri-urban areas in developing cities have skyrocketed and this may represent a major constraint. Additionally, cooperatives have a history of high failure rates especially in developing countries. Stringent and efficient measure need to be put in place to ensure its success.

Summary assessment – SWOT analysis

NAWACOM represents an initiative of a group of CBOs who successfully sustained their business following the exit of donor funding. The cooperative has been particularly successful by implementing an oversight committee, which has been essential in the smooth running of business operations. Assured high quality and affordability of Mazingira fertilizer has been instrumental for NAWACOM in increasing its market demand and exploring other market segments (Figure 137). The decentralization

FIGURE 137. SWOT ANALYSIS – NAWACOM

	HELPFUL TO ACHIEVING THE OBJECTIVES	HARMFUL TO ACHIEVING THE OBJECTIVES
INTERNAL ORIGIN ATTRIBUTES OF THE ENTERPRISE	**STRENGTHS** • High nutrient level organic fertilizer • High up-scaling potential • Low cost of technology • Availability and easy access to waste and production inputs (partially processed compost) • Self-branding has increased market share • Fortification has increased product marketability	**WEAKNESSES** • Financial instability • High transportation cost • Non-government-certified product
EXTERNAL ORIGIN ATTRIBUTES OF THE ENVIRONMENT	**OPPORTUNITIES** • Production of animal feed • Potential to produce granulated compost and access new markets • Increase in the scope of market from product certification • Acquiring KEBS certification increases market and revenues per unit	**THREATS** • High rental prices of land • Attitudinal problem – farmers see organic fertilizer as secondary input • Lack of certification from KEBS may disrupt business activity • Absent scheme for the promotion of organic materials vis-à-vis continued heavy subsidization of chemical fertilizers limits growth. • With continued governmental support, chemical fertilizer prices will continue to be more competitive than organic fertilizer

of NAWACOM's activities has ensured that: a) smaller-scale CBOs are still able to financially sustain their businesses by not having to put up significant capital investment for equipment and establishing sound marketing and distribution channels; and b) NAWACOM allocates its resource efficiently – i.e. waste collection and separation is outsourced to communities, reducing high transportation costs. The organic fertilizer produced by NAWACOM has not yet been approved by KEBS and this has limited its access to different and larger market segments. It is so far only serving about 3000 small-scale farmers per year, which is less than 2% of the market. A certification by KEBS and pelletization/granulation of its product will enable it to penetrate new market segments. Increasing governmental support along with growing demand for organic fertilizers will represent key opportunities for replication and up-scaling of the business.

Contributors

Jasper Buijs, Sustainnovate, Netherlands; formerly IWMI, Sri Lanka
Johannes Heeb, CEWAS, Switzerland
Josiane Nikiema, IWMI, Ghana

References and further readings

Environmental Protection Agency. 1985. Composting of municipal wastewater sludge. Seminar Publication. United States Environmental Protection Agency, Centre for Environmental Research Information.

Personal observations and interviews with NAWACOM personnel. 2015.

Scheinberg, A., Agathos, N., Gachugi, J.W., Kirai, P. Alumasa, V., Shah, B., Woods, M. and Waarts, Y. 2011. Sustainable valorization of organic urban wastes. Insights from African case studies. Wageningen UR.

Case descriptions are based on primary and secondary data provided by case operators, insiders or other stakeholders, and reflect our best knowledge at the time of the assessments 2015. As business operations are dynamic data can be subject to change. Plant operations were noted to have halted at time of latest edit (October 2017).

Note

1   Ksh is Kenyan shillings. 2015 Exchange rate: USD 1 = Ksh 87.85.

BUSINESS MODEL 11
Subsidy-free community-based composting

Miriam Otoo and Munir A. Hanjra

A. Key characteristics

Model name	Subsidy-free community-based composting
Waste stream	Municipal solid waste (including plant and animal waste)
Value-added Waste product	Provision of waste management services to communities, and provision of an affordable and safe compost for soil conditioning
Geography	Replicable in medium and large cities where land availability is limited; abundance and inexpensive labor
Scale of production	Small to medium, 20–30 tons of waste processed per day
Supporting cases in this book	Nakuru, Kenya
Objective of entity	Cost-recovery [X]; For profit [X]; Social enterprise [X]
Investment cost range	Capital cost about USD 3,500–5,500 excluding land costs, and O&M cost USD 7,50 –12,500 per year
Organization type	Cooperative
Socio-economic impact	Improved waste management service, creation of new jobs, provision of organic fertilizers for agriculture, improved soil productivity and a cleaner environment
Gender equity	Pro-gender model. Community based job opportunities for women

B. Business value chain

Community-based composting models have shown some success but can be limited by poor management, limited access to financing due to investors' reluctance in funding smaller-scale initiatives. The community-based cooperative model however offers opportunities to address these limitations as small communities are able to mobilize their own resources by encouraging members to join the cooperative on voluntary basis and raise their own funding through membership fees. This business model is initiated by a cooperative – a distinct form of enterprise that provides services and/or products to the members, by the members, and for the members at a cost and divides the profits, known as surpluses in a cooperative, among the members pro rata to the amount of business each member did with the cooperative (Figure 138). Community-based organizations (decentralized composting facilities) form the consortium of the cooperative. Membership is voluntary and based on mutual social, cultural and economic needs – waste management and composting in this case. Whilst this could be a cost recovery model of decentralized composting operations at individual member's level, the cooperative element transitions this model into a profit-making model.

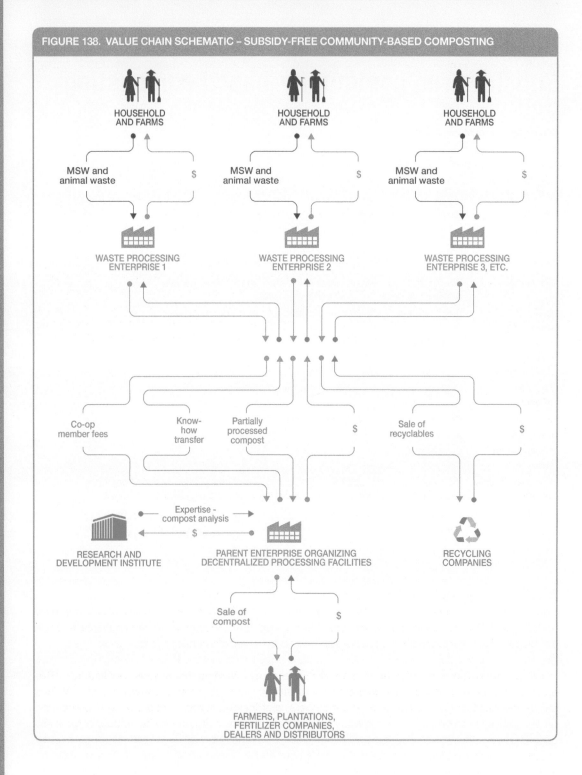

FIGURE 138. VALUE CHAIN SCHEMATIC – SUBSIDY-FREE COMMUNITY-BASED COMPOSTING

The CBOs collects waste from households and farms at a fee. Waste separation and its partial composting is done at the premises of each member, although depending on the scale for example, the local government may provide only land and infrastructure for plant operation. Outsourcing waste collection and separation implies land and transport cost savings to the parent enterprise that organizes the CBOs into a cooperative. The partially processed compost is sold to the parent enterprise. The members are incentivized to ensure high quality of the partially processed compost if the price they receive is dependent on product quality. The CBOs additionally generate revenue from the sale of recyclables. The parent enterprise that organizes the CBOs into a cooperative provide technical know-how to its members' composting. The parent enterprise can add value to the partially processed compost received from the CBOs by processing it further (i.e. fortification with nutrient minerals, pelletization), packaging, branding, marketing and distributing the final product. The outsourcing of specific activities to the CBOs by the parent enterprise ensures that an efficient allocation and use of resources. The parent enterprise generates revenue via membership fees and the sale of compost.

The unique features of this business model are: a) no recurrent governmental subsidies are required; b) assured monetary benefits accruing to all economic actors create incentives that underpin success; c) members of the cooperative circumvent the need for high capital investments for purchasing advanced equipment by producing a partially processed compost; d) by outsourcing waste collection, separation and partially composting the parent company reduces its operational costs and need for space, whilst on the other hand, CBOs have an assured market (parent enterprise) for their product; e) product quality and price dependency ensures a high quality product.

C. Business model

The basic value proposition of the model depends on the enterprise initiating the business model. Since this model can be initiated by a cooperative, unique value propositions that underpin this model are the ideals of cooperative movement – providing services for the members, by the members and to the members at cost and sharing the benefits. In that regard, the constituting value propositions are: a) provision of sustainable waste management services to communities; and b) increasing access to an affordable organic fertilizer to agricultural producers. The business model described here is from the perspective of a standalone private enterprise, operating as a cooperative (parent enterprise organizing the CBOs into a cooperative). Cooperative membership is open to all, provided that each member shares in the cooperative's vision and pays their annual membership and subscription fee. CBOs which form the core of the cooperative are contracted out for waste collection, separation and production of a partially processed compost, which is sold to the parent enterprise at a quality-determined price. A key partnership with a research institute is essential in developing a final compost product that is competitive on the fertilizer market. Third party product certification can help garner significant market demand and mitigate market competition effects from the often subsidized chemical fertilizer. The partially processed compost is further processed, packaged, branded and sold to farmers, fertilizer companies, dealers and distributors. The cooperative generates revenue from membership fees and compost sales.

The following elements aggregately ensure the success of this model: a) assured benefits to CBOs ensures commitment, output delivery and success of the cooperative; b) decentralized activities reduce transportation and land/ space costs; c) community involvement reduces waste segregation costs as they have a buy-in and awareness programs are more effective; d) CBOs are able to generate their own capital investment (which is modest given the decentralized nature and scale of operations); e) quality-determined pricing ensures a high quality product and invariably a greater market demand. This is a model that is not only financially self-sustaining (no recurrent governmental support) but also profitable, accruing significant benefits to society. This model can be extended to under-serviced areas such as new settlements and slums, under the scenario where community involvement can be encouraged and depending on scale of operations, land/ space provided by the municipalities to the CBOs. See Figure 139 below for the diagrammatic overview of the business model.

D. Potential risks and mitigation

The business model presented here was designed and optimized based on the analysis of the NAWACOM case (see previous section). In designing this optimized business model, risks related to safety, local acceptance by the community, and business attractiveness for investors were assessed.

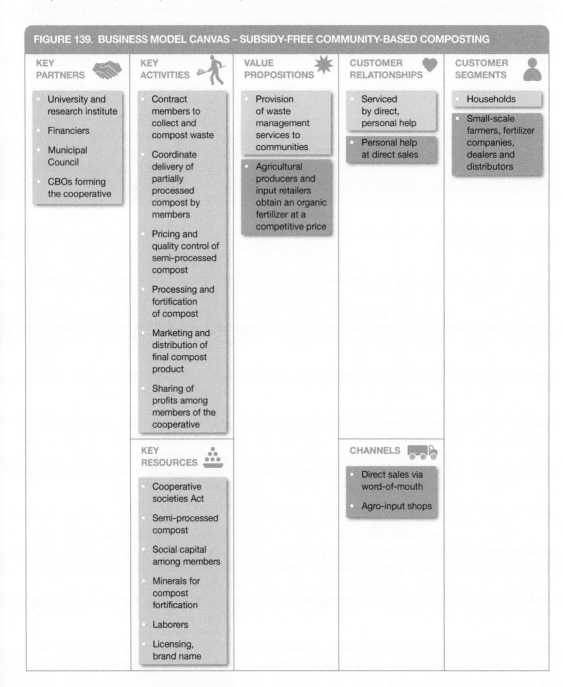

FIGURE 139. BUSINESS MODEL CANVAS – SUBSIDY-FREE COMMUNITY-BASED COMPOSTING

BUSINESS MODEL 11: SUBSIDY-FREE COMMUNITY-BASED COMPOSTING

COST STRUCTURE
- Fee based quality analysis
- Capital investment (land, machinery, licensing)
- Operation and maintenance (electricity, land rent, maintenance costs)
- Quality control fees payment to university and research institute
- Administrative costs (collection of member fees and selling the compost)

REVENUE STREAMS
- Sales of compost
- Membership subscription fees

SOCIAL & ENVIRONMENTAL COSTS
- Dispersed and decentralized processing poses greater health risk to members and neighbours
- Manual execution of activities such as sieving and packaging may be a source of occupational health risk

SOCIAL & ENVIRONMENTAL BENEFITS
- Model to collect waste free of charge
- Reduces existing waste management costs
- Reduces human exposure to untreated waste
- Creates jobs
- Income for low-income population
- Increased access to fertilizer alternatives for farmers
- Government savings from reduced expenditure on waste management
- Community empowerment by means of cooperative principles and sharing of profits among members

General risks: Lack of community awareness and interest. There is a need for a reliable leader among the community, which is a pre-requisite to prevent falling into the trap of a 'failed cooperative'. The management structure can be fairly complex and this can affect the sustainability of the enterprise.

Market risks: The model has a very low input supply risk as supply is assured from its members. On the other hand, there are potential risks in the output market and this can arise from policy instruments such as chemical fertilizer subsidies. Additionally, the scale of operations (if small) can imply that the cooperative cannot target large-scale agro-producers who often have large orders. Product certification and branding is imperative to permit greater market penetration.

Competition risks: Key market competition (fertilizer market) as noted above arises due to policy instruments that make substitute products more affordable to farmers than compost.

Technology performance risks: The composting technology typically used (windrow composting) is a relatively mature and simple technology. It can be more labor-intensive and less mechanized which implies that factors such as equipment breakdown, maintenance and repair costs will have a limiting effect on technology perforamance. Members' quest to reduce waste segregation costs and improve the quality of the partially processed compost can result in them being selective of the types of waste they collect, and thus reducing the waste collection coverage in the communities (and increased burning of waste).

Political and regulatory risks: Cooperative models, particularly in developing countries, have shown a mixed record of success even in cases where community involvement and support have been strong.

This has been mainly attributed to poor management. Moral hazard issues often arise, for instance, due to the misuse of funds (sometimes attributable to lack of financial management skills and due diligence) by the executives and influential members. Effects of these issues can however be mitigated via the establishment of an oversight committee (with cooperative members required as signatories in addition to the executives), regular audits, disclosure of financial performance to all the members. Policies and regulations related to waste-based compost sectors differ by country. The oftentimes stronger political support for chemical fertilizer use (slow phasing-out of fertilizer subsidies) and lack of specific government guidelines for the certification of compost and internationally accredited third-party certification entities can represent a significant risk to the sustainability of the business model.

Social equity related risks: There are no distinctive social inequity risks associated with this model. In contrast, the model generates opportunities for increased benefits to women as they are culturally noted to collectively engage in small-scale waste segregation and recycling initiatives. The model supports employment opportunites and additional revenue, suited particularly for the women.

Safety, environmental and health risks: Whilst the simplicity and labor-intensiveness of the technology implies low-level skills and greater job opportunities for the informal workers and people who would otherwise be unemployed, there is a higher risk of worker exposure to waste and related pathogens if the approapriate gear is not used. Additionally, given that the pre-composting process is dispersed and occurs in multiple locations, there may be a larger number of people exposed to waste-related pathogens, depending on their level of training on safety measures and use of safety gear. Similarly, manual execution of activities such as sieving and packaging could be a source of occupational health risk. Trainings on occupational health risk mitigation is imperative for all members of the cooperative, particularly the CBOs. To address the safety and health risks to workers, standard protection measures are also required as elaborated below in Table 36.

TABLE 36. POTENTIAL HEALTH AND ENVIRONMENTAL RISK AND SUGGESTED MITIGATION MEASURES FOR BUSINESS MODEL 11

RISK GROUP	EXPOSURE					REMARKS
	DIRECT CONTACT	AIR/DUST	INSECTS	WATER/ SOIL	FOOD	
Worker	HIGH					Risk of sharp objects in MSW and fecal contamination. Potential risk of dust, noise and chemical compost contaminants
Farmer/user				LOW		
Community			MEDIUM	MEDIUM		
Consumer					LOW	
Mitigation measures	🧤	😷 🎧	⚠️ 🧴	⛑️	🍴 Pb Hg Cd	

Key: ☐ NOT APPLICABLE ☐ LOW RISK ☐ MEDIUM RISK ☐ HIGH RISK

E. Business performance

This model is ranked highest on profitability due to the cooperative and cost-saving nature of the decentralized operations that produce a partially-processed compost product (Figure 140). The supplementary value-addition to the product via fortification and branding can represent an incremental price mark-up of the final compost product. The model also ranks high on scalability and replicability. This is because of the simplicity of the technology (low-level skill requirements), low capital costs requirements, relatively lower operational and maintenance costs and profits generated makes it attractive for communities with a cooperative vision to adopt and implement. Social impact and environmental impact rank next, whilst innovation is ranked the lowest which is attributable to the simplicity of the technologies and the word-of-mouth marketing strategy used.

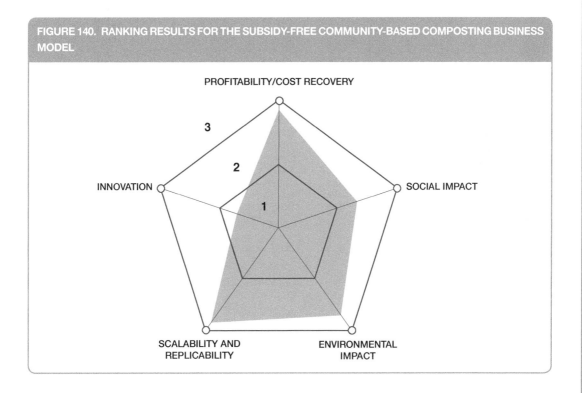

FIGURE 140. RANKING RESULTS FOR THE SUBSIDY-FREE COMMUNITY-BASED COMPOSTING BUSINESS MODEL

9. BUSINESS MODELS ON LARGE-SCALE COMPOSTING FOR REVENUE GENERATION

Introduction

Nutrient recovery from waste initiatives primarily aim to address the waste management challenge, and oftentimes geared towards only partial cost recovery, rarely full-cost recovery or profit maximization. Continuous dependence on external financial support from government grants, subsidies, tax credits and rebates is unsustainable, particularly in view of the ever-diminishing public budget allocations to waste management. Looking beyond cost recovery and aiming for profit-making models is imperative if sustainable financial and economic returns on investments are expected.

Multiple revenue generation streams (i.e. portfolio diversification) represents additional avenues for businesses to become financially viable. This business approach offers a way for businesses to mitigate risk associated with limited/seasonal market demand of certain services and products. A clear example is that of compost with highest demand around the planting season. Seasonal demand implies increased storage costs for compost plants with all year-round production. Additionally, oftentimes given the strong competition in the fertilizer market, compost demand may be low and not generate enough funds sufficient to cover the plant's operational and maintenance costs. In this instance, it will be important for the business to tap into other revenue streams with more stable returns such as sale of recyclables and energy (electricity). Under this model, the multiple-revenue stream approach translates into several value propositions that generate even greater benefits to actors in the sanitation and agricultural sectors. We consider the following value propositions: a) improved waste management services to communities and businesses; b) provision of an environmentally-friendly organic fertilizer at competitive market prices to agricultural producers; c) increased access to input resources for recycling companies; d) increased energy availability to communities and businesses; e) provision of tradable certified emission reduction to meet carbon emission commitments.

The ability for businesses to successfully implement the above value propositions and capture the greatest economic benefits will partly depend on **scale** and **strategic partnerships**. While the composting concept is applicable across scale, large-scale composting offers greater opportunities for capturing economies of scale benefits, revenue generation and market proliferation. Large-scale composting can generate significant environmental and socio-economic benefits as it offers an opportunity for municipalities to manage massive quantities of solid waste generated and collected in the cities. The scale element of the model presents an option to significantly reduce waste quantities transported to landfills (final disposal sites), thus reducing waste management costs. Large-scale operations can also offer access into markets that smaller-scale facilities are often excluded from. In considering the energy sector, for example, waste reuse facilities have to operate at a certain scale (large-scale) to meet the minimum wattage requirements for sale to the grid. This is also applicable to the sale of carbon credits to UNFCCC Annex I countries[1]. Studies show that 98% of all registered Clean Development Mechanism (CDM) composting projects fall in the category of medium- to large-scale composting plants (Fenhann, 2012). The need for strategic partnerships extends beyond those with NGOs for development of CDM projects, compost marketers and dealers to increase market share to include municipal authorities for exclusive rights/access to waste streams, research institutes for product and technology innovation, and informal workers for increased access to slums and waste segregation efficiency.

While a great potential exists for business viability (profitability) and significant accrual of economic benefits to other actors in the agricultural and sanitation value chains, the implementation of the noted value propositions does not come without challenges and risks. Price volatility in carbon credit market, strong buyer power (monopoly) in the electricity market and price distortions in the fertilizer market from policy instruments (subsidies) are among a few of the key factors to be taken into consideration and whose effects need to be mitigated.

This chapter describes the generic **large-scale composting for revenue generation** model and five supporting case examples. The presented examples are not exhaustive and some better cases could have been inadvertently omitted due to information and time constraints but cover a wide range of easily accessible cases at scales ranging from medium to large scale operations in selected settings in Bangladesh and India. It is interesting to note that whilst large-scale composting is a growing concept in Africa, particularly Sub-Saharan Africa – this model is more established in the Asian context.

References and further readings

Fenhann, J. 2012. CDM pipeline overview. UNEP DTU Partnership: www.cdmpipeline.org/ (accessed 19 August, 2016).

Note

1 Industrialized or transitional economies as listed in Annex I of the United Nations Framework Convention on Climate Change (UNFCCC). http://unfccc.int/parties_and_observers/parties/annex_i/items/2774.php (accessed November 8, 2017).

CASE

Inclusive, public-private partnership-based municipal solid waste composting for profit (A2Z Infrastructure Limited, India)

Miriam Otoo, Joginder Singh, Lesley Hope and Priyanie Amerasinghe

Supporting case for Business Model 12	
Location:	Ludhiana, India
Waste input type:	Municipal solid waste (MSW), High density inorganic material
Value offer:	Provision of waste management services, high quality compost and renewable energy
Organization type:	Private (with several public-private partnership projects)
Status of organization:	Operational since 2011
Scale of businesses:	900 tons of municipal solid waste / day
Major partners:	Ludhiana Municipal Corporation, Indian Potash Limited, Indian Farmers Fertilizer Corporation Limited, Krishak Bharti Cooperative Limited

Executive summary

A2Z Infrastructure Private Limited (A2Z-PL), established in 2011, is a subsidiary business of the A2Z Group – one of India's leading waste management companies. With a core mandate to provide sustainable waste management solutions to municipalities across India, A2Z-PL operated at the time of the assessment 21 integrated resource recovery facilities (IRRF) across India, processing in total 8,000 tons of municipal solid waste (MSW) per day. One of such projects, which has shown significant success is the 900-ton IRRF in Ludhiana, Punjab. With a partnership agreement with the Ludhiana Municipal Corporation (LMC), A2Z-PL is contracted to collect, transport, process and dispose the MSW in five jurisdictional zones in Ludhiana. Their activities have so far had an immense impact in addressing the health and environmental problems associated with the open dumping of waste. A2Z-PL's success is based on a solid business model grounded in five principles: 1) self-sustainability via a multi-revenue stream approach; 2) using an integrated and inclusive approach via synergies in business operational activities and a public-private partnership (PPP); 3) zero tolerance for compromise of product quality; 4) maximum resource derivation; and 5) strict compliance to regulations. The Ludhiana business generated an annual net profit of 25–30 million Indian Rupees[1] (Rs.) in 2012. This mainly came from the sale of recovered resources – compost, high density plastics and metals as the total cost of waste collection, provision of bins, transportation and processing is equivalent to the revenue made from the provision of such services at Rs.395 per ton, a cost borne by the municipality. With a business model that cuts across the entire MSW value chain, the Ludhiana business employs about 300 people of which 70% are unskilled laborers. This has improved the livelihoods of landfill ragpickers by ushering them into mainstream jobs. The acitivies of the Ludhiana IRRF have substantially reduced human direct exposure to waste, reduced the municipality's waste management costs and saved several acres of landfill area.

KEY PERFORMANCE INDICATORS (AS OF 2015)						
Land use:	20 ha					
Capital investment:	USD 1,114,620					
Labor:	300 (210 unskilled, 90 skilled)					
O&M cost:	USD 5,249/day					
Output:	150 tons of compost / day					
Potential social and/or enviornmental impact:	Creation of 300 jobs, reduction of GHG emissions, waste management cost savings, improved environmental health.					
Financial viability indicators:	Payback period:	3–3.5 years	Post-tax IRR:	N.A.	Gross margin:	N.A.

Context and background

Ludhiana is a centrally located city of Punjab situated between Delhi and Amritsar. It is the industrial hub of Punjab State and the district is agriculturally advanced as the granary of India. It is the most densely populated city of Punjab with a total population of about two million. About 20% of its population is comprised of migrant laborers from Bihar, Uttar Pradesh, Rajasthan and other states, and even from Nepal. As the industrial hub of Punjab State, Ludhiana has experienced a rapid and unplanned expansion of the city, creating an increase in waste generation disproportionate with its management. Amid increasing public criticism of limited and ineffective collection systems and poor disposal practices especially in slum areas, Ludhiana Municipal Corporation entered into a 25-year PPP contract with A2Z to collect and process waste generated from five zones in Ludhiana. A2Z has taken advantage of the deficiencies in the municipality's waste management approach, increasing demand for energy and chemical fertilizer prices, and established a sound and financially sustainable waste management and reuse business. The recovery of resources from the collected waste represents opportunities for A2Z to solidify its business approach. The city's acute power shortage has created a great demand for RDF generated power, suggesting a sustained revenue stream for A2Z. Additionally, considering that Ludhiana is agriculturally advanced, the need for affordable and environmentally sustainable agricultural input options is imperative. The availability of MSW-based compost in the market offers agricultural producers an environmentally safe and cheaper fertilizer alternative. A2Z-PL believes that its activities will help address the health and environmental problems associated with poor waste management and the nexus of energy and fertilizer deficiency in India.

Market environment

Ludhiana, as most cities in India, is facing an alarming energy shortage due to increasing urbanization and industrialization. With dwindling natural energy resources in India, the demand for renewable energy sources such as bio-energy is growing, which has resulted in a demand surge for related inputs such as RDF. Although A2Z Group has established profitable businesses in many cities in India (for example, Varanasi, Meerut, Jaunpur, Moradabad, Badaun, Fatehpur, Basti, Loni, Mirzapur and Ranchi), it is relatively new in the organic fertilizer market in Ludhiana and currently penetrates a very small share of the market. The market for compost is in its nascent stage while that for substitute goods such as chemical fertilizer has a well-established market and currently controls the largest share of the fertilizer market. Key drivers incentivizing farmers to use chemical fertilizer over more environmentally sustainable alternatives such as organic fertilizers–compost have been related to subsidy provision, and the high nutrient content and low application rate of the product. Although compost provides the dual advantage of price competitiveness and improved crop yields, these benefits typically occur on a long-term basis. For subsistence and smallholder farmers, additional incentives need to be put in place to encourage the use of compost. The Indian government has proposed phasing out the subsidy program to incentivize farmers to use chemical fertilizers more efficiently, lower related costs to the government and increase the adoption of environmentally sustainable alternatives – organic fertilizer.

Macro-economic environment

Chemical and synthetic fertilizers, particularly Nitrogen, Phosphorus and Potassium (NPK), are highly subsidized in India. The amount of subsidies on chemical fertilizer has grown in the last couple of decades from Rs.60 crore[2] during 1976–1977 to Rs. 349,980 crores in 2009–2010. Significant subsidy allocation has not only resulted in inefficient use by farmers and high costs to the government, but also significant soil degradation (NCOF, 2017). With a growing need to increase the availability and quality of bio-fertilizers and composts in the country for agricultural productivity improvement while still maintaining soil health and environmental safety, India set up a scheme to augment the infrastructure for the production of quality organic and biological inputs. As a result, the National Project on Organic Farming was birthed in 2004. This programme introduced the capital investment subsidy scheme for commercial production units for organic and biological agricultural inputs. Implemented by the Department of Agriculture and Cooperation through the National Centre of Organic Farming (NCOF), the scheme provides credit linked and back-ended capital investment subsidy equivalent to 33% of total financial outlay subject to the maximum of Rs. 60 lakh per unit, and 25% of total financial outlay subject to a maximum of Rs. 40 lakh per unit, whichever is less for bio-fertilizer/bio-pesticides production units (Ministry of Agriculture, 2011; NCOF, 2017). Policies to reduce the budget allocation for chemical fertilizers and provide capital investments for new and existing compost businesses such as these are important instruments that catalyze the business development in the RRR sector and the scaling-up of initiatives similar to that of A2Z.

Business model

Figure 141 represents A2Z-PL Ludhiana's business model canvas. A2Z Ludhiana's business model is centred around the provision of several value propositions with its success grounded in five principles: 1) using an integrated and inclusive approach via synergies in business operational activities and a public-private partnership (PPP); 2) self-sustainability via a multi-revenue stream approach; 3) zero tolerance for compromise of product quality; 4) maximum resource derivation; and 5) strict compliance to regulations. A2Z-PL Ludhiana has a 25-year partnership agreement with Ludhiana Municipal Corporation (LMC) to collect and process all solid waste generated within the municipality. This partnership gives A2Z sole ownership, i.e. continuous and unrestricted access to waste in five municipalities and provides land free of charge for all operations. With business operations cutting across the entire MSW value chain and increasing land prices, this PPP agreement allows: 1) A2Z to diversify its portfolio, mitigating risk associated with fluctuations in compost demand; and 2) alleviates it of high initial investment costs (optimizing its allocation of resources and activities), whilst the municipality gains from effective waste collection and processing systems. Strategic partnerships with chemical fertilizer companies such as Indian Potash Limited, Indian Farmers Fertilizer Corporation Limited and Krishak Bharti Cooperative Limited allows A2Z to use their established countrywide marketing and distribution system, providing A2Z with an assured and large market base for their compost product. This has proven to be a valuable business approach given that A2Z is a fairly new entrant in the fertilizer market. A2Z-Ludhiana is however gradually increasing its market share via the branding of its compost by ensuring to maintain a product quality surpassing the recommendations of the Fertilizer Control Order (FCO) board and selling at competitive market prices. Based on fertilizer application recommendations, A2Z's compost sold at USD 0.05/kilogram is comparatively cheaper than chemical fertilizer at a cost of USD 1. Another element to A2Z's pricing strategy is that it segments its compost market by selling to bulk buyers at USD 0.025/kg which is half of the price paid by retailers. Recovered non-degradable materials (high-density plastics and metals) are sold directly to plastic companies and industrial units. Additional revenue is earned from waste collection fees paid directly

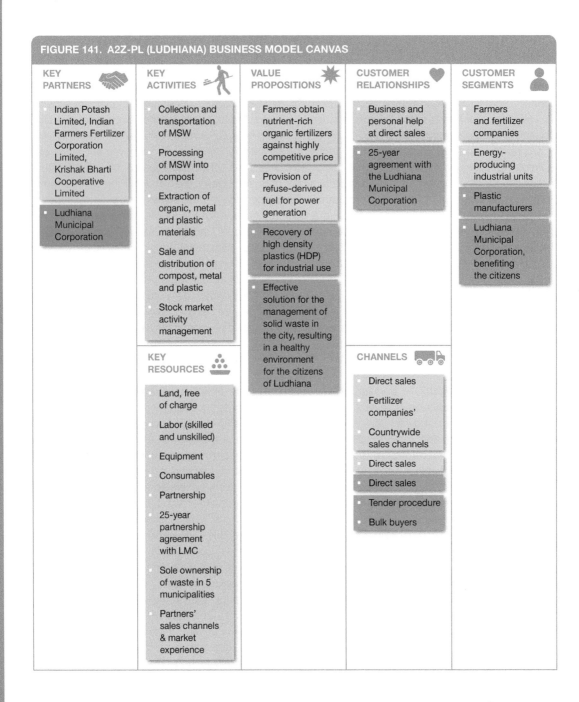

FIGURE 141. A2Z-PL (LUDHIANA) BUSINESS MODEL CANVAS

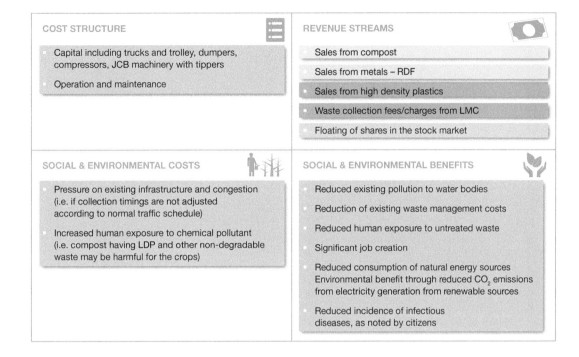

by LMC (recovered from household) at a rate of USD 7 per ton of waste collected. Also, essential to the model is the company's shares it floats in the stock market to generate additional revenue.

Value chain and position

A2Z's business operations cut across the entire MSW value chain – from collection and transportation of waste to processing and disposal. The value chain involves three key actors namely: waste suppliers – LMC and informal waste collectors; compost clients – fertilizer companies and farmers; inorganic material clients – plastic manufacturers and energy-producing industry units (Figure 142). A2Z is the focal point in the value chain. The raw material used by A2Z for compost production is municipal solid waste sourced directly from households and markets via informal waste collectors under permission from LMC. There is no competition from other entities in terms of input supply given the contractual agreement between A2Z and LMC, which ensures continuous and unlimited access to the waste from five zones in Ludhiana. A2Z contracts out some its waste collection activities to informal waste collectors. This has not only improved the livelihoods of landfill ragpickers by ushering them into mainstream jobs but has allowed A2Z to efficiently cover slum areas where poor road infrastructure make them less accessible. Compost produced by A2Z is sold mainly to chemical fertilizer companies who process the compost further or sell as is through their established distribution systems. With A2Z been fairly new in the organic fertilizer market and depending on others to access markets, they are also facing high price risk as the chemical fertilizer companies have a high buyer power. There is an increasing number of competitors – organic fertilizer businesses entering the market. Product branding strategies and field demonstrations to validate the product quality is been adopted by A2Z to gradually increase its market access and share. On the other hand, the demand for inorganic materials (i.e. RDF, high density plastics) is high and growing, although A2Z is not yet in a position where it can dictate the selling price.

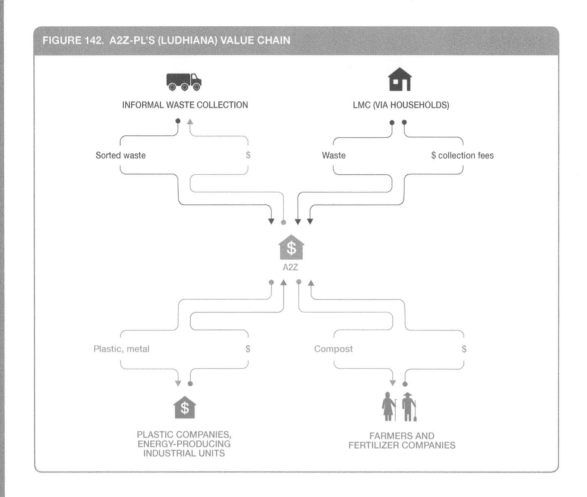

FIGURE 142. A2Z-PL'S (LUDHIANA) VALUE CHAIN

Institutional environment

The institutionalization of the Municipal Solid Waste (Management & Handling) Rules 2000 has resulted in the provision of bins for households by LMC which has facilitated the collection and reuse of MSW in Ludhiana and the resulting business activities of A2Z. In terms of production, there is currently a statutory guideline – the Fertilizer Control Order (FCO) instituted by the Ministry of Agriculture and Rural Development for the production and distribution of all fertilizers including organic fertilizer. Product quality recommendations are provided for different organic fertilizer types for which producers have to adhere to. This is particularly beneficial to farmers as they get what they are paying for, but also for compost businesses as they are able to build their product brand. Although yet to be fully implemented, the phasing out of the subsidy program for chemical fertilizers by the Indian government represents an opportunity for compost producers to gain an easier entry into the fertilizer market.

Technology and processes

Open-windrow composting system is the technology adopted by A2Z for processing MSW into compost (Figure 143). The technology has a high rate of recovery for the bulking material and thus suitable to composting large volumes of waste. Although this technology is not space efficient, it has low capital investment requirements as it is manufactured locally and has the capacity to handle large volumes of waste at a time. The first process includes collection and sorting of the waste. Sorting out waste into biodegradable and non-biodegradable portions is mainly a mechanized process although some level of segregation is manually done by the informal waste collectors serving mainly the slum areas. Waste of particle size greater than 50mm are separated, shredded, packaged and sold partly to electricity-generating units and cement and tile manufacturers. A precentage of the RDF material is sold and the remaining is burnt to generate electricity at one of A2Z's plants at Nakodar, where 15MW electricity is generated.

The organic component of separated waste (partical size <50mm) undergoes the composting process. The waste is piled into windrows. The additional aeration from the bottom of the pile allows microorganisms to decompose the organic waste efficiently through better oxygen supply and improved temperature control. Within 24 hours the micro-organisms within the waste start to multiply and generate heat. Pile temperature increases to 55–65°C, which is optimal for aerobic composting. To enable the micro-organisms to obtain sufficient oxygen, the pile is additionally aerated by turning the waste from time to time (approximately once a week depending on the temperature reached). High temperature leads to water losses through evaporation, so additional water must usually be added with each turning. After 40 days of composting the temperature has decreased, indicating a slowing

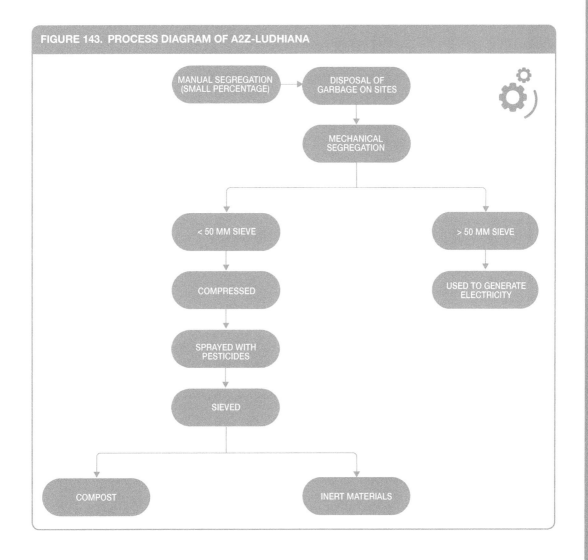

FIGURE 143. PROCESS DIAGRAM OF A2Z-LUDHIANA

down of the process. As less oxygen is demanded, the raw compost enters the maturation phase. For another 15 days, mesophilic micro-organisms further stabilize the compost leading to the final mature compost product. The final stage involves screening the piles for undecomposed materials and unwanted products. The compost product is then bagged into different weights for sale.

Funding and financial outlook

The investment cost at the start of the business is estimated at USD 1,114,620. Land for plant operations is provided for free and on a long-term lease basis from the Ludhiana Municipal Corporation. Operation and maintenance costs comprising of wages, salaries, fuel and other consumables are estimated at USD 5,248/day. A2Z receives financial support in the form of a 10% subsidy to cover operational and investment costs offered by the Jawaharlal Nehru National Urban Renewal Mission (a city-modernization scheme launched by the Government of India under Ministry of Urban Development). A2Z generates revenue from the sales of compost, non-degradable materials (plastic and metals) and waste collection fees. Collection fees of USD 7.4/ton of collected waste paid by LMC is sufficient to exactly cover the costs of waste collection and transportation and thus surplus revenue (i.e. profit) comes from compost and inorganic materials sales. On a yearly basis A2Z-Ludhiana makes a net profit ranging between USD 465,290 and 558,348, indicating a 3–3.5-year payback period.

Socio-economic, health and environmental impact

The simple idea of converting the high organic content of the waste into compost has brought about a valuable substitute for chemical fertilizers. Overuse of chemical fertilizers has been a serious problem in India, which has led to severe soil degradation and a costly venture for the government. Farmers now have real alternatives to chemical fertilizers and have the potential to increase their per hectare yield and soil health, which will improve agricultural productivity in the long term. A2Z's activities have so far had an immense impact in addressing the health and environmental problems associated with the unhygienic collection, open transportation and dumping of waste. Ludhiana citizens have noted that the waste management activities of A2Z has significantly reduced the risk of spreading of diseases (such as malaria, diarrhoea and cholera) through the proper collection and disposal of municipal solid waste. Additionally, improved collection systems have reduced water pollution and there is limited to no indiscriminate disposal of waste into nearby flowing *Budha Nala* (water bodies) and sewer pipes. A2Z's business activities has created 300 jobs (both skilled and unskilled) and counting along the entire MSW value chain – from informal waste collectors to plant workers, reducing the level of unemployment in Ludhiana.

Scalability and replicability considerations

The key drivers for the success of this business are:
- Increasing fertilizer prices and industrial demand for power supply, which suggest a foreseeable increase in the demand for the recovered resources – RDF, compost and high-density plastic.
- Strong industrial development and agriculturally advanced status in the area go hand-in-hand, requiring a solution that works both ways.
- Strong commitment of state government in providing an enabling environment for the implementation of the public-private-partnership.
- Positive reporting of A2Z's activities and potential benefits by media.
- Widespread public acceptance of A2Z activities has facilitated their waste collection activities.
- Policy initiatives to phase-out chemical fertilizer subsidies and capital investment subsidies to new and existing compost businesses.

A2Z's model has a high replication potential in cities of developing countries with the support from external support agencies as well as local entrepreneurs. Adopting a labor-intensive, cheap and low technological approach, the business does not require a large capital investment (except for land purchase) or state of the machinery, which removes one of the major constraints for business start-ups especially in the developing world context. But if scaling up can be achieved, then an advanced technology will have to be adopted. Public support is needed to dismantle the existing system of paving way for systematic disposal for which public awareness is needed. Additionally, field demonstrations to validate compost product quality are necessary to increase a business's entry into the fertilizer market as oftentimes compost sales constitute a fair share of the revenue generated and thus key factor for business sustainability.

Summary assessment – SWOT analysis

Figure 144 presents the SWOT analysis for A2Z-Ludhiana. Composting is a promising business in India, although a nascent market in Ludhiana. A2Z has been particularly successful by implementing innovative business partnerships with different actors across the entire value chain. Self-sustainability has been driven by a multi-revenue stream approach and gradually gaining market share via product branding. The use of a simple technology has been key – taking advantage of cheap labor; however with increasing wages, A2Z will have to consider other alternatives with future expansion plans. Increasing

FIGURE 144. SWOT ANALYSIS FOR A2Z-LUDHIANA

	HELPFUL TO ACHIEVING THE OBJECTIVES	**HARMFUL** TO ACHIEVING THE OBJECTIVES
INTERNAL ORIGIN ATTRIBUTES OF THE ENTERPRISE	**STRENGTHS** • Already in business and thus has experience, expertise and resources at their command • Technology has limited investment and energy requirements • Requires little technical skills or expertise to operate • Continuous and unrestricted access to waste	**WEAKNESSES** • Opposition from private sweepers and farmers close to dumping site • Technology is labor-intensive (costly amid rising wages) • High initial capital investment cost • High cost of maintenance and repairs • Dependency on door-to-door garbage collection by rag pickers • Up-scaling requires adapted technology and more skilled labor
EXTERNAL ORIGIN ATTRIBUTES OF THE ENVIRONMENT	**OPPORTUNITIES** • Government scheme set up for promotion of production of biological/organic fertilization products (set up a scheme to augment the infrastructure for production of quality organic and biological inputs) • Mechanization of activities to increase production and economies of scale • Up-scaling potential for a CDM project to earn carbon credits • Replicate activities in other cities given market entry opportunities with capital investment subsidies and phasing-out of chemical fertilizer subsidies • Possible phasing-out of subsidy on chemical fertilizers	**THREATS** • Increasing and high cost of labor • Well-established and subsidized chemical fertilizer market

governmental support along with growing demand for normal and enriched compost, spurred by the user awareness building programmes, will represent key opportunities for replication and up-scaling of the business. A2Z is an example of an innovative PPP utilizing a simple business approach to address some of the major waste management and environmental challenges in Ludhiana, India.

Contributors

Jasper Buijs, Sustainnovate, Netherlands; formerly IWMI, Sri Lanka
Alexandra Evans, Independent Consultant, London, United Kingdom
Michael Kropac, CEWAS, Switzerland
Josiane Nikiema, IWMI, Ghana

References and further readings

A2Z Group. Web: http://a2zgroup.co.in/index.html (accessed November 8, 2017).

Department of Local Government, Punjab Government. 2009. Request for qualification (RFQ) document for setting up an integrated municipal solid waste management project report.

Luthra, A. and Sareen, R. 2004. Municipal solid waste management – Current concerns in Ludhiana City. ITPI Journal: 62–72.

Ministry of Agriculture. 2011. Operational guidelines for capital investment subsidy scheme for vegetable and fruit market waste compost, and biofertilizers – biopesticides production units. http://ncof.dacnet.nic.in/Operational_Guidelines/Guidelines_for_Capital_Investment_Subsidy.pdf (accessed November 8, 2017).

Personal communication with: Sh. BPS Chauhan (Vice President), Sh. Parmod RM (Human Resources), Sh. Ravinder (Supervisor).

Case descriptions are based on primary and secondary data provided by case operators, insiders or other stakeholders, and reflect our best knowledge at the time of the assessments 2015/16. As business operations are dynamic data can be subject to change.

Notes

1　USD 1 = about INR (or Rs.) 65.62 in 2015.
2　Crore are 100 lakh, and lakh is a unit for 100,000. Rs. 60 lakh were in 2004–2012 about USD 120,000, and about USD 90,000 in May 2017.

CASE
Municipal solid waste composting with carbon credits for profit (IL&FS, Okhla, India)

Solomie Gebrezgabher, Sampath N. Kumar, Pushkar S. Vishwanath and Miriam Otoo

Supporting case for Business Model 12	
Location:	Okhla, India
Waste input type:	Municipal solid waste (MSW)
Value offer:	Provision of an affordable, organic compost and generation of carbon credits
Organization type:	Public-private partnership (PPP)
Status of organization:	Operational since 2008 (registered as Clean Development Mechanism (CDM) project since 2009)
Scale of businesses:	Processes 200 tons of MSW per day (73,000 ton/year)
Major partners:	Municipal Corporation of Delhi (MCD)

Executive summary

The Infrastructure Leasing and Financial Service Okhla composting plant (IL&FS Okhla) started composting operations in 1981 with the aim of avoiding methane (CH_4) emissions generated in the landfill site through the controlled aerobic decomposition of MSW in a windrow composting process. However, the plant was shut down in 2000, as the business was not viable due to insufficient revenues from the sale of the compost. In 2007 IL&FS Ltd. signed a Concession Agreement and a public-private partnership (PPP) with the Municipal Corporation of Delhi (MCD) to rehabilitate the Okhla compost plant on a build, operate and own (BOO) model with carbon finance support. This project demonstrates the significant role of CDM in ensuring sustainable operation of waste reuse businesses while contributing sustainable climate protection. As reported, the plant converts approximately 73,000 tons of MSW into compost every year. The plant has two brands for its compost, the Harit Lehar and the EcoSmart Home Garden, which are both FCO (Fertilizer Control Order) compliant composts sold to farmers and to urban residents. Around 1,600 tons of CH_4 (34,000 ton CO_2eq) emissions are avoided on average per year and it is estimated that 234,231 tons CO_2eq is likely to be achieved within the seven-year renewable crediting period[1]. Moreover, the compost is used as a replacement to chemical fertilizer and thus avoids greenhouse gas (GHG) emissions from the production of chemical fertilizer. Another environmental and economic benefit is that the compost is rich in organic carbon, which increases the soil fertility and farm productivity.

KEY PERFORMANCE INDICATORS (AS OF 2014)					
Land use:	3.27 ha				
Water use:	50,000 L/day				
Capital investment:	USD 1,454,250				
Labor:	10 skilled, 15 unskilled, 14 other adminstrative full time employees				
O&M:	USD 44.5/ton				
Output:	14,600 tons/year				
Potential social and/ or enviornmental impact:	Reduce pollution of water bodies, reduce waste management costs, reduce human exposure to untreated waste, enhance soil fertility and farm productivity, reduce GHG emissions, generate employment				
Financial viability indicators:	Payback period:	6–7 years	Post-tax IRR:	14.48%	Gross margin: 40%

Context and background

IL&FS Environmental Infrastructure & Services Ltd. (IL&FS Environment) is a 100% subsidiary of India's leading non-banking financial institution Infrastructure Leasing & Financial Services Ltd. (IL&FS). Its remit is to enhance the urban environmental infrastructure of Indian cities especially in terms of MSW management including new projects as well as the upgrading, operation and maintenance of non-functional compost plants all over India. The company has extensive experience in providing MSW consulting and advisory services to municipalities, and designing and implementing similar projects within the public-private partnership (PPP) framework in various parts of the country. It operates nearly 16 urban MSW processing facilities across the country, including the Okhla composting facility. The Okhla compost plant was constructed in 1981 and closed in 2000, as the operation was not cost effective due to insufficient revenues from the sale of compost. In May 2007, IL&FS signed a Concession Agreement with the Municipal Corporation of Delhi (MCD) to rehabilitate the Okhla compost plant with carbon finance support. IL&FS is responsible for financing, rebuilding, operating and maintaining the compost plant. The concession also provides exclusive rights and authority to retain, control, own, possess, collect and appropriate all possible revenue that can be generated from or in relation to the Project. The term of the concession is for 25 years from the date of agreement.

Market environment

The rapidly growing urbanization in Indian cities and the resulting increased need for good waste management practices has made MSW a top priority of most urban local bodies. Like the majority of landfills in India, the Okhla landfill was poorly managed and no precautions were taken to avoid the emission of methane. These have created a serious environmental and public health problem. Appropriate waste management is gaining priority with the government. This is evidenced by the fact that the MCD has signed a Concession Agreement with IL&FS to rehabilitate the Okhla compost plant. The Government of India is also supporting balanced nutrient management for agricultural soil in order to ensure that the productivity of agricultural land does not keep declining due to overuse of chemical fertilizers. The compost produced by IL&FS Ltd. is rich in organic carbon and increases soil fertility. The plant has two brands for its compost, the Harit Lehar and the EcoSmart Home Garden, which are both FCO (Fertilizer Control Order) compliant composts sold to farmers and to urban residents. Since the price of the compost is subsidized using revenue from carbon credit, marketing of compost is easier thus ensuring the sustainability of the project. The demand for the compost exceeds production but is highly seasonal. Demand is high from May to June and November to December. IL&FS sells its products through marketing alliances with fertilizer companies but is planning to be involved in direct sales of organic compost. There is competition from substitute products such as press mud, which is cheaper than the compost produced by the company.

IL&FS Okhla compost plant is also planning to produce and sell Refuse Derived Fuel (RDF), which is fuel produced from the combustible components of MSW such as plastics and other biodegradable waste. RDF is an alternative fuel to coal and IL&FS plans to sell RDF to cement industries.

Macro-economic environment

MSW management has become essential in India as there has been a significant increase in MSW generation in the last few decades due to rapid urbanization and high population growth rate. Around 90% of waste is landfilled, requiring around 1,200 hectare of land every year. With the growing population and urbanization, municipal bodies are facing financial pressures and challenges in coping with demands. The municipalities are therefore looking at alternative ways of handling waste by identifying activities that generate resources from waste. The government is encouraging reuse businesses. In addition to this, India signed the United Nations Framework Convention on Climate Change (UNFCCC) in 1992 and ratified the Kyoto Protocol on 26 August, 2002. The government has a very proactive approach to attract investors to develop CDM projects. India has been ranked first in the world in terms of approved CDM projects and it is considered as one of the countries with high potential for CDM projects. This is partly attributed to the proactive policies of the Indian government towards CDM.

Business model

IL&FS reconstructed the Okhla composting plant and signed a concession agreement with MCD to manage the plant. They obtain revenue from the sale of compost and through the CDM mechanism, by selling carbon credits to UNFCCC Annex I countries[2] (Figure 145). As per the concession agreement with MCD, 25% of the CER earning is shared with MCD for the first five years. The company has not started earning revenues from the by-products (RDF) yet, but it has a contract with cement factories to supply RDF as an alternative fuel to coal in the future. Strong partnership is required with the MCD and good relationships are needed with the customer base, farmers and urban household and organizations maintaining gardens. Sales of compost are either direct or through agreements with fertilizer companies.

Value chain and position

The compost plant receives the MSW from the urban local body, composts the waste, segregates the recyclables and sells the organic compost and recyclables to recover the costs. The MCD is a key partner as it not only supplies the raw materials but also it provided land to set up the facility (Figure 146). The compost is used in the agriculture fields. The company sells its Harit Lehar compost to farmers via fertilizer distributers and its EcoSmart Home Garden compost directly to urban residents and institutions with gardens. The company generates revenue from emission reduction credits and shares 25% of the CER revenue with MCD for the first five years.

Institutional environment

Since IL&FS Okhla composting plant is registered as a CDM project, both the UNFCCC/Kyoto protocol requirements and host country requirements apply. The Municipal Solid Waste Management and Handling Rules, 2000 directed the municipalities to supply only segregated waste to composting facilities but due to financial constraints, municipalities in India have still not implemented the rules. The organic compost is produced as per the Fertilizer Control Order (FCO) rules. MSW Rules 2000 for the overall management of the facility and the FCO rules for the compost quality are adhered to in the operation of the compost facility. The State Pollution Control Board does regular reviews of the facility and provides recommendations, which are to be followed.

FIGURE 145. IL&FS OKHLA COMPOST PLANT BUSINESS MODEL CANVAS

KEY PARTNERS
- Municipal corporation of Delhi (MCD)

KEY ACTIVITIES
- Receive MSW
- Production of compost and other recyclables (RDF)
- Marketing and sale of compost
- Managing CDM process & obtaining emission reduction credit

KEY RESOURCES
- Equipment (composting platform, segregation machinery, vehicles and other)
- Labor
- Consumables (MSW, energy, bio-culture)
- Intangibles (emission reduction)
- Subsidization resources from carbon trade income

VALUE PROPOSITIONS
- Farmers get high organic carbon content compost ('Harit Lehar' brand) against very low price due to subsidy
- Urban residents and institutions with gardens get high organic content compost (Brand EcoSmart Home Garden) against very low price due to subsidy
- Tradable certified emission reduction (CER) credits to meet carbon emission commitments
- Refuse derived fuel (RDF) at low price

CUSTOMER RELATIONSHIPS
- Farmer contact via fertilizer distributors (contract)
- Personal
- Registered as CDM at UNFCCC
- Personal (contract)

CHANNELS
- Marketing alliances
- Direct
- CDM certificate trading
- Direct

CUSTOMER SEGMENTS
- Farmers
- Urban residents and institutions
- Companies from UNFCCC Annex I countries
- Cement industries

COST STRUCTURE
- Investment cost:
  - Civil works
  - Plant cost
  - Equipment
- Annual operating cost
- Interest on borrowed funds
- MCD share payment
- Compost subsidization

REVENUE STREAMS
- Sales of compost Harit Lehar
- Sales of compost EcoSmart Home Garden
- Expected Sale of CDM benefits
- Sale of RDF (planned)

SOCIAL & ENVIRONMENTAL COSTS
- Possible human health risk while handling MSW

SOCIAL & ENVIRONMENTAL BENEFITS
- Reduced pollution of water bodies (about 1,000 litres of leachate is treated in the facility)
- Reduced existing waste management costs
- Reduced human exposure to untreated waste and chemical pollutants
- Enhanced soil fertility and productivity
- Reduced GHG emissions
- Employment generation

CASE: MUNICIPAL WASTE COMPOSTING WITH CARBON CREDITS

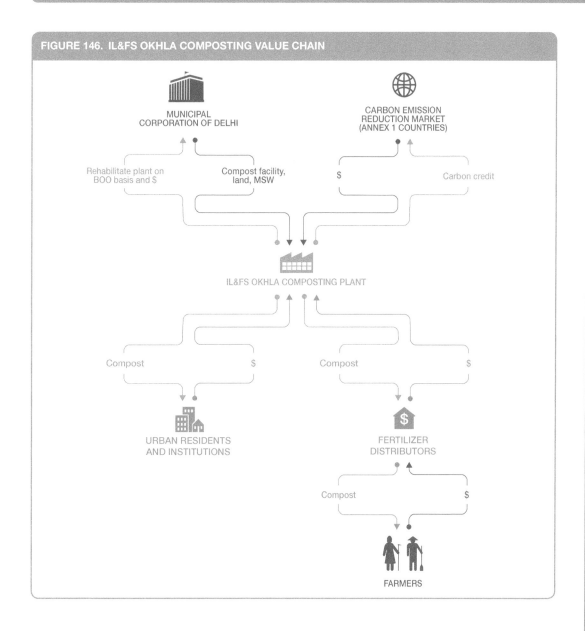

FIGURE 146. IL&FS OKHLA COMPOSTING VALUE CHAIN

Technology and processes

The technology used in the composting process is open windrow aerobic composting. Figure 147 depicts the composting process. The first step in the composting process is that the waste carried by the trucks is weighed and undergoes pre-sorting in which most of the large inorganic particles are separated out. The leachate is pumped to a separate treatment tank and the treated water is reused for the composting process. Inert materials and plastics are removed using sieving machines. The rejects are sent to landfill. The screened organic rich waste undergoes the process of composting. The duration of the composting process is about one month. During this period, the waste is sorted into windrows and undergoes turning and heaping. The compost pad is a concrete platform on which waste is allowed to undergo decomposition. The windrows are turned and shifted once a week using loaders for aeration and temperature control to enable aerobic decomposition of waste. A bio-culture is

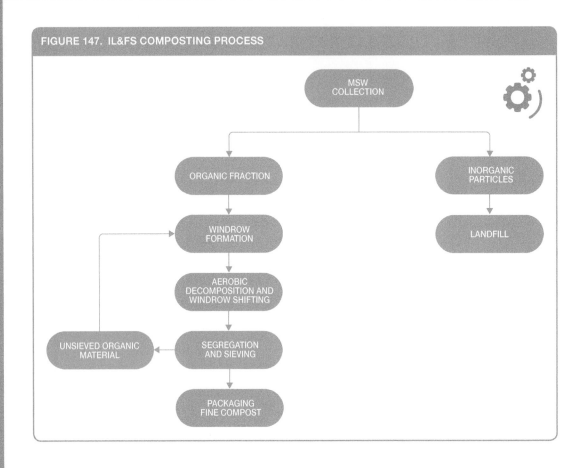

FIGURE 147. IL&FS COMPOSTING PROCESS

sprinkled on the waste heaps to aid growth of microorganisms and speed up the composting process. The temperature and oxygen of the waste heaps are measured and recorded every week. After four to five weeks, the composting heap is shifted to the "monsoon shed" for further stabilization. Next, it is sieved and the remaining inert and inorganic materials are separated out. To achieve maximum screening efficiency, one vibrating screen of 35mm and one trammel of 14mm are used. Cascading action inside the trammel ensures better screening of the waste. Screened material coming out after composting is uniform in texture and contains pure organic compost while the unsieved organic material is recycled back to the windrows for further degradation. The quantity of compost produced is about 15–26% of the quantity of MSW by weight. The NPK content of the compost is respectively 0.4%, 0.4% and 0.8%, organic matter of 50–60% and carbon content of 12%. The equipment used in the composting process is locally produced and spare parts can be easily purchased. However, the equipment needs frequent repairs. In terms of efficiency of the technology, there is a rapid composting technology which is more efficient than the one used by IL&FS but the cost is much higher.

Funding and financial outlook

The total investment cost of the project is USD 1,454,250. The civil works and plant costs account for more than 50% of the total project cost and equipment and other costs account for 42% of the project cost. Land was provided by the MCD. Financing was split between the owner's equity (24% of the total project cost) and debt (74%) at an interest rate of 14%. Table 37 gives the projected annual profits assuming that the first-year capacity utilization is 50% and the second year onwards, it is 100%. The plant has a capacity of producing 14,600 tons of compost and the selling price is 2,000 Rs./ton (USD 40/ton).

TABLE 37. FINANCIAL SUMMARY AND PROJECTED PROFITABILITY OF IL&FS COMPOST PLANT WITH CDM BENEFIT (USD)

ITEM	2008	2009	2010	2011	2012	2013	2014	2015	...
Investment cost:									
Land	0								
Civil works	425,250								
Plant cost	417,500								
Equipment	330,250								
Other cost	282,250								
Total investment	**1,454,250**								
Revenue:									
Compost sales	365,000	613,200	643,860	676,053	709,856	745,348	782,616	821,747	...
Sales of CER	49,850	121,937	203,111	278,173	347,606	411,850	471,312	0	...
Total revenue	414,850	735,137	846,971	954,226	1,057,461	1,157,198	1,253,928	825,747	...
Total expense	451,134	613,714	630,939	649,105	668,266	688,475	709,792	732,280	...
PBDIT	(36,284)	121,422	216,032	305,121	389,196	468,723	544,135	89,467	...
Interest	154,000	118,580	106,260	47,740	27,207	23,100	18,993	14,887	...
Depreciation	67,235	53,788	53,788	53,788	53,788	53,788	53,788	53,788	...
PBT	(257,519)	(50,946)	55,983	203,593	308,201	391,835	471,354	20,792	...
Income tax	0	0	4,714	17,143	25,951	70,089	158,192	10,119	...
Profit after tax	(257,519)	(50,946)	51,270	186,450	282,250	321,746	313,162	10,673	...
Projected IRR (%)	14.48								
NPV (USD)	482,398								
Payback period	6–7								

PBDIT = Profit before depreciation, interest and tax; PBT = Profit before tax

Assuming a discount rate of 10% and useful life of 25 years, with benefits from CDM, the project is viable and results in a positive net present value (NPV) and an internal rate of return (IRR) of 14.48% and payback period of six to seven years. Under the scenario where there is no revenue from CDM, the plant does not break even and results in a negative NPV and IRR of 7%.

Socio-economic, health and environmental impact

The business was set up to reduce the burden on the environment caused by untreated MSW waste. The compost plant treats biodegradable waste and on average it diverts approximately 73,000 tons of MSW per year (200 tons per day) and thus reduces the amount of waste disposed in landfill sites. The project avoids the emissions of methane that would be produced by landfill and thus contributes to

GHG emissions reduction. Around 1,600 tons of methane (34,000 ton CO_2eq) emissions are avoided on average per year and it is estimated that 234,231 tons CO_2eq is likely to be achieved within the seven years' crediting period. Moreover, the compost is used as a replacement to chemical fertilizer and thus avoids GHG emissions from the production of chemical fertilizer. About 1,000 litres of leachate is also treated in the facility which would otherwise get into the underground water. The organic compost is rich in organic carbon content and increases the soil fertility and farm productivity. The company had conducted field trials in the district of Agra, Uttar Pradesh state to check the yield gain using the organic compost, which was shown to be 25%–30% higher than the yield obtained using chemical fertilizers. In addition to its environmental benefit and contribution to better management of MSW, the project generates employment opportunities. The plant is semi-mechanized and created jobs for local people directly in the composting facility and indirectly through waste collection and transportation of compost to the end user. It also results in reduced human exposure to untreated waste and chemical pollutants.

Scalability and replicability considerations

The key drivers for the success of this business are:
- Strategic PPP model with the municipal corporation of Delhi (MCD).
- Government support and proactive policies towards CDM.
- Government encouragement of reuse businesses.
- Innovative financing scheme and sharing of benefits between municipality and IL&FS.
- Rapid urbanization combined with high population growth.
- Government support/priority to appropriate MSW management and sustainable soil (fertility) management.

The design and operation of this project, in conjunction with the avoidance of GHG emissions and production of compost as a soil amendment, will serve as an example to many other urban areas in countries that are facing similar waste management challenges. The IL&FS composting uses a holistic approach to processing waste where almost all waste types both degradable and non-degradable are used. The technology is semi-mechanized, simple and relatively inexpensive. In regards to scaling up or scaling out, IL&FS has developed and transferred similar waste management projects to other Indian regions. For example, RWE (German Power Supplier) and IL&FS are working in cooperation on two further composting projects close to Delhi and Varanasi, India. Both were registered as CDM at the UNFCCC in 2009. This project has a good potential to be replicated in other countries. Replicating this business in a locality close to landfill sites will reduce transportation cost and increase performance of the business. Receiving tipping fees for the MSW which does not exist in the case of IL&FS compost plant would also reduce production cost. However, a major limitation for setting up a composting plant of similar scale of operation and which would qualify to be considered as a CDM project, is the high capital requirement, especially in localities yet to be developed in terms of infrastructure. In order for this business to be replicated in other countries, strong partnerships with local authorities (municipalities) along with innovative financing mechanisms and good expertise in waste management practices are important.

Summary assessment – SWOT analysis

Figure 148 presents the SWOT analysis for IL&FS Okhla compost plant. Key strengths of the business are its strong partnership with the Municipal Corporation of Delhi (MCD) and its multiple revenue streams from sales of compost and CER credits. However, the carbon credit market is highly volatile, which puts the sustainability of the business under risk. This can be mitigated through additional revenues from by-products such as RDF, which can replace coal used in cement industries.

FIGURE 148. SWOT ANALYSIS FOR IL&FS OKHLA COMPOSTING PLANT

	HELPFUL TO ACHIEVING THE OBJECTIVES	**HARMFUL** TO ACHIEVING THE OBJECTIVES
INTERNAL ORIGIN ATTRIBUTES OF THE ENTERPRISE	**STRENGTHS** • Multiple revenue streams from compost sales and CDM • Availability of infrastructure • Concession agreement with MCD • Contract with major fertilizer companies • Professional management capability • Extensive experience in design and operation of composting plants • Local technology	**WEAKNESSES** • Less efficient technology • No tipping fee for MSW • Viability of business highly dependent on carbon credit sales
EXTERNAL ORIGIN ATTRIBUTES OF THE ENVIRONMENT	**OPPORTUNITIES** • Expected higher revenue from CDM • Availability of financing organizations • Revenue from by-products	**THREATS** • Competition from substitute products • High seasonality of the demand for compost may increase investment cost in storage facilities • Sales price of compost is low • The implementation of the FCO order on compost is stringent and uncertain

Contributors

Jasper Buijs, Sustainnovate, Netherlands; formerly IWMI, Sri Lanka
Alexandra Evans, Independent Consultant, London, United Kingdom
Michael Kropac, CEWAS, Switzerland

References and further readings

Clean Development Mechanism Project Design Document form (CDM-SSC-PDD). 2006. Version 3.

Loikala et al. 2006. Opportunities for Finnish environmental technology in India. ISBN 951-563-521-7. www.sitra.fi.

UNFCCC/CCNUCC. 2010. Monitoring report: Upgradation, operation and maintenance of 200 TPD composting facility at Okhla, Delhi.

Case descriptions are based on primary and secondary data provided by case operators, insiders, or other stakeholders, and reflect our best knowledge at the time of the assessments 2014/15. As business operations are dynamic data can be subject to change.

Notes

1 The crediting period for a CDM project is the period for which reductions from the baseline are verified and certified by a designated operational entity for the purpose of issuance of certified emission reduction (CERs). The crediting period for IL&FS is 7 years.

2 Industrialized or transitional economies as listed in Annex I of the United Nations Framework Convention on Climate Change (UNFCCC). http://unfccc.int/parties_and_observers/parties/annex_i/items/2774.php (accessed November 8, 2017).

CASE

Partnership-driven municipal solid waste composting at scale (KCDC, India)

Miriam Otoo, Sampath N. Kumar, Pushkar S. Vishwanath and Lesley Hope

Supporting case for Business Model 12	
Location:	Bangalore, Karnataka, India
Waste input type:	Municipal solid waste
Value offer:	Provision of waste management services and high quality compost for agricultural purposes; provision of consultancy services for waste management
Organization type:	Public entity (government-owned corporation)
Status of organization:	Operational since 1975
Scale of businesses:	Processes 300 tons of municipal solid waste per day
Major partners:	Bruhat Bengaluru Mahanagara Palike (BBMP), Karnataka Agro Industries Corporation (KAIC), Karnataka State Co-operative Marketing Federation (KSCMF)

Executive summary

Karnataka Compost Development Corporation Limited (KCDC) is one of the oldest public entities involved in the production of compost from municipal solid waste (MSW) for agricultural purposes in India. The business of compost production provides significant value to KCDC by offering viable options for cost recovery and ensuring sustainable sanitation services provision. KCDC has been particularly successful by using an innovative business partnership model. Its strategic partnerships with other local government entities and private enterprises have allowed it to optimize the allocation of resources and activities reduce risk associated with high capital investments and establish an assured market for their product. Another important success driver has been KCDC's ability to mold its business to local context elements. The use of a simple and labor-intensive technology not only gives KCDC a competitive advantage for production, but also generates employment particularly for low-income persons who would otherwise be unemployed. An additional socio-economic benefit from KCDC's businesses the reduction in chemical fertilizer imports from increased usage of organic compost. This in turn has significant ecological benefits, reducing residual chemical pollutants in soils and water bodies.

KEY PERFORMANCE INDICATORS (AS OF 2014)					
Land use:	6 ha				
Capital investments[1]:	USD 910,000				
Labor:	40 (13 skilled, 20 unskilled, 7 administrative)				
O&M cost:	USD 12,400/ day				
Output:	10,000–16,000 tons of compost per year				
Potential social and/or environmental impact:	40 direct jobs created with worker earnings higher than minimum wage; increased crop yield and reduced costs of fertilizer use, reduced waste management costs, reduced human exposure to untreated waste				
Financial viability indicators:[2]	Payback period:	7 years	Post-tax IRR:	N.A.	Gross margin:	N.A.

Context and background

Karnataka Compost Development Corporation Limited (KCDC) is a 39-year-old company based in Bangalore engaged in the business of hygienic disposal of solid wastes generated in Bangalore city through composting. The city of Bangalore, the capital of Karnataka, has a population of about 8,000,000 and generates about 3,500 to 5,000 tons of solid waste per day. With an ever-increasing urban population and limited waste management budgets, the local government invested in several integrated resource recovery facilities with the dual purpose of cost recovery and rehabilitating agricultural lands. Bangalore has a number of waste processing facilities at various locations, which are of larger processing capacity ranging from 200–1400 tons per day but KCDC remains one of the few still functioning. KCDC was incorporated in the year 1975 with an equity capital of USD 84,690[3] (in 2014 currency value) as equity infusion. The company started by setting up a composting plant using international technology along with 11 other similar plants across the country. The highly mechanized technology proposed was not sustainable for the Indian context and all the plants of similar technology closed down by 1980. KCDC was the only one who continued operations by doing incremental changes to its technology and by early 1990 transitioned completely to the use of an indigenized technology. Given its success, in 2000, KCDC received a subsidy of USD 34,000 from the Government of India to set-up a bio-fertilizer plant. KCDC is a state government entity with equity participation from Karnataka Agro Industries Corporation (KAIC), Bruhat Bengaluru Mahanagara Palike (BBMP) and Karnataka State Cooperative Marketing Federation (KSCMF). The principal shareholder is KAIC, which falls under the agricultural department of the Government of Karnataka. BBMP is the urban local body of Bangalore city and is responsible for municipal waste management in the city. The role of BBMP is to supply municipal solid waste to KCDC. Originally, the role of KSCMF was envisaged to support KCDC with marketing, however overtime KCDC has established its own marketing strategies for its products. KCDC is possibly the only government owned and longest operating municipal waste processing company in India. The waste processing facility of KCDC is located at Haralakunte, near Singsandra, about 13 km from the centre of Bangalore.

Market environment

Sanitization of waste is seen traditionally as a public sector obligation and consumes a large percentage of municipal budgets. A key challenge is managing the daily generation of millions of cubic meters of solid and liquid waste. The potential combinations of domestic, commercial and/or industrial waste streams are primarily viewed as a threat on which the public sector must spend resources to sanitize. Appropriate sanitation services to safeguard public health are however as expensive as they are crucial for exploding cities, consuming most of the municipal budget. Additionally, increasing chemical fertilizer prices, continuous degradation of agricultural soils from over-application of chemical fertilizer and subsequent reductions in crop yields have caused the government of India to shift to a soil nutrient based fertilizer plan and promoting organic agriculture. KCDC thus took advantage of the government's push for organic agriculture to convert readily available MSW into organic fertilizer for use in the agricultural sector. The size of the organic fertilizer market although fairly large and growing, is comprised of 90% of animal-manure based fertilizer producers. Of the remaining 10% that is non-animal manure-based; the majority of businesses is small-sized and found in the informal sector. These businesses generate demand for their product based on field demonstration, personal relationships and reputation. There have been many products that have been promoted and have not been found useful on the ground. The market acceptability especially for organic compost is based on proof by demonstration and product branding. The compost produced by KCDC competes with the numerous organic fertilizer products produced by private manufacturers as well as imported chemical fertilizers. KCDC, however, has a competitive advantage, as its product is priced lower than the average market price and is able to do this partly due subsidy receipts from the government.

Macro-economic environment

Significant increase in MSW generation in the last decades due to rapid urbanization and high population growth rate has put the identification of sustainable waste management systems at the forefront of local government issues. Around 90% of generated waste in Bangalore is currently landfilled, requiring around 1,200 hectares of land every year. The ever-increasing cost of waste management has limited public investment in other economic sectors. Additionally, chemical and synthetic fertilizers are highly subsidized in India, and this has not only led to inefficient use by farmers and high costs to the government; significant soil degradation has also been observed as a result. To curb public spending on waste management services and chemical fertilizer subsidies, the Indian government has implemented a number of schemes that support the reuse of waste. With a growing need to increase the availability and quality of bio-fertilizers and composts in the country to improve agricultural productivity while maintaining soil health and environmental safety, India has set up a scheme to augment the infrastructure for production of quality organic and biological inputs.

Accordingly, under the National Project on Organic Farming a capital investment subsidy scheme provides credit linked and back-ended capital investment subsidy equivalent to 33% of total financial outlay subject to the maximum of Rs. 60 lakh[4] per unit and 25% of total financial outlay subject to a maximum of Rs. 40 lakh per unit for commercial production units for organic and biological agricultural inputs has been introduced (see Case A2Z Infrastructure Limited in Chapter 9).

Business model

KCDC is a state government corporation that converts municipal solid waste into organic fertilizer for agricultural purposes. It also provides consultancy services (expertise on technology) to other waste processing companies. It partners with Karnataka Agro Industries Corporation (KAIC), Bruhat Bengaluru Mahanagara Palike (BBMP) and Karnataka State Co-operative Marketing Federation (KSCMF). All the partners contributed to the initial capital investment and are current shareholders in the company. The partnership with BBMP gives it access to municipal solid waste. Although the originally envisaged role of KSCMF was to support KCDC in marketing, overtime KCDC established its own marketing brand and has been successful in increasing its share of the organic fertilizer market. Essential aspects of KCDC's model are its marketing strategy and technology use. The major compost products are marketed through government institutions, dealers' network, KAIC retail outlets and direct selling to consumers. KCDC uses these intermediaries to sell its products to rural and urban farmers, plantation owners, nurseries, floriculturists, landlords and urban households. KCDC captures the large rural market through the well-organized distribution channels of government institutions with which farmers are familiar. The use of dealer networks has widened their market coverage, allowing them to capture most of the Karnataka state and some parts of the Tamil Nadu and Kerala market. KCDC sells to the marginal farmers through the state's agricultural department with a 50 per cent subsidy, under a scheme to promote organic farming. In addition, KCDC gives a discount on metric ton basis to private buyers. The promotion and marketing strategies adopted by KCDC have doubled its sales in the past one year. Another key sustainability factor of KCDC is its technology, which is simple, indigenous and has low-energy and investment requirements. KCDC has mastered the technology of aerobic windrow composting and vermicomposting and its expertise has been recognized by many municipalities who are now seeking their technical and managerial advice; for which KCDC now generates revenue from their consultancy services. See Figure 149 for diagrammatic representation of the business model for KCDC.

Value chain and position

Figure 150 opposite presents KCDC's compost value chain and position. KCDC was built with equity from three government entities: BBMP, KAIC and KSCMF to promote sustainable waste management and agricultural production practices. The City of Bangalore generates about 3,500 tons of solid waste per day. The capacity of existing facilities is insufficient to process all of the city's quantity of waste and is currently overloaded. KCDC has a contract with BBMP for the supply of 600 tons of MSW each day of which only 50% is being processed. Even with the entry of new organic fertilizer businesses in the market, there is adequate availability of waste for KCDC's operation and even for future scaling-up of operations. KCDC produces two types of compost, namely: (a) regular compost marketed as BIO AGRO; and (b) enriched compost marketed as BIO AGRO RICH (which is enriched with micro nutrients). KCDC's customers are mainly directorates of agriculture, horticulture and sericulture, estate plantations, smallholder farmers and households. As partners, KAIC and KSCMF are responsible for establishing a solid marketing and distribution network for the products. KCDC sells their products through dealer networks, KAIC retail outlets and the existing distribution channels of Karnataka state departments of agriculture. The compost is sold to marginal farmers with a 50 per cent subsidy under a scheme to promote organic farming. Pricing is based on cost of production and a profit mark-up. BIO AGRO is sold at Rs. 1,000/ton in loose form and Rs. 1,550/ton if bagged. BIOAGRO RICH is sold at Rs. 1,500/ton in loose form and Rs. 1,850/ton if bagged. All pricing includes transportation up to 100 kilometres and free loading charges. An additional government subsidy of Rs. 30/ton is provided if the user segment is farmers.

Chemical and other organic fertilizers found on the market are good substitutes for KCDC's BIO AGRO and BIO AGRO RICH. Terra Firma Biotechnologies Limited is a major competitor in the market and produces a variety of equally high quality compost products tailored to different customer segments. Additionally, Terra Firma implements a door-to-door sales strategy (direct sales) for urban households and uses HOPCOMS outlets to reach larger scale agricultural producers which have worked well to increase its market share. Terra Firma's products are however perceived to be an up-market product as they are not cost-effective for marginal farmers. Terra Firma thus has had to focus on the household and large-scale farmer segments. With government subsidies, KCDC's products are the most cost-effective product on the market given its quality (high nutrient levels and compliance with safety standards). KCDC seems to be the market leader compared to the main competitor, Terra Firma, mainly due to its long-standing existence; however, if government support and subsidies are withdrawn, the survival of the product is doubtful.

Institutional environment

There are no legal or regulatory policies that limit the processing of MSW to organic fertilizer products. The key regulation is that waste reuse businesses assure the safety of all actors involved in the business operations and the quality of the product as outlined in the Municipal Solid Waste (Management and Handling) Rules, 2000. In terms of production, there is a statutory guideline – the Fertilizer Control

Order (FCO) instituted by the Ministry of Agriculture and Rural Development for the production and distribution of all fertilizers including organic fertilizer. Product quality recommendations are provided for different organic fertilizer types for which producers have to adhere to. This is particularly beneficial

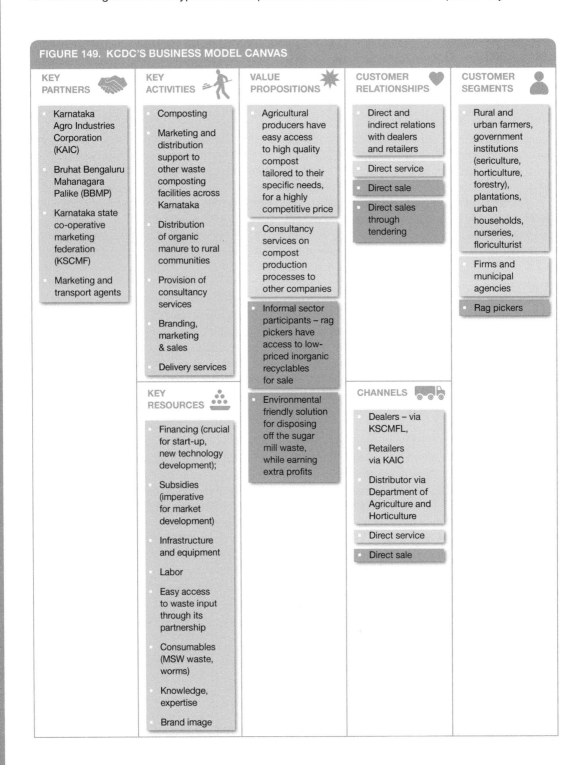

FIGURE 149. KCDC'S BUSINESS MODEL CANVAS

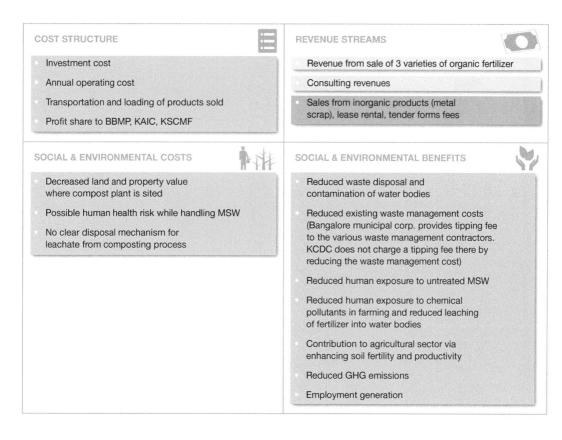

to farmers as they get what they are paying for, but also for compost businesses as they are able to build their product brand.

Technology and processes

KCDC had initially adopted the mechanical composting process that was developed essentially for western nations but after much experimentation, it adopted a simple, economical and rapid aerobic decomposition method which essentially consists of the rapid decomposition of organic materials in the presence of oxygen (Figure 151). KCDC's technology is found to be cost effective and simple. It implements two types of composting technologies: a) aerobic windrow composting; and b) vermin-composting. The aerobic windrow composting method can handle large quantities of waste as it is mostly mechanized in operation. The vermin-composting operation, on the other hand, has significant manual input requirement and is suitable for the processing of smaller quantities of organic waste. The quality of the compost and the associated price for vermin-compost is higher than that for aerobic windrow composting. The company produces both composts to meet customers demand and specification. KCDC does not conduct a detailed analysis of incoming waste. Visual assessment is done and unsuitable waste is not accepted. For the aerobic decomposition windrow method, the garbage received is arranged in windrows before segregation on the concrete platform. An inoculant is sprayed on the waste to speed up decomposition and reduce odors. The windrow is turned with augers and front end loaders once every seven days to ensure proper aeration and the aeration process continues uninterrupted. Water is sprayed as and when required depending on the moisture content of the mixture. The decomposition process is completed over a period of six to eight weeks. The decomposed mixture which has undergone sanitization and stabilization is taken up for processing by way of screening with different sized sieves. KCDC produces different intermediary and

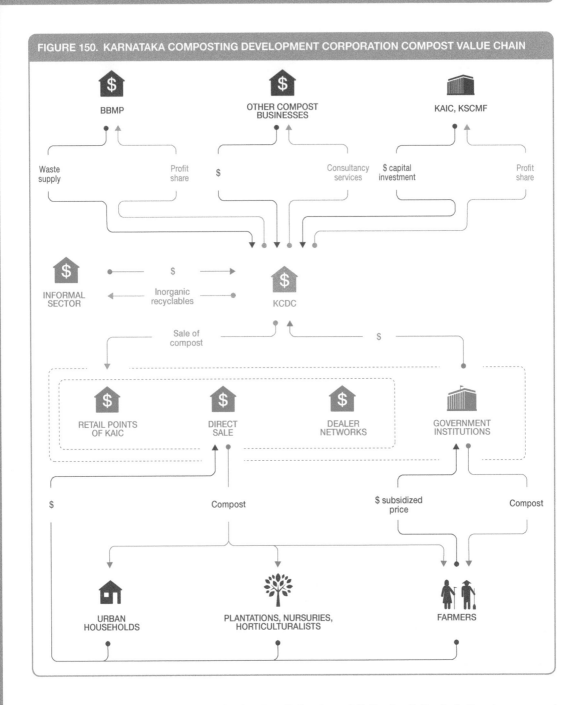

FIGURE 150. KARNATAKA COMPOSTING DEVELOPMENT CORPORATION COMPOST VALUE CHAIN

compost products: BIOAGRO, BIOAGRO RICH, B Grade and C Grade. B Grade is the decomposed matter after 25mm sieving and C Grade is decomposed matter without sieving (which is rarely sold). BIO-AGRO is the pure form of screened compost (particle size $\leq$ 4mm) without any additives, whereas BIOAGRO RICH is enriched with micro nutrients such as Neem, Gypsum, Cow dung, Rock Phosphate and Poultry Litter. The final product is a safe (free from harmful pathogens) and high nutrient product.

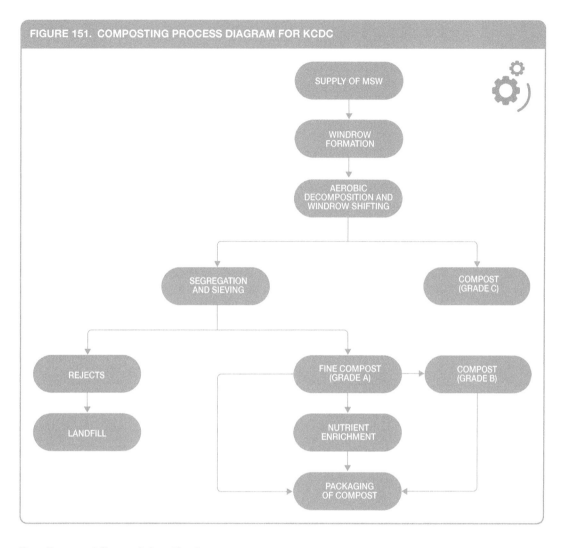

FIGURE 151. COMPOSTING PROCESS DIAGRAM FOR KCDC

Funding and financial outlook

KCDC was set up in 1975 with capital infusion in the amount of Rs. 5 million from KAIC (51%), BBMP (24.5%) and KSCMF (24.5%). These entities are government bodies and have invested in KCDC to promote effective waste handling and supporting usage of organic compost in agriculture. In 2000, KCDC received a grant subsidy of Rs. 2 million from the Government of India to further expand its activities to set up a bio-fertilizer plant. KCDC generates revenue by sale of compost and consultancies. Bangalore City Corporation does not pay any tipping fees to KCDC for processing the city's waste. KCDC manages its operations and maintenance on its own funds. The quantity of waste processed and sales have doubled in the last year. The quantity of sales was around 8,000 tons per year for last few years but has doubled up to about 15,000 tons from 2012 onwards. KCDC had revenues of about Rs. 51 million and an expenditure of Rs. 54 million. KCDC has been incurring losses from 2009–2012 due to the company having to adopt an aggressive pricing strategy to increase the quantity of compost sold. The quantity of compost sold has been significantly growing, doubling between 2010–2011 to 2011–2012 and with a similar trend in 2012–2013 (Table 38). The company reduced its losses in 2013 from Rs. 4.3 million to 0.6 million by increasing the quantity of processing and sales, and thereafter averaging annual profits of Rs. 1–3 million.

TABLE 38. FINANCIAL DATA FOR KCDC FROM 2009–2012

ITEMS	2009–10	2010–11	2011–12
Quantity of compost sold (in metric tons)	8,760	8,060	15,333
Total revenue (in millions of Rs.)	20.4	15.9	51.6
Total expenditure (in millions of Rs.)	19.1	20.4	54.5
Operating Profits (in millions of Rs.)	1.3	(4.4)	(2.9)
Profit after tax/(Losses)	(1.38)	(4.3)	(0.67)

Socio-economic, health and environmental impact

KCDC provides direct employment to about 40 personnel and indirectly about 60 personnel involved in the transportation and distribution of organic compost. In addition, KCDC is helping address the city's waste management problems and creating value out of the waste which was environmentally hazardous. KCDC started running as a profitable firm with average annual profits of Rs. 1–3 million and pays taxes for the consultancy services it renders to other waste reuse businesses. KCDC activities strongly support sustainable agriculture and provide advisory support to new companies and municipalities involved in waste reuse. The products of KCDC have been influential in adding value to farmers by enriching their farmland via increased microbial activity from compost use. The use of compost has also resulted in increase in crop yields. Table 39 below provides details about the economic value of organic compost considering requirement for banana crop. Typically by using organic compost, a farmer gains an economic advantage of about Rs. 6,600 per every hectare of crop. KCDC by serving about 20,000 customers per year by selling about 15,000 tons of organic compost in 2011–2012 added a total economic value of about Rs. 105 million to its consumers. The usage of organic compost in place of chemical fertilizers has also helped the country's economy by reducing imports through chemical fertilizers.

Scalability and replicability considerations

The key drivers for the success of this business are:
- Increasing need for alternative sustainable agricultural production inputs and waste management services.
- Strong business partnerships that reduced capital investment risk and eased entry into a highly competitive fertilizer market.
- Strong commitment of state government in providing an enabling environment for marketing and distribution of the compost products.
- Policy initiatives to phase-out chemical fertilizer subsidies and capital investment subsidies for new and existing compost businesses (government schemes to augment the infrastructure for the production of quality organic and biological inputs).

TABLE 39. ECONOMIC VALUE OF COMPOST USE FOR BANANA PRODUCTION

PARTICULARS	COST IF ORGANIC COMPOST IS NOT USED	COST WHEN ORGANIC COMPOST IS USED
Quantity of fertilizer required per Ha	2 tons of chemical fertilizer	1 ton of chemical fertilizer + 2 tons of organic compost
Cost of fertilizer per hectare	Rs. 40,000	Rs. 20,000+ Rs. 6,800
Total cost of fertilizer	Rs. 40,000	Rs. 26,800
Economic benefit per hectare	–	Rs. 13,200
Economic benefit per ton of compost	–	Rs. 6,600

The KCDC model has high replication potential especially for developing countries in need of sustainable waste management approaches and environmentally-safe agricultural input alternatives. The scale of KCDC's business model is applicable to cities with population size of 1.5 million or above. Strategic partnerships and governmental support are essential at both the start-up and business development phase to mitigate capital investment risk and gain access into new markets. With chemical fertilizer companies typically owning the greatest share of the market, governmental support via price subsidies, for example, will be important to ease the entry of new compost businesses into the fertilizer market. The adopted technology is semi-mechanized and offers opportunity to use unskilled and informal labor an abundant resource in developing countries. The use of a labor-intensive and inexpensive technology also implies that the business will not require large capital investment which mitigates one of the major constraints for business start-ups especially in developing countries.

Summary assessment – SWOT analysis

Figure 152 presents the SWOT analysis for KCDC. KCDC has been particularly successful in leveraging its business partnerships to mitigate capital investment risk and gain entry into a fiercely competitive fertilizer market. Increasing governmental support along with growing demand for organic fertilizers will represent key opportunities for replication and up-scaling of the business. KCDC implements a

FIGURE 152. SWOT ANALYSIS FOR KCDC

	HELPFUL TO ACHIEVING THE OBJECTIVES	HARMFUL TO ACHIEVING THE OBJECTIVES
INTERNAL ORIGIN ATTRIBUTES OF THE ENTERPRISE	**STRENGTHS** • Abundant and easy access to raw materials • Low capital investment requirements • Cost effective technology • Strategic partnerships for accruing capital investments and establishing strong distribution channels, as well as enabling competitive pricing • Aggressive pricing strategy • Business longevity • Strong brand image • Extensive experience in design and operation of composting plants	**WEAKNESSES** • No tipping fees for MSW • Viability of business dependent on price subsidies • High transportation costs given centralized operations
EXTERNAL ORIGIN ATTRIBUTES OF THE ENVIRONMENT	**OPPORTUNITIES** • Expanding to other markets – a market of about 100,000 tons of compost is achievable. • Availability of financing organizations and support • Government support • Proactive policies and acts of government, especially toward organic agriculture	**THREATS** • Competition from substitute products • High seasonality of demand for compost may increase investment cost for storage facilities

segmented pricing approach where it charges peri-urban and rural farmers less than its other clients. Its pricing strategy is however dependent on price subsidies provided by the government and its removal may expose KCDC to fierce competition in the fertilizer market, in which case it would have to rebrand its product to maintain its market share. KCDC is exploring the development of a high nutrient granulated compost. This new product retains its nutrient value over a period from production to actual use that can sometimes be between three to six months. Additionally, granulation would provide stability through transportation of the product. The use of a simple technology has also been essential to KCDC's success – taking advantage of cheap and abundant labor. However, with one of the most expensive operational components of the composting business being transportation, KCDC will need to explore a decentralized production unit approach and sourcing operation to reduce its transportation costs. KCDC is an example of an innovative business utilizing a simple partnership approach to address some of the major waste management and environmental challenges in Bangalore, India.

Contributors
Jasper Buijs, Sustainnovate, Netherlands; formerly IWMI, Sri Lanka
Alexandra Evans, Independent Consultant, London, United Kingdom
Michael Kropac, CEWAS, Switzerland

References and further readings
Personal communication with plant managers. 2014.

Case descriptions are based on primary and secondary data provided by case operators, insiders or other stakeholders, and reflect our best knowledge at the time of the assessments 2014/15. As business operations are dynamic data can be subject to change.

Notes
1. Based on estimates derived by authors from secondary data on the scale of operation and data provided by the business given that KCDC was incorporated (i.e. legally established) in 1975 and accurate details are unavailable.
2. Calculations were based on the assumption that if a new project were to be set up today to handle 600 tons per day of compost, the estimated project cost would be about Rs. 350 million. Based on some of the projections done by the company, the payback period would be in the range of about seven years.
3. Exchange rate: INR (Rs.)1 = USD 0.02
4. 1 lakh = 100,000; Rs. 60 lakh were in 2004–2012 about USD 120,000 and about USD 90,000 in May 2017.

CASE

Franchising approach to municipal solid waste composting for profit (Terra Firma, India)

Miriam Otoo, Lesley Hope, Sampath N. Kumar, Pushkar S. Vishwanath and Ishara Atukorala

Supporting case for Business Model 12	
Location:	Bangalore, Karnataka, India
Waste input type:	Municipal solid waste
Value offer:	Organic fertilizer, biogas, recyclable plastics
Organization type:	Public limited company
Status of organization:	Operational since 1994 (that plant was not receiving municipal waste at the time of final review in October 2017).
Scale of businesses:	Processes 1,400 tons of waste per day
Major partners:	Bruhat Bengaluru Mahanagara Palike (BBMP) – Bangalore municipality, Coromandel Fertilizer Limited, Karnataka Antibiotics and Pharmaceuticals Limited, Rallis India Limited

Executive summary

Terra Firma Biotechnologies Limited (Terra Firma) is one of the oldest operating municipal solid waste (MSW) processing companies in India. It is a public limited firm involved in the processing of MSW to organic compost, bio-methanation and the recycling of plastics and inert materials, with a processing capacity of up to 1,400 tons of municipal solid waste per day. With an increasing need for sustainable waste management options and agricultural inputs alternatives, nutrient recovery from different waste streams, particularly MSW is being promoted and showing promise in India. Terra Firma owns and operates several integrated resource recovery plants that receive MSW from the city of Bangalore. The success of Terra Firma's model rests on a multiple-revenue stream approach. Revenue is generated from five major streams: i) sales from organic fertilizer products; ii) service fees from the municipality and other private clients (townships and commercial establishment) for waste processing; iii) sales from recyclables; iv) consultancy fees; and v) franchising royalties. The diversification of their portfolio mitigates risk associated with fluctuations in demand for organic fertilizer products. Strategic partnerships have also contributed to the business' sustainability. The municipal corporation of Bangalore city and other commercial establishments and townships are contracted-out for the collection, separation and delivery of waste to Terra Firma for a fee – ensuring a consistent supply of high quality input (waste). It also partners with fertilizer companies and their network distributors to market and sell their compost whilst restricting its human resources to plant operations. These partnerships allow Terra Firma to sell compost under the fertilizer company's Coromandel brand name.

Terra Firma has also adopted a process of in-house technology development based on clear needs and locally appropriate solutions. The use of a simple and labor-intensive technology not only gives Terra Firma a competitive advantage for production, but also generates employment particularly for low-income persons who would otherwise be unemployed. Terra Firma's activities have helped to significantly reduce the city's waste management costs, reduce human exposure to untreated waste and contribute to the livelihoods of local communities.

KEY PERFORMANCE INDICATORS (AS OF 2015)	
Land use:	42 ha
Capital investment:	USD 527,996 (additional investments have been made with scaling-up of activities)
Labor:	215 (200 unskilled 15 skilled)
O&M cost:	USD 1,278,807 including cost of marketing
Output:	20,000–22,000 tons of compost per year
Potential social and/or environmental impact:	Significant job creation, reduced human exposure to untreated waste, reduced waste management costs
Financial viability indicators:	Payback period: 7–8 years / Post-tax IRR: N.A. / Gross margin: N.A.

Context and background

Established in 1994, Terra Firma set out with the goal of transforming the agricultural production landscape by promoting organic agriculture and substantially replacing chemical fertilizer use with more sustainable options such as organic fertilizers. Terra Firma noted that its business activities of converting MSW to organic fertilizers would additionally address the significant waste management challenges faced by the city. Although having the capacity to process up to about 1,400 tons per day of solid waste, Terra Firma processes about 600 tons/day of solid waste on an average and has been instrumental in reducing garbage disposal burden in Bangalore. The company was set up by a group of professionals in the area of chemical engineering and agriculture technology hailing from rural backgrounds. The activities of the company can be broadly classified into 3 parts: a) resource (nutrients and energy) recovery from the city's waste; b) consultation and design of turnkey projects; and c) franchising operations. The company set up a vermi-composting facility from municipal solid waste and has successfully operated it from 1995 to 2007. From 1998 to 2003, the company promoted franchisee operations for the processing of municipal solid waste across 38 locations in the country. In 2007, the company scaled up its operations to a new facility in a 42 hectare integrated waste management facility (ISWM). The company also undertakes training of agricultural graduates in the area of composting and other agricultural activities.

Market environment

The waste management service provided by Terra Firma to the city of Bangalore is unparalleled given the magnitude of MSW it processes daily. With increasing urban population growth and the resulting generation of significant amounts of waste, BBMP will continue to heavily depend on resource recovery businesses such as Terra Firma. Another key driver for the development of Terra Firma is related to the high demand for organic fertilizers for agriculture. The extensive use of chemical fertilizers has degraded the soil to a great extent and this has necessitated the demand for alternative agricultural inputs to replace synthetic fertilizers. With increasing farmers' awareness of benefits accruing from organic fertilizer use from government programs and increasing fertilizer prices, a surge in demand has been observed in Karnataka and neighbouring states. Although demand for organic fertilizer – compost is seasonal, with a few number of existing players in the organic compost sector, market demand in

Karnataka and neighbouring states still exceeds supply. Additionally, the recovery of biogas represents a revenue-generating opportunity for Terra Firma in the instance where it generates power in excess of its own energy and power requirements, which can be sold to the electricity grid. Given the shortage of electricity supply in Karnataka and India as a whole, there is a growing demand for alternative sources of energy production. Furthermore, with increasing national urban populations and limited waste management budgets, the demand for waste management solutions in other states and cities in India is growing as is the demand for consultation and design services for turn-key waste reuse projects by businesses like Terra Firma. The current market environments for Terra Firma's business activities are very supportive for its sustainability and indicates a foreseeable up-scaling their operations.

Macro-economic environment

Chemical and synthetic fertilizers are highly subsidized in India and this has not only led to inefficient use by farmers and high costs to the government; significant soil degradation has also been observed as a result. Even with the promotion of bio- and organic fertilizers via local research institutions and businesses, chemical fertilizer subsidies continue to be one of the key barriers for entry of organic fertilizer producers into the fertilizer market. With a growing need to increase the availability and quality of bio-fertilizers and composts in the country to improve agricultural productivity while maintaining soil health and environmental safety, the Indian government has set up a scheme to augment the infrastructure for production of quality organic and biological inputs.

A capital investment subsidy scheme for compost production has been introduced under the National Mission for Sustainable Agriculture (NMSA). The scheme provides 100% financial assistance to state governments and government agencies up to a maximum limit of about USD 300,000 per construction unit, and for individuals or private companies up to about USD 100,000 per unit (max 33% of project costs) through the National Bank for Agriculture and Rural Development (NABARD). Moreover, the Government of India is providing a Market Development Assistance of about USD 23.4 per metric ton to fertilizer companies for sale of city waste compost (Ministry of Agriculture, 2017).

Business model

Figure 153 provides an overview of Terra Firma's business model, which is centred on a multiple-revenue stream approach. Revenue is generated from five major streams: i) sales from organic fertilizer products; ii) service fees from the municipality and other private clients (townships and commercial establishment) for waste processing; iii) sales from recyclables; iv) consultancy fees; and v) franchising royalties. The diversification of their portfolio mitigates risk associated with fluctuations in demand for organic fertilizer products. The value proposition of provision of a nutrient-rich compost comes from the desirable social impact of providing an environmentally safe and cost-effective alternative agricultural input to local agricultural producers. It partners with Coromandel Fertilizer Limited and other retail distribution networks for sales and marketing of their organic fertilizer products whilst restricting its human resources to plant operations. Product demonstration, proof of concept farm fields and sustained interactions with agricultural producers was instrumental in garnering market demand for their products. Another key element of their model is the provision of waste management services to the municipality BBMP and other large scale generators via the processing of their wastes for a fee. This is a win-win partnership as all parties benefit: a) municipalities save on landfill costs; b) local businesses are to comply to waste management ordinances; and c) Terra Firma generates revenues, ensuring the sustainability of the partnership. Terra Firma also implements a franchise-based approach to increase revenue streams and capture additional markets. Terra Firma has entered into franchise agreements with several enterprises all over India. By this agreement, Terra Firma provides training on the composting technology at no cost but charges a cost-price fee for method trainings on bio-fertilizers production to the franchisee. Terra Firma markets the compost produced by the

FIGURE 153. TERRA FIRMA'S BUSINESS MODEL CANVAS

KEY PARTNERS
- Bangalore municipality (BBMP)
- Coromandel Fertilizer Limited
- Karnataka Antibiotics and Pharmaceuticals Limited
- Rallis India Limited
- Waste collection and separation partners

KEY ACTIVITIES
- Production and sale of enriched organic compost
- Promotional activities to catalyze compost market demand
- Sales of inorganic materials and other recyclables
- Consulting and design of waste reuse turn-key projects
- Training of agricultural professionals
- Franchising operations
- Production of biogas (bio-methanation) for internal use
- Research and development

KEY RESOURCES
- Knowledge and experience of 18 years in MSW business sector
- MSW
- Land, infrastructure, equipment, labor
- Relationships and contract agreement with partners
- Franchising contracts
- Composting technology development
- Partner-brands

VALUE PROPOSITIONS
- Agricultural producers obtain nutrient-rich compost at a competitive price
- Local government bodies and businesses obtain recyclables (plastics) as a substitute for bitumen
- Provision of sustainable waste management alternative for city and local businesses
- Provision of consulting and design services for the development of waste reuse turn-key projects and franchisees
- Reduction in power demand and expenditure due to biogas usage for internal power generation of the facility

CUSTOMER RELATIONSHIPS
- Product promotion activities through demonstration, proof of concept farm fields
- Indirect interaction with clients via network of distributors
- Direct interaction with clients
- Public-private partnerships with municipality and formal agreements with townships and commercial centres
- Recurrent partnerships for individual projects
- Dedicated one-to-one coaching and consulting services
- Self service

CHANNELS
- Distribution and retail networks (BBMP, Coromandel, KAPL)
- Direct communication and customer relationship
- Bidding (calling of tenders)
- Bidding and the word of mouth

CUSTOMER SEGMENTS
- Farmers, landlords, urban households
- Government institutions and plastic scrap dealers
- Municipal and private agencies in charge of the solid waste management, commercial institutions and townships
- Professionals in the SWM sector (potential franchisees and clients for consulting/design and turn-key projects)

COST STRUCTURE	REVENUE STREAMS
• Land and infrastructure: USD 4,589,675 to 4,956,849 • Material inputs: USD 17,624 • Marketing costs: USD 475,673.92 • Manufacturing costs: USD 392,325.42 • Administrative expenses: USD 206,351 • Interest and finance charges: USD 292,086.92 • Savings from in-house energy production • Waste collection and separation fees • R&D	• Sales of organic fertilizer products • Service fees from the municipality and other private clients (townships and commercial establishments) for waste management services • Sales from recyclables • Consultancy fees • Franchising royalties
SOCIAL & ENVIRONMENTAL COSTS	SOCIAL & ENVIRONMENTAL BENEFITS
• None noted based on information provided by business	• Reduced existing pollution of water bodies • Reduced existing waste management cost for the municipality • Reduced human exposure to untreated waste • Job creation for the poor without any gender discrimination • Contribution to restoring degraded soils and food security • Savings in landfill area

franchisees through a partner – Rallies India. In return, Terra Firma retains 24 percent of the equity with the franchisee and 10 to 20 per cent of the profit margin goes to the franchisee. Additionally, Terra Firma produces biogas, which is used internally to reduce production costs. Labor is employed on contractual terms to further reduce production cost especially during low production periods as well as when they are not in production. Terra Firma has established and demonstrated that it is possible to run a waste business sustainably over a long period based primarily on non-municipal tipping fee revenues. The company provides employment to about 200 people at its facility and about 15 people for management and administration. The indirect employment for transport, dealers, distributors and waste recycling industry supported is estimated to be at least twice these numbers.

Value chain and position

The City of Bangalore generates about 3,500 tons of waste per day. Like most cities in India, Bangalore faces a huge challenge in processing all the waste generated in the city. Terra Firma has a supply contract agreement with Bruhat Bengaluru Mahanagara Palike (BBMP) for supplying a minimum of 600 tons of waste daily. There is no short to medium term threat about availability of waste as a raw material to Terra Firma or to any other waste processing facilities in Bangalore. Terra Firma additionally has an independent collection system from large waste generators like commercial establishments (hotels, industries, institutions) and residential townships for their solid or organic waste. About 25–30% of the capacity is collected from these sources. This partnership ensures continuous waste (input) supply thus mitigating any production risk associated with input supply. Terra Firma also partners with Coromandel Fertilizer Limited and other network distributors to market its compost. This partnership allows Terra Firma to sell compost under Coromandel's brand name. Terra Firma has invested in product demonstrations and proof of concept farm fields to penetrate the fertilizer market and create a niche for itself. Currently, it serves between 35,000 farmers annually. It competes with other organic as well as chemical fertilizer producers. The organic fertilizer produced by Terra Firma is lower in price and thus has a competitive advantage over other organic fertilizers however highly subsidized chemical fertilizer still represents a great threat. Existing players in the organic fertilizer sector are currently unable to meet market demand in Karnataka and neighbouring states; and the demand is expected to grow in the near future; thus great opportunities exist for Terra Firma to increase its market share. Additionally, increasing urban populations and related waste management challenges along the promotion of integrated resource recovery facilities suggest that municipalities in India will continue to demand consultancy services of business like Terra Firma for the establishment of waste reuse turnkey to projects. Figure 154 above provides a diagrammatic overview of Terra Firma's value chain.

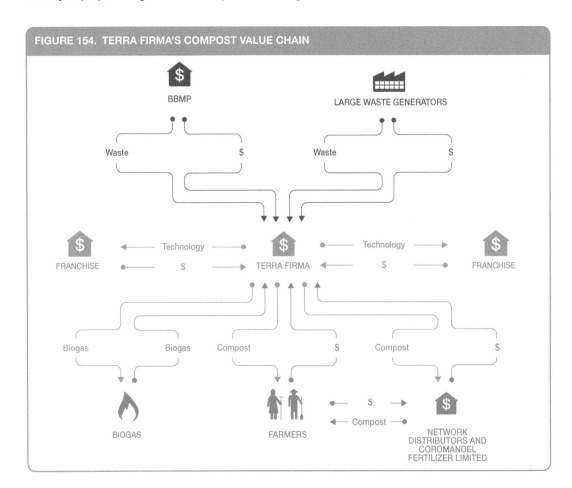

FIGURE 154. TERRA FIRMA'S COMPOST VALUE CHAIN

Institutional environment

Terra Firma is a public company registered under the Indian Companies' Act of 1956. The company has more than 50 shareholders. Waste processing facilities are usually not accepted by communities due to visual, odor and traffic pollution. This issue snowballed into a political, social and cultural resistance. It has been necessary for Terra Firma to manage the socio-political environment on an ongoing basis. It has garnered local support with its community works and corporate social responsibility. Terra Firma has also generated employment to large number of local residents near the facility.

Technology and processes

Terra Firma developed their technology for treating MSW in house (Figure 155). All developed technologies meet the requirements of environmentally, safe waste handling processes and being cost effective to ensure business viability. The company started with the vermi-composting of organic waste, then as it started receiving mixed waste, a microbial culture based waste composting method was adopted. As the waste included plastics, new techniques for plastic cleaning and conversion were incorporated. Terra Firma receives an average of 1,400 tons of MSW per day. About 650 tons of this waste is treated in the aerobic composting facility and 350 tons utilized for bio-methanation purposes daily. The remaining waste is sent to the bioreactor landfills.

1) **Composting** – An aerobic windrow composting process is used for the treatment of the organic waste. After the unloading of the waste, water is sprinkled on the waste to achieve a desired moisture level. The waste heap is then pushed by a tractor blade or front-end loader, which is used make high heaps of the waste, which is then sprayed with water and formed into a minimum of three meters heaps (maximum height five meters). A cow dung solution or bio-culture act as a catalyzing agent and accelerates the process. The heap is then turned by tractor blade or front-end loader into another windrow to allow aeration. This process is repeated after another seven to ten days. At the end of the three to four weeks period, the green or fresh compost may have fully decomposed but not the cellulosic content. The mixture is, therefore, stored in large sized windrows under a covered/roofed area for maturation for four to eight weeks. The mixture is then sieved to meet client requirements.

2) **Bio-methanation** – The bio-methanization of organic wastes is accomplished by a series of biochemical transformations. In the first step hydrolysis, acidification and liquefaction take place and in the second step acetate, hydrogen and carbon dioxide are transformed into methane. At Terra Firma, all these reactions take place simultaneously in a single reactor.

3) **Recycling of inorganic materials** – Metals, plastic, glass and paper separated, cleaned, packaged and sold.

4) **Bioreactor landfills** – The bioreactor landfill technology is an accelerated process of decomposition of municipal waste in the landfill. This technology involves placing the waste in specially designed cells. The cells have provision for leachate collection and recirculation. As one cell gets filled, it is covered and closed with an impervious liner. Gas extraction pipes are placed. The leachate from the landfill and the bio-methanation effluents are sprayed / injected to accelerate the bio-methanation process. The degradation time is reduced from about 10–15 years to about two to four years. Subsequent to the gas extraction the inert waste is removed from the landfill, compost and other recyclables are mined and the balance materials are sent for final disposal in a sanitary landfill.

5) **Final disposal** – The final disposal is proposed in a sanitary landfill. The incoming waste is spread in thin layers and compacted using landfill compactors to achieve high density of the wastes. The waste is covered immediately or at the end of each working day with a minimum of 10cm of soil, inert debris or construction material. Prior to the commencement of the monsoon season, an intermediate cover of 40–65cm thickness of soil is typically placed on the landfill to ensure proper compaction and prevent soil infiltration during monsoon.

The technologies developed by Terra Firma are focused on cost-effectiveness within the regulatory frameworks. The outputs of the company have been tested and approved by the regulatory authorities and are acceptable by the clients. But there are technological constraints with the changing nature of municipal solid waste; there is a significant quantity of waste which cannot be processed. The company is now facing issues on managing large amounts of inert materials, about 15% of the waste received (100 tons per day). The conventional method of sanitary landfill would exhaust the land

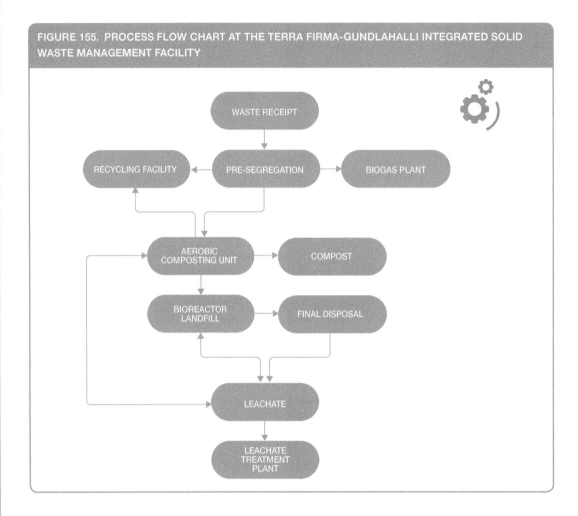

FIGURE 155. PROCESS FLOW CHART AT THE TERRA FIRMA-GUNDLAHALLI INTEGRATED SOLID WASTE MANAGEMENT FACILITY

very quickly, thus the company is exploring new technologies for processing this waste including converting them into refuse derived fuels and panel blocks.

Funding and financial outlook

The capital investment for the company was through equity infusion from diverse sources including private financiers, the municipality, in a total amount of USD 6,237,402. Of this amount, the municipality contributed USD 366,906 at no interest as an advance against tipping fees. Terra Firma uses working capital loans from banks and its own financial resources to cover operational and maintenance costs. Terra Firma generates revenue via five streams: a) sale of MSW-based organic fertilizer; b) waste management fees; c) sale of recyclables; d) consultancy fees; and e) franchise royalty fees. Of the total revenue, sales from compost contribute the most, that is, USD 770,000 per annum followed by sales from recyclables mainly plastics. Tipping fees of USD 280,320 per annum contribute the least. Terra Firma has been making profit since its inception, except the financial year 2009–2010 (a loss of about USD 352,435). The company was scaling up its operations at a new facility and incurring additional investment costs in the process. Table 40 below provides the percentage contributions of total revenue and expenditure. It is noted that administrative expenses including maintenance of the facility, utilities, rent, salaries to the staff is the highest contributor to the expenses.

TABLE 40. TERRA FIRMA'S REVENUES AND EXPENDITURES EXPRESSED IN PERCENTAGES

ITEM NUMBER	COMPONENT	
A	Revenues	Percentage of total revenue
1	Sale of compost	41%
2	Other revenues	59%
B	Expenditure	Percentage of total expenditures
1	Raw material consumed	20%
2	Manufacturing expenses (includes labor charges)	24%
3	Administrative expenses (includes rent and utilities)	36%
4	Selling and distribution	18%
5	Interest and financing charges	1%
6	Depreciation and other charges	1%

Socio-economic, health and environmental impact

Terra Firma has established and demonstrated that it is possible to run a waste business sustainably over a long period based on non-municipal tipping fee revenues. The company is a leading example in support of privatizing the municipal solid waste business to maximize recovery of resources from waste. The company influences a series of economic activities as part of its process. The polluter pays principle is implemented via direct payments to Terra Firma by large waste generators for the management of their waste. The transport sector is organized to collect and transport over larger distances in an efficient way thereby setting standards for similar operations. The concept of component-wise treatment of waste using smart segregation strategies has been exemplified by the company. The economic valuation of the enterprise can be assessed from its internal economic activities. Terra Firma set up the plant facility on its own land which in itself is trend setting in India, reducing the burden on the municipal agencies for provision of land to solid waste management operators. The company provides employment to about 200 people at its facility and about 15 people for management and administrative purposes. The indirect employment for transport operators, dealers, distributors and waste recycling industries supported is at least twice the current number of total employees. Terra Firma's activities additionally will in the long term reduce chemical fertilizer imports, resulting in foreign exchange savings. Averted greenhouse gas emissions and groundwater contamination from indiscriminate waste disposal are among the additional benefits of Terra Firma's activities.

Scalability and replicability considerations

The key drivers for the success of this business are:
- Strong business partnerships that reduced capital investment risk and eased entry into a highly competitive fertilizer market.
- Solid multi-revenue stream based business model that mitigated initial risk associated with fluctuations in market demand.
- Marketing strategies based on product demonstrations and proof of concept farm fields garnered the market demand for their organic fertilizer products.
- Increasing need for alternative sustainable agricultural production inputs and waste management services.
- Increased awareness among farmers about the advantages of organic fertilizers and in the face of increasing fertilizer prices.
- Strongly increasing urban populations and associated MSW problem.

Terra Firma's model has a high replication potential especially in large urban areas facing solid waste management challenges. High initial investment costs may represent barriers for entrepreneurs in developing countries where accessing capital investment is one of the key constraints for business development. Implementation of institutional policies such as the polluter pays principle, especially for large waste generators would be essential to ensure the viability of one of its revenue streams.

Summary assessment – SWOT analysis

Figure 156 presents the SWOT analysis for Terra Firma. Its model demonstrates that waste reuse businesses can be successful without government subsidies. By diversifying its portfolio, Terra Firma mitigates risk associated with fluctuations in demand for its organic fertilizer products. It has been particularly successful in leveraging its business partnerships to gain entry into a fiercely competitive fertilizer market via using the well-established marketing and distribution channels of other companies. This marketing strategy is however highly dependent on partners, exposing Terra Firma to some buyer power risk. Terra Firma has been conducting product demonstrations and proof of concept farm fields to establish its product brand and gain some market share. Increasing governmental support along with growing demand for organic fertilizers will represent key opportunities for replication and up-scaling of the business. Local community support programs may help dissipate occasional community

FIGURE 156. SWOT ANALYSIS FOR TERRA FIRMA

	HELPFUL TO ACHIEVING THE OBJECTIVES	HARMFUL TO ACHIEVING THE OBJECTIVES
INTERNAL ORIGIN ATTRIBUTES OF THE ENTERPRISE	**STRENGTHS** • Well-established multiple revenue streams • Good local buy-in for products due to establishment of brand name • Strong and strategic business partnerships • Independent decentralized franchisee operations • Reduced production costs due to generation of energy for internal use	**WEAKNESSES** • High capital investment • Operational pulls and pressures by municipal authority in waste supply which result in management issues of waste • Marketing strategy currently highly dependent on fertilizer companies
EXTERNAL ORIGIN ATTRIBUTES OF THE ENVIRONMENT	**OPPORTUNITIES** • Absence of other key competitive organic fertilizer suppliers in the market • Up-scaling potential of CDM project to earn carbon credits • Tax free policy for organic fertilizer in India • Production and sale of refuse-derived fuels • Sale of electricity to national grid from increased production	**THREATS** • Possible human health risk of being exposed to waste at collecting, sorting and processing • Village protest against frequent waste transportation across their village • Chemical fertilizer subsidies hamper organic fertilizer growth

protests against waste transportation through community neighbourhoods. Several opportunities exist for Terra Firma to further expand its operations. These include: a) the production and sale of refuse-derived fuels; b) sale of excess electricity to national grid; and c) the establishment of a CDM project for sale of carbon credits. Terra Firma is an example of an innovative business utilizing a multi-revenue approach and strategic partnerships to address some of the major waste management and environmental challenges in Bangalore, India whilst generating significant profits and benefits to society.

Contributors
Johannes Heeb, CEWAS, Switzerland
Jasper Buijs, Sustainnovate, Netherlands; formerly IWMI, Sri Lanka
Josiane Nikiema, IWMI, Ghana

Reference and further readings
Ministry of Agriculture. 2017. Press release on promoting the use of organic manure. (ID 159439; 17 March 2017). Government of India.

Personal communication with business owners and plant managers. 2015.

Terra Firma Biotechnologies Ltd. http://terrafirmabiotech.com/. Accessed May, 2015.

Case descriptions are based on primary and secondary data provided by case operators, insiders or other stakeholders, and reflect our best knowledge at the time of the assessments 2015. As business operations are dynamic data can be subject to change. At the time of final review (October 2017), for example, the plant was noted to not be receiving municipal solid waste.

CASE

Socially-driven municipal solid waste composting for profit (Waste Concern, Bangladesh)

Miriam Otoo and Lesley Hope

Supporting case for Business Model 12	
Location:	Dhaka, Bangladesh
Waste input type:	Municipal solid waste (MSW)
Value offer:	Efficient waste management service and provision of high quality compost
Organization type:	Private (Social Business Enterprise)
Status of organization:	Operational since 1995
Scale of businesses:	Small to medium scale: 3–20T of organic waste per day Large scale: 75–100T of organic waste per day
Major partners:	Dhaka City Corporation (DCC), Ministry of Environment and Forest (MoEF), Sustainable Environment Management Programme (SEMP) of the UNDP

Executive summary

Waste Concern Group, established in 1995, is a Social Business Enterprise (SBE) comprising both 'For Profit' and 'Not-for Profit' enterprises with the vision to contribute towards waste reuse, environmental improvement and poverty reduction through job creation and sustainable development. Waste Concern works in partnership with the government, private sector, local communities and international agencies. Amongst its various lines of business activities, the key ones are solid waste management and resource recovery where compost production plays an essential role. Waste Concern's compost business models implement both a small-to-medium decentralized community-based approach and large scale CDM (Clean Development Mechanism)/carbon trading approach. Waste Concern has been particularly successful by forging strategic partnerships with the local government, private enterprises and community-based organizations to optimize the allocation of resources and activities, reduce risk associated with high capital investments and establish an assured market for their product. The local government gave Waste Concern legal access to the city's waste and provided land for the plants. This is a win-win partnership as by alleviating Waste Concern of its high initial investment costs, the municipality gains from reduced waste collection and landfill costs. Waste Concern earns revenue through its established door-to-door collection service by means of rickshaw vans for which households pay a nominal amount of between USD 0.14 to 0.57 depending on income levels.

Additional revenue is generated from compost sales and carbon trading on international markets. Compost is sold in bulk to private chemical fertilizer companies who rebrand and sell through their own marketing and distribution networks. This sales strategy ensures an assured, large and growing market base for Waste Concern's compost. Waste Concern's extensive business activities has created a value chain generating thousands of jobs among the urban poor particularly women; and has also contributed to reducing greenhouse gas emissions by 62,200 tons between 2001 and 2006 (excluding the CDM project). This local business has reduced solid waste management expenditures and saved landfill area.

KEY PERFORMANCE INDICATORS (AS OF 2012)				
	SMALL SCALE		MEDIUM SCALE	LARGE SCALE
Scale of production (quantity of waste processed):	3 tons/day	10 tons/day	20 tons/ day	700 tons/day
Land use (square meter):	468	1,338	2,341	N.A.
Capital investment (USD):	14,609	41,739	73,043	16,500,000
O&M cost (USD):	4,348	14,493	28,986	N/A
Output (tonnes of compost/day)	0.75	2.5	5.0	130.0
Potential social and/or environmental impact:	Value chain generated approx. 1,000 jobs among urban poor; reduced GHG emissions by 62,200 tons between 2001 and 2006; 13.4 ha of savings in landfill area			800 jobs created; reduction of 89,000 tons of GHG emissions [as of 2012, 150 jobs created and reduction of 34,200 tons of GHG emissions]
Financial indicators:				
Pay Back period (years)	2	1.71	1.5	-
Post-tax IRR	N.A.	N.A.	N.A.	N.A.
Gross margin	8,696	28,986	57,971	N.A.

Context and background

The city of Dhaka, the capital of Bangladesh, produces about 4,700 tons of solid waste per day. The Dhaka City Corporation (DCC) is responsible for managing the waste; however with an ever-shrinking waste management budget and unavailability of landfill sites, it is only able to collect less than 40% of the total waste. As a result, waste is dumped in open areas and unmanaged landfill sites, creating many serious threats including diseases, intolerable odor, contamination of water sources, emission of greenhouse gases and exposing the rag-pickers to hazardous waste. In view of the then-prevailing problem, two young and dynamic urban planners, Iftekhar Enayetullah and Maqsood Sinha, founded Waste Concern, initially a research-based non-governmental organization (NGO) in the field of waste management and environment. Waste Concern is mainly involved in collection and processing of municipal solid waste (MSW) into compost and marketing thereafter. It began its composting operations in 1995 on an experimental basis in a small area of 1,000m² lent to it by the Lions Club for a period of three months. This demonstration project was to explore the technical and commercial feasibility of the labor-intensive aerobic composting technique. It also adopted door-to-door collection of waste with the help of rag pickers by providing them with rickshaw vans. This activity started by covering 100 households which subsequently increased to 600 households by 2004. At the time of the study, the service was extended to 1,400 households by partnering with community-based organizations. Waste Concern has set an example for a successful decentralized community-based waste management business. Using an appropriate composting technology in combination with sound financial management, as well as an appropriate marketing strategy ensures high quality compost and constant sales throughout the year. This model is already been replicated in 27 cities of Bangladesh and 10 cities of other developing countries with the support form external support agencies as well as local entrepreneurs. In 2005, to scale up its model with private investment, Waste Concern in partnership with a Dutch recycling company called World Wide Recycling BV initiated a project where carbon trading has been harnessed. This is the world's first compost plant using CDM opportunity.

Market environment

Huge amounts of waste are generated daily in the city of Dhaka, which the Dhaka City Cooperation has found difficult to manage. Indiscriminate waste disposal and unmanaged landfills spurred Waste Concern's desire to enter into a partnership with both private and public organizations to process MSW into organic fertilizers for agricultural purposes. While this initiative addressed the imminent environmental and social challenges, the production of compost represented a valuable agricultural input alternative for farmers. The extensive use of chemical fertilizers has degraded the soil to a great extent and an alternative to successfully replace synthetic fertilizers was a necessity and Waste Concern compost with value addition by MAP Agro filled this gap and made for the correct type of replacement for chemical fertilizers. Additionally, the growing popularity of industrial poultry farming in the country also created an increasing opportunity for compost as poultry feed. An approval from the Bangladesh Agriculture Research Council for the suitability of the compost product for agricultural purposes and policy support from Ministry of Agriculture was essential for market acceptance.

Macro-economic environment

With an estimated population of 291 million by 2050 in Bangladesh, total rice demand is expected to reach 68 million tons, which is more than twice that compared to 2007. To match this anticipated spike in agricultural production, chemical fertilizer application and demand are expected to reach an all-time high (Basak, 2014). Government provides subsidies on chemical fertilizer for agricultural producers which accounts for about 6% of total public expenditure. Farmers have generally been found to use chemical fertilizers indiscriminately without adequate information on actual soil and plant requirement. Over-application is common and this has resulted in depleted soils and a decline in crop yields. The use of organic fertilizers will play a vital role in restoring soil fertility and improving crop productivity. Policy instruments to address market price distortions created by the current subsidies on chemical fertilizers will be imperative to catalyze business development in the organic fertilizer market. A detailed analysis of the policy environment was provided by Matter et al. (2015).

Business model

Figure 157 represents Waste Concern's business model canvas. Using strategic partnerships that engage both public and private entities, Waste Concern's compost business models implement a small-to-medium decentralized community based approach and large scale CDM/carbon trading approach. This figure presents an aggregate of both the small-to-medium decentralized community based approach and large scale CDM/carbon trading approach. As a key characteristic of their business model, Waste Concern has forged strategic partnerships with the local government, private enterprises and community-based organizations (CBOs) to optimize the allocation of resources and activities; reduce risk associated with high capital investments and establish an assured market for their product. At the start-up, development agencies such as UNDP, UNICEF and CIDA provided both financial and expertise support for smooth operations of the business. Research institutes (universities) did and continue to provide periodic quality testing of the finished compost for which the services are paid for by Waste Concern. The local government provided land for the composting plants and gave Waste Concern legal access to the city waste. In alleviating Waste Concern's initial investment costs, the municipality gains from reduced waste collection and landfill costs. Whilst Waste Concern has a

CASE: SOCIALLY-DRIVEN MUNICIPAL WASTE COMPOSTING

FIGURE 157. WASTE CONCERN BUSINESS MODEL CANVAS (INCLUDING BOTH THE COMMUNITY-BASED AND CDM COMPOSTING MODELS)

KEY PARTNERS
- Dhaka City Corporation (DCC)
- Ministry of Environment and Forest (MoEF)
- UNDP
- Private chemical fertilizer companie
- Local communities and community-based organizations (rickshaw van operators)
- CDM Board

KEY ACTIVITIES
- Collection of municipal solid waste, also door-to-door
- Segregation of waste
- Processing of MSW to organic compost
- Sale of compost and carbon credits
- Quality control
- Awareness campaigns
- CER administration activities

KEY RESOURCES
- Right to collect city's MSW
- Land
- Capital
- Labor
- Contract partnership agreement with municipality
- CDM registration, certification etc.

VALUE PROPOSITIONS
- Provision of safe and nutrient-rich organic fertilizer (valuable substitute for chemical fertilizer)
- Provision of waste management services including door-to-door
- Provision of certified Reduction of greenhouse gas emissions

CUSTOMER RELATIONSHIPS
- Direct interaction and personal help with clients
- Contract agreement with municipality

CHANNELS
- Direct and bulk sales to fertilizer companies
- Contract agreement
- Direct sale

CUSTOMER SEGMENTS
- Both rural and urban farmers
- Fertilizer trading companies (MAP Agro industries, Alpha Agro Ltd.)
- Municipality (local government)
- International carbon-credit clients

COST STRUCTURE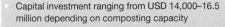
- Capital investment ranging from USD 14,000–16.5 million depending on composting capacity
- Operation and maintenance

REVENUE STREAMS
- Sale of compost
- Household waste collection fees
- Sale of Carbon emission reductions through CDM

SOCIAL & ENVIRONMENTAL COSTS
- None noted with available data

SOCIAL & ENVIRONMENTAL BENEFITS
- Reduced human exposure to untreated waste
- Possible reduction in human health cost in the locality due to reduction of pollution
- Significant job creation for the poor without any gender discrimination
- Reduction of greenhouse gas emissions
- Reduction in soil degradation with adoption of organic compost for agricultural production

legal permit from the DCC (main governing body in charge of managing waste) to access and process municipal waste in Dhaka, it does not have exclusive rights (own) to the waste and thus there remains the risk of facing competitors (e.g. compost producers) for the waste input. However, with over 4,700 tons of waste generated daily in the city and DCC limited capacity to properly manage only 40% of the waste, risk associated with input (waste) supply is relatively low. Community-based organizations are contracted for the collection and separation of waste, which ensures a consistent supply of high quality waste input for Waste Concern and income for the CBOs. Waste Concern earns a revenue through the established door-to-door collection service by means of rickshaw vans by the CBOs for which households pay a nominal amount of between USD 0.14 to 0.57 depending on income levels.

Municipal solid waste is processed into compost and sold directly in bulk through an established countrywide marketing and distribution system of private chemical fertilizer companies such as MAP Agro, providing an assured and large market base for their product. On the other hand, without established marketing and distribution channels, Waste Concern faces a strong buyer power as they mainly sell their compost to price-setting private chemical fertilizer companies who rebrand and sell the compost product. To reduce buyer power risk, Waste Concern launched an information campaign using farm demonstrations to raise consumer awareness and product demand. Waste Concern is negotiating with other large bulk compost users to limit their dependency on their main customer – MAP Agro Fertilizer. This has been an important strategy to also increase their direct market share as substitute products (e.g. other organic fertilizer products and chemical fertilizers) continue to flood the market. The threat of new business entrants is very high as there is an increasing availability and unlimited access to the waste input. Waste Concern, however, has an edge over new entrants given its strong partnerships across public and private sectors and communities, which is essential to mitigate many of the market risks it faces.

Value chain and position

Waste Concern's business operations cover the entire MSW value chain, providing services from collection to processing of the waste. Its activities have been implemented under two main business models namely:

a. Partnership model of community-based composting

Waste Concern's initiatives combine both public and community spheres with private sector involvement (Figure 158). Seed money from UNDP in partnership with the Ministry of Environment and Forests, Dhaka City Corporation (DCC) and Public Works Department (PWD) were utilized to implement community-based, solid waste management projects. A key characteristic of Waste Concern's community-based composting model is that it can be adapted to many contexts both in urban and rural areas. It has also shown great potential for implementation in slum areas at a small-, medium- or large-scale. The small-scale model processes three tons of organic waste daily, with the medium- and large-scale models processing three to 10 tons and more than 11 tons of organic waste per day, respectively. By focusing its efforts on the city's slums, an area where more than a third of the city's 11 million people live, Waste Concern has created a system that allows the community not only to dispose of trash effectively but also helps them to raise money. The organizational set-up of the composting scheme follows a business approach, which means that the community is seen as client who is paying for the service of waste collection. Waste Concern earns revenue through its established door-to-door collection service by means of rickshaw vans with capacity of 1.18m^3 for which households pay a nominal amount of between USD 0.14 to 0.57 depending on income levels. Waste Concern largely sells its compost in bulk to private chemical fertilizer companies such as MAP Agro Fertilizers, who rebrand and sell the compost through their established countrywide marketing

CASE: SOCIALLY-DRIVEN MUNICIPAL WASTE COMPOSTING

FIGURE 158. WASTE CONCERN VALUE CHAIN – PARTNERSHIP MODEL OF COMMUNITY-BASED COMPOSTING UNDER SEMP

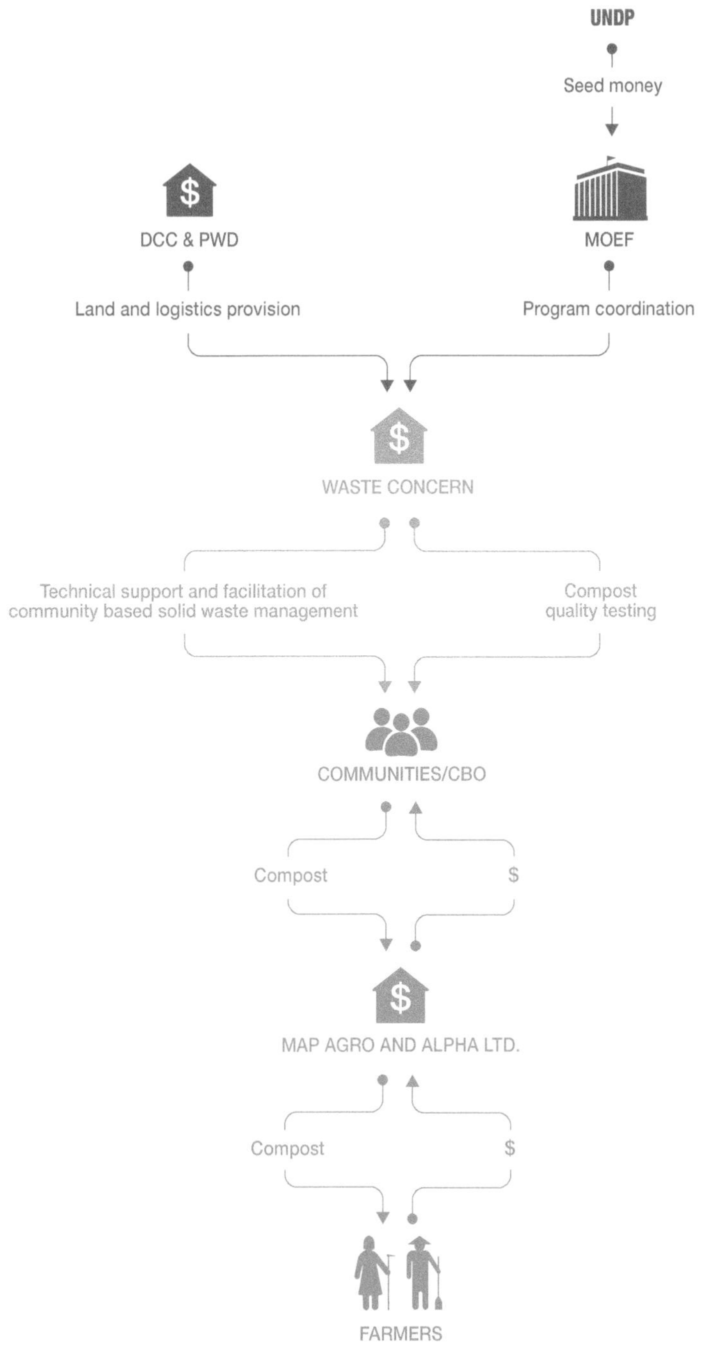

UNDP
Seed money

DCC & PWD
Land and logistics provision

MOEF
Program coordination

WASTE CONCERN

Technical support and facilitation of community based solid waste management

Compost quality testing

COMMUNITIES/CBO

Compost $

MAP AGRO AND ALPHA LTD.

Compost $

FARMERS

and distribution system. This partnership provides access to an assured, large and growing market base for Waste Concern's compost, selling about 10,000 tons of organic fertilizer per year (2010), which represents a significant portion of the market. This marketing strategy mitigates competition risk that they would otherwise face with chemical fertilizers. The community-based composting scheme has an added benefit for the communities of Dhaka in that they share in the profits made in selling the compost, earning USD 0.09 per kilogram. This model has improved the livelihoods of community members as the compost collectors come from the community and earn up to USD 52 per month[1]. The sustainability of this model is grounded in strong partnerships and the assured benefits accruing to each partner.

b. Composting under CDM/carbon trading model

Waste Concern has also established the world's first CDM compost plant in Bangladesh. This carbon trading-based business model is based on strong partnerships between the public, private and community spheres (Figure 159). Waste Concern partners with the Clean Development Mechanism's Board, which approves a compost plant project owned as a joint venture by Waste Concern and World Wide Recycling (WWR). Dhaka City Corporation (DCC) provides the approval for the collection and processing of the city's waste by Waste Concern. The compost plant obtains organic waste from the urban population through direct collection from vegetable markets. The resulting higher-yield, lower-cost compost is sold to rural farmers, and the carbon credits obtained are sold on the international market. A key characteristic of this model is that the municipality does not bear any cost with the setup of the project. Waste Concern collects all waste free of charge; and also bears

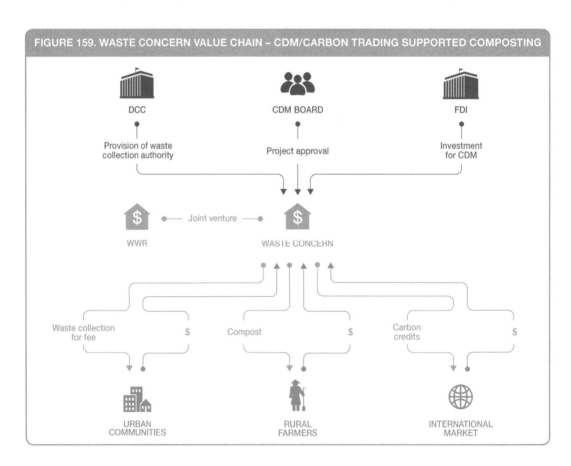

FIGURE 159. WASTE CONCERN VALUE CHAIN – CDM/CARBON TRADING SUPPORTED COMPOSTING

the cost for the land of the compost plant. This model saves the city numerous costs associated with waste collection, transportation, and disposal. The plant has two major sources of revenue: one is compost (organic fertilizer), and the other is Certified Emission Reduction (CERs)[2]. The compost plant processed between 75–100 tons of organic waste on a daily basis between 2009 and 2010. By 2012, the project had processed 76,697 tons of organic waste and generated 34,200 CERs. In addition to reducing greenhouse gas emissions, this model also generates valuable carbon credits on the international market. This project has improved livelihoods in the community, creating 150 direct jobs for the poor, with these jobs cutting across the entire MSW value chain from compost plant operation, transportation of waste and in the distribution of compost. This model is grounded in a win-win partnership between key players and has been instrumental in attracting large amounts of foreign direct investment (FDI) in the area of organic composting and carbon trading using the Clean Development Mechanism (CDM) of the Kyoto Protocol.

Institutional environment

Although the solid waste management system in Bangladesh is still not well organized, efforts are under way to improve the organizational structure for solid waste management in different cities. An example is Dhaka City Corporation which has established a Solid Waste Management Cell to improve the waste management services in the city. At the national level, the Urban Management Policy Statement, 1998 was enacted and implemented by the Government of Bangladesh, which recommends municipalities to privatize waste management services and give priority to slum areas. For more recent policy development see Matter et al. (2015). The special emphasis and encouragement of private sector participation in water supply and sanitation in urban areas is gradually resulting in the provision of efficient and reliable waste management services to the public, especially those in slum areas.

Technology and processes

A box-type composting technique was adopted because it is a low-cost process that needs less turning compared to the Indonesian Windrow Method, which was originally used (Figure 160). It has limited mechanization and is suitable for Bangladesh's climatic conditions. The composting process requires 40 days for decomposition and a maturing period of 10–15 days. Special measures are taken to reduce the odor. After maturing, the compost is screened and graded according to particle size and packed for marketing. Waste Concern has also developed two other types of composting methods apart from the Box Composting under the UNDP supported Sustainable Environment Management Programme (SEMP). These are the Aerobic Composting and Barrel Type Composting methods. All three techniques are simple, low cost and labor intensive methods which are suitable to the socioeconomic and climatic condition of Bangladesh.

Funding and financial outlook

For Waste Concern's decentralized business model, there is a range of plants across the city of different sizes and investment cost. These range from USD 14,000 to USD 73,000. The cost of maintaining and operating a plant also varies from USD 4,300 to USD 29,000 depending on the size of the plant. The company has benefited from the provision of land by the local government at no cost and financial support from Lion's Club, UNDP, UNICEF and CIDA, as well as technical guidance. Financial data was not accessible for the CDM business model. For both models, Waste Concern generally has two main revenue streams: a) compost sales; and b) carbon trading. About 31,100 metric tons of compost is sold on yearly basis yielding revenue of USD 998,621. There is the possibility of revenue generation from carbon credits for the decentralized business model. However, with a decline of the carbon market, these options have to be carefully analyzed.

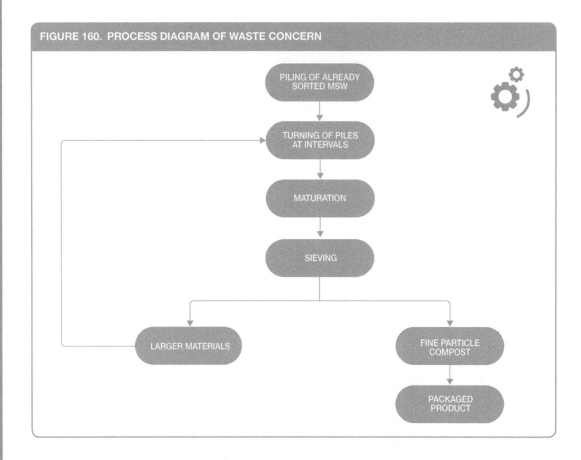

FIGURE 160. PROCESS DIAGRAM OF WASTE CONCERN

Socio-economic, health and environmental impact

Waste Concern's diverse projects have created numerous direct and indirect benefits for the economy and the environment. The simple idea of converting the high organic content of the waste, into compost brought about a valuable substitute for chemical fertilizers. Overuse of chemical fertilizers has been a serious problem in Bangladesh which has led to severe soil degradation. Farmers had no real alternatives in the absence of the organic fertilizers in the market prior to the entrance of compost from Waste Concern in the agricultural input market. Compost produced by Waste Concern has increased per hectare yield by 30–50% by adopters (potato farmers). The Waste Concern business, extending from collecting and processing waste produced in urban areas to selling compost to rural farmers, has created a value chain generating close to 1,000 jobs among the urban poor, especially women. The total value of the compost sold in the local market between 2001 and 2006 was USD 1.10 million. Close to 500,000 people are benefiting from household waste disposal system across the country. Waste Concern has also contributed in reducing greenhouse gas emissions by 62,200 tons of CO_2e between 2001 and 2006 (excluding the CDM project), and saved 13.4 ha of landfill area. The upcoming CDM project is also expected to reduce greenhouse gas emissions by 1 million tons over eight years, produce 50,000 tons of compost per year. At a global scale, this initiative has the potential to reduce transboundary impact of GHG and attract foreign direct investment. Waste Concern has also extended itself in the policy-making arena, steering environmentally appropriate governmental regulations, both existing and new. To date, they have been influential in the development of 27 governmental policies and spearheaded efforts at influencing the government to develop national policies and guidelines in issues in the like of CDM Project Approval Process for Government.

Scalability and replicability considerations
The key drivers for the success of this business are:
- Increasing need for sustainable waste management solutions.
- Strong, strategic partnerships with city municipality, Ministry of Environment and Forests, Dhaka City Corporation, Public Works Department, Community-Based Organizations, Private Fertilizer Companies and development agencies, gaining Waste Concern a.o. free or low-cost access to waste and to land.
- A perceived necessity to replace chemical fertilizers due to their effect in degrading soil and environment.
- Government (ministry of agriculture policy) that support/promoted use of compost for agricultural purposes.

The Waste Concern model has high replication potential and has already been replicated in 27 cities of Bangladesh and 10 cities of developing countries with the support from external support agencies as well as local entrepreneurs. Adopting a labor-intensive, cheap and low technological approach, the business does not require a large capital investment (except for land purchase) or state-of-the-art machinery, which removes one of the major constraints for business start-ups especially in the developing world context. The decentralized composting approach reduces transportation costs and makes use of low cost technologies based on manual labor and ensures waste is well sorted before it is composted. This minimizes many of the problems and difficulties that have led to the failure of large centralized composting plants in the past. There is great potential for the upscaling of this model due to its simplicity. Many decentralized units can be attached to the main business as long as raw material or the market demand does not become limiting factors. However, the decentralized approach to composting of waste work best for secondary cities and small towns where local government can allocate land. Similarly, the large-scale carbon trading model has a high replication potential. The technology adopted is semi-mechanized and offers opportunity to use unskilled and informal labor, indicating its suitability for developing countries.

Summary assessment – SWOT analysis
Figure 161 presents the SWOT analysis for Waste Concern. Composting has become a promising business in Bangladesh. Waste Concern has been particularly successful by using a suitable composting technology in combination with a sound financial management and an appropriate marketing strategy, which enables Waste Concern to produce high quality compost and ensure constant sales throughout the year. This business can hardly meet the demand for compost and processes several hundred tons of city waste daily since 2010 (Waste Concern, 2011). Increasing governmental and international support along with growing demand for normal and enriched compost, spurred by the user awareness building programmes, are seen as key opportunities for replication and up-scaling of the business. Waste Concern will, however, face increasing competition from new market entrants and increased buyer power if it does not explore new key customers or begin to establish its own marketing and distribution channels. Waste Concern is an example of an innovative social entity utilizing a simple business approach to address some of the major waste management and environmental challenges in Dhaka, Bangladesh and its model of organic composting is a clear demonstration of a successful business model that includes the poor, especially women both in the supply and the demand chain.

FIGURE 161. SWOT ANALYSIS FOR WASTE CONCERN

	HELPFUL TO ACHIEVING THE OBJECTIVES	**HARMFUL** TO ACHIEVING THE OBJECTIVES
INTERNAL ORIGIN ATTRIBUTES OF THE ENTERPRISE	**STRENGTHS** • Limitless and exclusive supply of municipal solid waste • Low O&M due to adoption of simple technology • Goodwill earned due to environmental stress relief • Good local buy-in due to establishment of brand for quality • Research and development work to strengthen the product • Excellent relationship with partner organizations • Decentralized composting units that do not depend on one another	**WEAKNESSES** • High dependence on few main customers – strong buyer power • Profit structure highly dependent on cheap labor
EXTERNAL ORIGIN ATTRIBUTES OF THE ENVIRONMENT	**OPPORTUNITIES** • Absence of strong competitive organic fertilizer suppliers to the market • Up-scaling potential of CDM project to earn carbon credits and set-up of additional decentralized units • Compost as poultry feed increases the market • Scale of production and transboundary GHG effects can potentially attract Foreign Direct Investment • Scale of impact and steering environmentally-appropriate governmental regulations provides positive market effect	**THREATS** • Increasing competition from other compost businesses • Increasing labor wages • Unstable carbon market

Contributors

Jasper Buijs, Sustainnovate, Netherlands; formerly IWMI, Sri Lanka
Alexandra Evans, Independent Consultant, London, United Kingdom
Michael Kropac, CEWAS, Switzerland
Josiane Nikiema, IWMI, Ghana

References

Alamgir, M. and Ahsan, A. 2007. Municipal solid waste and recovery potential: Bangladesh Perspective. Iran. J. Environ. Health. Sci. Eng. 2007 4 (2): 67–76.

Basak, J.K. 2014. www.unnayan.org/reports/Livelihood/future_fertilizer_demand.pdf (accessed November 8, 2017).

Bhuiya, G.M.J.A. 2007. Solid waste management: Issues and challenges in Asia. Bangladesh.

Enayetullah, I. 2006. Community based solid waste management through public-private-community partnerships: Experience of Waste Concern in Bangladesh. Paper presented at 3R Asia Conference, Toyko, Japan, 30 Oct–1 Nov, 2006.

Enayetullah, I., Khan, S.S.A. and Sinha, A.H. Md.M. 2005. Urban solid waste management. Scenario of Bangladesh: Problems and prospects. Waste Concern Technical Documentation.

Kwon, K. 2005. Financial feasibility of composting market waste in Vientiane, Lao PDR. University of Toronto, Master of Engineering Thesis, 2005.

Matter, A., Ahsan, M., Marbach, M. and Zurbrügg, C. 2015. Impacts of policy and market incentives for solid waste recycling in Dhaka, Bangladesh. Waste Management 39: 321–328.

Memon, M.A. 2002. Solid waste management in Dhaka, Bangladesh: Innovation in community driven composting. 19–20 September 2002, Kitakyushu.

Personal communication with plant managers. 2014.

Rahman, H. 2011. "Waste Concern: A decentralized community based composting through public-private community partnership." GIM Case Study No. B102. New York: United Nations Development Programme.

Seelos, C. 2008. Corporate strategy and market creation in the context of deep poverty. The Gold Prize Essay at the International Finance Corporation and the Financial Times Annual Essay Competition.

Sinha, A.H.M.M. and Enayetullah, I. 2000. Community based solid waste management: The Asian experience. Dhaka: Waste Concerns.

Waste Concern. 2011. www.wasteconcern.org. Accessed 5 March, 2016.

Waste Concern. 2004. Waste Concern, Department of Environment. Country Paper Bangladesh.

Zurbrügg, C. 2002. Decentralised composting in Dhaka, Bangladesh; Production of compost and its marketing. ISWA 2002 Annual Congress, Istanbul 8–12 July, 2002.

Zurbrügg, C., Dreschera, S., Rytza, I., Maqsood Sinhab, A.H. and Enayetullah, I. 2005. Decentralised composting in Bangladesh, a win-win situation for all stakeholders. Resource Conservation and Recycling 43 (2005): 281–292.

Case descriptions are based on primary and secondary data provided by case operators, insiders or other stakeholders, and reflect our best knowledge at the time of the assessments 2014/15. As business operations are dynamic data can be subject to change.

Notes

1 Based on 2014 exchange rates: USD 1 = 77.65 taka.
2 For carbon trading purposes, one CER is considered equivalent to one metric ton of CO_2 emissions.

BUSINESS MODEL 12
Large-scale composting for revenue generation

Miriam Otoo and Munir A. Hanjra

A. Key characteristics

Model name	Large-scale composting for revenue generation
Waste stream	Municipal solid waste (MSW), minor percentage of agro-waste
Value-added waste product	Recovered soil nutrients in the form of compost from MSW to address dual challenge of soil nutrient depletion and waste management
Geography	Any urban centre, assuming availability of land for plant construction
Scale of production	Medium to large scale; minimum plant size processes 60–100 tons of MSW per day, with a maximum size of 1,500 tons per day
Supporting cases in this book	Delhi, Ludhiana, Karnataka in India; Dhaka, Bangladesh
Objective of entity	Cost-recovery [ ]; For profit [X]; Social enterprise [X]
Investment cost range	USD 415,000–1.5 million depending on technology used and pay-back period of 2 to 7 years
Organization type	Public, private, public-private partnership, or social enterprise/entity
Socio-economic impact	Environmental benefits from reduced nutrient release into soils and waterbodies from reduced chemical fertilizer use, reduced GHG emissions via reduced production of chemical fertilizers and landfill emissions, reduced human exposure to untreated waste, improved waste management services, cost savings to municipalities from reduced land acreage for landfills and disposal costs
Gender equity	Employment generation for the urban poor, including women. Technology-wise no particular (dis)advantage for any gender

B. Business value chain

This business model rests on the notion that there is great potential for addressing the dual challenge of waste management and to some extent nutrient soil depletion via the recovery of nutrients from municipal solid waste (MSW) in large urban areas of developing countries. It is important to note that although the former may be the main driving force given the widening service gap between provision of waste management services and municipalities' budgets and infrastructural capacities an equally important driver is the increasing need for environmentally friendly and cost-effective fertilizer alternatives for agricultural producers. Thus, the opportunity of increased cost savings from reduced transportation costs and landfills as well as revenue generation and even profit making explicitly represents opportunities for different entities to engage in compost production from MSW.

A myriad number of constellations based on different scales of production, technologies, business strategies, partnerships, financing, among other factors, exist for this model. This business model can be initiated by a public, private entity, public-private partnership or a social enterprise to provide sustainable solutions for urban waste management issues and produce value-added products and services that generate significant benefits to several actors in both the sanitation and agricultural value chains. The goal of profit maximization via increased revenue generation drives the business strategies that the entities institute which is hinged on: a) portfolio diversification (multiple-revenue stream approach); and b) strategic partnerships.

Whilst the core business centres on provision of waste management services and fertilizer alternatives to agricultural producers and generates revenue from: a) waste collection fees charged to the municipality, households or commercial entities; b) sales of organic fertilizer products; c) sales of recyclables; by leveraging its scale, additional revenue streams that can be tapped into are sale of energy (electricity, biogas) and carbon credits. Businesses can also implement a franchise-based approach to increase their revenue streams and capture additional markets. A typical arrangement can include the following: a) the franchiser provides training on technology and management for compost production on a (discounted) fee basis; b) the franchisee sells the compost to the parent company who (can further add-value to the compost) markets and distributes the compost through its established distribution networks or those of its partners. Profits are shared between the franchiser and franchisee depending on agreed percentages outlined in contractual agreements.

Large-scale operations, whether through centralized or decentralized systems, offer the opportunity to capture benefits from economies of scale. Large-scale operations using efficient technologies along the entire compost production process can reduce production costs. This implies that the business can charge lower prices for the compost product and significantly increasing their market share and additionally gain access to new markets, such as the carbon credit market which has scale requirements. Additionally, efficient energy production whether for internal use to reduce production costs or for sale typically occurs at a larger-scale. Especially in the latter case, businesses can only connect to the grid if they are able to supply a certain wattage of electricity.

Strategic partnerships on different levels with the local government, private enterprises and community-based organizations to optimize the allocation of resources and activities, reduce risk associated with high capital investments, establish an assured market for their product, among others, will be imperative for the sustainability of the model, particularly given the multiple elements (activities) of the business. Central to this business model is the enterprise initiating and implementing the model for better waste management and revenue generation, as shown in a generic value chain schematic (Figure 162). Depending on the organizational structure of the model, the ownership, financing and operation of the enterprise transforming MSW to compost can take different forms. For example, management models can include: a) municipally owned – municipally operated; b) municipally owned – privately operated; and c) privately owned – privately operated. This often translates to the mode of financing of the initiative which can be through private equity, government or donor grants or a combination of these (Kaza et al., 2016).

Particularly for PPP initiatives (for example, in the cases of ILFS-Okhla and A2Z Infrastructure Limited-Ludhiana in India), the public entity typically provides the capital investment and outsources the overall management of the plant – to include sales and marketing of the compost products to the private entity. Additionally, from a private entity's perspective, partnership with government authorities in relevant sectors provides easy access to the city's waste streams and the often well-established fertilizer marketing and distribution networks. The former implies that there is no competition from other entities in terms of input supply, ensuring continuous and unlimited access to the waste, whilst the latter increases market access for the compost products. On another front, contracting-out some of the waste collection

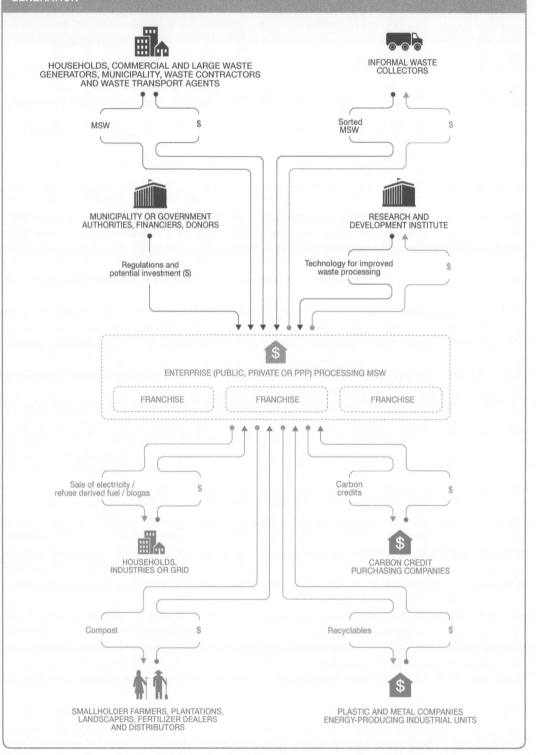

FIGURE 162. VALUE CHAIN SCHEMATIC – LARGE-SCALE COMPOSTING FOR REVENUE GENERATION

activities to informal waste collectors brings an inclusive element to a 'for-profit' model. This not only improves the livelihoods of landfill ragpickers by ushering them into mainstream jobs but it can allow the business to efficiently cover slum areas where poor road infrastructure make them less accessible.

C. Business model

The business model is hinged on a multiple-revenue stream approach which results in three value propositions: a) provision of sustainable and affordable waste management services to communities and businesses; b) increased supply of environmentally-friendly fertilizer alternatives to agricultural producers at affordable prices; and c) provision of recyclables to energy-producing industrial units at competitive market prices. The business model described here presumes the operation for a standalone private enterprise (Figure 163).

The provision of waste management services (i.e. waste collection) from households, commercial entities, institutions at a fee, can generate significant income. The business will however require a sound partnership agreement with the local authorities or municipality to ensure exclusive rights to the city's waste. The business additionally produces organic fertilizer products from MSW and minimally agro-waste. The main customer segments are agricultural producers who can be reached via direct sales or partner dealer networks. Given the large scale of operation, a secure market is needed for the compost. In that regard, the business has to consider innovative marketing and distribution strategies as well as product development. Strategies to be considered include: a) **partnerships** with government, agriculture departments and agro-industries, to take advantage the often well-established fertilizer distribution networks; b) **market segmentation** – different prices are charged to different customer segments to capture a larger share of the consumer surplus; c) **production innovation** – increase the accessibility and usability of compost via pelletization (as the bulky nature of compost often acts as a barrier to the transportation of the product to markets, increasing the distribution costs, which are borne by the end-users) and nutrient fortification to boost compost fertilizer value. For the latter strategy, partnership with a research institute is crucial to ensure continued product and process innovation.

This business model can also derive additional revenue from recovered non-degradable materials including high density plastics and metals that could be sold directly to the plastic and metal companies and the remaining solid materials to energy producing industrial firms for refused derived fuel (RDF). This business model adds two new stakeholders – inorganic material clients such as plastic manufacturing and energy producing commercial units using RDF, and informal waste collectors, adding value through collection and sorting of these materials, while also generating employment for these informal sectors workers including women. For large-scale operations, waste segregation into biodegradable and non-biodegradable portions is mainly a mechanized process but some level of sorting can be done by rag pickers. This model does not only improve the livelihoods of rag pickers (via assured and increased earning) but it increases coverage of slum areas where poor road infrastructure makes them inaccessible for mechanized operations. The demand for inorganic materials including refused derived fuel and plastics/metals is growing and collection costs could easily be covered through household fees. Wastes of particle size greater than 50mm can be sorted, shredded, packaged and sold partly to electricity units as well as cement, tile manufacturing and brick units. A portion of the remaining RDF material can be sold and the remaining quantities burnt to generate electricity for the business' internal use.

Alternate scenarios

The generic business model described above is to produce compost from MSW for agricultural purposes and provide waste management services. The business can be modelled along three different scenarios to include: a) a franchise system; b) energy (biogas and/or electricity) generation for internal use or sale to the grid; c) large-scale operations for carbon credits under the CDM.

Scenario I: Commercial establishment for composting through consultancy services and franchising royalties

This business model (Figure 164 on page 440) builds on the generic model described above. The business sets up a franchising system to its compost production component of the business to further increase its market access (in terms of provision of waste management services to communities and organic fertilizers to agricultural producers) and revenue. The multi-revenue approach adopted by the business will support its transition from a cost-recovery model to one of profit generation. In addition to

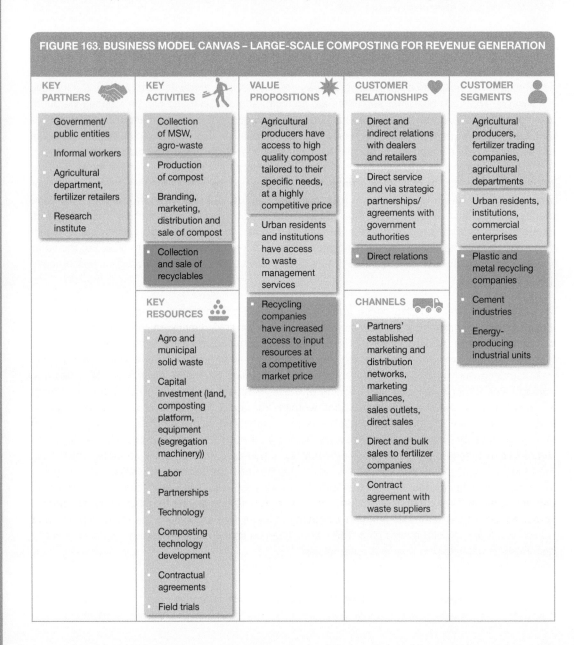

FIGURE 163. BUSINESS MODEL CANVAS – LARGE-SCALE COMPOSTING FOR REVENUE GENERATION

BUSINESS MODEL 12: LARGE-SCALE COMPOSTING FOR REVENUE GENERATION

COST STRUCTURE	REVENUE STREAMS
• Capital investment costs (land, infrastructure, equipment) • Operation and maintenance costs (labor, R&D, marketing and distribution costs, administrative costs) • Financing costs (interest on borrowed funds) • Payment of shared revenues (if any)	• Sale of compost products • Service fees from households, institutions, commercial establishments • Sale of recyclables • Government assistance (if any) for compost programs under sustainable agriculture programs
SOCIAL & ENVIRONMENTAL COSTS	SOCIAL & ENVIRONMENTAL BENEFITS
• Possible health risk to workers if appropriate protective gear is not used during production processes, especially when handling the raw waste • Possible water and soil contamination if leachate from the compost production is not captured and treated appropriately	• Reduced existing waste management costs • Reduced human exposure to untreated waste • Reduction of greenhouse gas emissions • Reduced acreage for landfills • Improved environmental health • Enhanced soil fertility and productivity from increased use of organic fertilizers for agricultural production • Capacity development of local farmers • Significant job creation for the poor without any gender discrimination

earning revenue from waste management service fees, sale of organic compost and fertilizer products and sale of inorganic recyclables, they can earn franchising royalties from their franchise network operating across different locations that provide waste management products and services. Depending on if the business is a public, private, PPP or social entity, the incremental revenue (if representative of a surplus) can be reinvested in technology innovation and new marketing strategies to further improve production efficiencies and dependence on partners' distribution networks, respectively. The franchise system also creates a greater opportunity for the parent business to enter into a CDM program. This is because the parent business may only be able to meet the scale of operation requirements for carbon credits sale upon inclusion of the franchisees' operations. The incorporation of a franchising system additionally builds inclusivity into the original business model as smaller-scale enterprises (such as CBOs) gain access into the waste management sector and generate jobs/income for individuals that would otherwise be unemployed. The parent business can further earn revenue via consultancy fees charged for the design and commissioning of waste management projects for townships and commercial clients, and the training of agricultural graduates and professionals in the field of waste management and compost production.

Scenario II: Energy generation and carbon credit sales

With the inherent large scale of operations (or derived from aggregate scale of franchises, the parent enterprise can efficiently produce energy for its own onsite use or enter into a partnership with the state electricity board and sell any surplus energy to the national grid. The business' ability to tap into the energy market is highly dependent on its scale given the minimum wattage requirements for electricity sale to the national grid. Cost-savings from use of internally-produced energy imply decreased production costs, and along with the sale of electricity increased revenue generation. This model can maximize resource recovery from municipal solid waste, diversify its portfolio beyond compost production, mitigating risk associated with seasonal compost demand and marketing, and

FIGURE 164. BUSINESS MODEL CANVAS – LARGE-SCALE COMPOSTING FOR REVENUE GENERATION WITH FRANCHISING SYSTEM

KEY PARTNERS	KEY ACTIVITIES	VALUE PROPOSITIONS	CUSTOMER RELATIONSHIPS	CUSTOMER SEGMENTS
• Government/public entities • Informal workers • Agricultural department, fertilizer retailers • Research institute	• Collection of MSW, agro-waste • Production of compost • Branding, marketing, distribution and sale of compost • Collection and sale of recyclables • Provision of consultancy and design services for turn-key waste management projects and franchisees operations • Marketing and distribution support to franchisees facilities	• Agricultural producers have access to high quality compost tailored to their specific needs, at a highly competitive price • Urban residents and institutions have access to waste management services • Recycling companies have increased access to input resources at a competitive market price • Increased creation of business operations for communities via franchises	• Direct and indirect relations with dealers and retailers • Direct service and via strategic partnerships/agreements with government authorities • Direct relations • Formal franchisee agreements	• Agricultural producers, fertilizer trading companies, agricultural departments • Urban residents, institutions, commercial enterprises • Plastic and metal recycling companies • Cement industries • Energy-producing industrial units • Clients of franchise network

KEY RESOURCES	CHANNELS
• Agro and municipal solid waste, • Capital investment (land, composting platform, equipment) • Labor • Partnerships • Technology • Composting technology development • Field trials • Contractual agreements (including franchising contracts)	• Partners' established marketing and distribution networks, sales outlets, direct sales • Direct and bulk sales to fertilizer companies • Contract agreement with waste suppliers • Direct sale • Open bidding, call for tenders

COST STRUCTURE	REVENUE STREAMS
• Capital investment costs (land, infrastructure, equipment) • Operation and maintenance costs (labor, R&D, marketing and distribution costs, administrative costs) • Financing costs (interest on borrowed funds) • Payment of shared revenues (if any)	• Sale of compost products • Service fees from households, institutions, commercial establishments • Sale of recyclables • Government assistance (if any) for compost programs under sustainable agriculture programs • Franchising royalties • Consultancy revenue
SOCIAL & ENVIRONMENTAL COSTS	SOCIAL & ENVIRONMENTAL BENEFITS
• Possible health risk to workers if appropriate protective gear is not used during production processes, especially when handling the raw waste • Possible water and soil contamination if leachate from the compost production is not captured and treated appropriately	• Reduced existing waste management costs • Reduced human exposure to untreated waste • Reduction of greenhouse gas emissions (improved environmental health) • Reduced acreage for landfills • Enhanced soil fertility and productivity from increased use of organic fertilizers for agricultural production • Capacity development of local farmers • Significant job creation for the poor without any gender discrimination

allow entry into the energy market. There is a great potential to improve the financial viability of the model from energy generation as there is generally a significant and growing demand for electricity in developing countries. Additionally, there are increasing opportunities for waste-to-energy entities to fill this gap based on the anticipated rapid rural electrification program; foreseeable increasing trend in electricity prices; structural and legal feasibility for private sector involvement (structural unbundling of the power sector, vertically integrated monopoly and privatization of the generation and distribution); a lesser vertically integrated market; and supportive renewable energy policies among others. It is noted however that particularly in developing countries, electricity producers are currently price takers and restricted to the price ceiling set by the state-owned transmission entity (limited negotiation ability – monopolistic market). Thus, the level of market concentration and market prices will determine whether investments in plant upgrades and equipment for energy production is worthwhile. The opportunity for waste-generated electricity can only materialize if the price offered in power purchase agreements (PPA) can substantially cover production costs and generate a net profit. The generation of energy, in addition to providing cost savings from internal use and generating sales revenues, can be accounted for carbon credit sales.

The business entity can also be registered as a CDM (Clean Development Mechanism) project to earn additional revenue from carbon credit sales to UNFCCC Annex I defined countries[1]. The composting of municipal solid waste offers opportunities for earning carbon credits through two main pathways: a) avoided GHG emissions from landfills; and b) reduced GHG emissions from reduced chemical fertilizer production and use. Carbon credits earned through avoided emissions over the base-case scenario can be sold in the global credit market to institutional and private investors (Figure 165). Carbon credits provide an additional value proposition that in most cases can help composting businesses on

FIGURE 165. BUSINESS MODEL CANVAS – LARGE-SCALE COMPOSTING WITH ENERGY GENERATION AND CARBON CREDITS SALE

KEY PARTNERS
- Government/public entities
- Informal workers
- Agricultural department, fertilizer retailers
- Research institute
- International development agencies (for CDM process)

KEY ACTIVITIES
- Production of compost
- Branding, marketing, distribution and sale of compost
- Research and development
- Collection of MSW, agro-waste
- Collection and sale of recyclables
- Provision of consultancy and design services for turn-key waste management projects and franchisees operations
- Marketing and distribution support to franchisees facilities
- Production of biogas (bio-methanation) for internal use
- Production and sale of electricity
- Managing CDM processes to obtain emission reduction credits

KEY RESOURCES
- Agro and municipal solid waste,
- Capital investment (land, composting platform, equipment)
- Labor
- Partnerships
- Technology
- Composting technology development
- Field trials
- Contractual agreements (including franchising contracts)
- Positive (image) reporting by media on Green Economy

VALUE PROPOSITIONS
- Agricultural producers have access to high quality compost tailored to their specific needs, at a highly competitive price
- Urban residents and institutions have access to waste management services
- Recycling companies have increased access to input resources at a competitive market price
- Increased creation of business operations for communities via franchises
- Reduction in energy demand and expenditure from internal energy generation and use
- Improved energy availability to communities and businesses
- Tradable certified emission reduction to meet carbon emission commitments

CUSTOMER RELATIONSHIPS
- Direct and indirect relations with dealers and retailers
- Direct service and via strategic partnerships/agreements with government authorities
- Direct relations
- Formal franchisee agreements
- Formal contractual agreement with government
- Registered as CDM at UNFCCC

CHANNELS
- Partners' established marketing and distribution networks, marketing alliances, sales outlets, direct sales
- Direct and bulk sales to fertilizer companies
- Contract agreement with waste suppliers
- Direct sale
- Open bidding, call for tenders
- Contract agreement with state electricity board
- CDM certificate trading on international market

CUSTOMER SEGMENTS
- Agricultural producers, fertilizer trading companies, agricultural departments
- Urban residents, institutions, commercial enterprises
- Plastic and metal recycling companies
- Cement industries
- Energy-producing industrial units
- Clients of franchise network
- State electricity board
- Investors from UNFCCC Annex I countries

BUSINESS MODEL 12: LARGE-SCALE COMPOSTING FOR REVENUE GENERATION

COST STRUCTURE	REVENUE STREAMS
- Capital investment costs (land, infrastructure, equipment) - Operation and maintenance costs (labor, R&D, marketing and distribution costs, administrative costs) - Financing costs (interest on borrowed funds) - Payment of shared revenues (if any) - Cost savings from in-house energy production	- Sale of compost products - Service fees from households, institutions, commercial establishments - Sale of recyclables - Government assistance (if any) for compost programs under sustainable agriculture program - Franchising royalties - Consultancy revenue - Sale of electricity to state electricity board (or relevant entity) - Sale of carbon credits
SOCIAL & ENVIRONMENTAL COSTS	**SOCIAL & ENVIRONMENTAL BENEFITS**
- Possible health risk to workers if appropriate protective gear is not used during production processes, especially when handling the raw waste - Possible water and soil contamination if leachate from the compost production is not captured and treated appropriately	- Reduced existing waste management costs - Reduced human exposure to untreated waste - Reduction of greenhouse gas emissions (improved environmental health) - Reduced acreage for landfills - Enhanced soil fertility and productivity from increased use of organic fertilizers for agricultural production - Capacity development of local farmers - Significant job creation for the poor without any gender discrimination - Supporting national efforts to tackle climate change - Attract FDI - Strong strategic partnerships and government support

a trajectory for profitability. However, the application process for a CDM project can be lengthy and complicated, involving certification, verification and accreditation to ensure compliance with various international standards, and often requiring additional investments for plant upgrade or retrofits. This thus requires support from international development agencies, government entities, and other private sector entities (consultancy support for formulation and submission of the application). In view of associated risks, the net returns on investment in the CDM project have to be carefully considered.

Scale plays an important role in this model given the related requirements for carbon credit sales. Additionally, waste-to-energy generation, which can contribute to improving the eligibility for a CDM project, requires a certain scale of operation for full efficiency. See Figure 165 for a diagrammatic representation of the business model.

Potential risks and mitigation

The business model presented here was designed and optimized based on the analysis of different case studies (see previous sections). In designing this optimized business model, risks related to safety, local acceptance by the community, and business attractiveness for investors were assessed.

Market risks: In developing countries, the composting business has the potential of being a burgeoning industry. However, there are oftentimes market entry barriers that may limit business development. The organic fertilizer market is typically less commercialized and the related market structure and business dynamics can be informal, while the inorganic fertilizer market, on the other hand, is more formal and commercialized. A market condition that would potentially affect the sustainability of compost businesses is the market power held by chemical fertilizer producers. This is because the fertilizer market can be traditionally highly concentrated – with few chemical fertilizers companies having the largest share of the market (characteristic of a strong oligopolistic market) – although a limited established distribution network represents an opportunity that organic fertilizer producers can capture.

Additionally, existing policies (e.g. price subsidies) supportive of chemical fertilizers distort market prices making compost comparatively more expensive; and making it difficult for compost producers to enter the market. New organic fertilizer businesses will need at the start-up a highly unique and differentiated product, and innovative marketing strategies to mitigate these competition effects. Furthermore, high seasonality in demand for compost may increase investment cost for storage facilities which may also imply increased operational costs. Risks related to the waste input market are relatively low for this model as it is assumed that depending on the type of entity operating the composting business (i.e. public, private, PPP or social entity), they have exclusive ownership or access (via partnership agreement) to the relevant waste streams. Another significant risk that the business needs to consider is the price volatility in the carbon market. If a business is highly dependent on carbon credit sales for its viability, then it puts its sustainability at an increased risk. As mentioned above, particularly in developing countries, electricity producers are price takers and restricted to the price ceiling set by state-owned transmission entities. Limited negotiation ability in a monopolistic/oligopolistic market puts the business' viability at risk if highly dependent on energy revenue sales.

Competition risks: Key market competition (fertilizer market) as noted above arises from policy instruments that make substitute products more affordable to farmers than compost. Additionally, competition for cheap labor will imply increasing labor wages which may imply increased operational costs for the business if the technologies/production processes are more labor-intensive than mechanized. A profit structure that is highly dependent on cheap labor exposes a business' viability to significant uncertainties.

Technology performance risks: The composting technology typically used (windrow composting) is a relatively mature and simple technology. For large-scale operations, it can be highly mechanized which implies increased investments in advanced technologies and labor costs for highly skilled labor. Additionally, given its high energy requirements, any shortage or infrequency in energy supply can significantly affect operations and in turn business viability. The option of energy generation for internal use can address this challenge. Although, it is worth noting that investments in the required technologies can be costly. Centralized operations may imply high transportation costs, however the adoption of a more decentralized operational system (e.g. via franchises) can reduce the resulting operational costs.

Political and regulatory risks: Policies and regulations related to waste-based compost sectors differ by country. The oftentimes stronger political support for chemical fertilizer use (slow phasing-out of fertilizer subsidies) and lack of specific government guidelines for the certification of compost and internationally accredited third-party certification entities can represent a significant risk to the sustainability of the business model. Furthermore, for the additional value proposition of energy generation, certain limiting factors to business development and sustainability have to be taken into consideration, particularly for developing countries: a) continued interest and large hydro-power potential; b) significant interest in small hydro-power projects; and c) waste-to-energy projects currently viewed as high-risk ventures by financial investors. While producer prices can be increased, additional market failures inherent in the energy sector can only be rectified with the institution of sound policies. Additionally, even with fairly easy entry into the energy market, transaction cost associated with long negotiation processes can be representative of a barrier to market entry. Additionally, high capital requirements and difficulty in accessing funds can be a disincentive for new businesses. By nature of the industry, the lead time for projects can be long and the cost of loan appraisal huge, especially for small projects. Lenders often tend to be concerned about government's interference in the tariff review process and which can increase the tariff risk (regulatory risk) and viewed as reducing businesses' repayment ability.

Social equity related risks: Similar to Business Model 11, this model does not result in any clear social inequity risks. On the other hand, with an extensive reach across the MSW value chain, it has the potential to generate thousands of jobs among the urban poor, particularly for women who are traditionally known to engage in waste segregation. On another front, contracting-out some of the waste collection activities to informal waste collectors brings an inclusive element to a 'for-profit' model. This not only improves the livelihoods of landfill ragpickers by ushering them into mainstream jobs but it can allow the business to efficiently cover slum areas where poor road infrastructure make them less accessible.

Safety, environmental and health risks: On one hand, the simplicity and labor-intensive technology of large scale MSW composting can offer many job opportunities for unskilled workers. On the other hand, MSW is usually contaminated by fecal matter ("flying toilets") and thus poses a higher risk of pathogenic exposure, aside physical hazards (glass, metal) for workers, as well as possible chemical contaminants which might enter the compost and food chain. The provision and use of protective gear for all production operations should thus be mandatory. From the consumer perspective, microbial testing should be a routine measure for quality assurance of MSW compost products. Additionally, farmers must be trained on the appropriate application methods for the waste-based fertilizer products. Recommendations of national agriculture agencies must also be implemented in tandem, in association with agricultural extension agents. To address safety and health risks to workers, standard protection measures are required as shown in Table 41.

TABLE 41. POTENTIAL HEALTH AND ENVIRONMENTAL RISK AND SUGGESTED MITIGATION MEASURES FOR BUSINESS MODEL 12

RISK GROUP	EXPOSURE					REMARKS
	DIRECT CONTACT	AIR/ DUST	INSECTS	WATER/ SOIL	FOOD	
Worker	HIGH RISK	MEDIUM RISK				Risk of sharp objects in MSW and fecal contamination. Potential risk of dust, noise, and chemical compost contaminants
Farmer/user					MEDIUM RISK	
Community						
Consumer						
Mitigation Measures	🧤	🥽 🎧	🦟 💧 🌳	⛰️	🍴 Pb Hg Cd	

Key: ☐ NOT APPLICABLE   LOW RISK   MEDIUM RISK   HIGH RISK

C. Business performance

This model ranks high on profitability and this is attributable to the multiple-revenue stream approach it implements (Figure 166). By diversifying its portfolio, the business is able to mitigate risks, for example, associated with seasonal compost demand, with a combination of revenue generation from sale of energy, carbon credits, recyclables, waste collection service fees and franchise royalties. This model is ranked high on social impacts due to benefits to the wider society in terms of providing sustainable waste management services and nutrient recovery as organic fertilizer for reuse to support more productive and sustainable farming, also generating new jobs for people. The model ranks high on environmental impacts due to its role in protecting public health and the environment by significantly reducing GHG emissions from landfilled waste, waste disposal costs and (large-scale operations) and contributing to soil health while restoring degraded and exhausted soils. The model also ranks high on innovation in terms of adaptation of technology to local conditions and innovative partnerships and pricing strategy, but lower on scalability and replicability due to large capital investment requirements.

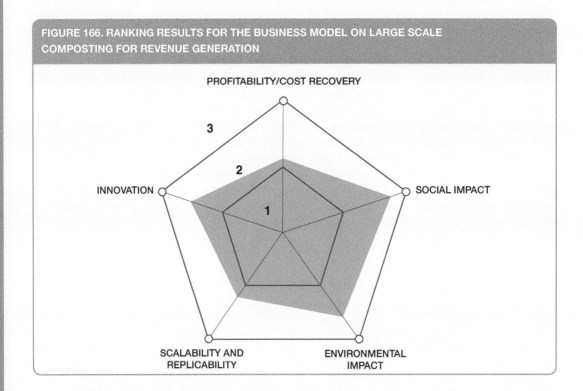

FIGURE 166. RANKING RESULTS FOR THE BUSINESS MODEL ON LARGE SCALE COMPOSTING FOR REVENUE GENERATION

References and further readings

Kaza, S., Yao, L., Stowell, A. 2016. Sustainable financing and policy: models for municipal composting. Urban development series Knowledge Papers 24. Washington, D.C.: World Bank Group.

Note

1 Industrialized or transitional economies as listed in Annex I of the United Nations Framework Convention on Climate Change (UNFCCC).

10. BUSINESS MODELS ON NUTRIENT RECOVERY FROM OWN AGRO-INDUSTRIAL WASTE

Introduction

With increasing scales of production, agro-industrial businesses (tea plantations, livestock producers, agro-processing businesses) are increasingly looking for sustainable treatment and disposal alternatives for the huge amounts of agro-waste (vegetative and livestock) that they produce. Livestock production has shown an accelerated growth in the past two decades, increasing by 62% in comparison with the 1990s, for example, in Latin America. As a result of this progressive increase in the agricultural sector, 84% of the total methane emissions were accounted to livestock production in 2002 (SEMARNAT, 2008). Additionally, livestock operations (swine, dairy cows, etc.) can generate serious environmental consequences such as greenhouse gas (GHG) emissions, odor and water/land contamination, all a result from storage and disposal of animal waste. Confined Animal Feeding Operations (CAFOs) use Animal Waste Management System (AWMS) options to store animal residues. These systems emit both methane (CH_4) and nitrous oxide (N_2O) resulting from aerobic and anaerobic decomposition processes (UNFCCC, 2012). Agricultural producers and food processors similarly face the challenge of sustainably treating and disposing off the waste generated. To ensure business sustainability (largely compliance with legislative mandates), these entities are increasingly implementing an additional arm to their main business to convert their waste into an organic fertilizer, especially given that the implicit cost of non-compliance can be significant, in view of their large scale operations and the resulting potential loss of up to several millions of dollars in annual revenue.

This business model – **onsite nutrient recovery** – is therefore hinged on the concept of the processing of a business' 'own' waste to organic fertilizer to reduce waste disposal costs and, generate revenue while ensuring the sustainability of the larger business entity as a whole. The model generates the double value proposition of:
- Provision of sustainable and environmentally friendly waste management options for agro-industrial entities (livestock producers, agricultural producers and agro-processors);
- Provision of affordable and high quality organic fertilizer for agricultural production.

This business model works for several reasons: a) it is built around harnessing economic value from agro-waste whilst ensuring business sustainability at a higher level and providing a highly-demanded, affordable and nutrient-rich organic fertilizer to farmers; b) the parent company typically provides the capital investment for the set-up of compost operations which mitigates capital investment risk; c) assured supply of key production input (livestock waste); and d) increasing global demand for organic foods and invariably organic farmers. The business also takes advantage of economies of scale and focuses on low cost, yet efficient technologies for compost production and improved waste management. By using value-addition technologies, high quality compost tailored to specific clients and agricultural purposes can be produced, and along with third party product certification can help garner significant market demand. Depending on the waste stream (e.g. livestock waste) and technology used, some health risks may ensue particularly to actors along the compost production chain. The exclusive focus and dependence on the launching customer (parent entity) can induce the business to lose touch with the larger 'agricultural' market and limit opportunities for business growth.

Several variants of this business model are possible as explained in the model description and case examples provided in this chapter. Our examples are not exhaustive and better cases could have been inadvertently omitted due to information and time constraints, but cover a wide range of easily accessible cases at scales in selected settings in India, Kenya and Mexico. Our case examples show that this business model can be technically feasible and financially viable.

References and further readings

SEMARNAT. 2008. Mexico profile. Animal waste management methane emissions. Prepared to be presented in the Methane to Markets, Agriculture Subcommittee. https://www.globalmethane.org/documents/ag_cap_mexico.pdf (accessed November 8, 2017).

UNFCCC. 2012. Benefits of the Clean Development Mechanism 2012. Bonn: UNFCC.

CASE

Agricultural waste to high quality compost (DuduTech, Kenya)

Miriam Otoo, Nancy Karanja, Jack Odero and Lesley Hope

Supporting case for Business Model 13	
Location:	Naivasha, Kenya
Waste input type:	Vegetative waste, livestock waste
Value offer:	Vermicompost
Organization type:	Private
Status of organization:	Operational since 2005
Scale of businesses:	Processes 125 tons of waste per month
Major partners:	Finlays Kenya Limited, Local livestock farmers

Executive summary

DuduTech is an autonomous division within the parent company Finlays Kenya Limited, producing and selling biological control organisms for Integrated Pest Management (IPM), together with the production and sales of vermicompost. Finlays – a wholly owned subsidiary of the Swire Group – is engaged in the production and processing of tea and horticultural products. With increasing scales of production, Finlays needed to identify sustainable treatment and disposal alternatives for their vegetative waste and dependence on synthetic pesticides – thus their motivation for the establishment of DuduTech. DuduTech's business model – onsite nutrient recovery – is hinged on the concept of the processing of a business' 'own' waste to organic fertilizer to reduce waste disposal costs, generate revenue via portfolio diversification and mitigate risk associated with fluctuations in compost while ensuring the sustainability of the larger business entity on a whole. Key success drivers for DuduTech's model are: a) portfolio diversification through the sale of biological control organisms and vermicompost; and b) market segmentation – sale of compost at USD 0.4/ton to Finlays (mother company) and USD 0.74/ton to other clients. Strategic partnerships have also contributed to DuduTech's sustainability. Animal manure is purchased on a contractual basis from local livestock producers for a fee as a corporate social responsibility gesture. Windrow and vermicomposting technology is used to process the livestock waste and vegetative waste from Finlay into a vermicompost – Vermitech. The use of a simple and labor-intensive technology not only gives DuduTech a competitive advantage for production, but also generates employment particularly for low-income persons who would otherwise be unemployed. The purchase of feedstock from local livestock farmers represents an added income-generation stream and implicit improvement of their livelihoods. DuduTech's activities have contributed to a reduction in water and soil pollution from reduced nitrate release attributed to chemical fertilizer use. DuduTech's long-term goals remain: a) to achieve good practices in sustainable and safe agriculture; b) to improve and sustain soil health; c) to up-scale its activities via production mechanization to satisfactorily serve other customer segments; and d) to develop versatile products for soil health improvement to carve its niche in the fertilizer market.

KEY PERFORMANCE INDICATORS (AS OF 2015)					
Land use:	0.5 ha				
Capital investment:	USD 46,460				
Labor:	11 people (2 skilled, 9 unskilled)				
O&M costs:	USD 103 per ton of vermicompost				
Output:	40 tons of vermicompost per month				
Potential social and/or enviornmental impact:	Creation of 11 jobs, reduction of water and land pollution, reduction of CO_2 emissions				
Financial viability indicators:	Payback period:	5 years	Post-tax IRR:	N.A.	Gross margin:	N.A.

Context and background

DuduTech is located in the outskirts of Naivasha, a market town in rift valley province of Kenya, lying North West of Nairobi. Naivasha is on the shore of Lake Naivasha and along the Nairobi-Nakuru highway and Uganda Railway. It is part of the Nakuru district and has an urban population of 14,563 (1999 census). The main industry is agriculture, especially floriculture. DuduTech was established in 2001 as an autonomous division within the parent company Finlays Limited which was founded in 1750. Finlays as a wholly owned subsidiary of the Swire Group, has extensive tea and horticultural interests in Kenya, South Africa, Sri Lanka and China. The motivation for the establishment of DuduTech was Finlays' vision for sustainable and safe agriculture. Apart from environmental conservation through reduction in the amount of nitrates released into the soil from the use of chemical fertilizers, the availability of safe vermicompost has enabled Finlays to produce certified organic products and obtain Fair Trade Certification. This certification brands products as those meeting internationally-set environmental and labor standards and thus receives higher market prices – from which Finlays has substantially benefited.

Market environment

Finlays – a major tea and horticultural production and processing business entity – generates approximately 125 tons of vegetative waste weekly. With plans for increasing their scale of production, Finlay faces a significant challenge with the management of their waste, which was disposed of in open spaces within their farms. The conversion of vegetative waste to compost represents a sustainable option for Finlays to reduce its current and future land requirements for waste disposal. Furthermore, the continuous use of chemical fertilizers has had a negative effect on soils and water bodies from the release high quantities of nitrates. This in addition to the increasing international demand for organic agricultural products has catalysed the promotion of organic farming and the demand for related agricultural inputs. Finlays' desire to tap into the international market segment requires their use of agricultural inputs that meet organic farming standards. It is in this regard that the development of DuduTech remains crucial for the sustainability of Finlays but also the growing agricultural sector in Kenya.

Macro-economic environment

In the early 1990s, fertilizer markets were liberalized, government price controls and import licensing quotas were eliminated, and fertilizer donations by external donor agencies were phased out. Fertilizer use then almost doubled over the 15-year period from 1992 to 2007, with much of the increase attributable to smallholder farmers. The liberalization of the foreign exchange regime in 1992, resulted in the convergence of what were then the official and the parallel market exchange rates, and effectively removed implicit taxation on fertilizers amongst other imports. While availability of fertilizers has been enhanced, these measures did not have the desired impact of lowering retail prices. This suggests that although businesses such as DuduTech may face fierce competition, organic fertilizer prices remain comparatively more cost-effective than those of chemical fertilizer. Additionally, increasing consumer preferences for organic foods and related local and global prices are representative of factors supportive for the development and sustenance of businesses such as DuduTech – given the related demand by farmers for organic agricultural inputs.

Business model

DuduTech's business is to process the waste of its parent company – Finlays Kenya Limited – into a valuable resource, vermicompost, and also produce and sell biological control organisms for agricultural purposes. Key success factors of DuduTech's business model have been: a) partnership with parent company to mitigate capital start-up risk and ensure continuous supply of vegetative waste; b) diversified portfolio through the sale of biological control organisms and vermicompost; c) segmented markets for its compost product. Vermitech, the brand name for the compost product is sold directly to Finlays and other local agricultural producers. Finlays' purchases represent 80% of all sales, with the remaining 20% by local farmers. The large purchase of the parent company represents an assured product demand and mitigates any risk associated with fluctuations in demand. Essential to DuduTech's business model is the market segmentation of its customer base. It sells compost at USD 0.4/ton to Finlays and USD 0.74/ton to other clients. It is thus able to recover the majority of its cost from the price differential. Additionally, DuduTech has invested a lot in developing high quality products, which has given it a competitive advantage in the fertilizer market, and has also enabled Finlays to produce certified organic products and obtain Fair Trade Certification. This certification brands products as those meeting internationally-set environmental and labor standards and thus receives higher market prices – from which Finlays has substantially benefited. For the production of the vermicompost, DuduTech sources its waste inputs – vegetative waste and animal manure – from Finlays and local livestock producers, respectively. These strategic partnerships have contributed to DuduTech's sustainability as they assure a consistent supply of inputs. Windrow and a vermicomposting technology is used to process the livestock and vegetative waste into a vermicompost. The use of a labor-intensive technology not only gives DuduTech a competitive advantage for production, but also generates employment particularly for low-income persons who would otherwise be unemployed. Although making use of an abundant input, labor, increasing wages have motivated DuduTech to explore the use of a more mechanized technology for labor-intensive activities such as heaping, turning and bagging, especially in light of foreseen production expansion. See Figure 167 for diagrammatic representation of the business model for Dudutech.

Value chain and position

Figure 168 provides an overview of DuduTech's value chain. The business sources its key inputs: vegetative waste and animal manure from Finlays and local livestock farmers (as part of its corporate social responsibility project), respectively. Access to and supply of vegetative waste is assured as Finlays currently produces more waste than DuduTech can actually process. On the other hand, however, DuduTech faces potential competition for animal manure given its demand for agricultural purposes. To mitigate this production risk, DuduTech plans to source this waste from larger scale livestock producers on a long-term contractual basis. DuduTech sells its products – compost and biological control organisms – to Finlays and other local farmers. The production capacity of DuduTech is approximately 10 tons per week of which about 80% is sold to Finlays. Vermicomposting gives Vermitech an edge over other compost products and chemical fertilizers in terms of its water retention capacity. Field trials have established a 30% reduction in irrigation when Vermitech was used in

CASE: AGRICULTURAL WASTE TO HIGH QUALITY COMPOST

FIGURE 167. DUDUTECH'S BUSINESS MODEL CANVAS

KEY PARTNERS	KEY ACTIVITIES	VALUE PROPOSITIONS	CUSTOMER RELATIONSHIPS	CUSTOMER SEGMENTS
• Finlay company (parent company) • Small scale local livestock farmers	• Vermicomposting of plant and animal manure • Quality control of compost products • Sales of vermicompost • Collection of plant waste from tea and flower plantations of mother company • Collection of animal manure from small farms (free of cost) • Produce and sell biological control organisms	• Provision of sustainable options for waste generated from production and processing of tea and flowers, providing space of productive areas. • Offer of a nutrient-rich, certification-grade quality vermicompost suitable for the production of 'Fair Trade Certified' agricultural products • Provision of biological control organisms for the control of pests and diseases	• Parent–daughter company • Personal at direct sales	• Finlays Kenya Limited • Small- and large-scale farmers; Finlays Limited
	KEY RESOURCES • Equipment • Labor (skilled and unskilled) • Capital • Land • Vegetative and animal waste • Partnership • Branding		**CHANNELS** • Direct channel with parent company • Direct sales	

COST STRUCTURE	REVENUE STREAMS
• Capital investment (equipment, land, infrastructure) • Operation and maintenance costs • Laboratory costs	• Sale of vermicompost to segmented markets: 80% of sales to parent company and 20% to large and small farms • Sale of biological control organisms

SOCIAL & ENVIRONMENTAL COSTS	SOCIAL & ENVIRONMENTAL BENEFITS
• Possible contamination of soil and groundwater from disposal of untreated leachate on-farm	• Job creation • Reduction in chemical fertilizer use • Reduction of carbon dioxide emissions • Improvement in structure and nutrient composition of soil • Management of animal and plant waste

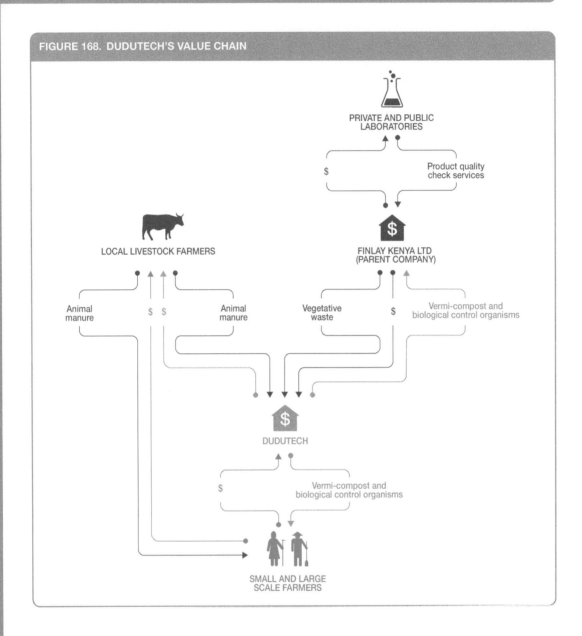

FIGURE 168. DUDUTECH'S VALUE CHAIN

replacement of some quantity of chemical fertilizer. However, Vermitech can be comparatively more expensive than chemical fertilizer given the relatively high application rates. Gaining additional share of the fertilizer market will require a more competitive product price. On the other hand, DuduTech's products are garnering great demand given the increase in global and local demand for organic products.

Institutional environment

Management of solid waste in Kenya in general is dealt with under several laws, by-laws, regulations and acts of parliament. As with DuduTech, in order to legally engage in composting activities on a business scale in Kenya, a waste management permit from the county council and waste recycler's permit from NEMA are a requirement and these are renewable on an annual basis. Additional regulations have been set in place including the Occupational Safety and Health risk Act and the Factories Act (cap 514 of the laws of Kenya) to protect plant workers and for which Dudutech has to comply to. The main policy and regulatory bodies that are responsible for overseeing the operations of composting activities in Kenya are: the City Council, Local Authorities in the Ministry of Local Government; Kenya Bureau of Standards (KEBS) in the Ministry of Industrialisation; and the National Environmental Management Authority in the Ministry of Environment and Mineral Resources (Onduru et al., 2009). The City Council provides guidance on waste management practices (collection, transportation and safe disposal), zoning and licensing. KEBS is mandated to develop standards (product quality certification) and ensure compliance with such standards. In collaboration with Kenya Organic Agriculture Network, KEBS has developed standards for the use and marketing of compost and other organic inputs (Onduru et al., 2009). The standards being developed recognize three categories of compost: liquid compost (e.g. leachates from vermicomposting), pelletized/granulated compost and natural/solid compost. KEBS' activities in particular will enable businesses like DuduTech to brand their product and increase their share of the fertilizer market.

Technology and processes

DuduTech employs a combination of windrow and vermicomposting for the production of compost (Figure 169). A tractor fitted with trailers transports the vegetative waste from Finlays and a 10–20 ton lorry transports the animal manure from livestock producers to the production site. DuduTech uses both manual and mechanical methods for the vermicomposting process; however, plans are underway to mechanize other activities for its future expansion plans. Activities that are done manually include heaping, turning, watering and bagging. The equipment used is locally manufactured and spare parts are obtained locally. Danish International Development Agency (DANIDA) trained staff on vermicomposting and quality monitoring at the onset of the business. For the processing activity, vegetative waste is mixed with animal manure in the ratio of 1:2 and the mixture is composted for eight to ten weeks after which it is spread on beds to form a layer of ten centimetres. The beds are 45 meters long. As the substrate is digested by the worms, the volume shrinks and so additional waste is added in intervals to maintain the 10 centimetres depth until the vermicompost is mature for harvest. Once mature, there are two ways of harvesting. One way is discontinuing moisturization/ watering so that the worms move to the lower parts of the compost in search of water. Upper parts are scooped until all matured vermicompost is harvested. This is a dry harvesting technique and bagging can be done without having to re-dry the compost. The second harvesting technique involves creating a layer of food substrate on top of the matured vermicompost, separated by a net. This allows for easy separation between the matured compost and added food substrate but also permits the worm to access the food. Moistening continues until almost all the worms have penetrated the net into the substrate. The worms are harvested along with the food substrate, leaving only the vermicompost which is then harvested and dried to attain 40% moisture content then bagged for sale.

Funding and financial outlook

Initial capital cost comprising of land, equipment and other infrastructure was financed by DuduTech at a cost of USD 46,457 (4 million Kenyan Shillings). Total operational costs amount to USD 4,126 per month of which wage and salaries is the largest component, constituting 64%. Cost of waste input (largely acquisition costs of animal manure) accounts for 18%; and water, fuel and repairs each representing 3% of all costs. DuduTech earns revenues from the production and sale of biological control organisms and vermicompost. An annual profit of USD 7,000–8,500 is made from sales of vermicompost. Revenue and profit data were not disclosed for the sale of biological organisms.

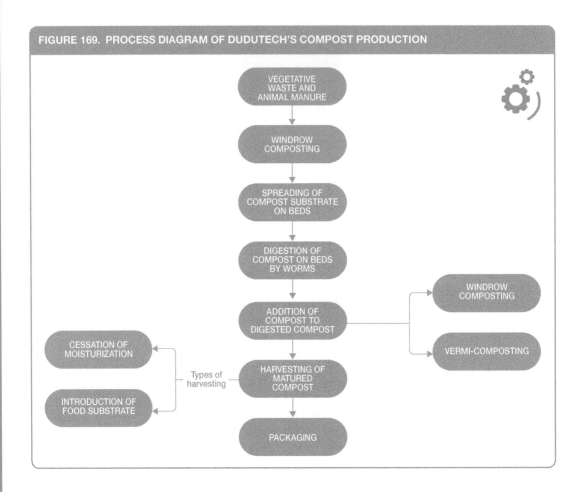

FIGURE 169. PROCESS DIAGRAM OF DUDUTECH'S COMPOST PRODUCTION

Socio-economic, health and environmental impact

DuduTech's activities have resulted in several socio-economic and environmental benefits. DuduTech's business activities, particularly compost production, provide employment to 11 people on a regular basis. Furthermore, the use of vermicompost has contributed to the reduction of nitrates released into the soil and water bodies within and around the Finlay's farms from reduced chemical fertilizer usage. Although actual nitrates reduction figures were not provided, evidence of good agricultural practices can be attested to through Finlay's attainment of a Fair Trade Certification and receipt of premium prices for its agricultural products. Additionally, monetary gains are represented by cost savings from the use of Vermitech instead of chemical fertilizers by Finlays. Available data indicates that Finlays saves up to 20% in fertilizer costs and up to 30% in reduction of water used for irrigation. The conversion of vegetative waste to compost has also made available productive space which was originally used from disposal purposes. Improved livelihoods beyond benefits from reduced CO_2 emissions and groundwater contamination include increased revenues to livestock farmers from the sale of animal manure to DuduTech. DuduTech's operations, however, release raw leachate into the soil and water bodies. Plans are underway to add value to the leachate also for agricultural purposes. Health risks to workers are very low as any likelihood of exposure to pathogens from waste handling, for example, is mitigated from workers use of protective gear.

Scalability and replicability considerations
The key drivers for the success of this business are:
- Provision of start-up capital by parent company, Finlays Limited – which mitigated capital investment risk.
- Assured supply of key production input (vegetative waste) at no cost.
- Diversified portfolio – which mitigates risk associated with fluctuations in market demand.
- Increasing international demand for (certified) organic produce.

DuduTech's model has a high replication potential especially in developing countries with increasing agro-processing businesses and related limited waste management options. An opportunity for the up-scaling of DuduTech's composting relates to the abundant vegetative waste produced by Finlays that is still being dumped untreated and used on farmlands. Increased production represents potential economies of scale that DuduTech can capture; which will help reduce its production costs and invariably lower product prices. This strategy will help capture a larger share of the fertilizer market. It is important to note however that adaptations to the production process may be necessary given increasing costs of labor and animal manure, in order to make the increase in scale of production monetarily worthwhile. The organic foods market in growing globally, suggesting a potential increase in demand for organic agricultural products for which DuduTech can additionally take advantage of.

Summary assessment – SWOT analysis
Figure 170 presents an overview of the SWOT analysis for DuduTech. Composting is a promising business in Kenya especially given the abundance of waste inputs and the growing need for environmentally sustainable agricultural input. DuduTech has been particularly successful in leveraging its business partnerships to mitigate capital investment risk and ensure consistent supply of waste inputs. Additionally, DuduTech implements a segmented pricing approach where it charges local farmers almost double the price its parent company, Finlays, pays. DuduTech produces a quality compost with high nutrient contents that is in high demand in spite of its comparatively higher market price. Its additional investment in quality assurance and monitoring by a third party has also enabled Finlays to produce certified organic products and obtain Fair Trade Certification. This certification brands products as those meeting internationally-set environmental and labor standards and thus receives higher market prices – from which Finlays has substantially benefited. The sustainability of DuduTech is however largely dependent on the parent company – Finlays Kenya Limited. Finlays provides raw materials at no cost and also buys 80% of the compost. Although unlikely, decreased demand from Finlays will significantly affect its profitability. Additionally, the technology currently in use is highly labor-intensive and any up-scaling initiatives without some changes to the technology process, exposes DuduTech to unpredictable labor costs. Despite these limitations, several opportunities exist for DuduTech to ensure sustainability: a) increase its scale of production to capture economies of scale; b) increase its market scope via the production and sale of leachate-based products; and c) sale of carbon credits through the establishment of a CDM project. DuduTech represents an example of an innovative business making use of its parent company's (Finlays) agricultural waste to ensure its sustainability whilst generating significant profits and benefits to society.

FIGURE 170. SWOT ANALYSIS FOR DUDUTECH

	HELPFUL TO ACHIEVING THE OBJECTIVES	HARMFUL TO ACHIEVING THE OBJECTIVES
INTERNAL ORIGIN ATTRIBUTES OF THE ENTERPRISE	**STRENGTHS** • Continuous and assured supply of vegetative waste • Produces nutrient-rich compost • Diversified portfolio • Segmented markets • Cost-effective technology • Good local buy-in for products due to establishment of brand name	**WEAKNESSES** • High capital investment required • High level of technical expertise required • Production process adaptations required for up-scaling
EXTERNAL ORIGIN ATTRIBUTES OF THE ENVIRONMENT	**OPPORTUNITIES** • Mechanization of labor-intensive production processes to increase scale of compost production • Increase in market scope – production of leachate-based products • Up-scaling potential of CDM project to earn carbon credits • Increasing organic food/ agriculture markets globally	**THREATS** • High dependency on parent company for sales of compost • Unpredictable labor cost • Competition for animal manure

Contributors
Jasper Buijs, Sustainnovate, The Netherlands; formerly IWMI, Sri Lanka
Johannes Heeb and Leonellha Barreto-Dillon, CEWAS, Switzerland

References and further readings
Onduru, D.D., Waarts, Y., de Jager, A. and Zwart, K. 2009. http://edepot.wur.nl/51329 (accessed November 8, 2017).

Personal interviews with management personnel of DuduTech. 2015.

Case descriptions are based on primary and secondary data provided by case operators, insiders or other stakeholders, and reflect our best knowledge at the time of the assessments 2015/16. As business operations are dynamic data can be subject to change.

CASE
Enriched compost production from sugar industry waste (PASIC, India)

Miriam Otoo, Marudhanayagam Nageswaran, Lesley Hope and Priyanie Amerasinghe

Supporting case for Business Model 13	
Location:	Pondicherry (Puducherry), India
Waste input type:	Sugar mill organic waste
Value offer:	Provision of enriched pressmud compost for agricultural production
Organization type:	Public
Status of organization:	Operational since 1996
Scale of businesses:	Processes 6,000–9,000 tons of waste/year
Major partners:	Puducherry Cooperative Sugar Mills (PCSM), Agricultural Department of the Government of Pondicherry; Government of India

Executive summary

The Pondicherry Agro Service and Industries Corporation Limited (PASIC) is a government-owned agricultural inputs producer and supplier. Seeing an opportunity with producing enriched pressmud compost from sugar mill waste and effluent water, PASIC set up a compost production arm to its business in partnership with the Pondicherry Cooperative Sugar Mill (PCSM) – the largest industrial unit in the cooperative sector under the Pondicherry government to process their waste. PCSM's inefficient disposal practices were adversely affecting groundwater quality and polluting surrounding areas. Thus, this partnership represented a win-win for both parties – PCSM was able to continue their operations according to legislative guidelines and PASIC produced and sold a nutrient-rich organic fertilizer to farming communities. The business arrangement is such that profits are split equally between both parties. PCSM provides the waste input to PASIC free of charge and provides the land for the processing of the waste. PASIC on the other hand covers all other capital and recurrent costs and has a budget of USD 45,600 per year. The corporation has so far created 25 jobs to benefit local workers and their families. The corporation deliberately keeps its annual net profit low at 5–7% given its social orientation. The compost, which is heavily subsidized by the agricultural department, is sold in agricultural depots and outlets. A 75% subsidy scheme is provided for farmers and 100% for Schedule Caste (SC) farmers. The project has significantly contributed to the peri-urban economy and safeguarded the health of local water bodies and environment in general. Beyond this, the increased adoption of organic fertilizer will contribute to the reduction of imported chemical fertilizer and related government subsidy expenditures.

KEY PERFORMANCE INDICATORS (AS OF 2015)						
Land use:	2.43 ha					
Capital investment:	USD 75,000 including cost of 2.43 ha of land					
Labor:	25 people (9 skilled and 16 unskilled)					
O&M cost:	USD 49 per metric ton					
Output:	3,000 tonnes of enriched pressmud compost / year					
Potential social and/or enviornmental impact:	Creation of 25 jobs, reduction in groundwater and land pollution, waste management cost savings and improved environmental health					
Financial viability indicators:	Payback period:	8 years	Post-tax IRR:	N.A.	Gross margin:	5–7%

Context and background

Pondicherry Agro Service and Industries Corporation Limited (PASIC) is located in the southern part of peninsular India, which is a Union Territory. It was incorporated in 1986 and is owned by the Government of Pondicherry. The main activity of the Corporation is to distribute agricultural inputs such as fertilizers, seeds, organic fertilizer (enriched pressmud and municipal solid waste-based compost), plant protection equipment, horticultural plants, implements, tools, bio-fertilizers etc., to the farming communities at a reasonable price. In 1996, PASIC and the Pondicherry Co-operative Sugar Mills Limited (PCSM), entered into a joint venture for the processing of sugar mill waste to an enriched pressmud compost. This became necessary due to the difficulty experienced by PCSM with the disposal of its sugar mill waste. Each processed ton of crushed sugarcane produces between 0.16 to 0.76 m^3 of wastewater. PASIC processes about 6,000 to 9,000 tons per annum of pressmud and effluent from PCSM units. The sugar mill's wastewater has excessive amounts of suspended solids, dissolved solids, BOD, COD, chloride, sulphate, nitrates, calcium and magnesium, creating significant deleterious effects to both water bodies and soil when disposed of untreated. PASIC also took advantage of the increasing chemical fertilizer prices and need for sustainable agricultural inputs alternatives and established a sound and viable reuse business.

Market environment

Government expenditures on chemical fertilizer imports for agricultural production are at an all-time high and on an increasing trend in India, in an effort to increase agricultural production. Government subsidies on chemical fertilizer have however resulted in inefficient use by agricultural producers. Over-application and extensive use of chemical fertilizers has had a dilapidating effect on agricultural soils and resulted in less productive yields. The demand for more sustainable agricultural input alternatives coupled with the increasing awareness of organic farming are some of the factors that PASIC capitalized on in setting up the business enterprise. In addition, there was the need to properly manage the waste generated by the sugar mill industry which had become a source of land and water pollution. India has a gross cropped area of 190 million hectares and would require about 627,000,000 tons/year of enriched pressmud compost to cover this agricultural production area. There are 600 sugar factories crushing 145 million tons of sugarcane annually in the country. The annual by-products generated through these industries are about 5 million tons of pressmud/year. This is indicative of a potential demand that will be greater than supply, assuming there are mechanisms in place to incentivize adoption by farmers. Organic fertilizer businesses face fierce competition in the fertilizer market from chemical fertilizer and other organic fertilizer businesses. The enriched pressmud compost produced by PASIC is heavily subsidized by the government – 100% subsidy for schedule caste farmers and 75% for general farmers. Additionally, although PASIC is socially-oriented, its profit margin remains positive and regulated between 5–7%. These measures have given PASIC a competitive advantage over other new market entrants (organic fertilizer producers) and chemical fertilizer. PASIC produces and sells about 3,000 tons of enriched pressmud compost, accounting for 90% and 15% of the compost and chemical fertilizer markets respectively in Pondicherry. Although PASIC's compost is fairly substitutable with other organic fertilizers, the relatively low price of USD 0.01/Kg and its high nutrient content (N: 1.24%, P: 2.77%, K: 1.68%, OC: 21.6%, Mg: 0.95% and Zn 0.012%) give it an edge over other products.

Macro-economic environment

The Indian government highly subsidizes chemical and synthetic fertilizers, particularly Nitrogen, Phosphorus and Potassium (NPK). The amount of subsidies on chemical fertilizer has grown exponentially in the last few decades and has been mainly attributed to inflation and price fluctuations in the international market (Mishra and Gopikrishna, 2010). Significant subsidy allocation has not only led to inefficient use by farmers and high costs to the government; substantial soil degradation has also been observed as a result. With a growing need to increase the availability and quality of bio-fertilizers and composts in the country to improve agricultural productivity while maintaining soil health and environmental safety, the Indian government has set up over the last few years new schemes to augment the infrastructure for production of quality organic and biological inputs, and also from organic municipal waste.

A capital investment subsidy scheme for compost production has been introduced under the National Mission for Sustainable Agriculture (NMSA). The scheme provides 100% financial assistance to state governments and government agencies up to a maximum limit of about USD 300,000 per construction unit, and for individuals or private companies up to about USD 100,000 per unit (max 33% of project costs) through the National Bank for Agriculture and Rural Development (NABARD). Moreover, the Government of India is providing a Market Development Assistance of about USD 23.4 per metric ton to Fertilizer Companies for sale of City Waste Compost (Ministry of Agriculture, 2017). Policies to reduce the budget allocation for chemical fertilizers and provide capital investments for new and existing compost businesses are important instruments that catalyze the business development in the RRR sector and the scaling-up of initiatives similar to that of PASIC.

Business model

PASIC undertook a long term (99-year) agreement with PCSM to process the sugar mills' waste into an enriched pressmud compost (Figure 171). PASIC is funded by the government of India; and produces and sells enriched pressmud compost to farmers directly through agricultural depots. It implements both a value-driven and a price-driven sales strategy, and a segmented market approach, selling enriched pressmud compost at a higher price to urban horticulturist than general farmers who represent 99% of its customer base. This is because, although PASIC's compost is fairly substitutable with other organic fertilizers, the relatively low price of USD 0.01/Kg and its high nutritive value (N: 1.24%, P: 2.77, K: 1.68%, OC: 21.6%, Mg: 0.95% and Zn 0.012%) gives it an edge over other products. Essential in its business model is PASIC's partnership with PCSM and the Indian Government via the agricultural department. It partners with PCSM for the continuous supply of waste at no cost. In addition, all production activities are executed on PSCM's production site to reduce investment costs (land purchase) and transportation costs thereby reducing overall production costs. PASIC manages and covers all costs associated with the production unit, technology, manpower, and production and marketing activities of the processed pressmud. PASIC does not compensate PCSM for the raw materials as it carries out the task of value addition of waste and disposal. Profits are shared on a 50:50 basis between PASIC and PCSM. The partnership with the government mainly is for easy marketing of products through price subsides provided to farmers. The government of Pondicherry through agricultural department annually allocates budget for the distribution of the pressmud compost

FIGURE 171. PASIC'S BUSINESS MODEL CANVAS

KEY PARTNERS
- Pondicherry Cooperative Sugar Mills
- Government of India
- Agricultural department of the Government of Pondicherry

KEY ACTIVITIES
- Treatment of organic solid of sugar industry and effluent water
- Processing of pressmud compost
- Transport and distribution to agricultural depots
- Marketing and sales

VALUE PROPOSITIONS
- Local farmers obtain a nutrient-rich, enriched pressmud compost at a reasonable price
- Provision of sustainable waste management alternative for agro-processing units

CUSTOMER RELATIONSHIPS
- Personal and direct sales
- Use of direct personal help for long-term agreement with PCSM

CUSTOMER SEGMENTS
- Farmers
- Urban horticulturists
- Pondicherry Cooperative Sugar Mills

KEY RESOURCES
- Free Sugar mill organic waste
- Land as included in the partnership deal
- Labor (25 workers), equipment
- Long-term agreement with PCSM

CHANNELS
- Direct sales through agricultural depots

COST STRUCTURE
- Capital investment (excluding cost of land)
- O&M cost
- Cost for quality control noted to be the lowest O&M cost; and input costs for micronutrients and enriched materials been the highest.
- Profit sharing with PCSM
- Transport cost (only for product delivery thanks to partnership)
- Marketing and sales

REVENUE STREAMS
- Sale of enriched pressmud compost

SOCIAL & ENVIRONMENTAL COSTS
- None noted based on information provided

SOCIAL & ENVIRONMENTAL BENEFITS
- Reduction of waste from sugar mill industry
- Job creation
- Reduced pollution of water bodies from sugar mill effluent
- Reduced existing waste management costs
- Reduced human exposure to untreated waste pollutants
- Enhance soil fertility and productivity
- Reduced GHG emissions

to farmers and also offers a 75% and 100% price subsidy to general and schedule caste farmers respectively. With the adoption of a social-oriented approach, profit margins are deliberately kept low and have been fixed at 5–7% by the government of Pondicherry. This in addition to the subsidies provided has made the product affordable to majority of farmers. These partnerships enable PASIC to maximize its profits in spite of a profit ceiling, obtain a regular supply of raw materials and also create an assured market for the enriched pressmud compost product.

Value chain and position

PASIC's key business activities are the production, marketing and sale of the pressmud compost (Figure 172). The value chain is very simplistic and has PASIC as the key player. PASIC sources its raw materials from PCSM and is the sole user of the 6,000–9,000 tons of sugar mill waste generated per year. Given the long-term agreement between these two parties, PASIC faces no competition with any other company for raw materials and has an assured supply of inputs. PCSM, in addition, provides the space and facilities for the processing operations of pressmud compost. PASIC in turn covers all remaining operational costs and the profits are split equally between the two parties. PASIC

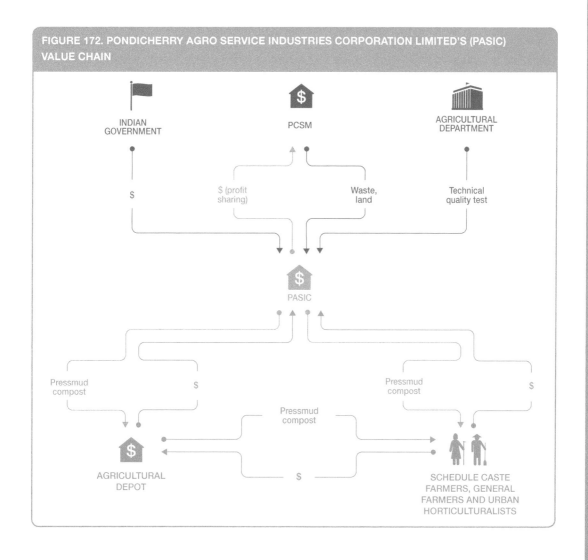

FIGURE 172. PONDICHERRY AGRO SERVICE INDUSTRIES CORPORATION LIMITED'S (PASIC) VALUE CHAIN

was funded by the Indian government at a cost of USD 75,000 excluding land costs, and provides significant subsidies to farmers. These subsidies have eased PASIC's entry into the fertilizer market in the face huge competitors such as chemical fertilizers who own a large share on the market. The agricultural department provides technical expertise for the laboratory analysis of compost to ensure that the pressmud compost is a safe and nutrient-rich product. PASIC is sold directly to farmers and also through agro-outlets and agricultural depots. PASIC has been able to capture a significant share of the organic fertilizer market in Pondicherry mainly due to using the agricultural depots via its partnerships with the agricultural department and government subsidies.

Institutional environment

At the local government level, the Pondicherry Government has been very supportive of the business activities of PASIC. In addition to putting up the start-up capital for the business, it annually makes a budgetary allocation for the distribution of the pressmud compost under a 75% subsidy scheme for general farmers and 100% subsidy for schedule caste farmers via the Department of Agriculture. The subsidy scheme has been essential for PASIC in gaining an easy enty into the fertilizer market. At the country level, there is a statutory guideline – the Fertilizer Control Order (FCO) instituted by the Ministry of Agriculture and Rural Development for the production and distribution of all fertilizers including organic fertilizer. Product quality recommendations are provided for different organic fertilizer types for which producers have to adhere to. This is particularly beneficial to farmers as they get what they are paying for, but also for compost businesses as they are able to build their product brand.

Technology and processes

Composting of pressmud is carried out using an aerobic decomposition of pressmud in windrows (Figure 173). Most of the processing equipments are simple and locally manufactured, making them more cost-efficient. The technology has a waste input–output conversion ratio of about 30%. Decomposition is accelerated by inoculation of microbial cultures and the provision of required fermentation optima (maintenance of optimum moisture, aeration and temperature). The composting process takes between 45 to 70 days, after which the decomposed material is mixed with other products listed in Table 42 to produce the enriched compost. The majority of these organic materials are produced by PASIC. For the aerobic composting process, raw pressmud is formed in windrows and dried for three to four days to reduce the moisture content. With an aero-tiller, the product is aero-tilled once in three days. The sugar mill effluent is sprayed on the product when the moisture level reaches 50%, and the process of aero-tilling is carried out again. This process is repeated for 60 days. The product is then enriched with bio-fertilizers and micronutrients through spraying over the windrows. This mixture undergoes the aero-tilling process to ensure a uniform mixture. The final enriched pressmud compost is then packed into 50kg high density polyethylene bags. The cost for

TABLE 42. TYPE AND QUANTITY OF PRODUCTS ADDED TO ENRICH THE PRESSMUD COMPOST PRODUCT

NAME OF THE NUTRIENTS	QUANTITY PER 10 TONS OF PROCESSED COMPOST
Rock phosphate	200 kg
Azospirillum broth	10 litres
Phosphobacterium broth	10 litres
Pseudomonas broth	10 litres
Magnesium sulphate	75 kg
Zinc sulphate	75 kg

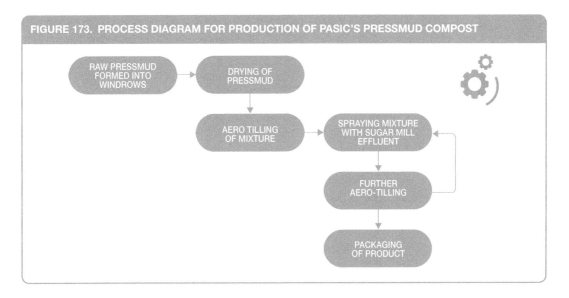

FIGURE 173. PROCESS DIAGRAM FOR PRODUCTION OF PASIC'S PRESSMUD COMPOST

quality control is noted to be the lowest O&M cost, with input costs for micronutrients and enriched materials been the highest. Micronutrients and enriched materials cost is about USD 8.81 per ton of enriched pressmud compost, accounting for almost a fifth of per unit operational cost.

Funding and financial outlook

PASIC is a public company established by the government of Pondicherry at a cost of USD 75,000 excluding land costs, with a payback period of eight years. There are no land costs to PASIC as all plant operations take place on the PCSM production site as part the established long-term agreement. PASIC has an average production capacity of 3,000 tons per annum. The average production cost of the enriched pressmud compost is USD 49 per ton, with labor costs comprised of wages, salaries and management cost accounting for 45% of the total operation cost. PASIC covers all costs related to technology, manpower, production and marketing of the enriched pressmud compost. Profit margins are estimated at 5–7% and with 50:50 sharing system between PASIC and PCSM – annual profit per entity of USD 7,900. Sales from enriched pressmud compost and waste management fees paid by PCSM are the revenue streams for PASIC. Twenty-five percent of the compost sale price is paid by farmers and the rest is paid for by the state government (i.e. Pondicherry government) through the agricultural department. Plans are underway to have enriched pressmud compost sold in other states.

Socio-economic, health and environmental impact

The business activities of PASIC have reduced the purchase of chemical fertilizer and subsequently led to enhanced sustainable crop production. In the last sixteen years, PASIC has processed about 1.46 million tons of sugar mill waste into about 444,350 tons enriched press mud compost. Applying a nominal value of USD 56 per ton to the waste, the project has generated approximately USD 2.56 million in "new waste to value" to the community. The project will continue to produce approximately 3,000 tons of packaged enriched pressmud compost annually, resulting in an increase in rice yields of 1,067 tons equivalent to about USD 0.25 million. This project has reduced environmental pollution due to unregulated disposal of untreated sugar mill waste which hitherto was a major problem. It has and continues to safeguard the health of local water bodies and soil health. It has also improved the livelihoods of the local community through the provision of jobs. The project supports 25 jobs and has a budget of USD 45,600 per year to benefit to local workers and their families. Additionally, PASIC ensures to safeguard the health of its workers through the provision of safety gear – hand gloves and rubber boots and annual medical check-ups.

Scalability and replicability considerations

The key drivers for the success of this business are:
- Strong commitment of the state government in providing an enabling environment for the implementation of the business via the provision of start-up capital and price subsidies.
- Strong partnerships with the agricultural department provided key technical expertise to produce a high quality product and easy access to customers via its agricultural depots.
- Long-term contractual agreement with PCSM (agro-processing unit) ensures continuous supply of waste input and premises for plant operations.
- Policy initiatives to phase-out chemical fertilizer subsidies and capital investment subsidies to new and existing compost businesses.
- Environmental legislation making waste treatment a requirement.
- Government scheme set up to augment the infrastructure for production of quality organic and biological inputs.
- Local government supportive of the business initiative.

PASIC's model has a high replication potential in agrarian developing countries with large agro-processing units. Initial governmental support will be required to mitigate capital investment risk and gain entry into an oligopolistic fertilizer market. The contractual agreement between PASIC and PCSM on use of all the sugar mill's waste and premises for processing activities, eliminates transportation costs and land rent (implying higher profits) which have been known to be substantial costs incurred by organic fertilizer producers. PASIC, however, faces a profit margin ceiling which prevents over-pricing but also the maximization of profits. This business has a social focus and its pricing model may not be applicable to a profit-oriented business. Out-scaling of PASIC's model will increase the costs of production proportionately more than the generated revenue, thus governmental support at least at the start-up stage will be required in replicating this model. It would be ideal for the sugar processing companies to contribute to the investment cost in addition to the land cost in the instance where government support is lacking.

Summary assessment – SWOT analysis

Figure 174 presents the SWOT analysis for PASIC. Composting is a promising business in India especially given the abundance of waste inputs and the growing need for environmentally sustainable agricultural input. PASIC has been particularly successful in leveraging its business partnerships to mitigate capital investment risk and gain entry into a fiercely competitive fertilizer market. Additionally, PASIC implements a segmented pricing approach where it charges urban horticulturists more than it does peri-urban and rural farmers. The sustainability of this business is however largely dependent on price subsidies provided by the government. The removal of these subsidies may expose PASIC to fierce competition in the fertilizer market, in which case it would have to rebrand its product to maintain its market share. Increasing governmental support along with growing demand for organic fertilizers will represent key opportunities for replication and up-scaling of the business. The use of a simple technology has been important to the business' success – taking advantage of cheap labor, however with increasing wages and energy prices, PASIC will have to consider other alternatives with future expansion plans. PASIC is an example of an innovative waste reuse business utilizing a simple partnership approach to address some of the major waste management and environmental challenges in Pondicherry, India.

FIGURE 174. SWOT ANALYSIS FOR PASIC

	HELPFUL TO ACHIEVING THE OBJECTIVES	**HARMFUL** TO ACHIEVING THE OBJECTIVES
INTERNAL ORIGIN ATTRIBUTES OF THE ENTERPRISE	**STRENGTHS** - Strong business partnerships - Ability to provide market advantage to farming community members - Limited transportation costs due to use of agro-processing site - Assured, continuous supply and easily accessible waste - Zero pollution processing - Strong marketing strategy	**WEAKNESSES** - Dependence on agricultural depots for marketing and sales of product - Energy-demanding technology - High capital investment requirements
EXTERNAL ORIGIN ATTRIBUTES OF THE ENVIRONMENT	**OPPORTUNITIES** - Increase demand for organic fertilizers thus potential to upscale operations and scope of market - Potential to enter into new agreements with other agro-processing entities to out-scale operations	**THREATS** - Dependence on financial support from government – subsidy removal could cause significant fluctuations in product demand - Increasing entry of other organic fertilizer businesses into fertilizer market - Rainy season interferes with production activities and may dampen business operations

Contributors

Alexandra Evans, Independent Consultant, London, United Kingdom
Johannes Heeb, CEWAS, Switzerland
Josiane Nikiema, IWMI, Ghana

References and further readings

Ministry of Agriculture. 2017. Press release on promoting the use of organic manure. (ID 159439; 17 March 2017). Government of India.

Mishra, S. and Gopikrishna, S.R. 2010. Nutrient Based Subsidy (NBS) & support systems for ecological fertilization in Indian agriculture. Policy Brief. Greenpeace India.

PASIC – Puducherry Agro Services and Industries Corporation. http://pasicpondy.com/ (accessed May, 2016).

Personal observations and interviews with PASIC personnel. 2015.

Case descriptions are based on primary and secondary data provided by case operators, insiders or other stakeholders, and reflect our best knowledge at the time of the assessments 2015/16. As business operations are dynamic data can be subject to change.

CASE

Livestock waste for compost production (ProBio/Viohache Mexico)

Javier Reynoso-Lobo, Miriam Otoo, Lars Schoebitz and Linda Strande

Supporting case for Business Model 13	
Location:	Culiacan, Sinaloa, Mexico
Waste input type:	Agro-waste (livestock waste)
Value offer:	Organic fertilizer – compost and nutrient-rich liquid fertilizer from processed leachate
Organization type:	Private
Status of organization:	Operational since 2003
Scale of businesses:	Large-scale processing 420,000 tons of animal waste per annum
Major partners:	SuKarne

Executive summary

Productos Bioorganicos (ProBio) is Mexico's largest compost and vermicompost producer with the well-known Humibac brand. Although recently, its name changed to Viohache, this presentation is still using "ProBio". ProBio is a private company created in 2003 to manage the animal waste generated by SuKarne – the largest beef producer and marketer in Mexico. Given the significant quantities of livestock waste produced by SuKarne, traditional waste disposal (i.e. landfilling) systems no longer seemed sustainable and the identification of viable and environmentally safe alternatives was imperative. ProBio maintains a strategic partnership with SuKarne by providing pen-cleaning services in return for their feedstock – animal waste. The business processes 420,000 tons of livestock waste per annum to produce a total of 231,000 tons of compost and 500,000 liters of nutrient-rich liquid fertilizer from processed leachate. It operates in five locations around the country, and supplies a low cost, high quality organic fertilizer to the vegetable, fruit and grain crop sectors. ProBio implements a commodity-value based business model by using simple, low-cost and innovative strategies for the production and branding of the products they offer. It has garnered significant market demand through third party certification and the tailoring of its products to specific clients and agricultural purposes. The business also takes advantage of economies of scale and focuses on low cost, yet efficient technologies for organic fertilizer production and improved waste management. ProBio's operations have had a strong impact on society and the environment as its activities contribute to the reduction of greenhouse gas emissions, on-site waste odor, groundwater and surface water contamination, agricultural crop burning, and local air and soil pollution, among a few.

KEY PERFORMANCE INDICATORS (AS OF 2015)	
Land use:	130 ha
Capital investment:	USD 6,410,000 (land – USD 600,000; infrastructure – USD 377,240; machinery – USD 5,130,000; R&D – USD 100,000)
Labor:	65 employees
O&M costs:	USD 2.5 million per year
Output:	231,000 tons of organic compost and vermicompost, 500,000 liters of nutrient-rich liquid fertilizer from processed leachate per annum
Potential social and/or enviornmental impact:	Reduction of methane and CO_2 emissions, waste odor, groundwater contamination, local air and soil pollution, fertilizer requirements and improvement of agro-industrial waste management systems
Financial viability indicators:	Payback period: 5 years — Post-tax IRR: N.A. — Gross margin: USD 1.9 million

Context and background

Grupo Viz is a family-owned business established in 1969 at Culiacan, Sinaloa, Mexico. Over the years, Grupo Viz has expanded its operations to other sectors of the cattle production value chain and now owns five subsidiary companies operating independently. The five subsidiaries of Grupo Viz are:

a) SuKarne, a beef, poultry and pork producer;
b) ProBio, dedicated to the production of organic compost and vermicompost from animal waste;
c) Rendimientos Proteicos (RenPro), specialized in the processing of tallow, meat and blood meals for livestock and animal feed production;
d) SuKuero, a leather commercialization business; and
e) Agrovizion, an agribusiness dedicated to the promotion and commercialization of agricultural products such as corn, wheat, oats and roughage.

At the time of assessment, SuKarne owned five production facilities around the country, located in the states of Nuevo Leon, Baja California, Michoacan, Durango and Sinaloa. These five facilities maintain a daily average of 425,000 animals confined in open feedlots through the year. As the largest beef producer in Mexico, it significantly contributes to the generation of animal waste both nationally and worldwide. The national and local state legislation prohibit the unlicensed disposal and/or uncontrolled burning of animal waste, which results in significant quantities of waste that are left to decay in open-air landfills. This contributes to the production of large amounts of methane from the anaerobic process of landfilling, and invariably contributing to greenhouse gas (GHG) emissions. The above situation triggered the creation of ProBio in 2003, an independent private company with the objective of incorporating an efficient waste management solution for SuKarne's feedlot operations. The animal waste is removed from the feedlots at their facilities once every 6 months by ProBio and is processed into compost and vermicompost, a total of 231,000 tons per annum (70 and 30%, respectively), and an additional 500,000 liters of nutrient-rich liquid fertilizer from processed leachate. As SuKarne is the company's waste provider, this makes ProBio by far the largest compost and vermicompost producer in the country.

Market environment

According to the Mexican Secretariat of Agriculture, Livestock, Rural Development, Fisheries and Food ("SAGARPA"), 58% of Mexico's land, a total of 113.8 million hectares, is used for beef production. There is a total of 31 million cattle livestock in Mexico owned by 1.13 million breeders: 2 million dairy cattle and 29 million beef cattle (SAGARPA, 2015). According to the Mexican Ministry of

Environment and Natural Resources "SEMARNAT", livestock production has shown an accelerated growth in the past two decades, increasing by 62% in comparison with the 90's (SEMARNAT, 2010). As a result of this progressive increase in the agricultural sector, 83% of its emissions were accounted to livestock production in 2002, equivalent to 8% of the total emissions in Mexico. Additionally, waste management systems currently adopted no longer seem sustainable. There is a growing need for environmentally sound waste management alternatives, particularly in the livestock sector, given increasing enforcement of legislative mandates related to environmental protection.

A key factor driving the development of businesses such as ProBio is increasing chemical fertilizer prices and a need for sustainable agricultural alternatives. Soils in Mexico have a high susceptibility to erosion especially in the high valleys, which are mostly formed from volcanic materials (with a high concentration of sand and silt). Farmers favor fertilizers that facilitate plant nutrient assimilation at soil level and promote the formation of mycorrhizae and root absorption. These factors are indicative of the increasing demand for organic fertilizers and in general the development of more waste reuse businesses in Mexico.

Macro-economic environment

The increasing demand in higher food safety standards and organic products has triggered an increased use of vermicompost as high quality soil conditioner in several regions across the world. Since the 90's, the global market for organic food products has grown rapidly, reaching US $63 billion worldwide in 2012. This demand has driven a similar increase in organic agricultural inputs, including fertilizers (Willer et al., 2013). Mexico is estimated to have more than 110,000 organic farmers, considered the greatest number in any country worldwide. As demand for organic food in the United States expands, Mexico's certified organic acreage has been growing at a rate of 32 percent per year. A 2009 study found an annual organic production value of more than $370 million with 80% destined for export (Agri-Food Trade Service, 2009). Nutrient management has also become increasingly relevant with the price increase of chemical fertilizers and their inherent accountability for human health issues and environmental contamination. To date, there are few organic fertilizer producers in Mexico with large-scale capabilities – most producers constitute small operations. Affordable organic fertilizers have strong market potential for Mexico in the agricultural sector.

Business model

Figure 175 summarizes ProBio's business model. By using simple and low-cost yet effective technologies, ProBio produces high quality organic fertilizers tailored to specific customer segments and agricultural purposes. This, in addition to third-party product certification has garnered significant market demand. Its three main products, compost, vermicompost and nutrient-rich liquid fertilizer from processed leachate are sold directly to vegetable, fruit and grain crop farmers. Product promotion is achieved through field demonstrations and pre-commercial tests and have been instrumental in creating greater market access. A key aspect of ProBio's model is its partnership with SuKarne, an important waste generator. Initially, ProBio established an agreement with SuKarne to provide pen-cleaning services in exchange for the waste and a small fee. Additionally, SuKarne aided ProBio financially in order to start up the business as establishing a waste management system was a pressing issue for the beef producer. Nowadays, ProBio is a well-established profitable business and no longer charges SuKarne pen cleaning fees. Close proximity of ProBio to SuKarne's plant operations eliminates significant transportation and labor costs associated with the acquisition of waste. Yet, transportation and waste collection costs constitute the largest share of all operational costs at 68%. ProBio has recently restructured its business model and made a significant investment in machinery and increasing operative personnel as most equipment and required resources for operative activities were initially outsourced. This will significantly reduce O&M costs and yield higher long-term margins.

FIGURE 175. PROBIO'S BUSINESS MODEL CANVAS

KEY PARTNERS
- SuKarne
- Certification partners

KEY ACTIVITIES
- Liaison with government and other entities
- Provide pen cleaning services
- Collection of animal waste
- Production and sale of compost, vermicompost and liquid fertilizer from processed leachate
- Product tailoring
- Brand management, marketing and sales

KEY RESOURCES
- Quality assurance laboratory
- Waste input
- Machinery and equipment
- Irrigation system
- Operators
- Technical sales personnel
- Pest free certificates
- Brand
- Contractual agreement

VALUE PROPOSITIONS
- Offer tailored, low price compost, nutrient-rich vermicompost and liquid fertilizer
- Provision of sustainable waste management options for livestock producers

CUSTOMER RELATIONSHIPS
- Sales staff advise farmers and promote the products via technical workshops, field demonstrations, semi commercial tests
- Long-standing business relationship

CHANNELS
- Direct and indirect (telephone) sales
- Radio and web advertisement

CUSTOMER SEGMENTS
- Vegetable and extensive crop farmers)
- SuKarne

COST STRUCTURE
- Capital investment of USD 6.4 million (land, infrastructure, equipment, machinery, R&D)
- O&M costs of USD 2.5 million per annum

REVENUE STREAMS
- Compost and vermicompost sales

SOCIAL & ENVIRONMENTAL COSTS
- Dust generation from daily activities

SOCIAL & ENVIRONMENTAL BENEFITS
- Job creation
- Reduction of methane and CO2 emissions, waste odor, groundwater contamination, local air and soil pollution
- Reduction of chemical fertilizer requirements for agriculture
- Improvement in livestock waste management
- Improved soil quality

ProBio has demonstrated that waste reuse businesses can be profitable with government support and generate significant benefits to both industry and society. The next step for ProBio is set to be a more technological and innovative-based business, already with available technologies being tested at pre-commercial stages.

Value chain and position

ProBio's value chain is depicted in Figure 176. It benefits from SuKarne's capability to provide constant and large volumes of animal waste (feedstock), which enables the company to produce significant quantities of organic fertilizer. ProBio also takes advantage of other waste streams such as leftover corn stover and paunch from SuKarne's feed mill and slaughterhouse to use them as nutrient additives into their process. Such scale allows ProBio to develop optimization strategies in order to maximize its efficiency and increase profit margins. Through its economies of scale, both compost and vermicompost are priced significantly lower than the competition's products, mainly chemical fertilizer and smaller organic fertilizer producers, and thus providing an important competitiveness factor. Product demand relies on two customer segments, the vegetable and extensive crop farmers; the latter particularly expected to grow given the increasing demand for organic food products. ProBio has a strong sales team that is strategically divided by regions with important agriculture operations, where they establish product promotion programs with local farmers.

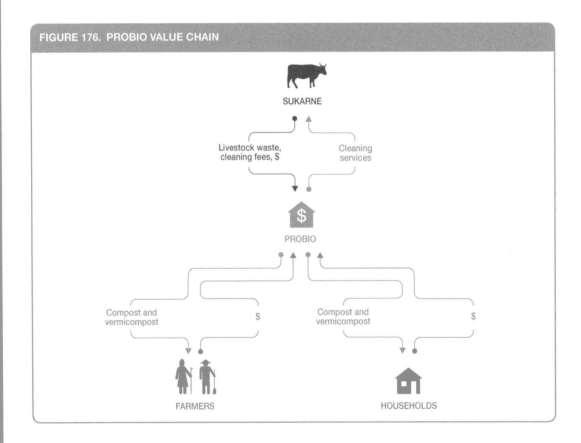

FIGURE 176. PROBIO VALUE CHAIN

Institutional environment

Livestock production units are bound by the Mexican Official Standard 001, which sets forth the maximum limits of solid and liquid waste allowed to be disposed of and discharged to federal water channels or bodies, respectively. This standard has forced livestock producers to develop waste management systems to meet those maximum limits, especially in the face of increasing production scales. This regulation implicitly incentivizes livestock companies to invest in businesses like ProBio to ensure their compliance and sustainability. Whilst there are no specific governmental guidelines for the certification of compost, several internationally accredited third party certification entities exist (e.g. Bioagricert and Metrocert) in Mexico. Product certification conveys a message of assured product quality to consumers (assuming they trust the certification body), which enables entities such as ProBio to increase their credibility and market share.

Technology and processes

Figure 177 provides an overview of the technological processes used by ProBio for the production of its organic fertilizers. The animal waste is collected from the feedlot pens every 6 months using a scraping system and stockpiled near their operations. Waste is constantly removed from this pile to enter the composting process. For such a process, windrows of 200 m length × 6 m wide × 3 m height of animal waste are formed, and corn stover and some paunch is added to the mixture to adjust for carbon and nitrogen requirements. Additionally, water is added to reach optimal humidity content for the fermentation process to start (this takes about a week). This part of the process undergoes an aerobic thermophilic fermentation stage for about 8 weeks, where temperatures of up to 70°C are reached and promote pathogen elimination. Further aerobic degradation is achieved throughout an approximate 14-week mesophilic stage where temperature drops to 25–30°C to enter a final compost maturation stage. Finally, the compost is screened to remove stones and other unwanted particles. The overall composting process lasts from 120–160 days.

Finished compost is utilized to feed the vermicomposting process. New windrows are formed and California Redworms (Eisenia fetida) are added. The redworms further contribute to the organic matter degradation, producing a compound called 'humus', or vermicompost, a nutrient-rich organic fertilizer with important soil conditioning properties. Once the worms are well established, additional compost is added weekly in order to "feed" them and increase the production of vermicompost. The windrows are watered every day through an automated irrigation system in order to maintain a humidity level between 60–70%. Windrows are placed over a sloped terrain to enable natural leachate collection throughout the process, where it is then pumped into large containers for further oxygenation and packaging. After a period of 5–6 months the worms are removed using a trommel and further reincorporated into a new vermicomposting process; the humus or vermicompost is finally screened and ready for sale. Both finished compost and vermicompost are analyzed to determine nutrients and other constituents. Overall, the whole process from waste to final product has a conversion efficiency of 55%. The final product contains a nitrogen, phosphorus and potassium content of 0.5–1%, 1–1.5%, and 1–1.5%, respectively, and provides a crop yield (tested e.g. for potatoes) increase of 15–30%.

Funding and financial outlook

The business required an initial capital of USD 2.2 million, for land, infrastructure, machinery and equipment. The payback period for such an investment is estimated at three years. Overall, the business has production costs of USD 5.5 million (Table 43), where 46% is accounted for operation and maintenance, which breaks down in the following way: 68% for transport and waste collection, 15% for machinery lease related to the composting and vermicomposting processes, 10% for equipment maintenance, 6% for fossil fuel, 1% for tools and equipment and the balance for final

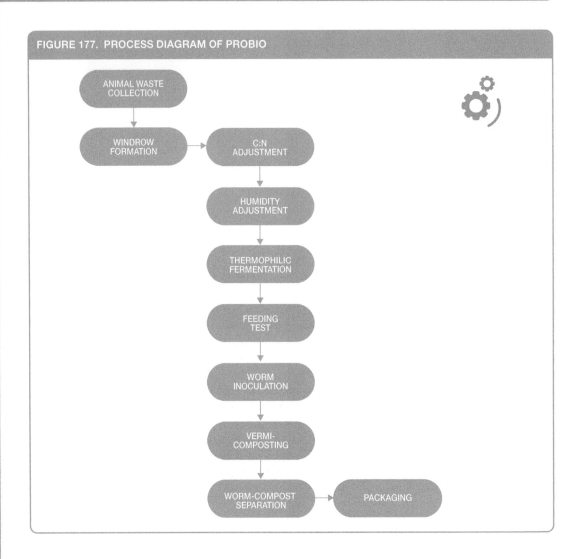

FIGURE 177. PROCESS DIAGRAM OF PROBIO

product packing. Land lease accounts for 20% of production costs, while labor constitutes 17%. Services, which account for 2%, comprise costs such as water, security, mail, etc. Quality control refers to laboratory analyses conducted by external entities and accounts for 2%. Finally, depreciation and administration costs comprise 1% and 12% of production costs, respectively.

ProBio has three key income streams. The main income streams are sales of compost and vermicompost. A minor income is acquired from sale of the nutrient-rich liquid fertilizer from processed leachate. In total, ProBio had revenues of USD 5.7 million in 2013, with a total net income of USD 1 million. The volume of sales for compost and vermicompost is estimated at 231,000 tons per year at a price of USD 30 per ton and USD 70 per ton, respectively. ProBio has been generating profit for several years; indicating that with increased production and demand, aside from incorporation of more innovative-oriented processes, the business stands to attain higher profits and benefits to its shareholders. ProBio restructured its business model in 2014 and made a USD 4.2 million investment in machinery and R&D infrastructure, which will significantly contribute to a cost reduction, particularly in transport and process maneuvers (over 50%) as well as in so far outsourced laboratory analyses (up to 100%).

TABLE 43. BREAK-DOWN OF OPERATIONAL COSTS (2013)

COST ITEM	TOTAL COST (PER ANNUM)
Land lease	USD 1,114,400
Labor	USD 970,700
Operation and maintenance	USD 2,565,500
– Fossil fuel	USD 158,000
– Equipment maintenance	USD 245,000
– Transport and waste collection	USD 1,750,000
– Machinery lease	USD 391,000
– Tools and equipment	USD 15,000
– Product packing	USD 6,500
Quality control	USD 98,000
Depreciation	USD 31,500
Administration costs	USD 798,600
Total	USD 5,578,700

Socio-economic, health and environmental impact

Agricultural operations have become increasingly more intensive to execute economies of production and scale around the world, as pressure to become more efficient continues to grow. This is especially true in livestock operations (swine, dairy cows, etc.), which can generate serious environmental consequences, such as GHG emissions, odor, and water/soil contamination, all a result from improper storage and disposal of animal waste. Confined Animal Feeding Operations (CAFOs) use similar Animal Waste Management System (AWMS) options to store animal residues. These systems emit both methane (CH_4) and nitrous oxide (N_2O) resulting from anaerobic decomposition processes (Clean Development Mechanism, 2007). Additionally, displacement of chemical fertilizers conveys a set of environmental and health benefits that may be achieved by production of organic fertilizers processed from agricultural waste. Businesses that incorporate cleaner waste management solutions such as ProBio have important environmental benefits such as:

- Reduction of CH_4 and CO_2 emissions by avoiding landfill anaerobic conditions;
- Reduction of waste odor, local air and soil pollution by accelerating the decomposition of organic matter present in waste streams;
- Reduction of groundwater contamination and health issues related to nitrogen accumulation derived from chemical fertilizer demand;
- Overall improvement in livestock waste management;
- Overall soil quality improvement from prolonged organic fertilizer application.

Scalability and replicability considerations

The key drivers for the success of this business are:
- Strong relationship and win-win partnership with SuKarne – main input supplier.
- Assured and continuous supply of large quantities of waste – free of charge, aiding economy of scale development.
- Guaranteed high quality product sold at a competitive market price.
- An effective market development strategy.
- Incorporation of efficient and innovative technologies across its operations.
- Increasing chemical fertilizer prices.
- Increasing demand for organic fertilizers due to soil stability issues.
- Fast-growing livestock markets and subsequent insufficient waste management capacity.

There is great potential for ProBio to expand its services to other livestock producers, however land availability for operation set-up close to the waste source may be a constraint. Regarding market share, one of its customer segments, the grain crop sector, is not fully aware and certain of the benefits of organic fertilizers, and considers them an additional cost rather than a long-term sustainable alternative. Further development of this segment will have a significant impact in market access as such crops represent the vast majority of cropland in Mexico. SuKarne's scale in terms of waste generation is probably one of the biggest success factors for ProBio since they are able to provide a constant and high amount of feedstock to the business. This model has a high potential for replication in agrarian countries with large-scale livestock production systems. It is important to note however that the implementation of such a model requires significant start-up capital investment – which is among the most cited barriers for business development in developing countries. In ProBio's case, SuKarne provided key initial financial support as it is obliged to comply with legislative mandates for waste disposal and the implicit cost of non-compliance would be significantly higher – so an incentive for the private sector to invest in such initiatives should exist if similar legislation applies.

Summary assessment – SWOT analysis

Figure 178 provides an overview of the SWOT analysis for ProBio. ProBio is a successful company that reuses the animal waste generated by the beef producer SuKarne to produce compost and vermicompost, and then sells it directly to farmers and households. Essential in its business model is the certification and branding of their organic fertilizer products. This in addition, strategic marketing and sales programs have increased ProBio's market share. Additionally, their agreement with SuKarne has ensured consistent supply of feedstock, mitigating production risk associated with fluctuation in input supply. The establishment of the compost facility in close proximity to the waste source significantly reduces related transportation and labor costs. Technology and related production efficiency, on the other hand, must be improved in order to increase the profit margin, since ProBio takes advantage of economies of scales to generate profit. Opportunities exist for ProBio to fully access the grain crop market segment. This would significantly increase its market share and profit margins due to its important cropland area in Mexico. The latter however requires a bold incentive program for farmers where they would be able to initially try out the product and experience tangible benefits prior to any investment, as uncertainty drives them to consider such fertilizers as an additional cost rather than a strategy to displace high-priced chemical fertilizers. ProBio is willing to bear this risk given its confidence in the quality of its products, as this practice has already proven to be effective. ProBio, however, solely relies on SuKarne to provide livestock waste. Although unlikely given their contractual agreement, in the event that SuKarne would decide to divert its waste supply to another purpose or business, ProBio would face a significant production risk. ProBio is an example of a novel business using a commodity-value approach and a solid partnership with an agro-waste generator to address some of the major waste management and environmental challenges in Mexico whilst generating significant profits and benefits to society.

Contributors

Jasper Buijs, Sustainnovate, Netherlands; formerly IWMI, Sri Lanka
Radheeka Jirasinha, Consultant, Colombo, Sri Lanka

FIGURE 178. SWOT ANALYSIS FOR PROBIO

	HELPFUL TO ACHIEVING THE OBJECTIVES	**HARMFUL** TO ACHIEVING THE OBJECTIVES
INTERNAL ORIGIN ATTRIBUTES OF THE ENTERPRISE	**STRENGTHS** • Strong business partnership • Low price and high nutrient value of products • Access to cost-free inputs • Creation of image through positive impact on the environment	**WEAKNESSES** • Technology improvement to increase profit margins • Customers not fully aware of benefits of products • Lack of an effective marketing and branding campaign
EXTERNAL ORIGIN ATTRIBUTES OF THE ENVIRONMENT	**OPPORTUNITIES** • Up-scaling of business operations to cater to other large-scale livestock producers • Increasing market share of extensive crop farmer segments • Market expansion to the United States	**THREATS** • Chemical fertilizer market mature and well-established • Waste provider diverting waste for different purposes

References and further readings

Agri-Food Trade Service. 2009. Mexico Organic Market Study, Comercio e Integración Agropecuaria, S.C., May 2009.

Personal interview with Mr. Horacio Hernández – General Manager of ProBio. 2012.

Personal interview with Mr. David Vizcarra – General Manager of ProBio (June 2014).

Personal interview with Mr. Marco Medina – Administration Manager of ProBio (June 2014).

SAGARPA. 2015. Ministry of Agriculture, Livestock, Rural Development, Fisheries and Food (SAGARPA) of Mexico. www.gob.mx/sagarpa. Accessed on 15 July, 2015.

SEMARNAT. 2008. Mexico profile. Animal waste management methane emissions. Report for the Methane to Markets, Agriculture Subcommittee. https://www.globalmethane.org/documents/ag_cap_mexico.pdf (accessed November 8, 2017).

Willer, H., Lernoud, J. and Home, R. 2013. The World of Organic Agriculture: Statistics and Emerging Trends 2013. Research Institute of Organic Agriculture (FiBL) and the International Federation of Organic Agriculture Movements (IFOAM, 2013).

Case descriptions are based on primary and secondary data provided by case operators, insiders or other stakeholders, and reflect our best knowledge at the time of the assessments 2015/2016. As business operations are dynamic data can be subject to change.

BUSINESS MODEL 13
Nutrient recovery from own agro-industrial waste

Miriam Otoo and Munir A. Hanjra

A. Key characteristics

Model name	Nutrient recovery from own agro-industrial waste
Waste stream	Vegetative waste, livestock waste
Value-Added Waste Products	'Regular' compost, enriched vermi-compost
Geography	Regions with significant livestock production, agro-processing enterprises
Scale of production	Medium: 5–40 tons/day; Large: 1,000–2,000 tons/day
Supporting cases in this book	Navaisha, Kenya; Pondicherry, India; Culiacan/Sinaloa, Mexico
Objective of entity	Cost-recovery [ ]; For profit [X]; Social enterprise [ ]
Investment cost range	USD 45,000–USD 2.5 million, depending on scale and technology
Organization type	Private, Public
Socio-economic impact	Cost savings, new revenue and income generation, job creation, reduction of water and land pollution, reduction of CO_2 emissions, averted human health risk
Gender equity	Where biogas is produced in addition to the agro-waste based compost, this can represent increased access to improved fuel options for women

B. Business value chain

Many agro-industrial entities continue to face the increasing challenge of managing their waste. To ensure business sustainability (typically in compliance to legislative mandates for environmentally friendly waste management practices), agro-industrial entities set up subsidiary businesses to the parent company to convert the agro-waste (tea, horticultural products, sugar mill waste, livestock waste) generated from operations of the latter into an organic fertilizer. The concept is primarily based on the notion that parent agro-businesses generate sufficient business such that its sustainability justifies new capital investments in an onsite nutrient recovery entity to support its own back-end agricultural operations. The concept is simple but the impacts are multi-fold, due to the forward and backward linkages between the parent agribusinesses entity and subsidiaries engaged in nutrient recovery for self-supply to the parent entity but also entry into the larger fertilizer market.

This business model can be initiated by a public, private or public-private partnership entity seeking to address an internal business waste management challenge and additionally generate revenue and diversify their business portfolio. Although this business model is typically geared towards cost-savings, the agro-waste processing entity can generate significant revenue from compost sales primarily to its parent company (usually if it is an agricultural producer) and local farmers. Investment in innovative technologies (e.g. inclusion of biologically active compounds that promote plant growth and health) can allow them to self-brand their compost product and invariably capture a share of the local fertilizer market. The agro-waste processing unit sources its waste input primarily from the parent company and its affiliates (contract farmers) thus ensuring a consistent supply of resources, oftentimes free of charge or at a lower cost. Quality monitoring activities can be performed by a local university/R&D institute at a fee or their own laboratory. The business concept involves a simple value chain schematic as depicted in Figure 179.

C. Business model

The business model is hinged on two value propositions: a) provision of sustainable waste management (collection and treatment) services and options (nutrient recovery) for 'primary' agro-industrial (parent company) business; and b) provision of affordable, high nutrient organic fertilizer

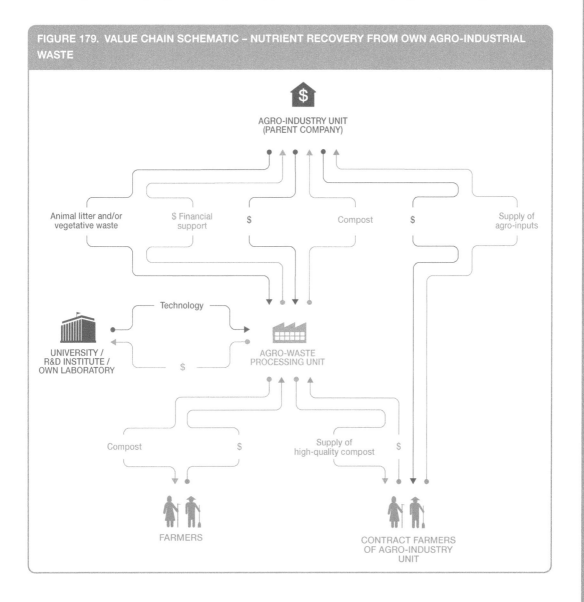

FIGURE 179. VALUE CHAIN SCHEMATIC – NUTRIENT RECOVERY FROM OWN AGRO-INDUSTRIAL WASTE

for agricultural production (Figure 180). Key success drivers of this business model are based on: a) mutually-beneficial partnership with its parent company – which ensures a consistent supply of waste input (vegetative and livestock waste) free of charge or at low cost and provision of capital investment which mitigates capital start-up risk; b) option of a diversified portfolio through the sale of biological control organisms and different grades of compost tailored to different markets; c) price differential gains from market segmentation for its compost product. Waste input used for compost production is sourced from the parent company. This is a win-win partnership as the latter has a reliable waste management system to ensure sustainability of its business and the former – a reliable source of waste input for production at a fee and start-up capital investment. The business model's main revenue generation streams are from: a) organic fertilizer sales to segmented markets; b) fees received from parent company for waste management.

The business typically sells its compost products primarily to its parent company (if it is an agricultural producer) and directly to local farmers often implementing a segmented-pricing approach with bulk sales to parent company and large-scale farmers at a lower price and a higher price to retailers and smallholder farmers. It is important to note however that depending on the contractual agreement between the parent company and the subsidiary (agro-waste processing) entity, the compost price may be adjusted to account for the cost of collecting and transporting the waste to the waste processing facilities. A competitive marketing strategy such as the provision of free samples of compost to first time users can help build the business' product brand and customer base. Also, by adopting a commodity-value (and using value-addition technologies) the agro-waste processing entity can produce high quality compost tailored to specific clients and agricultural purposes. The success of this approach is dependent on the partnership the business has with key research institutes that can provide support for the development of innovative technologies to produce high-quality products, and also provide product quality analysis services for certification. Third party product certification can help garner significant market demand and mitigate market competition effects from the often subsidized chemical fertilizer. Field demonstrations and semi-commercial tests (farmers, particularly crop farmers are able to initially try the product and observe actual benefits prior to payment) can be instrumental in creating greater market access.

The business can take advantage of economies of scale, depending on the scale of operations of the parent company, and focus on low cost, yet efficient technologies for compost production. Large-scale operations will permit the business to reduce its production costs and charge a lower price of its compost and help capture a larger share of the fertilizer market. The overall investment required for this type of business is relatively modest depending on the scale of operations and investments required at the start-up for R&D (development of innovative technologies), technologies and related equipment. This business model has the potential to generate significant socio-economic and environmental benefits including: job creation and reduced CO_2 emissions. Additionally, monetary gains to farmers are represented by increased crop yields and related incomes. This model has a high replication potential especially in developing countries with an increasing number of agro-processing businesses and related limited waste management options.

D. Potential risks and mitigation

The business model presented here was designed and optimized based on the analysis of different case studies (see previous sections). In designing this optimized business model, risks related to safety, local acceptance by the community, and business attractiveness for investors were assessed.

Market risks: The main market risk is related to the business' strong focus and dependence on the launching customer (parent). This can induce the business to lose touch with the market and limit its

BUSINESS MODEL 13: NUTRIENTS FROM OWN AGRO-INDUSTRIAL WASTE

FIGURE 180. BUSINESS MODEL CANVAS – NUTRIENT RECOVERY FROM OWN AGRO-INDUSTRIAL WASTE

KEY PARTNERS
- Agro-industrial company (parent company)
- Local university/R&D institute
- Small scale local livestock farmers

KEY ACTIVITIES
- Waste (vegetative and/or livestock waste) collection
- Production of organic fertilizer
- Brand management, marketing and sales
- Sales and distribution of organic fertilizer

KEY RESOURCES
- Partnership with parent company
- Equipment
- Labor (skilled and unskilled)
- Land and labor
- Vegetative and animal waste
- Partnership with university/R&D institute
- Branding and quality certification
- Contractual agreements

VALUE PROPOSITIONS
- Provision of sustainable waste management options for 'primary' agro-industrial (parent company) business
- Provision of affordable, high nutrient organic fertilizer for agricultural production

CUSTOMER RELATIONSHIPS
- Long-standing business relationship
- Direct sales between parent–sister company
- Personal at direct sales

CHANNELS
- Direct channel with parent company
- Direct sales to local farmers

CUSTOMER SEGMENTS
- Agro-industrial (parent) company
- Small- and large-scale farmers

COST STRUCTURE
- Capital investment (equipment, land, infrastructure)
- Operation and maintenance costs
- Laboratory and product certification costs

REVENUE STREAMS
- Sale of organic fertilizer to segmented markets:
  - Bulk sales to parent company and large-scale farmers (lower price)
  - Sales to smallholder farmers (higher price)
- Fees received from parent company for waste management

SOCIAL & ENVIRONMENTAL COSTS
- Potential contamination of soil and groundwater from disposal of untreated leachate on-farm

SOCIAL & ENVIRONMENTAL BENEFITS
- Job creation
- Reduction in chemical fertilizer use
- Reduction of carbon dioxide emissions
- Improvement in soil structure and nutrient composition
- Reduced nitrate contamination

opportunities for growth. Traditionally, farmers have a high acceptability of agro-waste based compost – especially given its high nutrient content. It is however important to consider quality testing by a third party to minimize market risks associated with consumers' negative perceptions. Whilst this approach can in turn allow the businesses to charge a higher price (from the 'branded' product), it may entail additional costs for which the compost producers have to take in account.

Competition risks: One of the key competition risks to be considered is supportive policies for chemical fertilizer use which may create a non-competitive market environment that negatively affects the sustainability of compost producers. This effect can be mitigated based on the scale of operation and targeted (assured) clientele – bulk purchases from government-owned agricultural department services and the parent firm. Innovative marketing strategies related to free samples and demonstration trials can be adopted to mitigate some of these effects. Resource/input (waste) supply risks are considered to be relatively low due to the assured supply of waste from the parent company.

Technology performance risks: The composting technologies (traditional windrow-composting and vermi-composting) considered under this model are relatively mature and freely available in the market. However, depending on the waste input and technology used, some residual risk may remain. For example, livestock waste-related diseases such as mad cow disease and foot-and-mouth infections need particular attention and quality monitoring and testing programs by a third party should be considered to reduce such risks.

Political and regulatory risks: Policies and regulations related to waste-based compost sectors differ by country. The oftentimes stronger political support for chemical fertilizer use (slow phasing-out of fertilizer subsidies) and lack of specific government guidelines for the certification of compost and internationally accredited third-party certification entities can represent a significant risk to the sustainability of the business model.

Social equity related risks: There are no distinctive social inequity risks anticipated for this business model in terms of poverty and gender. Smallholders could potentially benefit from improved agricultural productivity from increased access to comparatively inexpensive organic fertilizer, if the compost producers choose to sell the excess.

Safety, environmental and health risks: There are potential environmental and health risks that need to be considered under this model. Workers involved in all activities along the compost production value chain (waste collection, separation, compost production, etc.) can be potentially exposed to livestock waste-related diseases if technology performance is not up to par. To safeguard the health of workers, it is imperative that businesses provide and ensure the use of safety gear – hand gloves and rubber boots; conducts an annual medical check up. To address the safety and health risks to workers, standard protection measures are also required as elaborated below (Table 44).

E. Business performance

This model ranks high on scalability and replicability due to the increase in agro-industrial businesses and related limited waste management options especially in developing countries (Figure 181). Significant environmental benefits can be realized through nutrient recovery and improved waste management options, as the reduced release of nitrates and GHG emissions results in decreased environmental pollution. This business model however ranks low on social impacts, as aside from employment generation (and oftentimes labor is obtained from the parent company and used internally) and increased access to alternative fertilizers, leaner social benefits accrue to other economic actors along the value chain (e.g. waste collectors, compost retailers). It is noted that most entities either use

BUSINESS MODEL 13: NUTRIENTS FROM OWN AGRO-INDUSTRIAL WASTE

TABLE 44. POTENTIAL HEALTH AND ENVIRONMENTAL RISK AND SUGGESTED MITIGATION MEASURES FOR BUSINESS MODEL 13

RISK GROUP	EXPOSURE					REMARKS
	DIRECT CONTACT	AIR/ DUST	INSECTS	WATER/ SOIL	FOOD	
Worker	■					Potential risk of exposure to e.g. bovine parasites and diseases requires monitoring; Potential risk of dust, noise and chemical compost contaminants
Farmer/user						
Community						
Consumer						
Mitigation Measures	🧤	😷	🦟	⛰		

Key: ☐ NOT APPLICABLE   ☐ LOW RISK   ▨ MEDIUM RISK   ■ HIGH RISK

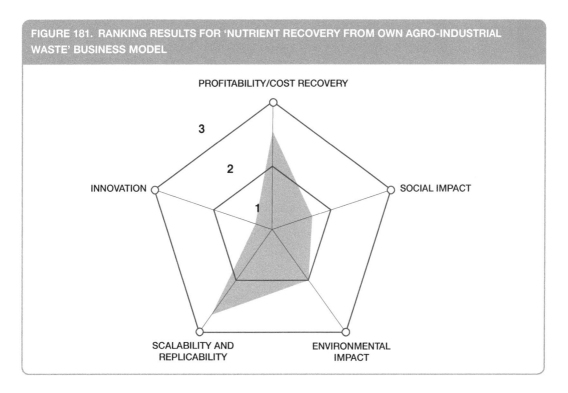

FIGURE 181. RANKING RESULTS FOR 'NUTRIENT RECOVERY FROM OWN AGRO-INDUSTRIAL WASTE' BUSINESS MODEL

the traditional open-windrow composting technology or vermicomposting or both, to produce regular compost and vermicompost. These technologies are simple, low cost and easily available (technical training) in the market such that the model ranks lowest on innovation. New technologies that help reduce energy costs could improve its rank on the innovation frontier.

11. BUSINESS MODELS ON COMPOST PRODUCTION FOR SUSTAINABLE SANITATION SERVICE DELIVERY

Introduction

Improved access to sanitation is one of the major policy goals throughout developing countries. An emphasis so far has been on the eradication of open defecation, hygiene and improved toilet facilities, ideally connected to sewer systems where urban centres are exploding. Global movements have to date increased access to basic sanitation products which has resulted in a significant percentage of rural and urban populations been connected to household-based latrines and septic tanks (CSE, 2011), however the majority of the population in developing countries still lack access to toilet facilities and substantial efforts are continuously being needed to close this gap. An increasing number of private businesses are setting up public toilet facilities to cater particularly to migratory populations and slum inhabitants who still have marginal access to sanitation products and services, however limited septage collection and treatment can undermine the sustainability of these services.

An effective and sustainable sanitation service delivery is one that provides products and services across the entire sanitation value chain, interlinks with the agricultural or other sectors to generate benefits to all economic actors in the respective value chains, and creates connectivity of resources among physical, and biological systems (Figure 182). Resource recovery and reuse of urban septage as peri-urban fertilizer has so far been largely an informal sector activity (Kvarnström et. al., 2012). But with the increasing interest in a green economy, and new technical innovations for fertilizer generation, there is scope for resource recovery to play an increasingly significant role (EAI, 2011). The business model on **sustainable sanitation service delivery** via nutrient recovery from fecal sludge presented here generates the double value proposition of:
- Provision of sanitation systems/ products (such as urine diversion dry toilets (UDDTs)), and reliable waste management (collection and treatment) services to poorer segments of society in greatest need of these services;
- Provision of an affordable, sanitized and nutrient-rich compost product for farmers.

The crux of the business model is hinged on the desirable social impact of providing hygienic sanitary facilities to society, particularly the masses at public places, whilst also providing an effective way to meet agricultural input needs of the farming community via compost production from human excreta. The business approach works because it is built around harnessing economic value from human waste whilst providing sanitation services to the poorer segments of society which represents the greatest percentage of population in need of such services, particularly in developing countries. By re-branding human waste as a needed input instead of a waste output, sanitation/waste reuse-based businesses can create both a physical and financial demand for waste, completely reinventing the economics of sanitation (Murray, Waste Enterprisers, pers. comm., 2014).

In this chapter, we describe a case from **Rwanda** which recognized the opportunities in human waste and is gradually playing an important role in leveraging private capital to help provide sustainable sanitation services and realize commercial the value in waste by shifting the focus from treatment for waste disposal to treatment of waste as a resource for reuse for the ultimate benefit of poor farmers and households (EAI, 2011; Murray and Buckley, 2010).

FIGURE 182. SUSTAINABLE SANITATION VALUE CHAIN WITH RESOURCE RECOVERY

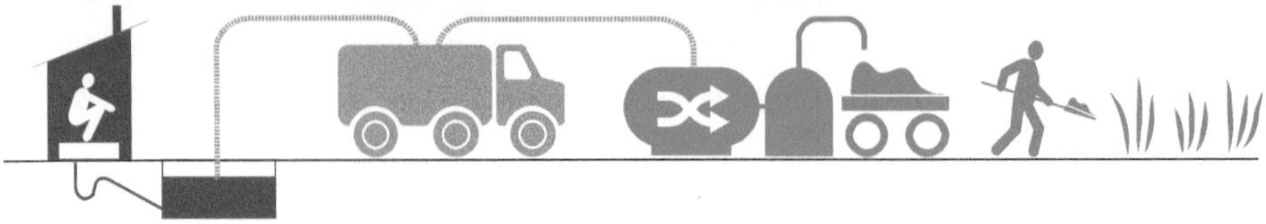

STAGE I
Access to toilet systems, capture and storage of fecal sludge

STAGE II
Collection, storage and transport

STAGE III
Treatment and conversion of fecal sludge into valuable resource for agricultural use

References and further readings

AECOM International Development, Inc. and SANDEC/Eawag. 2010. A rapid assessment of septage management in Asia: Policies and practices in India, Indonesia, Malaysia, the Philippines, Sri Lanka, Thailand and Vietnam. Report for USAID Regional Development Mission for Asia.

Chowdhry, S. and Koné, D. 2012. Business analysis of fecal sludge management: Emptying and Transportation Services in Africa and Asia. Seattle, WA, USA: Bill and Melinda Gates Foundation Report.

Centre for Science and Environment (CSE). 2011. Policy paper on septage management in India. Delhi.

Dodane P., Mbeguere, M., Sow O. and Strande, L. 2012. Capital and operating costs of full-scale fecal sludge management and wastewater treatment systems in Dakar, Senegal. Environmental Science and Technology 46 (7): 3705–3711.

EAI. 2011. Sustainable recovery of energy from fecal sludge in India. Report for BMGF. Energy Alternatives India. Chennai.

Kvarnström, E., Verhagen, J., Nilsson, M., Srikantaiah, V., Ramachandran S. and Singh, K. 2012. The business of the honey-suckers in Bengaluru (India): The potentials and limitations of commercial faecal sludge recycling – An explorative study. (Occasional Paper 48) [online] The Hague: IRC International Water and Sanitation Centre. www.irc.nl/op48 (accessed November 8, 2017).

Murray, A. Waste Enterprisers, personal communication. 2014.

Murray, A. and Buckley, C. 2010. Designing reuse-oriented sanitation infrastructure: The design for service planning approach. In Wastewater Irrigation and Health: Assessing and Mitigation Risks in Low-Income Countries, edited by Drechsel, P., Scott, C.A., Raschid-Sally, L., Redwood, M. and Bahri, A. UK: Earthscan-IDRC-IWMI, 303–318.

CASE

Fecal sludge to nutrient-rich compost from public toilets (Rwanda Environment Care, Rwanda)

Andrew Adam-Bradford, Miriam Otoo and Lesley Hope

Supporting case for Business Model 14	
Location:	Kigali, Rwanda
Waste input type:	Source-separated urine and feces from urine diversion dehydrating toilets (UDDT)
Value offer:	Provision of sanitation services and sanitized urine and feces as a safe organic fertilizer for agricultural production
Organization type:	Private
Status of organization:	Operational since 2009 (NGO since 2006); assessed in 2012-2014
Scale of businesses:	Production: 200 tons of fecal-based organic fertilizer per year
Major partners:	Kigali City Council (KCC), United Nations Development Programme (UNDP) and European Union (EU)

Executive summary

Rwanda Environment Care (REC) is a privately owned company engaged in the business of providing public toilet services and producing organic fertilizer from fecal sludge for sale to agricultural producers. With a mismatch between an ever-increasing urban population and the sanitation services provided by the municipalities, a significant number of inhabitants in Kigali have limited to no access to sanitation products such as toilets and when they do, there are virtually no collection systems in place. REC tapped into this gap in the sanitation value chain and has set up several public toilets at different locations in Kigali, Rwanda, using the ecological sanitation (eco-san) technology. The main goal of REC is to implement a sustainable sanitation services delivery system – which ensures that customers not only have access to services (i.e. toilets) but also mechanisms to ensure consistent and efficient waste collection and treatment systems are put in place. Its activities extend to the agricultural sector via the conversion of the collected fecal sludge from their public toilets into a valuable resource – urea-rich organic fertilizer (urine-enriched compost). REC implements a multiple revenue stream strategy comprised of: toilet fees amounting to USD 324 per day, kiosk and shop rentals (USD 334 per month), compost sales (USD 6,483/year) and consultancy service fees from the provision of technical assistance in the design and construction of eco-san latrines. The adopted technology – eco-san toilets – is simple and cost-effective and also ensures easy access to segregated waste inputs. REC's activities provide inhabitants, especially, the migrating population in Kigali with access to toilets which has significantly reduced the incidence of open defecation and 'flying toilets'. Additionally, reduced open-dumping of human excreta in the environment will reduce the risk of soil and groundwater contamination. Increased availability of environmentally safe fertilizer alternatives will contribute to reducing water and soil pollution from reduced nitrate release attributed to chemical fertilizer use. While the current scale of REC may not have a notable employment impact, with plans to out-scale their activities, it is expected that a significant number of jobs will created along the sanitation value chain.

KEY PERFORMANCE INDICATORS (AS OF 2013/14)						
Land use:	1.6 ha					
Capital investment:	USD 29,173 excluding land costs					
Labor:	2 unskilled full-time laborers					
O&M cost:	USD 188.39 per toilet block of 8 units and 2 kiosks					
Output:	200 tons of organic fertilizer per year					
Potential social and/or enviornmental impact:	Reduced risk of ground- and surface water pollution, reduced health cost associated with poor sanitation, reduced human exposure to untreated waste and chemical pollutants, enhanced soil fertility and productivity, increased food security					
Financial viability indicators:	Payback period:	2 years	Post-tax IRR:	N.A.	Gross margin:	N.A.

Context and background

Rwanda Environment Care (REC) was established in 2005. It received an award of USD 50,000 from a United Nations Development Programme (UNDP) Partnership Small Grant Programme in 2006 to establish fee-paying ecological sanitation services to residents in Kigali alongside rainwater harvesting. In 2007, an additional UNDP grant was awarded which allowed further development of public eco-san latrines in Kigali including the construction of public eco-san toilets in the main districts of Kigali. Rwanda Environment Care (REC) was first established as a pilot project but is now a profit-generating business. In 2009, they introduced a 'sanitation as a business' model which included an improved eco-san design along with additional adjoining units that were rented as kiosks, small shops and/or communication centres. REC's initiatives have been particularly important for Kigali as it has filled an important gap in the sanitation sector as the coverage of sanitation in urban areas is limited, particularly in the low-income areas (slum areas). It is equally important that revenue through fee-charging is generated from such facilities to cover routine repairs and staff salaries ensuring a level of sustainability. In addition to the high demand for public latrines in urban areas, there is an equally high demand for soil conditioners and fertilizer in farming systems throughout the country. Maintaining soil fertility through sustainable land management practices remains a major challenge which is compounded by poor agricultural practices and a lack of access to affordable fertilizers (Donovan et al., 2002).

Market environment

In Kigali, 80% of the population has access to latrines but only 8% of these latrines meet hygienic standards, hence improved access to hygienic and convenient public latrines is an important environmental sanitation and public health measure. Additionally, the significant migration population that characterizes this city makes this an even more important necessity. Furthermore, a continuously available supply of human effluent coupled with farmers' quest for an alternative to chemical fertilizer have been some of the driving forces for the establishment of this business. The maintenance of soil fertility through sustainable land management practices is a major challenge in the agricultural sector of Rwanda, and particularly for peri-urban agriculture in Kigali. REC thus processes fecal matter collected from its eco-san toilet to nutrient-rich organic fertilizer for sustainable agriculture.

Macro-economic environment

Given the relatively high global fertilizer prices, most farmers in Rwanda cannot afford to purchase fertilizers at the beginning of the season. Increasing oil prices and fuel costs have also greatly influenced fertilizer prices in landlocked Rwanda. Hence to make fertilizers more affordable for smallholder farmers, the government introduced the fertilizer subsidy programme for certain food crops. This measure will potentially have an undesirable impact on new businesses like REC who are entering the fertilizer market. They will be facing fierce competition if chemical fertilizer remains comparatively low in price and more cost-effective than organic fertilizers. Comparable incentives will have to be implemented for organic fertilizers to mitigate the effects of competition and facilitate entry of new waste reuse businesses in the fertilizer market.

Business model

Figure 183 below presents an overview of REC's business model. REC's business model is based on two main value propositions: a) provision of hygienic eco-san public toilets on a fee-for-use basis; and b) offer of affordable urea-rich, fecal sludge-based organic fertilizer (urine-enriched compost) which is sold directly to farmers. The high demand for public toilets in Kigali ensures a daily revenue through toilet fees. On average, the 4,000 daily users generate a total of 200,000 Rwandan Francs (RWF) (USD 324 per day). An essential part of this enterprise is the inclusion of other shops in the toilet complex, from which rent is derived, increasing the revenue stream available to the enterprise. In addition to the provision of public latrines, REC plans to provide an eco-san consultancy service through the provision of technical assistance in the design and construction of eco-san latrines which will include follow-up visits in the first six months of operation. An example of this consultancy work has included constructing eco-san toilets in over 18 schools over the last five years which were funded through the American NGO Water for People. The multiple revenue stream strategy ensures and secures funds for the composting component of the business and safeguards it from shocks such as delayed payments. REC received financial support from UNDP and the EU and land free of charge from the Kigali City Council. These grants were crucial at the start-up phase of the business given how traditionally difficult it is to access funds from formal financial institutions. REC's activities have resulted in several socio-economic and environmental benefits. Increased access to toilets especially in low-income areas have significantly reduced the incidence of open-defecation and 'flying toilets' and consequently environmental pollution. Increased access to environmentally safe and affordable fertilizer alternatives represent monetary gains for small-holder and large-scale farmers.

Value chain and position

Figure 184 provides an overview of REC's value chain. REC's business is composed of four main parts: a) provision of toilet facilities on a fee-per-use basis; b) provision of shops and kiosks to traders; c) sale of fecal sludge-based organic fertilizer to farmers; and d) provision of consultation services on technical assistance in the design and construction of eco-san latrines. From its early years, REC has constructed and managed five eco-san units in Kigali at the following locations: Kigali City Council (12-door toilet facility); Nyabugogo (12-door toilet facility); Kacyilry (four-door toilet facility); Kimironko (12-door toilet facility); and Kicukiro (eight-door toilet facility). The resulting 48 toilets in the city which on average receive 4,000 users on a daily basis are producing an estimated average of 0.6 tons of fecal matter per day. The high demand for public toilet use ensures a consistent waste supply stream. Quality factors such as moisture (i.e. eater use) can be regulated and monitored, ensuring high quality of the waste input. Currently there is no competition in this supply stream as new eco-san toilets are located where public toilets facilities are limited. The enterprise uses human effluents obtained from its toilet business and processes it into fertilizer, hence faces no competition for the waste input. The urine (urea)-enriched organic fertilizer is sold directly to large-scale farmers in the Northern Province who come to the site for purchase. Prior to collection the compost is stored at a central site in Kigali

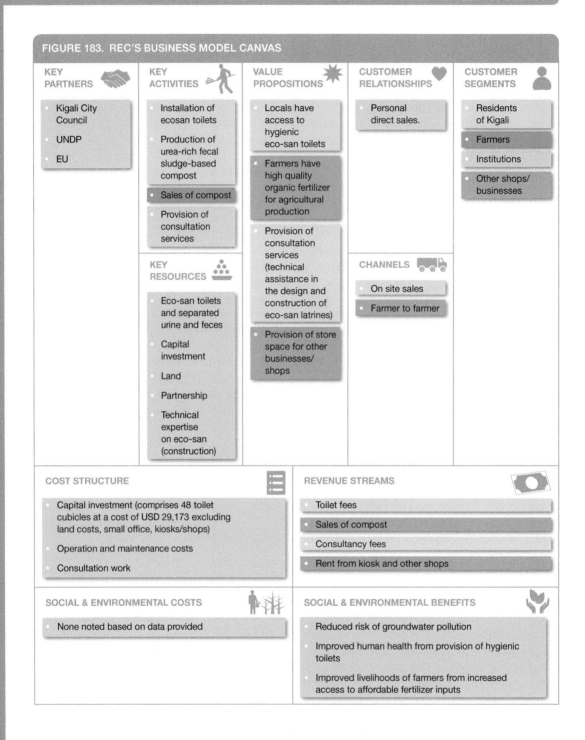

FIGURE 183. REC'S BUSINESS MODEL CANVAS

where it undergoes final decomposition before being bagged and stored ready for collection. During the assessment period, REC produced annually over 200 metric tons of compost generating over RWF 4 million, which in 2012/13 corresponded with about USD 6,483. The compost is bagged and stored at a centralized yard in Kigali ready for collection. Demand was from the start higher than production and this has remained constant. REC does recognize that the government subsidized chemical fertilizer

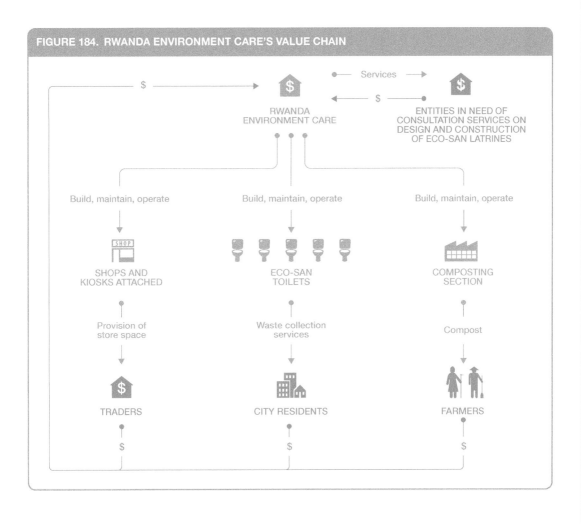

FIGURE 184. RWANDA ENVIRONMENT CARE'S VALUE CHAIN

programme could represent competition for their compost product and thus the need to implement a long-term marketing strategy to increase its share of the market. Additional revenue streams such as renting out shops and kiosks incorporated in the toilet building design has been important for REC in mitigating fluctuations in compost demand, thus invariably improving the sustainability of the business.

Institutional environment

Eco-san toilets were a relevant new introduction in Rwanda and while there are laws and regulations on the use of human waste issued by the Rwanda Utility Regulation Agency these did not have specific quality standards or guidelines for ecological sanitation. Consequently, REC has been working with the government agency to draft appropriate eco-san quality standards and guidelines. The Rwandan government is supportive of eco-san interventions as illustrated in the fact that urban land is provided by local authorities for projects such as eco-san toilets as they recognize this as an important contribution to service provision in urban environmental sanitation.

Technology and processes

Eco-san toilets are based on a very pragmatic principle of on-site treatment while separating the liquid and solid elements of human waste (Figure 185). In doing so, it brings several advantages such as

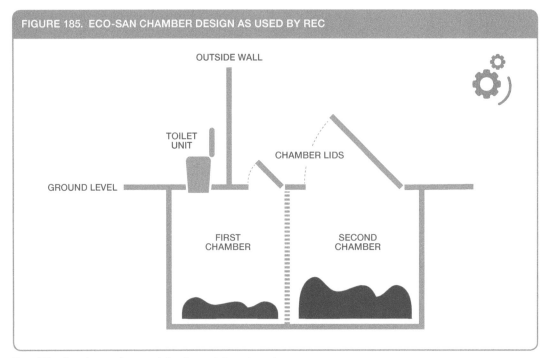

FIGURE 185. ECO-SAN CHAMBER DESIGN AS USED BY REC

Note: Urine diversion not shown and chambers not drawn to scale

removing the necessity for having flushing water in a toilet system, thus eliminating any wastewater that would normally flow into a septic tank or sewer. Also diverting urine from fecal matter and then keeping the fecal matter relatively dry eliminates the strong odors associated with the combination of urine and fecal sludge. Once urine is isolated and stored in a container the odor is reduced, moreover with usually no bacterial contamination the urine can be added to the latter stages of composting for compost enrichment or be diluted with water and instantly applied as a crop fertilizer. The fecal matter takes considerably more time to decompose into a state ready for crop application, consequently in the design of all eco-san toilets the separation of liquid and solid waste is a central feature. For the liquid element, urine is normally channelled into a receptor thus providing a safe method of harvesting and storing the urine, but in regards to managing the solid matter there is a degree of variation in how the solid element of human waste is collected, stored and treated, for example variations in chamber size, the use of chambers in series or in standard parallel arrangement and addition of solar heated chambers. The eco-san toilet systems have been designed in such a way that the physical structure fits the surrounding environment. One or two chamber systems can be used. In the latter, the smaller chamber is directly under the toilet unit while the adjoining larger chamber allows entry for a worker to shovel the dry waste from the first chamber to the second, and to empty the compost once it has matured. To increase heat in the large chamber and provide optimum decomposition conditions, the metal lid of the chamber is painted black to absorb solar radiation. Due to high number of users, the pits get full within a short time. Ideally, once the pits get full, the toilets are decommissioned for a period of at least three months during which the feces are left to compost. However, as the toilets are needed, the fecal matter is transferred to an external dry place to complete the composting which allows the vaults to be used again.

Funding and financial outlook

The project was funded by the UNDP and EU at an initial cost of USD 29,173 for the construction of an eight-door toilet complex with two kiosks. This amount is exclusive of land costs which was provided for free by the urban council. It is estimated that initial construction investment can be recovered in a two-year period. Operation and maintenance costs for a block unit is projected at USD 84 per day. REC has currently three revenue streams: toilet fees, kiosk/shop rentals and compost sales. On average 4,000 daily users of eco-san toilet generate USD 324 per day and the sale of 2,000 bags of compost generates USD 6,343 annually. Toilets fees anchor the compost business as noted from the significant difference in the revenues generated.

Socio-economic, health and environmental impact

REC's initiatives provide eco-san toilets of hygienic standards to the Kigali community and has reduced the intense pressure which hitherto existed on the available public toilets. It employed at the time of the study two people who work on a full time basis and are responsible for the collection of toilet fees as well as daily cleaning and maintenance of the facility. REC ensures to mitigate occupational health risks by providing protective gear (i.e. masks and gloves), which the staff are obliged to wear while working on site and particularly during chamber emptying operations. REC also ensures that it produces a safe compost product which is achieved from the long storage period of the decomposed substrate in the eco-san systems before collection. This ensures that most pathogens are eliminated before the product is used for any agricultural production. In the early phase of the project, microbial levels were tested and found to meet an acceptable level but it was noted that such testing has not become a routine measure and the results of the initial testing were not available. This however does not discount the significant quantities of nutrients recovered from the human effluent which is used for farming activities, thus improving the nutrient level of soils and increasing productivity.

Scalability and replicability considerations

The key drivers for the success of this business are:
- Significant migrating population that are in need of convenient public latrines.
- Strategic partnerships to mitigate capital investment risk at start-up phase; technology and product development/innovation.
- Assured supply of key production input (human excreta) at no cost.
- Increasing farmers' quest for a more affordable alternative to chemical fertilizer.
- Multiple revenue streams – which mitigates risk associated with fluctuations in demand of any of their products and services.

REC's model is replicable and can be scaled out and up especially in communities with no access to the flush toilet system. However, the replication and scaling up and out of this model is highly capital intensive. In Rwanda, eco-san toilets have proven to be successful and socially acceptable, however the main constraint in replicating such services is access to investment funds although the work of REC is now being recognized and the sector is attracting the interest of local development banks.

Summary assessment – SWOT analysis

Figure 186 presents an overview of the SWOT analysis for REC. By implementing a multiple revenue stream strategy, REC is able to safeguard the business from shocks such as delayed payment for compost or seasonal demand, or decreased demand in the provision of any of its products and services. This business has been particularly successful in leveraging its business partnerships to mitigate capital investment risk. Also importantly, it uses a technology that has a key advantage, i.e. there is no wastewater or sludge produced as in a flush-based toilet systems or pit latrines. The technology can be raised off the ground and is thus compatible with flood prone areas or in locations with high water tables as the risk of groundwater contamination is avoided. Currently, the use of urine-based fertilizers remains an underexploited resource in farming systems around Kigali, so demand

FIGURE 186. SWOT ANALYSIS FOR RWANDA ENVIRONMENT CARE

	HELPFUL TO ACHIEVING THE OBJECTIVES	HARMFUL TO ACHIEVING THE OBJECTIVES
INTERNAL ORIGIN ATTRIBUTES OF THE ENTERPRISE	**STRENGTHS** • Diversified revenue streams • Easy accessibility and availability of waste input • No wastewater or sludge is produced from process • Suitable in flood prone areas • Compost is rich in urea • Strong partnership with the municipality • Latrines meet hygienic standards	**WEAKNESSES** • Inadequate market strategy • Capital intensive at start-up phase • Site location may increase distribution cost • Technology requires expertise • Limited storage capacity • Under high usage, composting periods within the pit gets shorter
EXTERNAL ORIGIN ATTRIBUTES OF THE ENVIRONMENT	**OPPORTUNITIES** • Increase in toilet users • Increase in compost users • Sale of urine-based organic fertilizers	**THREATS** • Competition from continued subsidization of chemical fertilizers • Some cultural values prevent usage of public toilets • Regular monitoring necessary to ensure correct usage and effectiveness of composting

remains low mainly due to a lack of awareness in its benefits as a liquid fertilizer. As REC does not have the capacity to store and transport urine for on farm applications they have found an alternative use for the resource, which consists of using the urine to enrich the compost by adding quantities of urine to the compost heap during the later stages of decomposition. This is a common practice found in small-scale gardening as the urea feeds the bacterial action in the composting process. There is a great opportunity for REC to add value to the collected urine and with a sound marketing strategy increase its share of the fertilizer market. The compost from human excreta is sold directly to farmers and plans are underway to develop a market for the enriched urine. Although operating so far on a small-scale, the scaling-up and out of REC's initiatives supported by its partners, like SNV, has a high potential to generate significant impact.

Contributors
Alexandra Evans, Independent Consultant, London, United Kingdom
Josiane Nikiema, IWMI, Ghana
Valentin Mucyomwiza, Rwanda Environment Care (REC)

References and further readings

Donovan, C., Mpyisi, E. and Loveridge, S. 2002. Summary comments on forces driving change in Rwandan smallholder agriculture 1990–2001: Crops and livestock. Agricultural Policy Synthesis. Rwanda Food Security Research Project, Rwandan Ministry of Agriculture and Livestock.

Dusingizumuremyi, E., Ruzibiza P. and Nkurunziza, T. 2011. Effectiveness of eco-toilets management in public places, case of Kigali City. www.ircwash.org/sites/default/files/Dusingizumuremyi-2011-Effectiveness.pdf (accessed November 8, 2017).

Case descriptions are based on primary and secondary data provided by case operators, insiders or other stakeholders, and reflect our best knowledge at the time of the assessments 2012/14. As business operations are dynamic data can be subject to change.

BUSINESS MODEL 14
Compost production for sustainable sanitation service delivery

Miriam Otoo and Munir A. Hanjra

A. Key characteristics

Model name	Compost production for sustainable sanitation service delivery
Waste stream	Source-separated urine, feces from urine diversion dry toilets (UDDT) and pit/septic tanks
Value-added waste product	Urine-based fertilizer and fecal sludge-based soil conditioner
Geography	Suitable for slum areas/communities with limited provision of waste management service and/or no access to the flush toilet system. UDDT technology particularly suitable for flood prone areas or in locations with high water tables
Scale of production	Small to medium: 150–200 tons of fecal-based organic fertilizer
Supporting case in this book	Kigali, Rwanda
Objective of entity	Cost-recovery [ ]; For profit [X]; Social enterprise [ ]
Investment cost range	USD 25,000–32,000
Organization type	Private or business foundation
Socio-economic impact	Improved access to sanitation facilities, reduced health cost associated with poor sanitation, reduced human exposure to open waste dumping, enhanced soil fertility and agricultural productivity, jobs for unemployed
Gender equity	Toilet provision. Reduced practice of open defecation away from home, especially in the dark, reduces personal risk for women and girls

B. Business value chain

Many cities and towns across Africa and Asia have a huge gap in sanitation services and waste management – and are far below required international coverage standards. Open defecation continues to be a common practice in view of limited access to basic sanitation products such as toilets facilities suited to the local environment. Additionally, limited public funds to support waste management infrastructure and services has resulted in significant environmental pollution as the majority of the generated waste (e.g. human excreta), whether collected or uncollected is often disposed of untreated in unofficial and open spaces, water bodies and/or landfills (Kinobe et al., 2015). This situation is particularly exacerbated for large urban areas characterized by a growing population and rapid migration.

The business model – sustainable sanitation service delivery system – can be initiated by a private entity or a business-oriented foundation seeking to fill the gap in sanitation service delivery value chain by providing products and services particularly to poorer segments of society (e.g. slums) in greatest need of these services, and also converts collected fecal sludge from households and public toilets into a valuable resource: organic fertilizer for agricultural use (Rao et al., 2016). In the primary market, the business entity provides sanitation products (toilets) and services (i.e. public toilets, waste collection services) to two main customer segments: households and public masses at a fee (Figure 187). The value for customers in the primary market is increased access to toilet facilities, and reliable and clean removal of fecal sludge. In the secondary market, the collected septage is converted into a nutrient-rich organic fertilizer and is sold to peri-urban farmers. In the secondary market, increased availability of environmentally safe fertilizer alternatives will contribute to reducing water and soil pollution from reduced nitrate release attributed to chemical fertilizer, and also represent significant savings for farmers.

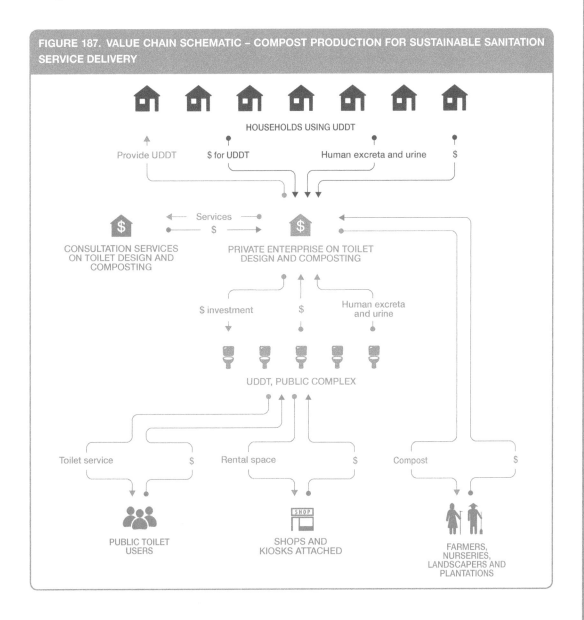

FIGURE 187. VALUE CHAIN SCHEMATIC – COMPOST PRODUCTION FOR SUSTAINABLE SANITATION SERVICE DELIVERY

A unique feature of this model is its viability potential which is driven by a multi-revenue stream and hinged on its primary market. The business generates the majority of its revenues from the sale of toilet facilities, provision of public toilets and waste collection services. It is able to generate sufficient funds to additionally cover the compost production costs, if needed. This is crucial as the sustainability of the primary market largely depends on the business been able to reuse or dispose of the human excreta safely.

C. Business model

The business model is hinged on three value propositions: a) supply and maintenance of ecological sanitation systems (such as urine diversion dry toilets (UDDTs)); b) provision of reliable waste management (collection and treatment) services to poorer segments of society in greatest need of these services; c) provision of affordable and high quality organic fertilizer for agricultural production. This translates into a multiple revenue stream strategy comprised of: sale of eco-san toilets, toilet user fees, kiosk and shop rentals, compost sales and consultancy service fees from the provision of and technical assistance in the maintenance of eco-san toilets and latrines, which ensures sustainability in business operations. This reflects the important success driver for the model which is the diversification of its portfolio which cuts across the entire sanitation value chain in the provision of toilets, waste collection services and organic fertilizers. Additionally, this business model adopts a service oriented approach in which it uses revenue generated from the provision of toilet facilities to run the composting section, which safeguards the business from shocks such as delayed payment for compost or seasonal demand, which could otherwise halt the smooth running of operations and affect the sustainability of the business.

The overall investment required for this type of business is relatively modest depending on the scale of operations, with major investments required at the start-up for the purchasing of toilet facilities and construction of the public toilets. Strategic partnerships with local government, municipalities, city councils, agriculture department and international financiers are instrumental not only for the purposes of gaining access to start-up financing but also customer segments for their compost product. The business model described in this chapter presumes the operation for a standalone private enterprise (Figure 188), and could also be useful for festivals and music events.

D. Alternate scenarios

Scenario I: Franchise model for safe and sustainable sanitation service delivery

An alternative to the generic business model of sustainable service delivery is the inclusion of a franchising system (Figure 189). It is assumed that at this scaling-up stage of the business, the private entity has sufficient private equity or collateral to obtain financing in order to set up the franchise system. The private/business entity (franchiser) creates a network of entrepreneur managed toilets and composting units. The network is organized within the framework of a franchise. The franchiser supplies the toilet and composting units on demand to its franchise partner network across several cities. The use of the franchiser's name brand and access to their business strategy comes at a cost to the franchisee. The franchisees deliver their composted material to the nearest franchise collection point which the franchise purchases. The franchisees have a sustainable system where they are able to earn revenue from toilet user fees and sale of composted materials without worrying about having a market for their product. The franchiser has the opportunity to sell to bulk buyers such as commercial farmers and large-scale organic food producers, given their increased scale of production. They are able to monitor the quality of the compost via their own product testing and occasional checks. Whilst the franchisor's success depends on the success of the franchisees, the franchisee has a greater incentive than the direct employee because they have a direct stake in the business. There is a risk

BUSINESS MODEL 14: COMPOSTING FOR SUSTAINABLE SANITATION SERVICES

FIGURE 188. BUSINESS MODEL CANVAS – COMPOST PRODUCTION FOR SUSTAINABLE SANITATION SERVICE DELIVERY

KEY PARTNERS	KEY ACTIVITIES	VALUE PROPOSITIONS	CUSTOMER RELATIONSHIPS	CUSTOMER SEGMENTS
- City Council - International development partners - Financiers	- Supply and maintenance of ecological sanitation toilets - Waste collection services - Production of fecal sludge-based fertilizer - Sale and marketing of compost product - Provision of consultation services	- Urban population have increased access to toilet facilities - Reliable provision of waste collection services - Provision of high quality and affordable fertilizer alternative - Provision of rental space to other businesses	- Personal and direct sale	- Urban population (households, institutions) - Commercial and medium-scale farmers, organic farmers (especially flower industry) - Other small-scale businesses
	KEY RESOURCES - Eco-san toilets and human excreta - Capital investment - Land - Partnership - Equipment (protective gear, desludging machinery)		**CHANNELS** - Direct sales of products and services to households, institutions and farmers	

COST STRUCTURE
- Capital investment (comprises small office, public toilets, kiosks/shops)
- Operation and maintenance costs

REVENUE STREAMS
- Sale of eco-san toilets
- Toilet user fees
- Sales of compost
- Rent from kiosk and other shops

SOCIAL & ENVIRONMENTAL COSTS
- Possible human health risk from handling of excreta if appropriate protective gear is not utilized

SOCIAL & ENVIRONMENTAL BENEFITS
- Reduced risk of water pollution and nitrate leaching
- Improved human health from increased access to hygienic toilets
- Improved yield due to application of affordable and nutrient-rich organic fertilizer
- Reduced incidence of open defecation
- Reduced open dumping of human excreta
- Increased access to toilets in low-income areas/slums

for the people that are buying the franchises as failure rates are noted to be higher for franchise businesses than independent business start-ups. Factors related to fair pricing of equipment and supplies from the franchisor, fees for training and advisory services charged by the franchisor, royalty fees, amongst others can influence the sustainability of the franchises. Overall, the franchising model has great potential to generate significant benefits to multiple economic actors in both the sanitation and agricultural value chains as it provides not only an opportunity for the franchiser to increase its profits but it also represents increased access to toilet facilities and waste management services for a greater number of households and improved fertilizer options for agricultural producers.

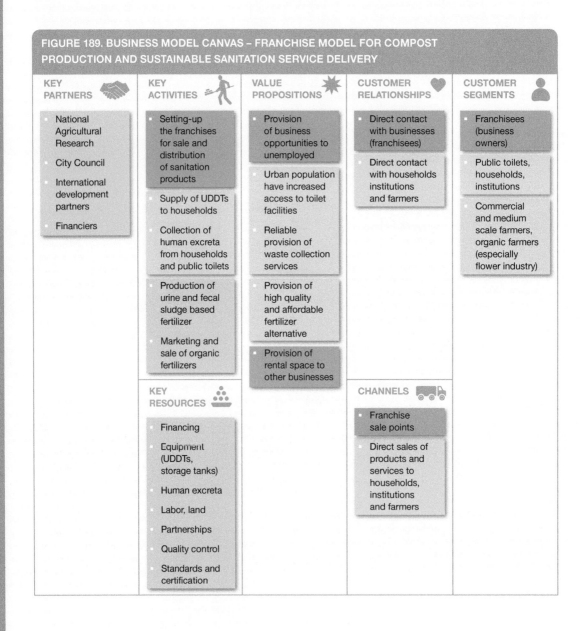

FIGURE 189. BUSINESS MODEL CANVAS – FRANCHISE MODEL FOR COMPOST PRODUCTION AND SUSTAINABLE SANITATION SERVICE DELIVERY

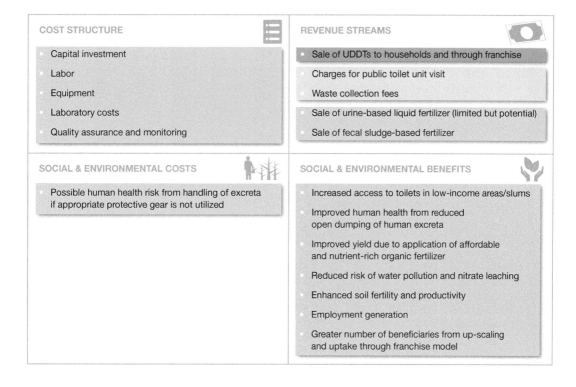

E. Potential risks and mitigation

The business model presented here was designed and optmized based on the analysis of different case studies and literature review. In designing this optimized business model the risks such as safety, local acceptance by the community and business attractiveness for investors were addressed.

Market risks: There is a huge imbalance between the demand and supply of sanitation products and services especially in fast growing cities in developing countries, such that open defecation and open dumping signals huge market potential but in some settings the affordability comes into question due to the very low income and socio-economic status of the communities. Households' low-ability to pay for sanitation products and services may pose a market risk for this model. This model has proven some initial success and social acceptability despite the stigma associated with waste-based fertilizers. Farmers' low willingness to pay for the compost in view of chemical fertilizer alternatives poses a risk to the sustainability of the model. This risk can however be mitigated from revenue generated from other streams. Additionally, storage and transportation challenges of the liquid-based urine fertilizer may also require an agricultural community nearby for reuse.

Competition risks: Competition risk could come from other suppliers of comparable sanitation products and services, more evidently from the chemical fertilizer sector. Policies and programs such as fertilizer subsidy programs make chemical fertilizer prices relatively lower than compost prices, and thus more cost-effective for farmers. Comparable incentives are needed to mitigate these effects for waste-based organic fertilzer businesses.

Technology performance risks: There are minimal to no technical performance risks associated with the composting technology. Whilst the technology is quite new in most developing country settings, it is relatively simple to implement. The sustainable sanitation technology design separates urine and keeps the fecal matter dry to elimate strong odor. Dry fecal matter can be processed into compost and directly used as fertilizer. After storage the separated urine can be directly used as liquid fertilizer in dilution with water, and after storage for two to six months for unrestricted application. The compost production is also low cost and flexible in terms of scale and has relatively simple quality assurance procedures and does not require a high-level of technical expertise.

Political and regulatory risks: National regulations on the reuse of human excreta for agricultural purposes differ, and this determines the scope within which sanitation businesses can engage in resource recovery. Even in cases, where reuse is permitted, the lack of regulations and standards on products and associated certification and quality minitoring pose significant risks for businesses. The provision of ecological sanitation facilities in cities is generally well-received by the governmental entities, in many developing countries.

Social equity related risks: This business model does not have any known social inequity risks. On the other hand, it significantly increases access to sanitation products and services, especially for migratory populations and slum inhabitants. From an agricultural perspective, farmers have improved livelihoods given their increased access to high nutrient organic fertilizers which contributes to improved agricultural productivity.

Safety, environmental and health risks: Also where UDDTs are used, potential pathogenic health risks to different actors along both the sanitation and agricultural value chains remain, associated with the collection, treatment, processing and use of human excreta (Table 45). In particular, workers that collect the (largely dried) fecal sludge and composted materials are at risk. The provision of protective gear for chamber emptying operations should be mandatory. For the compost buyer, microbial standards can provide trust, while from the food consumer perspective, careful washing and boiling should be a routine measure. Additionally, farmers must be trained on the appropriate application methods for the waste-based fertilizer products. Recommendations of national agriculture agencies must also be implemented in tandem, in association with agricultural extension agents.

TABLE 45. POTENTIAL HEALTH AND ENVIRONMENTAL RISK AND SUGGESTED MITIGATION MEASURES FOR BUSINESS MODEL 14

RISK GROUP	EXPOSURE					REMARKS
	DIRECT CONTACT	AIR/DUST	INSECTS	WATER/ SOIL	FOOD	
Worker	HIGH					Potential health risks to different actors along both the sanitation and agricultural value chains are associated with the collection, treatment, and processing of human excreta
Farmer/user			LOW			
Community			LOW			
Consumer					MEDIUM	
Mitigation measures	🧤	😷	🚫🦟 / 🧴	⛰️	🧺 / 🍲	

Key: ☐ NOT APPLICABLE ▨ LOW RISK ▨ MEDIUM RISK ■ HIGH RISK

F. Business performance

This model can be scaled up and decentralized through franchise operations across cities in Africa and Asia. A greater opportunity for scaling up and out the sanitation products (UDDTs) and services (waste collection, composting) exist particularly for slum areas due to limited provision of sanitation services. This model ranks highest on environmental impacts due to its catalytic role in protecting human and environmental health by reducing open defecation and unsafe disposal of human excreta. The model ranks second on scalability and can be replicated extensively in cities and neighbourhoods lacking toilet facilities (Figure 190). The model ranks next highest on profitability, because the model generates several revenue streams including toilet visit fees, sale of urine-based liquid fertilizer, sale of compost, fees for waste collection services and rental from shops in the toilet complex and even consultancy services where applicable. For the generic business model, the technology involved is simple, low cost and easy to use, and hence innovation rank is the lowest.

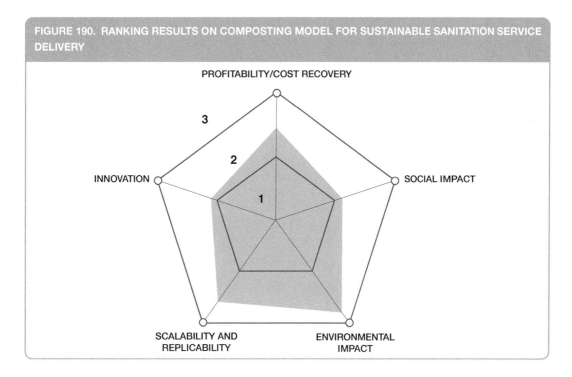

FIGURE 190. RANKING RESULTS ON COMPOSTING MODEL FOR SUSTAINABLE SANITATION SERVICE DELIVERY

References and further readings

Anand, C.K. and Apul, D.S. 2014. Composting toilets as a sustainable alternative to urban sanitation – A review. Waste Management, 34, 329–343.

Kinobe, J.R., Bosonaa, T., Gebresenbet, G., Niwagaba, C.B. and Vinnerås, B. 2015. Optimization of waste collection and disposal in Kampala city. Habitat International 49 (2015): 126–137.

Lalander, C.H., Hill, G.B. and Vinnerås, B. 2013. Hygienic quality of faeces treated in urine diverting vermicomposting toilets. Waste Management, 33, 2204–2210.

Mawioo, P.M., Hooijmans, C.M., Garcia, H.A. and Brdjanovic, D. 2016. Microwave treatment of faecal sludge from intensively used toilets in the slums of Nairobi, Kenya. Journal of Environmental Management, 184, Part 3, 575–584.

Rao, K.C., Kvarnstrom, E., Di Mario, L. and Drechsel, P. 2016. Business models for fecal sludge management. Colombo, Sri Lanka: International Water Management Institute (IWMI). CGIAR Research Program on Water, Land and Ecosystems (WLE). 80p. (Resource Recovery and Reuse Series 06).

12. BUSINESS MODELS FOR OUTSOURCING FECAL SLUDGE TREATMENT TO THE FARM

Introduction

With a limited number of septage treatment systems in many parts of the developing world, business entities that empty latrines or cesspits often discharge the sludge onto open lands, in landfills or into wetlands, instead of driving to remote official dumping sites. There is an urgent need to address this challenge through more fecal sludge treatment plants. Where this is not possible also farm based systems can offer safe treatment while directly recovering nutrients from fecal sludge for agricultural production.

Fecal sludge is an abundant and valuable resource as the dominating urban sanitation system in both South Asia and Sub-Saharan Africa are septic tanks and latrines (Chowdry and Koné, 2012; Dodane et al., 2012). Its low chemical and metal contamination in household based on-site treatment facilities makes the collected fecal sludge (septage) a valuable soil ameliorant similar to other organic manure such as farmyard manure with high application potential in farming and landscaping (Otoo et al., 2015). The reuse opportunity that lies in the fecal sludge waste stream is especially important where soils are poor and the availability of alternative inputs is expensive. In particular, in areas where affordable fertilizer production or its access is limited, smallholder farmers might use the fecal sludge for fodder, tree (crop) plantation or cereal production. Farmers in West Africa and South India, for example, re-direct cesspit truck operators to their fields to obtain the nutrient rich manure (Drechsel et al., 2011; Evans et al., 2013; Kvarnström et al., 2012). In Northern Ghana, this typically occurs after cereal harvest in the dry season (Cofie et al., 2009). Due to the aridity and heat, the sludge dries over several months and is then incorporated into the soil.

The observed reuse business model between farmers and truck operators reverses the cash flow, as farmers pay the drivers for farm-gate delivery, while otherwise the transporter must pay a tipping fee for desludging into a treatment pond. In an optimized business model, the revenue would ideally support the operation and maintenance costs of the cesspit operation, supplementing the fecal sludge household collection fee. However, an economic drawback to the sustainability of the system is the seasonality in demand for fertilizer, which are often only applied once or twice over the cropping cycle. Fecal sludge is applied as a basal fertilizer at the start of the dry season, allowing it sufficient time to dry over several months before it is incorporated into the soil, and cereals are planted. Sludge marketability is different with (tree) plantation crops, like in India, which can benefit from fecal sludge throughout the year. However, where farmers do not have spare land for the fecal sludge to be initially stored, the voluminous characteristic of the raw fecal sludge can become a constraint. This bottleneck has been bypassed in parts of Karnataka where sludge is collected and sun-dried by larger enterprises, for auctioning to farmers.

In most developing countries, fecal sludge as a source of fertilizer has not received much recognition, due to both the informal nature of reuse and possible cultural or perception barriers. Moreover, the disposal of fecal sludge onto land, particularly agricultural land, is often prohibited by law – or is, at least, a grey area governed by 'tacit approval'. In other words, 'culprits' have not been punished, especially where engineered, official dumping places are still an exception and the authorities are left with little choice. Where official dumping sites exist, cesspit truck owners pay to use them. Health concerns by authorities concerning the use of raw fecal matter in food production limit the extent of this activity, although with sufficient solar drying as observed in Ghana, and crop restrictions, the risks can be minimized (Seidu, 2010; Keraita et al., 2014), even where no other regulations govern the process. Most pathogens die during sun exposure, so health risks for consumers of cereals grown on this land are minimized (Seidu, 2010). To mitigate also health risks for farmers, they are required to use protective gear.

Other controlled resource recovery approaches can further reduce the potential health and environmental risks associated with fecal sludge use, and increase farmers' accessibility and usability. These steps and trajectories of increasing value proposition have been realized in different regions and are illustrated as shown in Figure 191. An observed pathway of value proposition for agricultural reuse is:

1) Direct land application of the raw fecal sludge for agricultural purposes – where value addition occurs in the form of sludge collection and transportation to the farm or plantation, usually followed by natural solar-treatment (sun drying) or incorporation in the soil as an alternative treatment and risk reducing option (Keraita et al., 2014).
2) To limit the risks for farmers, the fecal sludge can also be dumped on designated unplanted drying beds followed by composting (or co-composting with other organic waste to improve the carbon–nitrogen ratio) before sale. The value addition lies in removing pathogens, reducing the volume and concentrating the nutrients. Moreover, co-composting is an approved Clean Development Mechanism (CDM) activity. The bulky nature of composted fecal sludge can however act as a barrier to the transportation of the product to markets, increasing the distribution costs, which are borne by the end-users.
3) To increase the accessibility and usability of the composted product, pelletization and blending of fecal sludge-based compost with rock-phosphate, urea/struvite or any industrial fertilizer will allow the product to have nutrient levels specific for target crops and soils, and a product structure improvement (pellets) to improve its competitive advantage, marketability and field use. Several business cases have been identified in Nigeria, Ghana[1], Sri Lanka and South Africa which offer related value proposition (Rao et al., 2016). While farmers generally show a positive perception, for those who already use raw sludge (for free or a low fee), they may require field demonstrations to appreciate any other form of sludge with a higher price tag.

This chapter presents the business model on **Outsourcing fecal sludge treatment to the farm** and a supporting case from India, demonstrating how the informal business sector can support the sanitation value chain for the benefit of agricultural production.

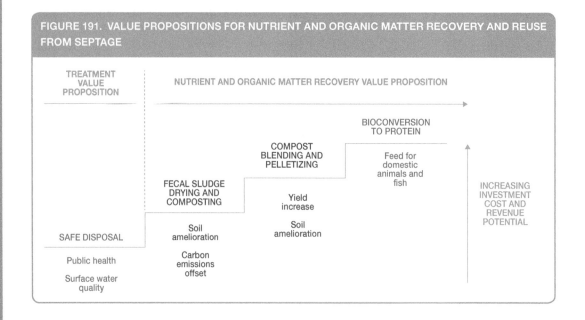

FIGURE 191. VALUE PROPOSITIONS FOR NUTRIENT AND ORGANIC MATTER RECOVERY AND REUSE FROM SEPTAGE

References and further readings

Chowdhry, S. and D. Koné. 2012. Business analysis of fecal sludge management: Emptying and transportation services in Africa and Asia. Seattle, WA, USA: Bill and Melinda Gates Foundation Report.

Cofie O.O., Drechsel P., Agbottah, S. and van Veenhuizen, R. 2009. Resource recovery from urban waste: Options and challenges for community based composting in Sub-Saharan Africa. Desalination 248 (2009): 256–261.

Drechsel P., Cofie, O.O., Keraita, B., Amoah, P., Evans, A. and Amerasinghe, P. 2011. Recovery and reuse of resources: Enhancing urban resilience in low-income countries. Urban Agriculture Magazine 25 (September 2011): 66–69.

Dodane P., Mbeguere, M., Sow, O. and Strande, L. 2012. Capital and operating costs of full-scale fecal sludge management and wastewater treatment systems in Dakar, Senegal. Environmental Science and Technology 46 (7): 3705–3711.

Evans, A., Otoo, M., Drechsel, P. and Danso, G. 2013. Developing typologies for resource recovery businesses. Urban Agriculture Magazine 26: 24–30.

Keraita, B., Drechsel, P., Klutse, A. and Cofie, O. 2014. On-farm treatment options for wastewater, greywater and fecal sludge with special reference to West Africa. Colombo, Sri Lanka: International Water Management Institute (IWMI). CGIAR Research Program on Water, Land and Ecosystems (WLE). 32p. (Resource Recovery and Reuse Series 1).

Kjelle, N.M., Pensulo C., Nordqvist P. and Fogde, M. 2011. Global review of sanitation system trends and interactions with menstrual management practices. Stockholm Environment Institute.

Kvarnström, E., Verhagen, J., Nilsson, M., Srikantaiah, V., Ramachandran S. and Singh, K. 2012. The business of the honey-suckers in Bengaluru (India): The potentials and limitations of commercial faecal sludge recycling – An explorative study. (Occasional Paper 48) [online] The Hague: IRC International Water and Sanitation Centre. www.irc.nl/op48 (accessed November 8, 2017).

Otoo, M., Drechsel, P. and Hanjra, M.A. 2015. Business models and economic approaches for nutrient recovery from wastewater and fecal sludge. In: Drechsel, P., Qadir, M., Wichelns, D. (eds) Wastewater: Economic asset in an urbanizing world. Springer, Chapter 13, pp. 247–270.

Rao, K.C., Kvarnström, E., Di Mario, L. and Drechsel, P. 2016. Business models for fecal sludge management. Colombo, Sri Lanka: International Water Management Institute (IWMI). CGIAR Research Program on Water, Land and Ecosystems (WLE). 80p. (Resource Recovery and Reuse Series 06).

Seidu, R. 2010. Disentangling the risk factors and health risks associated with fecal sludge and wastewater reuse in Ghana. PhD thesis. Norwegian University of Life Sciences.

Note

1. https://goo.gl/wfCksE (accessed November 8, 2017).

CASE
Fecal sludge for on-farm use (Bangalore Honey Suckers, India)

Jasper Buijs, Heiko Gebauer, Miriam Otoo and Alexandra Evans

Supporting case for Business Model 15	
Location:	Bangalore, India
Waste input type:	Fecal sludge
Value offer:	Provision of waste removal and collection services, and fecal sludge as organic fertilizer to farmers
Organization type:	Small and medium enterprise (SME), private entity
Status of organization:	Currently in operation
Scale of businesses:	Number of businesses (fecal sludge collection trucks) operating in Bangalore is estimated to be up to 300
Major partners:	Truck and pump system supply and repair sector; municipality

Executive summary

Due to shortcomings in sewage treatment systems and the availability of a large number of cement pit latrines without good maintenance and service planning, an informal sector of micro business ventures named "honey suckers" has emerged to fulfil the market need for on-site sanitation services. "Honey suckers" is the term given to the businesses that pump the waste out of pit-latrines, septic tanks and other types of on-site wastewater treatment plants. These businesses have been successful in exploiting this opportunity for the past few years and Bangalore now has an estimated 300 of such businesses. The primary market is where honey suckers collect fecal sludge from pit latrines for a fee. The sludge is then disposed of either at an approved site (rarely) or more typically it is dumped illegally on open lands or into drains. A secondary market has emerged in which the honey suckers deposit the sludge on farmlands at the farmer's demand, either in pits or directly on designated sites. There is usually no fee but the farmer may tip the driver. The sludge is used as a fertilizer and in some cases for the water content. The value for household is the clean removal of fecal sludge to ensure a working water closet and a clean property and environment. The value for farmers resides in obtaining nutrient-rich manure for free or for a very low fee. This model works well, because no other fast, reliable, high-quality pit cleaning service is available. The model works best when the cleaning service is easily combined with the dumping service, for which a smart network with farmers is required. The socio-economic and environmental benefits can be significant, with the creation of jobs, reduction of wild sludge dumping and associated health and environmental problems, improved sanitation and living comfort. However, risks have to be controlled and the informal nature of honey suckers and 'illegal' aspect of the business (i.e. the supply of collected fecal sludge to the farmers) prevents monitoring of the practice.

KEY PERFORMANCE INDICATORS (AS OF 2015)					
Land use:	Limited (car park). On farm for drying and reuse				
Capital investment:	Variable depending on fleet size; cost per truck is USD 24,000 for new trucks				
Labor:	Variable, depending on fleet size, 3 people per truck				
O&M cost:	USD 7,500 year, excluding legal dumping fees				
Output:	20,000 people reached per truck per year (single homes and apartment buildings)				
Potential social and/or enviornmental impact:	3 jobs per truck, possible reduction of open-dumping of fecal sludge[1], improved sanitation and resulting waste build-up reduction				
Financial viability indicators:	Payback period:	Ca. 9 months	IRR:	98%	Gross margin: 81%

Context and background

In India, 46% of the urban population uses a septic tank, a pit or vault latrine. This population that is not connected to the sewerage network relies on different forms of self/hired services to cover their basic needs. The common services combine on-site containment such as latrines or septic tanks, with removal and off-site disposal. In the best cases, the fecal sludge is emptied at a designated site where sludge dewatering and treatment takes place. However, more often the collected fecal sludge is disposed of haphazardly and illegally, like in wetlands, thereby creating health and environmental risks. Opportunities to change this practice lie in the reuse value of the sludge, i.e. in productively utilizing this waste by capturing and using resources such as nutrients, organic matter, energy and water. Fecal sludge thus presents – like farmyard manure – a value in particular to farmers, which has been recognized by on-site sanitation entrepreneurs. Additionally, the drying of fecal sludge on farm, and incorporation in the soil represents an 'outsourcing of fecal sludge treatment' to the farm which can help mitigate the challenge of open-dumping and the related health and environmental risk. However, reuse of fecal sludge or night soil, without taking precautionary measures can pose health risks to workers, farmers and consumers.

Market environment

Many people in urban areas in Bangalore do not have access to sewage systems, or even basic sanitary services. The current sewerage network in Bangalore only serves 37% of the city's population. Moreover 53% of the total generated sewage goes untreated in the environment. Sanitation deficiency is largely prevalent in the conurbation and green belt of Bangalore. In conurbations, only 47% of households have toilets, 19% share toilets and 35% defecate in the open. In the green belt areas, only 26% of the households have toilets while 4% share toilets and 70% defecate in the open. Bangalore, like India in general, has invested majorly in the development of septic tanks, pit latrines and eco-san toilets, however, a sound plan for maintenance and services has been lacking, creating multiple problems. Waste is often disposed of haphazardly, with all the associated health and environmental consequences. A relatively large number of houses and apartment complexes have pit latrines. The existence of these circumstances and the fact that no appropriate pit cleaning management exists has created a strong market opportunity for the evolution of the informal honey sucker businesses. Another market driver is fertilizer demand, which has tended to far exceed fertilizer supply. In areas where urban dwelling is in relatively close proximity to farmland, an opportunity arose for honey sucker businesses to dispose of fecal sludge on farmlands, especially where farmers are asking for it in view of declining soil fertility. A honey sucker business of average size serves about 20,000 people per year. Bangalore has 1.9 million households, of which 63.4% have no access to the sewage systems, and of those, 46% do have a tank or pit. With an average household size of 4.5 in Bangalore, the total serviced available market (SAM) in number of people is 2.49 million. Thus, with an average fleet of three trucks per smaller honey sucker business, and 20,000 people reached per truck per year, the market penetration (or, share of market – SOM) is 2.4% per honey sucker business. There is thus a large portion of the market that is yet untapped. On the other hand, with urban spread the transport distances and costs to reach farms around the city is increasing. Thus, the business will be most interesting for truck operators in new (unsewered) settlement areas towards the city outskirts than in its centre.

Business model

The business climate for honey sucker operations in Bangalore is different in various city areas. In the Northern part of Bangalore, there has been an intensive, but healthy competition between the honey sucker business ventures. Here, honey suckers have access to farmers and farmland that can be used as composting sites. In the Eastern part of Bangalore, this access to farmers and farmland is missing, which makes transportation distances long and expensive.

Fundamentally, the honey sucker business operates in two markets. The primary market is payment for the collection of fecal sludge from pit latrines or other onsite storage/treatment facilities. The secondary market is the 'sale' of the sludge to farmers[2]. The value for customers in the primary market is clean removal of fecal sludge to ensure a working water closet and a clean property. The value for farmers is the provision of low cost nutrients. This model works well, because no other fast, reliable, high-quality pit cleaning service is available. The model works best when the cleaning service is easily combined with the dumping service, for which a relative proximity to farm land, and a smart network with farmers is required. The socio-economic and environmental benefits can be significant, with job creation, reduction in pathogenic pressure from waste build-up and associated health and environmental problems, thus improved urban sanitation in general. However, the illegal character of the business creates problems with illegal networks, and uncontrolled dumping and land-use which may give rise to possible health risks for farmers and consumers of farm produce. See Figure 192 for the diagrammatic overview of the business model.

Value chain and position

Honey sucker businesses operate in a relatively simple value chain (Figure 193). The business has two different markets that rely on each other. The primary market, and the driving force of the business, is people who need their pit latrines emptied. The secondary market is formed by farmers who wish to make use of the sludge. The business relies on the availability of trucks adapted to the job and specialized equipment, which is available in the country. However, also secondary value chains have been observed where larger farmers dry sludge for resale to fellow farmers. The farm market might be seasonal, depending on the type of crops grown.

Institutional environment

In Bangalore, the Environmental Protection Rules and Acts of 1986 requires honey suckers to dispose of the sludge in designated areas, these being Bangalore Water Supply and Sewerage Board (BWSSB) sewage treatment plants. The reality is that few exist, which means long journeys for the truck operators, high fuel costs and a disposal fee of Rs. 50/kilolitre (0.82 USD/kilolitre). Instead truck operators dispose of the waste into open drains or onto wasteland. In some cases the truck operators have made arrangements with farmers who receive the waste and either use it directly on their fields, thereby making use of the water content, or store it and compost it over a period of time. The

CASE: FECAL SLUDGE FOR ON-FARM USE

FIGURE 192. HONEY SUCKER BUSINESS MODEL CANVAS

KEY PARTNERS	KEY ACTIVITIES	VALUE PROPOSITIONS	CUSTOMER RELATIONSHIPS	CUSTOMER SEGMENTS
• Truck and pump systems suppliers and repair services • Municipality	• Collecting sludge • Transporting and disposing of sludge • Marketing (customer generation)	A fast and affordable fecal sludge removal service for citizens, improving the household and city environment Free/low-cost 'human manure' for farmers, who will save on costs for fertilizer.	Personal help at contractual agreements (apartment blocks) Personal help (ad-hoc)	Households, apartment blocks and institutions with septic tanks or pits Farmers
	KEY RESOURCES • Truck(s) (capital) • Labor, 3 people per truck • Driver–farmers network • Household and apartment base • Loyalty of truck drivers, fortified by right to claim tipping fees		**CHANNELS** Telephone marketing Trucks, leaflets and banners advertising Informal, word of mouth	

COST STRUCTURE	REVENUE STREAMS
• Truck maintenance • Truck fuel • Labor wages • Legal dumping/tipping fees (not always done)	Pit latrine emptying fees Informal fees for manure plus possible savings on transport costs and tipping (desludging) fees

SOCIAL & ENVIRONMENTAL COSTS	SOCIAL & ENVIRONMENTAL BENEFITS
Human health risk arising from use of fecal sludge on farms without proper regulation and training in safety options Operating ex-legally strengthens illegal networks	Improved sanitary situation citizens Reduced city-internal fecal waste pressure and related health and environmental problems Increased land fertility where fertilizer prices are prohibitively expensive for poor farmers

business of honey suckers supplying collected fecal sludge to farmers suggest that it is a desirable commodity, which acts as a means of effectively and cheaply dealing with the sludge. However this activity is not supported by legislation (although some government officials state that fecal sludge is implied in the Fertilizer Control Order which permits the use of animal dung). There are no effective policies and regulations in place for either pit emptying or reuse on agricultural land. Standards would however be important to reduce the risks to workers, farmers and consumers of farm produce which may be contaminated with pathogens. The urban governance structure in India is highly complex

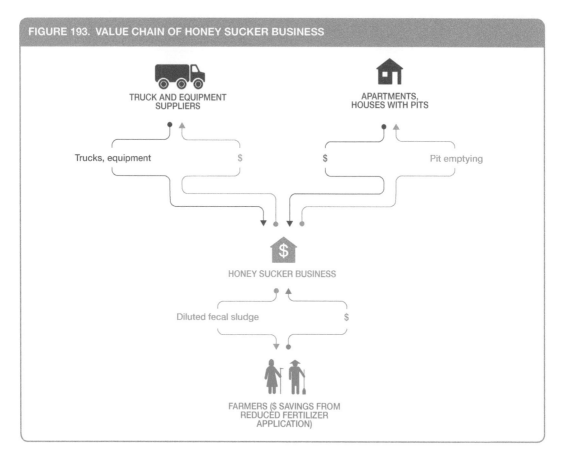

FIGURE 193. VALUE CHAIN OF HONEY SUCKER BUSINESS

with overlapping as well as weak mandates. The result of institutional complexities combined with a lack of funds is described as 'local governments operate in an implementation muddle', demanding improvisation, flexible interpretation and inviting the bending of rules and corruption.

Technology and processes

Honey suckers operate with dedicated trucks with a storage tank, which have a (vacuum) pumping system to suck up the sludge and an opening for desludging of their load. An increasing number of trucks are being manufactured in the country. Besides normal maintenance of the trucks and their equipment, there is little requirement for specialized maintenance services or training. Depending on the age of the sludge in the pit, and its hardness, truck operators might need access to water for sludge dilution and removal. On farm, the sludge might be stored and dried in larger pits (usually over about three months) before it is applied to the crops, e.g., to coconut trees. Wet fecal sludge can also be directly applied to the farm land. This is done either through trenches (for instance, in between banana trees), or on vacant farmland that will be farmed later in the season. Some farmers also sell dried sludge to other farmers (Figure 194).

Funding and financial outlook

An enterprise typically starts with an entrepreneur initiating a honey sucker business until it reaches about three to four trucks. The initial investment requirements for starting a honey sucker business venture are relatively low. It needs only a telephone number, a dedicated and registered truck, and a driver and two assistants per truck. Capital injection is required for establishing a truck fleet. Costs

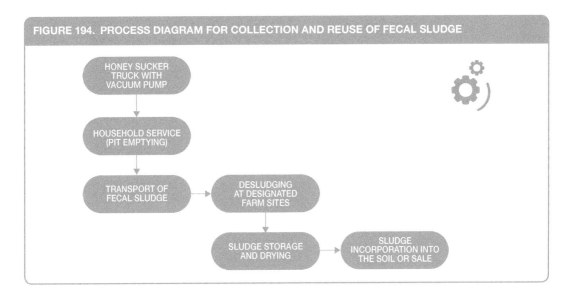

FIGURE 194. PROCESS DIAGRAM FOR COLLECTION AND REUSE OF FECAL SLUDGE

for one (new) truck are about USD 24,000. Major variable costs are related to truck maintenance and operation, labor wages and fuel for the trucks. These costs accrue to about USD 7,500 per year. Legal dumping fees are an additional cost, but turn into revenue if farms are the target. Because most of the businesses are not registered, considerable costs are incurred avoiding fines, and an opportunity loss is incurred due to business scaling limitations. In the current setting, the only, but profitable, revenue stream is from pit emptying fees, which amounts to a maximum USD 27 per pit emptied. With four services per day, 30 days per month, the revenues accrue to about USD 39,000 per truck per year and profit is estimated at about USD 31,500 per year. Thus, the payback period is nine months, with an IRR of 98% and a gross margin of 81%. Revenues are highest in the monsoon season, when servicing is required more often due to rainfall and overflowing pits. Drivers take tips from farmers for delivering sludge to their farms. However, the larger benefit can be savings on petrol (and desludging fees) if the farm is closer to the pit than the official dumping site. In more conducive legal-institutional settings, revenues could increase based on contractual customer relationships. Also specialized services such as ensured environmentally friendly dumping or guaranteed-time collection could be offered. Moreover, an official and larger customer base would allow businesses to perform more sophisticated services.

Socio-economic, health and environmental impact

Sewers are expensive and water to flush them increasingly rare. This gives on-site sanitation system an important place on the urban sanitation agenda. Due to the booming business of emptying pit latrines and holding tanks with honey sucker trucks, less fecal sludge finds its way into city drains and waterways, and household facilities function better. Disposal to farmlands outside the city offers the advantage of controlled drying and soil application, and improved crop production, but it needs oversight and risk reduction measures. The risks to farmers and potentially consumers are manageable without particular costs as long as the sludge can be well dried, crop restrictions are in place, and farmers wear protective gear (Keraita et al., 2014; WHO 2006). In this case, several social and environmental benefits could be attributed to honey sucker businesses as a valuable component of the sanitation service chain.

Scalability and replicability considerations

Honey sucker businesses thrive in places where sewage service is minimal and where people require affordable, fast and reliable sanitation services (Rao et al., 2016). The business requires a high density of easily accessible pits. There must be dedicated trucks available, with suction pumps. If the waste is to be provided to farmers, they must be within an economically viable radius (i.e. closer than official dumping sites or alternative illegal dumping sites currently are). While sludge supply is year-round, agricultural demand depends on cropping systems and might be seasonal. Another major restriction to honey sucker business growth in Bangalore is the lack of a supportive legal framework, which also links to the availability of farmers interested in the sludge. Currently businesses operate on a small scale, avoiding official marketing systems such as yellow pages and websites, and avoiding penalties. A legal standing would reduce the cost of acquiring new customers and improve access to finance. In such a situation, honey sucker businesses could follow multiple avenues to expand their operations: use of their specialized knowledge in advisory roles; offering improved services, e.g. time-guarantee arrival and emptying, eco-friendly processes (customers explicitly mention their willingness to pay for guaranteed environmentally safe handling and disposal); production of safe compost and information services to farmers.

Summary assessment – SWOT analysis

Figure 195 presents an overview of the SWOT analysis for the honey sucker business model. Due to shortcomings in the sewage systems, and the availability of a large number of cement pit latrines

FIGURE 195. SWOT ANALYSIS FOR INFORMAL FECAL SLUDGE REUSE IN BANGALORE

	HELPFUL TO ACHIEVING THE OBJECTIVES	HARMFUL TO ACHIEVING THE OBJECTIVES

INTERNAL ORIGIN ATTRIBUTES OF THE ENTERPRISE	**STRENGTHS**
	• Low capital investment
	• Only affordable, reliable emptying service available for people without sewage systems or with sub-standard ones
	• Positive influence on sanitary situation in urban areas
	• Virtually unlimited supply
		WEAKNESSES
		• No legal standing which creates other weaknesses:
		  • Limited access to affordable finance
		  • Must make additional efforts to avoid penalties
		  • Marketing difficulties including use of economies of scale
		  • No support in view of health protection
		• Requires climates with sufficiently long dry periods for sludge drying
		• Agricultural demand might be seasonally limited for some crops

EXTERNAL ORIGIN ATTRIBUTES OF THE ENVIRONMENT	**OPPORTUNITIES** • Legalization leading to opportunities to increase scale • Higher health standards and sustainability offer avenues for qualitative service offers • Decreasing land fertility and rising fertilizer prices drive growth of business • Increasing urban populations drive growth of business	**THREATS** • New sewage lines developed in underserved areas destroy honey sucker market • Legalization incurs additional costs for waste disposal, administration

without good maintenance and service planning, an informal sector of micro-business ventures named 'honey suckers' has emerged to fulfil the market need for on-site sanitation services. This model works well, because no other fast, reliable, high-quality pit cleaning service is available in the city. With very limited capital investment requirements and a strong revenue stream from pit-emptying services, this model offers entrepreneurs an opportunity for recouping their investment in a very short time period and with a relatively high gross margin. Although profitable, the honey sucker business is a highly risky investment option as their activities occur in a legally restrictive environment with significant uncertainty. This has implications for business sustainability and any scaling-up opportunities. Legalization of these initiatives may positively influence the honey sucker sector although there is some concern, especially among NGOs, honey suckers and farmers, that legalization and regulation may reduce its viability.

Contributors
Vishwanath Srikantaiah, Biome Environmental Solutions, India
Michael Kropac and Leonellha Barreto-Dillon, CEWAS, Switzerland
Sharada Prasad C.S., UC Berkeley, USA

References and further readings

Keraita, B., Drechsel, P., Klutse, A. and Cofie, O. 2014. On-farm treatment options for wastewater, greywater and fecal sludge with special reference to West Africa. Colombo, Sri Lanka: International Water Management Institute (IWMI). CGIAR Research Program on Water, Land and Ecosystems (WLE). 32p. (Resource Recovery and Reuse Series 1).

Kvarnström, E., Verhagen, J., Nilsson, M., Srikantaiah, V., Ramachandran, S. and Singh, K. 2012. The business of the honey-suckers in Bengaluru (India): The potentials and limitations of commercial faecal sludge recycling – An explorative study. (Occasional Paper 48) [online] The Hague: IRC International Water and Sanitation Centre. www.irc.nl/op48.

Rao, K.C., Kvarnström, E., Di Mario, L. and Drechsel, P. 2016. Business models for fecal sludge management. Colombo, Sri Lanka: International Water Management Institute (IWMI). CGIAR Research Program on Water, Land and Ecosystems (WLE). 80p. (Resource Recovery and Reuse Series 06).

WHO. 2006. Guidelines for the safe use of wastewater, greywater and excreta in agriculture and aquaculture. Vol. IV. Excreta and Greywater use in Agriculture. Geneva, Switzerland: World Health Organization (WHO).

For a photographic journey please see: http://arghyam.org/wp-content/uploads/2013/07/Honeysuckers-S-Vishwanath.pdf and www.flickr.com/photos/sharadaprasad (both accessed November 8, 2017).

See also: www.downtoearth.org.in/coverage/shit-its-profitable-47389 (accessed November 8, 2017).

Case descriptions are based on primary and secondary data provided by case operators, insiders or other stakeholders, and reflect our best knowledge at the time of the assessments 2015/16. As business operations are dynamic data can be subject to change.

Notes
1 While sludge disposal on farmland can reduce wild dumping of fecal sludge, the actual contribution has not been quantified as many farms might be too far away (transport costs) or their demand seasonally limited.
2 These fees are important as they reverse the normal process where drivers pays a tipping fee at a formal treatment pond. Thus, even if the token does not necessarily enter the business' revenue stream, there are savings, and it is a means for creating a trusted relationship with the driver. However, while earlier, farmers were approaching vehicle owners to have the sludge dumped into their fields, there is today much competition among trucks, and drivers are increasingly seeking farmers willing to accept sludge. The situation is different e.g. in Dharwad where larger farmers organize interim sludge storage and after drying auction the material.

BUSINESS MODEL 15
Outsourcing fecal sludge treatment to the farm

Jasper Buijs, Pay Drechsel and Miriam Otoo

Key characteristics

Model name	Outsourcing fecal sludge treatment to the farm
Waste stream	Fecal sludge (FS)
Value-added waste product	Organic fertilizer, waste removal and collection services
Geography	Urban population with no connection to sewerage network and use on-site containment such as latrines or septic tanks with off-site disposal. Dry climate over 3+ months for on-farm sludge drying before application.
Scale of production	Small to medium sized service operation; 20,000 people reached per truck per year (single homes and apartment blocks)
Supporting case in this book	Bangalore, India (with additional lessons learnt from Northern Ghana)
Objective of entity	Cost-recovery [ ]; For profit [X]; Social enterprise [ ]
Investment cost range	Variable but low; depending on fleet size, per truck ca. USD 24,000 (new)
Organization type	Private
Socio-economic impact	Jobs (3 people per truck), reduced disposal costs, agricultural production increase, sanitation improvement, living comfort increase
Gender equity	Primarily more benefits accrue to men who farm crop plantations and male drivers of vacuum trucks who gain from improved desludging and disposal measures

Business value chain

This business model can be used by private enterprises in smaller and larger towns and cities with a significant share of on-site sanitation facilities like septic tanks and cement pit latrines at households or office/apartment blocks in need of servicing (desludging). In the primary market, the business will collect fees from the household for collecting the fecal sludge (septage). In the secondary market the sludge is sold to peri-urban farms or plantations where the material is treated on-site, potentially composted and used as manure (Figure 196). The value for customers in the primary market is clean removal of fecal sludge to ensure a working water closet and a clean property. On the secondary market, the sludge supports crop growth on even unfertile soils, easily replacing commercial fertilizer, which can represent significant savings for the farmer while reducing the disposal/pollution costs for the city. The truck operator gains significantly economically if the farm is closer than the official dumping site and due to a reversed cash flow: instead of paying a tipping fee, the farmers pay the drivers. This model works best where farmers have no objection to the use of fecal sludge, know how

BUSINESS MODEL 15: FECAL SLUDGE TREATMENT ON FARM

to treat it safely and official dumping sites are far out of town. As farm demand might be seasonal, sludge that cannot be sold to farmers must be legally dumped.

An alternative scenario in the secondary market is that a farmer has multiple partnerships with different truck operators to deliver sludge to the farm. The farmer treats the sludge through sun drying (e.g. over 6 months like in Dwarward, Karnataka) and sells/auctions the treated dried sludge as fertilizer to other farmers. Compared to conventional septage collection from households and disposal in treatment ponds, the model has increased safety issues due to sludge disposal on farm and its possible link to the food chain. On-farm treatment, hygiene and crop restrictions must be strictly managed in this model, unless the fecal sludge is professionally dried and sanitized in a dedicated facility before being sold to farmers.

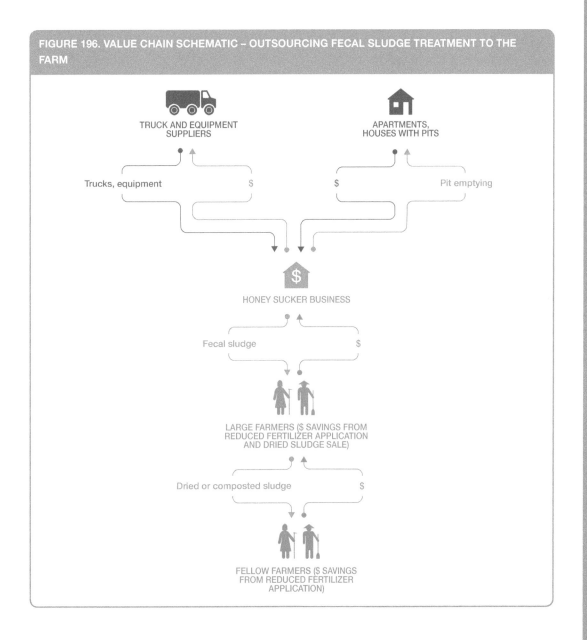

FIGURE 196. VALUE CHAIN SCHEMATIC – OUTSOURCING FECAL SLUDGE TREATMENT TO THE FARM

This type of business operates in a relatively simple value chain, and has two different markets that rely on each other. The primary market, and the driving force of the business, is the one where people are in need of on-site sanitation service, to clean out pit latrines where houses, apartment blocks, etc. in urban areas have no connection to sewage systems. The secondary market is formed by farmers who are interested in buying the fecal sludge for use on their land, thus saving on fertilizer costs, or for drying/composting and resale to fellow farmers. The business relies on availability of adapted trucks and specialized equipment.

Business model

The primary concept of the business model is to provide on-site sanitary cleaning services to households in the city by collecting fecal sludge from households' pit latrines, and provide nutrient-rich sludge to peri-urban farmers as a form of cheap 'manure' (Figure 197). A private enterprise operates throughout (parts of) the city, providing pit latrine emptying services to households and apartment blocks that have no connection to sewer systems or any other effective on-site sanitation treatment service. The service is based on the operation of fecal sludge emptying trucks that have specialized equipment on board to flush, suck up and store fecal sludge. The overall investment required for this type of business is relatively modest, with major investments required only for buying trucks (ca. USD 24,000 for each new truck, not counting for variation per country). The business makes a contribution to improvements in the environment through reduction of fecal sludge-based pollution in the city and related possible contamination of water bodies. It provides an important sanitation service where sewer systems are not available, and offers an opportunity for farming communities to improve soil quality with minimal investment. The business, however, may be prone to seasonality unless perennial crops are grown, and suffer from ex-legal status. The best business conditions arise where the use of fecal sludge on farms is legal, like the use of manure, but also, where the safety of such business systems is thoroughly investigated and where regulation compliance is monitored and incentivized.

Alternate scenarios

In an alternative, legal (but hypothetical) model the enterprise will be operating with a larger fleet of trucks. This model builds on the possibilities that arise when raw fecal sludge reuse on farms is permitted and regulated. The enterprise is a public-private-partnership in which the private partner, having the materials and equipment as well as operative expertise, gains operational freedom leveraged through its public partner (e.g. a municipal sanitation body or a government-owned operation). Operations will be bound to strict selection of complying farms (monitored), but the enterprise also gains the advantage of economies of scale, enabling the transition to improved value offerings, such as 'eco-friendly fecal sludge removal' or 'guaranteed time of pick-up of fecal sludge'. The enterprise invests in and gains from extensive expertise on fecal sludge removal and pit latrine construction and cleaning knowledge. Competition from micro and small enterprises of the same sort is minimal because of value proposal superiority and operational freedom arrangements. Costs are incurred for monitoring of compliance, also at farming sites in the network. This model strongly reduces negative externalities such as health risk to consumers of farm products, and illegal networks.

Potential risks and mitigation

The business model presented here was designed and optimized based on the analysis of previous studies and a case study. In designing this optimized business model, risks decribed below were addressed. However, risks defined below would continue to remain and are hence acknowledged.

Market risks: Market risks in terms of accessing fecal sludge are minimal, unless there are plans to extend the coverage of the sewer system. Market risks in terms of accessing farm land can occur

BUSINESS MODEL 15: FECAL SLUDGE TREATMENT ON FARM

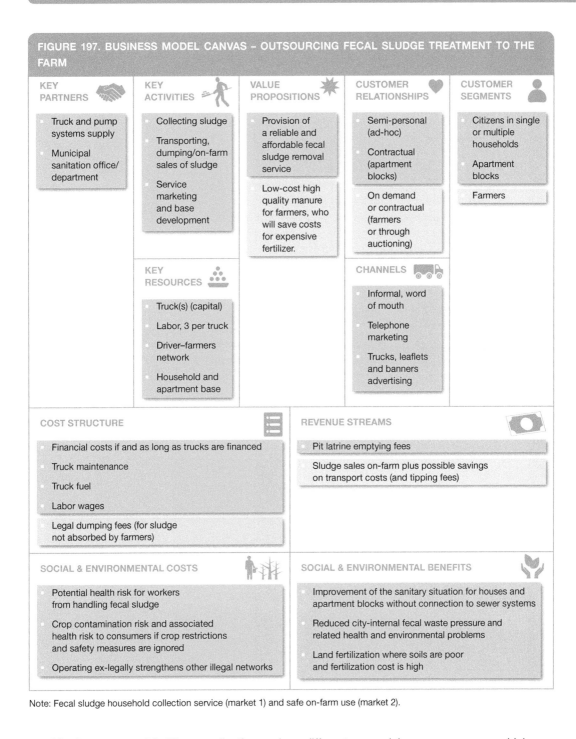

FIGURE 197. BUSINESS MODEL CANVAS – OUTSOURCING FECAL SLUDGE TREATMENT TO THE FARM

Note: Fecal sludge household collection service (market 1) and safe on-farm use (market 2).

outside the season of fertilizer application, unless different perennial crops are grown which can absorb fecal sludge throughout the year. Market risk in terms of consumer acceptance could become a factor where crops are not mixed in markets.

Competition risks: Competition risk for small-scale business is high, with low new entry barriers.

Technological performance risk: The business relies on availability of specialized trucks and equipment, as well as parts and repair expertise for the same. If such are imported, a real technological risk exists.

Political and regulatory risks: Regulatory risks exist for the business as long as they operate in an ex-legal manner (which is common practice rather than exception). The ex-legal character forms a barrier to enterprise growth and maturation. Legalization of the business and associated regulation and compliance forms a further complexity to this type of business.

Social equity related risks: This business model does per se not create any particular social inequity, but this depends on the type of crops used and the associated gender. As ideally perennial plantation crops are preferred, the model might in many cultures favour men who have better access to land and capital. Also, most truck drivers will be male. Otherwise, the model rather contributes to ensuring that households using non-sewered systems have access to waste collection services. This is because cesspit operators now have 'informal' designated disposal sites and are thus incentivized to provide services to a larger proportion of the population.

Safety, environmental and health risks: Health risks exist for personnel operating latrine emptying trucks. Serious health risks to consumers of farm products exist where the model is employed ex-legally, and sludge handling practices on farms do not follow basic safety recommendations. Risk mitigation options are known and should be sought, like protective clothing for workers and farmers, and monitored farming practices such as crop restrictions, sufficient time for sludge drying and safe sludge application (Table 46).

TABLE 46. POTENTIAL HEALTH AND ENVIRONMENTAL RISK AND SUGGESTED MITIGATION MEASURES FOR BUSINESS MODEL 15

RISK GROUP	EXPOSURE					REMARKS
	DIRECT CONTACT	AIR/DUST	INSECTS	WATER/SOIL	FOOD	
Worker	High	Low	Medium	Low	Low	At farm level, sufficient drying time for the sludge and crop restrictions are recommended, as well as personal protection (gear and hygiene) from sludge collection to farm work. See Stenström et al. (2011) and Keraita et al. (2014) for more details on risks and risk mitigation
Farmer/user	High	Low	Low	Low	Low	
Community		Low	Medium			
Consumer				Low	Low	
Mitigation Measures	🥾🧤	😷🌳	✖🧴	💧 Pb Hg Cd	🧽🍲	

Key: ☐ NOT APPLICABLE   ▨ LOW RISK   ▨ MEDIUM RISK   ▨ HIGH RISK

Business performance

The business model scores high on scalability, environmental impact and profitability (Figure 198). This business model may thrive in places where sewage services are minimal and where people require affordable, speedy, on-the-spot sanitation services. A strong driver for the business is the large availability of pit latrines that are accessible by truck, and the availability of local vacuum truck

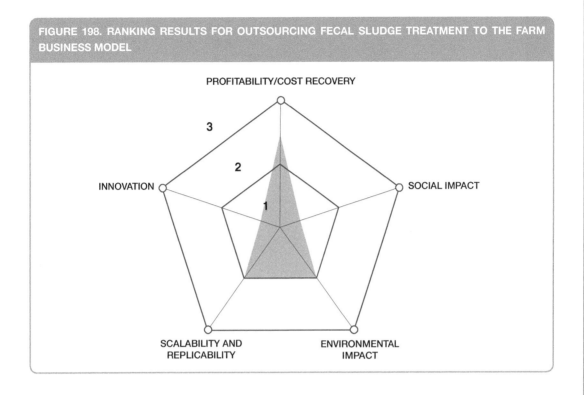

FIGURE 198. RANKING RESULTS FOR OUTSOURCING FECAL SLUDGE TREATMENT TO THE FARM BUSINESS MODEL

manufacturers. There is need for farming activities in proximity with ample (ideally year-round) demand for sludge, like via perennial (tree) crops. Although driving the ease of entry, an ex-legal climate for the business operation also forms the major restriction of business growth, because official marketing systems such as yellow pages and websites are avoided to steer away from penalties. The development of a conducive legal-institutional framework would benefit the industry greatly.

Under the right circumstances, i.e. a legally conducive framework, companies will be able to grow and make use of economy of scale principles. The cost of acquiring new customers then are lower, as well as the cost of accessing and buying finance. In such a situation, this type of business could follow multiple avenues to vertically scale their operations: 1) through the exploitation of their growing specialized knowledge (e.g. of construction details and cleaning ease, efficacy) towards 'smart' sanitary solutions advice and consulting; 2) by bringing customers new quality offers services (e.g. time-guarantee arrival and emptying, guarantee towards eco-friendliness – customers explicitly mention their willingness to pay extra if they can be sure the entrepreneur's handling of the waste is guaranteed to be environmentally safe); 3) by development toward production of safe compost that will allow sales of compost to farmers and companies, and information services to farmers for safe handling and use of sludge for composting and crop growing; and finally 4) by offering maintenance service for 'smart' latrines that are built to fit the housing and offer higher safety, e.g. monsoon times or are easier to clean and empty. For scaling towards eco-friendliness, more emphasis would have to be put on the secondary market, the services to farmers. Stronger relationships, built on solid trust, would have to be developed. In the long run, these businesses would need to spend more effort in design and cleanliness of trucks, the appearance and training of personnel, and increasingly good handling of sludge, to sell their services to increasingly developed and richer communities.

References and further readings

Keraita, B., Drechsel, P., Klutse, A. and Cofie, O. 2014. On-farm treatment options for wastewater, greywater and fecal sludge with special reference to West Africa. Colombo, Sri Lanka: International Water Management Institute (IWMI). CGIAR Research Program on Water, Land and Ecosystems (WLE). 32p. (Resource Recovery and Reuse Series 1).

Sharada Prasad CS, University of California at Berkeley, personal communication, September 2017.

Stenström, T.A., Seidu, R., Ekane, N. and Zurbrügg, C. 2011. Microbial exposure and health assessments in sanitation technologies and systems. Stockholm, Sweden: Stockholm Environment Institute (SEI).

13. BUSINESS MODEL ON PHOSPHORUS RECOVERY FROM EXCRETA AND WASTEWATER

Introduction

Among the essential plant food nutrients, phosphorus (P) is of particular interest as it is a non-renewable (finite) resource and means of its production other than mining are unavailable. With about 90% of known phosphate rock reserves found in only a few countries, the slowly declining reserves have stimulated a lively discussion ("Peak phosphorus") on sustainable P management and P recovery before it ends in waterways (Cordell et al., 2011; Edixhoven et al., 2014; Sartorius et al., 2012).

According to Latimer et al. (2016), phosphorus (P) recovery in the form of struvite is for now the most established technology for facilitating extractive nutrient recovery at scale during wastewater treatment. Nitrogen-only recovery is also feasible but has not been implemented extensively. Taking this into account, Latimer et al. (2016) estimated that the existing domestic wastewater treatment industry can optimistically bring between 100,000 and 210,000 metric tonnes of P_2O_5/yr (as struvite) and up to 220,000 metric tonnes N/yr to the fertilizer market. Although this corresponds only to 2–5% of the global P_2O_5 and N fertilizer demand, the sector is expected to grow. Moreover, in financially more rewarding niche markets, like fertilizer for ornamental plants, already between 30% and over 100% could be covered.

A particular interesting source for P recovery is human excreta[1]. Each year, the average human excretes up to 500 litres of urine and 50–180 kg (wet weight) of feces depending on water and food intake. Comparing feces and urine, most of the nutrients, i.e. 88% of the nitrogen, 67% of the phosphorus, and 71% of the potassium are found in the urine (Drangert, 1998). For low-income countries, there are three broader options for accessing and recovering P from human excreta, which are in order of increasing scale:

a) Collecting separated urine and feces at source (toilet), for urine use as liquid or crystal mineral fertilizer;
b) Collecting mixed excreta (septage) from unsewered systems, for use as organic fertilizer (fecal sludge composting);
c) Extracting P crystals during or after sewage treatment, for use as inorganic P fertilizer.

a) Collecting excreta before they are mixed with other potentially harmful waste streams appears most straight forward. Given the different nutrient amounts in feces and urine, and also the differences in pathogen loads, an ideal system collects both fractions separated, like in urine diverting dry toilets (UDDTs). The separated products can be safely treated and reused in agriculture ideally directly at household level (gardens). However, where households have no space, means or interest in reusing the produced excreta, collection services can be set up, where – depending on available alternatives – households either pay a fee for being served or receive payment for the provided waste resource. Different models are possible:
- **Decentralized excreta collection from households with UDDTs**. This has been tried at scale, e.g. in Ouagadougou (see **case example** following) with resale of the recovered and treated resources to farmers. There are very few similar examples yet to promote a particular business model. From a financial perspective, success is so far mixed, especially when the provision of the UDDTs is included (WSP, 2009). Additional challenges, like in the case of Ouagadougou, are the high management overheads to organize excreta collection and distribution as well as the related (urine) transport costs.
- A related business model is to focus on the **collection of urine from large one-point supply sources** such as sport arenas, youth hostels, prisons, industrial fares, music-, business- or entertainment-parks, universities and colleges, research institutes, etc. which are (or can be

INTRODUCTION

temporarily) equipped with normal urinals or UDDTs. This model avoids the costs of dealing with multiple clients as well as expenditures related to transport and logistics. The Dutch GMB[2] Bioenergy company in the Netherlands runs such a business using the SaNiPhos® process for urine treatment. The plant has been operating since 2010 and sourcing urine from music festivals, treating about 1300m^3 of urine per year. Each cubic meter of urine yields 3–4kg struvite (solid fertilizer) and about 60kg ammonium sulphate (liquid fertilizer). In another Dutch example, Amsterdam's water company (Waternet) and water authority Amstel, Gooi and Vecht recover phosphate in a special phosphate factory since 2013. They targeted Amsterdam's five-day maritime festival in 2015 to harvest about 100m^3 of urine. The expected 140 kg of struvite will be used in three innovative urban greening projects by the Amsterdam Rainproof platform[3]. A significant disadvantage of urine collection is its large water content and related volume and weight. The most common method for reducing the urine volume is through P precipitation (Pronk and Koné, 2010) as used in the examples above which can be catalysed through the addition of magnesium and results in "struvite" which is a soft P-crystal ($NH_4MgPO_4 \cdot 6H_2O$). The process has been piloted in many countries, like in Nepal, and can be financially viable unless magnesium access becomes too expensive (Tilley et al., 2009; Etter et al., 2011). An alternative option could be membrane filtration. Urine collected during a music festival in Ghent, Belgium, has been heated in larger (e.g. solar powered) tanks before passing it through a membrane which separates the nutrients and recovers the water in the urine[4].

b) Where urine and feces are not separated, and collected in latrines or septic tanks, resource recovery can transform the generated and collected septage during treatment into a safe organo-mineral compound fertilizer for example through drying and composting or co-composting (Nikiema et al., 2014). The compound nature of the material with different macro- and micro-nutrients has its own value proposition and business models (Rao et al., 2016). In many developing countries where treatment plants are too expensive, the agricultural use of nutrient rich (composted) sludge from septic tanks can be the most cost-effective option. This does not apply to sewage sludge, which with increasing industrialization has a growing risk of chemical contamination limiting its direct reuse potential. For sewage sludge, other P extraction options exist (see next point).

c) Where households are connected to sewers, and the excreta are flushed away, the process of extracting at this stage nutrients is increasingly complex and costly. However, to protect water bodies from eutrophication and treatment plants from unwanted phosphorus crystallization (valve and pipe damage), a large array of technical options is available to not only remove but recover different percentages of reusable P from wastewater and sludge during or after the treatment process (Egle et al., 2014; Latimer et al., 2016). These technologies have different requirements on the treatment process and energy and not all might be suitable for developing countries. However, especially in larger plants, they offer an important value proposition for saving maintenance costs, next to the generation of high quality Ca or Mg based P crystals with potential for use as fertilizer.

In this section we will describe two examples from the spectrum of opportunities listed above, one as a **case study** (Ouagadougou) and the other as **a model** (**P extraction from sewage treatment**). The Ouagadougou case was selected as a promising but also highly subsidised attempt for going at scale without qualifying yet as a model recommended for replication. The other case is based on P extraction from sewage treatment using the approach of Ostara (Canada) as an example. Given the success of the Ostara model, the example was chosen as a ***business case*-cum-*business model*** based on data from Ostara's operations in Canada, USA, and Europe and application potential also in middle-income countries. It is however important to add that there exist a wide array of other companies, processes and technologies for P recovery with different advantages for different situations and recovery targets (Sartorius et al. 2012; Egle et al., 2014).

References and further readings

Cordell, D., Rosemarin, A., Schröder, J.J. and Smit A.L. 2011. Towards global phosphorus security: A systems framework for phosphorus recovery and reuse options. Chemosphere 84: 747–758.

Drangert, J.-O. 1998. Fighting the urine blindness to provide more sanitation options. Water SA 24 (2): 157–164.

Edixhoven, J.D., Gupta, J. and Savenije, H.H.G. 2014. Recent revisions of phosphate rock reserves and resources: A critique. Earth Syst. Dynam., 5: 491–507.

Egle, L., Rechberger, H. and Zessner, M. 2014. Phosphorus recovery from wastewater: Recovery potential, heavy metal depollution and costs. IWA Specialist Conference on Global Challenges for Sustainable Wastewater Treatment and Resource Recovery, Kathmandu, Nepal; Paper-Nr. 0206D.

Etter, B., Tilley, E., Khadka, R. and Udert, K.M. 2011. Low-cost struvite production using source-separated urine in Nepal. Water Research 45(2): 852–862.

Latimer, R., Rohrbacher, J., Nguyen, V., Khunjar, W.O., Jeyanayagam, S., Alexander, R., Mehta, C. and Batstone, D. 2016. Towards a renewable future: Assessing resource recovery as a viable treatment alternative state of the science and market assessment. London, United Kingdom: IWA.

Nikiema, J., Cofie, O. and Impraim, R. 2014. Technological options for safe resource recovery from fecal sludge. CGIAR WLE (Resource Recovery and Reuse Series 2), 47p. Colombo, Sri Lanka: IWMI.

Pronk, W. and Koné, D. 2010. Options for urine treatment in developing countries. Desalination 251 (2010): 360–368.

Rao, K.C., Kvarnström, E., Di Mario, L. and Drechsel, P. 2016. Business models for fecal sludge management. Colombo, Sri Lanka: International Water Management Institute (IWMI). CGIAR Research Program on Water, Land and Ecosystems (WLE). 80p. (Resource Recovery and Reuse Series 06).

Sartorius, C., Horn, J. Von and Tettenborn, F. 2012. Phosphorus recovery from wastewater – Expert survey on present use and future potential. Water Environ. Res. 84: 313–322.

Tilley, E., Gantenbein, B., Khadka, R., Zurbrügg, C. and Udert, K.M. 2009. Social and economic feasibility of struvite recovery from urine at the community level in Nepal. Proceedings of the International Conference on Nutrient Recovery from Wastewater Streams. Vancouver, Canada, May 10–13, 2009.

WSP. 2009. Study for financial and economic analysis of ecological sanitation in sub-Saharan Africa. Nairobi: WSP.

Notes

1 Urine and feces together are called excreta.
2 http://www.gmb-international.eu (accessed November 8, 2017).
3 https://amsterdamsmartcity.com/projects/amsterdam-rainproof (accessed November 8, 2017).
4 http://firstwefeast.com/drink/2016/07/scientists-discover-way-to-turn-urine-into-beer (accessed November 8, 2017).

CASE
Urine and fecal matter collection for reuse (Ouagadougou, Burkina Faso)

Miriam Otoo and Linus Dagerskog

Supporting case for Business Model 16	
Location:	Ouagadougou, Burkina Faso
Waste input type:	Urine and feces
Value offer:	Provision of sanitation services and sanitized urine and feces as a safe organic fertilizer for agricultural production
Organization type:	Public-private partnership
Status of organization:	Project started in 2006, full system operational in 2008/2009
Scale of businesses:	Collection, treatment and reuse: 75,000 litres of urine and 11 tons of fecal sludge/year
Major partners:	European Union (EU), Water and Sanitation for Africa (WSA, formerly known as CREPA), Deutsche Gesellschaft für Internationale Zusammenarbeit (GIZ), National Water and Sanitation Authority (ONEA), Municipality of Ouagadougou

Executive Summary

The ECOSAN-EU initiated project was selected as a unique example of a large-scale household based resource recovery venture, while providing urban farmers with a reliable nutrient source for agricultural production. As with many other rapidly growing cities in the developing world, Ouagadougou is representative of a huge nutrient sink – where massive amounts of nutrients brought into the city with food are not recycled back to productive land. Coupled with poor waste management practices, especially the risk of groundwater contamination from the accumulation of human excreta in deep-pit latrines and septic tanks, the current waste management approach has dire effects in terms of soil fertility loss, increased disease burden and eutrophication. The project's activities which cuts across the entire sanitation value chain via the provision of sanitation products and waste collection services, whilst having a direct linkage to the agricultural sector via the conversion of human excreta into organic fertilizers for supply to local farmers, represents a sustainable market-driven solution especially in the absence of political pressure. The initial pilot phase of the project, from June 2006–December 2009, was set up with funding from the EU with contributions from the implementing organizations, GIZ, CREPA and ONEA. The EcoSan system was implemented in four of Ouagadougou's 30 urban sectors and the project was engaged in the provision of household urine diverting latrines, decentralized collection and treatment of urine and feces and the sale/delivery of the treated excreta as fertilizers for crop production. A key characteristic of the project has been its transfer of ownership to the municipality of Ouagadougou in 2010 and strong engagement of community-based organizations (CBOs) in different business activities along its value chain. The ECOSAN-EU business model is based on a CBO approach where in each urban sector, one group association or community-based organization (CBO) has a contract with the municipality to ensure the collection, treatment and delivery of sanitation products from households to farmers. A key success factor for this model has been the diversification of their portfolio as represented by the multiple products and services they provide. The variable income for the associations include monthly collection fees of USD 0.69 per UDDT (urine diversion dehydrating toilet), income from sales of EcoSan fertilizers (sanitized urine and feces sold at USD 10.37/m^3 and USD 5.34/50kg bag, respectively).

ECOSAN-EU has contributed to improved health and hygiene of households with installed UDDTs and offers a monthly collection service comparatively cheaper than having a one-off pit emptying service. Improved excreta management practices has resulted in a reduction of environmental pollution. Additionally, the activities of this project have created a significant number of jobs along the entire sanitation value chain and provided a low-cost and sustainable agricultural input alternative for farmers.

KEY PERFORMANCE INDICATORS (AS OF 2014)						
Land use:	Data not available					
Capital investment:	USD 20,145 per year					
Labor:	Data not available					
O&M cost:	USD 3,319–3,651 per sector per year					
Output:	223,760 litres of sanitized urine and 21 tons of solid fertilizer over a 3-year period					
Potential social and/or environmental impact:	Improvement in health and hygiene of households with installed UDDTs, creation of jobs, reduction in environmental pollution, low-cost fertilizer for farmers					
Financial viability indicators:	Payback period:	N.A.	Post-tax IRR:	N.A.	Gross margin:	N.A.

Context and background

Only 19% of the population in Ouagadougou, Burkina Faso, had access to improved sanitation (i.e. increased waste collection and treatment services) in 2006. With an annual population growth rate of around 5%, it has become increasingly difficult for municipalities to keep up with that with the provision of sanitation services. Large quantities of human excreta accumulating in deep-pit latrines and septic tanks not only represent a potential risk for groundwater contamination but are also wasted nutrient resources. An integrated ecological sanitation (EcoSan) system was implemented in 2006–2009 by the EU-funded ECOSAN-EU project led by WSA (Water and Sanitation for Africa), GIZ[1] (Deutsche Gesellschaft für Internationale Zusammenarbeit) and ONEA (National Water and Sanitation Authority). The key activities of this project were to support 1,000 households in obtaining appropriate and affordable, urine diverting dry toilets (UDDTs) with an associated collection service followed by treatment and reuse demonstrating novel excreta management systems that protect human health, contribute to food security and enhance the protection of natural resource and promote small and medium size enterprises. The project was implemented in four of Ouagadougou's urban sectors – "arrondissement" 17, 19, 27 and 30. Public UDDTs were initially installed at the central prison of Ouagadougou, the Bangrweogo Park, town hall and the zoo. Subsequently, households were willing to install UDDTs after the subsidies were increased, and within six months, 400 double vault UDDTs were built. By June 2009, 922 homes were using UDDTs and some 800 gardeners and small-scale farmers were trained on the application of treated urine and feces for their crops. The Ouagadougou municipality took over the coordinating role from January 2010 when the project was officially completed, after a transition phase of six months. The municipal waste department (or "Department for Cleanliness" - Direction de la Propreté) set up an EcoSan committee, which has a chairman, one rapporteur and one focal point. A municipal budget line was dedicated for continued support to the associations. The total investment for the three-year project (2006–2009) was USD 2,070,218. In 2010, the municipality of Ouagadougou allocated USD 14,735 of its budget for continued support to the service providers (local CBOs), and took over the coordinating role of the project. The waste management regulations of Burkina Faso are such that the municipalities organise the collection, treatment and disposal of waste, which can be carried out in partnership with private organisations and Decree 95 indicates the setting up of a fee for household waste collection. The ECOSAN-EU is based on a concept where in each urban sector, a community-based organization (CBO) has a contract with the project to ensure the collection, treatment and delivery of sanitation products from households to farmers.

Market environment

With increasing waste management costs but ever-dwindling budgets, municipalities are in dire need of sustainable alternatives such as integrated ecological sanitation solutions involving the reuse of waste in cities like Ouagadougou. Additionally, Burkina Faso is a landlocked country, affected by droughts and desertification, overgrazing, soil degradation and deforestation, with only 14.43 % of its land being arable. Around 90% of the population is engaged in subsistence agriculture and with unpredictable chemical fertilizer prices, exemplified by the price hike in 2008, reuse of treated human excreta can be a reliable nutrient recovery strategy for agriculture. This represents opportunities for business development in both the sanitation and agricultural value chains. It is important to note that although the demand for sanitation infrastructure (i.e. UDDTs) and services will demonstrate an increasing trend for the next decade, human fertilizer demand at least in the city may not reflect a similar trend. Factors related to transportation constraints especially for sanitized urine and the current area of urban agriculture within city limits may potentially limit the amount of excreta that can be absorbed in the agricultural sector, suggesting an excess supply. Based on 2012 data, the present farming activities in the city can potentially only absorb the excreta from approximately 50,000 people, whereas there are 1.5 million inhabitants in Ouagadougou. New technologies to add-value to urine and feces such as pelletized fecal sludge-based compost will allow businesses to access new markets beyond the city limits, as realized for example in Accra, Ghana.[2]

Macro-economic environment

The Government of Burkina Faso does not have an officially recognized chemical fertilizer subsidy program (IFDC, 2013). However, financial difficulties experienced by cotton companies in the country in 2005 and the food crisis of 2008 influenced the government to undertake actions to support the production of cotton and staple food crops by facilitating access to fertilizers. The goal of the fertilizer support operation in Burkina Faso was "to increase the current level of fertilizer use by reducing its cost and facilitating farmers' access to quality fertilizers". There is no prescribed fertilizer package for farmers under this program, but it covers two types of fertilizers: a combined nitrogen, phosphate and potassium (NPK) fertilizer and urea. The fertilizer support program was first introduced in 2008–2009 with exclusive funding from the national budget, and subsequently from 2010 through 2012 with support from the African Development Bank in addition to government funds. So far, the government does not have an exit strategy for the fertilizer support program. Subsidized fertilizers account for approximately 17 percent of all fertilizer products consumed in Burkina Faso. While the availability of chemical fertilizers has been enhanced, these measures will have an undesirable impact on new organic fertilizer businesses which have to compete with the subsidized market prices of chemical fertilizer. Similar incentives may be required to be put in place to enable new 'Resource Recovery and Reuse' businesses producing pelletized fecal sludge-based compost, for example, to penetrate the fertilizer market.

Business model

The ECOSAN-EU project's main goal was to facilitate access to sustainable, safe and affordable sanitation systems for the residents of Ouagadougou, support 1,000 households in obtaining appropriate and affordable closed-loop sanitation systems, provision of sanitation infrastructure (toilets) and waste collection services and contribute to food security via the conversion of human excreta into organic fertilizers for supply to local farmers (Figure 199). A notable aspect of this initiative has been the transfer of ownership to the municipality and the engagement of local community-based organizations. Although the initiative runs today as reduced level, the implementation of the initiative till this step was a success on its own. In that regard, the business model is to be viewed from the perspective of the CBO that operates, manages and owns the business entity. There are several factors that have driven

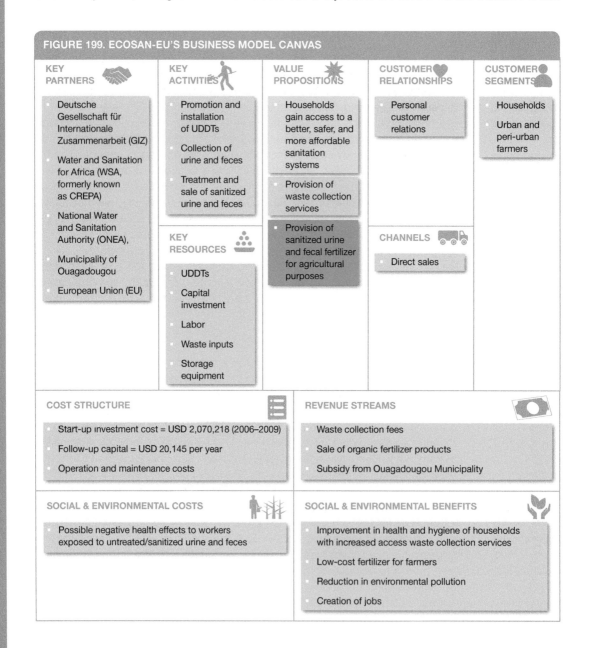

FIGURE 199. ECOSAN-EU'S BUSINESS MODEL CANVAS

the sustainability of this initiative: a) key partnerships for financial support at the start-up stage; and b) portfolio diversification/multiple revenue streams to mitigate fluctuations in market demand of certain products and/or services (waste collection services and sale of organic fertilizers). Financial support from the municipality in the form of price subsidies on UDDTs incentivized the rapid adoption by households. This has a direct implication for the production side of the organic fertilizer products as the use of UDDTs represents easy access and availability to high quality waste inputs. With a business model that cuts across the entire sanitation value chain and also links in with the agricultural sector, the benefits from this initiative are multi-fold. The value proposition of increased access to safe and affordable sanitation systems translates into improved health of society, especially for low-income urban households in slum areas which are typically characterized by limited to no access to sanitation infrastructure and services. This notion can be extended through the second value proposition of provision of waste collection services. It is important to note that the CBOs are not directly engaged in the sale of UDDTs but the project provided subsidies to households for the construction which was done by local masons who in turn were contracted by ONEA. Benefits to the agricultural sector from the availability of organic fertilizers are noteworthy especially given the agro-ecological conditions (i.e. droughts, poor soil fertility) in Burkina Faso. Additionally, access to affordable agricultural inputs is crucial as most urban and peri-urban farmers are budget-constrained.

Value chain and position

Figure 200 below provides an overview of the value chain for a community-based organization in each urban sector. The CBO provides waste collection services to households for which it has total market control as the municipality gives them sole responsibility for this activity and thus faces no competition for provision of this service or access to the waste as an input. The CBOs in all the four sectors however noted experiencing low levels of waste supply. This has been attributed to a significant decrease (41%) in the number of households using UDDTs from 2009 to 2014 and also the supply of excreta from each household being extremely low. Only 16% of urine and 25% of feces of the expected quantity from each household was collected. Broken and non-functioning UDDTs due to rains and inundations and misinformation about collection fees led to discontinued use by households. This suggests the need for CBOs to invest in and provide repair and maintenance services for the household toilets or at the least partner with an entity to provide such services as this component of their business has significant implications for their entire business value chain. Other possible reasons include other toilet alternatives, overestimation of expected volume of excreta and open-dumping by households if collection services were irregular. Despite the fact that approximately only 1.6% of households in the four sectors were connected to the project system, the demand for the fertilizer products is fairly low as not all the produced fertilizer (both sanitized urine and feces) had been sold. The CBOs currently face stiff competition from subsidized chemical fertilizer and other factors related to stigma of using excreta-based fertilizers, strong smell of urine, transportation challenges and additional labor costs due to bulkiness of urine and feces. The businesses subsequently have rebranded their products with labelling to dispel the negative perceptions of waste-based products. Sanitized urine is sold in green 20L cans labelled "birg-koom", which means liquid fertilizer; and sanitized dried feces are sold in bags labelled "birg-koenga" which means solid fertilizer. Field demonstrations have also been key to show the efficiency and use of the fertilizer products and this has significantly increased demand especially for the dried feces in the past year. The main clientele are farmers and nurseries, with a few large-scale buyers – plantation owners from outside Ouagadougou. From 2009–2012, 21 tons of dried feces (424 bags of 50 kg) were sold, which represents 48% of the total quantity collected. The CBOs continue to face challenges with the sale of urine – which amounted to 11,188 20L jerry cans, which represents 74% of the total quantity collected. Additional awareness programs are being planned.

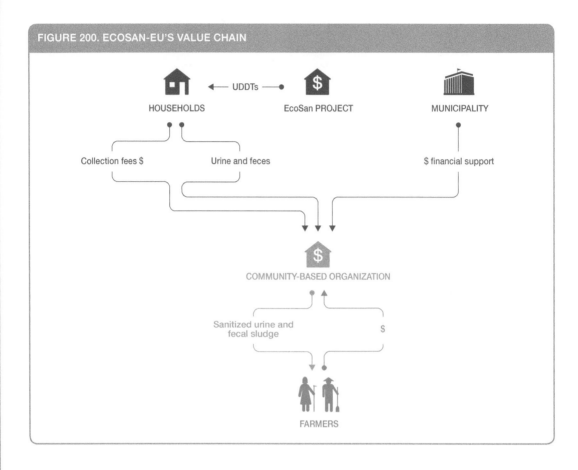

FIGURE 200. ECOSAN-EU'S VALUE CHAIN

Institutional environment

The management of waste in Burkina Faso in general is dealt with under several laws and regulations. As with the CBOs, in order to legally engage in any waste management activities, a clearance must be provided by the municipality. The sole assignment of the CBOs to excreta management in the different sectors by the municipality has enabled the CBOs to ward off any competition for the provision of waste collection services but also access to the waste input. The municipality additionally provides financial support to the CBOs by paying the salaries of all staff for the four associations. Approximately, CFA 7 million (USD 14,735 – using 2014 conversion rates) is set aside annually in the municipal budget for the system.

Technology and processes

The process of production of the sanitized urine and feces is very simple and involves a low-level technology (Figure 201). There were originally three types of UDDTs used for the collection and separation of feces and urine at the household level: double-vault toilets, single-vault toilets and box toilets. Households are advised to add ash after each defecation to enhance pathogen die-off and drying. In the double vault toilet, the vaults are used in alternation and the full vault is kept closed for at least 6 months to sanitize the excreta. The vaults are then emptied by the collection service workers and brought to an eco-station for further drying and storage of at least two months before final packaging and sale. The sanitization of urine occurs once transferred to the eco-stations via storage in closed 1m³ plastic tanks for at least one month. Feces from single vault and box UDDTs were directly collected in lined containers (using rice-bags). After the trial period, it was however

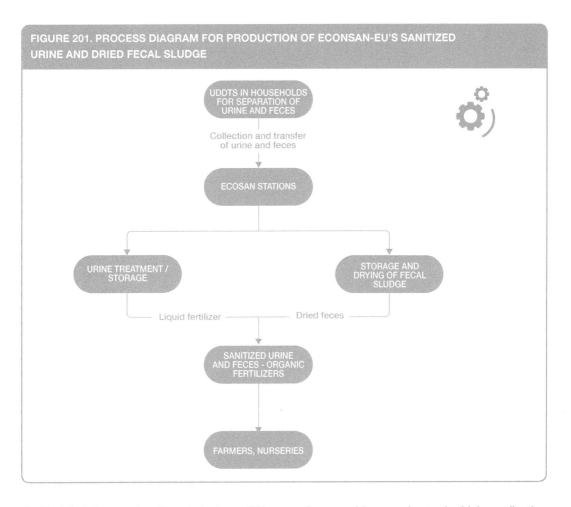

FIGURE 201. PROCESS DIAGRAM FOR PRODUCTION OF ECONSAN-EU'S SANITIZED URINE AND DRIED FECAL SLUDGE

decided that the construction of single vault/ box versions would cease due to the higher collection rates needed and challenges associated with providing adequate lining for the containers. During the period of 2006–2009, more than 300,000 litres of urine and 44,000 kg of feces in total were collected from the four sectors. This amounts to 27 20L jerry cans of urine and 80 kg of feces per household, which suggests that there are many households who are currently not using the UDDTs on a regular basis. The associations stated that collection services to households are provided on a weekly basis although cases of limited use were attributed to irregular provision of collection services and the malfunctioning of UDDTs. While this technology is simple and cost-effective for the CBO – in regards to easy access to waste inputs and income generation from waste collection, it is imperative that the CBOs pay particular attention to efficiently providing consistent collection and maintenance services.

Funding and financial outlook

Initial capital cost for the project was provided by the following institutions in the amounts of: EU = EUR1.11 million (USD 1,534,908), CREPA = EUR207,120 (USD 286,405.54) and GIZ = EUR180,000 (USD 248,904). The involvement of households in the construction process of the toilets via the provision of building materials and assistance for the construction workers significantly reduced the start-up costs. Since 2010, when the municipality took over the management role of this initiative, it invests USD 14,735 yearly in the four sector associations which cover the staff salaries for all associations.

The support the associations receive varies from USD 157.64–324.96 per month depending on size of each sector and quantity of UDDTs. Based on the information of the associations, this means a subsidy of USD 1.78 (CFA850) per household per month. There were two CBOs appointed per urban sector, and these form one association in each of the sectors to reduce management costs. The associations are trained and involved in project management and operation, which includes collection, transport, treatment, management, delivery. The expenditure of an association consists primarily of salaries, maintenance work at the eco-stations, transport and communication expenses and this amount varies from USD 277–304 per month. The monthly income for each association consists of a fixed sum of about USD 415 for associations in sectors 17 and 30, and USD 318 for associations in sectors 19 and 27. This fixed amount was taken over in 2010 by the Ouagadougou municipality after the EU project was completed. The variable income for the associations include monthly collection fees of USD 0.69 per UDDT (dependent on households that are able to pay), income from selling the EcoSan fertilizers (sanitized urine sold at USD 0.21 for 20-litre jerry can or USD 10.37/m^3, and sanitized feces at USD 5.34 for a 50kg bag). The total income received from all sectors from sales and collection fee, is about USD 451 (CFA214,400) per month and this goes toward maintenance of equipment. The income stream from current sales of sanitized urine and feces is fairly low compared to the revenue from waste collection fees at a ratio of about 70/30. The generated revenue only constitutes about 24–43% of the total revenue for the associations, with the rest been subsidies from the municipality. The associations could potentially become independent with increased demand and sales of organic fertilizers from increased product awareness, branding and product differentiation, to name a few.

Socio-economic, health and environmental impact

This initiative has had noteworthy impacts on the communities in Ouagadougou. With a business model that cuts across the entire sanitation value chain, this initiative has created jobs especially for low-income persons who would otherwise be unemployed. Additionally, smallholder farmers who are typically budget-constrained have access to comparably cheaper fertilizer alternatives. The introduction and incentives put in place to facilitate household adoption of UDDTs have significantly improved the health and hygiene of households with installed UDDTs. Communities have also noted a reduction in air pollution and flies from reduced open dumping of human excreta. In total, approximately 224,000 litres of urine were sold from 2009–2012 for all four sectors, which represent 74% of the collected urine, and 21 tons of sanitized feces sold, representing 48% of the collected feces. Another advantage from the adoption of UDDTs by households is that the monthly collection service is cheaper than having a one-off pit emptying service and the lower risk of inundation of the latter toilet types compare to the former. Households, however, tend to empty jerry cans filled with urine into street gutters and the environment if collection services are irregular. Additionally, environmental pollution could also potentially occur at the eco-stations from leakages of aging 1m^3 urine tanks or from flooding of fecal storage vaults during extreme rains, which happened in 2009.

Scalability and replicability considerations

The key drivers for the success of this initiative are:
- Strong partnerships for provision of start-up and working capital.
- Diversified portfolio – which mitigates risk associated with fluctuations in market demand for any one product or service.
- Assured supply of waste input at limited operational cost.

This initiative has a good potential for replication especially in low-income developing towns and cities with well developed urban and peri-urban market farming able to absorb the recovered resources. The strategy of close cooperation with communal authorities, community based organisations in peri-urban areas, and the local private sector was adopted throughout the project and this brought positive results with a high degree of engagement from all stakeholders involved. This focus has helped to increase the capacities of actors to engage in a programme of sustainable sanitation systems aiming at ensuring that activities will be integrated into ongoing work when the initial project ended – an important strategy for any plans for out-scaling. Monitoring activities throughout the project phase were an integral part of the project cycle. This allowed improving the design, mitigating construction errors, ensuring that the households maintained their new toilet facilities properly, and to encourage safe reuse practices. The study was carried out for 2.5 years after which the municipality took over. Results indicate that the number of toilets had decreased from 938 in 2009 to 551 in 2012. The drastic decrease is due to reasons such as abandonment of toilets that were broken and not functioning, destroyed latrines by rains and inundations and households not using or removing toilets as a result of misinformation about waste collection fees. This suggests the need for CBOs to invest in and provide repair and maintenance services for the infrastructure (toilets) or at the least partner with an entity to provide such services as this component of their business has significant implications for their entire business value chain. The present farming activities in the city can absorb the excreta from approximately 50,000 people, compared to 1.5 million inhabitants in Ouagadougou. Both land and water resources may limit urban agricultural expansion. Therefore, any up-scaling of reuse of sanitation products has to connect with the hinterland of the city, and in the case of Ouagadougou, applied in rain-fed farming. This requires the use of new technologies to add-value to urine and feces such as pelletized fecal sludge-based compost, which will allow businesses to increase demand by accessing new markets beyond the city limits. Product differentiation will: 1) increase the competitiveness of the products; and 2) eliminate the transportation challenges and additional labor costs associated with the bulkiness of urine and feces.

Summary assessment – SWOT analysis

Figure 202 presents an overview of the SWOT analysis for the EcoSan system in Ouagadougou. This initiative has been particularly successful in leveraging strategic partnerships to mitigate capital investment risk. The strategy of close cooperation with communal authorities, community-based organizations and the local private sector resulted in positive results with a high degree of engagement from all stakeholders involved, facilitating the transition phase from a project to a 'business'. The implementation of a multiple revenue stream strategy has been crucial in sustaining the viability of the initiative as it is noted that income generation from the sale of organic fertilizer products contributes only 30% of the overall revenue generated. One of the key weaknesses of this initiative is that it is highly subsidized, with municipal support covering 65–75% of the associations' income. The present system is not working in an optimal and efficient way, and it is clear that a subsidy that was close to CFA 10,000 per household per year would not be sustainable in the case of up scaling. In 2001 there were 154,000 households in Ouagadougou (SUSANA, 2012), which most likely is around 200,000 households today. Such a subsidy per household city wide would amount to approximately CFA 2 billion (equivalent to about Euro 3 million) per year for the municipality. There is an apparent gap in the business' value chain of activities – that is, a lack of provision of maintenance services for UDDTs and irregular waste collection services. This is negatively affecting the supply of waste inputs and directly affects profit levels. This represents an opportunity for the CBOs to invest in and provide repair and maintenance services for the UDDTs or at the least partner with an entity to provide such services as this component of their business has significant implications for their entire business value chain. The EcoSan system is also facing stiff competition from chemical fertilizers which are easily accessible and are now subsidized in Burkina Faso. Thus the sale of organic fertilizers will be difficult as long as chemical fertilizers are reasonably affordable. In the long run however, it is likely that chemical fertilizers will become more expensive as energy prices increase and resources become scarcer. On the other hand, these challenges present opportunities for the business to reinvent its product innovation and marketing strategy. Adoption of new technologies to add-value to feces such as pelletized fecal sludge-based compost will increase the business' access to new markets beyond the city limits – reducing transportation challenges and additional labor costs due to bulkiness of feces, while supporting higher market prices for its products. Additionally, extending its business value chain to include provision of repair and maintenance services would be a new revenue source but also increase the number of households to which waste collection services can be provided and the amount of waste actually collected. This represents additional income and ensures an incremental quantity in the waste input available. The new EcoSan system in Ouagadougou is by no means ideal, but it has taken some innovative steps to go to scale in urban waste and nutrient management. Public funding is needed for investments in and control of the system and to a certain extent for running costs, at least in the short term. It is always difficult to mobilize scarce public funds but if the gain in health and environmental protection can be evaluated in addition to agricultural benefits, it can prove to be an economically sound public investment. Additionally, several opportunities exist for this initiative to become financially self-sufficient.

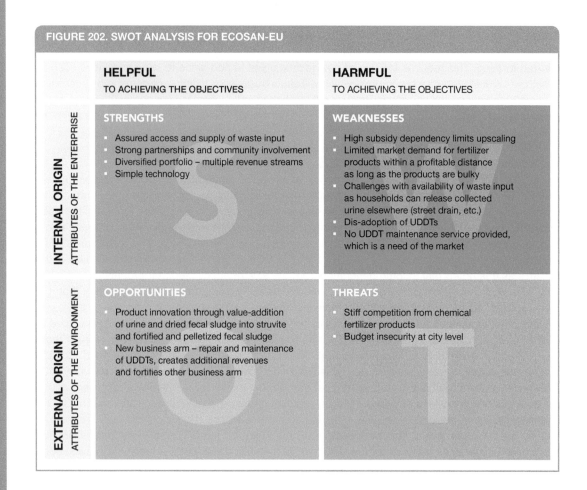

FIGURE 202. SWOT ANALYSIS FOR ECOSAN-EU

Update of the Ecosan system in Ouagadougou (Oct. 2017):
The international NGO Action Contre la Faim (ACF) coordinated a follow-up EU-funded project 2013–2016 in Ouagadougou to improve sanitation and hygiene in peri-urban sectors of the city. Part of the efforts included support to two of the existing CBOs involved in the EU EcoSan project. The

CBOs received help to develop business plans in addition to receiving improved equipment such as motorized tricycles for waste collection. Demonstration gardens were developed next to the eco-stations, enabling a supplementary source of income. The project subsidized 403 new urine diverting toilets (mainly constructed in 2016) and rehabilitated 37 old ones. During 2013–2016, the two supported CBOs sold 35m^3 of urine (43% of collected) and 17.5 tons of feces (86% of collected), which can be compared to the period 2009–2012 when 145m^3 urine was sold (82% of collected) and 12 tons of feces (60% of collected) in the same two sectors.

Challenges to sustain the operations remained, especially since the municipal subsidy for the CBOs was removed in 2013 during a turbulent period in the local administration while also households willingness to pay for collection decreased. Apart from variable demand of the fertilizer products, transport distances for input collection and product delivery is the main cost factor. To reduce costs, collection is today only 'on demand'. Technical innovations to transform urine, reducing volume and odor in a cost efficient way, will be necessary to sustain the business and enable further scaling in view of fertilizer demand and transport costs.

Contributors
Alexander Evans, University of Loughborough, United Kingdom
Josiane Nikiema, IWMI, Ghana

References and further readings
About.com. 2014. Geography: Burkina Faso [Online]. http://geography.about.com/library/cia/blcburkinafaso.htm.

Dagerskog, L. 2008. Financial and economic evaluation of sanitation systems in Ouagadougou in Burkina Faso, with focus on ecological sanitation. CREPA's case study report to Hydrophil as part of the WSP-commissioned "Study for Financial and Economic Analysis of Ecological Sanitation in Sub-Saharan Africa". 2009. CREPA.

Dagerskog, L., Coulibaly, C. and Ouandaogo, I. 2010. The emerging market of treated human excreta in Ouagadougou. Urban Agriculture Magazine 23 (April 2010).

IFDC. 2013. Practices and policy options for the improved design and implementation of fertilizer subsidy programs in Sub-Saharan Africa. NEPAD Agency Policy Study. https://ifdcorg.files.wordpress.com/2015/01/sp-41_rev.pdf. Accessed 15 August, 2017.

Sawadogo, H. 2008. Approche GIRE et expansion de l'agriculture urbaine à Ouagadougou. Master's Thesis, 2iE, Ouagadougou.

SUSANA. 2012. Compilation of case studies on sustainable sanitation projects from Africa: Urban urine diversion dehydration toilets and reuse, Ouagadougou, Burkina Faso.

Case descriptions are based on primary and secondary data provided by case operators, insiders or other stakeholders, and reflect our best knowledge at the time of the assessments in 2014 and 2017. As business operations are dynamic data can be subject to change.

Note
1. It is important to note that GIZ only funded the start-up of the initiative and does not have a continuous role in the business model. This is also applicable to the case of WSA (Water and Sanitation for Africa) and the EU (European Union).
2. www.iwmi.cgiar.org/tag/fortifier/ (accessed 18 January 2018).

BUSINESS MODEL 16
Phosphorus recovery from wastewater at scale

Pay Drechsel, George K. Danso and Munir A. Hanjra

Key characteristics

Model name	Phosphorus recovery from wastewater at scale
Locations	Tested so far in 14 commercial installations worldwide (status January 2017)
Waste stream	Wastewater (sewage)
Value-added waste product	Recovery of phosphorus for reuse as clean-green fertilizer with environmental benefits
Geography	Any urban centre, applicable to a wide range of sewage treatment plants
Scale of production	Medium to very large; minimum plant size of 19 MLD sewage
Supporting case in this book	None (the case of urine collection in Ouagadougou, Burkina Faso is unrelated)
Objective of entity	Cost-recovery [ ]; For profit [X]; Social enterprise [ ]
Investment cost range	USD 2–5 million with a capex pay-back time of 3 to 7 years
Organization type	Public, private
Socio-economic impact	Enhanced compliance with environmental regulations, reduction in eutrophication and environmental pollution, cost savings for municipalities, reduced damage to public/municipal infrastructure, reduced financial costs for the society and potentially cost-efficient fertilizer reuse
Gender equity	Technology-wise no particular (dis)advantage for any gender

Business value chain

After food digestion, our ultimate 'food waste' is discharged as excreta into toilets and where toilets are connected to a sewer, sewage treatment plants become vast nutrient transformation hubs where depending on the technology significant amounts of nutrients can be extracted from the waste stream, ranging in the case of phosphorus (P) from 20% to over 80% of the P in the wastewater. The cost per unit of P recovered varies with the wastewater volume and P concentration and are significantly higher for smaller plants and for lower discharge effluent P concentrations. So far, the cost of recovered P exceeds the cost of natural rock-phosphate (Petzet and Cornel, 2013; Mayer et al., 2016) making P recovery financially not viable. As it is uncertain when rock-phosphate prices will change, and if the fertilizer industry will accept the new product[1], the double value proposition offered for example by Ostara is interesting. The Ostara technology, like similar ones, aims at P removal from the liquid generated from sludge dewatering. As the liquid (sludge liquor) feeds back into the treatment

BUSINESS MODEL 16: PHOSPHORUS RECOVERY FROM WASTEWATER

process, and contains a significant share of the overall P load, the removal of P in the return flow improves the biological nutrient removal performance of the treatment plant and prevents unplanned P crystallizing in the form of struvite[2]. The business concept is based on a PPP where Ostara is assisting the treatment provider in reducing its maintenance and disposal costs for P removal after its unplanned crystallization, while generating a high-quality slow-release fertilizer. The process offered by Ostara does not replace traditional sewage treatment, but can be (retro)fitted into the facility's existing treatment process (see http://ostara.com).

The benefits from the concept are multiple: the treatment plant saves costs, high enough to finance the investment, the captured phosphorus is of high quality (no contaminants) and can be marketed as fertilizer raw material, the functionality of the treatment plants is extended while its effluent meets (even better) environmental standards (Figure 203).

Business model

This business model has a double value proposition. The first (and most important one) offers savings in treatment maintenance through an alternative P removal process; the second, a high-quality P

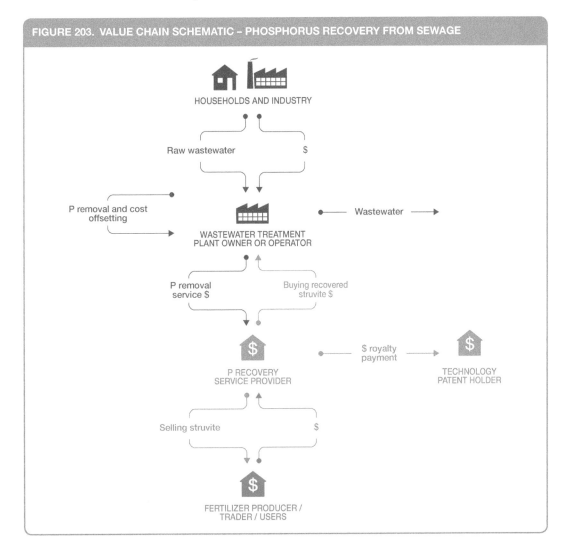

FIGURE 203. VALUE CHAIN SCHEMATIC – PHOSPHORUS RECOVERY FROM SEWAGE

crystal with potential use as fertilizer. The model is as such cost-driven for utility clients, and value-driven for resource sales. There are two models for financing the capital investment required for the P removal/recovery. Ostara offers its PEARL™ process based on either a traditional **capital purchase business model**, or through a **treatment fee model**. In the treatment fee model, Ostara pays for the installation and keeps ownership while the municipality or treatment plant operator (the client) runs the nutrient recovery process. Using a long-term contract, the client pays a monthly treatment fee based on agreed performance on phosphate removal. The treatment fee is lower than the costs of conventional phosphorus removal leading to immediate savings on operational costs. In the capital purchase model, the client pays for the installation and recovers the costs through maintenance savings usually over three to seven or max. ten years.

In both models, Ostara has a multi-year **purchase agreement** with the client to buy back the generated P crystals which are technically for the treatment plant a 'waste' product, while Ostara offers struvite marketing under the brand name Crystal Green™. In other cases than Ostara, the municipality might engage itself in fertilizer sales, be it for (green) image marketing or revenues. This, however, requires additional investments to enter the fertilizer market. In the case of Ostara's PEARL™ process, the struvite is generated as a side product which gives Ostara flexibility in its pricing and makes it relatively independent from the current rock phosphate price.

In alternative processes where P is, for example, extracted from the **ash** of mono-incinerated[3] sewage sludge, the P recovery can be much larger than from sludge liquor, but does not reduce the cost for the treatment plant, and has to be largely financed through P sales unless the recovery process is subsidized due to the environmental benefits. This dependency on the global rock-P price remains a challenge for the acceptance of several P recovery technologies, and most companies target premium (niche) markets with higher than usual willingness to pay. In general, ecological and economic benefits of closed loop concepts are not (yet) the driving force for the implementation of P recovery technologies, but financial advantages.

The business model described in this chapter presumes the operation for a standalone **private enterprise** (Figure 204). A largely complementary description of the Ostara case has been provided by P-Rex (2015).

Potential risks and mitigation

Market risks: From the recovery enterprise perspective, the number of wastewater treatment plants currently being built, or already set up, without P recovery units is larger than what the suppliers of P recovery technology could satisfy. In this sense there is limited risk, especially as with increasing emphasis on the SDGs, environmental sustainability and a circular economy, the recycling market will certainly grow. There remains a risk of missing out on prestigious projects.

In view of the market for recovered P, there can be a variety of challenges which differ from country to country and are still limiting the potential of P recovery despite its obvious benefits:

1) In many countries a range of markets might not be accessible due to prohibitive legislations or missing legislation on the reuse of waste derived resources.
2) The volumes of the recovered P are still too small compared with the market size, which increases the costs of entering the current mainstream value chain.
3) Although many studies showed that recovered P crystals are of high quality, and show often even less micro contamination, e.g. with metals than natural rock phosphate, not only legislations but also the fertilizer industry is hesitant to accept the product, be it for blending of other P sources or as stand-alone slowly-soluble fertilizer.

BUSINESS MODEL 16: PHOSPHORUS RECOVERY FROM WASTEWATER

FIGURE 204. BUSINESS MODEL CANVAS – STRUVITE RECOVERY INTO PREMIUM GRADE P FERTILIZER (OSTARA TYPE)

KEY PARTNERS	KEY ACTIVITIES	VALUE PROPOSITIONS	CUSTOMER RELATIONSHIPS	CUSTOMER SEGMENTS
• Treatment plant operator • Patent holder • Fertilizer industry	• Capturing P from the treatment process • Marketing and sales of recovered struvite • Obtaining permits and certification for struvite marketing	• A modular P removal systems and financing model to recover a non-renewable resource with a potentially high fertilizer value. • Savings in M&O cost for chemically controlling unwanted P crystallization during the treatment process	• Direct contact with plant operators • Technical support • Direct contact with potential struvite traders	• Treatment plant operator/ municipality • Fertilizer market (so far mostly niche markets)
	KEY RESOURCES • Technology for P recovery • Ability to obtain legal permits for struvite sale		**CHANNELS** • Direct sales of P technology • Direct sales of fertilizer to traders	

COST STRUCTURE	REVENUE STREAMS
• Investment cost in P recovery unit • Operational cost if run by enterprise • Struvite collection/storage/marketing cost. Transaction costs related to the penetration of the fertilizer value chains with small P volumes • Research & Development/validation/ licensing/certification	• Sales of P technology • Monthly treatment fees based on P removal • Sales of premium grade P fertilizer

SOCIAL & ENVIRONMENTAL COSTS	SOCIAL & ENVIRONMENTAL BENEFITS
• Uncertain acceptance of the product by traders and customers	• Environment benefits from preventing eutrophication • Supporting circular economy jobs and added-value via P (and N) recovery • Extended life time of a finite resource • Potentially a cheaper P source once rock-phosphate prices go up

4) More progressive legislation in support of a circular economy could help penetrate the conventional P market by demanding for a certain ratio of recovered to natural P; an example is one of the Indian Government which requires the fertilizer industry to co-sell bags of industrial fertilizer with a number of bags of waste-based compost.

5) To avoid perception related risks, marketing strategies normally avoid any connection between the name of the P product and its source.

6) With the never-ending generation of wastewater, also the supply of recovered P will be continuous irrespective of agricultural seasons. This will pose storage challenges unless multiple market segments next to seasonal crops are available (e.g. parks and gardens, forest or fruit plantations, year-round home gardens).
7) It is a significant advantage if like in the Ostara case the cost of P recovery can be (more than) absorbed by savings in conventional P removal, as the price of rock-phosphate is still too low compared with the break even price of recovered P, pushing recovered P into premium or niche markets which are able to pay higher-than-average prices.

P recovery from wastewater should be complemented with source separation. Capturing urine for example at large point sources (e.g. festivals) for nutrient recovery gives more flexibility to balance supply and demand, requires however similar to the case above legal support to enter established markets.

Competition risks: The number of providers of P recovery technology (and related patents) is increasing, and so is the diversity of processes supporting different treatment technologies, recovered amounts of P, and scales (WERF, 2010). Several companies have moved beyond technical pilots and are now competing on the market. However, compared to conventional suppliers of wastewater treatment technology, and demand for new plants, the internationally competitive group specialized on P recovery is still small. Where the enterprise partner has obtained a license from the patent holder, it needs to be understood how stringently the license is restricting similar business and upscaling. Patenting might open business avenues, while new technologies will continue to evolve. Competition risk is highest from the conventional P market where rock-P dominates in quantity, price-wise and is favoured also in view of some physical properties. Moreover, conventional P fertilizer might be subsidised, a benefit which is not easily applicable to a waste-derived product. Over time, it is anticipated that a higher rock-P price will help to stimulate P recovery.

Technology performance risks: Most P recovery technologies on the market have been repeatedly tested and produce a high quality final products. As the recovery potential between the technologies varies significantly (see Figure 280 in chapter 19) as does the cost-effectiveness (Sartorius et al., 2012; Petzet and Cornel, 2013), the municipality has to choose the one most appropriate for its plan, be it preventing unplanned struvite crystallization and/or compliance with P recovery targets. Where urine is collected with UDDTs their maintenance requires attention. Logistical challenges for urine storage and transport could be solved through low-cost innovations in urine dehydration (e.g. Senecal and Vinneras, 2017).

Political and regulatory risks: The regulatory context is in many countries not yet supporting 'secondary' phosphorus containing fertilizers and their producers as it is often classified as waste (P-Rex, 2015). While stringent environmental regulations on the discharge of P effluents into water bodies are on the increase and provide an opportunity to promote recovery and reuse, and so do SDG 12.4 and 12.5, these regulations mostly favour P removal, but not yet recovery and reuse. In fact, in Europe, regulations on the reuse of waste derived resources, including urine and struvite, are often very restrictive (Winkler et al. 2013). On the other hand, in many developing countries, regulations and standards might be lacking which can place resource recovery and reuse in a grey area where entrepreneurs might have an easy go, but quality control and legal security remain risk factors. However, with increasing attention to the SDGs and a circular economy the situation is changing, especially in Europe (Box 6).

Social equity related risks: There are no social risks with the model or technology, unless urine diverting toilets are targeted and household urine collection which might add to the workload of those

Box 6. P-recovery regulations and obstacles in Europe

Switzerland was the first European country to make phosphorus recovery and recycling from sewage sludge and slaughterhouse waste obligatory. The new regulation entered into force on 1.1.2016 with a transition period of 10 years. Switzerland banned direct use of sewage sludge on land in 2006, so that the new regulation will lead to obligatory technical recovery and recycling in the form of inorganic P products. Swiss sludge and slaughterhouse waste together represent an annual flow of 9100t of phosphorus.

In Germany, a new sewage sludge ordinance (AbfKlärV) is expected to enter into force early 2018, making phosphorus recovery obligatory for larger sewage works within 12 years (> 100 000 p.e.) or 15 years (> 50 000 p.e.), under certain conditions. P-recovery will thus be required for around 500 sewage plants, treating around 2/3 of German sewage. Following the legislative developments in Switzerland and Germany, Austria is now also opting for mandatory P recovery from municipal sewage sludge. The draft Federal Waste Plan 2017 (Bundes-Abfallwirtschaftsplan) includes a ban of direct land application or composting for sewage sludge generated at Wastewater Treatment Plants (WWTP) with capacities of 20,000 p.e. or above within a transition phase of 10 years. Alternatively, these WWTP will have to recover the P from sludge or its ash. This regulation will cover 90% of the P contained in the Austrian municipal wastewater.

However, P recovery within a Circular Economy requires reuse. Until now, struvite recovered from wastewater is only authorised for use as a fertilizer for some producers in some countries (e.g. the Netherlands, Denmark and Japan), or only on a case-by-case (e.g. Ostara plant by plant) authorization. Even in a country like the Netherlands, approval as a fertilizer does not ensure for struvite the **End-of-Waste** status. End-of-waste criteria specify when certain waste ceases to be waste and obtains a status of a product (or a secondary raw material). This current lack of clarity and disparities even between EU Member States poses a significant obstacle also to investments in the technology as long as it cannot necessarily be sold in another country, because the resulting product cannot be sold as a fertilizer.

The currently (2017) discussed new EU Fertilisers Regulation will enable recycled nutrient products to be sold in any Member State, when the new Regulation comes into force. Recognised products will also be granted de-facto End-of-Waste status. Composts and digestates are already included in the proposed Regulation text, but struvite is not. The EU's Joint Research Centre (JRC) has been mandated to make an impact assessment and (if this concludes positively) to propose criteria to add struvite, biochars and ash-based recycled nutrient products to the new Regulation annexes.

Source: http://phosphorusplatform.eu/

culturally in charge of household waste and sanitation. At larger scale, mineral fertilizer recycling not only saves jobs in the long term, but also creates additional green jobs and industries. As further increases in the price of rock-phosphate (based fertilizer) will hurt poorer countries first, the suggested resource recovery options – especially those with guaranteed cost recovery – could provide a low-cost alternative.

Safety, environmental and health risks: The industrial production of struvite shows good safety records, and the final product is usually of high purity for direct application in agriculture (Table 47).

There can however be variations in the heavy metal content with some of the technologies (Egle et al., 2014). Urine-based fertilizer is P and N rich and requires as a liquid fertilizer precaution. Although urine is per se sterile, there is a limited risk if it is collected from unhealthy people or if there is cross-contamination with fecal material. A higher risk from farmers' point of view is its unpleasant smell, and high pH which can damage crops if applied undiluted or too often. Guidelines for handling urine related risks, also in farming have been presented by Richert et al. (2010) and Stenström et al. (2011).

TABLE 47. POTENTIAL HEALTH AND ENVIRONMENTAL RISK AND SUGGESTED MITIGATION MEASURES FOR BUSINESS MODEL 16

RISK GROUP	EXPOSURE					REMARKS
	DIRECT CONTACT	AIR/ ODOR	INSECTS	WATER/ SOIL	FOOD	
Worker						Independently of the struvite recovery, workers at sewage plants face the relatively highest risk
Farmer/user						
Community						
Consumer						
Mitigation measures	🧤	😷	🚫🦟	⚠️		

Key: ☐ NOT APPLICABLE   ☐ LOW RISK   ☐ MEDIUM RISK   ☐ HIGH RISK

Business performance

P recovery technologies are on the increase. Currently, technologies with the highest economic viability for P removal during the treatment process has a cost recovery pay-back time of up to seven years. Other technologies, where P is recovered at the end of the treatment process, are financially struggling, although the P recovery percentage can be much higher. The reason is that their revenues depend – if not subsidized – on the P market price which is so far too low to compete and break even (Cornel and Schaum, 2009; Molinos-Senante et al., 2010). Thus, from the perspective of resource recovery, some of the best recovery rates are only viable when all aspects are considered, including economic, environmental and social (Balmer, 2004). In industrialized countries, a push for circular economics are expected to drive the establishment of P recovery (Sartorius et al., 2012), while the tipping point when the price of rock P exceeds the cost of P recovery remains uncertain (Horn and Sartorius, 2009). As in addition the legal framework for the reuse of resources recovered from waste remains a challenge, business models like the one of Ostara have significant advantages as their viability is independent of the P market. In general, the PPP model as run by Ostara has, except for smaller treatment plans, high replication potential. The prospects of cost recovery for the public partner and the win-win perspectives for both partners outshine the possible challenges of entering the fertilizer value chain for the generated struvite. The model ranks high on innovation, profitability and positive environmental impacts, but low on social impact (Figure 205).

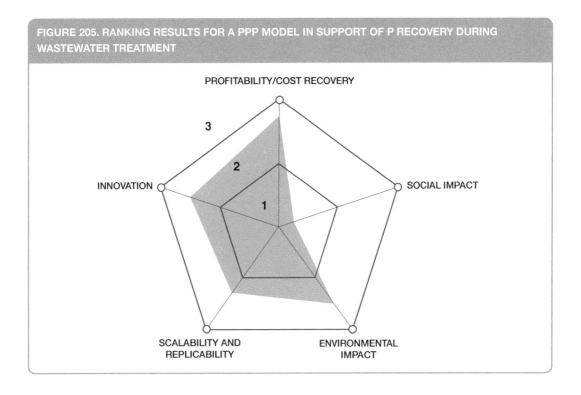

FIGURE 205. RANKING RESULTS FOR A PPP MODEL IN SUPPORT OF P RECOVERY DURING WASTEWATER TREATMENT

References and further readings

Balmer, P. 2004. Phosphorus recovery – an overview of potentials and possibilities. Water Sci. Technol. 49: 185–190.

Cornel, P. and Schaum, C. 2009. Phosphorus recovery from wastewater: Needs, technologies and costs. Water Sci. Technol. 59: 1069–1076.

Egle, L., Rechberger, H. and Zessner, M. 2014. Phosphorus recovery from wastewater: Recovery potential, heavy metal depollution and costs. Vienna: Vienna University of Technology.

ESPP. 2015. European Sustainable Phosphorus Platform. www.phosphorusplatform.eu.

Foley, J., de Haas, D., Hartley, K. and Lant, P. 2010. Comprehensive life cycle inventories of alternative wastewater treatment systems. Water Res. 44: 1654–1666.

Gaterell, M.R., Gay, R., Wilson, R.J., Gochin, R.J. and Lester, J.N. 2000. An economic and environmental evaluation of the opportunities for substituting phosphorus recovered from wastewater treatment works in Existing UK fertiliser markets. Environ. Technol. 21: 1173–1180.

Horn, J.V. and Sartorius, C. 2009. Impact of supply and demand on the price development of phosphate (fertilizer). In: Ashley, K., Mavinic, D., Koch, F. (eds), International Conference on Nutrient Recovery from Wastewater Streams. Vancouver: IWA Publishing.

Mayer, B.K., Baker, L.A., Boyer, T.H., Drechsel, P., Gifford, M., Hanjra, M.A., Parameswaran, P., Stoltzfus, J., Westerhoff, P. and Rittmann, B.E. 2016. Total Value of Phosphorus Recovery. Environmental Science & Technology, 10.1021/acs.est.6b01239.

Mihelcic, J.R., Fry, L.M. and Shaw, R. 2011. Global potential of phosphorus recovery from human urine and feces. Chemosphere 84: 832–839.

Molinos-Senante, M., Hernández-Sancho, F., Sala-Garrido, R. and Garrido-Baserba, M. 2010. Economic feasibility study for phosphorus recovery processes. Ambio 40: 408–416.

Ostara. 2017. Company information. www.ostara.com/.

Otoo, M., Drechsel, P. and Hanjra, M.A. 2015. Business models and economic approaches for nutrient recovery from wastewater and fecal sludge. In: Drechsel et al. (eds). Wastewater: Economic asset in an urbanizing world. Springer.

Petzet, S. and Cornel, P. 2013. Phosphorus recovery from Wastewater. In: Hester, R.E. and Harrison, R.M. (eds), Waste as a resource. Royal Society of Chemistry, p. 110–143.

P-Rex. 2015. Phosphorus recycling – Now! Building on full-scale practical experiences to tap the potential in European municipal wastewater. P-Rex Policy Brief. www.p-rex.eu. Accessed 22 April 2015.

Pronk, W. and Koné, D. 2010. Options for urine treatment in developing countries. Desalination 251 (2010): 360–368.

Richert, A., Gensch, R., Joensson, H., Stenstroem, T. and Dagerskog, L. 2010. Practical guidance on the use of urine in crop production. Sweden: Stockholm Environment Institute (SEI).

Sartorius, C., Horn, J. Von, Tettenborn, F. 2012. Phosphorus recovery from wastewater – Expert survey on present use and future potential. Water Environ. Res. 84: 313–322.

Senecal, J. and Vinneras, B. 2017. Urea stabilisation and concentration for urine-diverting dry toilets: Urine dehydration in ash. Science of the Total Environment. 586: 650–657.

Shu, L., Schneider, P., Jegatheesan, V. and Johnson, J., 2006. An economic evaluation of phosphorus recovery as struvite from digester supernatant. Bioresour. Technol. 97: 2211–2216.

Stenström, T.A., Seidu, R., Ekane, N. and Zurbrügg, C. 2011. Microbial exposure and health assessments in sanitation technologies and systems. Stockholm, Sweden: Stockholm Environment Institute (SEI).

Verstraete, W. and Vlaeminck, S.E. 2011. Zero Wastewater: Short-cycling of wastewater resources for sustainable cities of the future. Int. J. Sustain. Dev. World Ecol. 18: 253–264.

WERF. 2010. Nutrient recovery state of the knowledge. Water Environment Research Foundation, USA. www.werf.org/c/2011Challenges/Nutrient_Recovery.aspx.

Winkler, M-K.H., Rossum, F. van and Wicherink, B. 2013. Approaches to urban mining: Recovery of ammonium and phosphate from human urine. GIT Laboratory Journal Nov 11, 2013. www.laboratory-journal.com/science/chemistry-physics/approaches-urban-mining (accessed November 9, 2017).

Case descriptions are based on primary and secondary data provided by case operators, insiders or other stakeholders, and reflect our best knowledge at the time of the assessments 2014/15. As business operations are dynamic, data can be subject to change. More recently, for example, Ostara added the WASSTRIP (Waste Activated Sludge Stripping to Remove Internal Phosphorus) process to its technology solutions. See http://ostara.com/nutrient-management-solutions/.

Notes

1 Some resistance had been explained with the characteristics of recovered P crystals, like their slow solubility as well as regulatory challenges (see box 6 and chapter 19).
2 The spontaneous and unplanned formation of struvite in treatment plants affects pipes and other inner surfaces of the treatment process, making operation of the plant inefficient and costly because the struvite must be dissolved with sulphuric acid or broken down manually.
3 "Mono-incineration" means that the sewage sludge is incinerated separately, not mixed with municipal solid waste or other waste, and the ash contains high phosphorus levels (up to 7% P).

SECTION IV – WASTEWATER FOR AGRICULTURE, FORESTRY AND AQUACULTURE

Edited by Pay Drechsel and Munir A. Hanjra

Wastewater for agriculture, forestry and aquaculture:
An overview of presented business cases and models

Between now and 2030, the sourcing of water for human needs is expected to change, as the pressure on natural freshwater resources becomes more intense. This pressure is likely to come primarily from agriculture, as increasing demands for higher protein diets and biofuels will require a significant increase in agricultural output, which can only be met through greater water use. This will accelerate the over-exploitation of our freshwater resources, including a 66% increase in non-renewable groundwater withdrawals which is likely to affect millions of people by 2030, and billions by the end of the century (GWI, 2014). Under these circumstances, there will be limited alternatives to water reuse and desalination, especially where long-distance transfer is not cost-competitive. As public agencies seek economically and socially acceptable solutions to cope with increasing water demands, matching waters of different qualities with appropriate uses and implementing helpful reuse incentives will become essential for achieving the Sustainable Development Goals 6.3, 7.2 and 12.5, which directly address resource recycling, recovery and reuse. Unfortunately, the wastewater sector has long been a neglected utility, driven by regulation rather than economics or business thinking. But the situation is changing and water reuse is gaining significant momentum in discussions around green economies, urban resilience and enhancing urban food security. The awareness is growing that wastewater is in fact the only source of additional water that is increasing with population growth and higher water consumption, offering a range of opportunities for transforming wastewater and bio-solids into value propositions (Figure 206).

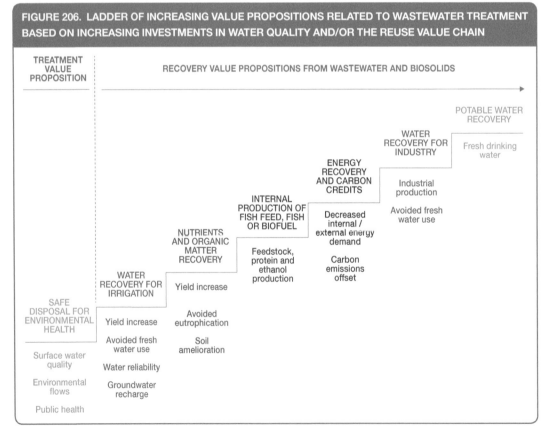

FIGURE 206. LADDER OF INCREASING VALUE PROPOSITIONS RELATED TO WASTEWATER TREATMENT BASED ON INCREASING INVESTMENTS IN WATER QUALITY AND/OR THE REUSE VALUE CHAIN

Source: Drechsel et al., 2015.

The highest market growth has been forecasted by Global Water Intelligence (GWI, 2010) for advanced water treatment supporting high value industrial and potable use. GWI is predicting that despite higher treatment costs the returns on investment will be rewarding. Already today we see many examples, also in developing countries, where up to 100 % of the operational and maintenance (O&M) requirements can be met from the sale of treated wastewater to local industries for uses such as cooling, power generation or air-conditioning (WSP, 2016). Cost recovery is usually less promising in view of agricultural reuse. Although the agricultural demand might be much higher than from particular industries, the sector's willingness and ability to pay are usually much lower, especially in low-income countries, while demand is often seasonally limited. Also in regions with highly subsidized freshwater tariffs or free groundwater access, cost recovery potential is low (Hanjra et al., 2015).

Thus, many wastewater business models are mainly **social models**, which are economically strong but fall short in view of financial sustainability unless the societal benefits are internalized. A survey conducted, for example, by the Water Environmental Research Foundation showed that only 12 out of 79 projects setting reclaimed water rates, aimed at full cost recovery (GWI, 2010). In these cases, motivating the use of reclaimed water takes precedence over cost recovery. In a report by the Tunisian Ministry of Agriculture, cost recovery rates in different irrigation projects with treated wastewater ranged between 13% and 76% of operational expenses for the agricultural supply component only (Chenini et al., 2003), not including the operational costs of the treatment facility. This is because financing water reuse projects can be challenging in that it is often expensive to build and operate an additional set of pipes and pumps to reach the final users, unless investors take over the responsibility. A more interesting point is why do some projects manage 76% while others only 13%? Such differences can derive from the choice of technology, institutional set-up, value proposition and targeted investment in cost reduction and recovery as our examples will show.

The first example of **Business Model 16** was presented in the previous section, and showed how wastewater treatment plants can reduce their operational costs of unwanted phosphorus (P) removal by investing in its recovery. The model by the company Ostara has therefore been presented in the Nutrient Recovery section. The model can be combined with energy recovery as shown recently in Amersfoort, the Netherlands where a 12,000-ton sludge treatment installation at the local wastewater treatment plant was commissioned in 2016, that will annually produce 900 ton (t) P-rich struvite and has an energy surplus of 2 million kilowatt hours (kWh) that will be delivered to the national grid.

This wastewater section of the Resource Recovery from Waste catalogue starts by describing three water reuse projects from **Egypt, Tunisia** and **Morocco** in Chapter 14. They represent different plant sizes, and institutional and regulatory challenges, and can therefore stimulate discussion on how to best maximize social and environmental benefits while targeting cost savings and recovery through closed loop processes and the sale of, for example, forest products. The three cases are located in a region where water is a scarce resource and reclaimed water can be of high importance for different sectors including farming and afforestation, therefore providing the basis for **Business Model 17: Wastewater for greening the desert (Institutional and regulatory pathways to cost recovery).** A fourth related case is Jordan's As Samra plant. However, due to its interesting financial set-up, the case is presented separately (see below).

The potential for cost recovery or even profit is multiplied when costs are minimized and returns maximized. This is possible where treatment systems are low in energy consumption and the resource recovery and reuse (RRR) value proposition goes beyond simply recovering water to incorporate the next steps of the value chain by selling products (e.g. fish fed with fodder) grown with the nutrients wastewater offers. In such cases, the likelihood of recovering both the fixed and variable costs of the (added) reuse component as well as the operational and maintenance costs of the treatment process can be substantial as the analyzed cases show. Technology choice is important, particularly in developing countries. Wastewater use in agriculture or aquaculture, can be supported through pond based treatment processes, with low investment costs and affordable operation and maintenance. Such processes are particularly suited to countries with warm climates where biological processes perform well. The investment costs for such locally 'appropriate' technologies are in the range of 20% to 50% of more conventional treatment plants, while the operation and maintenance costs are in the range of 5% to 25% of conventional (activated sludge) treatment plants (Libhaber and Orozco-Jaramillo, 2013). Chapter 15 presents two cases from **Bangladesh** and **Ghana** which show community based low-cost treatment systems in combination with the establishment of a fish-based value chain, taking advantage of the nutrient content of the wastewater. In the case of Bangladesh, the pond operator-cum-entrepreneur even recovered the construction costs of the treatment system. This is followed by an explanation of the overall model derived from the cases, which could potentially be applied to other settings – **Business Model 18: Leapfrogging the value chain**.

However, this business model does not imply that only smaller community-based pond systems can build on RRR to achieve high cost recovery. Many of the largest wastewater treatment plants in the world minimize their operational costs through highly efficient energy recovery mechanisms. As described in Chapter 16, **Jordan**'s extended As Samra wastewater treatment plant which was inaugurated in October 2015 is able to generate up to 95% of its energy needs, supported in part by a favorable topography. With minimized operational costs and an innovative overall cost sharing model, it contributes significantly to Jordan's entire renewable water resources, freeing up fresh water for more valuable uses. The main sources of finance for capital expenditure are public spending, external aid (loans, grants) and revenues from potable and industrial water use. The set-up of funding sources and guarantees can be of high complexity as shown in case of As Samra, but also much simpler as described for example by the Water and Sanitation Program (WSP) (2016) for Tamil Nadu. **Business Model 19: Enabling private sector investment in large-scale wastewater treatment** explains the institutional arrangements and overall characteristics for this type of model.

Energy recovery has also in smaller treatment plants the most significant potential for cost savings. While water and nutrient recovery can provide a certain contribution to offset the costs of sanitation and wastewater management systems, it is mostly the recovery of bioenergy that supports more substantial O&M savings. A survey carried out by WSP in India, for example, showed that energy recovery rates of 80–95% allowed to cut O&M cost of the studied wastewater treatment plants by half. The addition of a biogas plant, which costs about 15% of the wastewater treatment plants own capital cost, showed a pay-back period of only two to three years with an Internal Rate of Return of 33%. To support on-site electricity generation, contracts with private plant operators can be designed so that twice the amount for the power is charged whenever power is drawn from the grid to meet the plant's energy need and this is deducted from the payment made to the contractor (WSP, 2016). According to a 50-country analysis by Wang et al. (2015) bioenergy recovery has a high potential to realize environmental sustainability in developing countries where approaches should be customized, rather than attempting to replicate the successful models of developed countries.

Another set of interesting business models are emerging in the rural–urban interface. Growing urban water demands are placing substantial pressure on urban and peri-urban areas, leading to increasing calls for water reuse and integrated inter-sectoral water management and transfers. Chapter 17 covers four cases in this important interface: the rural-urban water-wastewater swaps in **Spain** and **Iran** (**Business Model 20: Inter-sectoral water exchange**), and the cases of **India** and **Mexico** where urban wastewater refills peri-urban and rural aquifers. In these cases, peri-urban areas function increasingly as 'kidneys' for the urban metabolism (**Business Model 21: Cities as their own downstream user**), which can be a promising model as long as possible environmental and human health risks are controlled, a statement which of course applies to all waste-based RRR models.

The last set of 'business models' differ from the others and are intended to stimulate further discussion. This is needed as wastewater reuse in agriculture is actually much more common than any official statistics so far have shown. The latest estimates indicate 36 million hectares of irrigated cropland depend on untreated or partially treated wastewater, used directly or indirectly after dilution (Thebo et al., 2017), in areas where urban treatment capacities are not keeping pace with population growth. The widespread use of unsafe water in these areas has prompted the World Health Organization (WHO) to test and recommend alternative on- and off-farm options for safeguarding farmers and public health, such as the multi-barrier approach (Drechsel et al., 2010). However, the adoption – or more precisely the provision of incentives for the adoption – of such safety measures remains a major challenge, and is urgently needed where regulations are not able to enforce measures such as crop restrictions in the informal irrigation sector. In support of WHO's sanitation safety planning concept (WHO, 2015) this catalogue presents three "business models" based on empirical cases from **Pakistan, India** and **Ghana**, supported by similar observations from other countries. The models are not mutually exclusive and show entry points and opportunities for increasing the safety of informal wastewater irrigation (**Business Model 22–24**) based on corporate social responsibility, the marketing of wastewater as a commodity and farmers' own investments in infrastructure, respectively. A model related to No. 23 and 24 with focus on improving the safety of informal (sludge) reuse was presented in the Nutrient Recovery section (Business Model 15).

In summary, most examples presented in the wastewater section of the *Resource Recovery from Waste* catalogue address the more common, but still complex and financially challenging situation of water reuse for agriculture, forestry and aquaculture, covering cases from Latin America, Africa, Asia and the Middle East and North African region. Several of the examples recover more than one resource and/or support more than the agricultural market. Further wastewater reuse examples from other sectors than agriculture have been covered elsewhere (e.g. 2030 WRG, 2013; Lazarova et al., 2013; USEPA, 2012).

A significant **weakness** throughout large parts of the wastewater section of the catalogue is the lack of reliable data on infrastructure financing or financial performance, as well as economic benefits. Extracting financial data from authorities or their publications posed a significant challenge, while economic impact assessments are generally rare. This is unfortunate, as internalizing the social and environmental benefits of wastewater treatment would probably well justify any public investments.

The presented cases and models – although by far not exhaustive – show a tremendous potential for RRR and private sector participation, where the enabling environment is in place (Chapter 19). If the well-known health and environmental risks can be controlled appropriately, there are many options to go beyond the social benefits of wastewater treatment and monetize the reuse value in ways that enable public and private sectors to achieve higher degrees of savings as well as cost recovery or even to generate profit. This 'double value proposition' will hopefully pave the way for a better delivery of wastewater services, and a more 'circular economy' for overall system sustainability (Andersson et al., 2016; Drechsel et al., 2015).

References and further readings

Andersson, K., Rosemarin, A., Lamizana, B., Kvarnström, E., McConville, J., Seidu, R., Dickin, S. and Trimmer, C. 2016. Sanitation, wastewater management and sustainability: From waste disposal to resource recovery. Nairobi and Stockholm: United Nations Environment Programme and Stockholm Environment Institute.

Chenini F., Huibers, F.P., Agodzo, S.K., van Lier, J.B. and Duran, A. 2003. Use of wastewater in irrigated agriculture. Country studies from Bolivia, Ghana and Tunisia. Volume 3: Tunisia. Wageningen: WUR.

Drechsel, P., Scott, C.A., Raschid-Sally, L., Redwood M. and Bahri A. (eds) 2010. Wastewater irrigation and health: Assessing and mitigation risks in low-income countries. London, UK: Earthscan-IDRC-IWMI, 404 pp.

Drechsel, P., Qadir, M. and Wichelns, D. (eds) 2015. Wastewater: An economic asset in an urbanizing world. Springer, 282 pp.

Global Water Intelligence (GWI). 2010. Municipal water reuse markets 2010. Oxford, UK: Media Analytics Ltd.

Global Water Intelligence (GWI). 2014. Global Water Market 2014. Volume 1. Introduction and the Americas. Oxford, UK: Media Analytics Ltd.

Hanjra, M. A., Drechsel, P., Mateo-Sagasta, J., Otoo, M. and Hernandez-Sancho, F. 2015. Assessing the finance and economics of resource recovery and reuse solutions across scales. In: Drechsel, P., Qadir, M. and Wichelns, D. (eds.) Wastewater: Economic Asset in an Urbanizing World. Springer Dordrecht.

Lazarova, V., Asano, T., Bahri, A. and Anderson, J. 2013. Milestones in water reuse: The best success stories. IWA, 408 pp.

Libhaber M. and Orozco-Jaramillo, A. 2013. Sustainable treatment of municipal wastewater. IWA's Water 21 (October 2013): 25–28.

Thebo, A.L., Drechsel, P., Nelson, K. and Lambin, E.F. 2017. A global, spatially-explicit assessment of irrigated croplands influenced by urban wastewater flows. Environmental Research Letters. 12: 074008.

USEPA. 2012. Guidelines for water reuse. EPA/600/R-12/618. Washington, DC: United States Environmental Protection Agency (USEPA). http://nepis.epa.gov/Adobe/PDF/P100FS7K.pdf (accessed 5 Nov. 2017).

Wang, X., McCarty, P.L., Liu, J., Ren, N.-Q., Lee, D.-J., Yu, H.-Q., Qian, Y. and Qu, .. 2015. Probabilistic evaluation of integrating resource recovery into wastewater treatment to improve environmental sustainability. Proc. Natl. Acad. Sci. U. S. A., 112 (5): 1630–1635.

WHO. 2015. Sanitation safety planning manual for safe use and disposal of wastewater, greywater and excreta. Geneva: World Health Organization.

WSP. 2016. Approaches to capital financing and cost recovery in Sewerage Schemes implemented in India: Lessons learned and approaches for future schemes. Water and Sanitation Program: Guidance note. New Delhi: WSP.

2030 Water Resources Group (WRG). 2013. Managing water use in scarce environments. A catalogue of case studies. Washington, DC: 2030 WRG.

14. BUSINESS MODELS ON INSTITUTIONAL AND REGULATORY PATHWAYS TO COST RECOVERY

Introduction

Most countries in the Middle East and Northern Africa (MENA) are severely affected by deforestation, or are simply too dry to sustain forests. Building green infrastructure (orchards, parks, green belts, forests, farms) in such a harsh environment can have substantial benefits for the ecosystem and society, especially if the investment does not compete for limited freshwater reserves but can build on 'waste' resources and even help avoiding disposal costs. In this context, there is no question about the multiple values wastewater treatment can offer society in dry climates on top of safeguarding public health. The three examples from Egypt, Tunisia and Morocco presented in this chapter were selected from a wide variety of similar cases. All three are located in water-scarce Northern Africa and show similar patterns and typical challenges of the region, as well as complementary features and innovations on the trajectory towards successful resource recovery and reuse (RRR). All three cases are aiming at **cost savings and cost recovery**, using RRR to create new revenue streams. Several other cases were explored but data availability did not allow adding more within the study period. A fourth case, however, could be Jordan's As Samra plant, which will be introduced in Chapter 16 as a model on its own (Business Model 19) due to its interesting financial set-up.

While in **Egypt**, the implementation of wastewater reuse is struggling with its institutional and regulatory set-up and missing incentives, significant progress can be reported from water reuse in afforestation and also in view of value creation from sludge, both with a huge potential for scaling up. The **Tunisian** example, on the other hand, showed the advantages of shared institutional responsibilities, private sector participation and a more flexible regulatory framework based on a strong political will to achieve environmental targets. The Tunisian example appears some steps ahead on the trajectory towards cost recovery although the case is struggling with its reuse percentages as many farmers can access alternative water sources with less stigma, risks and crop restrictions. In order to catch up, Egypt will have to revise its regulations and choice of crops to attract private sector participation for large-scale investment. Finally, the case study in **Morocco** shows how smart planning could allow achieving full cost recovery via decentralized, smaller systems for peri-urban communities. The setup of the case combined par excellence an applied low-cost technology, stakeholder participation, local resource recovery demands and a business plan for replication across towns and suburbs, with a dedicated accounting system to support full financial cost recovery. However, the potential of this setup received less attention after plant ownership and operations were transferred to the national sanitation agency. The progress and challenges in all these situations allow for the identification of possible bottlenecks and opportunities for new projects, and can help to steer the reuse agenda in view of SDG 6.3.

Following these case studies, the chapter presents **Business Model 17: Wastewater for greening the desert**, based on the country examples. It is relative flexible in its design and technical options as the main challenges appear to be vested in the (non-) supporting environment. After setup of a wastewater treatment facility, which follows in most cases the Build, Operate, Transfer (BOT) model supported by external loans, the plants could be managed by a public or private entity, with the same applying to the irrigation system. The Egyptian model of all components under one public sector company (in charge of sanitation) could allow to cut on transaction costs and improve cost recovery through the sale of wood, but can also be challenged by public sector inefficiencies and constraints, like overstaffing and limited entrepreneurial ambition, marketing knowledge and capacity. The Tunisian model with two governmental entities (sanitation and agriculture) working hand in hand for wastewater treatment and delivery to independent private water user associations combines complementary strength and expertise. If accompanied with a stakeholder dialog for participatory reuse planning, the model could be well positioned to thrive under different local conditions and crop demands (fruits, cotton, flowers, wood, etc.). The potential of such an approach is shown in the Moroccan case. Depending on local needs and social acceptance, an alternative reuse model could be aquifer recharge as is increasingly supported in Tunisia. However, the key determinants in the analyzed cases are often in the regulatory context, institutional capacity and interests, and in the fiscal policy of the respective national government.

While the running costs of the treatment plants can be covered by household connection fees, especially if energy costs are kept low, appropriate freshwater pricing is needed to value wastewater. The value chain for farmers can be enhanced where reuse goes beyond primary production and supports for example protein generation via fish or fodder production for the dairy industry (see Business Model 18 on Leapfrogging the value chain).

CASE

Wastewater for fruit and wood production (Egypt)

Pay Drechsel and Munir A. Hanjra

Supporting case for Business Model 17	
Location:	El-Gabal El-Asfar, northeast Cairo, Egypt
Waste input type:	Domestic and small industrial wastewater
Value offer:	Secondary treated wastewater reuse for cactus fruits (70%), lemon trees, and wood production; sludge sale for composting and construction (cement mix)
Organization type:	Public
Status of organization:	Secondary treatment level operational since 1998, commercial reuse of lemons and cactus fruits since 2007 with breaks
Scale of businesses:	Treatment: medium (450,000m^3/day); Reuse: small 10,000 to 30,000m^3/day on max. 147 ha.
Major partners:	Holding Company for Water and Wastewater (HCWW) through (the Greater Cairo Sewage Water Company); Ministry of Water and Wastewater Utilities; Undersecretariat for Afforestation and Environment of the Ministry of Agriculture and Land Reclamation (MALR), Ministry of Water Resources and Irrigation; Other Ministries (Housing, Health), Desert Research Center

Executive summary

The Greater Cairo Sewage Water Company (GCSWC) operates the El Berka wastewater treatment plant in the north-eastern part of Greater Cairo. Although the bulk of its wastewater is discharged back into the environment, about 5% of its secondary treated wastewater is used for irrigating lemon trees, cactus and trees for wood production, such as Khaya senegalensis, and, on pilot basis, industrial oilseeds including Jojoba and Jatropha. In addition, about 1,500 tenant farmers renting government land use approximately another 12% of the treated wastewater to irrigate about 1,000 hectares (ha) to support their livelihoods. This activity is informal and no fees are charged. The majority of the entity's revenue comes from household wastewater fees levied on around 1 million connected households, helping achieve a high cost recovery for the treatment of the wastewater. However, only about half of the households pay regularly resulting in USD 3.6 million revenues. The plant also raises revenue of about USD 0.6 million from selling one third of the generated sludge for composting and to the construction sector. There is significant potential for expansion into the agroforestry sector which is underused due to different challenges typical for the wastewater irrigation sector (not only) in Egypt. Therefore, compared with its potential, cost recovery through wastewater reuse is low, and the overall plant revenues subsidize the reuse system, in particular via household fees. This situation could be improved significantly with a change in the regulatory framework to support more progressive commercialization opportunities (choices of plants) and reuse standards, which is likely catalyzing private sector engagement.

KEY PERFORMANCE INDICATORS (AS OF 2014)					
Land use:	42 ha for treatment plant, 210 ha available for afforestation of which so far only up to 30% were used				
Wastewater treated/reused:	0.4–0.5 million m³/day of which 10,000–30,000m³ are formally and 49,000m³ informally reused; 50–60,000t sludge produced per year of which 20,000t are sold.				
Capital investment:	48 million (discounted to 1990 prices) for treatment plant; USD 1.6 million for plantation and irrigation system				
Labor:	About 270 persons at treatment plant; 110 at the plantation				
O&M costs:	USD 3 million/year for the treatment plant; USD 0.6 million/year for the plantation (due to overstaffing) (2013)				
Output (revenue):	USD 3.65 million/year from household sanitation fees; USD 11,700–28,000/year from agroforestry system using 10,000–30,000 m³/day; USD 609,000/year from sludge sale (unpacked, packed, largely for cement mix)				
Potential social and/or environmental impact:	Employment creation through afforestation programs; public health and environment protection; forest (fruits), wood, oilseeds products; benefit of research and outreach in wastewater reuse in agroforestry systems				
Financial viability indicators:	Payback period:	Depends on tree growth rate	Post-tax IRR:	N.A.	Gross margin:	N.A.

Context and background

Egypt has an arid climate with an annual precipitation in Cairo of only 26mm. The El Berka wastewater treatment plant and its wastewater reuse scheme is one of the smaller wastewater irrigated agroforestry plantations in Egypt. The total area allocated to the Holding Company for Water and Wastewater (HCWW) across Egypt for reuse is about 37,000 ha of which in 2013 about 4,622 ha were used. The El Berka wastewater treatment plant is managed by the Greater Cairo Sewage Water Company (GCSWC), a subsidiary company of HCWW. The plant is located in El-Gabal El-Asfar, in the north-east of Greater Cairo, in the Cairo Governorate, and employs about 270 permanent staff. Outputs from the secondary treatment (activated sludge) are sludge and water. While about 30% of the sludge is used for composting and cement production, only a minor part of the generated wastewater is formally used to irrigate fruits (lemons, cactus) and different wood producing trees (e.g. *Cupressus sempervirens*, *Kaya senegalensis*). At pilot scale oilseed/energy crops, like Jatropha and Jojoba, are being tested with promising results. The El Berka forest and horticulture plantation was established in 1998 by GCSWC and covers about 210 ha, of which so far 147 ha have been designed for irrigation offering jobs to 110 permanent employees at the plantation, 20 of them in support of irrigation. Since 2007, when lemon and cactus were commercialized for the first time, the actual area under irrigation varied between 21 and 60 ha.

The water is pumped from the treatment plant to the land parcels and a drip irrigation system is installed for the wood trees (wastewater only), whereas lemon and cactus receive both wastewater and freshwater via flood irrigation. The daily consumption of treated wastewater within the plantation varied over the last years between approximately 10,000 and 30,000m³, while farmers outside the plantation use informally about 49,000m³. The variation in land area and water consumption were due to restriction placed on the sale of crops. After a successful production in the first years (2007–2010) the Ministry of Agriculture prohibited the commercialization of (already produced) products irrigated with treated wastewater in 2011 and 2012, while in 2013 commercialization was allowed with restrictions which led to a decrease in the managed production area. Efforts to harmonize and standardize regulations on wastewater use in agriculture culminated in the Egyptian Code for the Reuse of treated Wastewater in Agriculture (ECP 501/2005) by the Ministry of Housing, Utilities, and New Communities. These standards set in 2005 reference the 1989 WHO guidelines, not the updated 2006 revision and are considered as too strict especially in terms of crop choices for commercialization. Additional difficulties occur through lacking standards for laboratory analyses (different methods result in different values). Thus, although the reuse potential for land reclamation is high and there are many profitable cropping options, so far the legal framework and its dynamic is not attracting investors and requires substantial improvements (Soulie, 2013).

Next to land reclamation and productive reuse, another driver for water reuse is operational risk reduction. Discharging wastewater to the Nile, canals or drains are controlled by law through licensing which requires compliance with set discharge standards. Failure to comply can mean withdrawal of the licence; however, there is hardly any source control.

Market environment

The public Holding Company of Water and Wastewater (HCWW), established in 2004, owns all water and sanitation infrastructure in Egypt. It works through its 26 affiliated subsidiaries companies across all Egyptian governorates where its 126,000 employees serve 85 million citizens. In 2013, HCWW operated 2,690 water treatment plants, and 357 wastewater treatment plants in the country, with 80% of the latter providing secondary treatment. Today, Egypt produces about 7.6 billion m³ wastewater per year, of which 3.8 billion are treated and about 0.7 billion formally reused (Abdel Wahaab, 2014). As regulations are difficult to enforce in the informal sector, direct and indirect use of (treated and untreated) wastewater is common.

Reuse in forestry systems is permitted by law and has been widely promoted by the government. According to HCWW around 63 man-made forests irrigated with treated wastewater occupy 4,622 ha. The total allocated land to HCWW (only) for reuse is about 37,000 ha which is about half of the size of all public and private forest plantations in Egypt. So far most plantations involving wastewater reuse have been government-driven. The government's support for private sector participation in water supply and sanitation did not go much beyond build-operate-transfer (BOT) arrangements for wastewater treatment plants. To stimulate wastewater use, Egypt and other countries in Middle East and North Africa (MENA) adopted a low-pricing policy for reclaimed water. As in addition freshwater use is subsidized, also for irrigation, it is most common to set a price for treated wastewater below the price of freshwater, in order to increase its market share. Thus, cost recovery via the sale of wastewater is far from being a viable option. In fact, thus far the rule is that water is provided for free to the plantations. The generally low water tariff rates lead to overconsumption and wasteful practices. Water consumed by Egyptian citizens, as measured by litres per capita, exceeds international rates, e.g. in the EU by a wide margin (USAID, 2013).

Free supply of treated wastewater is a significant loss for those treatment plants where the plantation is run by a different entity like the Undersecretariat for Afforestation and Environmental Affairs. In plantations run by the same operator as the treatment plant, like in the case of El Berka, reuse offers at least some value creation to extend the revenue stream beyond household fees and sludge sale. In

2009, dried sewage sludge produced in El Berka was directly sold to farmers with a gate price of USD 8.20/m^3. In other plants, HCWW sells the produced sludge to contractors for (on average) USD 6.1/m^3 and the contractors sell it to farmers with a profit margin. Other organic fertilizer in the Egyptian market are sold at about USD 17.76/m^3 (Ghazy et al., 2009).

Little information is available about demand for plantation products. Market assessments and marketing strategies are urgently required. Rotational forest production and harvesting schemes are so far missing, but it is assumed that the market for fruits, industrial oil and wood is significant. However, several plantations show very inhomogeneous wood production and commercialisation to major wood manufacturing companies for wood chips, wood fibre or board production is doubtable, unless wood quality (i.e. plantation management) is improved and overall production is increased. Sale of carbon credits generated due to the increased carbon sink effect in biomass and accumulation of organic matter in the soil have not yet been explored (Becker et al., 2013).

Macro-economic environment

Egypt, like other MENA countries, offers great opportunities for large-scale afforestation projects due to the availability of significant amounts of sewage water and wide areas of desert land. Given the lack of any substantial natural forests, aggressive desertification and the dependency of the national wood industry on imported raw materials, the productive reuse would serve multiple benefits for society and nature, and help the national wood industry. Following basic treatment, sewage water can be efficiently used as a resource for the production of wood, woody biomass and biofuel crops. The HCWW supports this vision through its 25 subsidiary companies, plans to stronger encourage private sector investments in reuse projects via tenders and to establish an affiliated company dedicated to the management and operation of wastewater reuse projects. While the production of edible and non-edible crops is in line with the Egyptian Code for the Reuse of treated Wastewater in Agriculture (ECP 501/2005), adjustments are in discussion to support stronger the cultivation of industrial crops, like cotton, and selected edible crops that are not eaten raw (Abdel Wahaab, 2014).

The Egyptian Water Regulatory Agency estimates the degree of overall cost recovery in 2012–2013 at 62%, and the recovery of operation and maintenance costs excluding depreciation at 76%. Although low water fees deprive treated water from its potential value, the free water supply to plantations supports their cost recovery potential. However, until now private sector participation in plantations is missing. Challenges are complex institutional arrangements with inadequate communication and coordination among authorities; unclear regulations for commercialization, land ownership issues and limited initiation of public participation in reuse to promote its value. Efforts have been made to establish a new policy to sell or lease desert land adjacent to wastewater treatment plants to private investors for forest plantations (Loutfy, 2011).

Business model

The value proposition is to create commercial and amenity value by turning desert soils with the help of secondary treated wastewater into a plantation for the commercial production of wood and fruits (Figure 207). This transformation entails significant economic benefits for nature and society if it can be replicated across all 350 to 370 wastewater treatment plants operated under the umbrella of HCWW (reduced wastewater discharge into other water bodies, reduced dependency on wood import, wind breaks/microclimate improvement, carbon sink, fresh water savings, employment, etc.) given the lack of any natural forests in the country.

Key factors in support of this proposition are full government support, the advantages of a central coordination (HCWW) and that the required land and water inputs are free. However, the already

FIGURE 207. BUSINESS MODEL CANVAS FROM THE OPERATORS' PERSPECTIVE (HCWW/GCSWC IN EL BERKA)

KEY PARTNERS
- Ministry of Water and Wastewater Utilities
- Under Secretariat for Afforestation
- Ministry of Water Resources and Irrigation
- Egyptian Environmental Affairs Agency
- Farmers (informal use)
- Connected urban households

KEY ACTIVITIES
- Collection and treatment of wastewater for safe water and sludge reuse or disposal
- Plantation management and sale of fruits and wood
- Collection of fees

KEY RESOURCES
- Wastewater and sludge, land and labor
- Expertise in irrigation and plantation management
- Capital investment

VALUE PROPOSITIONS
- To transform wastewater and sludge into safe products for reuse in agro-forestry, cement production and safe environmental disposal

CUSTOMER RELATIONSHIPS
- Government to government contracts
- Government to private sector contracts
- Automated system for tariff payment to the Gov./HCWW

CHANNELS
- Tender for reuse
- Water via pipeline
- Wood via direct delivery or collection at plant/plantation
- Fees via online payment (Cairo)

CUSTOMER SEGMENTS
- Wood and cement industry
- Fruit market

Under discussion:
- Cotton market
- Industrial oil market
- Biofuel market
- Export market

COST STRUCTURE
- Treatment plant OPEX, CAPEX
- Agroforestry system labor and machinery
- Opportunity costs of idle reuse capacity
- Customer service and fee collection

REVENUE STREAMS
- Household wastewater fees (via water bill)
- Sale of sludge for compost and construction
- Sale of fruits, wood and potentially other plant products
- Sale of wastewater to informal reuse sector (potential)
- Governmental subsidy; carbon credits (potential)

SOCIAL & ENVIRONMENTAL COSTS
- Possible long-term risk for groundwater but limited costs if safety protocols are followed

SOCIAL & ENVIRONMENTAL BENEFITS
- Reduced water pollution from wastewater disposal into streams
- Creating forest and farmland from desert
- Microclimate benefits; carbon sequestration
- Reduced dependence on wood imports and freshwater resources
- Employment and amenity values

installed reuse systems, like at El Berka, operate significantly under capacity in terms of planted land, used water, marketing and actual sale volumes, while staffing reflects design capacity, resulting in high operational costs and negligible revenues. A key reason for the mismatch relates to insecurity in the choice of crops which can be commercialized under the governing regulatory framework and other, mostly institutional challenges. This insecurity translates into scaling back in the planted area, limited investments in forest management (sustainable planting-harvest rotations) and across similar locations lack of private sector engagement.

The revenues to cover the expenditures of the afforestation efforts come from the sewer surcharges on the water bill, with some additional revenues from the sale of sludge. The overall El Berka treatment plant including the plantation achieved according to FAO (2014) a 119% operational cost recovery despite the fact that only half of all connected households pay regularly the billed fees. USAID (2013) estimated for the operating GCSWC a more conservative 79% on O&M.

The business concept would gain momentum by revisiting the regulatory framework and institutions in charge, to avoid that whole harvests get lost, engagement in an annual planting/harvest cycle, increasing the cropped plantation area (and returns per paid staff), improving collection rates from households and the consideration of charging those 1,500 tenant farmers who are informally abstracting a large volume of treated wastewater from the El Berka drainage channels. The charges could be levied as part of the Governmental land rent while offering farmers extension services, e.g. on how to comply with the safety code 501/2005.

Other revenue streams, once available, could be carbon or biodiversity credits. Given the social dimension of this business model the level of governmental support could be supported based on an evaluation of the provided economic benefits in terms of ecosystem services.

Value chain and position

The main revenue streams are wastewater fees and additional governmental support (Figure 208). While the wood value chain in Egypt depends on import from Northern Europe and sub-Saharan Africa, this does not automatically make irrigated local forest plantations an attractive venture, especially not for private sector engagement due to the long growing time needed before the first harvest and returns on investments. A major initiative of HCWW is therefore to support the revision of the Egyptian Code for reuse to allow for the cultivation of industrial crops and some edible crops that are not eaten raw but have a significant market value, like cotton, industrial oil plants or biofuel and allow quick returns on investment (Abdel Wahaab, 2014). Growing such plants can reduce private sector risks, allows to diversify production and bridge till the first tree rotation after 13 years is on (FAO, 2014). The long initial waiting period is also risky for private sector investments considering the reform-friendly institutional landscape and related insecurities that policies might change over time to their disadvantage.

Companies in charge of drinking water and sanitation (like HCWW) are not mandated to set the tariff structure for the services they provide. It is the State which approves rates according to socio-economic and political criteria. This results in low prices that do not cover the cost of the service or the operation of these organizations in the majority of cases. There are continuing efforts to work towards a tariff policy and reform package, in support of an improved financial performance of the sector.

Institutional environment

The water and sanitation sector of Egypt went over the last decades through a series of institutional reforms.[1] Given the common water scarcity and the fact that the agricultural sector is the highest freshwater consumer, utilizing about 86% of the available supplies, water reuse, especially in

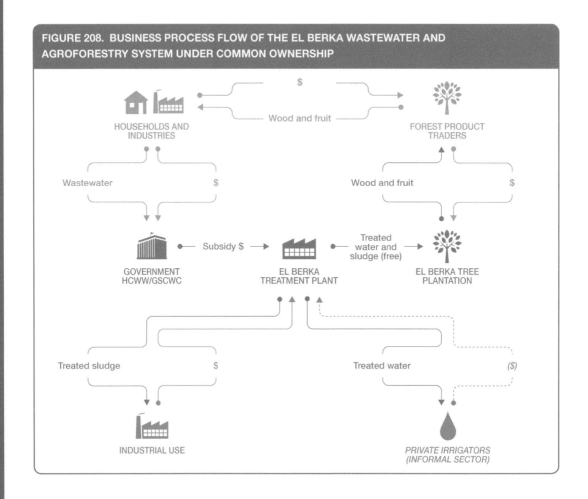

FIGURE 208. BUSINESS PROCESS FLOW OF THE EL BERKA WASTEWATER AND AGROFORESTRY SYSTEM UNDER COMMON OWNERSHIP

agriculture, was always part of the agenda. Laws and decrees have been issued including guidelines for mixing drainage water with fresh water, regulations for sewage and industrial effluents, wastewater reuse, cropping patterns, and health protection measures and standards specifications. The most important one is the Egyptian Code for the Reuse of Treated Wastewater in Agriculture (501/2005) by the Ministry of Housing Utilities and New Communities (Abdel Wahaab, 2014). There are at least five to six ministries with different roles involved in the wastewater management and reuse in Egypt. To streamline the institutional landscape the 2004 established HCWW owns, manages and operates all wastewater treatment plants across Egypt through its about 25 subsidiary companies. Other public companies under the 2012 created Ministry of Water and Wastewater Utilities (MWWU)[1] are the Cairo and Alexandria Potable Water Authority (CAPWO), which is responsible for the execution of water and wastewater projects in Cairo and Alexandria, the National Organization for Potable Water and Sanitary Drainage (NOPWASD) in charge of the execution of water and wastewater projects in other Governorates, and the Egyptian Water Regulatory Agency (EWRA) as an independent body of the others in charge of monitoring, inspection and customer satisfaction (Figure 209). However, due to overlapping responsibilities the regulatory agency remains so far relatively weak.

Besides the MWWU, several other ministries and institutes are involved in the wastewater activities. The Ministry of Agriculture and Land Reclamation (MALR) manages agricultural aspects, especially it operates forest plantations on reclaimed desert lands via the Undersecretariat of Afforestation and

CASE: WASTEWATER FOR FRUIT AND WOOD PRODUCTION

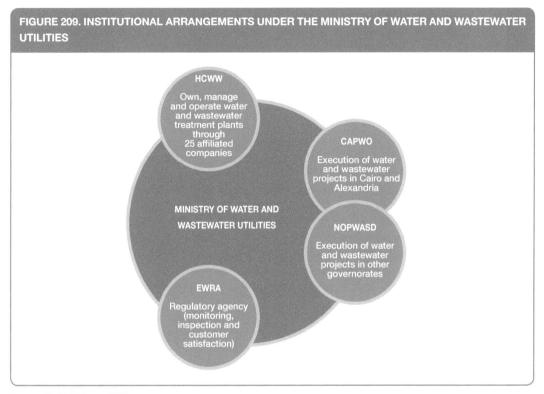

FIGURE 209. INSTITUTIONAL ARRANGEMENTS UNDER THE MINISTRY OF WATER AND WASTEWATER UTILITIES

Source: Abdel Wahaab, 2013.

Environment (UAE) as its subsidiary body. MWWU took over function from the Ministry of Housing Utilities and Urban Communities (MHUUC), which was concerned with the planning and construction of municipal wastewater treatment plants.

The Ministry of Health and Population (MHP) assumes responsibility for sampling and analysis of all wastewater effluents. It is also responsible for setting water and wastewater quality standards and regulations in addition to its central role as the custodian of public health. The Ministry of Water Resources and Irrigation (MWRI) allocates water for reclamation areas and is responsible for the Nile. The Ministry of State for Environmental Affairs (MSEA) and the Egyptian Environmental Affairs Agency (EEAA) cater to environmental aspects (FAO, 2014).[2] Other stakeholders are scientific institutions and universities conducting basic and applied research activities and international partners (USAID, AfDB, GIZ, EU, etc.) supporting the sector (Abdel Wahaab and Mohy El-Din, 2013).

However, lack of communication and coordination among the authorities, overstaffing and overlapping institutional responsibilities, strict regulations for reuse, but only enforced in formal, not informal, systems, are some of the recurrent issues challenging progress (Abdel Wahaab and Mohy El-Din, 2013). While the ongoing reforms addressed major issues, others remain unresolved. For example, sector fragmentation was not actually reduced. No organization was dissolved; instead several new organizations were created. Cost recovery is still very low; overstaffing has apparently even increased, and the institutional separation of responsibilities for investment and operation remains a challenge, also for foreign assistance.

Technology and processes

Water: The wastewater treatment plant at El Berka was realized in two steps. The primary treatment was constructed in the year 1990. Facilities for secondary treatment were established in 1998. The plant has a total area of 42 ha and receives the wastewater from the 5 million people (1 million households) in the northeast Cairo through a specific sewer system. The treatment is biological and activated sludge in the aerated basins. An additional chlorine treatment is used to limit microbial contamination and potential disease risks for people and animals. While chemical characteristics of the treated wastewater were reported within the acceptable range for reuse with beneficial crop nutrient levels, microbial and parasitic data indicate that chlorination levels might be too low and do not reduce viable nematode numbers (Abd El Lateef et al., 2006). Consequently, additional safety measures are recommended where the water is used informally for crops eaten unwashed or raw. Groundwater levels are between 15–17m in the study area and an impact from irrigation difficult to verify (Abd El Lateef et al., 2006).

The treated wastewater arrives at the plantation from the treatment plant by gravity and by electric pumps. The water is pumped to the parcels and a drip irrigation system is installed for the wood trees, whereas flood irrigation is used for lemon trees and for cactus. The lemon trees and the cactus plants are irrigated both with treated wastewater and fresh water, whereas the wood trees are irrigated only with treated wastewater.

The plantation size is very small given the treatment capacity and there is a strong call for better matched systems, where decentralized, smaller wastewater treatment facilities allow to reuse a larger proportion of the treated water for agro-forestry than in large-scale facilities where the majority is discharged into receiving water bodies.

A particular challenge is the lack of controls to monitor wastewater discharge. This situation is untenable to a public private partnership (PPP) investor/contractor who is subject to a significant risk due to the practices of upstream dischargers that could easily compromise with toxic effluents the ability of a treatment plant to satisfy contractual obligations related to the quality of the plant effluent (USAID, 2013). There is a need to treat industrial wastewater separately and/or before discharge onto public sewer networks.

Sludge from the activate sludge treatment is dewatered in a gravity thickener and then sun dried. While its chemical characteristics were found acceptable (Ghazy et al., 2009), pathogen levels are natural high. To destroy pathogens a mixture of the sludge and agricultural waste (e.g. rice straw) are air composted where the temperature reaches about 65°C. The composted product is then sold as organic fertilizer for landscaping or for construction (to be mixed with cement). A limitation of the project is the expensive cost of rice straw (Massoud, 2010). The untreated sludge is discharged in desert areas.

Energy: For the replication of the system lessons from the neighbouring El Gabal El Asfar treatment plant will be useful where methane from the anaerobic digestion of sludge allows to produce 37–68% of the total power consumption for the treatment plant (Ghazy et al., 2009; Massoud, 2010).

Funding and financial outlook

Both the reliable supply of **wastewater** of suitable quality as well as vast areas of land are freely available for reuse. Although soil quality is poor, there are large volumes of nutrient rich organic (sludge) compost in direct proximity. Several trees and agro-industrial crops species showed good performance under the given climatic conditions. Thus, there should not be any biophysical problem to establish agro-forestry plantations. That so far most agroforestry schemes in Egypt operate sub-optimally and current irrigated areas are far below the areas actually planned and/or equipped for irrigation has man-made reasons which can be addressed. So far only a small fraction of treated wastewater is reused, also in the El Berka agroforestry systems and the bulk is being discharged amid some informal reuse by crop farmers, yet no wastewater reuse charges are levied either on land rent or as reuse fees, which could offer additional revenues.

The main source of cost recovery are the household sanitation fees charged with the water bill. Given that only every second household pays as required, and the tariffs are far too low, adjusting the tariffs and increasing fee collection provide the largest opportunity for exceeding cost recovery while subsidizing any further expansion of the plantation (FAO, 2014). Higher freshwater tariffs could also stimulate demand for lower priced reclaimed water. On the other hand, if the plantation is supposed to be run by a third party, low wastewater tariffs will support investments. More important is in this case that the regulatory framework supports the commercialization of short-rotation crops with high market value, including export markets, allowing the operator to have diverse income sources and returns on investment before the first tree rotation is due. With such measures and annual planting/harvesting cycles the prospects of business viability are high.

The use of **sewage sludge** in landscaping and forestry should be part of the model as plants need organic matter and nutrients. According to the Egyptian Government future plan, there is the possibility that the El Berka composting project may be expanded to a full-scale project to produce a compost of 720 tons per day from the dried sewage sludge accumulated from El Berka, Shobera and Al Gabel Asfer WWTPs (Ghazy et al., 2009). Such a significant sludge production supports the Egyptian policy to reclaim land lost due to desertification. The extensive sunshine exposure, high temperature, and dry conditions provide aggressive and unfavorable conditions for the survival of microbial pathogens. Chemical risks can be limited by industrial source treatment and sludge reuse for non-edible crops. Moreover, the high pH of most soils limits crop uptake of heavy metals. Indeed, most soils in Egypt would benefit from sludge compost, as reclaimed land is usually poor in micro-nutrients, such as zinc and copper which are required for plant growth and present in sludge (Ghazy et al., 2009). The theoretical calculated monetary value of the dried sewage sludge in Egypt is about USD 53/ton (USD 28.5/m^3). This value probably indicates the maximum price of the dried sewage sludge that can be paid by farmers, including the transport costs in the Egyptian market (Ghazy et al., 2009). Where sludge quality does not match safety standards, other reuse options exist. The El Amria Cement Company in Alexandria has been granted EEAA approval for use of substitute fuel to natural gas in the cement kiln including hazardous waste. The proposed project is to use part of the dried and dewatered sludge produced from the wastewater treatment plants in Alexandria as substitute fuel in the cement kilns. This will reduce GHG emissions generated from the anaerobic conditions if sludge is disposed in the landfill. Moreover, incineration of this type of bio-fuel will produce less CO_2 emissions if compared to fossil fuel. Therefore, there is a possibility that this project can be considered as a potential clean development mechanism (CDM) project; which offers interesting perspectives for other plants in Egypt (Massoud, 2010).

Socio-economic, health and environmental impact
An environmental and social impact assessment carried out by the African Development Bank for the extension of the El-Gabal El-Asfar wastewater treatment plant with comparable treatment quality and draining into the same water resources as El Berka confirmed an overall positive impact of the plant on its social and ecological environment (AfDB, 2008). The wastewater and sludge reuse activities, if done at scale, are reducing in addition their unproductive discharge into the environment while creating employment opportunities. These like other benefits for land reclamation, the support of the local wood or cement industry, micro-climatic improvements and carbon sequestration will depend on the scale

of the water reuse and afforestation activities. The largest socio-economic and environmental benefit of reuse is its contribution to addressing water scarcity of communities across Egypt in the face of rising demand and shrinking freshwater volumes. The strict reuse standards provide less flexibility than the current WHO guidelines and Sanitation Safety Plans and could be adjusted, especially in formal reuse schemes where compliance monitoring is feasible. Efforts have to be increased, however, to address informal wastewater reuse by farmers outside the plantations where crop restrictions are not enforced. There are signs of microbial soil and groundwater contamination and a need for monitoring if and how far the irrigation is affecting the 15m deep groundwater table (Abd El Lateef et al., 2006).

Scalability and replicability considerations

The key drivers for the success of the reuse model are:
- Government's financial support for water, land and sludge use.
- Vast amounts of available resources, including a reliable water supply.
- Political will to further transform the sector.

The key obstacles are of institutional and regulatory nature:
- Limited private sector interest due to a too firm wastewater reuse code especially in terms of crop restriction.
- Overlapping responsibilities and limited cooperation among ministries and agencies in view of reuse and crop commercialization.
- Operational risk due to continuing sector reforms and reorganization with changing mandates and responsibilities.

Once the obstacles have been addressed, the El Berka model has significant potential for replication at decentralized level where land is available and forest plantations have a notable economic and social value for the local communities. Governmental support for initial capital cost will remain instrumental, both for the treatment systems and agroforestry system. However, cost recovery for the plantation is feasible. It can be run by the treatment facility or a third party. Product diversity with different payback intervals will be crucial while tapping into the emerging CDM market could be an additional option (Becker et al., 2013).

To support investments in reuse, HCWW embarked on an initiative to revise the Egyptian code for reuse to allow use of treated wastewater for cultivation of industrial crops and some edible crops that are not eaten raw, taking into consideration the required health protection measures. Ideally, a National Plan of Wastewater Reuse has to be established making freshwater use an exception where reclaimed wastewater is available, and where no reclaimed water is available, the freshwater tariffs have to be increased to stimulate wastewater (treatment for) reuse. However, since the Arab Spring residential tariff increases have become even more difficult to approve.

Summary assessment – SWOT analysis

Figure 210 presents the SWOT analysis of this business case from the reuse in agroforestry perspective. The case shows a high potential where resource supply and demand are in place but the institutional and regulatory environment prevents public and private sectors to make optimal use of the given opportunity.

FIGURE 210. SWOT ANALYSIS FOR EL BERKA WASTEWATER AND AGROFORESTRY SYSTEM

INTERNAL ORIGIN — ATTRIBUTES OF THE ENTERPRISE

HELPFUL TO ACHIEVING THE OBJECTIVES

STRENGTHS
- Vast land and wastewater resources at low cost or free
- Strong vision of HCWW to support reforms for private sector engagement and increased cost recovery
- Wastewater quality matching reuse for forestry and species mix
- Location with limited risk of negative trade-offs
- Highly trained staff in place

HARMFUL TO ACHIEVING THE OBJECTIVES

WEAKNESSES
- Severe overstaffing due to scaling back of production
- Unclear and complex decision making regarding what can be planted and commercialized
- Unclear standards for laboratory analyses to comply with water and sludge standards
- Treatment quality can vary and monitoring is weak
- Also groundwater quality needs monitoring

EXTERNAL ORIGIN — ATTRIBUTES OF THE ENVIRONMENT

OPPORTUNITIES
- Revised regulatory framework allowing more non-edible products for commercialization
- Up-scaling of El Berka to design capacity and replication of similar plantations via decentralized smaller treatment plants
- WHO 2015 Sanitation Safety Planning concept facilitating compliance monitoring
- Export market
- Large national and export demand for wood which will support cost recovery

THREATS
- Mixed messages from authorities due to overlapping competencies
- Too restrictive regulations for cash crops which allow fast returns on investment
- Delay of reforms of regulatory framework and standards
- Negative public attitude and concerns related to product safety
- Fear of second Arab Spring preventing water tariff increase

Contributors

Mochamud Mantamuna, El Berka treatment plant
Mahmud Mohamed Alhanafy, El Berka treatment plant
Emad Salah, El Berka treatment plant
Ahmed K. Mohamed, Desert Research Centre, Egypt
Achim Kress, FAO
Michele Baldasso, FAO
Simone Targetti, University of Basilicata
Javier Mateo-Sagasta, IWMI

References and further readings

Abd El Lateef, E.M., Hall, J.E., Lawrence, P.C. and Negm, M.S. 2006. Cairo-East Bank effluent re-use study 3 – Effect of field crop irrigation with secondary treated wastewater on biological and chemical properties of soil and groundwater. Biologia, Bratislava, 61/Suppl. 19: S240–S245.

Abdel Wahaab, R. 2014. Wastewater reuse code in Egypt. SWIM-Sustain Water MED 12–16 May 2014, Djerba Plaza, Tunisia. http://www.swim-sustain-water.net/fileadmin/resources/regional_training-Tunis/Egypt.Reuse_Code_Presentation.Djerba.pdf (or https://goo.gl/nQpKra; accessed 5 Nov. 2017).

Abdel Wahaab, R. 2013. Wastewater reuse in Egypt: Opportunities and challenges. SusWaTec-Workshop, 18–20 Feb. 2013, Cairo; http://www.suswatec.de/download/presentations/Abdelwahaab.pdf (accessed 19 Oct. 2017).

Abdel Wahaab, R. and Mohy El-Din, O. 2013. Wastewater reuse in Egypt: Opportunities and challenges. Arab Water Council. www.scribd.com/document/169795860/Egypt-Country-Report.

AfDB. 2008. Egypt: Gabal El Asfar wastewater treatment plant. Environmental and social impact assessment summary. https://www.afdb.org/en/documents/document/egypt-gabal-el-asfar-wastewater-treatment-plant-esia-summary-12893/ (or https://goo.gl/RTKnfL; accessed 5 Nov. 2017).

Becker, K., Wulfmeyer, V., Berger, T., Gebel, J. and Münch, W. 2013. Carbon farming in hot, dry coastal areas: An option for climate change mitigation. Earth Syst. Dynam. 4: 237–251.

FAO. 2014. Cost recovery of wastewater use in forestry and agroforestry systems: case studies from Egypt, Tunisia and Algeria. Unpublished final report completed for IWMI, November 2014.

Ghazy, M., Dockhorn, T. and Dichtl, N. 2009. Sewage sludge management in Egypt: Current status and perspectives towards a sustainable agricultural use. World Academy of Science, Engineering and Technology 57: 492–500. http://citeseerx.ist.psu.edu/viewdoc/download?doi=10.1.1.193.4445&rep=rep1&type=pdf (accessed 5 Nov. 2017).

Loutfy, N.M. 2011. Reuse of wastewater in Mediterranean region, Egyptian experience. In: Barceló, D., Petrovic, M. (eds.) Waste water treatment and reuse in the Mediterranean region. Springer, p. 183–214.

Massoud, S. 2010. A study on potential of CDM projects in municipal water & wastewater utilities sector. First interim report prepared for the Clean Development Mechanism Awareness & Promotion Unit (CDM/APU), Ministry of State for Environmental Affairs, Egypt.

Soulie, M. 2013 Review and analysis of status of implementation of wastewater strategies and/or action plans; National report Egypt. Sustainable Water Integrated Management (SWIM) – report. www.swim-sm.eu/files/National_Report_WW_strategies-EGYPT.pdf (accessed 5 Nov. 2017).

USAID. 2013. Clean water for Egypt. Egypt water policy and regulatory reform program: Final report prepared by Chemonics International Inc. under contract No. EPP-I-00-04-00020-00.

WHO. 2015. Sanitation safety planning: Manual for safe use and disposal of wastewater, greywater and excreta. Geneva, Switzerland: World Health Organization (WHO).

Case descriptions are based on primary and secondary data provided by case operators, insiders or other stakeholders, and reflect our best knowledge at the time of the assessments 2014/15. As business operations and institutional environments are dynamic data can be subject to change.

Notes

1. Also called Ministry of Water Supply and Sanitation Facilities.
2. The names of institutions and ministries and their responsibilities changed frequently in the past and can change again.

CASE
Wastewater and biosolids for fruit trees (Tunisia)

Pay Drechsel and Munir A. Hanjra

Supporting case for Business Model 17	
Location:	Ouardanine near Monastir, Tunisia
Waste input type:	Domestic wastewater
Value offer:	Secondary treated wastewater sold for reuse in farmer-managed tree crop system (peaches, olives, grapes, grenades)
Organization type:	Public-private
Status of organization:	Plant set up in 1993, irrigation scheme operational since 1997
Scale of businesses:	Small: 1,590m³/day treatment; reuse on 65–75 ha
Major partners:	National Sanitation Utility (ONAS); Regional Offices of Agriculture Development (CRDA); Groups of Agricultural Development (GDA) and local farmers

Executive summary

The National Sanitation Utility (ONAS) is a public institution in charge of the Tunisian sanitation sector and operates the small Ouardanine wastewater treatment plant near Monastir city (328mm annual rainfall), treating mostly domestic (non-industrial) wastewater from about 3,400 households. About a quarter of the reclaimed water is used by the nearby Ouardanine tree plantations (65–75 ha), managed by about 40–46 private farmers to produce olives, peaches and pomegranates to sell at the local market. In the case of irrigation, the downstream infrastructure is managed by the Governmental Commissariat Régional de Développement Agricole (CRDA), which receives the water from ONAS free of charge and is responsible for distributing and billing reclaimed water to end users (farmers' collectives or Water User Associations called Groupement de Développement Agricole; GDA). CRDAs charge the GDA a subsidized water price which is fixed by the government as an incentive for reuse and low compared to the value it is creating. The Water User Associations then distribute the water among their members/ farmers, while collecting an annual subscription fee, and also charging a mark-up on the water price to undertake routine repairs of the distribution network. The Ouardanine plant also supplies biosolids (sludge) on-demand as soil conditioner free of charge. ONAS recovers 40% of operation and management (O&M) costs from wastewater transfer to the irrigated plots (with the balance buffered by the CRDA), and has an overall operational cost recovery of 56% for the total wastewater treatment system when adding sewage taxes levied on households (FAO, 2014).

KEY PERFORMANCE INDICATORS (AS OF 2013)	
Land use:	65–75 ha under fruit trees
Wastewater treated and reused:	Treated: 1,590m³/day wastewater; 2050t/year sludge produced; of this reused: up to 410m³/day wastewater; about 105t/year of treated sludge
Capital investment:	USD 1.2 million for the treatment plant (1993) USD 337,000 for agroforestry system (1997)
Labor :	About 46 farmers at the plantation; additional seasonal harvester
O&M cost:	USD 30,500/year for treatment plant USD 11,700/year for wastewater transportation to the irrigated 65 ha
Output:	USD 17,000/year from household sanitation fees (USD 5 per household and year) USD 1,950/year from CRDA selling water to GDA; sludge valued at USD 1,270 (data from Egypt) but uncounted as free USD 2,780/year from GDA selling water to farmers USD 817,000/year from fruit sales by farmers, not counting further gains along the value chain
Potential social and /or environmental impact:	Water savings, public health and marine environment protection, nutritious food, carbon sequestration
Financial viability indicators:	Payback period: Depends on tree growth rate — Post-tax IRR: N.A. — Gross margin: N.A.

Context and background

In Tunisia about 84% of the generated wastewater is collected and treated at least 109 wastewater treatment plants. Nearly all (95%) of this water is treated at secondary level. A key motivation is the preservation of Tunisia's marine environment and coastal resorts, given the national importance of the tourist sector as part of Tunisia's overall commitment to prevent pollution in the Mediterranean Sea. As a semi-arid country, Tunisia is also aware of the pressure on its existing water resources and as many resources are saline, authorities are determined to increase water reuse.

In 2009, about 63 million m³ (i.e. 26% of the annually treated 238 million m³ wastewater) have been reused directly (agriculture, landscaping) or indirectly (aquifer recharge, etc.). The total agricultural area equipped for irrigation with treated wastewater was about 8,065 hectares (ONAS, 2009), although not all of this land might be actually irrigated. In addition, wastewater use has been reported for landscape irrigation such as golf courses (1,040 ha) and green areas (450 ha). The main crops irrigated across the country with treated wastewater are fruit trees (29%), fodder crops (45%), cereals (22%) and industrial crops (4%) (Bahri, 2002; Abid, 2010). In 2021, the plans are that 172 million m³ would be made available for reuse on 40,500 ha farmland, 50 million m³ for landscaping of 3,500 ha, and 25 million m³ for aquifer recharge (ONAS, 2009).

Reuse has been regulated under the 1975 Code des Eaux (Water Code) and several more detailed decrees which are setting norms for chemical and biological loads in reclaimed water, prohibit the use of untreated effluents for irrigation and stipulated that reclaimed water could be used on a range of crops except vegetables or fruits that are consumed raw, such as tomatoes, lettuces, carrots and berries. The list of crops which could benefit from treated wastewater remains valid and includes industrial crops (e.g. cotton, tobacco, flax, jojoba and castor oil plant), grain crops (e.g. wheat, barley, oat), fodder crops (e.g. clover, corn, alfalfa), fruit trees (e.g. date palms, citrus trees, olive trees, vines), forest trees, flowers and herbs (e.g. rose, lily, jasmine, marjoram and rosemary).

The Ouardanine case is an example of a small and decentralized wastewater treatment plant which is serving about 3,400 households. Until 1993, the town of Ouardanine had to cope with the impacts of untreated sewage discharge. Environmental degradation combined with limited employment opportunities contributed to many local youth leaving this rural town. Called to remediate the untreated discharge situation, ONAS was met with pressing demands by local farmers to reclaim the water for irrigation. While ONAS implemented the treatment system, the CRDA elaborated the irrigation scheme with the farmers regrouped in a formal water user association, responsible for site selection, land rights resolutions and plant culture selection. This has allowed to ease use restrictions and avoid rejection of reclaimed water by users.

About 26 % of the secondary treated wastewater is reused by about 46 farmers for different fruit trees in an irrigation scheme set up in 1997 at a cost of USD 337,000, as part of the national water reuse program. Of the 65 (max 75) ha allocated for reuse, 34–45 ha were in recent years under peaches, 20–21 ha under olives and a small area under grapes, barley, alfalfa, cut roses and pomegranates. Drip and furrow irrigation are used. The wastewater treatment plant also produces sludge, but so far only a small percentage gets composted and recycled in agriculture in a free of charge pilot program. The application of biosolids on agricultural land is by law limited to experimental plots conducted as demonstration pilot projects.

Market environment

With some geographic variability, water scarcity is the defining feature of the agricultural economy in Tunisia. Against this backdrop Tunisia has since the mid-1960s increasing experience in wastewater reuse with a strong supporting legal framework and political commitment that has led to continuous expansion of wastewater treatment and reuse in the country. Perception studies show a reasonably high level of farmers' hypothetical acceptance to use reclaimed wastewater (80%), preferably without restrictions, and public acceptance (71%) to consume crops irrigated with treated wastewater (Abu-Madi et al., 2008). However, despite increasing water shortages and substantial economic incentives, actual demand for reclaimed water between 2001 and 2009 plateaued at around 25–30% of treated wastewater. According to Abu-Madi et al. (2008) and GWI (2010), factors that fuel the farmers' hesitation are: (i) availability of or accessibility to freshwater; (ii) distrusted water quality; and (iii) worries about crop/fruit marketing and acceptance. Less important are however concern for public criticism, concern for health impacts, religious prohibition, or psychological aversion. Reasons for water quality concerns which led farmers to fall back on conventional resources include (GWI, 2010):

- Plant saturation (particularly in coastal areas and in summer when tourist numbers put a strain on capacities) and ageing.
- Industrial pollution due to poor upstream pre-treatment (a legal requirement for industrials but often poorly observed in practice) which refers in particular to salinity in central and southern Tunisia.

The existing system for crop marketing in which crops produced with reclaimed-water crops are on offer together with freshwater irrigated crops 'facilitates' marketing although some consumers seem to be able to distinguish between the crops. However, there are calls for more transparency and monitoring, also to increase the confidence of the consumer. So far only the national market is targeted. According to Abu-Mari et al. (2008) Tunisia has not yet reached a stage where the crops irrigated with reclaimed wastewater can be exported.

Sludge reuse has been tested on pilot farms (about 300 ha) for several years, in line with the national standards (Normes Tunisiennes (NT) 106.20 – 2002) (MAERH, 2003; ONAS, 2009). Also at Ouardanine soils were amended under the regular monitoring of the Ministry of Agriculture as one of the demonstration projects. The estimated amount of 6 t sludge/ha is expected to be spread over five years.

Macro-economic environment

All of Tunisia's infrastructure has been financed by the state, usually with a combination of loans and grants from state finance and international lenders. Tunisia has good links, e.g. with the European Investment Bank, the Agence Française de Dévloppement, the German KfW, the Japan International Cooperation Agency, etc. Most new wastewater treatment plants are medium-sized plants (15,000m³/day), built on a turnkey contract basis and financed by ONAS with international loans. But also different financing and procurement avenues are being explored, including a 25-year BOT contract for the construction of two large WWTPs.

Tunisia is determined to reduce the discharge of the wastewater to the sea, and to develop water reuse. The government policy strongly supports wastewater treatment and incentivizes wastewater reuse. Sanitation charges for domestic users, industry and tourist establishments vary according to water consumption and the principle of 'polluter pays' and do so far only cover about 60–65% of ONAS's operational costs, which comprise personnel salaries (about 30%), energy (60%) and equipment repairs and replacements (10%). The rest is financed by the state.

Cost recovery via wastewater use is constrained by water pricing. The tariff set by the Government in 1997 demanded Tunisian dinar (TND) 0.02/m³ (ca. USD 0.02/m³ at that time, or USD 0.015/m³ in 2010). The target is to keep the price for reclaimed water significantly below the one of the subsidized freshwater[1] which was about 3 to 4 times higher for irrigation, and 7–40 times higher for domestic and industrial use in the year 2000 (Bazza and Ahmad, 2002; GWI, 2010). The price for reclaimed water has remained unchanged since 1997 and covers only a fraction of the real cost of wastewater treatment, estimated at TND0.3–0.7/m³ (GWI, 2010).

Despite the low price charged to Tunisian farmers for reclaimed water compared to conventional water supply, the demand for reclaimed water remained so far modest (Qadir et al., 2010). The mismatch has (i) in part geographical reasons with most wastewater being produced in the Greater Tunis area and along the coast, i.e. not where it is mostly needed; (ii) is supported by the availability of alternative water sources like shallow groundwater which only attracts pumping costs until a depth of 50m (FAO, 2009); and (iii) is also driven by seasonal demand–supply gaps.

Business model

The treated wastewater coming for free from the wastewater treatment plant of Ouardanine is pumped to a ground reservoir which is under the supervision of the Regional Offices of Agriculture Development (Commissariat Régional de Développement Agricole, CRDA). The CRDA also operates the pumps and wastewater distribution network connecting to the irrigation scheme. CRDA sells the water in bulk to the Water User Associations (Groupement de Développement Agricole, GDA), at a price of TND 0.02/m³ (2013: USD 0.012/m³) which recovers about 17% of the costs of CRDA to convey the water to the irrigation scheme; the balance is covered by the Ministry of Agriculture.

The GDA then sells the water to the farmers at TND 0.035/m³ (USD 0.022/m³), thus earning a mark-up of about USD 0.01/m³. Besides wastewater sales to the farmers, the GDA also raises revenue (about USD 1,250) from the annual subscription fees paid by farmer (USD 32 per farmer) which allows them to support CRDA with minor maintenance of the irrigation network at farmers' end. According to FAO (2014) farmers are the main beneficiary of the irrigation system with an annual income from their production sale of about USD 5/m³ or USD 12,570/ha. Moreover farmers engaging in reuse are entitled to purchase irrigation equipment at a 30% discount, or use for free treated sludge (Figure 211).

CASE: WASTEWATER AND BIOSOLIDS FOR FRUIT TREES

FIGURE 211. BUSINESS MODEL CANVAS OF ONAS SUPPORTED WATER REUSE FOR GROWING FRUIT TREES IN OUARDANINE, TUNISIA

KEY PARTNERS
- Regional Offices of the Agriculture Department (CRDA)
- Water User Associations (GDA)
- Fruit market and consumers (public acceptance)
- Sewer connected households
- ANPE and DHMPE as monitoring agents

KEY ACTIVITIES
- Collection and treatment of wastewater
- Sludge drying and composting
- Maintenance of irrigation supply network (via CRDA)
- Collection of water fees and charges

KEY RESOURCES
- Treatment plant and treated wastewater
- In and outflow network
- Interested farmers in vicinity
- Land (or machinery) for sludge drying and composting

VALUE PROPOSITIONS
- To transform wastewater and sludge into safe products for reuse in peach and olive orchards

CUSTOMER RELATIONSHIPS
- Government to government contract (ONAS/CRDA)
- Automated system for tariff payment (ONAS/Households)
- Government to private sector contract (via CRDA to GDA)

CHANNELS
- Roundtable (water reuse)
- Online payment (house-hold sanitation fees)

CUSTOMER SEGMENTS
- Regional Offices of the Agriculture Department (CRDA)
- Water user association (GDA) and its farmers
- Private sector (golf resorts)

COST STRUCTURE
- Treatment plant, capital cost and operational costs including in-house monitoring of water and sludge quality
- Water conveyance to CRDA reservoir (or water user association)
- Opportunity costs of idle reuse capacity
- Customer service and fee collection

REVENUE STREAMS
- Household wastewater fees (via water bill)
- Bulk sale of wastewater (and potentially sludge)
- Governmental subsidy; carbon credits (potential)

SOCIAL & ENVIRONMENTAL COSTS
- Possible long-term risk for groundwater and food safety
- Occupational risks
- Monitoring failure can lead to loss of trust with high costs across all reuse schemes

SOCIAL & ENVIRONMENTAL BENEFITS
- Year-round water reuse resulting in less discharge into water bodies
- Nutritious fruits from wastewater irrigation
- Orchard amenity value and ecosystem functions
- Less human exposure to untreated waste
- Created a new resource and employment

Available data compiled by FAO (2014) indicate that the treatment system in Ouardanine is recovering about half of its operational costs from household fees, while the CRDA as intermediary recovers about 17% via the internal contribution of the water user associations. Other sources mention 25%. Governmental subsidies remain crucial for the remaining pumping and maintenance work. The Water User Associations (GDA) itself is in a better position to charge its members enough to break even, whereas the individual farmers make profit.

Value chain and position

There are four main 'business' segments in the value chain. These include the wastewater treatment plant; bulk sale of treated wastewater by the local agricultural authorities to the water user association; and distribution and resale of wastewater to its members. The resource recovery cycle gets closed with the irrigated fruits entering the market. Each segment also has responsibilities for the operational aspects of the transformation from wastewater to fruit. The key business activities for the wastewater treatment plant are the treatment of wastewater to obtain its environmental sustainability objectives. A secondary objective is recovering costs. The involvement of intermediaries between ONAS and farmers makes much sense as the treatment plant has neither capacity nor expertise in dealing with farmers. Water sale to farmers generates revenue for CRDA and the water user association. CRDA is in charge of water transfer (2.5 km pipeline), pumps and routine maintenance work. Farmer and traders up the value chain make net profits (Figure 212). The business activity also involves production and treatment of sludge for composting and fertilizer yet generates no revenue but saves disposal costs.

The value chain and market position could be elevated through better collection and rationalization of sewage taxes, water pricing to achieve full cost recovery, sludge sale to outside buyers (potential) and channelling a larger part of farmer revenues to investment in the maintenance of the pumps and water transport. Additional revenue from forest carbon sequestration would depend on the size of the plantation and could be explored. With households being the source of the water and recipient of the fruits, it is obvious that an important component is the compliance with health standards, i.e. the monitoring responsibilities of the involved actors.

Institutional environment

Under the Ministry of Environment and Sustainable Development, ONAS is the central authority charged with protecting water resources and for this with managing the wastewater systems from collection to treatment and disposal. ONAS is in charge of sanitation planning to operation and maintenance throughout Tunisia. However, the monitoring of wastewater treatment plants, i.e. treatment standards and discharge quality, is with the National Environmental Protection Agency (ANPE) and the Department of Hygiene and Environmental Protection (DHMPE). Several other ministries (see Box 7) are also involved in wastewater reuse. The private sector plays a role where wastewater is reused, e.g. for golf courses, which for their part, own and operate their infrastructure ensuring the transfer of treated effluents from the treatment plant to the field.[2] All investment and operational costs must be met by the golf course operators, but they do not pay for the water provided by ONAS.

From a scale perspective, water used for irrigation is managed at three levels: at the national level are the Ministry of Environment and Sustainable Development and its sanitation utility ONAS; at the regional level the 24 Regional Offices of Agriculture Development (CRDA); and at the local level the Water User Associations (GDA) whose objective is to ensure self-management of the hydraulic systems established by the state for irrigation.

Operation and maintenance costs are covered by the governmental budget as well as by the farmers. CRDA, under the supervision of Ministry of Agriculture, Hydraulic Resources and Fisheries, is the main

CASE: WASTEWATER AND BIOSOLIDS FOR FRUIT TREES

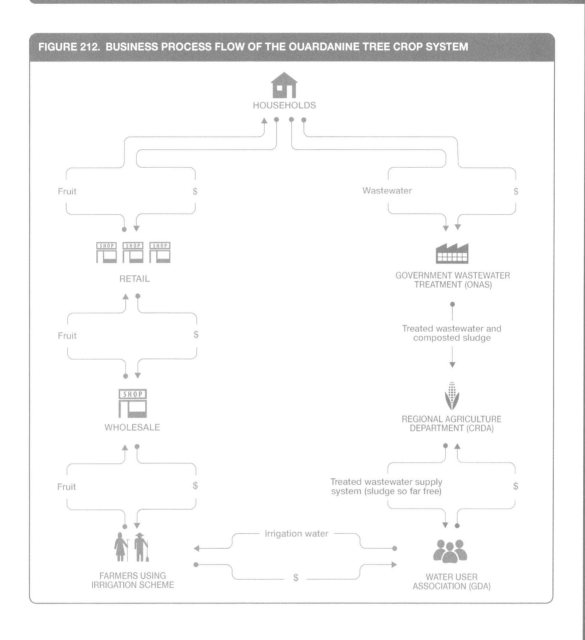

FIGURE 212. BUSINESS PROCESS FLOW OF THE OUARDANINE TREE CROP SYSTEM

responsible entities for operation and maintenance of the wastewater distribution network. The GDA are essentially associations of farmers that take responsibility for minor maintenance of the irrigation systems, as well as selling water, collecting water fees and keeping accounts. The GDA also sets fees and users' contributions to cover all the costs of running the association. This kind of regulatory structure and the devolution of management responsibility for wastewater treatment and supply to the user level has enabled significant improvements in terms of a participatory reuse planning and management.

Considering the strong Governmental support of both public entities, ONAS and CRDA in this context and the lack of financial transaction between both, it might be justifiable to say that both entities co-convene the business model.

> **Box 7. Main government bodies and institutions for wastewater treatment and reuse in agriculture in Tunisia (updated from GWI 2012)**
>
> **Ministry of Agriculture, Water Resources and Fisheries**: formulates regulations on water resources, including irrigation and water reuse in agriculture.
>
> **Regional Offices of Agriculture Development (CRDA)**: under Ministry of Agriculture, and responsible for the distribution of treated wastewater from the plants to irrigated perimeters through pumping stations and a supply network while coordinating the monitoring of water quality.
>
> **Ministry of Public Health**: sets standards for drinking water and effluent discharge to the environment with the focus on human health protection.
>
> **Department of Hygiene and Environmental Protection (DHMPE)**: a division of the Ministry of Health which controls the sewage system and purification stations as well as irrigation water to ensure compliance with public health standards.
>
> **Ministry of the Environment and Sustainable Development**: formulates regulation for environmental protection and the prevention of pollution, including effluent discharge standards and reuse standards.
>
> **National Environmental Protection Agency (ANPE)**: in charge of preventing and controlling pollution and sole body controlling direct discharge of effluents from an environmental health perspective, including monitoring of ONAS's treatment plants. Like ONAS, ANPE works under the Ministry of the Environment and Sustainable Development.
>
> **National Sanitation Utility (ONAS)**: responsible for the country's wastewater infrastructure. It collects, treats and discharges municipal (and some industrial) effluents and sells (heavily subsidised) treated wastewater for reuse.
>
> **National Water Supply and Distribution Company (SONEDE)**: Tunisia's bulk water supplier and main drinking water utility, which serves all urban areas and about half the country's rural areas.
>
> **Ministry of Tourism and Handicrafts:** supervision of societies in charge of golf courses including irrigation.

Wastewater reuse is covered under Tunisia's Water Code from 1975, which is the overarching legislation covering the water sector. Different supporting decrees define norms for chemical and biological loads in reclaimed water (based on FAO and WHO recommendations), those crops that could benefit from reclaimed water and the terms and conditions and precautions required for using reclaimed water in agriculture (such as cattle not grazing on land that has been irrigated with reclaimed water, or sprinklers not being used for the irrigation of fruit trees). Excluded from reuse are vegetables or fruits that are consumed raw. However, as exactly vegetables are a key cash crop, there is a strong call for high quality treatment to include also vegetables. As there is also a call to extend aquifer recharge with treated wastewater, Tunisia started recently revising its reuse norms to reflect quality norms for different applications (irrigation, landscaping, aquifer recharge, industrial use, etc.).

CASE: WASTEWATER AND BIOSOLIDS FOR FRUIT TREES

While effluent quality is monitored by ONAS and ANPE, an explicit risk management system is missing and enforcement of corrective measures remains limited given that one Governmental body is monitoring another, in part under the same ministry (ANPE and ONAS). Theoretically, if treated effluents fail to comply with the standard NT106.03, CRDA has to notify ONAS to turn down the treated effluents. If other water supplies are scarce (notably in summer) it is not unusual for CRDAs to accept below par treated effluents. On the other hand, it is interesting to note that sector performance seems to be motivated by the 'carrot' rather than the 'stick': the impact that poor water quality, environmental pollution and scarce water resources have on the economy and society seem to be enough to keep the sector on its toes (GWI 2012).

Sludge reuse as a fertilizer for agricultural purposes is permitted as long as it derives from urban wastewater treatment plants, i.e. not from pre-treatment by commercial and industrial facilities to remove harmful pollutants, or is recovered from cleaning of wastewater infrastructure. Sludge cannot be applied to land used for the cultivation of vegetables (GWI 2012).

Technology and processes

Reclaimed water receives in 95% of all cases secondary treatment in Tunisia, mostly via activated sludge systems. Tertiary treatment is seldom (5%) and was so far only considered in exceptional circumstances for specific uses because of its cost. Treatment technologies comprise low (56%) and medium load (30%) activated sludge plants, stabilization ponds (lagoons 14%) and in a few cases trickling filter and others systems making treated wastewater sufficiently safe for reuse as permitted in the Water Code. In general, ONAS's compliance with environmental discharge standards is with 80–90% high (GWI, 2012; Table 48).

In Ouardanine, the treatment process consists of preliminary, primary and secondary treatment (activated sludge process). The Ouardanine wastewater treatment plant supplies secondary treated wastewater to the Ouardanine peach, olive and pomegranates orchards. The plant receives raw wastewater from 3,400 households (around 17,000 citizens) and the daily treatment capacity is around 1,000 to 1,590m^3. The collected wastewater is to 91% of rural and domestic origin (residential, commercial and institutional), to 9% of industrial sources. The following Table 48 reports some key indicators of raw and reclaimed water discharged from the Ouardanine wastewater treatment plant compared to the agricultural reuse standard.

After primary treatment, the secondary step consists of an aeration tank with activated sludge in which the organic content of the sewage is digested by microorganisms. The remaining wastewater is subsequently pumped to a final clarifier which allows the sludge to settle. Parts of the secondary

TABLE 48. AVERAGE ELEMENT CONCENTRATION FOR TREATMENT PLANT INFLOW AND OUTFLOW

PARAMETER	BEFORE TREATMENT	AFTER TREATMENT	TUNISIAN STANDARD*
pH	7	8	6.5 – 8.5
Total Suspended Solids (TSS)	386	28	30 mg/L
Chemical Oxygen Demand (COD)	1,131	80	90 mg/L
Biological Oxygen Demand (BOD)	472	31	30 mg/L
Chloride	622	426	2,000 mg/L

* Tunisian Standards NT 106.03, 1989

Source: Salem et al., 2011

sludge is (usually mechanically) dried and composted to be used as a soil conditioner at rates of 6–11t/ha/year.

The irrigation scheme of Ouardanine was set in 1997 as a part of the national water reuse program and covers 65–75 ha of irrigated land and orchards cultivated by about 40–46 independent farmers. The CRDA operates the wastewater distribution network and is responsible for the organization of water quality monitoring. About 70% of the total irrigated area are cultivated with peaches and the remaining area with olives, grapes and pomegranates. A young peach tree begins fruiting by the third year after planting, which keeps the investment period much lower than for wood plantations (e.g. Egypt). Mature crop yields can reach up to 35t/ha for peaches and 7t/ha for olives.

The treated wastewater coming from the wastewater treatment plant is pumped during 20 hours/day to feed a ground reservoir with a capacity of 1,000m^3. According to Decree No 89–1047, CRDAs must test the quality of the treated effluents before using them, with regular controls from ANPE and DHMPE. The water must be tested for bacteriological load fortnightly. Tests for the water's pH, BOD, COD, TSS, chloride, sodium, ammonia, nitrogen and electrical conductivity must be carried out at least monthly. And tests for a broad range of heavy metals and other potential contaminants must be carried out at least once every six months (GWI, 2012). From the reservoir the water is then pumped to the distribution system of the orchard passing a battery of sand and gravel filters. The pumping station is not supplied with potable water. The total amount of treated wastewater used to irrigate the field is estimated at 2,300m^3/ha/year. Drip irrigation is the most frequent irrigation systems adopted in Ouardanine: about 60% of the field is irrigated by drip irrigation and the rest is irrigated using furrow irrigation. As the water still contains pathogens (Salem et al., 2011), irrigation remains restricted to certain crops. Some slight restrictions also derive from water salinity, which is moderate.

Funding and financial outlook

The funding and financial outlook for the Ouardanine wastewater treatment plant and agroforestry systems is positive due to clearly defined institutional responsibilities and opportunities for cost recovery within a regulatory framework which supports commercial reuse. Although public sector subsidies are well justified given the strong policy support and environmental and economic benefits (tourist sector) of wastewater treatment and reuse, charging for wastewater is an important step towards cost recovery. While cost recovery for water treatment and transport is still sub-optimal, there are options for improvement.

A higher cost recovery rate will be possible, e.g. via the sewage tax paid by the households, which could reflect more on the treatment costs rather than just the connecting fees. Wastewater charges paid by the farmers could also be adjusted to further support cost recovery for water conveyance. This could be at the end of the CRDA if it is allowed to revise the 1997 fixed rate of TND0.02/m^3 or at the end of the water user associations. A preliminary analysis based on the data reported in FAO (2014) shows significant scope for improving cost recovery. In general, it would be useful to learn from irrigation systems with higher than average cost recovery rate. Between different reuse schemes cost recovery can vary in wide margins (e.g. 13–63%) (Chenini et al., 2003). A third important step is to cut costs. ONAS has launched a comprehensive programme to rehabilitate and extend 19 of its treatment plants (including Ouardanine) in a bid to improve their compliance with standard NT106.02. Aside increasing the plants' capacity, ONAS plans to retrofit the plants with fine bubble aeration systems and/or biogas co-generation facilities to cut back on energy costs while improving water quality. A fourth opportunity is to start selling the treated and composted sludge to farmers and for landscaping.

The willingness to pay for treated wastewater is mostly undermined by the ability of many farmers to fall back on groundwater use, which is free of charge if found above a depth of 50m (FAO, 2009). However, extraction is increasingly unsustainable and there are options to regulate this, e.g. via pumping (electricity) charges. On the other hand, farmers' willingness to pay is increasing if water quality could allow growing vegetables which are the most appreciated cash crop (Abu-Madi et al., 2008).

Socio-economic, health and environmental impact

The Ouardanine project has eliminated raw wastewater discharges to the environment/ocean, while sustaining a strong new economic activity for a local farmer association. This is an example how investments in comprehensive wastewater collection, treatment and reuse can lead to positive impacts, locally and beyond, as marine pollution gets reduced which positively affect the overall economy given the importance of the tourist sector. About 70% of all treatment plants are located in towns and cities along the urban coast. Other positive impacts in the Ouardanine case relate to savings of freshwater, including the stressed groundwater reserves, local employment and support of economic activities and gains along the fruit value chain.

The trust in reclaimed water in Tunisia is based on comprehensive research on the possible impact of irrigation on crops, soils, groundwater and human health, which showed that in general the concentrations of almost all regulated elements in reclaimed water are below the maximum concentration recommended for agricultural reuse by the Tunisian standards (Bahri, 2002; Berglund and Claesson, 2010). However, different treatment processes show different results, with stabilization ponds performing best for microbial indicators, offering opportunities for unrestricted irrigation. Other treatment systems only support restricted irrigation unless other options (e.g. the multi-barrier approach) for pathogen reduction are put in place as promoted by WHO (2006) and IWMI (Amoah et al., 2011). Where crop choice is restricted, like in the example of citrus fruits, care has to be taken that the fruits are not in contact with the soil. To minimize risks, a non-irrigation period of 10–14 days is used in Ouardanine before crop harvest to support natural die-off (Berglund and Claesson, 2010), which is however not always easy as some fruits need regular watering. Occupational risk mitigation options and fencing against third parties will be routine measures to control related impacts. ONAS embarks also on a program to mitigate bad odor (via filtering, spraying, treatment plant coverage) and treats possible mosquito breeding grounds within its treatment premises and canals (MAERH, 2003).

Application of reclaimed water on different soils showed little modifications of their physical and chemical properties, except for a normal increase of salinity as also observed under irrigation with freshwater. However, there are regional differences in salinity level (see above) which can also be influenced by the treatment process making treated water less preferred than groundwater. Aside from salinity, also nutrient supply can be higher in reclaimed water resulting in better annual and perennial crop yields, but might also affect the balance of vegetative growth vs. fruit development. Therefore, irrigation with reclaimed water (and also sludge application) has to be considered as a complementary fertilization that has to be taken into account when calculating fertilizer application rates (Bahri, 2002; Mahjoub, 2016).

Scalability and replicability considerations

Tunisia is setting an exceptional example with higher investments in sanitation than drinking water, which is normally the opposite in the region. It is thus no surprise that it is the most advanced country in North Africa with regards to water and wastewater infrastructure, including regulation. The Ouardanine case, also small in scale, is in this context an excellent example of a decentralized treatment for reuse scheme. Key drivers for the success of the business are:

- A clear regulatory framework permitting reuse for a wide variety of seasonal and perennial crops against the payment of a water fee.
- Governmental will, financial support and inter-institutional cooperation down to water user associations.
- Early participation of the users.

In view of a more or less stagnant reuse rate, Tunisia's plans to multiply the volume of reclaimed water it uses in the years to come, targeting a 60% reuse rate, appears very ambitious, but not totally unrealistic since Tunisia is addressing head-on some of the main challenges that have delayed the development of the reuse sector until now – geographic imbalance, water tariffs, treatment quality and related reuse standards: (i) The geographic challenge is that most water needs are inland while most wastewater generation and easy disposal is along the coast. A major project for 2016–2021 is the planned transfer of treated wastewater from Tunis to the country's arid interior which will include irrigation of 25,000 ha as well as aquifer recharge of 30 million m^3 (World Bank, 2011). This builds on recharge experiments which started already in 1992. (ii) Compared to Egypt, where in the aftermath of the Arab spring authorities are thinking twice about any changes of tariffs, Tunisia used the wind of change to address chronic deficits in its national utilities by raising tariffs. The benefits of a rational increase of freshwater tariffs are threefold: first, it would make reclaimed wastewater more attractive. Second, it may help in saving water. Third, it could be used to recover part of the costs of conveyance and distribution of reclaimed wastewater. (iii) The quality challenge ONAS tries to address through the rehabilitation and extension of its treatment plants. ONAS has identified 48 plants (including Ouardanine) that it wants to equip with tertiary treatment facilities. The plants are located in areas with significant irrigation needs and the programme's objective is to produce 150 million m^3 of effluents treated at tertiary level (GWI, 2012) which would support unrestricted irrigation if the newly revised (but not yet published) reuse standards provide space for this option. For about 96% of the surveyed farmers by Abu-Madi et al. (2008) improving the quality of treated wastewater and allowing unrestricted irrigation have the power to change the negative attitudes of farmers with respect to reuse.

However, these measures might not be sufficient and attention will also be required to address other reasons for low reuse demand. Farmers complain for example about a mismatch in seasonal supply and demand which requires more investments in inter-seasonal storage facilities. Another key challenge is that compared to, for example, Jordan or Israel, many irrigators in Tunisia have more choices about which type of water to use than wastewater. In distinct contrast to, e.g. Israel and Jordan, reclaimed wastewater in Tunisia has not been mixed into reservoirs or aquifers or is by law replacing freshwater, thus many farmers can simply avoid using it, and opt for shallow groundwater which only costs pumping (FAO, 2009; Kfouri et al., 2009). To allow reuse to boom, the use of alternative water sources has to be restricted, like through higher electricity or diesel charges for pumping or aquifer protection by delineating perimeters where the quantity and quality of groundwater is compromised. Shallow groundwater accounts for 40% of groundwater use. This is now used almost exclusively for agriculture and it is being over-exploited nationally as demand exceeds supply. Faced with this situation, the government already decreed that a number of aquifers would be protected and drilling would be subjected to prior approval. The government also subsidizes water saving techniques up to 60% of the investment costs in irrigation systems when switching from traditional furrow irrigation to more water-saving methods like sprinklers or drip irrigation (Mahjoub, 2016).

Summary assessment – SWOT analysis

The strength of this business case is in its inter-departmental institutional setup with representation from the sanitation sector, environmental protection, health, agriculture and water users, a clear regulatory framework, charges for reclaimed water and promising options for increasing cost recovery aside a strong governmental will to support treatment and reuse. The regulatory framework is offering a variety of crop options although it could be extended in line with WHO (2006). Figure 213 presents the SWOT analysis of the Ouardanine case within its larger context.

FIGURE 213. SWOT ANALYSIS FOR OUARDANINE WASTEWATER AND AGROFORESTRY SYSTEM

	HELPFUL TO ACHIEVING THE OBJECTIVES	HARMFUL TO ACHIEVING THE OBJECTIVES
INTERNAL ORIGIN ATTRIBUTES OF THE ENTERPRISE	**STRENGTHS** • Sound institutional arrangements across agriculture–sanitation sectors • Multiple choices of crops and high economic benefits to farmers • Early farmer participation and sale of treated wastewater at reduced price • Reasonable O&M cost recovery with options for improvement • Continuous availability of wastewater, especially if inter-seasonal storage is supported • Investments in cutting energy costs • High compliance with safety standards	**WEAKNESSES** • Use of reclaimed water remains under potential • Tertiary treatment needed to support water acceptance and unrestricted reuse • Sludge use undervalued • No legal provision for risk management to allow unrestricted irrigation of vegetables • Monitoring of heavy metals should not be underestimated
EXTERNAL ORIGIN ATTRIBUTES OF THE ENVIRONMENT	**OPPORTUNITIES** • Political commitment with increasing private sector support • Current revision of reuse standards, planned investments in tertiary treatment, and transfer of reclaimed water to high demand areas • Willingness to increase freshwater tariffs to make reclaimed water more attractive • Opportunity to sell treated sludge • Cost recovery for reclaimed water supply would increase by restricting groundwater access	**THREATS** • Mismatch between governmental push for reuse and public perception of reuse • Monitoring failure can lead to loss of trust with high costs beyond this reuse scheme • Availability of alternative water sources with less risk (seasonal variation in quantity and quality, public perception)

Contributors

Abdelkarim Ben Ticha, Regional Director, ONAS Monastir
Fathi Chatti, CRDA
Mohamed Mkada, GDA
Imed Chakroun, Mechanical Engineer, CRDA
Hacib Amami, National Research Institute for Rural Engineering, Water and Forestry, INRGREF, Tunisia
Rafet Ataoui, University of Basilicata (UNIBAS), Italy
Simone Targetti, University of Basilicata (UNIBAS), Italy
Achim Kress, FAO
Michele Baldasso, FAO
Javier Mateo-Sagasta, IWMI

References and further readings

Abid, N. 2010. Sanitation sector in Tunisia. Follow-up Conference of the International Year Sanitation; Tokyo 26–27 January, 2010.

Abu-Madi, M., Al-Sa'ed, R., Braadbaart, O. and Alaerts, G. 2008. Perceptions of farmers and public towards irrigation with reclaimed wastewater in Jordan and Tunisia. Arab Water Council Journal, Volume 1 (II): 18–32.

Amoah, P., Keraita, B., Akple, M., Drechsel, P., Abaidoo, R.C. and Konradsen, F. 2011. Low cost options for health risk reduction where crops are irrigated with polluted water in West Africa. IWMI Research Report 141. Colombo.

Bahri, A. 2002. Water reuse in Tunisia: Stakes and prospects. Marlet, S. and Ruelle, P (eds). Montpellier: CEMAGREF, CIRAD, IRD, Cédérom du CIRAD. (Vers une maîtrise des impacts environnementaux de l'irrigation. Actes de l'atelier du PCSI, 28–29 Mai 2002). https://hal.inria.fr/file/index/docid/180335/filename/Bahri.pdf.

Bazza, M. and Ahmad, M. 2002. A comparative assessment of links between irrigation water pricing and irrigation performance in the Near East. In: Irrigation water policies: Micro and macro considerations. Agadir, Morocco: FAO. ftp://ftp.fao.org/docrep/fao/008/ag006e/ag006e00.pdf.

Berglund, K. and Claesson, H. 2010. A risk assessment of reusing wastewater on agricultural soils – A case study on heavy metal contamination of peach trees in Ouardanine, Tunisia. Report TVVR - 10/5004. Lund University, Sweden: Department of Water Resources Engineering.

Chenini, F., Huibers, F.P., Agodzo, S.K., van Lier, J.B. and Duran, A. 2003. Use of wastewater in irrigated agriculture. Country studies from Bolivia, Ghana and Tunisia. Volume 3: Tunisia. Wageningen University: WUR, 2003. (W4F - Wastewater).

FAO. 2009. Groundwater management in Tunisia. Draft Synthesis Report. Rome: Groundwater Governance. www.groundwatergovernance.org/fileadmin/user_upload/groundwatergovernance/docs/Country_studies/Tunisia_Synthesis_Report_Final_Groundwater_Management.pdf.

FAO. 2014. Cost recovery of wastewater use in forestry and agroforestry systems: Case studies from Egypt, Tunisia and Algeria. Unpublished final report completed for IWMI, November 2014.

GWI. 2010. Municipal water reuse markets 2010. Oxford, UK: Global Water Intelligence. Media Analytics Ltd.

GWI. 2012. Global water and wastewater quality regulations 2012: The essential guide to compliance and developing trends. Oxford, UK: Global Water Intelligence.

Kfouri, C., Mantovanic, P. and Jeuland, M. 2009. Water reuse in the MNA region: Constraints, experiences, and policy recommendations. In: Water in the Arab world. Management perspectives and innovations. Middle East and North Africa Region. The World Bank.

MAERH. 2003. National report: The state of the environment 2003. Tunis: Ministry of Agriculture, the Environment and Water Resources.

Mahjoub, O. 2016. Good practices in wastewater reuse for agricultural irrigation. Case study of Ouardanine, Tunisia. Paper presented at the UNU-FLORES co-organized Scientific and Capacity Development Workshop on Safe Use of Wastewater in Agriculture in Tehran, Iran, 5–7 Dec 2016.

ONAS. 2009. Annual report. Tunis: Ministry of Environment and Sustainable Development.

Qadir, M., Bahri, A., Sato, T. and Al-Karadsheh, E. 2010. Wastewater production, treatment, and irrigation in Middle East and North Africa. Irrigation Drainage Systems 24: 37–51.

Salem, I.B., Ouardani, I., Hassine, M. and Aouni, M. 2011. Bacteriological and physico-chemical assessment of wastewater in different region of Tunisia: Impact on human health. BMC Research Notes 4: 144 (11 pp).

WHO. 2006. Guidelines for the safe use of wastewater, excreta and greywater, volume 2: Wastewater use in agriculture. Geneva: World Health Organization.

World Bank. 2011. Water reuse in the Arab world: From principle to practice – Voices from the field. A Summary of proceedings. Expert Consultation – Wastewater Management in the Arab World. 22–24 May 2011, Dubai-UAE, Arab Water Council and World Bank.

Case descriptions are based on primary and secondary data provided by case operators, insiders or other stakeholders, and reflect our best knowledge at the time of the assessments 2014/16. As business operations are dynamic data can be subject to change.

Notes

1 In 2007 average public irrigation costs with freshwater were USD 0.097/m^3 and the average tariff applied was USD 0.084/m^3 – a national average cost recovery rate of 87%. Total cost recovery, however, based on infrastructure and operating cost remains low at 25% (FAO, 2009).
2 As at end of June 2009, the private sector operated 2,206 km of sewers and 17 wastewater treatment plants. It is also worth mentioning that the new regulations stipulate the adoption of concession contracts that can extend the contracts up to 30 years (ONAS, 2009).

CASE

Suburban wastewater treatment designed for reuse and replication (Morocco)

George K. Danso, Munir A. Hanjra and Pay Drechsel

Supporting case for Business Model 17	
Location:	Drarga, suburban Agadir; Morocco
Waste input type:	Domestic wastewater from Drarga town
Value offer:	Tertiary treated wastewater for irrigation, with capacity to produce organic fertilizer, reed grass and energy
Organization type:	Public
Status of organization:	Plant operations started in 2000; the reuse operation inaugurated in 2001
Scale of businesses:	Small with up to 1,800 to 2,700m^3/day (design capacity 600–1,000m^3/day) and 6–16 irrigated ha
Major partners:	National electricity and water company (ONEE); Ministry of Energy, Mining, Water and Environment, Drarga town and Prefecture of Agadir, Al Amal water users association, local farmers

Executive summary

The wastewater treatment plant in the town of Drarga (ca. 17,000 inhabitants in 2004) has attracted international attention as an example of (i) an applied low-cost technology designed and managed in close consultation with local stakeholders; (ii) a system able to support local resource recovery demands for revenue generation; (iii) a system with marketing plan for replication across towns and suburbs; and with (iv) a dedicated accounting system to support full financial cost recovery. The treatment technology involves screening, anaerobic basins, denitrification, a water recirculating sand filter system and reed beds. The effluent meets the World Health Organization standards for unrestricted use in irrigation. The RRR options the plant offers are internal energy recovery, and the possible sale of tertiary treated water, reed, and sludge based co-compost. Although the demand for resource recovery remained optional, the Drarga plant achieved its objective of operational cost recovery while eliminating soil and aquifer pollution from raw sewage. Controlled trials verified that farmers using the water could save significantly on pumping and fertilizer costs while gaining higher yields and profits. However, in 2004, the plant's operations were centralized under ONEP (now ONEE)[1] which deemphasized the exploration of resource recovery and reuse as revenue streams. This might change again as the use of treated wastewater is strongly supported in Morocco due to scarcity of water resources and recurring droughts.

KEY PERFORMANCE INDICATORS (AS OF 2012)					
Land use:	Plant: 2 ha; up to 16 ha irrigated (under potential)				
Water treated:	1,800 to 2,700m^3 per day (design capacity 600–1,000m^3 per day)				
Capital investment:	Total investment USD 1.7 million				
Labor:	About 5; ca. 27% of the O&M costs				
O&M cost:	USD 2,300 to 3,600 per month				
Output:	Tertiary treated wastewater				
Potential social and/or environmental impact:	As there was no treatment before, inhabitants in Drarga gained most of all from health risks reduction and an improved living environment				
Financial viability indicators:	Payback period:	N.A.	Post-tax IRR:	N.A.	Gross margin:	N.A.

Context and background

Morocco is facing severe water shortage with less than 800m^3 water per capita. Frequent and recurring droughts, rising demand for water, and pollution of freshwater threaten water security in Morocco, also affecting the tourist sector like in the Agadir region where Drarga is located. The Drarga treatment plant was constructed as one component of the Morocco Water Resources Sustainability (WRS) project (1996–2003) co-funded by the Moroccan Government (Ministry of Environment[2]) and the United States Agency for International Development (USAID). The area around Drarga is semi-arid with annually 236mm of rain in the winter months. Agriculture around the town depends on the limited water resources from the Souss-Massa River Basin (SRB). The Souss river is most of the year dry, and the aquifers in the region which are already to 95% supporting agriculture cannot cope with further withdrawal, making treated wastewater a promising alternative.

Today, many Moroccan towns have sewer systems, and the number of (functional) treatment facilities is on the increase. Also in Drarga, about 80% of local households are connected to a sewage system. However, before the treatment plant was constructed, Drarga's raw sewage was discharged through four outfalls into the environment, contaminating the aquifer and creating unhealthy sanitary conditions. This uncontrolled release of wastewater into the environment is still a common situation in many smaller and larger towns in and outside Morocco.

The Drarga treatment plant was inspired by a similar technological setup piloted in the 10-km distant Ben Sargeo and designed in consultation and partnership with the local community in Drarga and institutional stakeholders across administrative scales using a participatory approach. The feasibility study analyzed various options for selection of the site and of the technology for the plant, a detailed financial and economic analysis based on different water reuse scenarios, following an assessment of the community's willingness to accept crops irrigated with treated wastewater (EAU, 2004).

Market environment

For the Drarga wastewater treatment plant's O&M costs to break even, it is essential to combine low operational costs and sufficiently high revenue streams. Aside the sewage fee paid by households, the sale of treated wastewater to farmers is one of the most prominent design revenue streams. A wastewater tariff of USD 0.05/m^3 was suggested which is half the fresh water price. However, although the initial feasibility studies confirmed consumers' acceptance of the concept, the study fell short to predict farmers' refusal to pay for the treated water, arguing that the water will anyway be released after treatment (Dadi, 2010). Given the sufficiently high revenues through the water bill (see "Funding and financial outlook" below), the market for compost or reed was not explored by the operators.

Finding an acceptable and competitive price for wastewater compared to the freshwater tariff and aquifer pumping is a common challenge across the MENA region. In some regions where the level of the groundwater has witnessed a considerable decrease, like around Agadir, the pumping costs have however become very expensive (up to USD 0.14/m^3) which is increasing the financial competitiveness of the treated wastewater, but not improve its stigma, even with tertiary treatment as other farmers reported who were concerned about their image in view of crop exports (Dadi, 2010; Salama et al., 2014).

Macro-economic environment

Although Morocco is a water-scarce country, 46% of the active population works in the agricultural sector (80% in rural areas) contributing 14% of the gross domestic product (GDP). The agricultural sector's exposure to water stress and climate variability causes fluctuations in its economic contribution: its share of GDP ranges from 11% in water-scarce years to over 20% in years when the climate is favorable (Houdret, 2012). Recent estimates indicate an average water availability of around 730m^3 per person per year, which is significantly lower than the often cited 1,000m benchmark[3], and might further decline to 580m^3 by 2020, which poses a significant challenge to the government. In addition, the quality of water resources is deteriorating at an alarming rate as only 25% of the collected wastewater is actually treated (Hirich and Choukr-Allah, 2013). In an attempt to rectify these problems, the Government of Morocco is heavily investing in new treatment plants and recommends to make use of non-conventional water sources such as treated wastewater for extending irrigated areas, exploiting arid lands, improving public health, controlling environment pollution and managing the quality of water resources at the level of hydrographic basins (Salama et al., 2014). The *Liquid Sanitation and National Wastewater Treatment Programme* (2005), the *Green Morocco Plan* (2008) and the *National Water Strategy* (2010) support the agricultural reuse of treated wastewater (Salama et al., 2014). As of 2011, only 13% of the 32.38 million cubic meters (MCM) of treated wastewater was reused in agriculture across the country, a share which is expected to reach 50% by 2020 (MEMEE, 2013)[4]. The Drarga plant offers in this context the double value proposition of safe treatment and water for reuse at a favorable benefit–cost ratio which is tailored to smaller towns and suburbs targeting agricultural water reuse.

Business model

This is a cost-recovery business model which combines low investment and running costs with multiple cost recovery options supported by a special account to manage costs and revenues of the plant. Aside the use of household fees for the sewer connection, the plant can generate parts of its energy needs and obtain revenues from selling reed grass, highly treated wastewater and organic fertilizer to farmers, depending on demand; see Figure 214 on the following page.

Value chain and position

The plant treats wastewater from the Drarga commune against a fee (charge with the water bill) and sells depending on demand tertiary treated water to farmers which is of increasing interest where groundwater availability and pumping costs become challenges. Farmers benefit through guaranteed all-year access to low-priced water, and savings on fertilizer. A number of local field trials showed that farmers can gain through the use of the treated wastewater between USD 80 and more than USD 500 per ha with variations between crops (EAU, 2004; Choukr-Allah et al., 2005; Choukr-Allah and Hamdy, 2005; Mohamed and Young, 2013). Common crops in the areas, irrigated via surface, micro jet or drip systems are for example wheat, maize, tomatoes, zucchini, alfalfa and clover. With on average doubled yields using wastewater compared to irrigation with other water sources (Hirich and Choukr-Allah, 2013), price advantages could also be extended to consumers (Figure 215).

The volumes of actually realized water reuse and irrigated hectares vary between sources. The volume of treated effluent increased from 170m^3 per day (in the year 2000) to 400m^3 in 2010, irrigating initially

CASE: SUBURBAN WASTEWATER TREATMENT FOR REUSE

FIGURE 214. BUSINESS MODEL CANVAS – DRARGA DESIGN FOR REUSE

KEY PARTNERS	KEY ACTIVITIES	VALUE PROPOSITIONS	CUSTOMER RELATIONSHIPS	CUSTOMER SEGMENTS
• Former 'Water Resources Sustainability (WRS) project' • Ministry of Environment • Al Amal Water User Association • ONEP (initially as partner, later as operator and owner) • Town and provincial authorities	• Wastewater collection and treatment • Provision of treated water to potential users • Possible supply of reed grass and organic fertilizer	• Operational cost recovery, enhanced public and environmental health, combined with optional water, nutrient, biomass and energy recovery and reuse	• Indirect to farmers via water user association • Computer based household billing • Direct sale (reed) • Collective agreement (contract)	• Connected households in Drarga • Farmers • Buyers of reed grass
	KEY RESOURCES		CHANNELS	
	• Treatment plant (capital investment), wastewater • Dedicated bank account • Partnership across scales • Expertise (ONEP)		• Water User Association • Water bill • On-site sale (reed)	

COST STRUCTURE	REVENUE STREAMS
• Original capital investment • O&M • Opportunity costs (missed RRR revenues)	• Household sewage fees • Sale of wastewater, reed, co-compost (optional) • Energy recovery (optional)

SOCIAL & ENVIRONMENTAL COSTS	SOCIAL & ENVIRONMENTAL BENEFITS
• Potential health costs for workers and consumers of irrigated crops if safety guidelines are ignored or plant performs below standard • Potential environmental harm through plant flooding (wadi proximity) • Potential of reduced farm income based on non-acceptance of wastewater use for food production	• Much cleaner environment supporting town growth • Likely reduction of human health-related costs due to reduction of pollution of streams and the aquifer • Increased water and fertilizer savings if treated wastewater gets accepted, resulting in benefits for farmers • Reduction in energy use (optional)

an area of about 6 ha of crops and about 2.5 ha of green spaces from 2005 on. However, these were mostly demonstration farms for accompanying research. Hirich and Choukr-Allah (2013) mention with reference to data from 2003 an area of about 16 ha under wastewater irrigation. From 2010 onwards, some water was also routed to neighboring crops under greenhouses. However, the actual reuse remained far below its potential (see overleaf).

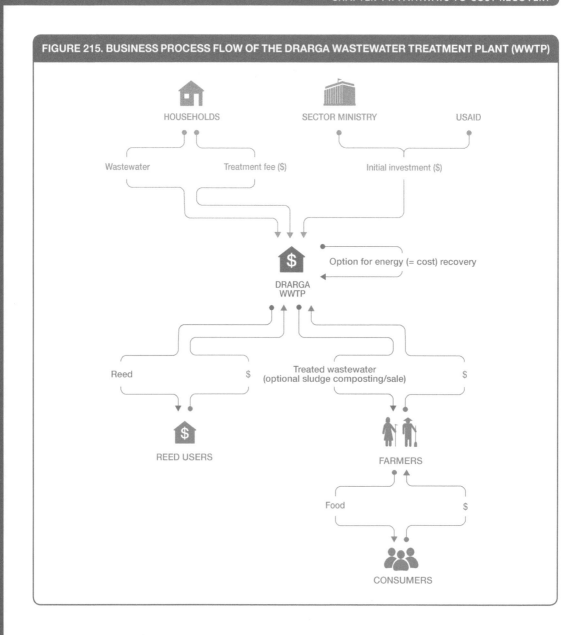

FIGURE 215. BUSINESS PROCESS FLOW OF THE DRARGA WASTEWATER TREATMENT PLANT (WWTP)

Institutional environment

The original set-up of the plant was based on dialog and effective institutional partnerships with key stakeholders from the town and the Agadir region. Each stakeholder was responsible for some aspect of project implementation: the commune of Drarga provided the land for construction and initially owned the wastewater treatment facility; the Province of Agadir facilitated administrative procedures; the Al Amal Water User Association managed the plant; the Regional Agency for Planning and Construction (ERAC-Sud) financed construction of the main sewage collector, etc. This partnership was sealed through a collective agreement signed in 1998, under the patronage of the Ministry of Environment. A steering committee of stakeholders from different sectors supported the implementation process, and a technical oversight was formed to oversee the plant's operation, as well as the quality of agricultural products irrigated with treated effluents. The committee has the authority to stop the delivery of treated water to farmers if monitoring analysis shows that the water fails to meet adequate reuse standards. The last institutional set up for the business was the establishment of an association of treated wastewater users in charge of maintaining the irrigation network, collecting fees, and distributing treated wastewater to its members. The project also assisted the Ministry of Environment to develop norms and standards for wastewater reuse, thus helped to support the enabling environment for replication and out-scaling (EAU, 2004).

Because of the limited financial and technical capacities across many smaller municipalities their wastewater treatment plants ceased over the years functioning which triggered a Governmental decision to gradually transfer from 2000 on the responsibility for sanitation in small- and medium-sized towns to the National Potable Water Agency (ONEP) whose mandate was amended to include sanitation (sewerage and wastewater treatment). According to Dadi (2010), lack of capacity was also a risk factor in the original setup of Drarga's WWTP, and in 2004, its ownership and operations were transferred to ONEP. ONEP was already involved in the Drarga project, including presiding over its technical committee. The commune of Drarga then requested ONEP to take over the management of the plant. This was a natural transition as also for replicating the Drarga model, the Drarga marketing plan had already recommended that ONEP becomes the "facilitator" or "dealmaker" (EAU, 2002). On the other hand, since agricultural water reuse is not within ONEP's mandate, interactions with farmers decreased, and so also efforts in the other resource recovery options. In September 2011, the National Electricity and Water Company (Office National de l'Electricité et de l'Eau Potable; ONEE), was created, with ONEP becoming its "water branch".

Technology and processes

The plant provides advanced wastewater treatment with limited energy demands. After initial wastewater screening and grit removal, the wastewater is treated in two 918m^3 anaerobic basins with an average hydraulic retention time (HRT) of about three days. The flow is then sent to two 736m^3 denitrification ponds (HRT of 2.4 days) and finally to ten recirculating sand filters, each with a surface of 893m^2. After passing again the denitrification ponds, the effluent is treated in two 2,900m^2 planted wetlands (reed beds) before being assembled in storage basins (Young et al., 2011). The treated wastewater meets the WHO standards for unrestricted irrigation. When required, the water is pumped to irrigate farms, or drained into a local wadi. No chemicals or mechanical equipment are used in the treatment process; however, all equipment parts like valves and pumps were imported from USA which could make local replacement difficult (Dadi, 2010).

The Drarga plant was designed for the production of co-compost and energy: the residual sludge from the anaerobic basins can be pumped, dried (on three drying beds of 337m^2 each), and combined with organic wastes from the town to produce compost. Also the biogas from the anaerobic basins could be captured and converted into energy to run the pumps at the plant, thereby reducing its electricity costs (Figure 216). While the station started to collect methane gas in the anaerobic stage, and a generator was put in place, electricity generation was not realized (Dadi, 2010; Mohamed and Young, 2013). The generated sludge has been sent for drying beds and disposed on the local landfill without any reuse (100–120m^3 annually). Actual flow to the facility has been much higher (1,800–2,700m^3/day) than originally thought (600–1,000m^3/day). However, the influent has been more dilute than the plant was designed for and the plant continued to perform as expected (Young et al., 2011; Dadi, 2010).

Funding and financial outlook

According to the project, total investment in the Drarga wastewater treatment plant in 2001 was about USD 1.7 million with the equipment and construction constituting about 70% of total investment cost. Given the technology chosen, the annual O&M costs were estimated at USD 22,000 to 30,000 with

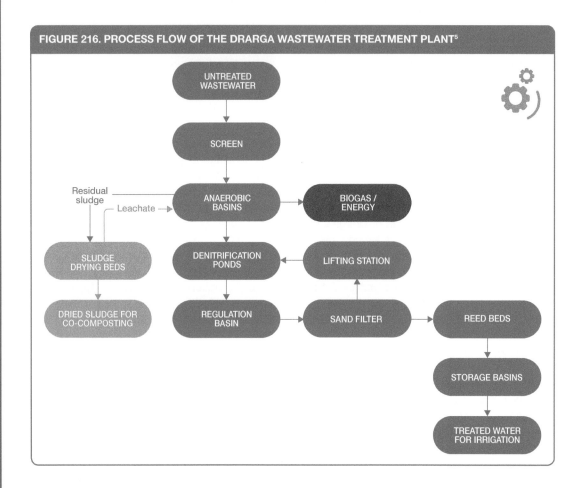

FIGURE 216. PROCESS FLOW OF THE DRARGA WASTEWATER TREATMENT PLANT[5]

electricity taking approximately 23% (Table 49). Considering also the data presented by Mohamed and Young (2013) a different assessment puts the annual O&M cost at USD 28,000 to 43,000.

Based on a treatment of 1,800m³ per day (Young et al., 2011), the operator pays around USD 0.06/m³. Households are provided with potable water consumption meters. At the end of each month, citizens pay an invoice that shows both potable water consumption and associated wastewater costs. The wastewater fee is about 20% of the water fee. The two lowest wastewater tariffs in 2010 were Morocaan dirham (DH) 0.51 and 1.28/m³ and are the most common charged (Dadi, 2010). Using an average tariff of DH0.9 (USD 0.11/m³ in 2010) the operator generates more revenues via the wastewater tariff than it has operational costs, even if not all households pay or an administrative overhead will be deducted.

Farmers were initially charged USD 0.05/m³ for the treated wastewater. The other revenue streams which are included in the design, i.e. the one-time household sewerage connection fee, revenues from reed and compost sales, plus cost saving from internal energy production show the potential of this type of plant to achieve cost recovery even if some of the revenue stream did never crystallize (USEPA and USAID 2004). Before the plant's finance became part of ONEP's operations in 2004, the running costs appeared to be fully covered (Table 50).

Based on this simple accounting system, the combined revenues from the plant were at least initially deposited into a **special wastewater treatment plant account** that is independent of the city's

TABLE 49. ESTIMATED COSTS OF THE WASTEWATER TREATMENT PLANT IN DRARGA, MOROCCO

ITEMS	COST IN 1,000 DH	COST (%)
(a) Investment costs		
Research and Feasibility Study	3,000	14.8
Equipment	6,900	34.0
Construction	7,600	37.4
Monitoring	1,800	8.9
Reuse component	1,000	4.9
Sub-total	20,300	100
(b) Annual Operating Costs		
Electricity	60	23.1
Salaries	70	26.9
Laboratory Analysis	80	30.8
Miscellaneous	50	19.2
Sub-total	260	100

Note: USD 1= DH 11.4 in Dec 2001, and DH 8.4 in Nov 2004

Source: EAU, 2004.

TABLE 50. OPERATIONAL COSTS AND ANNUAL REVENUES AT DRARGA

USD	2001 (SECOND HALF)	2002	2003
Total income	49,820	59,760	61,560
Total expenses	18,250	28,180	43,180
Balance (net income)	31,570	31,580	18,380

Source: Mohamed and Young, 2013.

general budget. This dedicated account was further divided into two parts: the first part deals with **current account expenses** and the second part deals with the **extension and renewal account** in which money is saved to pay for the replacement of equipment and any future expansion of the wastewater treatment plant. This special arrangement was a response to common bottlenecks in public financing of O&M costs which contributed among other factors to the breakdown of about 70% of the wastewater treatment plants in the country (Choukr-Allah et al., 2005).

It is unfortunate that the full potential of the plant as a regional demonstration project for RRR remains to be verified (Dadi, 2010): neither the reed harvest and sale, sludge composting nor the biogas production took off. However, depending on local demand, all these options could be activated without any major additional investment. The main material to be recovered was water, and farmers' reservations showed a clear gap in the feasibility study. Especially to farmers whose products are exported to foreign markets, even "treated wastewater" still appeared 'unclean' and not good for their business, while raw wastewater that went through the ground or river before reaching the farm appeared acceptable for use even if this water is highly polluted (Aomar and Abdelmjid, 2002). Other farmers were unwilling to pay for wastewater that will anyway be discharged into the environment after treatment. Both factors undermined the generation of revenue from irrigation.[6]

Socio-economic, health and environmental impact

By eliminating the discharge of raw wastewater, the plant has significantly improved environmental and living conditions in Drarga, and reduced potential risks to aquifers and human health. Especially high nitrogen levels entering the groundwater were of concern given the sandy nature of the soils. Construction of the plant has improved the living standards and value of local communities by eliminating problems associated with foul odors and mosquitoes. It has also supported water savings by promoting drip irrigation. Results from agricultural demonstration plots showed that the additional benefits for farmers (in particular savings on fertilizer) and the environment can be significant and could easily cover the irrigation water fee (EAU, 2004; Dadi, 2010). A note of caution refers (i) to the manual raking of the sand beds, which can pose an exposure risk to the employees; and (ii) the location of the plant next to a Wadi that is dry all year round but fills up to high levels during the rainy period and could potentially flood the plant (Dadi, 2010).

Scalability and replicability considerations

The Drarga wastewater treatment plant was designed as a demonstration plant for replication in small towns, with a strong emphasis on RRR and financial sustainability. Its planning and setup was based on strong stakeholder participation and included a dedicated self-marketing strategy for national replication of the model under the facilitation of ONEP (EAU, 2002). The strategy included demonstrations, also of financial viability, capacity development as well as various communication components. Key drivers which supported calls for replication were:

- High treatment standard based on applied technologies with a favorable cost effectiveness.
- Multiple opportunities to achieve O&M cost recovery.
- High environmental and social benefits.

FIGURE 217. SWOT ANALYSIS OF DRARGA BUSINESS CASE, MOROCCO

	HELPFUL TO ACHIEVING THE OBJECTIVES	HARMFUL TO ACHIEVING THE OBJECTIVES
INTERNAL ORIGIN ATTRIBUTES OF THE ENTERPRISE	**STRENGTHS** • Full O&M cost recovery based on: 　• Low cost treatment 　• Multiple cost recovery mechanisms • Participatory planning and operation • Special account for costs and revenues • Visible environmental benefits • Significant profit opportunities for farmers using wastewater	**WEAKNESSES** • Insufficient awareness creation to address farmers' unwillingness to pay • Limited local expertise in running of the plant and its RRR options • Centralized management de-emphasizing RRR given cost recovery via the water bill
EXTERNAL ORIGIN ATTRIBUTES OF THE ENVIRONMENT	**OPPORTUNITIES** • Awareness creation for the reuse of tertiary treated water for the creation of financial and economic value • High potential for (so far neglected) energy recovery and organic fertilizer production • Development of reuse policies based on the example • Indirectly charging farmers for the treatment via the land rent in locations where farmers would not accept paying for the treated water	**THREATS** • Backlash from consumers and importers of wastewater irrigated crops • Replication requires towns with sewer systems • National promotion of water reuse remains lip service without awareness and demand creation

The business case demonstrated that high cost recovery could be achieved where demand allows to capitalize on the different revenue streams the plant offers. While the environmental and social benefits of the plant were fully achieved, it might require more water stress or higher (pumping) electricity prices to see water reuse and energy recovery going to scale.

Summary assessment – SWOT analysis

The inspirational setup of the Drarga plant was featured in the 2004 *US EPA - USAID Guidelines for Water Reuse*. The close stakeholder involvement and joint design with water users has been praised and can be considered an excellent example of "Design for Reuse" as demanded by Murray and Buckley (2010). Although resource recovery faced in Drarga some challenges, the case main strength remains the combination of low setup and operational costs with multiple options for operational cost recovery. The challenges Drarga is facing are common also in other regions, which again makes it a good example. Farmers asked, like also in Pakistan's Faisalabad, why to pay for a product which will anyway be released. Others feared less (export) demand for their produce, based on the term 'wastewater' while highly polluted stream water would be without this terminology stigma.

The local demand for compost, reed, as well as plant-internal electricity generation remained underexplored, partly related to the change in plant ownership and operation which resulted in less emphasis in the demonstration of alternative cost recovery options via RRR. Figure 217 illustrates the full SWOT analysis of this business case based on the available information.

Contributors

Prof. Dr. Redouane Choukr-Allah, Institut Agronomique et Vétérinaire Hassan II, Agadir

References and further readings

Aomar, J. and Abdelmjid, K. 2002. Wastewater reuse in Morocco. Paper presented at the Water Demand Management Forum. Ministry of Agriculture, Rural Development and Waters and Forests. www.hitpages.com/doc/4664231974666240/1.

Bahri, A. 2008. Water reuse in Middle Eastern and North African countries. In: Water reuse: An international survey of current practice, issues and needs. Jimenez, B. and Asano, T. (eds). ISBN: 9781843390893. London, UK: IWA Publishing. Scientific and Technical Report. No. 20. pp. 27–47.

Choukr-Allah, R. and Hamdy, A. 2005 Best management practices for sustainable reuse of treated wastewater. In: Hamdy A. et al. (eds.), Non-conventional water use: WASAMED project. Bari: CIHEAM / EU DG Research, pp. 191–200 (Options Méditerranéennes: Série B. Etudes et Recherches; n. 53).

Choukr-Allah, R., Thor, A. and Young, P.E. 2005. Domestic wastewater treatment and agricultural reuse in Drarga, Morocco. In: Hamdy A. et al. (eds). Non-conventional water use: WASAMED project. Bari: CIHEAM /EU DG Research, pp. 147–155 (Options Méditerranéennes: Série A. Séminaires Méditerranéens; n. 66).

Dadi, E.-M. 2010. L'évaluation de la possibilité de réutiliser en agriculture l'effluent traité de la commune de Drarga. Master thesis at Université de Sherbrooke. http://savoirs.usherbrooke.ca/handle/11143/7134 (accessed 5 Nov. 2017).

Environmental Alternatives Unlimited (EAU). 2002. Water resources sustainability project (WRS). Marketing strategy for Drarga wastewater and reuse pilot project. Report for USAID. Contract No. 608-0222-C-00-6007-00. http://pdf.usaid.gov/pdf_docs/PNACW910.pdf (accessed 5 Nov. 2017).

Environmental Alternatives Unlimited (EAU). 2004. Morocco water resources sustainability project final report. www.ircwash.org/sites/default/files/Morocco-2004-Morocco.pdf (accessed 5 Nov. 2017).

Hirich, A. and Choukr-Allah, R. 2013. Wastewater reuse in the Mediterranean region: Case of Morocco. Daniel Thevenot (ed). 13th edition of the World Wide Workshop for Young Environmental Scientists (WWW-YES-2013) – Urban waters: Resource or risks? Jun 2013, Arcueil, France. https://hal.archives-ouvertes.fr/hal-00843370/document (accessed 5 Nov. 2017).

Houdret, A. 2012. The water connection: Irrigation, Water grabbing and politics in southern Morocco. Water Alternatives 5 (2): 284–303.

MEMEE. 2013. Stratégie Nationale de l'Eau en matière de protection et de développement de l'offre: Réutilisation des eaux usées épurées. Ministere de l'energie, des mines, de l'eau et de l'environnement. www.agire-maroc.org/fileadmin/user_files/2013-02-04-pnarr-rabat/1/2013-02-04-Strategie-Nationale-Eau-Makhokh-Jaoucher.pdf (or https://goo.gl/96tWf1; accessed 5 Nov. 2017).

Mohamed, H.O. and Young, R. 2013. Environmental and economic impacts of using effluent of Drarga WTP for irrigating potato. International Journal of Chemical & Environmental Engineering 4 (1): 75–80.

Murray, A. and Buckley, C. 2010. Designing reuse-oriented sanitation infrastructure: The design for service planning approach. In: Drechsel, P., Scott, C.A., Raschid-Sally, L., Redwood, M. and Bahri, A. (eds) Wastewater Irrigation and health: Assessing and mitigation risks in low-income countries. UK: Earthscan-IDRC-IWMI. pp. 303–318.

Rijsberman, F.R. 2006. Water Scarcity: Fact or Fiction? Agricultural Water Management 80 (1–3): 5–22.

Salama, Y., Chennaoui, M., Sylla, A., Mountadar, M., Rihani, M. and Assobhei, O. 2014. Review of wastewater treatment and reuse in the Morocco: Aspects and perspectives. International Journal of Environment and Pollution Research 2 (1): 9–25. www.eajournals.org/wp-content/uploads/Review-of-Wastewater-Treatment-and-Reuse-in-the-Morocco-Aspects-and-Perspectives.pdf (accessed 5 Nov. 2017).

US EPA/USAID. 2004. Guidelines for water reuse. Washington, DC: U.S. Environmental Protection Agency (EPA/625/R-04/108).

Van der Hoek, W., Ul Hassan, M., Ensink, J.H.J., Feenstra, S., Raschid-Sally, L., Munir, S., Aslam, R., Ali, N., Hussain, R. and Matsuno, Y. 2002. Urban wastewater: A valuable resource for agriculture. A case study from Haroonabad, Pakistan. Research Report 63. Colombo, Sri Lanka: IWMI.

Young, T., Copithorn, R., Karam, J. and Abu-Rayyan, O. 2011. Reliable low technology for pollution control in semi-rural Morocco and Jordan. Water (Journal of the Australian Water Association) Dec 2011: 77–81.

Case descriptions are based on primary and secondary data provided by case operators, insiders or other stakeholders, and reflect our best knowledge at the time of the assessments 2014/16. As business operations are dynamic, data might change.

Notes

1. In 2011, the National Potable Water Agency (ONEP) was regrouped with the National Electric Utility (ONE) to become the National Electricity and Water Company (Office National de l'Electricité et de l'Eau Potable or ONEE).
2. Since 2002, Ministry of Environment and Water, and since 2007, Ministry of Energy, Mining, Water and Environment.
3. A renewable water supply below 1,000m^3 per capita per year has been suggested as a threshold for water sarcity, based on estimates of water requirements in the household, agricultural, industrial and energy sectors, and the needs of the environment (Rijsberman, 2006).
4. The majority of treated wastewater is used on golf courses (66%) and for industrial reuse (20%). About one percent supports groundwater recharge. While the area under wastewater irrigated farming varies between sources (550 ha–max. 2,000 ha), there are estimates of additional 6,000–7,000 ha under informal irrigation with untreated wastewater (Bahri, 2008; MEMEE, 2013; Salama et al., 2014; www.fao.org/nr/aquastat/).
5. Energy and biomass (compost) recovery optional and not realized so far.
6. An opportunity to charge farmers indirectly for the treated water could be through owing and renting out farmland along the effluent channel or stream. As shown in Pakistan, the availability of wastewater can significantly increase the land value even above the one next to freshwater canals (van der Hoek et al., 2002).

BUSINESS MODEL 17
Wastewater for greening the desert

Pay Drechsel and Munir A. Hanjra

Key characteristics

Model name	Wastewater for greening the desert
Waste stream	Domestic wastewater from decentralized sewer systems
Value-added waste products	Treated wastewater and sludge (biosolids); wood and other tree products
Geography	Arid and semi-arid regions (e.g. MENA)
Scale of production	Small to medium (300 to 30,000m^3/day reused)
Supporting cases in this book	Cairo, Egypt; Ouardanine, Tunisia; Drarga, Morocco
Objective of entity	Cost-recovery [X]; For profit [ ]; Social/environmental enterprise [X]
Investment cost range	Treatment plants: up to USD 50 million Agroforestry system: USD 300,000 to 1.6 million
Organization type	Public or public-private or for the reuse component also only private
Socio-economic impact	Green infrastructure like urban and peri-urban tree plantations have multiple financial and economic benefits from wood and fruit production to water retention, pollution combatement, job creation along the value chain and locally increasing property values
Gender equity	Gender specific advantages vary along the water reuse value chain

Business value chain

The basic business concept is to recover in arid and semi-arid regions as much treated wastewater as possible for landscaping and productive reuse, like afforestation for timber, fuel or fruit production, while minimizing the unproductive or environmental harmful discharge of water and sludge. Given that treated wastewater of suitable quality for tree plantations will anyway be produced, or is already available, the additional value proposition for the creation of green infrastructure in a desert environment will have multiple social, environmental and economic benefits including improved overall living conditions while having the potential for recovering its own costs through the creation of opportunities for economic growth along the reuse value chain.

The treatment plant might be run by the public and/or private sector and has to be located at the border of a town or city with sufficiently available land for afforestation, recreation or agriculture in the vicinity. The high value for environment and society will help to sustain public subsidies, allowing the business to focus on the recovery of the additional reuse-related costs. For a high reuse rate and limited water conveyance (pumping), decentralized small to medium-sized wastewater treatment plants serving towns, peri-urban communities, suburbs and emerging cities would be most favourable. The institutional set-up across the sanitation-agriculture interface is important as all three case studies showed, and requires a high level of participatory planning and trust building for the recipients of the treated wastewater as well as their customers in its safety. The business model is most promising where no alternative water sources are available and the technology and safety standards permit the production of crops or produce in high demand. The model is at risk of limited impact where a) regulations are too weak; or b) do not match locally feasible technologies; and/or c) alternative water sources are available at a lower or even slightly higher cost.

Next to the sale of treated wastewater, also treated sludge (biosolids) can generate revenues as soils in dry areas are generally poor in organic matter and sludge could be an excellent soil ameliorant supporting soil fertility management and its water holding capacity. However, sewage sludge, even more than the treated water, requires a very reliable monitoring of potential contaminants. If these are too high, sludge can still be an asset but for other uses than food production. With the right institutional set up, market research, sales strategy and a pricing policy, net profits from the reuse scheme are possible.

The business concept involves a simple value chain schematic as shown in Figure 218. This treatment-reuse scheme can result in a public private partnership or remain in the public sector.

Business model

The basic value proposition of the model depends on the business goals and social objectives of the entity initiating the business model – government or private entity operating the wastewater treatment system, and government or farmers/ private enterprise operating the reuse system. Eventually the model will have several value propositions, but with different emphasis for cost recovery. **Next to the treatment of wastewater for safeguarding public health and other water resources, the second value proposition is to establish green infrastructure by offering water, crop nutrients and soil organic matter.** This will result in amenity values and other ecosystem services. Improved soil productivity can for example support tree or fruit plantations, wood and cotton production, biofuel, fodder or also vegetables as long as possible health risks can be minimized and controlled.

There are many institutional options for running the model. Two examples are:

a) Treatment plant and tree plantation are managed by the same public company. With free water and land allocation, cost recovery for the reuse component will largely depend on the efficiency of reducing operational (e.g. electricity) costs and possible overstaffing. Extending the privileges of free land and water to the private sector, would certainly constitute a strong incentive for its engagement assuming trees/crops with high market value and short turnover can be grown.
b) Alternatively, the responsibilities between treatment, water transport and reuse are shared between different stakeholders, which can be public or private like in the Tunisian case where water is sold along its pathway and each entity is using different strategies for cost recovery.

The key revenue sources for the treatment plant are (i) households via sanitation fees, usually collected as part of the water bill; (ii) governmental subsidies reflecting the treatment service for society and nature; and (iii) direct or indirect income from the sales of forest/tree crop products (Figure 219). For the conveyance of the treated water to the plantations, both the treatment entity and the government (saving directly/indirectly water disposal costs), and the benefitting water user association should contribute. A target could be to align wastewater selling rates with the operational cost of the water transfer and the market value of the irrigated product. Another possible revenue stream could derive

BUSINESS MODEL 17: WASTEWATER FOR GREENING THE DESERT

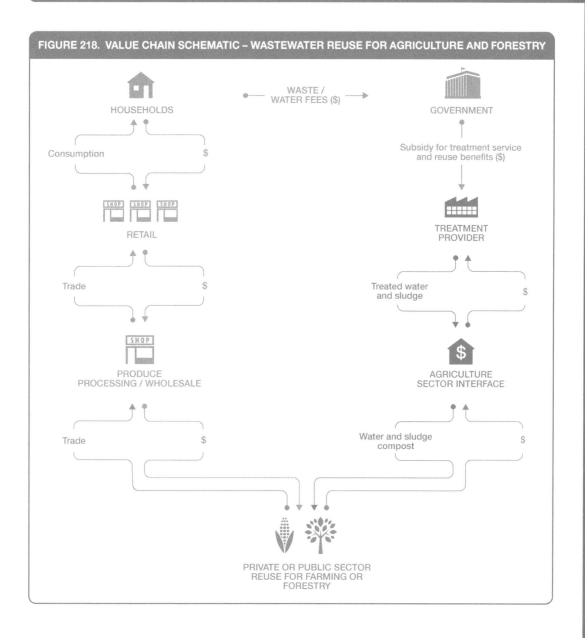

FIGURE 218. VALUE CHAIN SCHEMATIC – WASTEWATER REUSE FOR AGRICULTURE AND FORESTRY

from carbon sequestration in forest plantations or orchards (Box 8). While investments in perennial plants, like trees for wood production or evergreen citrus trees can absorb water year-round, their payback period till the first harvest (for fruits at least three to five years, for wood production twice as long) does not support quick returns on investment. In such cases, such trees might best be combined with other crops allowing earlier revenues. Many farmers call in particular for advanced treatment to grow highly profitable cash crops like vegetables.

CHAPTER 14. PATHWAYS TO COST RECOVERY

FIGURE 219. BUSINESS MODEL CANVAS – WASTEWATER REUSE FOR GREEN INFRASTRUCTURE IN (SEMI)ARID REGIONS

KEY PARTNERS
- Sector ministry
- Agroforestry systems (managed by public or private sector)
- Served households
- Authorities in charge of safety monitoring
- Larger public (trust building)
- Environmental ministry/authority

KEY ACTIVITIES
- Treatment and sale of wastewater and sludge
- Collection of household fees (sanitation tariff)
- Social marketing to support perceptions on reuse safety

VALUE PROPOSITIONS
- To build green infrastructure and value chains from treated wastewater and sludge (biosolids)

CUSTOMER RELATIONSHIPS
- Government to government
- Government to water user associations/private sector
- Online household billing
- Direct sale of biosolids to traders (or farmers)

CUSTOMER SEGMENTS
- Water user associations
- Golf courses
- Forestry projects
- Cotton plantations
- Orchards
- Biofuel and energy market
- Soil input dealers

KEY RESOURCES
- Treatment plant
- Wastewater with no or limited industrial contamination
- Land, labor and capital investment
- Partnerships with agricultural sector

CHANNELS
- Direct dialog and contract (water/biosolids reuse)
- Wholesale contracts (if treatment plant operator also runs plantation)
- Online communication with served households

COST STRUCTURE
- Capital investment (treatment plant, water to farm conveyance)
- Operation and maintenance costs (incl. in-house quality monitoring and costs of risk prevention)
- Customer interface and social marketing/promotional costs

REVENUE STREAMS
- Sewage tax for connected households
- Government subsidy (environmental and social benefits)
- Sale of forest products, wastewater and sludge
- Energy savings (via internal (treatment plant) energy recovery)
- Carbon sequestration (potentially, see Box 8)

SOCIAL & ENVIRONMENTAL COSTS
- Limited costs if treatment plant can avoid receiving untreated industrial effluents, treatment level is appropriate and monitored, and compliance with safety protocols strictly enforced.

SOCIAL & ENVIRONMENTAL BENEFITS
- Green infrastructure benefits from ecosystem service support, like water and nutrient cycling, micro-climate improvement, amenity values, jobs along the value chain, and editable or non-editable (e.g. firewood) produce
- It reduces the disposal of treated but nutrient rich wastewater into natural water bodies
- Reduction in energy use (optional)

Alternate scenario

Greater cost recovery through better accounting, pricing and market extension

As seen in the example of the Drarga plant near Agadir in Morocco, an advantage of decentralized plant management can be that for each plant's service area, all sales revenues and revenues from the water or sewage tariff are deposited into a special account, independently of others accounts, to serve solely the cost-recovery and maintenance of each individual treatment plant. This system can prevent that community revenues are redirected to other needs, and could also provide incentives for benchmarking where management is centralized if transaction costs can be minimized.

Greater cost recovery could come through improved pricing of the services, resources and products. For instance, household could be charged block rates prices for wastewater treatment based on actual water usage, instead of a flat sewage tax as it is common in some countries. Treated wastewater could be sold in bulk to the water user association at a price that reflects more on the costs of water treatment rather than just the additional cost of the water conveyance between treatment plant and irrigation system. The farmer body could then resell the water to its members charging them a markup to recover additional costs of operations including routine maintenance and repairs within the irrigation system. However, all this requires that farmers have limited access to other water sources.

Increasing the freshwater tariffs would make agricultural irrigation with freshwater unfeasible and might force farmers to shift to using reclaimed wastewater if its tariffs are maintained low and if its supply and quality are reliable. This incentive might be constrained by the fact that many farmers control their own facilities for meeting their needs from groundwater resources; thus, energy tariffs should also be considered to steer pumping costs.

For further income, new market segments are needed, like industrial demand for dried sludge as fuel. To reduce the industrial carbon footprint, especially in the cement industry, or where conventional fuel sources are in irregular supply or expensive, sewage sludge derived kiln fuels can be an alternative which the industry might favor as it will in addition qualify under the Clean Development Mechanism (CDM, Box 8).

Box 8. Forest carbon offset: An additional revenue stream?

Converting 'no forest' desert land into a 'forest' absorbs carbon in the growing wood which can be sold on the carbon offset market to carbon emitters, and add a revenue stream to the wastewater reuse project. The gain depends on the total carbon offset which is estimated in 'million tonnes equivalent' (mt CO_2-eq.) stored in living tree biomas. In 2012, a cumulative 134Mt CO_2e of offsets have globally been transacted from 26.5 million hectares of forests. Two out of every three offsets were sold to multinational corporations. Businesses were motivated by offset-inclusive corporate social responsibility (CSR) activities, or to "demonstrate climate leadership" in their industry or to send signals to regulators. Demand for offsets from afforestation or reforestation projects were in 2012 with 8.6 MtCO_2e at a similar level as demand based on *reduced emissions from deforestation and forest degradation* (REDD).

The issuing of carbon credits for afforestation activities has to meet a set of strict guidelines. The amount of carbon sequestered by forests has to be assessed and depends upon many factors including type of tree, tree age and local growth rate, which again depends on climate, irrigation and soil quality.

If a forest owner sells his forest then (s)he is committing to maintain the CO_2 stock. If wood gets lost, like to climate events, disease or unplanned instead of planned harvest, the owner would have to buy back offset credits to cover the loss. An 'ideal' carbon sequestration forest is one where the owner is able to sell carbon credits each year until tree growth and the carbon sequestration rate plateaus, at which time the forest could be harvested and the harvest revenue is higher than what is needed to pay for the lost (above ground) carbon stock. This requires close monitoring of the wood and carbon markets. An alternative target would be to establish a sustainable rotation with regular planting and harvesting, where the stock and growth rate of sold forest biomass could be maintained despite harvests.

Obviously, this type of management and certification has costs and the question is if the returns make them worthwhile. From a purely financial perspective, revenues from offsets in today's still-developing offset market are limited. The price per ton of CO_2e varies significantly but is commonly in the range of USD 4–10, although higher and lower prices can be found. Trees might bind five to ten metric tons of CO_2 per ha per year which translates on average into an annual gain of about USD 30–80 per ha. Thus a 50 ha irrigated wood plantation could generate a gross annual income from carbon sequestration of about USD 1,500–4,000 which has to be compared with the transaction costs of registration and alternative commercial options (timber, firewood, fruits, etc.). Orchards are in the carbon business less prominently as they are usually less densely planted and also pruned, i.e. their carbon accumulation rate will be lower than of many forest species. The plantation sizes as reported in our two case studies are rather small and as offset credits are often traded in units of 10,000t CO_2e or more, which might only be achieved on about 1,000 ha, forest owners need to pay an Offset Aggregator who functions like a broker between woodland owners and the carbon market. A possible alternative for the future could be other offset markets, such as BioBanking where plantation owners can sell Bioversity credits to the market as seen e.g. in Australia (NSW 2007) or payments for watershed services (The Rockefeller Foundation 2015).

Additional sources: http://www.rogerdickie.co.nz/Forestry.aspx; www.forestcarbonportal.com/; www.ecofys.com/files/files/world-bank-ecofys-2014-state-trends-carbon-pricing.pdf.

Potential risks and mitigation

The business model presented here was designed based on the analysis of three case studies in Tunisia, Egypt and Morocco, and other cases and references. There can be a variety of business risks affecting the successful implementation of such a model, most of them being more generic than model specific. For example, as reuse projects involving wastewater are potentially harmful to human and environmental health, particular health risk (mitigation) options are obvious and have to be highlighted.

Market risks: There is no risk related to the need for treating the wastewater, which is a necessity for safeguarding public health. Market risks exists however for the reuse part of the system, which can derive e.g. from (i) competing water/fertilizer sources; (ii) competing final products; and (iii) lack of trust in product quality.

i) Lower costs for accessing alternative water sources (e.g. groundwater) or organic fertilizer can reduce demand for reclaimed water or sludge as fertilizer.
ii) If imported fruits or timber have an established local market, market penetration will require extra efforts or highly competitive pricing which reduces the likelihood of cost coverage.

iii) Different kinds of reuse like irrigating trees, orchards, fodder or vegetables will require different water quality standards. It is mandatory that treatment and post-treatment options will meet these standards to maintain trust in the reclaimed water. Monitoring compliance with safety measures and final water effluent are part of the risk management protocol, as outlined in the WHO *Sanitation Safety Planning Manual* (WHO, 2015). However, technical capacity alone might not be sufficient to address negative consumer perceptions. Any reuse project requires active stakeholder engagement, transparency and feedback from the start on. The role of social marketing and awareness raising can be critical in reducing opposition to water reuse especially in agriculture.

Technology performance risks: A large variety of treatment technology and irrigation systems are available. In low-income countries, common reasons for low or decreasing technical performance in wastewater treatment are poor maintenance practices often due to lack of incentives, lack of electricity, or e.g. lack of sufficient water to flush the sewers. Poor maintenance can result in non-compliance with set treatment standards, which can translate into health risks and loss of customers. Also irrigation technology can have shortcomings, especially where wastewater has to pass in small tubes, like in drip irrigation, where clogging is more common than with fresh water.

A mismatch between imported treatment technologies and local requirements, possibilities and capacities has also been described, for example by Nhapi and Gijzen (2004) from Zimbabwe and supports the call for low-cost applied technologies (Libhaber and Orozco-Jaramillo, 2013) with any additional treatment levels matching cost-effectively the intended reuse or disposal.

Political and regulatory risks: These risks vary from country to country and can be high where the regulatory frameworks, like reuse standards, are under discussion or managed by different authorities with overlapping responsibilities.

Social equity related risks: The model is considered in general as neutral in view of particular gender advantages or disadvantages from the operational or business perspective. As the percentage of women graduating in both, agriculture/forestry and engineering in MENA countries is comparable to or higher than in more developed countries, the foundation for women employment in treatment plants or forestry is increasing. However, women's increasing enrollment in engineering and the sciences is not (yet) reflected in a higher female labor force participation or lower female unemployment (World Bank 2009). There are significant variations between countries, and there can be more permanent employment opportunities for men in forestry and wood processing, while the forest might provide firewood as primary or secondary objective, which could be a significant social benefit in an environment where women struggle finding fuel. There is also evidence of seasonal employment opportunities for women (e.g. olive harvest in Tunisia), although in many countries female workers receive lower wages than male.

Safety, environmental and health risks: Wherever wastewater is used there can be a health risk for different stakeholders and the environment, including occupational risks for workers, discomfort (odor) affecting communities in plant vicinity, and depending on what is produced also risk for buyers/consumers. Mitigation measures are ideally installed along the wastewater treatment to reuse value chain (WHO multiple barrier approach). To minimize safety and health risks to workers and other stakeholders, standard protection measures are required as elaborated below (Table 51). Among various reuse options, growing trees is considered one of the safest. However, where trees are harvested for editable products for the market (e.g. citrus, olives), care has to be taken that pathogens do not get in contact with the harvested product. A particular challenge derives from the use of wastewater and sludge (even if composted, i.e. sanitized) where the wastewater includes industrial effluent due to the possibility of heavy metal entering the food chain. As in all cases of industrial effluent, local pretreatment is required before the water enters the public sewer network and eventually the treatment plant. The risk would matter less for wood than fruit production where it requires monitoring. Although in the target areas aquifers are usually only found in considerable depth, regular groundwater monitoring is also required.

TABLE 51. POTENTIAL HEALTH AND ENVIRONMENTAL RISKS AND SUGGESTED MITIGATION MEASURES FOR BUSINESS MODEL 17

RISK GROUP	EXPOSURE						REMARKS
	DIRECT CONTACT	AIR/ DUST	INSECTS	WATER & SOIL	FRUIT	WOOD	
Workers							Sanitation Safety Planning (WHO 2015) recommended for entire value chain. Elevated risk if business opts for sewage sludge composting/sale.
Farmers							
Community							
Consumer							
Mitigation measures	🧤	😷	🚫🐛	Pb Hg Cd	Pb Hg Cd		

Key: ☐ NOT APPLICABLE   ☐ LOW RISK   ☐ MEDIUM RISK   ■ HIGH RISK

Business performance

Wastewater reuse to produce green infrastructure in human vicinity like tree plantations, parks or orchards can have significant social and ecosystem benefits in MENA region (Figure 220), although the overall social impact varies to some extend with the generated employment opportunities.

Using wastewater for wood production is one of the safest and financially promising reuse options. Thus from an investment perspective, stigma might be less an issue, and the main challenges of the model are more related to the time span between investment and payback, not the water itself. However, there are various options from fast growing trees to agro-forestry which can allow faster returns if supported by treatment quality and regulations.

Different ownership models are possible with cost recovery for the treatment plant largely depending on the freshwater and wastewater tariffs and prices. As long as freshwater is sold under value, the business model ranks low in view of recovering treatment costs although it reduces the water bill of the plantation, and can create significant financial value in form of wood and other forest products. The model ranks high in its adaptability to various bio-physical conditions and in terms of scalability and replicability wherever land is available and freshwater sufficiently scarce that farmers have no alternative. The model can work with plants of any size providing secondary treatment although the cost recovery share might be highest at the scale of smaller towns or suburbs. The right institutional setup to balance financial and economic benefits to the satisfaction of all involved parties is the challenge.

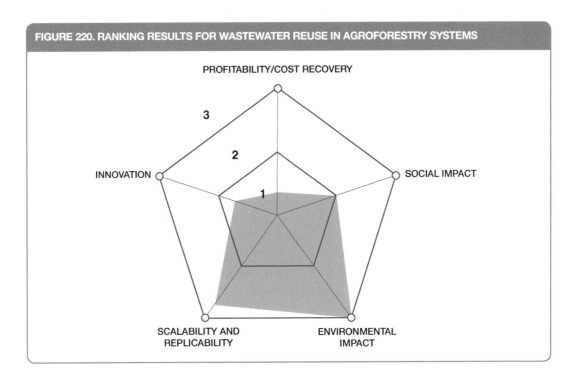

FIGURE 220. RANKING RESULTS FOR WASTEWATER REUSE IN AGROFORESTRY SYSTEMS

References and further readings

Abu-Madi, M.O.R. 2004. Incentive systems for wastewater treatment and reuse in irrigated agriculture in the MENA region, evidence from Jordan and Tunisia. CRC Press, 248p.

Kfouri, C., Mantovanic, P. and Jeuland, M. 2009. Water reuse in the MNA region: Constraints, experiences, and policy recommendations. In: Water in the Arab World. Management perspectives and innovations. Middle East and North Africa Region. The World Bank.

Libhaber, M. and Orozco-Jaramillo, A. 2013. Sustainable treatment of municipal wastewater. IWA's Water 21 (October 2013): 25–28.

Nhapi, I. and Gijzen, H.J. 2004. Wastewater management in Zimbabwe in the context of sustainability. Water Policy 6: 501–517.

NSW (New South Wales). 2007. BioBanking – Biodiversity banking and offset scheme. Scheme overview. Sydney, Australia: Department of Environment and Climate Change.

The Rockefeller Foundation. 2015. Incentive-based instruments for water management. Synthesis review. Pacific Institute, Foundation Center, Rockefeller Foundation.

WHO. 2006. Guidelines for the safe use of wastewater, excreta and greywater, volume 2: Wastewater use in agriculture. Geneva: World Health Organization.

WHO. 2015. Sanitation safety planning manual for safe use and disposal of wastewater, greywater and excreta. Geneva: World Health Organization.

World Bank. 2009. The Status and Progress of Women in the Middle East and North Africa. World Bank, Washington, DC. https://openknowledge.worldbank.org/handle/10986/28425 (accessed 5 Nov. 2017).

15. BUSINESS MODELS BEYOND COST RECOVERY: LEAPFROGGING THE VALUE CHAIN THROUGH AQUACULTURE

Introduction

According to Global Water Intelligence (GWI 2010) the market for water reuse is on the verge of major expansion, outpacing desalination. Especially capital expenditures on advanced water reuse are expected to grow significantly. The market will migrate away from irrigation, towards the production of water, which passes the quality requirement of industrial clients, and for potable reuse based on microfiltration, reverse osmosis and advanced disinfection. Despite being a more expensive process, this water reuse can provide better returns on investment.

Less sophisticated treatment options can also provide high-value products, especially if the reclaimed water is used for the production of more than crops such as animal protein. If such a value proposition can be based on low-cost applied technology, it would have a significant replication potential across low-income countries and/or where advanced treatment facilities have a limited lifetime. It would also be able to attract the local private sector where limited investment costs can be combined with high returns on investments. The examples presented in this chapter are based on aquaculture. Aquaculture-based models recycle water and also assimilate nutrients into the food chain. While in some models, like in the large-scale case of Calcutta (Bunting et al., 2010), fish are produced within the (natural) treatment system, in other cases fish are grown in the last pond of a constructed treatment system, or aside the treatment system which is producing fish feed. The feed consists of fast growing plants which are extracting nutrients from the water and contribute to its treatment (phyto-remediation). Although wastewater-fed aquaculture is according to Bunting et al. (2012) on the decline due to factors such as reduced availability of peri-urban land and increasing water contamination, aquaculture in general is considered as the fastest growing agricultural sector in the world (World Fish Center, 2011). It can be particularly attractive where fish is in high demand, land available and water sources do not pose particular health risks.

This chapter describes two cases of wastewater reuse in aquaculture in **Bangladesh** and **Ghana**. The first case looks at a wastewater system in the town of Mirzapur, Bangladesh, which generated over 20 years profit until the treatment system was phased out. The second case reviews the system pioneered by Waste Enterprises Ltd. in Kumasi, Ghana. In Mirzapur, protein-rich duckweed was produced in wastewater treatment ponds and fed to fish in adjacent ponds, while in Kumasi, the treated wastewater was used directly for fish production. The two examples are followed by **Business Model 18: Leapfrogging the value chain through aquaculture**, which showcases the possibility of a win-win situation for public-private partnerships that are able to cover operational costs as well as recover capital costs within an acceptable time frame.

As with the other chapters, these examples do not claim to be comprehensive and some better cases could have been missed due to information and time constraints. However, they show significant opportunities for moving reuse solutions beyond cost recovery to net profits for business by combining a relatively low-cost but highly efficient technology with an advanced value proposition. This is a remarkable achievement in the usually highly subsidized wastewater treatment sector.

References and further readings

Bunting, S.W., Pretty, J.N. and Edwards, P. 2010. Wastewater-fed aquaculture in the East Kolkata Wetlands, India: Anachronism or archetype for resilient ecocultures? Reviews in Aquaculture 2 (3): 138–153.

Global Water Intelligence (GWI). 2010. Municipal water reuse markets 2010. Oxford, UK: Media Analytics Ltd.

World Fish Center. 2011. Gender and aquaculture: Sharing the benefits equitably. Issues Brief 2011-32. http://pubs.iclarm.net/resource_centre/WF_2832.pdf (accessed 5 Nov. 2017).

CASE

Wastewater for the production of fish feed (Bangladesh)

Pay Drechsel, Paul Skillicorn, Jasper Buijs and Munir A. Hanjra

Supporting case for Business Model 18	
Location:	Mirzapur, Bangladesh
Waste input type:	Hospital complex-derived raw wastewater
Value offer:	Protein-rich feed to cultivate whole, fresh fish – carp species, and treated wastewater
Organization type:	Partnership of private trust and NGO
Status of organization:	Fully operational since 1993; phased out in 2013–2015
Scale of businesses:	Medium
Major partners:	PRISM Bangladesh / Kumudini Welfare Trust (KWT)/ Kumudini Hospital Complex (KHC)

Excutive summary

The for-profit business case describes the experience in Bangladesh to locally treat wastewater for fish production and crop cultivation, generating over 20 years net profits and improvements in environmental quality. The business known as 'Agriquatics' started full operations in about 1993 and run till about 2015 when the treatment system was decommissioned and replaced. The system at the town of Mirzapur received raw sewage and grey water from the local Kumudini Hospital Complex (KHC), water which would otherwise flow untreated to a nearby river. The treatment involved duckweed-based phytoremediation on a 0.6-hectare zig-zag plug flow. No fees were charged for the treatment, no subsidies received from the government and no water sold, but fish was reared on the harvested duckweed in adjacent tanks fed by groundwater and topped up with treated wastewater. Perennial crops such as papaya and bananas were grown along the pond perimeter providing additional income. The fish and crops produced were sold on-site and the income received did not only cover operational and maintenance costs of the combined system, but also recovered several times the original capital investments.

KEY PERFORMANCE INDICATORS (AS OF 2012)	
Land use:	1.6 ha
Wastewater treated:	ca. 300m^3/day
Capital investment:	USD 20,000 for the plug flow treatment system, of which 32% as loan for land development and equipment; and 68% long-term land lease
Labor:	4-persons for 1 hour each day – 7 days per week (0.7 full-time equivalent)
O&M cost:	The major O&M costs were harvesting and feeding the duckweed to fish, fish harvest, and seasonal cleaning of the fish tanks. No chemicals were required

Output:	About 7.5 tonnes/yr of mixed carp species fish sold on-site at an average price of USD 1/kg, earning USD 7,500 from fish (an equal amount possible pilfered) and about USD 1,000 from crops. With costs deducted the annual net revenue was around USD 2,000–3,000					
Potential social and/ or enviornmental impact:	Several part time jobs, inexpensive source of fish and a non-chlorinated treated effluent that meets US advanced tertiary standards (Alaerts et al., 1996)					
Financial viability indicators:	Payback period:	6 years (loan); less than 10 years all	Post-tax IRR:	26%	Gross margin:	20%

Context and background

Mirzapur town (ca. 28,600 inhabitants) in central Bangladesh is well known to the community for the Kumudini Welfare Trust (KWT) and its hospital complex with college and schools. This is also where the *Shobuj Shona* system – continuous duckweed farming and feeding to mixed carp species – for wastewater treatment was first developed. Initially, the local hospital had a four-cell facultative wastewater treatment system but this proved over time inadequate. The KWT contacted PRISM[1], an NGO that had a rural development and healthcare project in the area, and in a collaborated effort it was agreed to build, operate and manage a *Lemnaceae*[2] (duckweed)-based wastewater treatment system which supports fish farming on the condition that the operating entity would keep any profits that the system might generate. The development of duckweed-based, conventional wastewater treatment began in the 1980s with the finally installed plug flow system for the hospital complex starting full operations in 1993 (Gijzen and Ikramullah, 1999; UNEP, 2002). The interlinked aquaculture system continued over the years to supply the local Mirzapur population with a reliable, twice per week harvest of carp and free of charge wastewater treatment service for the local hospital, schools and staff housing complex.

Market environment

Situated on the banks of a largely perennial river, and with water still being relatively abundant in the Mirzapur area, there is no demand for (treated) water, but fruits and in particular fish which provides in Bangladesh more than 50% of total animal protein intake (FAO, 2014a). Agriquatics therefore adopted the Shobuj Shona system of duckweed farming to produce a protein-rich fish feed for its own aquaculture system and revenue generation. Despite a boom of aquaculture in the country, the large Dhaka city market is absorbing a huge share of what gets produced by formal aquaculture operations, allowing Agriquatics to focus on local demand. Fish sale was complemented by the production of fruit and vegetables including bananas and taro around the ponds. According to Gijzen and Ikramullah (1999) a substantial portion of the fish produced was bought by the Kumudini Hospital Complex (KHC), which reduced costs for distribution and marketing, and pressure from competitors in Mirzapur. The opportunity that Agriquatics exploited was the combination of the need for the treatment of wastewater, and the locally strong demand for fish, combined with the low-cost availability of land and potential fish tanks.

Macro-economic environment

Bangladesh ranks for many years globally among the top five countries in view of aquaculture production (FAO, 2014a). Aquaculture has been one of the fastest-growing economic subsectors of the economy, providing high-protein food, income and employment and earning foreign exchange. More than 4 million fish farmers, mostly small-scale, and more than 8.5 million other people derive a livelihood from it directly or indirectly. In 2012, farmed fish contributed some 1.73 million tons to the country's total fish production of 3.26 million tons (FAO, 2014a). This is an almost 19-fold increase from the 1980 aquaculture production of about 91,000t, and for example ten times the reported production in the USA. Export revenue in 2012 was estimated at USD 450 million (FAO, 2014b).

The macro-economic situation reflects a positive business driven investment climate for aquaculture in Bangladesh. However, Edwards (2005) and Parkinson (2005) stated that direct governmental support, institutional assistance and a lack of a national funding mechanism to support, e.g. the capital investments in aquaculture in general, or duckweed-based systems in particular are missing. This might be changing under the National Aquaculture Development Strategy and Action Plan 2013–2020 which is aligned with and draws guidance from the National Fisheries Policy, Country Investment Plans, the National Fisheries and Livestock Sector Development Plan and the preceding national fisheries strategy and action plan of 2006–2012. The new plan is however not addressing linkages between sanitation and aquaculture and leaves the model Agriquatics pioneered in a grey area, even more as also wastewater management and reuse are typically not acknowledged as a major element of water management in existing laws and policies in Bangladesh. The sector is hampered, in addition, from a considerable complexity with regard to the power of implementing authorities from both the agricultural and urban wastewater management perspectives.

Business model
The overall value proposition is high quality wastewater treatment paid through the production of fish feed, crops and fish at competitive market prices, making the system independent of fees and tariffs. The enterprise employs a value-driven and for profit, end-sales model whereby an even larger value derives from environmental and social responsibility impacts beyond sales revenues (Figure 221). Essential for the business model start-up was the partnership of the Hospital (via Kumudini Welfare Trust) and PRISM Bangladesh, enabling expertise-supported and cost-effective implementation of the duckweed water treatment and fish rearing system. This ensured that two important economic values were created: (i) wastewater that is treated to an advanced tertiary level at no extra cost to the hospital and thus adding value for the hospital in terms of avoided costs for financing an additional treatment level; (ii) a reliable and guaranteed supply of wastewater generated fish feed at no extra costs, and high quality water supporting crop and fish farming. The symbiosis between the non-profitable wastewater treatment and the highly profitable fish production made the Agriquatics model financially viable, not only to break even, but to pay back the initial loan taken for the setup of the treatment system.

PRISM inherited a defunct pond system which was redesigned for fish production while its capital investment went into the duckweed zig-zag treatment system (see below). Land, fish tanks, water and nutrients were effectively free. Since conventional fish feed is scarce and (consequently) prices are high, the use of alternative sources of quality fish feed remains until today very attractive.

Unlike conventional wastewater treatment systems in more developed countries, where treatment quality is enforced by regulatory agencies, the revenue generation of Agriquatics provided sufficient incentive for the highest quality of treatment found in Bangladesh.

Value chain and position
The Agriquatics initiative was developed under the Kumudini Welfare Trust-PRISM Bangladesh partnership. These two partners provided the business with its most critical resources (wastewater, treatment ponds, technology and expertise). Having these in place, the business was positioned to buy its other inputs such as fingerlings and seeds from up-chain suppliers and sell its products (fish and crops) directly to end-users (local fish consumers; Figure 222).

CASE: WASTEWATER FOR THE PRODUCTION OF FISH FEED

FIGURE 221. BUSINESS MODEL CANVAS FROM THE PERSPECTIVE OF PRISM BANGLADESH (AGRIQUATICS)

KEY PARTNERS	KEY ACTIVITIES	VALUE PROPOSITIONS	CUSTOMER RELATIONSHIPS	CUSTOMER SEGMENTS
• Kumudini hospital complex (KHC) • Kumudini Welfare Trust (KWT) • PRISM HQ • Local community	• Treatment of wastewater • Growing and harvest of crops, duckweed, fish • Fish and crops sales • Technical advice	• High quality wastewater treatment paid through the production of fish feed, crops and fish at competitive market prices, making the system free of fees or tariffs	• Recurrent purchase based on customer satisfaction (low price and availability) • Contractual relation • Strong (non-financial) public support	• Fish buyers (incl. Kumudini hospital complex) • Crop buyers • KWT (demanding wastewater treatment)
	KEY RESOURCES • Land use rights; ready ponds • PRISM technical expertise • Wastewater, duckweed • Capital access • Fingerlings		CHANNELS • Direct selling on-site • Contracts and direct interaction of partners at hospital site	

COST STRUCTURE	REVENUE STREAMS
• Capital investment (loan and land lease) • Regular fingerlings purchase/breeding costs • O&M (mostly labor employed for duckweed farming, fish feeding, harvest and sale; and crop irrigation, harvest, sale); debt repay	• Sales of fish • Sales of crops

SOCIAL & ENVIRONMENTAL COSTS	SOCIAL & ENVIRONMENTAL BENEFITS
• Laborers' health risk due to contact with wastewater • Possible human health hazard from consumption of fish if contaminants are transported via duckweed to the fish and not destroyed by fish cooking	• Wastewater efflux from hospital is treated, which results in reduced environmental pollution • Support of jobs and protein supply for the local community • Cheap food supply to the hospital supporting its free service to the poor

A notable portion of the fish and crops produced was bought by the hospital complex. Additional profits from water sales were not realistic in the local context as there is no market for the treated water due to the availability of adequate fresh water for agriculture, even in the dry season.

Institutional environment

The Kumudini Welfare Trust is a not-for-profit family trust managed by an external board of directors – one member of which is nominated by the Government of Bangladesh. PRISM Bangladesh is a not-for-

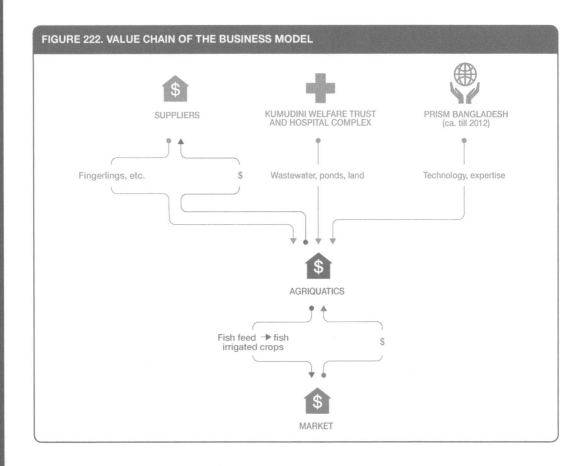

FIGURE 222. VALUE CHAIN OF THE BUSINESS MODEL

profit Bangladeshi NGO. The relationship between the two entities was specified under a succession of mutual agreements. At a later stage, PRISM's involvement phased out, while the treatment system continued to operate until 2013 when the Indian Government financed a new treatment plant for the hospital complex which was inaugurated on 7 June, 2015.

The wastewater fed aquaculture system received significant scientific interest. Public support was also strong, but involved no direct financial transaction beyond a continuing willingness by the local public to purchase fish. Agriquatics provided in-house training to the locals working as laborers. Linking between the sanitation and agricultural sector, the project fell under different policies and strategies without any direct support (see section on the Macro-economic environment above).

Technology and processes

The project inherited a defunct four-cell, single hectare facultative ponds complex and added to it a 0.6-ha plug flow duckweed wastewater treatment system. Only the first of the four ponds remains connected to the wastewater treatment system serving as a primary wastewater receiving and settling tank (Figure 223). The other three ponds were converted to fish production tanks, fed by groundwater and by the final effluent of the plug-flow (Iqbal, 1999).

Except for an initial lift pump, the wastewater moves by gravity to and through the whole treatment system from the initial 0.25-ha pond with a hydraulic retention time of two to four days, and followed by the duckweed-covered, 0.6 ha plug flow lagoon constructed as a 500m long non-aerated serpentine channel with seven bends. For this, depth of the lagoon increases gradually from 0.4 to 0.9m. The system was fed with a mixture of hospital, school and domestic (staff residencies) wastewater from a population of about 3,000–4,000 people with per capita production of wastewater estimated at around 100L/day. The hydraulic retention time in the plug flow wastewater-fed duckweed lagoon was estimated by different authors as 15–22 days, with parts of the water in the zig-zag been lost as seepage to the nearby canal. The lagoon was covered by a floating bamboo grid to contain the standing (100% cover) duckweed mat, at least in the first part of the system which is naturally the richest in nutrients. Early data suggest that the system produced 220–400t fresh duckweed/ha/year (about 17 to 31t dry weight/ha/year) (UNEP, 2002). Duckweed was harvested manually with nets, drained in bamboo baskets, weighed and then placed in one of 12 floating feeding stations distributed evenly across the surface of the originally three 0.25 ha fish tanks. Fish were fed in addition with rice bran and oil cake (Edwards, 2005).

Part of the treated water was eventually used to top up the fish tanks. Analysis by the International Center for Diarrheal Disease Research, Dhaka, Bangladesh, verified that indicator pathogen transmission to fish or workers was similar to control groups and within safety margins (Gijzen and Ikramullah, 1999; Islam et al., 2004). This might however not apply to all possible pathogens and heavy metals (see below).

The fish tanks were stocked with around 10,000 to 14,000 fingerlings at the onset of the monsoon season. The polyculture includes Indian major carp (Mrigal 25%, Catla 20%, Rohu 15%) and Chinese carps (Silver Carp 10%, Mirror Carp 20%, Grass Carp 10%). Tilapia was not stocked but fingerlings entered the tanks incidentally (UNEP, 2002). Fish were usually harvested twice a week. The production numbers varied between reports from on average of 7.5 to max. 15t/ha/year (of which usually a share got stolen).

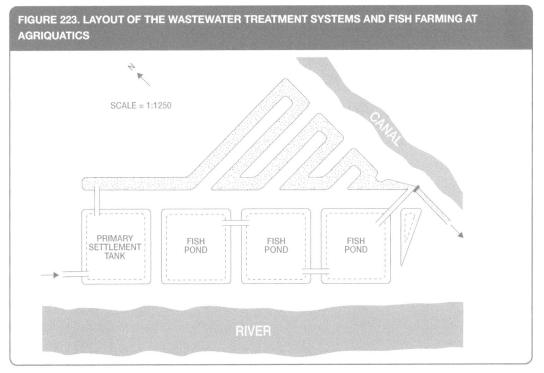

FIGURE 223. LAYOUT OF THE WASTEWATER TREATMENT SYSTEMS AND FISH FARMING AT AGRIQUATICS

Source: After Iqbal, 1999.

Movement of wind across the surface was mitigated by strategic placement of crops such as bananas, taro, papaya and lentils along the perimeter. These also contributed to the income of the system.

Funding and financial outlook

Agriquatics had the advantage that wastewater collection and channeling were already in place and so the defunct pond system was redesigned for fish production. The land was leased on favorable terms, and capital investments for the labor intensive construction of the plug flow system were limited. Financial support was provided by United Nations Capital Development Fund (UNCDF).

In view of operational cost recovery, a portion of the fish produced was bought by the hospital which provided financial security. Both initial partners (KWT and PRISM) had obvious interests in the effective operation of the system: KWT to achieve the effective treatment and proper disposal of its wastewater; PRISM to promote the duckweed technology while generating financial returns. Based on audited records from the first eight years (Table 52), revenues allowed a pay back of the initial loan from PRISM in about six years. Since then the wastewater-fed duckweed-fish system generated an annual net profit of about USD 2,000–3,000, which is larger per hectare than e.g. that of rice, the major agricultural crop in the area. The internal rate of return was calculated as about 25.9% (Gijzen and Ikramullah, 1999; UNEP, 2002; Patwary, 2013).

TABLE 52. AVERAGE ANNUAL INCOME AND EXPENDITURES 1993–2000 IN TAKA (USD 1 = 40–50 TAKA IN THIS PERIOD)

DESCRIPTION	YEAR 1	YEAR 2	YEAR 3	YEAR 4	YEAR 5	YEAR 6	YEAR 7	YEAR 8	8 YEARS AVERAGE
1. Recurring operational Cost									
Land rental (2 ha)	26,000	26,000	26,000	26,000	26,000	26,000	26,000	26,000	26,000
Staff salary and wages	85,600	92,020	98,922	106,341	114,317	122,891	129,036	136,480	110,701
Field supplies (duckweed)	10,000	12,000	13,500	14,300	15,200	15,960	15,678	16,512	14,144
Field supplies for agriculture & fish	28,000	29,000	30,000	31,000	33,000	32,300	34,000	33,600	31,363
Energy/fuel cost (pump)	43,500	45,500	47,900	50,430	55,720	58,500	62,400	63,100	53,381
Maintenance	13,700	14,000	14,500	15,200	16,720	17,556	18,375	18,500	16,069
Miscellaneous	6,285	6,580	7,000	7,350	7,700	7,900	7,500	7,720	7,254
Subtotal annual operation cost	213,085	225,100	237,822	250,621	268,657	281,107	292,989	301,912	258,912
Depreciation of loan (10 years)	25,000	25,000	25,000	25,000	25,000	25,000	25,000	25,000	25,000
Management overhead (7.5%)	15,981	16,833	17,837	18,797	20,149	21,083	21,974	22,643	19,412
Financial costs (9.5% on work capital)	10,450	10,925	11,590	12,350	13,300	13,352	13,916	14,340	12,528
Subtotal admin & finance costs	51,431	52,758	54,427	56,147	58,449	59,435	60,890	61,983	56,940

DESCRIPTION	YEAR 1	YEAR 2	YEAR 3	YEAR 4	YEAR 5	YEAR 6	YEAR 7	YEAR 8	8 YEARS AVERAGE
Total annual recurring costs	264,516	277,858	292,249	306,768	327,106	340,542	353,879	363,895	315,852
2. Income from farm revenue									
Sale proceed from duckweed-fed fish	128,778	253,800	316,509	402,231	404,982	445,702	419,440	413,354	348,100
Sale proceed from agriculture & fruits	25,000	30,000	34,000	44,000	65,000	58,250	56,667	60,223	46,643
Miscellaneous sales	3,600	4,400	4,600	5,200	5,400	5,200	5,100	5,600	4,888
Total income from sales	157,378	288,200	355,109	451,431	475,382	509,152	481,207	479,177	399,631
3. Operational profit	-55,707	63,100	117,287	200,810	206,725	228,045	188,218	177,265	140,719
4. Net profit before taxes*	-107,138	10,342	62,860	144,663	148,276	168,610	127,328	115,282	83,779

* No tax on agro-production (tax holiday)
Source: Patwary, 2013; modified.

Socio-economic, health and environmental impact

Local studies showed that duckweed recovered a significant portion of the nutrient value inherent in the wastewater, so much that in the last part of the zig-zag system it hardly grew due to low nutrient content. The nutrient removal had a positive impact on the effluent receiving water body and its water quality, reducing potentially human health-related costs in the vicinity. But nitrogen as ammonium and nitrate was not only efficiently captured through phytoremediation, but also transformed into protein rich biomass. Based on water quality data (oxygen demand, nitrogen, phosphorus) by Alaerts et al. (1996) and fecal coliform analysis by Islam et al. (2004), treated wastewater discharged to the adjacent river could be considered among the highest quality of treated wastewater in the country attainable without use of reverse osmosis and fit for unrestricted irrigation of vegetables according to WHO standards for wastewater reuse (UNEP, 2002). Further disinfection of the treated effluent prior to its discharge into the river had been considered, but found to be prohibitive on the basis of cost.

While the harvest of duckweed significantly exposed workers to wastewater and its pathogens, scientific monitoring could not determine a cause-effect relationship between incidences of worker diarrheal disease infection and their working at the site (Gijzen and Ikramullah, 1999). Also fish was tested to be safe for consumption. However, while duckweed absorbs nutrients, it also absorbs heavy metals, and if it used as herbivorous fish feed, the metals can be bio-accumulated as it was locally verified (Parven et al., 2009). There can also be gastroenteritis-causing bacteria which persist in the treatment system and might spread to fish (Rahman et al., 2007). An impact from such a pathogen transfer on human consumers was however considered low as fish is generally not eaten raw in Bangladesh (Gijzen and Ikramullah, 1999). Data on other potential contaminants such as estrogen or pharmaceutical residues do not exist. The recommendation was made that related research be included also in any replication of the system.

Entry into aquaculture appears to have fewer gender barriers, as this sector developed outside cultural traditions. According to FAO, Bangladeshi women make up about 60% of fish farmers, and many are successful entrepreneurs[3]. And while women's involvement in aquaculture has importantly improved the economic, nutritional and social benefits for their family, their work goes largely unrecognized in official statistics.

Scalability and replicability considerations

Over its lifetime, the Agriquatics system recovered several times its investment costs, which is unique in the domain of wastewater treatment. The key drivers for the success of the business were:
- Availability of land.
- Limited capital cost with several profitable revenue streams for high-value products resulting in fast payback.
- Low-tech and -cost treatment system supported by a mutually beneficial partnership ensuring availability of water, expertise and system maintenance.
- Profit incentive for treatment of wastewater that obviates requirement for external supervision and controls.

It is important to note that the positive financial performance of the wastewater treatment and aquaculture system was a product of a mutually beneficial partnership which created favourable conditions, such as no major costs for wastewater collection and channelling, and favourable terms for capital investment, land lease and cost recovery.

A pillar of the success was the value creation in terms of fish, i.e. to capitalize on increasing revenues with moving up the value chain, compared to treatment plants only providing treated water. On the other hand, the requirement for a suitably large land area for the combined treatment – aquaculture system will be a common constraint within towns and cities. This is especially true for Bangladesh with its very high population density, land speculations and rising opportunity cost of land, in particular within urbanizing areas (Edwards, 2005). An opportunity in drier areas could be to link such systems with inner-urban or peri-urban green belts, as realized in Parque Huascar in Lima, which can create significant social value[4], or biodiversity reserves. From a health perspective, it has to be added that although the system in Mirzapur was set up at a hospital, its replication potential will be highest where the wastewater derives only from domestic settings with minimal risk of chemical contamination.

Aside its benefits of nutrient accumulation and high crude protein production, also duckweed has some biological constraints which can limit its use in other regions: its growth is adversely affected by both low and high temperatures, and high light intensity; occasional insect infestation; and rapid decomposition following harvest, i.e. the fish ponds have to be in proximity.

Summary assessment – SWOT analysis

The success story builds on a win-win situation of treatment quality and revenue generation combined with favourable low capital and O&M costs, and a high-value product allowing the recovery of both operational and investment costs. The system requires a relatively large land investment for the spatial combination of aquaculture and treatment systems. Figure 224 shows the SWOT analysis of this business case.

FIGURE 224. SWOT ANALYSIS OF AGRIQUATICS BUSINESS CASE

	HELPFUL TO ACHIEVING THE OBJECTIVES	**HARMFUL** TO ACHIEVING THE OBJECTIVES
INTERNAL ORIGIN ATTRIBUTES OF THE ENTERPRISE	**STRENGTHS** - Viable business model based on multiple revenue streams, cost recovery and net profit - Limited capital and low O&M costs - Green infrastructure image with positive environmental impacts - Mutually beneficial partnership enabling free supply of wastewater and land including lagoons at favorable terms - Expertise in duckweed-fed fish farming	**WEAKNESSES** - Significant land-size requirement - No direct governmental support - Monitoring of system safety remained fragmentary - Sensitive to flooding - Fish farming not possible with increasing chemically contaminated wastewater
EXTERNAL ORIGIN ATTRIBUTES OF THE ENVIRONMENT	**OPPORTUNITIES** - Rising fish demand - Increasing demand for green infrastructure - Proof of concept for replication - Short pay-back period - Grey legal area	**THREATS** - Possible industrial effluent - Undetected human health risks - Local aquaculture competition - Changing government policy toward wastewater reuse in aquaculture - Area wide flooding (happened twice)

References and further readings

Alaerts, G.J., Mahbubar, M.R. and Kelderman, P. 1996. Performance analysis of a full scale duckweed sewage lagoon. Water Research 30 (4): 843–852.

Edwards, P. 2005. Demise of periurban wastewater-fed aquaculture? Urban Agriculture Magazine 14: 27–29.

FAO. 2014a. The state of world fisheries and aquaculture 2014: Opportunities and challenges. Rome: Food and Agriculture Organization of the United Nations. www.fao.org/3/a-i3720e/i3720e01.pdf (accessed 5 Nov. 2017).

FAO. 2014b. National aquaculture development strategy and action plan of Bangladesh 2013–2020. Ministry of Fisheries and Livestock, Government of the People's Republic of Bangladesh, and FAO. Rome: FAO. www.fao.org/3/a-i3903e.pdf (accessed 5 Nov. 2017).

Gijzen, H.J. and Ikramullah, M. 1999. Pre-feasibility of duckweed based wastewater treatment and resource recovery in Bangladesh. Final Report. Washington, DC: World Bank. www.ircwash.org/sites/default/files/341.9-15750.pdf (accessed 5 Nov. 2017).

Iqbal, S. 1999. Duckweed aquaculture. Potentials, possibilities and limitations for combined wastewater treatment and animal feed production in developing countries. SANDEC Report No. 6/99. www.coebbe.nl/sites/default/files/documenten/Duckweed-aquaculture.pdf (accessed 5 Nov. 2017).

Islam, M.S., Kabir, M.S., Khan, S.I., Ekramullah, M., Nair, G.B., Sack, R.B. and Sack, D.A. 2004. Wastewater-grown duckweed may be safely used as fish feed. Canadian Journal of Microbiology 50 (1): 51–56. See also www.ircwash.org/sites/default/files/Islam-2000-Faecal.pdf (accessed 5 Nov. 2017).

Leng, R.A. 1999. Duckweed: A tiny aquatic plant with enormous potential for agriculture and environment. FAO, Rome, 108 pp.

Parkinson, J. 2005. Decentralised domestic wastewater and faecal sludge management in Bangladesh. An output from a DFID funded research project (ENG KaR 8056). DFID. http://r4d.dfid.gov.uk/PDF/Outputs/Water/R8056-Bangladesh_Case_Study.pdf (accessed 5 Nov. 2017).

Patwary, M.A. 2013. Powerpoint presentation: Wastewater for aquaculture: The case of Mirzapur, Bangladesh. ADB, Manila, 29–31 January 2013. http://k-learn.adb.org/system/files/materials/2013/01/201301-wastewater-aquaculture-case-mirzapur-bangladesh.pdf (accessed 5 Nov. 2017).

Parven, N., Bashar, M.A. and Quraishi, S.B. 2009. Bioaccumulation of heavy and essential metals in trophic levels of pond ecosystem. Journal of Bangladesh Academy of Sciences 33 (1): 131–137.

Rahman, M., Huys, G., Rahman, M., Albert, M.J., Kuhn, I. and Mollby, R. 2007. Persistence, transmission, and virulence characteristics of Aeromonas strains in a duckweed aquaculture-based hospital sewage water recycling plant in Bangladesh. Appl. Environ. Microbiol. 73: 1444–1451.

Skillicorn, P. 2008. Mirzapur agriquatics system. (6:37 min) www.youtube.com/watch?v=M93HZDoqhsE (accessed 5 Nov. 2017).

Skillicorn, P., Spira, W. and Journey, W. 1993. Duckweed aquaculture: A new aquatic farming system for developing countries. EMENA, The World Bank.

Torres, R. 1993. Shobuj Shona evaluation. Enterprise. Asset accumulation and income generation in Bangladesh: A new model for women in development. The University of California, Davis. www.mobot.org/jwcross/duckweed/Shobuj_Shona_Evaluation.htm (accessed 5 Nov. 2017).

UNEP. 2002. International source book on environmentally sound technologies for wastewater and stormwater management. London: IWA Publishing. www.unep.or.jp/ietc/Publications/TechPublications/TechPub-15/2-9/9-3-3.asp (accessed 5 Nov. 2017).

See also (all accessed 5 Nov. 2017):
- http://genderandwater.org/en/bangladesh/gwapb-products/knowledge-development/policy-brief-gender-in-aquaculture (or https://goo.gl/kPqq3y)
- www.thefishsite.com/articles/1073/marketing-lowvalue-cultured-fish-in-bangladesh/
- www.thefishsite.com/articles/1447/fao-state-of-world-fisheries-aquaculture-report-fish-consumption/#sthash.9uYMKtfp.dpuf
- http://www.kumudini.org.bd/Environmental-participation1
- www.adb.org/sites/default/files/publication/31230/toilets-river.pdf (p. 75–76)

Case descriptions are based on primary and secondary data provided by case operators, insiders or other stakeholders, and reflect our best knowledge at the time of the assessments 2016. As business operations are dynamic data can be subject to change.

Notes

1 PRISM: Project in Agriculture, Rural Industry Science and Medicine. The PRISM group was founded in the 1980s as an international non-profit organization focusing on the support of local and family enterprise within rural communities in developing countries. PRISM Bangladesh was created as an affiliate of the PRISM Group in 1990 (Torres 1993).
2 Lemnaceae ("duckweed"), a family of aquatic macrophytes converts nutrients from the wastewater into protein rich biomass, that can be used as poultry and fish feed. According to Leng (1999), on average 40–50 tons of dry matter can be produced per year per hectare under optimal conditions, allowing the production of more protein per ha and year than via soybean or groundnut (Patwary, 2013).
3 www.fao.org/gender/gender-home/gender-programme/gender-fisheries/en/ (accessed 5 Nov. 2017).
4 https://wle.cgiar.org/thrive/2013/09/02/wastewater-reuse-benefits-beyond-food-production (accessed 5 Nov. 2017).

CASE
A public-private partnership linking wastewater treatment and aquaculture (Ghana)

Philip Amoah, Ashley Muspratt, Pay Drechsel and Miriam Otoo

Supporting case for Business Model 18	
Location:	Kumasi, Ghana
Waste input type:	Municipal wastewater
Value offer:	African catfish, treated water
Organization type:	Public-private partnership (PPP)
Status of organization:	Operational 2010–2012 (later transformed into a research project)
Scale of businesses:	Small-medium
Major partners:	Waste Enterprisers Ltd. (now: Waste Enterprisers Holding) Kumasi Metropolitan Assembly (KMA), Ghana Kwame Nkrumah University of Science and Technology (KNUST), Department of Fisheries and Watershed Management, Kumasi International Water Management Institute (IWMI), Accra

Executive summary

In Kumasi, Ghana, a public-private partnership was established between the Kumasi Metropolitan Assembly (KMA) and the private company Waste Enterprisers Ltd. (WE) to use aquaculture as a source of revenue for sustaining the sanitation services. As part of the agreement, WE is allowed to stock catfish in the final maturation pond(s) of governmental owned wastewater treatment plants, while in return WE uses half of its fish-sale profit to facilitate regular plant maintenance. This arrangement helps WE to access water and infrastructure for fish farming without related capital expenditures, while KMA gets its treatment plants well maintained which was so far more than challenging.

The business was co-funded by both parties without external support. Further beneficiaries are the low-income households charged for maintenance of the Waste Stabilization Ponds (WSP) and the maintenance subcontractor who is entitled of collecting the household fees.

Selling smoked catfish which is in high demand can make already the management of one treatment plant viable. For (unsmoked) fresh fish, with optimized production, break-even can be achieved from two managed plants upwards although only from three systems up the economic indicators will be positive. With full compliance with safety regulations and policy support, the model is easily transferable to other locations, as pond based treatment systems are very common in the tropics.

The case is an example of an innovative pro-poor PPP that helps to ensure the sustainability of a wastewater facility whilst providing benefits to the community. During its engagement in Kumasi, WE rehabilitated two WSP, built rearing infrastructure for its fingerlings, and increased stock survival rates from less than 10% to 80% over the course of four cultivations. This case attracted international donor funding for accompanying research.

KEY PERFORMANCE INDICATORS (AS OF 2012)					
Land use:	230–266m^2 (per fish pond); about 1 ha (total WSP)				
Water use:	225m^3/day				
Capital investments:	Limited to fish hedging as ponds were in place. From a PPP perspective, less than 30% borne by WE and over 70% by KMA				
Labor	2 staff (part-time), 2 workers				
O&M costs	USD 3,429 /year/WSP (for 5 WSP systems), to USD 11,440 /year/WSP (for 1 WSP)				
Output:	Per hectare (water): 40 tons/year of fish; Per actual area: 2 tons/year from two ponds				
Potential social and/or environmental impact:	Reduction in public sanitation and health expenditures, improvement in food supply and job creation; poor households exempted from treatment plant maintenance fees				
Financial viability indicators:	Payback period:	N.A.	Post-tax IRR:	45%	Gross margin:	N.A.

Context and background

Kumasi is the capital of Ghana's Ashanti region and the second largest city in the country with a 2013 population of over 2 million and an annual growth rate of about 4–4.5%. The increasing population is challenging urban water and sanitation services. Like across Ghana, also the wastewater treatment facilities in the Kumasi metropolis are not or only partially functioning due to constrained institutional and financial resources (Murray and Drechsel, 2011). The resulting pollution of water bodies remains unchallenged as the enforcement of environmental regulations is especially weak for governmental infrastructure. Innovative partnerships and financing mechanisms are needed for sustainable wastewater management.

Waste Enterprisers (WE) is a non-profit organization, which focuses on building business models that incentivize waste collection and treatment services without further burden for poor households (tariff-independency). WE was set up with the goal to create demand for value-added waste products whilst providing an avenue for investing a portion of profits back into the sanitation sector, generating cycles of local investment, sustainable sanitation and healthier communities (Murray and Buckley, 2010).

In early 2010, WE approached KMA with its PPP proposal. The business locations of WE in Kumasi were the Ahinsan and Chirapatre housing estates and their wastewater treatment systems. Both were built in the late 1970s by the now-defunct State Housing Corporation of Ghana. Over 200 houses in each community (with ca. 1,500 inhabitants in Ahinsan and ca. 1,800 in Chirapatre) are connected to a communal sewerage network, which, along with storm-water runoff, is channelled to the respective WSP for treatment (Tenkorang et al., 2012). Like most sanitation facilities in Ghana, both WSP systems have chronically lacked reliable maintenance. In theory, a KMA subcontractor is responsible for raising the necessary fees from the served households for undertaking the maintenance of the plant. However, as households are poor and consider this a task of the municipality, the effort of collecting the fees erases any incentive to do the job and ponds were hardly maintained over years (Tenkorang et al., 2012).

The aquaculture production by WE was accompanied by an extensive testing of fish quality and safety. Studies targeted pathogenic contamination, heavy metals and pharmaceutical residues (Amoah and Yeboah-Agyepong, 2015a; Asem-Hiablie et al., 2013). Also the cultivated species, African Catfish (*Clarias gariepinus*), was chosen for safety reasons as in the study region it is normally smoked and not consumed fresh, but cooked.

Market environment

Traditionally, fish is the preferred and cheapest source of animal protein in Ghana with about 75% of total annual production being consumed locally. Tilapia constitutes about 80% of aquaculture production, while catfish accounts for the remaining 20%. According to Cobbina and Eiriksdottir (2010), fish trading is an important occupation in Ghana with an estimated 10% of the population engaged in it, on a full time or part time basis, both in rural and urban communities. Commercial farms mostly deal with wholesale buyers who buy the bulk of the harvested product and go on to sell to retailers or fish processors while fish harvested by the non-commercial farmers is mostly retailed by themselves or their spouses. Only a few non-commercial farmers sell their product to wholesale buyers. Unsold fish is either frozen or processed via smoking, salting and/or fermentation. Fish availability and marketing is most common in the southern and the middle zone of Ghana (GLSS, 2014). Ghana's Ashanti region is currently the leading region in pond-based fish farming in Ghana, with about 1,205 fish ponds, involving over 500 fish farmers. Available water surface area in Ashanti for fisheries development is about 151 ha producing about 585 metric tons of fish annually. Ashanti also leads in the production, supply and export of catfish in Ghana (Rurangwa et al., 2015).

In Kumasi, the majority of people consume catfish at home and in street restaurants which offer traditional stews. About 68% of the interviewees indicated to eat fish eight times per month or all three to four days (Amoah et al. 2015b). The 2014 Ghana Living Standard Survey recorded an annual food budget share of 15.8% for fish and seafood, which is nationwide the second most important food consumption subgroup, after cereals (e.g. rice and bread) (17.7%), and twice as high as meat (GLSS, 2014). Product attributes that influence consumers' decisions prior to purchasing fish are price, size and quality of the fish. Source of fish is among the least important product attributes influencing consumers' decision. In surveys which explained the wastewater use, consumers in Kumasi reconfirmed that they are more likely to choose fish farmed in treated wastewater if it was less expensive and larger than fish from other sources (Amoah and Yeboah-Agyepong, 2015b), which mirrors consumers' behaviour in view of wastewater irrigated vegetables (Keraita and Drechsel, 2015). An indicator of demand and tolerance of the source of water is the frequent theft of fish directly from the wastewater treatment ponds.

Macro-economic environment

Ghana's total annual fish requirement has been estimated to be 880,000 t while the nation's annual fish production average is 420,000 t, leaving a significant deficit. This deficit is partly made up for through fish imports which were estimated at 213,000 t in the year 2007 and valued at USD 262 million (Cobbina and Eiriksdottir, 2010). However, import of farmed fish is not allowed so as to ensure good prices for local fish farmers. However, illegal import, especially of Tilapia, is a growing concern. The deficit between fish demand and production has been a main driving force for pushing the agenda of developing aquaculture (Awity, 2005). Studies conducted by Asmah (2008) reported a 16% mean annual growth rate in the number of aquaculture farms since the year 2000. Fish production in ponds range from about 35 kg to over 35t/ha/year (Asmah, 2008), with less than 10% of commercial farmers exceeding production levels of 20t/ha/year. Common production cycles range between seven and 12 months (Asmah, 2008).

Business model

The partnership arrangement between the WE and KMA offers an interlinked double value proposition: maintaining the treatment capacity financed through waste valorization via fish production (Figure 225). The model ensures that WE gets nutrient rich water at no cost and KMA derives benefits from cost savings, as a more reliable WSP maintenance will lead to lower public health expenditures

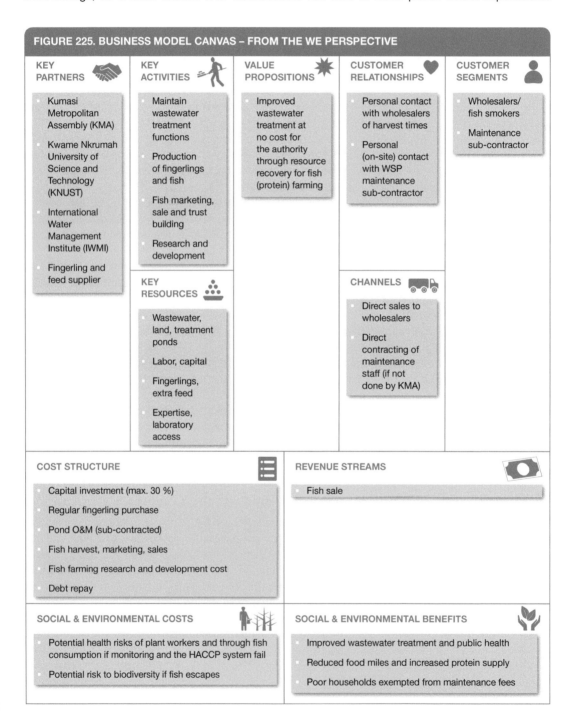

FIGURE 225. BUSINESS MODEL CANVAS – FROM THE WE PERSPECTIVE

CASE: PPP LINKING WASTEWATER AND AQUACULTURE

from insufficiently treated wastewater entering the environment. Other beneficiaries are (i) the WSP-connected households which were so far asked to pay the maintenance contractor (see Figure 226, option 1); and (ii) the contractor who faced significant opportunity costs trying to collect the household fees. While KMA provides the land and pond system, WE cultivates the fish under strict safety monitoring standards. KMA as public partner is not paying the WE for the expected service; in contrary, any profit WE achieves is shared 1:1 with the public utility allowing it to improve sanitation services, like to fully pay for pond maintenance (see Figure 226, option 1)[1], i.e. without need for the subcontractor to collect fees from the served households which appeared difficult as both estates were set up for low-income groups.

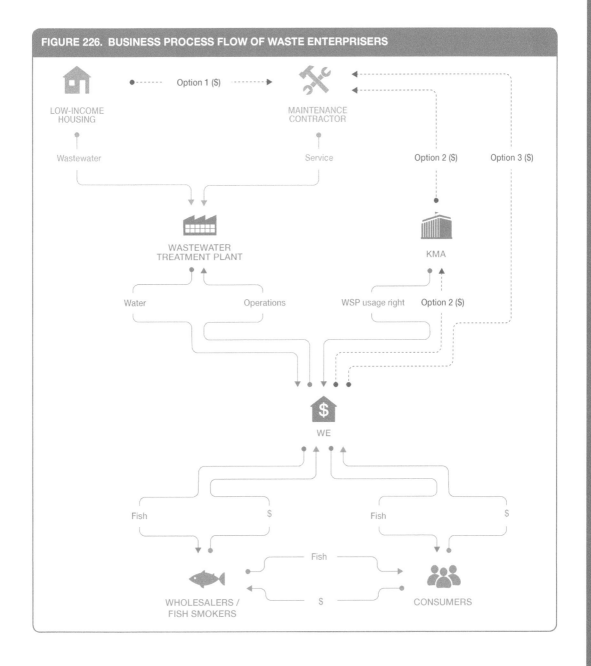

FIGURE 226. BUSINESS PROCESS FLOW OF WASTE ENTERPRISERS

So far, WE and its operational 'successors'[2] sold the produced catfish very easily to wholesalers who smoke the fish or sell it to fish smokers. Wholesalers are typically contacted and notified of harvest times. WE sold initially their product at a competitive price (USD 3/kg) equivalent to local market prices but could achieve far higher revenue by smoking its fish before sale, which would also help to control pathogenic health risks. One of the key strengths of the aquaculture business model is that once the WSP is in place, the additional start-up costs are low, and the operating costs (in particular staff salaries) become bearable with more than two WSP to manage. However, the fish production needs a pre-run to optimize fish stocking, feeding and survival (Amoah and Yeboah-Agyepong, 2015b). The key elements of the business model canvas are presented above.

Value chain and position

The fish produce is up to 80% directly supplied to wholesale/fish smokers, while 20% goes to consumers who roast or smoke the fish before it gets cooked (Figure 226). Up till now, demand for catfish remains higher than supply, and all fish brought offered gets also sold. Parts of the revenues from fish sale are used to maintain the treatment quality of the WSPs, without charging the low-income neighborhood. Although the "source of fish" is so far among the least important product attributes influencing consumers' purchasing decisions, a potential threat to the viability of the business could be that despite safety controls, traders or consumers start rejecting the fish.

Institutional environment

This is a public-private partnership business between WE and the city of Kumasi (KMA), where WE controls the operation and management of the WSP and KMA supplies the land with its treatment infrastructure and wastewater. While there are no legislations in Ghana that explicitly promote or ban the use of wastewater for aquaculture, an environmental impact assessment is required for commercial aquaculture[3]. With the permit from the Ghana Environmental Protection Agency (EPA) and a permit for (fresh) water usage from the Water Resources Commission (Act 522, 1996) the Fisheries Commission will approve the business. The WE-KMA public private partnership did not fit into the common scheme and was authorized through the agreement of the city to enter into contract with WE to support environmental sanitation in the city. Since then, the National Aquaculture Development Plan of 2012 was developed, which calls among others for more attention to fish health, and the 2013 established Ministry of Fisheries and Aquaculture Development published in 2014 through the Fisheries Commission, "National Aquaculture Guidelines and Code of Practice" to set minimum standards for operators in the aquaculture value chain and also prevent any possible negative impact of aquaculture on the environment in line with the Fisheries Regulations 2010 (L.I. 1968) and Fisheries Act, 2002 (Act 625).

WE's business was from 2011-on accompanied by research, e.g. on feeding, stocking and food safety by the Department of Fisheries and Watershed Management, Kwame Nkrumah University of Science and Technology (KNUST) and the International Water Management institute (IWMI). This was supported by a grant from the African Water Facility to Ghana's Water Resources Commission.

Technology and processes

In both project locations, the WSPs were overgrown and dysfunctional when WE arrived. The setup of the WSP systems is shown in Figure 227 on the example of Ahinsan. The system is made up of five sludge chambers: a grit, screening, influent, two inspection chambers and four treatment ponds, which were overgrown before WE took the WSP over. The four treatment ponds are: anaerobic pond (AP), facultative pond (FP), and first and second maturation ponds (MP I, MP II). Given the fixed number of connected households, the series of ponds make up an effective and low-cost means of treating wastewater, if well maintained. The last pond (MP II) or depending on water quality also MP I

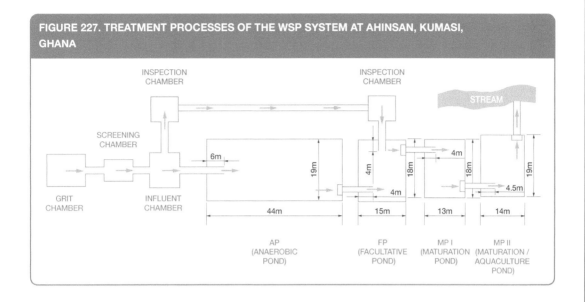

FIGURE 227. TREATMENT PROCESSES OF THE WSP SYSTEM AT AHINSAN, KUMASI, GHANA

and MP II are used to cultivate catfish, which has a relatively high tolerance for low levels of dissolved oxygen. Phosphorus and nitrogen provided with the wastewater are essential to facilitate production of natural microscopic plants and plankton which are food for the fish. There are two growing seasons per year and three fingerlings per m^2 are stocked in both maturation ponds per season, targeting an average annual production of about one ton per pond or 2t of fish per treatment plant with a survival rate of about 80%[4]. WE holds no inventory of fish at harvest and sells its product easily to wholesalers to be resold in the local markets to consumers and fish smokers for processing. Wholesalers are typically contacted and notified of harvest times.

Funding and financial outlook

Aquaculture, in general, appears to be a good business option in Ghana. A feasibility study by the Ministry of Food and Agriculture (MoFA) for Tilapia, indicated a positive Net Present Value (NPV) and Internal Rate of Return (IRR) of 32%, a Benefit Cost Ratio (BCR) of 1.18, and a payback period which is slightly longer than four years (Cobbina and Eiriksdottir, 2010). Aside labor and management costs, the cost of feed forms the bulk of the variable cost. Sensitivity analysis showed that the cost of feed, the fish survival rate as well as the farm gate price of fish are the main factors affecting profitability, while the most constraining factor for commercial aquaculture are the high start-up cost of which about 68% are fixed costs.

In the presented business case of WE, the possibility to use existing infrastructure, provided a huge cost saving (covering nearly all fixed cost except rearing infrastructure for fingerlings). Although wastewater was expected to support the development of a significant amount of feed for the fish, the experience of WE showed that this is not sufficient (or sufficiently balanced) and feeding remains recommended. This feeding pays off as catfish grown with wastewater eventually grew much larger than fish in freshwater control ponds (Amoah and Yeboah-Agyepong, 2015b).

Table 53 presents financial projection based on WE data for the management of one to five WSP systems, using a ten-year planning horizon. Data show that although with two systems, the business can break even, with three or more WSPs, staff costs are most efficiently used, resulting in a viable business with NPV and IRR positive.

SECTION IV: WASTEWATER AS A RESOURCE

CHAPTER 15. BEYOND COST RECOVERY

TABLE 53. INCOME STATEMENT OF WASTE ENTERPRISES

NUMBER OF WSP SYSTEMS				1	2	3	4	5
COST ITEMS								
A) Capital investment			GHC	4,495	8,990	13,485	17,980	22,475
B) Production Costs								
	Cost/stocking	Stocking/year						
Fingerlings	225	2	GHC	450	900	1,350	1,800	2,250
Fish Feed	70	2	GHC	140	280	420	560	700
Pond/Tank Maintenance	Cost/stocking	Stocking/year						
Patching Cement	10	2	GHC	20	40	60	80	100
Chlorine	10	2	GHC	20	40	60	80	100
Pond Liming	120	2	GHC	240	480	720	960	1,200
Hand Sanitizer	25	2	GHC	50	100	150	200	250
Total production costs			GHC	920	1,840	2,760	3,680	4,600
C) Administrative Costs								
Employees	Annual Salary							
Manager	GHC	15,000	GHC	15,000	15,000	15,000	15,000	15,000
Grounds Keeper	GHC	1,200	GHC	1,200	2,400	3,600	4,800	6,000
National Health Plan	GHC	20	GHC	40	60	80	100	120
Total Administration Costs			GHC	16,240	17,460	18,680	19,900	21,120
D) Total Operating Costs			GHC	17,160	19,300	21,440	23,580	25,720
Operating cost/system			GHC	17,160	9,650	7,147	5,895	5,144
REVENUE ITEMS								
E) Total revenue per number of systems			GHC	9,720	19,440	29,160	38,880	48,600
Administrative Cost as percentage of Total Revenues				167.0%	89.8%	64.0%	51.1%	43.4%

CASE: PPP LINKING WASTEWATER AND AQUACULTURE

Production cost as percentage of Total Revenues		9.5%	9.5%	9.5%	9.5%	9.5%
Total Cost as percentage of Total Revenues		176.5%	99.3%	73.5%	60.6%	52.9%
PROFIT ITEMS						
F) Total Profits	GHC	(7,440)	140	7,720	15,300	22,880
Interest Income/ (Expense)	GHC	–	–	–	–	–
Taxes	GHC	–	–	–	–	–
Total profits before Interest & Taxes	GHC	(7,440)	140	7,720	15,300	22,880
G) Net Profits	GHC	(7,440)	140	7,720	15,300	22,880
Net profits as percentage of Total Revenues		–76.5%	0.7%	26.5%	39.4%	47.1%
Net Present Value (calculated over a 10-year period)	GHC	(36,516)	(5,118)	25,030	55,178	85,325
Internal Rate of Return (calculated over a 10-year period)	negative net cash flows over 10 years	–17%	82.0%	119%	141%
Break-even year (calculated over a 10-year period)	not in ten years	year 1	year 1	year 1	year 1

Source: Waste Enterprises; updated.

1 Reported financials for Year 1 of business with two annual harvests of two 250m² ponds per system. 20% mortality rate. Fish sold fresh.
2 Aquaculture business has tax exemption for the first 5 years and thereafter an income tax of 10% is assumed. Capital costs after depreciation over 10 years.
3 Inflation rate assumed at 8% as in 2011/12; Exchange rate in mid-2011: USD 1.0 = GHC (GHS) 1.50.

SECTION IV: WASTEWATER AS A RESOURCE

Another option to make already a one-WSP system viable is to sell fish **smoked** and not fresh which allows a much higher sales price and return. For such a case, WE internal projections estimated for catfish an IRR of up to 45% (at 20% discount rate). With an estimated profitability index of 2.1, and a BCR of 1.13, a payback period of three years was estimated under favorable stocking and sales conditions (Amoah et al., 2015b).

As experience in wastewater aquaculture had first to be gained, time was lost with optimizing production on both WE sites, and after two years, revenues hardly covered daily operations. As indicated in the sensitivity analysis the profitability improved with increased fish survival, which was supported by the accompanying research. For the research, WE in association with IWMI and Ghana's Water Resources Commission attracted external funding from the African Water Facility. After its successful proof of concept, Waste Enterprisers planned initially to expand its aquaculture business across Africa with a Technical Director in charge of fish-farming, but then received funding to engage in another resource recovery challenge and discontinued fish farming (IWMI, 2012) while the accompanying research continued at the WSP sites until 2015.

Socio-economic, health and environmental impact

At the aggregate level, the business will help with the reduction in public health expenditures through avoided cost of diseases associated with untreated or only partially treated sewage entering surface water bodies, thereby leading to their improvements. On the other hand, health risks of workers at the WSPs, fish traders and consumers have to be assessed, monitored and minimized. This objective was supported through studies addressing pathogenic contamination, as well as the accumulation of heavy metals and pharmaceutical residues (Amoah and Yeboah-Agyepong, 2015a; Asem-Hiablie et al., 2013). Also the type of fish (African Catfish; *Clarias gariepinus*) was chosen for safety reason as in the study region it is normally smoked and cooked before consumption. As the Fisheries Act does not address fish health, quality assurance or product safety, a WHO recommended Hazard Analysis and Critical Control Point (HACCP) system was developed which allows to monitor a number of critical control points where compliance with safety procedures interventions is required to reduce or eliminate potential health risks (Yeboah-Agyepong et al. 2017).

In view of environmental impacts, the WSP rehabilitation and maintenance will improve the environmental situation. As wastewater aquaculture is so far not addressed in any legislation, the Ghana Environmental Protection Agency (EPA) became member of the steering committee of the business accompanying research.

Although fish meat analysis did so far not point at actual risks, in the Ahinsan or Chirapatre system, critical control-points are the smoking of the fish directly after harvest as well its well cooked consumption to remove pathogens from fish surface. An additional safety option would be to purify the fish in a fresh water pond after harvest and prior to sale, i.e. to clean as far as possible also the fish's digestive tract. Smoking of fish on-site, would also increase its market value, i.e. sales price.

Scalability and replicability considerations

In general, the investment climate for aquaculture is across and beyond Africa very positive. To promote and encourage new aquaculture enterprises in Ghana, they are granted for example a five-year tax free period (Cobbina and Eiriksdottir, 2010).

The use of (treated) wastewater for fish farming is more challenging. It has a long tradition, especially in Asia, and although it is supported by WHO (2006) with an own set of guidelines, many authorities might not agree with the idea especially where risk monitoring is weak. On the other hand, pond-based treatment systems are very common in many tropical countries, supporting housing estates, towns, military camps, universities, boarding schools, etc. Moreover, the majority of these systems are on a similar trajectory to failure as observed in Kumasi (Murray and Drechsel, 2011). Thus, the general drivers for the success of the business are:

- Supportive (or at least non-restrictive) regulations and policies, and positive perceptions.
- High local demand for catfish, allowing to share profits.
- Win-win public-private partnership resulting in low capital cost investment by the private partner.
- Research partnership to monitor and optimize system safety and productivity.

The implemented model has a significant potential for replication and scaling up if compliance with national or international safety guidelines such as WHO (2006) can be assured. The accompanying research in Kumasi resulted in fish farming manual and implementation plan summarizing the lessons learnt from wastewater aquaculture (Amoah et al., 2015a; Amoah and Yeboah-Agyepong, 2015b). But even with full compliance, market demand remains also a function of risk awareness and consumer perceptions, which has to be considered in local feasibility studies. Where wastewater treatment systems are to be newly set up for aquaculture, land requirements for pond-based systems have to be considered. The maintenance of the ponds can eventually be outsourced, or become part of the business.

Summary assessment – SWOT analysis

The model WE developed was intended to inspire opportunities that exist for using the resource value of human waste to the economic benefit of the sanitation sector. The aquaculture business supports via the productive use of treated wastewater the maintenance of otherwise dysfunctional wastewater treatment plants without charging poor households. With fish being nation-wide the second most important food consumption subgroup, market demand, especially for catfish is high. The strength of the business (Figure 228) is its ability to negotiate for the supply of free wastewater and land, which helps reduce fixed cost by 70%. The benefits are equally important for the municipality which is lacking funds to maintain environmental and human health. The HACCP system, fish smoking and boiling minimizes risks, and make the fish acceptable to traders. However, changing public perception remains a potential threat. Day to day challenges were more of technical nature, like optimizing fish survival and feeding.

FIGURE 228. SWOT ANALYSIS OF WASTE ENTERPRISERS AQUACULTURE BUSINESS

	HELPFUL TO ACHIEVING THE OBJECTIVES	**HARMFUL** TO ACHIEVING THE OBJECTIVES
INTERNAL ORIGIN ATTRIBUTES OF THE ENTERPRISE	**STRENGTHS** • Continuous free water supply • Low investment costs • Contractual agreement with the authority • Direct contact with wholesalers • Safety monitoring • Economies of scale in the long-run due to low average cost and gained expertise • Savings for authority and no charge to poor households	**WEAKNESSES** • Requires space for pond based treatment plants • Requires uncovered demand for cultivated fish species • Needs efficient systems to handle possible on- and off-site health risks (e.g. high risk where fish is eaten uncooked) • Requires aquaculture expertise (feeding, survival, predator and duckweed control, etc.)
EXTERNAL ORIGIN ATTRIBUTES OF THE ENVIRONMENT	**OPPORTUNITIES** • Easily replicable • Significant market demand for fish • Missing regulations and supportive investment climate	**THREATS** • With city growth, more compact treatment plants appear • Competition in alternative fish production may cause fish price to decrease • Health risks reduction and monitoring costs for fish consumption may be high for the business • Changing regulation and/or consumer perception of wastewater-reared fish

Contributors

George K. Danso, Government of Alberta
Michael Kropac, CEWAS
Mark Yeboah-Agyepong, KNUST, Kumasi
Nelson Agbo, KNUST, Kumasi

References and further readings

Amoah, P., Bahri, A. and Aban, E.K. 2015a. Treated wastewater aquaculture. Manual for production of African Catfish (Clarias gariepinus). A Report for AWF/AfDB Project no. 5600155002451 on Design for reuse submitted on 28 February 2015. 72pp.

Amoah, P., Esseku, H., Gebrezgabher, S., Yeboah-Agyepong, M. and Agbo, N. 2015b. Implementation plan. The production of African Catfish (Clarias gariepinus) in waste stabilization ponds in Kumasi and Tema, Ghana. A Report for AWF/AfDB Project no. 5600155002451 on Design for reuse submitted on 28 February 2015. 49pp.

Amoah, P. and Yeboah-Agyepong, M. 2015a. Public health considerations of African Catfish (Clarias gariepinus, Burchell 1822) cultured in waste stabilization pond system in Kumasi, Ghana. A Report for AWF/AfDB Project no. 5600155002451 on Design for Reuse. Submitted 28 February 2015. 42pp.

Amoah, P. and Yeboah-Agyepong, M. 2015b. Growth, survival and production economics of African Catfish (Clarias gariepinus) fed at different feed application rate in wastewater from waste stabilization pond system in Kumasi, Ghana. A Report for AWF/AfDB Project no. 5600155002451 on Design for Reuse. Submitted 28 February 2015. 35pp.

Asem-Hiablie, S., Church, C.D., Elliott, H.A., Shappell, N.W., Schoenfuss, H.L., Drechsel, P., Williams, C.F., Knopf, A.L. and Dabie, M.Y. 2013. Serum estrogenicity and biological responses in African Catfish raised in wastewater ponds in Ghana. Science of the Total Environment 463–464: 1182–1191.

Asmah, R. 2008. Development potential and financial viability of fish farming in Ghana. Institute of Aquaculture, University of Sterling, Scotland. www.storre.stir.ac.uk/bitstream/1893/461/1/PhD%20Document.pdf (accessed 5 Nov. 2017).

Awity, L. 2005. National aquaculture sector overview. Ghana. In: FAO fisheries and aquaculture department [online]. Rome. Updated 10 October 2005. www.fao.org/fishery/countrysector/naso_ghana/en. Accessed 26 April 2016.

Cobbina, R. and Eiriksdottir, K. 2010. Aquaculture in Ghana: Economic perspectives of Ghanaian aquaculture for policy development. Final Project report. UNU Fisheries Training Programme. www.unuftp.is/static/fellows/document/rosina_2010prf.pdf (accessed 5 Nov. 2017).

GLSS. 2014. Ghana living standards survey (GLSS) Round 6. Government of Ghana. www.statsghana.gov.gh/glss6.html (accessed 5 Nov. 2017).

Henriksen, J. 2009. Investment profiles based on intervention opportunities for Danida business instruments in the Ghanaian aquaculture sub-sector. Danish Ministry of Foreign Affairs. Cited in Cobbina and Eiriksdottir, 2010.

IWMI. 2012. Hitting pause on aquaculture. IWMI News 5 September, 2012 [online]. www.iwmi.cgiar.org/News_Room/pdf/Waste-enterprisers_com-Hitting_pause_on_aquaculture.pdf (accessed 5 Nov. 2017).

Keraita, B. and Drechsel, P. 2015. Consumer perceptions of fruit and vegetable quality: Certification and other options for safeguarding public health in West Africa. Colombo, Sri Lanka: International Water Management Institute (IWMI). 32pp.

Murray, A. and Buckley, C. 2010. Designing reuse-oriented sanitation infrastructure: the design for service planning approach. In: Drechsel, P., Scott, C.A., Raschid-Sally, L., Redwood, M. and Bahri A. (eds) Wastewater Irrigation and health: Assessing and mitigation risks in low-income countries. UK: Earthscan-IDRC-IWMI. p. 303–318.

Murray, A. and Drechsel, P. 2011. Why do some wastewater treatment facilities work when the majority fail? Case study from the sanitation sector in Ghana. Waterlines 30 (2), April 2011: 135–149.

Owusu-Sekyere, E., Harris, E. and Bonyah, E. 2013. Forecasting and planning for solid waste generation in the Kumasi metropolitan area of Ghana: An ARIMA time series approach. International Journal of Sciences 2.

Rurangwa, E., Agyakwah, S.K., Boon, H. and Bolman, B.C. 2015. Development of Aquaculture in Ghana. Analysis of the fish value chain and potential business cases. IMARES report C021/15 for the Embassy of the Netherlands. https://www.rvo.nl/sites/default/files/2015/04/Development%20of%20Aquaculture%20in%20Ghana.pdf (accessed 5 Nov. 2017).

Tenkorang, A., Yeboah-Agyepong, M., Buamah, R., Agbo, N.W., Chaudhry, R. and Murray, A. 2012. Promoting sustainable sanitation through wastewater-fed aquaculture: A case study from Ghana. Water International 37 (7): 831–842.

WHO. 2006. WHO guidelines for the safe use of wastewater, excreta and greywater. Vol III. Wastewater and excreta use in aquaculture. Geneva: UNEP, WHO.

Yeboah-Agyepong, M., Amoah, P., Agbo, W.N., Muspratt, A. and Aikins, S. (2017): Safety assessment on microbial and heavy metal concentration in Clarias gariepinus (African catfish) cultured in treated wastewater pond in Kumasi, Ghana, Environmental Technology http://dx.doi.org/10.1080/09593330.2017.1388851

See also: www.flickr.com/photos/waste-enterprisers/sets/72157627841508651/.

Case descriptions are based on primary and secondary data provided by case operators, insiders or other stakeholders, and reflect our best knowledge at the time of the assessments 2014/16. As business operations are dynamic data can be subject to change.

Notes

1 After a test period the arrangement was changed to accelerate the maintenance process, and WE organized directly full-time plant maintenance, i.e. without need for KMA to organize this (see Figure 226, option 3).
2 The pond systems were till 2015 maintained by the local university (KNUST) and IWMI for research purposes. One of the ponds is currently (2017) used as a fish hatchery.
3 So far, mostly commercial private sector operators undertook environmental impact assessment, but not small-scale operators (Awity, 2005).
4 High survival rates were achieved with longer feeding periods (rearing fingerlings to at least 20g) and after successful removal of a large numbers of predators (snakes) from the ponds.

BUSINESS MODEL 18
Leapfrogging the value chain through aquaculture

Pay Drechsel and Munir A. Hanjra

Key characteristics

Model name	Leapfrogging the value chain through aquaculture
Waste stream	Domestic wastewater
Value-added waste product	Reclaimed water, fish feed, fresh fish and/or packaged fish, irrigated crops
Geography	Regions where inland fish is in higher demand than supply
Scale of production	Small–medium scale; 200–1,000m^3 wastewater intake per day
Supporting cases in this book	Kumasi, Ghana; Mirzapur, Bangladesh
Objective of entity	Cost-recovery [ ]; For profit [X]; Social enterprise [ ]
Investment cost range	USD 20,000 to 100,000 plus cost of suitable land/lagoons of about 1–5 ha
Organization type	Mostly public-private partnership, but also other options
Socio-economic impact	Environmental pollution reduction, health risk reduction, job creation, food security
Gender equity	Inequity likely on farm in view of access to land, knowledge and capital, while gender roles in fish marketing vary between countries

Business value chain

Wastewater-fed aquaculture has a long tradition, especially in South and Southeast Asia, and is being recognized as an innovative business-oriented reuse system where sufficient land is available and possible health risks can be controlled, e.g. by avoiding mixed wastewater, which contains industrial effluent.

There are two different conceptual variations possible. From a safety perspective, a model as used in the presented case of Bangladesh is being preferred where the treatment process includes duckweed to absorb large amounts of nutrients, transforming them into high quality protein. The harvested duckweed is then used to feed fish grown, e.g. with groundwater in vicinity. Possible chemical contamination of the food is being monitored.

In a variation of the model, fish receives its food directly in the treatment system, where it is cultivated in the last maturation pond of multiple treatment pond set-up. To reduce health risks in this case, WHO guidelines are strictly to be observed. The treated water can be released safely in the environment, or reused for crop production in areas where irrigation water is scarce. The business model adds economic value to an existing pond-based treatment infrastructure by offering with limited additional investments different revenue options linking into high revenue value chains. The model is suitable for small- to medium-scale operations at community or institutional level where land is available, water quality is known and fish and irrigated crops have an assured local market demand (Figure 229).

The dotted short cut can further reduce capital costs but although increases public health risks, thus can only be recommended where strict water quality monitoring is possible.

In both studied cases, the institution in charge of safe wastewater disposal teamed up with an entity experienced in wastewater treatment and aquaculture. This could be a public-private partnership (PPP), but also private-private partnerships, e.g. where a private university or hospital is teaming up with an enterprise or NGO, or only public operation. In the public-private case, the public entity provides wastewater and [a budget to set up] infrastructure for wastewater treatment and safe disposal, while the private partner offers treatment expertise and invests either in additional fish ponds and/or fish fingerlings, and assures the O&M of the overall treatment system.

The interesting aspect of the PPP is the realization of a multiple win-win situation: while the public partner gets the treatment and waste disposal done without paying for the O&M service, the private partner benefits from the – in large – already existing/budgeted infrastructure and can with very limited own capital investment produce a high-value product for revenue generation. Depending on demand and supply, the contractual agreement for using the land and/or pond system can also include a profit sharing arrangement like in the Ghana case, which allows the public entity recover some of its own operational costs. Finally, the generated revenues can allow the authorities to 'pro-poor' waive sanitation fees for the served wastewater generating households or entities.

The key players in the business set-up are the aquaculture business entity, if needed with (access to) expertise in phyto-remediation, the local municipality and/or local organization in need of wastewater

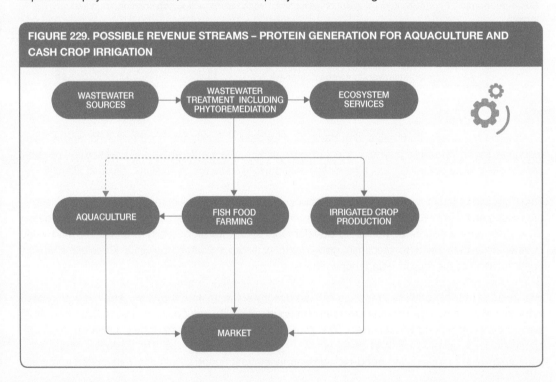

FIGURE 229. POSSIBLE REVENUE STREAMS – PROTEIN GENERATION FOR AQUACULTURE AND CASH CROP IRRIGATION

BUSINESS MODEL 18: LEAPFROGGING THE VALUE CHAIN

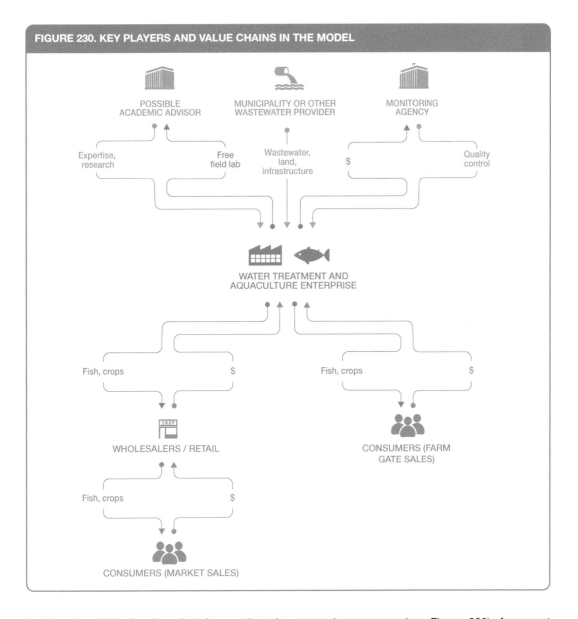

FIGURE 230. KEY PLAYERS AND VALUE CHAINS IN THE MODEL

treatment, and the local market, i.e. produce buyers and consumers (see Figure 230). An expert partner, able to carry out locally applied research in fish or duckweed farming, like a local university, could add value. Finally, an important stakeholder is the one in charge of monitoring water, crop and fish quality. Although the fish farming business will give highest priority to maintaining consumer trust, independent quality control is recommended. This could be the local agency in charge of food safety.

Given the limited capital investment needs for the enterprise, financing should be possible in many countries through a bank loan at a term of five years, best at a subsidized lending rate given the public sector support. The business has the potential to impact local residents through the production of inland fish and the creation of employment opportunities along the aquaculture value chain.

Business model

The wastewater from the community is brought to the treatment ponds through an existing sewerage network (in case of municipal wastewater treatment). Fish farming can be integrated in the treatment system or preferably be indirectly linked via the harvested phyto-biomass (e.g. duckweed). The recommended business model uses wastewater to produce on-site fish feed and with the feed off-site (i.e. not within the treatment system) fish. It offers through the sale of fish to end-users and/ or intermediate traders a value proposition with a much higher revenue stream than the sale of the reclaimed water would allow (Figure 231). The business usually relies on a (public-private) partnership, which acts on an opportunity that derives from a need for both wastewater treatment and a market which can absorb more fish than on offer. The business is cost-driven, and can offer cheap produce through minimal capital costs for infrastructure, and low-cost operation. Low cost operation is enabled through the free provision of nutrient rich water and the duckweed technology which allows

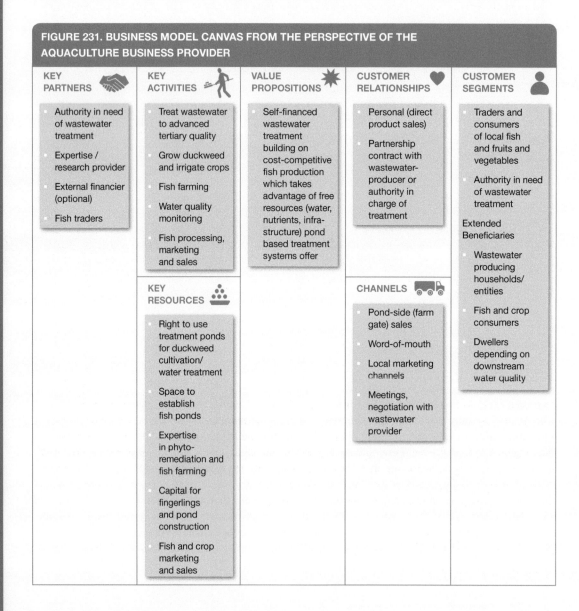

FIGURE 231. BUSINESS MODEL CANVAS FROM THE PERSPECTIVE OF THE AQUACULTURE BUSINESS PROVIDER

BUSINESS MODEL 18: LEAPFROGGING THE VALUE CHAIN

COST STRUCTURE	REVENUE STREAMS
• Capital investment in fish and fish ponds (unless part of final treatment system) • O&M of ponds, mostly labor incl. security against illegal fish harvest; debt repay • Fish and crop marketing and sales • Research collaboration (fish growth and produce safety) • Benefit sharing with public partner (optional)	• Pond-side sales of fish to customers, retail, whole sale • Pond-side sales of crops and fish feed (if in excess) • Payment for water treatment service (optional)
SOCIAL & ENVIRONMENTAL COSTS	SOCIAL & ENVIRONMENTAL BENEFITS
• Possible human health hazard from consumption of fish grown in (proximity to) treated wastewater, where human pathogens may still exist and may be carried into human • Laborers' health risk due to contact with wastewater, especially when harvesting duckweed	• Wastewater efflux of organization(s) is treated, which results in reduced environmental pollution pressure to water bodies and downstream water users • More fish protein and crops on the market at low cost • Job creation

to produce most of the required fish feedstuff within the treatment system. Costs are also kept low due to farm gate marketing and no need for storage. Irrigated crop production offers a secondary revenue stream. Labor required for duckweed management and feeding to the fish and for fish harvesting is locally available and manpower can be trained on-site. Although the manual operations are simple which helps to save costs and to move the business towards net profits, aquaculture and even more wastewater-aquaculture requires significant management experience and skills to maintain a high fish survival rate and manage the right feeding for optimal fish growth. Partnering with an expertise provider/research institution on phyto-remediation and fish rearing will be useful unless the expertise is internally available to avoid high startup costs through 'learning by doing'. This type of enterprise may flourish at small to medium scale wherever sufficient land for both, pond based treatment and fish farming can be set up in proximity or interlinked, and where water for fishing or fish farming is generally limited. In coastal regions, possible competition from saltwater fish has to be explored. In any situation, either if fish is grown with reclaimed water (or fish is fed with plants produced in wastewater) the business requires a conducive legal-regulatory setting and quality monitoring given potential consumption as well as occupational health risks.

Potential risks and mitigation

The business model presented here was designed based on a detailed analysis of the two case studies from Ghana and Bangladesh, as well as other cases and references. There can be a variety of business risks affecting the successful implementation of such a model, most of them being more generic than model specific. For example, as reuse projects involving wastewater are potentially harmful to human and environmental health, particular health risk (mitigation) options are obvious and have to be addressed, like also community acceptance. However, also other risks such as those defined below have to be addressed, although there will be location specific differences.

Market risks: Fish is a protein-rich, nutritious source of human food and the assumption is that a strong market exists for onsite direct sale to consumers and/or sale through retail. Where the source of fish on the market is known, some consumers might not like to eat fish raised with duckweed grown in wastewater. However, it is unlikely that traders will brand their produce in a way that could jeopardize their business.

Competition risks: Fish produced on wastewater competes directly with local freshwater (or also sea water) fish and indirectly with frozen product from oversea markets, which at times could be cheaper than the local produce in some countries. Therefore, the advantage of low-cost production (using free feed) have to be used to sell the fish at a competitive price.

Technology and performance risks: Natural water quality remediation measures are usually low-cost. The technology of duckweed production for fish farming is straight forward and mature, and can build on decades of research and development. Local workforce can be trained in the operations. Fish farming itself requires more expertise than the water treatment as well as quality monitoring.

Political and regulatory risks: Fish farming in general is a supported agricultural practice, and there are no known political and regulatory risks in most settings. If the water used for the fish is part of the treatment chain, the business requires a legal-regulatory setting that is conducive to this situation, and thus a threat to the business might come from particular or changing safety regulations.

Social equity related risks: The model is considered to have more advantages for male entrepreneurs (farmers) although in many places cultural tradition steers if more men or women are involved in fish farming. However, in many regions, women have comparatively to men less access to land, education or capital, which are crucial for entering aquaculture. Still, there can be regionally more women working in the sector than men. In Asian countries such as Cambodia, Bangladesh, Indonesia and Vietnam, for example, women carry out 42–80% of all aquaculture activities, with equally large variations along the value chain. See also World Fish Center (2011).

Safety, environmental and health risks: The model can be very safe but requires significant attention to risk monitoring and control (Table 54). There can be specific health concerns for workers harvesting the duckweed from the wastewater, which can however be addressed with protective gear, harvesting equipment and good hygiene. In the less preferred variation of the model where fish is grown with reclaimed water, the risks extend also to the fish and thus the consumer. For this situation, the WHO (2006) guidelines for wastewater use in aquaculture apply. A common way to reduce consumer microbial

TABLE 54. POTENTIAL HEALTH AND ENVIRONMENTAL RISK AND SUGGESTED MITIGATION MEASURES FOR BUSINESS MODEL 18

RISK GROUP	EXPOSURE					REMARKS
	DIRECT CONTACT	AIR/ ODOR	INSECTS	WATER/ SOIL	FOOD (FISH)	
Fish farmer/operator	HIGH	LOW	MEDIUM	MEDIUM		Consumer awareness and information has to be supported on the source of the traded fish
Community		LOW				
Fish consumer					MEDIUM	
Mitigation measures	🧤	😷	🦟	🚱 Pb Hg Cd	🥬 🍲 Pb Hg Cd	

Key: NOT APPLICABLE | LOW RISK | MEDIUM RISK | HIGH RISK

risks is through fish smoking or grilling, although contaminants might survive, in particular within the fish. Purification in clean water ponds could address this challenge to some degree, as well as careful separation of meat and the digestive tract during slaughtering, and cooking. Regular monitoring of the inflowing wastewater and fish could help detect possible chemical risks, although laboratory capacity for so-called "emerging contaminants" is still missing in many developing countries.

Business performance

The business model supports the move beyond cost recovery towards profitability. The combination of ponds or zig-zag flow systems with a phyto-remediation step are applied low-cost technologies which treat wastewater to an advanced tertiary state. Using the phyto-(plant) biomass as 'in-house' production of fish feed, such as duckweed, and the low labor requirements of the system significantly reduces operational costs for nearby fish farming while the free use of land reduces capital cost. Where fish has a market, the system can make profits even where no subsidies are received and no wastewater treatment fees are charged. Capital costs could be further reduced where fish is grown within the last part of a pond based system. However, this variation of the model is significantly increasing health risks and can only be considered where water quality and risk mitigation measures fully correspond with safety recommendations. The model ranks also high in terms of environmental impacts due to the wastewater treatment, in particular nutrient removal, and on social impact due to protection of public health, plus the additional supply of nutritious fish and local jobs (Figure 232). The model ranks lowest on scalability and replicability criteria due to its land requirements. Yet, the business model highlights strong potential for replication for a developing country setup with limited institutional capacities and its applicability to peri-urban areas and towns where land is not yet in short supply. The model is thus attuned to the needs of small- and medium-size communities where high tech wastewater treatment plants will not achieve cost-economies and/or might not be affordable. The system can be scaled to the needs of the local communities as the inputs are as simple although

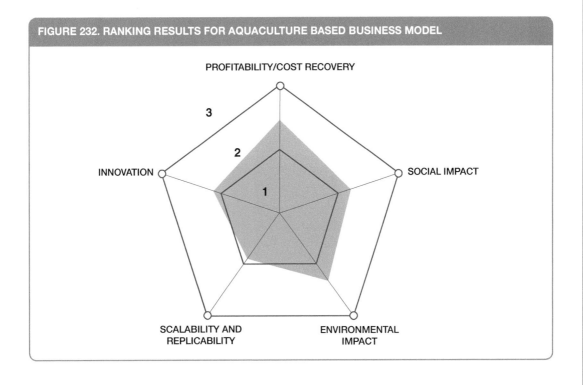

FIGURE 232. RANKING RESULTS FOR AQUACULTURE BASED BUSINESS MODEL

the rearing of fish should not be underestimated and requires an experienced partner. Further, the regulatory setup should support production and sale of fish from such a system, even if fish is only indirectly in contact with the water. There exists a greater potential for this model in countries that are land-locked (no sea food), have limited surface water resources while fish is a welcomed staple food in the local diet.

References and further readings

World Fish Center. 2011. Gender and aquaculture: Sharing the benefits equitably. Issues Brief 2011-32 http://pubs.iclarm.net/resource_centre/WF_2832.pdf (accessed 5 Nov. 2017).

WHO. 2006. Guidelines for the safe use of wastewater excreta and greywater. Volume III. Wastewater and excreta use in aquaculture. Geneva: World Health Organization.

16. BUSINESS MODELS FOR COST SHARING AND RISK MINIMIZATION

Introduction

Public resources, including Official Development Assistance (ODA) are not sufficient for achieving the Sustainable Development Goals (SDGs) targets in particular in the waste management and sanitation sectors with strong regional funding gaps. Private finance appears critical and the sector is increasingly looking to emerging and frontier markets for investment opportunities. However, current levels of private investment in sectors related to the SDGs are relatively low. Only a fraction of the worldwide invested assets of banks, pension funds, insurers, foundations and transnational corporations is in sectors critical to the SDGs (World Economic Forum, 2015). Translating these assets into SDG-compatible investments will be key, with the potential being greater in sectors related to the circular economy, including infrastructure (power, renewable energy, water and sanitation). Yet, despite growing interest, significant barriers to private sector engagement remain at all levels, including inefficient financial markets, weak institutions, regulatory frameworks and enabling environments, and macroeconomic and political instability (see also Chapter 19). All these barriers contribute to a more risky, challenging, and uncertain environment for investors, particularly when compared to more developed markets where beneficiaries do not face affordability constraints (and governments can set tariffs at cost recovery level) to balance the risk and reward of investments.

Investment guarantees are designed to mitigate risk for private and/or public sector financing. These can be guarantees for public projects (Partial Credit Guarantees) and for private projects (Partial Risk Guarantees; PRG) with counterguarantee from the member government. There can also be political risk insurance as offered by the World Bank's Multilateral Investment Guarantee Agency (Box 9).

Especially important for private investors are PRGs which cover private lenders against the risk of a public/governmental entity failing to perform its obligations with respect to a private project. Eligible projects are public-private partnerships (PPP) such as Build-Operate-Transfer (BOT) projects. PRGs can cover a range of risks including, changes in law, failure to meet contractual payment obligations, obstruction of an arbitration process, expropriation and nationalization, foreign currency availability and convertibility, failure to issue licenses, approvals and consents in a timely manner, etc.

Also, public investors can strategically use their funds to mitigate investment risk and/or enhance returns for private investors by supporting blended finance transactions. Blended Finance is the strategic use of development finance and philanthropic funds to mobilize private capital flows to

Box 9. Multilateral Investment Guarantee Agency (MIGA)

MIGA offers political risk insurance and can cover equity, shareholder loans and loan guarantees issued by equity holders; it can also cover loans by third party institutions, usually commercial banks, provided that a shareholder's investment in the project is also being insured by MIGA. Like other investment insurers, MIGA can provide broad coverage to investors against such risks as currency transfer, war and civil disturbance and expropriation; it can customize these coverages to suit the particular needs of investors. MIGA can normally issue coverage within a few months of an investor's application since it does not enter into counter-guarantee arrangements with the host country government of the project. MIGA is also a key partner in the presented case study from Jordan.

Source: http://siteresources.worldbank.org/INTGUARANTEES/Resources/Guarantees_Q&A_03172009.pdf

emerging and frontier markets to do more with limited public funds. Blended Finance enhances the impact of limited philanthropic and development resources by using those funds to tap into the dollars of private capital available in global markets. It offers promising potential as an ecosystem solution to close the development funding gap. Estimates suggest that public capital deployed through Blended Finance transactions can attract one to ten times the initial amount in private investment.

The public-private partnership setup of the As Samra wastewater treatment plant in Jordan was hailed for its innovative financing model, using government funds and donor grants to leverage private sector investments. Taking the business model of As Samra as example, the described model highlights the key components of the financial set up rather than the operational side of the plant as the approach could prove beneficial in other contexts. How the model was implemented in the case of As Samra is presented as case study.

While the financial set up can be applied to many large-scale treatment plants, independently of their efforts towards resource recovery and reuse, the model is particularly relevant for realizing social and environmental benefits which "treatment for reuse" projects offer in water-scarce regions, in particular if they fall short of financial viability.

The most common cost and risk sharing mechanism to support projects that are economically justified but not financially viable, is **viability gap funding**. Through targeted investment of public or donor funds in infrastructure development costs, private sector can be enticed to assume responsibility for construction, operation and maintenance of the facility, provided the venture becomes profitable and bankable. Especially in cases with fixed or capped water user tariffs, cost recovery options require intensive analysis and negotiation. The viability of the project is supported through a highly efficient energy recovery mechanism, which is significantly reducing operational costs.

For the model to work, several framework conditions are indispensable. A stable regulatory and political environment is prerequisite for partners to engage, especially for large-scale and long-term ventures. The combination of multiple public and private funding channels creates an interdependence of payment streams because each contribution will only pay off if the other parties fully comply with their commitments. All partners need to negotiate comprehensive contractual and risk mitigating agreements to provide necessary guarantees and remedies.

If these conditions are met, the public-private partnership setup can lead to state-of-the-art wastewater treatment facilities and management processes with large efficiency gains compared to traditional models. The public investment, albeit a grant contribution, leverages additional funds from private investors and can thereby deliver wastewater treatment services more efficiently and at larger scale. The provision of safe, treated water for reuse in agriculture and industrial operations contributes to economic development and environmental protection, especially in water-scarce regions.

References and further readings

Humphrey, C. and Prizzon, A. 2014. Guarantees for development: A review of multilateral development bank operations. London: Overseas Development Institute.

World Bank. 2010. The World Bank guarantees: Leveraging private finance for emerging markets. Washington DC: World Bank. http://siteresources.worldbank.org/INTGUARANTEES/Resources/GuaranteeBrochureEnglishApril2010Final.pdf (or https://goo.gl/Ht2EHY; accessed 6 Nov. 2017).

World Economic Forum. 2015. Blended finance Vol. 1: A primer for development finance and philanthropic funders. An overview of the strategic use of development finance and philanthropic funds to mobilize private capital for development. Geneva.

CASE
Viability gap funding (As Samra, Jordan)

P. Drechsel, G.K. Danso and M.A. Hanjra

Supporting case for Business Model 19	
Location:	As Samra, Amman, Jordan
Waste input type:	Wastewater
Value offer:	Treated wastewater, hydropower, biogas, carbon offsetting
Organization type:	Public-private partnership (PPP)
Status of organization:	New treatment plant completed in 2008 and extended between 2012 and 2015
Scale of businesses:	Large scale
Major partners:	Samra Wastewater Treatment Plant Company Limited; Millennium Challenge Corporation; Government of Jordan; Consortium of banks

Executive summary

Water scarcity puts water reuse very high on Jordan's development agenda. The As Samra wastewater treatment plant (WWTP) which is the largest in the country was purposely designed to support agricultural production in the Jordan Valley that relies increasingly on treated wastewater for irrigation purposes. Set up as a public-private partnership (25-year BOT contract) the WWTP is located near Amman. Building on an older pond-based treatment plant, a new WWTP was constructed between 2003 and 2008 (phase 1) and expanded from 2012 to 2015 (phase 2) with financial support from the United States Agency for International Development (USAID; phase 1) and a Viability Gap Funding by the Millennium Challenge Corporation (MCC, phase 2), to reach a capacity of 364,000m³ per day. Under the coordination of the Ministry of Water and Irrigation, the construction was facilitated by a 20-year commercial loan, the longest maturity that Jordanian banks have ever offered so far, and a comprehensive risk sharing arrangement. The contractual structure developed for the As Samra expansion (2012–2015) has a high replication potential elsewhere in the world, to allow projects that are economically and environmentally beneficial to be implemented and operated by the private sector also where such projects would otherwise be unaffordable to the public sector. The expanded As Samra Wastewater Treatment Plant was inaugurated in October 2015 to provide Jordan with up to 133 million cubic meters of treated water per year. Already today, treated wastewater is representing 13 percent of Jordan's entire renewable water resources, freeing up fresh water for more valuable uses. Ten percent of the country's agricultural water consumption comes from the As Samra plant. In addition, the As Samra plant is able to generate up to 95% of its energy needs, supported in part by a favorable topography. The production of renewable energies allows the plant to reduce its carbon footprint by about 300,000t of carbon dioxide (CO_2e) per year.

CASE: VIABILITY GAP FUNDING

KEY PERFORMANCE INDICATORS (2015/16)	
Land use:	About 400 ha owned by the Ministry of Water and Irrigation (MWI)
Water treated:	A design capacity of 364,000m^3 per day able to serve about 3.5 million capita
Capital investment:	Phase-1 (2003–2008) USD 169 million; Phase-2 (2012–2015) about USD 223 million[1]
Labor employment:	About 180–210 permanent local employees, of which about 70 are skilled workers; plus up to 2,500 during the construction phases
Operation and maintenance cost:	Full cost recovery (at the time of study USD 1.3 million per month)
Outputs:	364,000m^3 per day wastewater treatment capacity 90–95% energy self-sufficiency; 300,000t CO_2e per year carbon savings 118t of dry sludge (DS) per day in 2011, to increase to 194t in 2025
Potential social and/or environmental impact:	Significantly improved water quality, less contamination of soil and groundwater, reduced carbon foot print; treated water for irrigation; livelihoods support for irrigating farmers, plus 180–210 new jobs at the WWTP
Financial viability indicators:	Bank loan back period: 13–20 years Post-tax IRR: 10–18% (t.b.c.) Gross margin: undisclosed

Context and background

The Hashemite Kingdom of Jordan covers a territory of about 90,000 km^2. Rainfall is confined largely to the winter season and ranges from around 660mm in the north-west of the country to less than 130mm in the eastern and southern deserts, which form about 90% of the surface area. Under low rainfall, high evaporation and increasing crop intensification, Jordan is since long over-exploiting its available water resources with severe consequences for the Lower Jordan River Basin and the Dead Sea where over the last decades decreasing amounts of water arrived (Courcier et al., 2005).

Wastewater collection and treatment services were provided to about 63% of the Jordanian population in 2013, producing about 137 million cubic meters (MCM) of treated wastewater annually that is being reused primarily in agriculture. The remaining population uses septic tanks and cesspits in rural and dispersed settlements. With the increasing population and the country's social and economic development, the amount of treated wastewater is growing. It is estimated that by 2030, the volume of treated wastewater will be 240 MCM. Currently, more than 70% of the wastewater treated in Jordan comes from the As-Samra wastewater treatment plant which underwent between 2003–2008 and 2012–2015 major construction work. The plant replaced an overburdened stabilization pond system which was despite some extension work no longer able to maintain effluent water quality at acceptable levels. Its treated effluent is collected in the King Talal Reservoir (KTR) which is supporting most of the farming in the Jordan valley. In the KTR, the wastewater gets mixed with rain/freshwater from the Zarqa river basin. The mixed water irrigates about 20,000 ha in the middle and lower Jordan Valley, replacing its dwindling freshwater flow (Seder and Abdel-Jabbar, 2011). The wastewater flow is facilitated by a favorable topographical situation, allowing a low-cost transfer of urban wastewater via As Samra to the irrigation areas (McCornick et al., 2004; Courcier et al., 2005). Amman, the capital of Jordan, produces the bulk of the wastewater treated in As Samra.

About 80% of the agricultural water consumption and production in the lower and middle Jordan valley depends on blended wastewater (World Bank, 2016). Fruits and other cash crops form the major component of reuse in the Jordan valley. Aside indirect wastewater reuse of treated wastewater mixed with fresh water, also direct use (i.e. of unmixed wastewater) exists to a smaller extent in the vicinity of As Samra.

Market environment

While globally many WWTP have smaller reuse activities, As Samra is an example of a WWTP with a strong double value proposition (Wichelns et al., 2015), where the national water scarcity makes the production of water 'fit for reuse' of equal if not larger importance than the provided sanitation service.

The demand for the As Samra plant stimulated a range of institutional, financial and regulatory innovations to make the project happen. The plant represents the first private sector co-financed BOT project in Jordan, as well as the first public-private partnership in financing and management of a public infrastructure project in the country, using a mixed financing model that accommodates that neither water reuse nor the water tariff will be major revenue streams.

The market acceptance and penetration of mixed fresh/wastewater is high and competition is almost none as fresh water resources are fully exploited. Given population growth, which is expected to exceed 7.8 million by 2022, increasing fresh water abstraction or reallocation for domestic needs implies also more available wastewater for irrigation. The benefits of safely treated wastewater are well recognized by most stakeholder, especially in the public sector (Carr and Potter, 2013). In summary, the Jordanian market for further reuse-oriented WWTPs is very positive, and Jordan is not the only water-scarce country in the subregion.

Macro-economic environment

The inclusion of wastewater reuse in the country's National Water Strategy since 1998 was an important signal of placing high priority on the value of reclaimed water. The 2016–2030 National Water Strategy and the national substitution policy consider treated wastewater effluent as a core water resource that has been added to the water budget, with priority given to agriculture for unrestricted irrigation. The main pillars of the national substitution policy are public acceptance, suitability and adequacy of high-quality water, sustainability and enforcement of laws. As a result, treated wastewater has been used in place of fresh water (recommended in the National Wastewater Management Strategy) in accordance with the quality guidelines and standards of the World Health Organization (WHO) and Food and Agriculture Organization (FAO), to produce an effluent fit for reuse in irrigation (MWI, 2016). Table 55 shows the estimated value of water in different sectors. MWI strategy is to increase the use of unconventional and reclaimed water for industry and agriculture as much as possible in order to save fresh water for domestic use (which includes the tourist sector).

Jordan has also taken significant steps to encourage foreign investment. Several sectors have experienced key reforms in recent years. Foreign and domestic investment laws grant specific incentives to industry, agriculture, tourism, hospitals, transportation, energy and water distribution. The Public Private Partnership Law from 2014 aims to encourage the participation of the private sector in the Kingdom's economic development and provides a legislative environment for joint projects (U.S. Department of State, 2015). Following sector reforms, agriculture in Jordan is now virtually free of

TABLE 55. ECONOMIC BENEFIT FROM WATER USED, BY SECTOR

SECTOR	FINANCIAL RETURN USD/M^3 OF WATER	JOB OPPORTUNITIES PERSON/MCM OF WATER
Agriculture	0.36	148
Tourism	25	1,693
Industry	40	3,777

Source: MWI, 2016; Closson et al., 2010.

restrictions and all direct subsidies have been removed. Credit to agriculture at low interest rates is the single most important conduit for government subsidies to agriculture.

Critical challenges to agricultural development are water scarcity and the need for increasing water use efficiency as Jordan is among the world's most water deficit-countries. Its per capita share of renewable water resources is according to different sources between 106 and 156m^3 per year, which is even lower than the "absolute water scarcity" threshold of 500m^3 per person per year (Rijsberman, 2006). Despite limited arable land (2.4%), the agricultural sector is the largest water user (65–75% of the country's water resources) absorbing almost all treated wastewater. Although the agricultural contribution to Jordan's GDP appears with about 4% small, an estimated 28% of the national GDP is considered agriculture-dependent due to strong upstream and downstream linkages. The arrival of the Arab Spring in early 2011 had a profound effect on market confidence in the region. While the events of the Arab Spring did not directly impact Jordan, they inevitably raised the risk bar and prolonged completion of the transaction.

Business model

A public-private partnership (PPP) model was developed to finance the construction and operation of the As-Samra plant, with funding provided initially by USAID (construction phase I: 2003–2008), and for further expansion and technological upgrade by MCC (construction phase II: 2012–2016). The PPP is based on a 25-year Build-Operate-Transfer (BOT) contract signed in 2003 which was extended in 2012. Through this PPP, the government (MWI) delegates responsibilities to a private sector entity to finance, design, build, operate and maintain the facility for a 25-year period. The private sector entity is the Samra Wastewater Treatment Plant Company Limited (SPC), a private company whose investors include Morganti, an American affiliate of the Consolidated Contractors Group, Suez Environment, a Paris based utility company, and Infilco Degremont, an American company, since mid-2015 a subsidiary of Suez Environment. The Jordan-based Arab Bank arranged a consortium of nine local and international financial institutions to provide a commercial loan in local currency with a term of up to 20 years, the longest maturity in Jordan to date, with an initially fixed, then floating interest rate. Under this public-private partnership, the government of Jordan benefits from having the private sector both (i) raise the financing for and (ii) guarantee the high-quality construction, operation and maintenance of the facility. At the end of the concession period, in 2037, the facility will be transferred back to the government of Jordan in good working order and at no additional cost.

MCC funded USD 93 million of the USD 275[2] million cost of the As-Samra phase 2 expansion project, the Government of Jordan at least USD 19.8 million, the private sector sponsors contributed an equity injection of USD 8.6m (brownfield investment based on reinvesting phase 1 cash flows into the expansion) and the association of banks about $148 million. The MCC support is leveraged through the lenders and private sector's co-financing of more than 50% of the expansion cost. By bringing down the capital costs, the MCC grant enabled the project to be financially viable, thus benefiting the government and local rate-payers, while making the project attractive for SPC and local Jordanian banks. However, MCC's grant does not subsidize the private sector, as the private investors earn a return only on their portion of invested capital. The As Samra WWTP was the first in the Middle East to use a combination of private, local government and donor financing, using a **Viability Gap Funding** scheme (see related Box in Business Model 19) to bring down the capital costs via the MCC contribution. Closing the financing of the expansion supported its feasibility and demonstrated the significant benefits of combining private sector financing with viability-gap grant funding.

As unique as the template is, it has its challenges. The setup of the blended finance was complicated by MCC's inability to enter into any direct contractual relationships with the project sponsors (private sector) or the lenders (banks). Moreover, both MCC and the lenders were reluctant to fund ahead of each other; as a result, financial close and satisfaction of the initial conditions to the MCC disbursement had to occur on the same day (Keenan and Norman, 2012). This situation is indicative of another cornerstone of the business model, which is **risk sharing** as a necessity to attract investors (Figure 233; adapted from SPC, 2014).

Given the size of the plant and the current water tariff and fee structure, the finance model does not rely to any significant degree on revenues from the wastewater-generating households, e.g. in Amman, Zarqa and Russeifa, or fees from wastewater using farmers. In contrast, the applied finance model allows to keep the treatment tariffs very affordable (stated objective). This is supported by a significant measure to keep the WWTP energy efficient and in large self-sustainable and in this way the largest operational cost factor within limits.

Jordan's water tariff includes a wastewater levy which is based on the freshwater consumption. However, this is not sufficient to cover O&M cost of wastewater treatment, also if farmers water reuse fees are added. Farmers are charged differently depending on the scheme they are connected. Some pay per cubic meter consumed, others have an allocated amount of water and pay a lump sum. However, the fee for reclaimed water cannot exceed the one paid for the preferred freshwater (Rothenberger, 2010). According to Bahri (2008), farmers in the vicinity of WWTPs pay the MWI USD

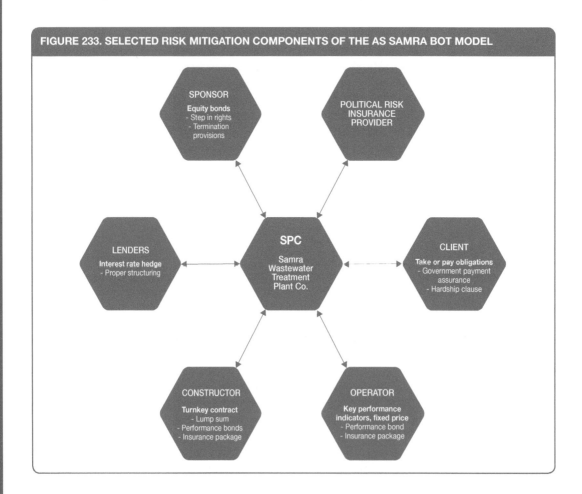

FIGURE 233. SELECTED RISK MITIGATION COMPONENTS OF THE AS SAMRA BOT MODEL

143–286 per ha and year, while those using mixed water pay the Jordan Valley Authority (JVA) USD 0.07 per m^3. During the rainy winter season, water is provided for free for salt leaching. However, the revenues of the JVA are so far not recorded at As Samra (or the Government) as the Water Authority of Jordan (WAJ) provides JVA with free supply of wastewater services from As Samra (OECD, 2014). Both authorities are reporting to the Ministry of Water and Irrigation.

Value propositions

In the water stressed situation of Jordan, the essence of the business model is the double value proposition of wastewater treatment and the recovery of as much reclaimed water as possible for further usage, especially crop irrigation (an increase from 61 to 83%), that high quality freshwater can be reserved for domestic (including potable) purposes benefitting 2 million people. This objective has been achieved with an innovative blended finance and risk sharing mechanism which makes the capital investment attractive and bearable for all parties, and covers in addition the operational costs (through the BOT arrangement, supported by a high level of energy recovery and potentially carbon credits). The model allows the WWTP to achieve financial viability despite low user tariffs (Figure 234).

Asides the main objective of supporting irrigation in the Jordan Valley, the plant also offers its direct proximity job opportunities and water for irrigation. There are about 300–500 ha within and around the As Samra plant premises planted with forage crops (clover), olive trees and, for example, sorghum. Most farmers have irrigation water rights and contracts with the Ministry of Irrigation. The irrigation method applied is surface and drip irrigation, often gravity based. The amount of irrigated water used is open and there is no particular system in place to regulate use. In addition, many farmers pump water directly out of the Zarqa river without any formal arrangement with the MWI. There are periodical field inspections to prevent the cultivation of leafy vegetables.

Also livestock owners benefit. According to farmers, the availability of wastewater irrigated forage has simplified the production of sheep and goats instead of relying on natural grazing in the surrounding areas. This is a significant advantage given that the local area has a poor natural vegetation cover due to the scarcity of rainfall (Seder and Abdel-Jabbar, 2011).

The plant also produces sludge and (biogas) slurry with a high potential for soil amelioration (e.g. for forestry) or the cement industry, once the regulatory framework becomes supportive. Given the significant amount of sludge the WWTP will generate, the MCC considers local storage only a temporary solution. The plant operator and the Government of Jordan have agreed to work together to provide alternative solutions including related policies, procedures and standards for an environmentally and socially sound permanent disposal and/or re-use of sludge[3]. A viable market for sludge produced by the plant is yet to be found, given the restrictions that apply. Until this happens the parties will continue to store and dispose of sludge in accordance with the terms of the concession agreement.

Institutional environment

The Ministry of Water and Irrigation (MWI) has overall responsibility for policies and strategies in the water sector, including water and wastewater supply and related projects, planning and management. Under MWI operate, among others, the (i) Water Authority of Jordan (WAJ) which is responsible for water supply and wastewater services, as well as for water resource planning and monitoring, construction, and operations; (ii) the Project Management Unit (PMU) within WAJ, which regulates water supply and wastewater utilities, promotes private sector participation in the water sector and carries tasks related to project planning and execution; and (iii) the Jordan Valley Authority (JVA) which manages water resources and provides bulk water in the Jordan Valley. The main institutions involved in the As Samra WWTP (Figure 235) and their roles are (SPC, 2014):

FIGURE 234. BUSINESS MODEL CANVAS OF AS SAMRA WASTEWATER TREATMENT PLANT FROM THE PRIVATE SECTOR PERSPECTIVE (SPC)

KEY PARTNERS
- Government of Jordan
- MCC: grant funding
- Bank consortium: loans
- MoF
- MIGA: credit risk insurance

KEY ACTIVITIES
- Design and construction of plant
- Treatment of wastewater
- Provide safe, treated water for agriculture / industry
- Hydropower and biogas generation
- Operation of plant for fixed period, then handover to government

VALUE PROPOSITIONS
- Treat wastewater and provide safe, treated water for reuse in agriculture and industry, freeing up freshwater for domestic use
- Internal energy recovery (95%)

CUSTOMER RELATIONSHIPS
- Government pays SPC per unit of treated water
- Government collects fees through water user tariffs
- Users of treated water in agriculture and industry pay fees to government

CUSTOMER SEGMENTS
- SPC provides water treatment service to government
- Wastewater from households and industry treated
- Wastewater reused in agriculture / industry

KEY RESOURCES
- Funding from various (MCC grant, bank loans, private equity, MoF fund)
- Private sector technology and expertise in construction, operation and maintenance

CHANNELS
- Direct / brokered negotiation with GoJ
- Existing conveyor pipeline to transport water from cities, and stream directing the treated waters to King Talal Dam

COST STRUCTURE
- Capital cost and upfront infrastructure investment
- Operating costs and maintenance

REVENUE STREAMS
- SPC is paid by government per unit of treated water
- Energy recovery allows significant operational cost savings

SOCIAL & ENVIRONMENTAL COSTS (cf. Consolidated Consultants, 2012)
- Infrastructure requires land and affects eco-system on site
- Health risk due to laborers' possible contact with wastewater and (during construction) impacts on air quality and noise
- Possible ecological impact from mixing freshwater and treated wastewater
- Potential health effects of using diluted wastewater to produced vegetables

SOCIAL & ENVIRONMENTAL BENEFITS (cf. Consolidated Consultants, 2012)
- Increased resource efficiency through water reuse
- Reduced pollution of receiving waters, reduction in public health expenditure associated with disease outbreak
- Improvement in groundwater level because of the additional water sources, improved irrigation technology and protection
- Job creation at the plant and downstream
- Use of treated water in agriculture and industry supports economic development
- Government steps in for cost recovery and can maintain low water tariffs for inclusive access to services

- Client: Government of Jordan; represented by the Ministry of Water and Irrigation (MWI).
- Donor (Phase 2): Millennium Challenge Corporation (MCC); U.S. foreign aid agency.
- Grant Fund Manager: Millennium Challenge Account (MCA-Jordan).
- Authorities Engineer: Fichtner (+ local consultant Eco Consult), also in charge of compliance monitoring with the health, safety and environment management plan.
- Project Companies: Samra Wastewater Treatment Plant Company Ltd. (SPC) and Samra Plant Operation and Maintenance Co. Ltd. (O&M).
- Sponsors: Suez Environment / Infilco-Degremont and Morganti- Consolidated Contractors Group.
- Lenders: Lender Syndicate led by Arab Bank; Lenders technical advisor: Mott MacDonald.
- Political risk insurance: Multilateral Investment Guarantee Agency (MIGA) of the World Bank.
- Beneficiaries: Mainly Amman, Russeifa and Zarqa populations as well as local towns in plant vicinity (e.g. Hashimiyya) and farmers irrigating crops with treated wastewater in the vicinity of the plant and across the Jordan valley.

An overview about relevant laws and bylaws, standards and regulations as well as the requirements of the funding agencies, of relevance for the WWTP, has been presented by Seder and Abdel-Jabbar (2011) and Consolidated Consultants (2012). Of particular relevance in the **agricultural** context are the 2006 standards for safe water reuse (JS 893/2006) which allow for a wide range of water reuse activities for highly treated reclaimed water for landscapes, cut flowers and high-value crops (except crops eaten uncooked), and for smaller scale treatment reuse activities with restricted cropping patterns. Reuse categories for treated wastewater are:

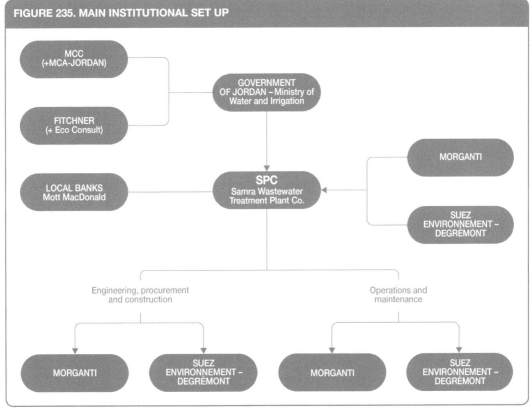

FIGURE 235. MAIN INSTITUTIONAL SET UP

Source: SPC, 2014

- Recycling of water for irrigation of vegetables that are normally cooked.
- Recycling of water used for tree crops, forestry and industrial processes.
- Discharges to receiving water such as wadis and catchment's areas.
- Use in artificial recharge to aquifers not used for drinking purposes.
- Discharge to public parks or recreational areas.
- Use in irrigation of animal fodder.
- Use of reclaimed water for cut flowers.

Although the 2006 standards were a big step forward (McCornick et al., 2004), Abdel-Jabbar (2009) argues that the existing water thresholds are often too stringent and less suitable than the multi-barrier risk reduction options promoted by WHO (2006). The author recommends updating JS 893/2006 towards a more accommodating model, supported by on- and off-farm risk mitigation measures. Although treated wastewater mixed with freshwater might no longer be labeled 'treated wastewater' (Carr and Potter, 2013) the government recommends that all crops irrigated with treated or mixed waters shall be analyzed and monitored periodically (MWI, 2010).

Technology and processes

The wastewater generated for example in Amman, where 80% of the households are connected to a sewage network, is transported over about 40 km to As Samra by gravity through a conveyor pipeline. During the year 2010, the maximum inflow ranged between 210,000 and 230,000m^3/day. Wastewater is under high pressure when arriving at the plant due to difference in elevation, and turbines have been installed to run on upstream wastewater flow, thereby generating renewable energy that is used on site. The same process is repeated after treatment where the effluent is used to power discharge hydraulic turbines generating additional energy before the water is released towards the KTR with its 86 MCM storage capacity.

The activated sludge treatment process consists of pretreatment and primary settling tanks, aerobe and anaerobe biological treatment, biomass settling and chlorination. Water quality changes between in- and outlets are shown in Table 56 (Consolidated Consultants, 2012; Suez, 2015).

Sludge from primary treatment and the aeration tanks undergoes thickening and anaerobic digestion, dewatering (target 18% dry solids) and sun drying (target 50% dry solids) (Suez, 2015). The daily sludge generation was in 2011 about 118 tons of dry sludge or 393 tons of sludge (at 30% dry solids). Given the current legal limitations for sludge reuse, MCC and SPC are given the exploration of alternative sludge disposal/reuse options, such as cement kiln or land application, highest priority (Consolidated Consultants, 2012) as space for future storage is declining and the potential negative environmental impact unacceptable for the WWTP's staff and people living in the area.

The company has implemented an energy management system as per ISO 50001 to evaluate and control its energy consumption. Between 80 and 95% of the plant's energy requirements are met using the in- and outflow turbines (1.7 and 2.5 megawatts, respectively) and the biogas generation from sludge (9.5 megawatts). An innovation was the use of hydraulic turbines on raw sewage water.

TABLE 56. WATER TREATMENT QUALITY AS SAMRA

WATER QUALITY INLET		WATER QUALITY OUTLET	
BOD$_5$	637–708 mg/l	BOD$_5$	5–30 mg/l
TSS	649–682 mg/l	TSS	15–30 mg/l
Total Nitrogen	100–107 mg/l	Total Nitrogen	15–30 mg/l

The expected increase of the wastewater inflows from the city of Zarqa will pose some challenges as its location is lower than the plant which will affect the power recovery ratio. This can in part be compensated by increasing the capacity of the biogas power generation system and a reduction of the power consumption by the aeration units.

Funding and financial outlook

Like across the region, Jordanian water tariffs do neither cover the water production cost nor the wastewater treatment costs. While MWI (2010) suggests that wastewater charges, connection fees, sewerage taxes and treatment fees shall be set to cover at least the operation and maintenance costs (ultimately aiming at full cost recovery), the As Samra BOT blended finance model allows to keep the plant also under the current (social) tariff structure viable over the 25-year contract period. To achieve this, the government pays for SPC's provision of wastewater treatment services about USD 0.17/m^3 (pers. communication with the plant manager, 2014).[4] Running at the targeted capacity of 133MCM per year, this would result in an annual governmental subsidy of USD 22.6m. This can be partially recovered in various ways. Household (waste)water tariffs contribute the largest share of about 60% on average over all WWTPs (MWI, 2013). If As Samra would have its own account, it could probably break even as its O&M costs are much lower than of other WWTPs, given its energy efficiency (MWI, 2015). Lower contributions could be expected from the agricultural sector (see above) and potentially through the carbon market. The UNFCCC (2010) application for registration under the Clean Development Mechanism (GHG reduction of 296,704t of CO_2e per year) is at the validation step.[5] Another (more lucrative) revenue source planned for 2021 is the possible sale of wastewater to Power Plants in the order of 22.5 MCM per year at USD 0.63/m^3 resulting in an estimated annual cost recovery of about USD 14million. Tariff adjustments would help reducing the governmental share. This applies more to As Samra (if budgeted separately) than other WWTPs in Jordan as in other cases energy tariff increases would undermine possible savings (MHI, 2013).

Socio-economic, health and environmental impact

An environmental and social impact assessment (ESIA) was prepared in January 2012 for the Samra WWTP expansion. The project sponsors' consortium then prepared a health, safety and environment (HSE) management plan based on the standards of all (national and international) partners to mitigate potential environmental and social risks and impacts during the construction period, while during operations environmental and social risks and impacts are managed by the SPC based on their "Quality, occupational health, safety and environment" (QHSE) management system.

Positive impacts of the As Samra wastewater treatment plant largely accrue as a result of improved quality of domestic and industrial sewage effluents entering ultimately surface water bodies. The treatment plant reduces disposal of raw sewage, risks of groundwater pollution and the spread of excreta-related diseases. Since the commissioning of As Samra, water quality in the King Talal Reservoir and the Zarqa river have significantly improved despite some recontamination (Al-Omari et al., 2013; Abdel-Jabbar, 2009) allowing fish to return. The plant is providing directly about 170–180 new jobs, nearly exclusively used by national staff. As so far only 3% of all employees are female, women's associations were contacted to encourage the participation of women in public consultations about job opportunities, and to analyze and address the barriers of women employment at the Samra WWTP. Finally, the treated wastewater is supporting about 10,000 jobs in agriculture. At the aggregate level, the treatment plant has significant indirect benefits for the whole country as improvements in wastewater use deliver fresh water savings for domestic use by an estimated 2 million people, reduce aquifer extractions, support the tourist sector and related jobs, food security, and adaptation to the risks of climate change and migration. As Samra is also producing 103,000 kwh green energy per day, making the plant 90–95% energy self-sustainable.

A challenge is sludge management. The drying lagoons and bio-solid storage lagoons provide a favourable environment for mosquito, fly and insect growth. The ESIA states that 15% of flies that originate at the project site can reach the nearest residential areas. Mitigation measures like fumigation have been put in place, but an extension of sludge drying could reach acceptability limits. The As Samra plant is in general designed to ensure that no odor nuisance occurs and the plant obtained highest certificates for health and safety as well as environmental protection (Suez, 2015). Risks and impacts related to groundwater infiltration were considered as low due to the physical characteristics of the sludge and 80m deep groundwater table.

Scalability and replicability considerations

The finance of water recovery and use becomes more favorable when treatment costs are low and the value proposition goes beyond recovering water from wastewater and includes for example the recovery of nutrients and energy (see below). In such cases, the likelihood of recovering both the fixed and variable costs of wastewater use, and parts of the operational and maintenance costs of the treatment process is improved. Technology choice is important, particularly in developing countries. Wastewater use, especially in agriculture, can be supported by relatively simple treatment processes of proven technology, with low investment costs and affordable operation and maintenance. Such processes are particularly suited to countries with warm climates, as biological processes perform better at higher temperatures. The investment costs for such simple or 'appropriate' treatment facilities are in the range of 20% to 50% of conventional treatment plants, and more importantly, the operation and maintenance costs are in the range of 5% to 25% of conventional activated sludge treatment plants. These cost differentials are substantial from a financial point of view (Libhaber and Orozco-Jaramillo, 2013). Appropriate technology processes include (but are not limited to) the following: lagoon treatment, upflow anaerobic sludge blanket (UASB) reactors, anaerobic baffled reactors (ABRs), constructed wetlands or stabilization reservoirs for wastewater use. Various combinations of these processes can be set up. In the context of fully exploited freshwater resources, the economic gain from treated wastewater can be significant. The business template developed in Jordan – namely, grant financing coupled with private finance from sponsors under a debt-to-equity ratio of 80:20, and debt finance raised on a limited-recourse basis with shared risks – offers significant potential for the development of much-needed infrastructure projects in developing countries in the future. The additional savings on operational costs through a high level of energy self-supply makes the model even more interesting. There is significant potential for its transfer to similar locations if a donor, such as USAID and MCC in this case, is ready to contribute to the overall costs. MCC expects to adapt the contractual structure developed for the As Samra expansion for use in upcoming infrastructure projects elsewhere in the world, thereby allowing projects that are economically and environmentally beneficial to be implemented and operated by the private sector where such projects would otherwise be unaffordable to the public sector (Keenan and Norman, 2012). The MHI capital investment program makes also reference to a possible third As Samra expansion phase for handling extra amounts of wastewater, budgeted with USD 324million (2020–2024).

Summary assessment – SWOT analysis

The As Samra business case presents a multi-partner model to transform urban wastewater into several benefits for the society. The case required large initial capital investment which was managed through an innovative and multiple award-winning finance model using Viability Gap Funding and risks sharing model. However, the case points asides strength and opportunities also at weaknesses and potential threats for its future and replication (Figure 236).

FIGURE 236. SWOT ANALYSIS OF AS SAMRA BUSINESS CASE

	HELPFUL TO ACHIEVING THE OBJECTIVES	**HARMFUL** TO ACHIEVING THE OBJECTIVES
INTERNAL ORIGIN ATTRIBUTES OF THE ENTERPRISE	**STRENGTHS** • Strong institutional and regulatory support • High profile partners • Excellent financing package offered by Arab Bank • No foreign exchange risk for Gov. • Innovative business model for infrastructure set up and O&M cost reduction • Financial risk mitigation instruments facilitate multi-party investments • Favorable topography reduces pumping needs while allowing to generate energy	**WEAKNESSES** • Plant of significant size with high O&M needs after BOT hand over • First co-financed BOT and PPP (for public infrastructure) in Jordan; i.e. it cannot build on lessons learnt • A similar favorable topography will be seldom to repeat the energy balance • Cost recovery from tariff system marginal, not to mention farmers • Long-term sludge/slurry disposal • Low job attraction for female workers
EXTERNAL ORIGIN ATTRIBUTES OF THE ENVIRONMENT	**OPPORTUNITIES** • Continuous supply of wastewater and demand for treated water • Virtually zero competition from fresh water due to water scarcity • Stable political environment • Leader in the regional Clean Development Programs • Job creation across sectors • Revision of national reuse guidelines based on WHO (2006)	**THREATS** • Political crisis • Non-compliance with safety plans resulting in human health risks and loss of image/trust/support • Increase in industrial effluent (salinity, metals) • Reuse restrictions

Contributors and resource persons
Bernard Bon, General Manager SPC, Samra Project Company
Rami Al-Zomor, Process Engineer, As Samra
Dr. Esmat Al-Karadsheh, Jordan
Dr. Bassim Eid Abbassi, Jordan
Manzoor Qadir, United Nations University, Canada
Miriam Otoo, IWMI
Javier Mateo-Sagasta, IWMI

References and further readings
Abdel-Jabbar, S. 2009. Assessment of use of reclaimed water in unrestricted agriculture in Jordan Valley in the light of the new WHO Guidelines. GIZ document. http://www2.giz.de/Dokumente/oe44/ecosan/en-assessment-of-use-of-reclaimed-water-in-unrestricted-agriculture-2009.pdf (accessed 6 Nov. 2017).

Al-Omari, A., Al-houri, Z. and Al-Weshah, R. 2013. Impact of the As Samra wastewater treatment plant upgrade on the water quality (COD, electrical conductivity, TP, TN) of the Zarqa River. Water Science & Technology 67 (7): 1455–1464.

Bahri, A. 2008. Case studies in Middle Eastern and North African countries. In: Jimenez, B. and Asano, T. (eds) Water reuse. An international survey of current practices, issues and needs. IWA, p. 558–591.

Carr, G. and Potter, R.B. 2013. Towards effective water reuse: Drivers, challenges and strategies shaping the organisational management of reclaimed water in Jordan. The Geographical Journal 179 (1): 61–73.

Closson D., Hansen H., Halgand F., Milisavljevic N., Hallot F. and Acheroy M. 2010. The Red Sea–Dead Sea Canal: Its Origin and the Challenges it Faces. In: Badescu V., Cathcart R. (eds) Macro-engineering Seawater in Unique Environments. Environmental Science and Engineering (Environmental Engineering). Springer, Berlin, Heidelberg, p. 106–124.

Consolidated Consultants. 2012. ESIA for the expansion of As-Samra wastewater treatment plant. MCA. www.mca-jordan.gov.jo/systemfiles/pages/file_635053196963205703.pdf (accessed 6 Nov. 2017).

Courcier, R., Venot, J.P. and Molle, F. 2005. Historical transformations of the Lower Jordan River Basin in Jordan: Changes in water use and projections (1950–2025). Colombo, Sri Lanka: International Water Management Institute (IWMI). vi, 85p. www.iwmi.cgiar.org/assessment/files_new/publications/CA%20Research%20Reports/CARR9.pdf (accessed 6 Nov. 2017).

Keenan, R. and Norman, M. 2012. As-Samra sets expansion template. Project Finance International, October 17: 48–51.

Libhaber, M. and Orozco-Jaramillo, A. 2013. Sustainable treatment of municipal wastewater. IWA's Water 21 (October 2013): 25–28.

McCornick, P.G., Hijazi, A. and Sheikh, B. 2004. From wastewater reuse to water reclamation: Progression of water reuse standards in Jordan. In: Scott, C., Faruqui, N.I. and Raschid-Sally, L. (eds) Wastewater use in irrigated agriculture: Confronting livelihood and environmental realities. Wallingford/Colombo/Ottawa: CABI/IWMI/IDRC, p. 153–162.

MHI. 2013. Structural benchmark – Action plan to reduce water sector losses. August 2013. MWI. https://goo.gl/v4xE8g (accessed 6 Nov. 2017).

Molle, F., Vernot, J.P. and Hassan, Y. 2008. Irrigation in the Jordan Valley: Are water pricing policies overly optimistic? Agricultural Water Management 95: 427–438.

MWI. 2010. Wastewater policy. www.mwi.gov.jo/sites/en-us/SitePages/Water%20Policies/Waste%20Water%20Policy.aspx (accessed 6 Nov. 2017).

MWI. 2015. Wastewater treatment. National plan for operation and maintenance. Ministry of Water and Irrigation of the Hashemite Kingdom of Jordan, Amman. www.mwi.gov.jo/sites/en-us/Hot%20Issues/Wastewater%20Treatment%20National%20Plan.pdf (accessed 6 Nov. 2017).

MWI. 2016. National Water Strategy 2016–2025 (v. 25. Feb 2016). Ministry of Water and Irrigation, Hashemite Kingdom of Jordan. http://www.mwi.gov.jo/sites/en-us/ and https://goo.gl/L1dNyP (accessed 6 Nov. 2017).

OECD. 2014. Jordan: Overcoming the governance challenges to private sector participation in the water sector. 10th meeting of the Regulatory Policy Committee. GOV/RPC/NER(2014)5.

Rijsberman, F.R. 2006. Water Scarcity: Fact or Fiction? Agricultural Water Management 80 (1–3): 5–22.

Rothenberger, S. 2010. ACWUA report: Wastewater reuse in Arab countries. Amman. www.ais.unwater.org/ais/pluginfile.php/356/mod_page/content/128/Jordan_Summary-Report-CountryCasestudies_final.pdf (accessed 6 Nov. 2017).

Seder, N. and Abdel-Jabbar, S. 2011. Safe use of treated wastewater in agriculture – Jordan case study. Prepared for ACWUA. http://www.ais.unwater.org/ais/pluginfile.php/356/mod_page/content/128/Jordan_-_Case_Study%28new%29.pdf (accessed 6 Nov. 2017).

SPC. 2014. Samra: Wastewater treatment plant – A major asset for Jordan. Presentation 21 Oct 2014. https://www.unece.org/fileadmin/DAM/ceci/documents/2014/water_and_sanitation_October/Jordan_the_Amman_wastewater_treatment_plant.pdf (or https://goo.gl/mi3VNY; accessed 6 Nov. 2017).

Suez Environment. 2015. As Samra wastewater treatment plant. A major asset for Jordan. www.degremont.com/document/?f=engagements/en/as-samra-r-er-020-en-1510-web.pdf (accessed 6 Nov. 2017).

U.S. Department of State. 2015. Investment climate statement Jordan, June 2015. Washington, DC. www.state.gov/documents/organization/241822.pdf (accessed 6 Nov. 2017).

UNFCCC. 2010. As-Samra Wastewater Treatment Plant project (SWTP), Jordan. Clean development mechanism project design document form (CDM-PDD); Version 03. https://cdm.unfccc.int/Projects/Validation/DB/DNBGISCCSJ6174719S0D97R4DBWHVS/view.html (accessed 6 Nov. 2017).

Wichelns, D., Drechsel, P. and Qadir, M. 2015. Wastewater: Economic asset in an urbanizing world. In: Drechsel, P., Qadir, M. and Wichelns, D. (eds) Wastewater: Economic asset in an urbanizing world. Dordrecht, Netherlands: Springer. pp. 3–14.

World Bank. 2016. Blended financing for the expansion of the As-Samra wastewater treatment plant in Jordan. Case studies in blended finance for water and sanitation. Washington, D.C.: World Bank Group.

Case descriptions are based on primary and secondary data provided by case operators, insiders, literature and other stakeholders, and reflect our best knowledge at the time of the assessments 2014/16. Possible errors are solely of the authors, not their resource persons. As business operations are dynamic, data can be subject to change.

Notes

1 Numbers vary between sources, depending e.g. on the inclusion of a Phase I loan.
2 Approximately USD 42 million of the debt package was used to refinance the outstanding loan on Phase 1. All numbers vary to some degree from source to source, also as the bank contribution is in Jordanian Dinar.
3 https://assets.mcc.gov/content/uploads/2017/05/action-2012-002-1136-01-first-major-build-operate-transfer-project.pdf (accessed 6 Nov. 2017).
4 According to MWI (2015) the treatment cost per cubic meter of wastewater in As Samra is USD 0.28 with capital cost and depreciation included. The average treatment costs of other plants are USD 0.76–2.8/m^3.
5 http://cdm.unfccc.int/Projects/Validation/DB/DNBGISCCSJ6174719S0D97R4DBWHVS/view.html (accessed 6 Nov. 2017).

BUSINESS MODEL 19
Enabling private sector investment in large scale wastewater treatment

Katharina Felgenhauer

Key characteristics

Model name	Enabling private sector investment in large-scale wastewater treatment
Waste stream	Wastewater treatment for reuse
Value-Added Waste Products	Treated wastewater for irrigation and a healthy environment
Geography	Water-scarce regions
Scale of production	Medium- to very large-scale
Supporting cases in the book	As Samra, Jordan
Objective of entity	Cost-recovery [ ]; For profit [X]; Social/Environmental enterprise [X]
Investment cost range	USD 100–400 million
Organization type	Public-private
Socio-economic impact	High-technology setup and efficient, nearly energy neutral operation of wastewater treatment facility while maintaining affordable tariffs for water users
Gender equity	Socio-economic benefits for male and female population. All users benefit from affordable water tariffs

Business value chain

Investments which are economically and socially desirable, like large-scale wastewater treatment for reuse, often lack financial viability. The upfront capital investment is too high for public or private sector to assume alone, and long gestation periods and the inability to increase user charges to commercial levels, decrease the likelihood of private sector buy-in. Especially larger plants with significant resource recovery potential often struggle with an appropriate finance plan. To share investment burden, investors are invited to cover the design and construction of the facility, coupled with a time-bound operation agreement, such as the Build-Operate-Transfer (BOT) model applied in the case of As Samra in Jordan. Private sector investment, however, can only be expected if the project is profitable and bankable.

Normally, revenue from such an investment is generated from user fees paid for wastewater treatment, public subsidies and to a minor degree, revenues from water reuse. In some cases, public sector services are configured with fixed or capped end-user fees. This may be useful to ensure broad and inclusive access to the service, such as in the framework of pro-poor policies. If fees are low and inflexible, the costs for infrastructure installation plus operation and maintenance is hardly recovered fully through user fees, let alone can a profit be made from the operation.

To address this common situation, the business concept applied in As Samra suggests ways to provide an attractive investment opportunity to private sector despite inflexible user fees and high capital costs. Government or donor funds can be used to cover up-front capital expenditure in infrastructure, thereby setting the stage for private investors. Such targeted investment of public sector funds can secure private sector resources in the forms of funding, material assets, technology and management expertise.

Public investment thereby achieves higher impact at a faster and more efficient rate compared to a solely public intervention. After a defined period of operation, the facility can be handed back to government, providing a return in kind on the initial public expenditure. Private sector management ensures a resource-efficient setup and running of the operation, giving the public sector opportunity to continue efficient service provision after the end of the public-private partnership (PPP) agreement. High degrees of energy recovery for system internal reuse is supporting the feasibility of the model.

To achieve this leverage, the upfront investment costs of the overall undertaking must be reduced to a level that makes the venture interesting and viable for private sector investors, including banks. A comprehensive risk management and reassurance scheme has to accompany and guide the partnership to ensure adherence to resource commitment by all parties throughout the duration of the PPP term.

Business model

This business model looks at blended finance options for the up-front investment of medium- to large-scale wastewater treatment plants. The model seeks to attract private sector co-funding and is applicable to situations in which the water user fees cannot fully recover investment, operating and maintenance costs. By reducing the up-front investment needs, the venture becomes financially interesting for private investors.

Public sector funds have to be available for this model to close the funding gap, either through domestic government budget or other sources, e.g. international development partners. Funds should be disbursed as grants to reduce financial liability. These funds are used to cover all or some initial infrastructure investment costs to reduce the up-front investment hurdle (Viability Gap Funding, see Box 10).

The funds should not subsidize the companies themselves nor their operations but the infrastructure development at hand; companies will earn a return only on their share of investment. Investors can create a project company, in which different sponsors can hold shares, to ease transaction management and tracking. Benefits can be combined with existing measures which attract foreign direct investment, such as tax breaks or reduction of duties and levies.

The private sector co-funding can only be secured if the viability gap funding has been fully committed to. Guarantee mechanisms have to be in place to back the commitments, e.g. through comprehensive contracts (see example in Figure 237) and guarantees from government or multilateral bodies. Backing by the Ministry of Finance (e.g. through a reserve fund) as well as international reinsurance and dispute resolution services help build trust among the partners and lower the investment risk. In the case of As Samra, Jordan, the Multilateral Investment Guarantee Agency (MIGA) provided guarantees against breach of contract for the expansion of the plant and its operation during the 20 year PPP term (MIGA, 2015). Failure to comply with any commitment should lead to strict and clearly spelled out penalties,

Box 10. Viability Gap Funding

Viability Gap Funding (VGF) refers to a grant, one-time or deferred, provided to support infrastructure projects that are **economically justified but fall short of financial viability**. The lack of financial viability usually arises from long gestation periods and the inability to increase user charges to commercial levels, making it unattractive for private sector investments. Viability Gap Funding (VGF) reduces the upfront capital costs of pro-poor private infrastructure investments by providing grant funding at the time of financial close, which can be used during construction. The VGF 'gap' is between the revenues needed to make a project commercially viable and the revenues likely to be generated by user fees paid mostly by poor customers. Although the economic benefits of a private investment project may be high, in situations where the incomes of end users are low, it may not be possible to collect sufficient user fees to cover costs. VGF is designed to make projects that are economically viable over the long term, commercially viable for investors. It helps mobilize private sector investment for development projects, while ensuring that the private sector accepts a share in the risks of infrastructure delivery and operation. Recognized by several international financial organizations the As Samra innovative financing has set up a new template for Viability Gap Financing. This new mechanism provides a significant leverage to the financial assistance of international donors and will allow new projects to materialize.

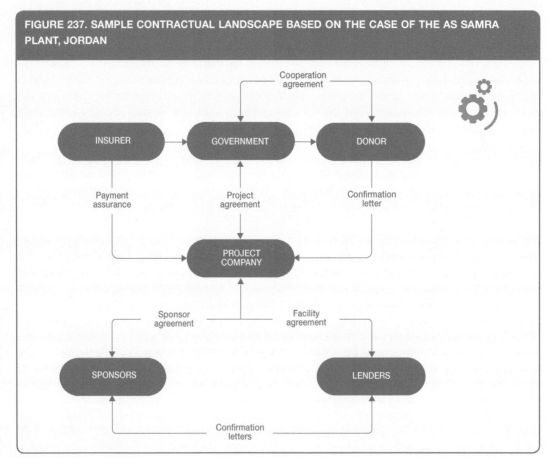

FIGURE 237. SAMPLE CONTRACTUAL LANDSCAPE BASED ON THE CASE OF THE AS SAMRA PLANT, JORDAN

Source: Adapted from SPC, 2014.

compensation or other rectification measures for all negligent parties. Banks are more likely to avail credit to private sector partners with a substantive risk-sharing mechanism in place.

For such a setup with multiple actors and a high level of interdependency to work, a number of framework conditions need to be fulfilled (OECD, 2014). Government requires strong and stable institutions with growing capacity to manage private sector partners. In Jordan, the Millennium Challenge Corporation (MCC, 2012) funded transaction advisors who would help broker the multi-party agreement system on behalf of the government. Unclear roles and responsibilities, ongoing reforms and policy gaps all contribute to a higher level of uncertainty, i.e. investment risk. The less flexible the water tariffs, the more reliable the government commitment to maintain minimum prices must be. Otherwise, cost recovery risks become difficult for the investor to hedge. Partners need to be aware that negotiations are likely to take considerable time before completion; project implementation will not commence before closure. These transaction costs add to the overall financial burden of the investment opportunity.

Once operational, the treatment plant can generate revenue from government payment or user fees for both, wastewater treatment and reuse of treated wastewater (Figure 238). If government steps in, expenses can partly be recovered through water fees or taxes at household or entity level. Farmers and companies which use treated water can be charged, however, fees will likely remain below the level of fresh water. A differentiated assessment of the clients' willingness and capacity to pay will estimate the cost recovery potential of this revenue stream. Ideally, tariffs should be calculated to cover at least operation and maintenance of the wastewater treatment facility to ensure long-term viability even after the end of the PPP agreement. Flexible tariff structures reduce the economic risk of the investment.

In return, government investment leverages private co-funding for a timely setup and operation of wastewater treatment to benefit large portions of society. Making additional water resources available for use in agriculture and industry supports economic development while maintaining affordable water user tariffs. At the end of the PPP agreement, government will receive the wastewater treatment facility at no additional cost. Efficient management processes will be in place, spurred by private sector interest in efficiency gains during the PPP term.

Alternate scenario

Lower viability gap funding through tender

Difficulties might arise when calculating the dimension of viability gap funding needed to make the venture interesting to private investors. Cost recovery alone will be insufficient to entice investors who are looking to make maximum profit. Investors, for the same reason, are motivated to predict inflated cost estimates when asked for advice in calculating the appropriate viability gap funding.

One way to limit the risk of overspending at the onset is to include the viability gap funding as element in a public tender. Expressions of interest from private sector partners should include an assessment of the amount of grant funding needed. The tender can then be allocated to the best bidder in terms of service provision and viability gap funding necessary to ensure maximum return on the public sector grant. The competitive nature of the bidding process encourages minimum gap funding requests. Service delivery quality, however, should not be compromised.

FIGURE 238. BUSINESS MODEL CANVAS – ENABLING PRIVATE SECTOR INVESTMENT IN WASTEWATER TREATMENT

KEY PARTNERS	KEY ACTIVITIES	VALUE PROPOSITIONS	CUSTOMER RELATIONSHIPS	CUSTOMER SEGMENTS

• Sector ministry
• Ministry of Finance
• Donor(s)
• Private investor(s) / operator(s)
• Commercial banks
• Risk mitigation broker / insurer
	• Secure viability gap funding and revenue guarantee
	• Negotiate effective risk management and insurance mechanisms
	• Moderate multi-partner platform
		• To create attractive business opportunities for private sector investment in wastewater treatment infrastructure	• Public-private partnership agreement
			• Price guarantee
			• Risk management and mitigation system
				• Private sector with strong track record in wastewater treatment, incl.
				  • Infrastructure companies
				  • Operators
				  • Commercial banks / investors

	KEY RESOURCES • Grant funding (government or donor funds) • Budget for payment to operators per m³ of treated water • Risk insurance		**CHANNELS** • Multilateral negotiations • Mutual risk management guarantees • Public-private partnership agreement	

COST STRUCTURE	REVENUE STREAMS
• Capital investment (one-time grant) • Payment to operators per m³ of treated water (ongoing expenses during PPP term)	• Water user fees for wastewater and water reuse • Government budget allocation and/or donor funds • High degree of energy recovery as cost saving measure • P-recovery as cost saving measure and possible revenue

SOCIAL & ENVIRONMENTAL COSTS	SOCIAL & ENVIRONMENTAL BENEFITS
• Large-scale treatment plant to occupy land and eco-system • Sludge storage	• Business-run operation will generate efficiency gains, i.e. save resources at high output, like low external energy requirements • Private sector funds increase scope and impact of operation, i.e. more water is treated and available for reuse, reducing e.g. groundwater abstraction

Potential risks and mitigation

This business model has been derived from the successful and acclaimed example of the As Samra Plant in Jordan. In addition to general risks related to reuse projects involving wastewater, such as harm to human and environmental health, the following risk mitigation options are particularly relevant to the financing model at hand.

Market risks: The viability gap funding requires a careful analysis of the business case for wastewater treatment in the region. Without reliable calculations of cost recovery and attractive profit margins, public overspent is likely. The risk can be partly mitigated by including an assessment of necessary viability gap funding in tender selection criteria (see alternate scenario above).

The careful assessment of the business case for wastewater treatment will also help to ensure long-term sustainability of the operation, in particular upon handover of the facility back to government at the end of the PPP. Water users' fees, as sole income to refinance the service, must cover operation and maintenance costs of the facility to avoid continuous subsidy. A differentiated fee structure for users of treated water, e.g. in agriculture or industry, can expedite cost recovery.

Private sector investors will only buy into the venture if viability gap funding is fully committed. A comprehensive risk-sharing and mitigation mechanism has to be negotiated for all parties to agree. This, in return, also provides security to government that public funds will effectively leverage additional investment and result in efficient wastewater service delivery. Sufficient time and resources need to be spent on the partnership negotiations and the establishment of a reliable contractual framework.

Technology performance risks: Leveraging private sector investment supports high-end technology because companies will operate at competitive levels to sustain their own business and generate profit, e.g. through efficiency gains. At the end of the PPP agreement, public sector is likely to receive state-of-the-art facilities. However, private sector partners must be selected competitively, considering track records of service delivery, to avoid technology and funding pitfalls. Quality of service should be guaranteed in unambiguous commitments (contracts) with clear remedy processes in case of non-compliance.

Political and regulatory risks: The model's dependency on reliable funding commitments and risk-sharing entails heightened relevance of political and regulatory stability. Reinsurance guarantees have to be given by stable, legitimate partners that are very likely to remain unchanged throughout the duration of the PPP agreement. A multi-layer support system which includes, for example, national and international partners alike, can be beneficial.

Social equity related risks: The model enables social benefits independent from gender differentiation, such as increased water resources for agricultural and industrial production. Additional jobs will be created at the plant (likely to favour male over female employees) as well as in irrigated agriculture benefitting both gender. The model facilitates the preservation of low water user fees, thus supporting broad and inclusive access to wastewater treatment services across social layers and income groups.

Safety, environmental and health risks: The model is about balancing financial risks for large-scale investments and as such not associated with any technology or particular environmental and health risks. In fact the financial volume is so high that it allows advanced treatment and risk mitigation. Naturally, the construction of a large-scale wastewater treatment plant will impact the site itself and its immediate surroundings, including eco-systems and communities. However, the downstream environmental benefits are significant in terms of preventing pollution, and providing large amounts of reclaimed water. The involvement of private companies in setup and operation of the wastewater treatment plant will support resource-efficient technology and management practices, e.g. covering the energy needs of production from own operation, and phosphorus recovery for reuse. In case of non-compliance with safety measures, potential health hazards will remain possible and demand risk mitigation measures as shown in Table 51 of Business Model 17. However, as this model is about the institutional–financial set-up, independently of the technology, **a separate table on potential risks and risk mitigation has been omitted.**

Business performance

Targeted viability gap funding by public sector helps leap ahead in wastewater treatment and water service delivery. Government and donor grants can leverage funding from private investors while tapping into business technology and expertise in wastewater treatment and management. Overall efficiency gains in water treatment (e.g. via energy recovery) coupled with the provision of additional water resources for agricultural or industrial consumption make the investment model attractive to government. While private sector partners exploit a profitable business opportunity, returns in economic development and environmental protection benefit society at large. Figure 239 shows the ranking of the model with its considerable strength to secure the anticipated positive environmental and social impacts as well as long-term viability.

That being said, the model can be challenging to set up with high transaction costs before operations can begin. Commitments need to be reliably secured through contracts and effective remedy mechanisms. Risk management and mitigation are of great importance, especially in large-scale and long-term ventures, as the model is vulnerable to economic, political and regulatory instability. If the capacity to effectively broker powerful public-private partnerships is further developed, substantial gains can be achieved in public service delivery.

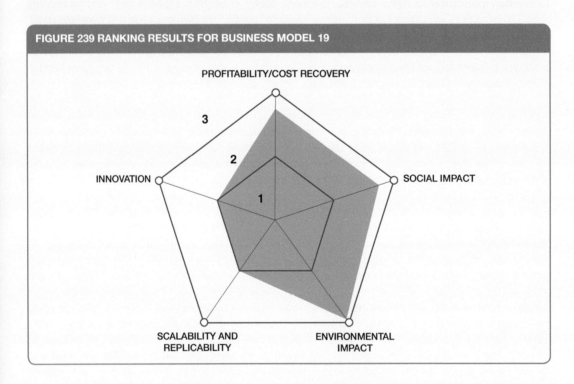

FIGURE 239 RANKING RESULTS FOR BUSINESS MODEL 19

References and further readings

Keenan, R. and Norman, M. 2012. As-Samra sets expansion template. Project Finance International October 17: 48–51.

MCC. 2012. MCC completes first major build-operate-transfer project financing in Jordan. 11 September, 2012. MCC. https://assets.mcc.gov/content/uploads/2017/05/action-2012-002-1136-01-first-major-build-operate-transfer-project.pdf (or https://goo.gl/LyRoH8; accessed 6 Nov. 2017).

MIGA. 2015. Guaranteeing investments in water and sanitation projects. MIGABrief Water July 2015.

OECD. 2014. Jordan: Overcoming the governance challenges to private sector participation in the water sector. 10th meeting of the Regulatory Policy Committee, 14–15 April 2014, GOV/RPC/NER(2014)5.

SPC. 2014. Samra: Wastewater treatment plant – A major asset for Jordan. Presentation 21 Oct, 2014. https://www.unece.org/fileadmin/DAM/ceci/documents/2014/water_and_sanitation_October/Jordan_the_Amman_wastewater_treatment_plant.pdf (or https://goo.gl/mi3VNY; accessed 6 Nov. 2017).

Wichelns, D., Drechsel, P. and Qadir, M. 2015. Wastewater: Economic asset in an urbanizing world. In: Drechsel, P., Qadir, M. and Wichelns, D. (eds. Wastewater: Economic asset in an urbanizing world. Dordrecht, Netherlands: Springer, pp. 3–14.

17. BUSINESS MODELS ON RURAL–URBAN WATER TRADING

Introduction

To sustain increasing urban water demands different strategies are common, such as a combination of long-distance water transfer and advanced wastewater treatment for reuse. Where possible also seawater desalination is being considered. Commonly referenced examples of technical excellence are the production of potable water from wastewater in Singapore and Namibia, based on a business model that is largely depending on reliable technology and positive public perceptions (Lazarova et al., 2013).

In this section, two business models (20 and 21) are presented which use a different approach of exchanging wastewater and freshwater, based on rural-urban water trading. Compared with inter-basin water transfers[1], the here presented models target **inter-sectoral transfers of water** to uses of higher economic value:

i) Water relocation takes place within the same basin or even the same watershed, moving water originally allocated to agriculture to domestic use, in particular drinking water.
ii) The models involve a two-way flow, i.e. freshwater release and transfer are based on the availability of a return flow of (treated) wastewater able to replace the created water gap and support if needed also other ecosystem service functions.
iii) Aquifer recharge is a common element complementing the available treatment capacity to produce water suitable for agricultural and/or domestic reuse also where treatment capacities are limited.

Given the young age of the presented cases and complexity of their setup, financial performance indicators as well as estimates of the social and/or environmental benefits or costs are largely missing, except for managed aquifer recharge, e.g. in USA or Australia (Maliva, 2014; Megdal et al., 2014; Gao et al., 2014).

Model 20: Inter-sectoral water exchange

Water exchange is driven by social and economic values. Not all uses of water are equally valued. Water for drinking has much high **social value** than for agriculture, yet the quantities involved are smaller. Water for irrigation has a lower **economic value** but the quantities involved are vast; on a global average about 70% of all the world's freshwater withdrawals go towards irrigation. Further, the quality requirements for drinking and agriculture are quite distinct. Therefore, taking a small volume of good quality water away from agriculture could make a sterling contribution to urban drinking water needs, while the resulting reduction to agriculture could be offset by substituting the lost amount with reclaimed water of lower but still appropriate quality, and this independent of seasons, i.e. throughout the year (Figure 240). In instances where farmers can get volumetrically more reclaimed water for irrigation than they release freshwater, and where a water-short municipality gets in a cost-competitive way a reliable supply of quality water for drinking, all partners benefit. Although such water exchange is in theory optimizing the value of the available water within a system, in support of greater environmental sustainability and climate change adaptation, it requires incentive systems and well-formulated contracts to secure the buy-in of a sufficiently large number of farmers who release freshwater for a mutually beneficial and thus sustainable business model. This is no easy endeavour with a range of possible gains but also conflicts (Molle and Berkoff, 2006; GWI, 2010), and might not recover its costs as long as swapped water volumes are low, but will greatly pay off in comparison with the direct and indirect costs of any extended drought period (Martin-Ortega et al., 2012).

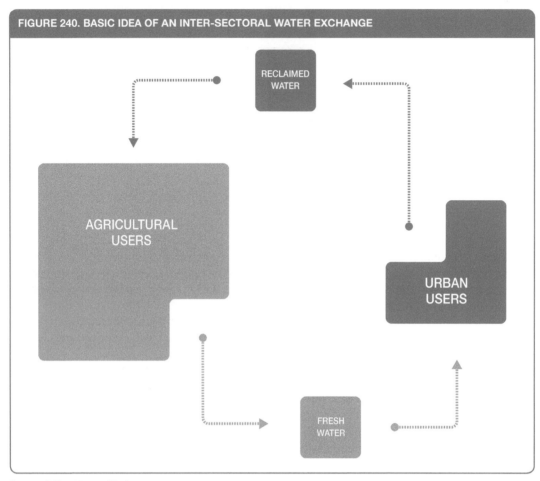

FIGURE 240. BASIC IDEA OF AN INTER-SECTORAL WATER EXCHANGE

Source: GWI, 2010, modified

The two case studies, which informed Business Model 20, are from **Iran** and **Spain** and based on the most recent experience with inter-sectoral water exchange. In the case of the Llobregat delta in Spain, a severe drought in 2007–2008 catalyzed significant investments into infrastructure able to produce high-quality reclaimed water to secure farmers' acceptance of a water swap in prolonged periods of drought. For this, the water swap contract remained flexible to allow transfers as needed. In Iran, on the other hand, the urban water deficit of the city of Mashhad is common reality and farmers received incentives to transfer their (entire) freshwater rights to the city in exchange of treated wastewater. Both cases face challenges which provide valuable lessons.

The model offers **several related value propositions**:
- Mitigating drought and related economic costs through reallocating freshwater from agriculture to urban use in exchange for reclaimed water allowing to realign water supply and demand from various sectors based on sector specific water quality requirements.
- Improved crop production and food security across seasons, the support of ecosystem services, aquifer recharge and increased resilience against drought and climate variability.
- Opportunities to raise revenue from sale of freshwater for high-value use and enhancing cost-effectiveness of the overall rural-urban water systems.

> INTRODUCTION

Although a water exchange could be approached from the perspective of both main parties, the reality is that in most cases the urban end is the driver of the business. In the case of Iran, for example, an initial survey showed that all city dwellers supported the planned exchange while about 97% of water right holding farmers opposed the plan (Yazdi, 2011). While in this case the political power of the urban sector determined the negotiations, the opposite could be possible, like in the case of Faisalabad, Pakistan (Business Model 23) where farmers strongly prefer (untreated) wastewater instead of (the only temporarily available and nutrient-poor) freshwater.

Model 21: Cities as their own downstream user

The rapid growth in urban population in countries like India is putting immense pressure on urban water supply and wastewater management. This has led to large-scale water transaction between urban and peri-urban areas. On one hand, urban water authorities and informal water traders are increasingly importing water from the urban periphery to meet the urban water need, while on other hand, farmers in the hinterland are using wastewater disposed by urban centers for irrigation (Londhe et al., 2004; Van Rooijen et al., 2005; Jampani et al., 2015; Hanjra et al., 2018). This rural-urban water exchange is a common situation today, and becomes more 'interesting' in water scarce areas, where the imported freshwater is actually the exported wastewater. Model 21 thus brings a developing country perspective to what is commonly referred to as managed aquifer recharge (MAR), looking at the increasingly common phenomenon of a closed water loop where the city is tapping into its own return flow. Aquifer recharge happens in this context on a trajectory from unplanned to planned, with limited wastewater treatment and differently developed formal and informal water markets closing the loop (Foster et al., 2010; Londhe et al., 2004; Jiménez, 2014). This makes the models rather complex and unsafe in contrast to the more commonly described experiences from Australia or USA (Dillon, 2009; Megdal et al., 2014) where in part dedicated agencies manage the underground water banking program under well-defined regulations and monitoring.

The chosen examples in this book are thus not success stories per se (Lazarova et al., 2013) with already documented, positive benefit cost ratio (e.g. Vanderzalm et al., 2015; Perrone and Merri Rohde, 2016), but reflecting situations and challenges observed on the trajectory to a more planned and managed RRR program, which have a significant potential for upscaling, if appropriately addressed.

Common related challenges in developing countries are weak institutional linkages for integrated surface and groundwater management across rural-urban borders, as well as missing regulations and monitoring of water quality (Bahri, 2012; Foster and Vairavamoorthy, 2013; Yuan et al., 2016). Without enabling environment, related business models struggle although the economic benefits appear worth the investment. The two cases, which informed Model 21, are from **Mexico** and **India**. In the example from Bangalore, India, largely untreated wastewater is transferred out of town to replenish peri-urban water tanks (reservoirs) and aquifers with multiple benefits for society, farming and ecosystem services. Some of the water returns through informal water markets back to the city, often at prices unaffordable for poorer households. Such rural–urban water transactions are increasingly common around Bangalore and many other cities in India, and need much stronger official acknowledgement to address likely externalities (Londhe et al., 2004).

The second case is the Mezquital Valley of Mexico, which is well-known for its enormous scale of wastewater reuse (Jiménez, 2009). With the recent inauguration of the Atotonilco treatment plant, the recovery of 'freshwater' from the replenished aquifer can become for Mexico City an increasingly important business model with lower pumping costs than any alternative option. The two business cases offer:

- Turning wastewater into a commodity for all-year irrigation and potable reuse through tank revival and/or groundwater recharge.
- Savings in land, disposal and treatment costs while supporting the delivery of ecosystem services.

The resulting water loop from both cases appears to reflect an increasing reality of the circular economy between urban and rural areas, where the urban hinterland functions as a **'kidney'** for urban water reuse (Figure 241).

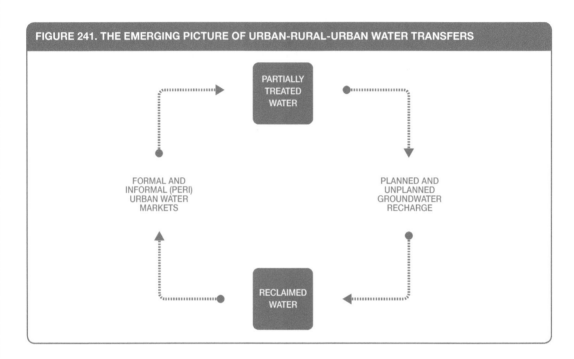

FIGURE 241. THE EMERGING PICTURE OF URBAN-RURAL-URBAN WATER TRANSFERS

References and further readings

Bahri, A. 2012. Integrated urban water management. GWP (Global Water Partnership). TEC Background Paper 16.

Dillon, P. 2009. Water recycling via managed aquifer recharge in Australia. Boletín Geológico y Minero 120 (2): 121–130. ISSN: 0366-0176.

Foster, S., Hirata, R., Misra, S. and Garduno, H. 2010. Urban groundwater use policy – Balancing the benefits and risks in developing nations. GW•MATE Strategic Overview Series 3. Washington DC, USA: World Bank.

Foster, S. and Vairavamoorthy, K. 2013. Urban groundwater. Policies and institutions for integrated management. Global Water Partnership Perspectives Paper. 20pp.

Gao, L., Connor, J.D. and Dillon, P. 2014. The economics of groundwater replenishment for reliable urban water supply. Water 6: 1662–1670.

GWI. 2010. Municipal water reuse markets 2010. Oxford: Global Water Intelligence.

Hanjra, M. A., Wichelns, D. and Drechsel, P. 2018. Investing in water management in rural and urban landscapes to achieve and sustain global food security. In: Zeunert, J. and Waterman, T. (eds.) The Routledge Handbook of Landscape and Food. Routledge: London, UK.

Jampani, M., Amerasinghe, P. and Pavelic, P. 2015. An integrated approach to assess the dynamics of a peri-urban watershed influenced by wastewater irrigation. Journal of Hydrology 523: 427–440.

Jiménez, B. 2009. Water and wastewater management in Mexico City. In: Larry Mays (ed) Integrated urban water management: Arid and semi-arid regions. UNESCO-IHP, CRC Press, p. 81–112.

Jiménez, B. 2014. The unintentional and intentional recharge of aquifers in the Tula and the Mexico Valleys: The megalopolis needs mega solutions. In: Garrido, A. and Shechter, M. (eds) Water for the Americas: Challenges and opportunities. London and New York: Routledge, p. 156–174.

Lazarova, V., Asano, T., Bahri, A. and Anderson, J. 2013. Milestones in water reuse: The best success stories. IWA, 408pp.

Londhe, A., Talati, J., Singh, L.K., Vilayasseril, M., Dhaunta, S., Rawlley, B., Ganapathy, K.K. and Mathew, R.P. 2004. Urban-hinterland water transactions: A scoping study of six class I Indian cities. Draft paper of the IWMI-Tata Water Policy Program Annual Partner's Meet 2004. 21pp.

Maliva, R.G. 2014. Economics of managed aquifer recharge. Water 6: 1257–1279.

Martin-Ortega J., González-Eguino, M. and Markandya, A. 2012. The costs of drought: The 2007/2008 case of Barcelona. Water Policy 14: 539–560.

Megdal, S.B., Dillon, P. and Seasholes, K. 2014. Water banks: Using managed aquifer recharge to meet water policy objectives. Water 6: 1500–1514.

Molle, F. and Berkoff, J. 2006. Cities versus agriculture: Revisiting intersectoral water transfers, potential gains and conflicts. Comprehensive Assessment of Water Management in Agriculture Research Report 10. Colombo, Sri Lanka: International Water Management Institute (IWMI), Comprehensive Assessment, 70pp.

Perrone, D. and Merri Rohde, M. 2016. Benefits and economic costs of managed aquifer recharge in California. San Francisco Estuary and Watershed Science 14 (2): Article 4.

Vanderzalm, J.L., Dillon, P.J., Tapsuwan, S., Pickering, P., Arold, N., Bekele, E.B., Barry, K.E., Donn, M.J. and McFarlane, D. 2015, Economics and experiences of managed aquifer recharge (MAR) with recycled water in Australia. Australian Water Recycling Centre of Excellence.

Van Rooijen, D.J., Turral, H. and Biggs, T.W. 2005. Sponge city: Water balance of mega-city water use and wastewater use in Hyderabad, India. Irrig Drain 54: 81–91.

Yazdi, P. 2011. A survey on social issues in replacing effluent with agricultural water rights (Case study: replacing Mashhad effluent with water rights from Karder River). International Congress on Irrigation and Drainage, 15–23 October, Tehran, Iran.

Yuan, J., Van Dyke, M.I. and Huck, P.M. 2016. Water reuse through managed aquifer recharge (MAR): Assessment of regulations/guidelines and case studies. Water Quality Research Journal of Canada 51 (4): 357–376.

Note

1 Inter-basin water transfer schemes attempt to reduce regional imbalance in the availability of water, in particular for agricultural or domestic use, by constructing elaborate canals and pipelines over long distances to convey surplus water from one river basin to another which shows a water deficit.

CASE

Fixed wastewater-freshwater swap (Mashhad Plain, Iran)

George K. Danso, Munir A. Hanjra and Pay Drechsel

Supporting case for Business Model 20	
Location:	Mashhad plain/city, Iran
Waste input type:	Treated wastewater
Value offer:	Treated wastewater for farmers in exchange for freshwater for urban use
Organization type:	Public and private (farmer associations)
Status of organization:	Operational (since 2005–2008)
Scale of businesses:	Medium
Major partners:	Khorasan Razavi Regional Water Company, Regional Agricultural Authority, farmer associations downstream of the Kardeh and Torogh dams

Executive summary

This is an inter-sectoral business case whereby treated wastewater from Mashhad city is exchanged for freshwater from farmers in Mashhad plain, Iran. In this business case, the regional water company negotiated the exchange of freshwater rights from farmer associations against access to treated wastewater. The main objective is to mitigate the impact of water scarcity in the urban area and to improve farmers' continuous access to water, also in view of the declining groundwater table in the Mashhad plain. The exchange of reclaimed water against reservoir water rights is one of two parts of a larger water swap project. It involves a number of villages downstream of two dams with the aim of exchanging annually fixed volumes of water: 15.7 and 9.4 million cubic meters (MCM) of treated wastewater for 13 and 7.8 MCM water rights from the Kardeh and Torogh dams, respectively. The project started in 2005 to 2008, and successfully replaced with treated wastewater the fresh water relocation to the city. In the other part of the exchange program, 192 MCM of wastewater are planned to replace farmers' rights to withdraw groundwater and to replenish the declining groundwater table. This part of the exchange was in late 2016, while studying the case, still work in progress.

Farmers' cooperation was facilitated by providing 1.2 times more replacement water than what was withdrawn. In contrast to the Spanish exchange model described in this book, the water volumes were defined and fixed. Major still ongoing challenges relate to wastewater treatment and low effluent quality which does not correspond with local standards and farmers' risk management capacity.

CASE: FIXED WASTEWATER-FRESHWATER SWAP

KEY INDICATORS (AS OF 2011)	
Land use	Up to 3,000 ha under irrigation
Water use:	About 25 MCM treated effluent used for irrigation (15.7 MCM for Kardeh area)
Capital investment:	USD 6 million (Kardeh dam area only)
Labor:	-
O&M cost:	USD 650,000 (Kardeh dam area only)
Output:	Release of ca. 21 MCM of freshwater for municipal use (13 MCM from Kardeh area)
Potential social and/or environmental impact:	Cost savings in water extraction, improvements in living standard and economic development (incl. tourism) because of additional freshwater for Mashhad, reduced overexploitation of aquifers, rivers and lakes. Benefits for ecosystem services.
Financial indicators:	Payback period: N.A. Post-tax IRR: N.A. Gross margin: N.A.

Context and background

Iran is a country facing significant water related challenges. The Mashhad plain in the Northeast of the country is a sub-basin of Kashafrud catchment and an example of a region with extended and increasing water crisis. While all surface water resources have been allocated, the only buffer for increasing demands has been groundwater. However, the groundwater table is declining rapidly (about 1.2m/yr) with an annual groundwater deficit of about 200 MCM in the Kashafrud basin. This development is strongly linked to increasing agricultural water needs to match the growing demands from the city of Mashhad. Mashhad is the second most populous city in Iran, with today about 3 million capita, and capital of Razavi Khorasan Province. Every year, about 30 million tourists and pilgrims visit the city for the Imam Reza shrine, which multiply urban food and water needs.

The Mashhad plain has a semi-arid climate with about 250mm of precipitation per year, mostly between December and May. In an attempt to rectify these interlinked issues, the city authorities decided to exchange treated wastewater for freshwater rights of farmers. Based on this objective, a total of about 25 MCM of wastewater have been allocated annually to various purposes. There are two sub-projects of the water swap model in Mashhad plain, one targeting surface water, the other groundwater. The first sub-project on surface water targeted two dams and is running since 2005 and 2008, while the groundwater exchange was in Dec. 2016 still work in progress or under reevaluation (Monem, 2013; Nairizi, pers. comm.):

Sub-project 1: Exchange treated effluent with water rights of the farmers from (a) 15 villages downstream of the Kardeh dam, (b) several villages downstream of the Torogh dam.

Sub-project 2: Exchange of treated wastewater with the right of groundwater exploitation from (a) the wells in the west of Mashhad, (b) the agricultural lands (sample farms) of the Astan Quds Razavi which owns the majority of the arable land in Khorasan Province.

The plan for the second sub-project was that a part of the groundwater will be supplied to meet Mashhad drinking needs and a part will remain in the aquifer to stabilize the groundwater table. Mashhad City's estimated water supply in 2016 of nearly 350 MCM would depend without water swap to over 90% on groundwater.

Market environment

Mashhad, like any other city in the Middle East, has been confronted with several challenges over the years. Most notable one being the explosive population growth and annual tourist inflow and related food demand making irrigated food production essential for urban food supply. The most common types of crops are cereals (55%), vegetables (21%), orchards (19%) and industrial crops (5%). The bulk of available water (77%) is allocated to agriculture. Substitution of the treated wastewater for farmers' right to use water from the reservoirs and allocation of the reservoir water to the citizens helps to assure water availability to the city with less impact on groundwater resources, while providing a reliable water source for farmer. In a study prior to the exchange, the large majority of the farmers who are water-right holders opposed the swap, while the urban stakeholders unanimously welcomed it (Yazdi, 2011). Aside possible quality concerns, farmers expressed lack of trust in governmental promises regarding water quantities and timing, a lesson learnt from what was promised by the construction of the local dams.

Macro-economic environment

Among the recent decisions taken by Iran's Expediency Council were the adoption and implementation of general plans for recycling water nationwide. The proposed policies and strategies flag prominently that to guarantee future urban water demands, agricultural water rights should be switched from the use of freshwater (from rivers, springs, wells, etc.) to treated effluents. According to Tajrishy (2011), about one-third of the municipal wastewater generated in Iran gets collected, of which 70% gets treated. Forty percent of the treated municipal wastewater (or ca. 10% of the generated wastewater) is already formally reused. A much larger share of (mostly untreated) wastewater is indirectly reused after entering freshwater bodies.

Another pillar of Iran's water resources policy is to improve water productivity by increasing water use efficiency, control the overexploitation of groundwater and avoid the use of high quality urban water for irrigating green spaces, and instead use low quality water for this purpose. Finally, the government also plans to cut off water supply to industries, which do not take practical measures for treating and reusing their wastewater. These government policies provide the legal support for reuse of wastewater for irrigation in Mashhad plain with the aim of improving the environment.

Business model

Like in other inter-sectorial water swaps, the larger economic and social benefits constitute also in this case the main objective. In Mashhad city, a part of the generated wastewater is collected, and treated, and so far released into the next stream. Transferring this water further to support villages downstream of two freshwater dams requires limited extra investments. The reclaimed water is replacing freshwater farmers are entitled to from the water reservoirs. To facilitate this exchange, a contract was signed between farmer associations and the regional water company. While the urban sector gets high quality freshwater from the reservoirs for high-value use, farmers in the two regions receive nutrient-rich reclaimed water at a 20% higher volumetric allocation than their original water entitlement supports. This was an important incentive for closing the contracts (Figure 242).

The other parts of the water swap which targets groundwater would add a significant benefit for the overall ecosystem as the majority of the reclaimed water would be used for aquifer recharge, not 1:1 exchange. This part is still work in progress and will hopefully have an appropriate water quality monitoring mechanism in place.

Before implementing the surface water swap, wastewater user associations were formed. This strategy enhanced cooperation and facilitated the contracting, especially as most farmers did not agree with an irrevocable contract, and the contracts eventually signed with farmers were in two categories (Yazdi, 2011):

A) Contracts between the Regional Water Company and representatives of the association of water right owners from a village based on the total water right of the village.

CASE: FIXED WASTEWATER-FRESHWATER SWAP

FIGURE 242. WATER EXCHANGE BUSINESS CASE IN MASHHAD, IRAN

KEY PARTNERS (cum customers in a swap model)	KEY ACTIVITIES	VALUE PROPOSITIONS	CUSTOMER RELATIONSHIPS	CUSTOMER SEGMENTS (cum partners in a swap model)
• Khorasan Regional Water Authority/ Company • Wastewater user association • Larger farmers • Khorasan Regional Agricultural Authority	• Establish wastewater reuse associations • Negotiations for water rights • Treat wastewater and distribute to farmers • Chanel freshwater from dams to urban users • Communication and awareness raising	• Mitigating drought through reallocating freshwater from agriculture to urban use in exchange for reclaimed water	• Formal contracts between water company and farmers associations or larger individual farmers • Automated services for urban households	• Khorasan Regional Water Authority/ Company • Wastewater user association • Larger farmers • Indirectly: Urban water users
	KEY RESOURCES • Financial resources • Legal and institutional framework • Water rights and rights exchange agreements • Wastewater treatment facilities		**CHANNELS** • Water distribution canals • Piped household water supply and automated billing	

COST STRUCTURE	REVENUE STREAMS
• Investment cost in wastewater conveyance/distribution • Operational cost (mostly wastewater pumping) • Cost of awareness creation and farmer safety training (so far this cost item is underdeveloped)	• Urban households, industry paying for freshwater • Water usage fee paid by farmers (if accepted) • Indirect (reduced groundwater pumping costs) • Indirect and direct cost savings from avoided inability to supply enough water to the city of Mashhad

SOCIAL & ENVIRONMENTAL COSTS	SOCIAL & ENVIRONMENTAL BENEFITS
• Ongoing challenge to meet the legal requirements for reclaimed water quality and application, leading potentially to costs related to health impacts on farmers and consumers, and groundwater and soil contamination	• Preventing agriculture production losses inflicted due to drought, and related social and economic benefits • City's larger benefits (domestic and industrial growth) • Improved ecosystem services once aquifer recharge takes off

B) Individual contracts between the Regional Water Company and the (larger) individual water right owner based on the right of every single water user.

Value chain and position

The water exchange in Mashhad supports the agricultural and urban value chains. Although there is no direct monetary exchange between the parties, there are environmental and social benefits associated with this business case for both sides. The actual exchange is of a higher water quantity against a higher water quality than what is available without swap (Yazdi, 2011). As arable land is not a limiting factor, in contrast to water, there will be an increase in cropping supporting the related industry. The city authorities, on the other hand, obtain freshwater and supply it to the urban dwellers to fulfill their mandate. The gains cover the costs of pumping the treated wastewater to the farms while wastewater treatment is anyway taking place, with or without swap. The strength of the business case is the possibility of a win-win situation, if the water quality matches the expectations at both ends. The authorities have the opportunity to sell the released water to households and industries at an affordable price, thus increasing their water sales revenue. So far, water quality delivered to farmers only partially matched national reuse standards and water quality adjustments have been demanded. Figure 243 illustrates the basics of the water exchange used by the business to generate value for all.

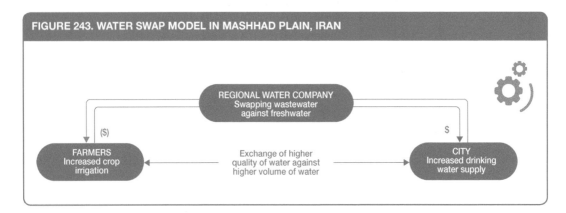

FIGURE 243. WATER SWAP MODEL IN MASHHAD PLAIN, IRAN

Institutional environment

According to national law all water bodies (rivers, lakes, aquifers) are public property and the government is responsible for their management. Allocating and issuing permits to use the water for domestic, agricultural and industrial purposes is the responsibility of the Ministry of Energy (MOE) which supervises the construction of large hydraulic works, including dams and primary and secondary irrigation and drainage canals. Within the MOE, the Water Affairs Department (WAD) is responsible for overseeing the development and management of water resources via the Water Resources Management Company (WRMC), provincial Water Authorities/Companies and provincial Water and Wastewater Engineering Companies (WWEC). They are supported by the National Water and Wastewater Engineering Company (NWWEC) which provides oversight and assistance to service providers.

Other direct and indirect stakeholders are the Ministry of Jahad-e-Agriculture (MOA), the Environmental Protection Organization, the Department of the Environment, as well as the National Economic Council and the Supreme Council for Environmental Protection. The amendment of wastewater effluent standards was published in 1994, and in 2010 the national guidelines for use of reclaimed water were published (IVPSPS, 2010; Tajrishy, 2011).

As the provincial Water Authorities now act like companies, water swap contracts were signed in most cases between district branches of the Khorasan Razavi Water Company (like the Mashahad water company) and the farmer associations, in part also with individual farmers irrigating larger land. The regional agricultural authority supported the cooperation with training and capacity development. The Khorasan Regional Water Authority is responsible for the quality of the treated wastewater, and the farmer cooperative handles water right compensation and collection of wastewater distribution revenue and transfers to the Regional Water Authority.

Technology and processes

The largest volume of wastewater comes from domestic sources. For the support of the water swap, the Olang and Parkandabad wastewater treatment plants (stabilization pond systems with anaerobic, facultative and in part maturation ponds) were constructed/adjusted along with distribution networks to transfer the treated wastewater to the farmers' fields. The transfer started operations in 2006 (Parkandabad) and 2008 (Olang). The Olang system receives sewage from east of Mashhad where most of the city hotels and commercial centers are located, while the amount of industrial flow coming to the system is negligible. The Parkandabad plant receives a combined domestic/industrial inflow and like the Olang plant is running over capacity and in need of a significant upgrade. Due to financial constraints both, the treated quantity and effluent quality remain therefore under discussion. The treated wastewater is pumped uplands to the agricultural fields, while the reservoir water is now channeled to the city, no longer to farmers. Treatment capacity upgrades would not only serve sanitation and public health but also farmers who are asking for more reclaimed water given dwindling groundwater reserves.

Funding and financial outlook

A cost analysis of the water swap was attempted for the villages at the Kardeh dam based on 2005–2006 prices when the transfer started. As the wastewater treatment is an independent investment in public sanitation, the major additional costs of the water exchange relate to water conveyance and pumping. The costs were evaluated based on the contract price adjusted to 2005–2006, using a 7% interest rate and 0.5% of the investment towards operation and maintenance costs for power transmission lines. The pump stations and treatment plants operation and maintenance costs are assumed at 2% of total investment. The total volume of reclaimed water exchanged in this sub-project is about 15 MCM per year. The estimated capital cost for conveyance pipelines, pump and power stations were in 2005–2006 about USD 6 million and annual O&M costs (mostly electricity) of around USD 650,000. Direct revenues accrue from farmers and urban water users. However, due to low tariffs and low bill collection, the water service providers do not recover their operation and maintenance costs. The same applies to the running costs of the wastewater transfer as farmers pay very little for the water they are receiving (1 to 3% of the produced crop value), which undermines efforts to increase water productivity and irrigation efficiency. Although water prices have gone up from time to time during recent decades, they have never risen as fast as the prices received for agricultural commodities. Using wastewater, farmers in the Kardeh area reported wheat, maize and barley yield increases by 20–30% and 50–68% for hay production for livestock feeding. Yields of leafy vegetable (lettuce) increased even more (82%) but also soil and crop contamination (Monem, 2013). The water company has as additional benefit savings on groundwater pumping based on the increased access to upstream surface water.

Socio-economic, health and environmental impact

Although farmers were initially skeptical about the transfer, mistrusting the regional water authority based on their past promises on allocation of reservoir water, the formation of associations for risk sharing and possibility to revoke the contracts if parties fail to deliver on their promises, facilitated their buy-in. The hierarchic institutional setup will have contributed, too. The formation of associations also had advantages for the water company. There were about 920 water right owners in the two sub-project areas although entitlements were not always clear (Alaei, 2011). The formation of associations significantly reduced the contractual transaction costs. As reported in December 2016, farmers appear satisfied with the model and are asking for more reclaimed water, especially as groundwater reserves continue to decline. Farmers also appeared more 'incentivized' to undertake water conservation practices.

Care has to be taken that any change in water flows and directions will not affect other water users and environmental flow requirements. Then the project has the potential to contribute significant aggregate economic benefits that could accrue in particular to municipal households and industry in terms of securing additional freshwater at an affordable price. If the additionally planned aquifer recharge-cum-wastewater/groundwater swap could be realized, also ecosystem services depending on the aquifer would gain. However, the transfer can only become a sustainable success if wastewater treatment capacities are increased and farmers (and potentially the aquifer) receives well treated wastewater. At the current stage, especially leafy vegetables like lettuce showed non-acceptable pathogen levels and also soils are affected. Without close monitoring and implemented risk reduction measures, farmers and consumer are at risk. Several stakeholders expressed concern that training for farmers in support of risk awareness and risk mitigation is missing, while facilities could adopt the WHO (2015) Sanitation Safety Planning which is operationalizing the WHO (2006) wastewater reuse guidelines. Authorities are well aware of the challenges and the Government of the Kashafrud basin has, for example, guaranteed a loan for vegetable farmers who like to shift to non-fruit trees instead of vegetables. The authorities also promised further supports in order to find a market for tree based products which might however be difficult, less profitable and for sure not providing returns on investment as fast as vegetables. Thus more thoughts and initiatives are needed. This also applies to those vegetable farmers in the suburbs of Mashhad who use untreated wastewater.

Scalability and replicability considerations

The water swap represents as a social business model an innovative approach of mitigating the impact of water scarcity, trading water between low and high value users in the society. The key drivers for the documented success of the business were the political will to:
- Address the growing water demands on surface and groundwater resources in an integrated way.
- Decrease high value water losses and inefficiencies in the agricultural sector.
- Consider reclaimed water as far as possible.
- Engage with farmers to work on a mutually acceptable solution.

It is possible to scale as well as to transfer this business case to other geographical areas with similar challenges and institutional set up. However, safety issues, capacity development in risk mitigation as well as issues around well-defined water rights, appropriate compensation schemes for water right holders, proper training and effective institutional coordination have to be fully addressed.

Summary assessment – SWOT analysis

The case presents a rural–urban water exchange (reallocation) to better support high value water needs of the booming city of Mashhad. The project offers interesting lessons on the need to provide farmers with incentives, in particular in comparison with the voluntary water swap in the Llobregat delta of Spain. Farmers' agreement to exchange their fresh water rights against reclaimed water allowed the Iranian water company to use the additional freshwater for domestic purposes while farmers gained additional volumes for increasing their crop production. In an apparent win-win situation, farmers in the Mashhad plain are asking today for even more reclaimed water, catalyzed by dwindling groundwater resources and drought. While in Mashhad the existing wastewater treatment capacity has reached its limit, the reuse-directed extra treatment facilities in the Llobregat case continue to run below capacity, as long as the Spanish farmers can access any alternative water source.

The SWOT analysis for water exchange in Mashhad plain is presented in Figure 244. The major strength of this business case is that farmers, regional water and agricultural authorities were involved in the negotiations from the start of the project. Farmers were given the needed recognition and incentives as the more obvious advantages of the water swap are at the urban end. While the model appears like a win-win for all parties, the economic benefits have not been quantified. This could however help the argumentation for further investments, e.g. in treatment capacity.

The challenges of the case are the cost of wastewater supply to the farmers, low cost recovery and the low treatment capacity within the city resulting in the release of reclaimed water for irrigation of in part low quality. Aside treatment upgrades, capacity development of farmers on possible risks and options for the safe use of wastewater have been strongly recommended.

FIGURE 244. SWOT ANALYSIS FOR MASHHAD PLAIN, IRAN

	HELPFUL TO ACHIEVING THE OBJECTIVES	**HARMFUL** TO ACHIEVING THE OBJECTIVES
INTERNAL ORIGIN ATTRIBUTES OF THE ENTERPRISE	**STRENGTHS** • Incentivized agreement with farmers via associations which reduces transaction costs • Win-win situation for city and farmers • Model can be up-scaled and repeated • Economic benefits likely high but so far not quantified	**WEAKNESSES** • Gaps in water quality monitoring and insufficient wastewater treatment • The cost of conveying treated water to farmers • Low education of farmers on waste water use and water conservation • Water rights partly unclear as some official title holders left the region
EXTERNAL ORIGIN ATTRIBUTES OF THE ENVIRONMENT	**OPPORTUNITIES** • Strong governmental support of water reuse and water swaps with surface and groundwater • Potential for aquifer recharge • Sufficient arable land for irrigation and increasing wastewater volumes	**THREATS** • Sustainability of the project without further investments in wastewater quality • Farmers perception on the use of treated wastewater could change if water of inferior quality is delivered • Public acceptability of wastewater for irrigation could change if potential risks are not controlled • The swap does not stop informal wastewater reuse and risk of epidemics which could also affect the exchange

Contributors

Dr. Saeed Nairizi, International Commission on Irrigation and Drainage (ICID)
Sayed Hamid Reza Kashfi, NWWEC, Ministry of Energy
Eghbal Rostami, Department of the Environment
Dr. Massoud Tajrishy, Sharif University, Iran

References and further readings

Alaei, M. 2011. Water recycling in Mashhad Plain (Effluent management: Opportunities and threats). ICID 21st International Congress on Irrigation and Drainage, 15–23 October, Tehran, Iran.

IVPSPS (Iran Vice Presidency for Strategic Planning and Supervision). 2010. Environmental criteria of treated waste water and return flow reuse, No. 535. Iranian Ministry of Energy. [In Persian]

Mahvi, A.H., Mesdaghinia, A.R. and Saeedi, R. 2007. Upgrading an existing wastewater treatment plant based on an upflow anaerobic packed-bed reactor. Iran. J. Environ. Health. Sci. Eng. 4 (4): 229–234.

Mateo-Sagasta Dávila, J. 2013. Agreement for creating a water users association for wastewater allocation of Karde Dam. Personal communication.

Monem, M.J. 2013. Best practice: Wastewater reuse in the Mashhad Plain. Presented at the wrap-up meeting of the project 'Safe Use of Wastewater in Agriculture', 26–28 June 2013, Tehran. www.ais.unwater.org/ais/pluginfile.php/550/mod_page/content/84/Iran_Wastewater%20reuse%20in%20Mash-had%20Plain_Monem.pdf.

Orumieh, H.R. and Mazaheri, R. 2015. Efficiency of stabilization ponds under different climate conditions in Iran. Indian Journal of Fundamental and Applied Life Sciences 5 (S1): 794–805.

Rahmatiyar, H., Salmani, E.R., Alipour, M.R., Alidadi, H. and Peiravi, R. 2015. Wastewater treatment efficiency in stabilization ponds, Olang treatment plant, Mashhad, 2011–13. Iranian Journal of Health, Safety & Environment 2 (1): 217–223.

Tajrishy, M. 2011. Wastewater production, treatment and use in Iran. Presented at the International Kick-Off Workshop on Safe Use of Wastewater in Agriculture, November 2011, Bonn, Germany. www.ais.unwater.org/ais/pluginfile.php/356/mod_page/content/128/Iran_Paper%20Bonn%20%20version%202.1.pdf. Accessed 02/12/16.

WHO. 2006. Guidelines for the safe use of wastewater, greywater and excreta in agriculture and aquaculture. Geneva, Switzerland: World Health Organization (WHO).

WHO. 2015. Sanitation safety planning: Manual for safe use and disposal of wastewater, greywater and excreta. Geneva, Switzerland: World Health Organization (WHO).

Yazdi, P. 2011. A survey on social issues in replacing effluent with agricultural water rights (Case study: Replacing Mashhad effluent with water rights from Karder River). International Congress on Irrigation and Drainage, 15–23 October, Tehran, Iran.

Case descriptions are based on primary and secondary data provided by case operators, insiders or other stakeholders, and reflect our best knowledge at the time of the assessments 2013–2016. As business operations are dynamic data can be subject to change.

CASE
Flexible wastewater-freshwater swap (Llobregat delta, Spain)

Pay Drechsel, George K. Danso and Munir A. Hanjra

Supporting case for Business Model 20	
Location:	Llobregat delta, Barcelona, Spain
Waste input type:	Treated wastewater
Value offer:	Treated wastewater for farmers to release in times of drought freshwater for domestic (and industrial) purposes
Organization type:	Public and private
Status of organization:	El Prat WWTP operational since 2004, with several upgrades since then; Sant Feliu WWTP operational since 2010
Scale of businesses:	Medium
Major partners:	Farmers, Catalonian Water Agency (ACA), City of Barcelona, European Union (EU)

Executive summary

This business case presents an example of integrated water resources management (IWRM) in support of a voluntary water exchange between local farmers and the Catalonian Water Agency (ACA) in the Llobregate River basin delta. The inter-sectoral water transfer builds on a flexible approach which allows negotiation between the parties involved to adapt to the intensity of seasonal drought and priority water needs. In this European Union co-funded project, the ACA treats urban wastewater to different, reuse defined levels. The main clients are farmers who are obliged to stop using surface water in times of drought. In exchange for accepting treated wastewater the city obtains the protected freshwater for aquifer recharge. This is in large a social responsibility business model, which allows on one hand (i) ACA to deliver on its water supply mandate also in times of extreme water shortage; and on the other hand (ii) gives farmers a reliable water supply to cope with drought or to go beyond (low revenue) rainfed farming; while (iii) the city gains in terms of drinking water, environmental health, aquifer protection and more resilient short food supply chains. From an economic perspective, the investment costs are marginal compared to the direct and indirect costs of a severe drought as experienced in 2007–2008 (Martin-Ortega et al., 2012). The case also realizes an often demanded paradigm shift where the degree of water treatment and allocation differ between types of reuse to optimize the overall returns on investment.

COMBINED KEY INDICATORS FOR THE EL PRAT AND SANT FELIU WWTPS (2012)						
Land use:	1076 ha (maximum irrigation area potentially served)					
Wastewater treated:	Up to 146MCM per year with about 20MCM for agriculture (water swap)					
Capital investment:	EUR 15.12 million (treatment upgrades)					
O&M:	EUR 3.11 million per year (treatment); EUR 2.56 million per year (water conveyance)					
Output:	Among others, the possible release of up to 20MCM freshwater per year					
Potential social and/or environmental impact:	Improvements in economic development because of additional freshwater, for domestic use and environmental flow, and reduced overexploitation and protection of the local aquifer.					
Financial indicators (for both plants assuming annual water swaps; FAO 2010):	Payback period:	Depending on the volumes actually reused/ swapped	Net Present Value	70–115 million Euro	Benefit-cost ratio:	3–5 to 1

Context and background

Eastern Spain has been experiencing severe droughts in its recent past and is expected to experience even more in the coming years. To support Barcelona, the government is using multiple strategies, including long distance transfer and seawater desalination. Another measure to reduce the water deficit is reallocation matching water needs and water quality. Reuse of treated wastewater is part of this approach. Already today, about 13% of Spain's total wastewater volume is reused, which is far above the European average[1]. The Lloberegat delta region presents an example of Spain's reuse efforts applying an IWRM approach to deal with the complexity of surface and groundwater resources under stress within a basin cutting across rural and urban boundaries. This stress has qualitative and quantitative dimensions. By the end of the 1980s, the Llobregat River, which runs through parts of Barcelona was one of the most degraded rivers in Western Europe, putting increasing pressure on water users and the aquifer (Sabater et al., 2012). Supported by the 1991 European directive on urban wastewater treatment, a comprehensive rehabilitation programme has been implemented along the river allowing the situation to improve dramatically.

The Llobregat River's lower valley and delta, located in Barcelona's province, consist of about 30 km² of alluvial valley, up to 1 km wide, and a delta of 80 km². In spite of the delta's very close proximity to the city, it constitutes a wetland of international importance for wildlife, especially migrating birds. Its fertile farmland supports intensive agriculture (fruits, vegetables) for the urban market, and as a protected green belt, the delta helps restricting urban sprawl. The delta aquifer is one of the most important fresh water resources of the Barcelona area, forming an underground source with a capacity of 100 million cubic meters (MCM) of water,[2] which is however under pressure from seawater intrusion. With an average annual precipitation in the Lloberegat delta around 620mm/yr (2015: only 346mm), spread over two to six rainy days per month, not only the city and local industries but also the delta farmers rely on the aquifer for supplementary irrigation, resulting increasingly in over-exploitation and water salinization. The need to optimize water allocations across sectors was highlighted during the severe drought of 2007–2008 in Northeast Spain, which caused very high societal, economic and environmental cost of an estimated EUR 1605 million (Martin-Ortega et al., 2012). Aside supporting human needs, a significant part of the EU supported effort targeted ecosystem services of the

Llobregat River and delta by reducing water loss to the sea, and pumping it upstream over 15 km to re-support the natural river flow.

Market environment

In a region suffering regularly from very low rainfall, access to water is fundamental to many economic sectors, including agriculture, as well as environmental needs. Based on a participatory stakeholder dialog, the treatment of the wastewater in the Llobregat delta follows a step-wise approach to meet the particular water quality requirement of each reuse purpose, considering that any additional treatment will cost extra and should only be activated on demand. Wastewater leaving the plant for the sea undergoes secondary treatment, while for aquifer recharge tertiary treatment including reverse osmosis can be used, while farmers demanded in addition the demineralization of the reclaimed water as water salinity prevented them from using it. As a result, the two wastewater treatment plants (El Prat and Sant Feliu) in the district of Baix Llobregat were designed to support directly or via water exchange a range of demands (agriculture, environmental flow, wetland ecosystem services, seawater barrier through managed aquifer recharge, urban water supply, recreation and industry) (Table 57).

About 20MCM/year of treated effluent from the two plants could support seasonal irrigation of up to about 1,000 ha (Heinz et al., 2011a, 2011b). As drought conditions vary, the water exchange was set up on voluntary base without specific quantitative targets. In general, most farmers prefer the usually less saline river or groundwater. Only when these sources get scarce, and farmers are no longer allowed to abstract water, reclaimed water was used. The efforts by the authorities to install additional treatment capacity for halving the salinity level of the reclaimed water to about 1.4 millisiemens per centimeter (mS/cm) responded directly to farmers' water quality concerns.

The water exchange can build in this case on an efficient water distribution system, where farmers are in relatively close proximity to the wastewater treatment system and freshwater users, limiting upstream pumping costs of the treated water.

Macro-economic environment

The government of Spain is giving high priority to the improvement of water use efficiency across sectors, especially in the drought affected Eastern region around Barcelona. While different coping strategies are being implemented, inter-sectoral water transfer based on wastewater treatment for reuse was described as the least costly option (EUR $0.34/m^3$) compared with desalination of sea water (EUR$0.45–1.00/m^3$) and water transfer from other areas (EUR $8.38/m^3$) (Hernández-Sancho et al., 2011).

TABLE 57. MULTI-PURPOSE USE POTENTIAL OF RECLAIMED WATER IN THE LLOBREGAT DELTA

	EL PRAT DE LLOBREGAT WWTP (MCM/YR)	SANT FELIU DE LLOBREGAT WWTP (MCM/YR)
Agriculture	13.09	7.36
River stream flow	10.37	–
Wetlands	6.31	–
Seawater barrier	0.91	–
Municipalities	–	0.11
Recreation	–	0.37
Industry	5.48	–
Total	**36.2**	**7.84**

Source: Hernández-Sancho et al., 2011.

To assess the economics of water exchange between farmers (releasing freshwater) and cities (providing reclaimed water) a broader perspective at watershed level is needed. The IWRM concept offers an appropriate framework which allows to consider water-related sectors, services and their interdependencies. The first analysis showed that water reclamation (treatment and conveyance) costs would be more than offset by the value the exchange offers urban water supply, not to mention the direct and indirect costs of the next prolonged drought. The macro-economic benefits will increase with more water transferred to high-value usage. While farmers' financial advantages are limited, the urban water sector is best positioned to absorb the costs for the exchange (Figure 245) unless the investment is considered an insurance against the possibility of significant loss.

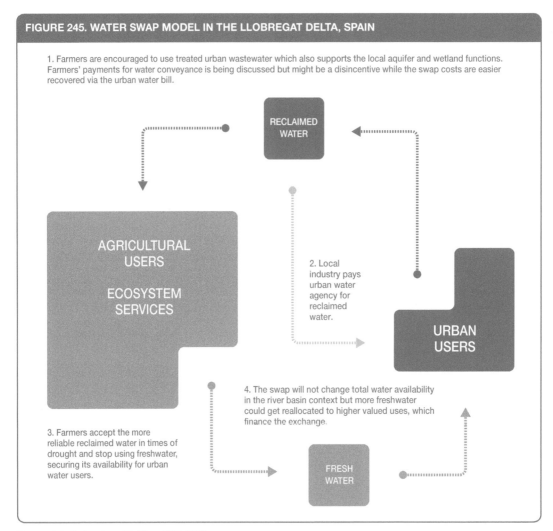

FIGURE 245. WATER SWAP MODEL IN THE LLOBREGAT DELTA, SPAIN

Source: Adapted and modified from GWI, 2009.

Business model

The business model offers multiple value propositions through need-based wastewater treatment for different water reuse purposes. Aside the support of ecosystem services, irrigated crop production will be an important water user in periods of drought when farmers are asked to withdraw from surface water use. Through freshwater savings and additional aquifer recharge, ACA can continue its freshwater supply for the urban population. The volume of the business transaction depends on the duration of the drought and related negotiations between ACA and farmers. While urban users would be the main source of finance for the costs, there will be a range of environmental benefits (Figure 246). While farmers can save in pumping costs and fertilizer application, the benefits for the city are large, and can provide the exchange with a net benefit depending on the traded water volume (see Finances below).

Value chain and position

Table 57 shows the technically possible volumetric benefits of the exchange for different usage of the water released by the two mentioned treatment plants in the Llobregat delta. While the numbers show the potential, the majority of the treated wastewater is used so far to maintain or re-establish the Llobregat River's flow while farmers shifted to treated wastewater so far only in those periods when there was no other (equally reliable) alternative left to maintain crop yields and/or to avoid shifting to low value rain fed crops. The city gains in this situation by securing additional freshwater for domestic and industrial purposes with a higher water value than what it can offer agriculture. While the exchange is so far of voluntary nature, farmers could gain higher bargaining power and opt for a formal exchange of water rights with other buyers once they have better information on the nature of the water market.

Institutional environment

The main stakeholders in the project are farmers, the water company of the metropolitan area of Barcelona, the water administrations (at regional and local level), and the environmental administration. Because the inter-sectoral water transfer relies on farmers and the city, a cooperation and negotiation process between farmers and the water supply company ACA was essential. Being part of the decision-making process, has been described as an important pillar for farmers' support of the model.

The European Water Framework Directive (WFD) and the Catalonian Water Reuse Program were key for the development and financial support of the water swap model, and also the regulations for reuse to be considered.

Since the swap became operational, farmers are making use of the reclaimed water, however, to a smaller extent than what could be made available based on treatment capacity. Farmers view the reclaimed water only as a last resort to be used when freshwater use is no longer permitted, reliable or salinity of the freshwater exceeds the one of the reclaimed water. As each swap is a response to a particular drought, negotiation between farmers and the water administration remain dynamic and prevented so far contractual commitments. To increase farmers' use of reclaimed water also under normal seasonal water stress, there are different instruments and incentives possible which have however to be aligned with farmers water rights (concessions), especially in view of groundwater abstraction.

Technology and processes

By generating a reliable flow of high quality reclaimed water, the options available for integrated water resources management have widely expanded to allow in-stream river water substitution, restoration of natural wetland areas, agricultural irrigation and aquifer recharge to block seawater intrusion. Those management options have been possible thanks to the implementation of an extensive water distribution system that allows distribution of reclaimed water to a point 15 km upstream of the reclamation facility, and to a seawater intrusion barrier within a few kilometers of the plant. The water distribution network has 18.8 km of main pipes.

The wastewater treatment plant of El Prat de Llobregat has been operating since 2004 and has a capacity of up to 420,000m³/day. It includes an activated sludge treatment process that was upgraded in 2006 to achieve nutrient removal, using biological nitrification-denitrification, plus biological and chemical phosphorus removal. About two-thirds of the secondary treated water is discharged into the Mediterranean Sea, while one-third could undergo depending on demand tertiary treatment for reuse, with a smaller part of it also reverse osmosis (RO). An additional desalination plant which is using membranes for electrodialysis reversal (EDR), is able to produce for farmers up to 57m³ of improved irrigation water per day (18.8MCM/yr).

FIGURE 246. BUSINESS MODEL CANVAS FOR INTER-SECTORAL WATER EXCHANGE IN SPAIN

KEY PARTNERS (cum customers in a swap model)	KEY ACTIVITIES	VALUE PROPOSITIONS	CUSTOMER RELATIONSHIPS	CUSTOMER SEGMENTS (cum partners in a swap model)
• Farmers irrigating in the Llobregat delta willing to accept reclaimed water • Catalonian Water Agency (ACA) • Others: EU-WFD	• Treat wastewater to an acceptable level for farmers and wetland • Make reclaimed water accessible for reuse • Negotiate water swap with farmers • Obtain freshwater and sell to households and industries • Awareness creation for water savings and reuse • Maintenance of treatment facilities	• Mitigating drought (related costs) through reallocating freshwater from agriculture to urban use in exchange for reclaimed water allowing to realign water supply and demand from various sectors based on sector specific water quality requirements	• Negotiations between ACA and farmers considering expected drought duration and sectoral water needs • ACA services for urban households	• Farmers in need of water of acceptable quality • Catalonian Water Agency (ACA) Indirectly: • Urban water users

KEY RESOURCES		CHANNELS	
• Financial resources for tertiary wastewater treatment and desalination units • Legal and institutional framework for collaboration • Awareness campaigns • Farmers' consent		• Roundtables for negotiation • Distribution canals for irrigation with reclaimed water • Piped water supply for households and automated billing	

COST STRUCTURE	**REVENUE STREAMS**
• Investment cost as well as O&M costs unless the swap allows sufficient urban revenues • Water conveyance and distribution cost • Cost of awareness campaigns	• Urban households pay ACA for extra (released) freshwater, and farmers for high-value crops • Farmers have been asked to pay for water conveyance (only) • Indirect revenues (cost savings) in view of socio-economic damage during drought from interrupted or reduced water supply
SOCIAL & ENVIRONMENTAL COSTS	**SOCIAL & ENVIRONMENTAL BENEFITS**
• Potential health impact on consumers from the consumption of crops irrigated with reclaimed wastewater not meeting all possible risk factors	• Avoidance of production losses inflicted due to drought • Urban consumers continue to have fresh fruits and vegetables • Improved water allocations for the Llobregat aquifer, river and wetlands and related ecosystem services • Hydraulic barrier against sea water intrusion

Box 11. Treatment for nature

A third WWTP operates since 2010 on the western edge of the delta at **Gavà-Viladecans** with a capacity of about 23MCM/yr. The treated effluent is sent to the headwaters of the system of canals and corridors feeding into the Murtra lagoon, with the goal of protecting water quality in the nature reserves and preventing eutrophication. One of the lines, which treats 50% of the total flow, has a membrane bioreactor system (MBR). This process gives high quality reclaimed water which can be reused. However, the water is usually not used directly for irrigation, but for stabilizing the hydrological balance and to recharge wetlands.

Funding and financial outlook

The overall project had an initial budget of EUR102 million; 85% of that amount has been covered by European Union Cohesion Funds, through the Spanish Ministry of the Environment, and the remaining 15% has been covered by the Catalan Water Agency. Comparing costs and benefits of the water swap, including discounted capital costs, the projected net profit of water transfer when considering agriculture and the city is around EUR16 million per annum (Table 58), without counting environmental benefits. The water swap could lead to savings as well as gains for farmers and the city. In an ideal situation, the investment of one euro in the use of reclaimed water creates an income increase in agriculture of approximately EUR1.6 (Hernández-Sancho et al., 2011). Farmers face less groundwater and surface water pumping costs as well as costs of fertilizing, while they can maintain high value crops or expand irrigation. The magnitude of the benefits increases with the duration of the swap.

In general, the cost of the additional wastewater treatment is paid by the urban water users and the cost of conveying irrigation water by farmers. However, with the largest share of benefits accruing at the city level, and the fact that the system depends on farmers' voluntary contribution, they would need

TABLE 58. COSTS AND BENEFITS OF WATER REUSE AT THE LLOBREGAT DELTA

CHARACTERISTICS	EL PRAT	SANT FELIU
Irrigated farmland (ha)	801	275
Effluent volume applicable for irrigated agriculture (MCM/yr)	13.0	7.3
ANNUAL COSTS...	MILLION EURO/YR	MILLION EURO/YR
Cost of new treatment units	1.09	0.08
Operation and maintenance cost of treatment	2.6	0.51
Cost of conveying effluents	0.12	0.20
Cost of conveying water released for urban use	1.43	0.81
Total cost of water reuse and exchange (A)	**5.24**	**1.60**
...AND ANNUAL BENEFITS		
Value added to agriculture	0.35	0.46
Value of water exchanged for city use	14.43	8.12
Total economic benefit of water reuse and exchange (B)	**14.78**	**8.58**
Total value added of water reuse and exchange (B-A)	**9.54**	**6.98**
UNIT COSTS AND BENEFITS	EUR/M^3	EUR/M^3
Unit cost of water reuse and exchange	0.40	0.22
Unit total economic benefit for agriculture and city	1.14	1.17
Unit cost/benefit ratio	**2.85**	**5.3**

Source: Heinz et al., 2011a.

to be convinced of the value of the exchange for themselves (reliability of the water supply, savings of pumping, nutrient value) and depending on urban needs be supported by additional incentives to engage in the exchange. If farmers' buy-in can be augmented, the urban benefits could be sufficiently high to carry the exchange, also if farmers do not pay for water conveyance.

It should also be considered that aside the stigma of wastewater use, farmers expressed concerns how the [European] market and legislations would perceive the use of reclaimed water.

Based on the first evaluation (Hernández-Sancho et al., 2011) the water swap model started successfully as farmers accepted the reclaimed water in times of water stress. In the first 1.5 years, 35.5MCM were reused to re-establish the Llobregat River flow, 2.4MCM for agricultural irrigation, 4.8MCM to stabilize wetland ecology and 0.4MCM to reduce salt water intrusion in the aquifer. Since then agricultural reuse (and water release) remained at a similar level although details on actual volumes during the drought of 2012 and 2015–2016 could not be accessed (Santos and Marcos, 2009).

If a sensitivity analysis were to be done, it would show that the overall NPV would be highly sensitive to the size of released water and resulting urban water benefits (FAO, 2010), which were so far much lower due to sufficient precipitation. Urban cost recovery remains also challenged due to low water tariffs combined with difficulties to accurately determine the cost of wholesale water services in a complex situation when the infrastructure is shared among different uses, e.g. regulation and transport of raw water for populations, energy uses and irrigation (García-Rubio et al., 2015).

Socio-economic, health and environmental impact

The anticipated main impact is based on the reduction of the direct and indirect costs of any forthcoming severe drought as in 2007–2008. The exchange of water towards higher value water use allows economic gains for different sectors without that the overall amount of water is changing. The project appears to succeed because farmers started to use the reclaimed water and freshwater has been released to other sectors, such that the overall availability of water in the metropolitan area of Barcelona has improved. The income of the farmers has increased to some extent and the availability of reclaimed water for irrigation has been improved in times of low freshwater supply.

An interesting side-effect is that water consumption for domestic use has decreased and the water quality of the Llobregat aquifer has improved widely. Although this was not a direct objective of the business case, the water crisis in 2007–2008 and implementation of the inter-sectoral water exchange and related educational efforts increased public awareness for water savings. Energy savings associated with the reduction of pumping groundwater were quantified at around 4m kWh/yr which translates approximately into 1,440t of CO_2 equivalent per year. The use of reclaimed water has also led to cost savings in chemical fertilizers and related energy quantified as 2,170t/yr, including the avoided use of phosphorus (Hernández-Sancho et al., 2011).

Also an improvement in the Delta aquifer for all parameters related to seawater intrusion has been verified (Hernández et al., 2011), and even the wastewater which is with less treatment discarded into the sea, still serves a purpose: brine produced at Barcelona's Desalinization plant (which support 20% of Barcelona with potable water) is blended with treated water from the El Prat WWTP in a ratio lower than 1:1 before it enters the sea.

Scalability and replicability considerations

The key drivers for the success of this business model are common in many water-stressed regions and replicable:
- Water scarcity combined with growing urban water needs made water reclamation and innovative water allocations for reuse important and necessary for the region.
- Early stakeholder consultation leading to the adaptation of treatment quality to farmers' needs and their voluntary acceptance of the seasonal water swap (which can also be key risk factor as long as the exchange remains voluntary).
- Single agency (ACA) with mandate for wastewater treatment and providing drinking water to the city, thus providing greater flexibility and ease for negotiating with farmers on the inter-sectoral water exchange.
- Economic analysis showed an overall positive economic balance, not counting improved ecosystem services.
- Support from the Government of Spain and European Commission to improve inter-sectoral water use efficiency.

Replication of the case is recommended as it represents an interesting example of the often demanded paradigm shift (e.g. Huibers et al., 2010; Murray and Buckley, 2010) where different water uses are matched with their required water quality, which includes that (i) wastewater treatment is designed for the planned type of reuse; and (ii) water is allocated to the type of use which allows the highest returns for the respective water quality. It is also a case where the IWRM framework was successfully applied across sectors including the urban one. However, monitoring crop and water quality will be needed to prevent that produce markets, also in other EU countries may reject crops irrigated with reclaimed water.

For a full success of the swap, the city might prefer to plan with a released minimum water volume, while farmers should not see the reclaimed water as additional water to increase their irrigated area, which would prevent any release of freshwater for the city. GWI (2009) stressed that voluntary water swap models can be flawed due to the potentially unlimited agricultural water demand and no direct benefit for farmers from the release of their water. Thus the swap needs regulatory support, for example in form of seasonal surface or groundwater abstraction limits (volumes, time periods) which farmers have to adhere to, in exchange of a reliable supply with reclaimed quality water. In the case of the Llobregat delta, extraction from the common irrigation channels by farmers is prohibited in drought periods and, at such times, farmers are obliged to use reclaimed wastewater from the El Prat de Llobregat WWTP. The same applies to the Sant Feliu de Llobregat WWTP where the limit for agricultural use of water from the Llobregat river is 1.5m^3/s, but in periods of water shortage this use is reduced to 0.8m^3/s, and farmers are obliged to use treated wastewater or to switch to less demanding crops (FAO, 2010).

Summary assessment – SWOT analysis

In this case significant investments went into infrastructure able to produce high-quality reclaimed water to secure farmers' acceptance of a water swap in prolonged periods of drought. Thus the water swap contract remained like an insurance policy flexible, given the, in large, unpredictable nature of the extent of a possible drought period and actual need for farmers to seek alternative water sources. Despite harsh conditions in 2007–2008, 2012 and 2015–2016, the installed infrastructure (reverse osmosis, desalination) was so far hardly used for serving agricultural demand. Financial considerations/limitations might have contributed to the underutilization.

FIGURE 247. SPAIN WATER SWAP SWOT ANALYSIS

	HELPFUL TO ACHIEVING THE OBJECTIVES	**HARMFUL** TO ACHIEVING THE OBJECTIVES
INTERNAL ORIGIN ATTRIBUTES OF THE ENTERPRISE	**STRENGTHS** • Governmental support to invest in infrastructure to mitigate possible risks from climate change • Dialog with farmers and offer of reliable water supply • Flexible targets and execution allow adaptation to extreme climate events and water savings • Multi-purpose reuse program with aligned treatment levels supporting urban and ecosystem needs • High economic benefits for society covering all investment costs	**WEAKNESSES** • The cost of temporarily unused infrastructure (RO, EDR) in times of sufficient freshwater supply • Missing incentive systems for farmers to use reclaimed water more, and more frequently • Water salinity challenge undermining farmers' acceptance • Farmers market reservations related to wastewater use • Limited cost recovery without urban users paying for released water
EXTERNAL ORIGIN ATTRIBUTES OF THE ENVIRONMENT	**OPPORTUNITIES** • Flexibility allows farmers to react to different drought situations, while the option to swap a fixed minimum volume could be an alternative. • Educational options to improve farmers' acceptance of water reuse • Similar locations and challenges exist in various countries for replication of concept and strategy	**THREATS** • Changing public perception on the use of treated wastewater • Financial and economic crisis affecting plant operations • Alternative freshwater sources (desalinization, long-distance transfer) appear more reliable than a voluntary agreement and are already in place or in construction

While farmers prefer to use the aquifer as their main water source, supplemented by the Llobregat River water, they complied with the swap although to a lower extent than anticipated. Without set targets, it is difficult to assess the difference between any intended and actual outcome or to predict if the swap will remain an option of choice once Barcelona can rely on sea water desalinization. This also poses questions how far the presented cost-benefit analysis (e.g. FAO, 2010; Heinz et al., 2011a, 2011b; Hernández-Sancho et al., 2011) for a regular water exchange remains valid. On the other hand, in view of the possible damage an extended drought period could cause, any of the current investments in risk mitigation (water swap, desalination, water transfer) would have significantly higher returns on investments already with the next drought (Martin-Ortega et al., 2012).

Figure 247 presents the SWOT analysis for water exchange in Llobregate. As the success of the water exchange depends mostly on farmers' willingness to accept reclaimed water, while stopping the use of other sources, tax and/or regulatory incentives should be discussed in support of the process. For a detailed risk analysis see FAO (2010).

Contributors

Francesco Hernández-Sancho, University of Valencia, Spain
Miquel Salgot, University of Barcelona

References and further readings

Cazurra, T. 2008. Water reuse of south Barcelona's wastewater reclamation plant. Desalination 218: 43–51.

FAO. 2010. The wealth of waste: The economics of wastewater use in agriculture. Roma: Food and Agriculture Organization of the United Nations.

García-Rubio, M.A., Ruiz-Villaverde, A. and González-Gómez, F. 2015. Urban water tariffs in Spain: What needs to be done? Water 7 (4): 1456–1479.

GWI. 2009. Municipal water reuse markets 2010. Global Water Intelligence.

Hernández, M., Barahona-Palomo, M., Pedretti, D., Queralt, E. Massana, J. and Colomer, V. 2011. Enhancement of soil aquifer treatment to improve the quality of recharge water in the Llobregat aquifer. 8 International Conference on Water Reclamation and Reuse, 26–29 September, 2011, Barcelona.

Hernández-Sancho, F., Molinos-Senante, M. and Sala-Garrido, R. 2011. WP3 Ex-post case studies. Voluntary intersectoral water transfer at Llobregat River Basin. EPI Water. Grant Agreement no. 265213. http://www.feem-project.net/epiwater/docs/d32-d6-1/CS9_Llobregat.pdf (accessed 4 Nov. 2017).

Heinz, I., Salgot, M. and Mateo-Sagasta Davila, J. 2011a. Evaluating the costs and benefits of water reuse and exchange projects involving cities and farmers. Water International 36(4): 455–466.

Heinz, I., Salgot, M. and Koo-Oshima, S. 2011b. Water reclamation and intersectoral water transfer between agriculture and cities – A FAO economic wastewater study. Water Science & Technology 63 (5): 1067–1073.

Huibers, F., Redwood, M. and Raschid-Sally, L. 2010. Challenging conventional approaches to managing wastewater use in agriculture. In: Drechsel, P., Scott, C.A., Raschid-Sally, L., Redwood, M. and Bahri, A. (eds) Wastewater irrigation and health: Assessing and mitigation risks in low-income countries. UK: Earthscan-IDRC-IWMI, pp. 278–301.

Martin-Ortega, J., González-Eguino, M. and Markandya, A. 2012. The costs of drought: The 2007/2008 case of Barcelona. Water Policy 14: 539–560.

Mujeriego, R., Compte, J., Cazurra, T. and M. Gullon. 2008. The water reclamation and reuse project of El Prat de Llobregat, Barcelona, Spain. Water Science & Technology 57 (4): 567–574.

Murray, A. and Buckley, C. 2010. Designing reuse-oriented sanitation infrastructure: The design for service planning approach. In: Drechsel, P., Scott, C.A., Raschid-Sally, L., Redwood, M. and Bahri, A. (eds) Wastewater irrigation and health: Assessing and mitigation risks in low-income countries. UK: Earthscan-IDRC-IWMI, pp. 303–318.

Sabater, S., Ginebreda, A. and Barceló, D. (eds) 2012. The Llobregat: The story of a polluted Mediterranean river. Springer.

Santos M.G. and Martos, P.A. 2009. Presentation on the Metropolitan Area of Barcelona Environmental Authority and the Baix Llobregat regeneration plant. http://www.eesc.europa.eu/resources/docs/presentation-aguilo-en.pdf (accessed 4 Nov 2017).

Case descriptions are based on primary and secondary data provided by case operators, insiders or other stakeholders, and reflect our best knowledge at the time of the assessments 2014–2016. As drought periods vary in frequency and extent, so will the voluntarily swapped water volumes and related costs and benefits.

Notes

1 http://ec.europa.eu/environment/water/reuse.htm (accessed 4 Nov. 2017).
2 http://geographyfieldwork.com/LlobregatWaterReclamation.htm (accessed 4 Nov. 2017).

BUSINESS MODEL 20
Inter-sectoral water exchange

Pay Drechsel and Munir Hanjra

Key characteristics

Model name	Inter-sectoral water exchange
Waste stream	Urban wastewater
Value-added waste product	Reclaimed water for domestic and industrial use
Geography	Seasonally or continuously water short areas where urban and agricultural water demands could be better aligned
Scale of production	Medium- to large-scale (no defined range)
Supporting cases in the book	Mashhad, Iran; Llobregat delta, Spain
Objective of entity	Cost-recovery [ ]; For profit [ ]; Social enterprise [X]; Insurance [X]
Investment cost range	Can vary in large margins depending on (i) how far existing treatment infrastructure meets standards for irrigation, and (ii) distance for water transport
Organization type	Public, public-private, private
Socio-economic impact	Increased freshwater supply for urban households in periods of drought; guaranteed agricultural supply with reclaimed water in all seasons
Gender equity	Beneficial in particular to urban women and children due to time savings in water access; improvement in hygiene and living conditions

Business value chain

To address increasing urban water demands in basins with limited water resources, or to cope with severe periods of drought, water reallocation within and across basins can be important adaptation strategies. Even without increasing the overall water volume, reallocating freshwater from agriculture to urban use in exchange for reclaimed water can help within the same basin urban needs, and help optimizing water allocations with sector specific water quality requirements. Such a water swap requires investments in appropriate treatment as well as incentive systems that farmers actually release their surface- or groundwater for urban use. This water can then be sold at a higher price to urban consumers than farmers would ever pay. The obtained revenues can support cost recovery of water transport and treatment, with an increasing probability of a positive benefit-cost ratio the larger the water volumes exchanged (Figure 248). The situation looks even better from an economic perspective: In the case of Spain, the direct and indirect costs of the affected regional economy due to multiple months of water scarcity in 2007–2008 were estimated at EUR1605 million or 0.48% of the regional GDP. The order of magnitude of these estimates is similar to others reported in the USA and Australia in recent years and easily outweighs the total investment costs in climate change adaptation measures in the region, including wastewater conveyance and treatment for reuse (Martin-Ortega et al., 2012).

FIGURE 248. VALUE CHAIN SCHEMATIC – INTER-SECTORAL WATER EXCHANGE

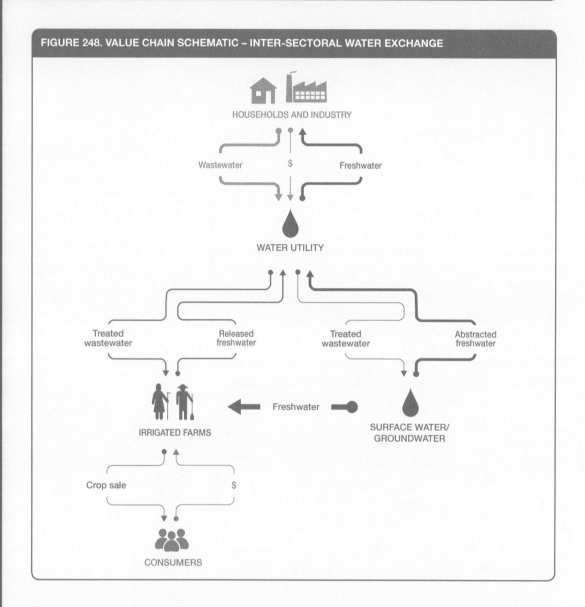

The business concept depends strongly on the incentives offered to (and accepted by) farmers, i.e. the contractual agreement (such as transfer of water rights) as otherwise farmers might absorb the wastewater to expand their operations without releasing freshwater. The exchange might only work where defined water rights exist, freshwater can be transferred to urban consumers without allowing access by third parties, and wastewater has to be redirected to farmers, e.g. pumped upstream (from city to farmers) as otherwise at least some downstream farmers will be able to access the urban return flow without contract.

Business model
This business model transfers freshwater from agricultural use to urban areas for domestic use in exchange for treated wastewater. This model is complex as it can entail many partners across the agricultural – water supply – wastewater and health sectors, different time horizons and mechanisms to support farmers' buy-in.

The main contract is between the public or private water utility and the farmers or their water users associations. The urban partner has to invest in additional treatment capacity as conventional treatment might result in water with too high in contaminants or salinity for crop irrigation. In addition, investments in water conveyance are needed although in many situations one of the flows might follow gravity.

Contracts can span the whole year where urban areas face a permanent supply deficit or be seasonal. If seasonal, the water swap can be limited to certain months or only be activated in times of severe drought. Water volumes can be defined or remain flexible according to the supply gap. Obviously, the pay-back period for treatment infrastructure and water conveyance increases when actual water swaps remain seldom, and/or volumes are low, like in the case of the Llobregat delta. However, in this case, the investment is more like a water supply insurance for parts of the 1.6 million city of Barcelona, aside other, and often more expensive, risk reduction and mitigation measures (desalination, long-distance water transfer).

Farmers, who have to give up on parts or all of their freshwater rights, need to understand the reasons and incentives to accept what looks per se as a disadvantage. These investments in awareness creation and incentives, and the contract, which builds on them, are the heart of the business model. The incentives can have pull and push factors. Depending on the local context, the authorities might limit farmers' freshwater withdrawal through regulations for times of drought while offering reclaimed water as substitute. To support farmers acceptance, the volume of supplied wastewater, can, like in the case from Iran, be higher than the released freshwater. Obviously, options to charge farmers for the water could be counterproductive. In contrary, wastewater acceptance could be bundled with financial incentives, such as access to micro-credit. Accompanying training in its safe application, protective gear and awareness creation on reduced fertilizer needs should be part of the package. Most important, as the studied cases stress, is the reliability of the supply and an acceptable water quality for plant growth. Social and economic benefits of the water exchange will be very high as the case from Iran shows where households and the local (tourist) economy depend on the additional freshwater year-round (Figure 249). On the other hand, the economic damage can be very high if a city is not prepared to adapt to such climatic extremes as the case from Spain shows.

Potential risks and mitigation from the urban perspective

In designing any optimized business model based on case studies, it is assumed that generic business risks are known and will be taken care of. However, some risks might be more model specific and will be acknowledged in the following:

Market risks: The market could be characterized as fragile as business success depends on willingness and availability of enough farmer or farmer associations to exchange freshwater against reclaimed water, which appears on the first view as a 'bad deal'. The business thus requires awareness creation on the reasons for the exchange, education on the advantages of wastewater (nutrients, reliability) and in addition tangible incentives for the farmers to accept the swap; all under the assumption of a water supply gap, thus a market for the released freshwater. Where urban water is constantly in short supply, long-term contracts would have an advantage; where the exchange is more an instrument for time of water crisis, also flexible contracts are possible. Market risks might be lower in societies where farmers have limited political power to negotiate agreements in their interests.

Competition risks: Different perspectives of competition are possible in a water swap:

a) farmers continue using freshwater;
b) the city receives freshwater through desalination or long-distance transfer at lower costs or less (human) risks (as the water exchange requires negotiations with farmers, reliance on behaviour change, etc.); and

FIGURE 249. BUSINESS MODEL CANVAS – INTER-SECTORAL WATER EXCHANGE FROM THE PERSPECTIVE OF THE WATER UTILITY

KEY PARTNERS
- Farmers or their associations
- If available, private entity managing water supply and sewerage
- Authorities (e.g. health, agriculture)

KEY ACTIVITIES
- Awareness creation for water swap
- Negotiation of water swap
- Treat wastewater fit for irrigation
- Convey wastewater to farmers/ freshwater to city
- Monitoring of water exchange agreement
- Households water supply
- Selling fresh water

KEY RESOURCES
- Treatment technology
- Capital
- Legal entitlement to exchange water rights

VALUE PROPOSITIONS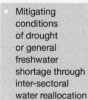
- Mitigating conditions of drought or general freshwater shortage through inter-sectoral water reallocation

CUSTOMER RELATIONSHIPS
- Negotiations to agree on formal contracts
- Inter-institutional collaboration
- Automated billing services to urban households

CHANNELS
- Negotiation roundtables
- Piped supply, direct or bulk sales
- Automatic billing through internet or supermarkets

CUSTOMER SEGMENTS
- Irrigating farmers
- Water demanding industry
- Urban residents

COST STRUCTURE
- Capital investment in treatment and water conveyance
- O&M (in particular water pumping, quality monitoring)
- Cost of awareness campaign for exchange, and safe wastewater reuse training

REVENUE STREAMS
- Sales of wastewater (optional)
- Sales of gained fresh water
- Indirect revenues (saving on socio-economic costs and damage claims from inability to supply water in (prolonged) periods of drought)

SOCIAL & ENVIRONMENTAL COSTS
- Possible health risks for farmers and consumers if wastewater quality does not meet agreed standards, or gets mixed with untreated wastewater before use, and insufficient risk reduction options have been put in place

SOCIAL & ENVIRONMENTAL BENEFITS
- Climate change adaptation measure to reduce the impact of extended water scarcity on agriculture and society
- Contribution to food security and continuing social welfare

c) technical advances allow to treat wastewater to potable quality making the swap redundant (assuming the water consumer accepts the reclaimed water).

Technology performance risks: The technology needed to upgrade existing treatment plants to meet the WHO guidelines for wastewater reuse in agriculture are common and in general not at risk of failure. However, the technology depends on political will and investments to meet the contractual quality and quantity targets the farmers are expecting. A severe performance risk concerns the limitations of the swap. In times of prolonged drought, also farmers' freshwater supply might decrease, reaching a limit where there is no more water to swap.

Political and regulatory risks: The business requires that farmers have well defined water rights or entitlements, which can be transferred, and regulations that allow the use of (partially) treated wastewater on farms serving local markets. Particular challenges relate to the regulation of groundwater usage and rights, e.g. where urban and rural users share the same aquifer. This also applies to the need of defining the ownership of raw wastewater as well as reclaimed water.

Social equity related risks: The model links different interest groups in need of water: farmers and urban dwellers/industry. This requires an inclusive process of planning and implementation where all parties can express their interests during fair contract negotiations. The reality might look different depending on the political power farmers have compared with the significant power of urban centers.

Where women farmers had no water rights before the swap, the model will not improve their situation unless the contract with the local community earmarks additional entitlements to reclaimed water for women. The swap is considered to have more advantages of social nature for women in the urban sector which vary with the scale of the prevented water shortage in terms of time and cost savings in water access, maintaining standards of hygiene and general living conditions.

Safety, environmental and health risks: Foreseeable health risks arise from the use of partially treated wastewater on farms, to farmers themselves, or, depending on the produce and the way it is consumed (e.g. cooked or uncooked) also other stakeholders along the value chain (Table 59). Perfectly treated wastewater which takes care of all pathogens, as well as inorganic and organic contaminants is still seldom, especially in low-income countries. Risks may be mitigated by following the WHO Sanitation Safety Planning process, including quality control measures or by regulation on the type of crops allowed to receive wastewater. As the Iran case showed, not only the quality of the replacement water matters, but also if the treated effluent is mixed with untreated wastewater before farmers access it. Therefore, this model should include the adoption of the multi-barrier approach for health risk reduction along the farm-to-fork value chain (WHO, 2006, 2015; see the introduction to Chapter 18).

Business performance

The model ranks high on the innovation criteria as it involves diverse actors across sectors and extends the value chain beyond cost recovery to social gains. The **scalability** of the model is contractually defined by the water volumes which have been negotiated between the parties, but ultimately by the physically available wastewater volume (and quality) which the city can offer to farmers as freshwater replacement. The essential building blocks for scaling are the existence of additional capacity to treat wastewater, latent irrigation demand in the area, and cooperation among farmers, industry partner, and municipality.

Where alternative adaptation measures to drought are not feasible, like seawater desalination, water swaps with farmers are possible if farmers can be convinced and incentivized to release their freshwater

TABLE 59. POTENTIAL HEALTH AND ENVIRONMENTAL RISK AND SUGGESTED MITIGATION MEASURES FOR BUSINESS MODEL 20

RISK GROUP	EXPOSURE					REMARKS
	DIRECT CONTACT	AIR/ ODOR	INSECTS	WATER/ SOIL	FOOD	
Farmer				MEDIUM		Higher risk possible where reclaimed water offered to farmers is poorly treated. WHO's multi-barrier approach highly recommended along food chain
Community	LOW	LOW	LOW	LOW		
Food consumer					HIGH	
Mitigation measures	🧤	😷 🌳💨	🦟	Pb Hg Cd	🚿 🍲 Pb Hg Cd	

Key: ☐ NOT APPLICABLE ☐ LOW RISK ☐ MEDIUM RISK ■ HIGH RISK

rights, or do not have sufficient political power to resist. There are different options to facilitate farmers' buy-in, of which the receiving water quality ranks highest. Surplus water allocations appear as another strong factor for decision support. The actual amounts to be exchanged, and the timing, depend on the local freshwater deficit and regularity of supply.

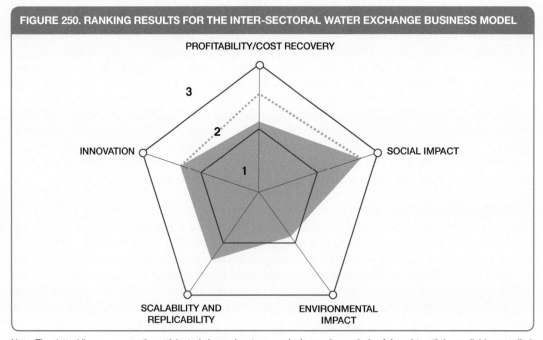

FIGURE 250. RANKING RESULTS FOR THE INTER-SECTORAL WATER EXCHANGE BUSINESS MODEL

Note: The dotted line represents the anticipated change in returns under increasing periods of drought until the available water limit has been reached.

The water swap has a high potential for **replication** wherever cities outgrow local water supply. Cost recovery (from the urban sector) depends on the frequency and volume of the exchange. However, like with any insurance scheme, this is foremost a social responsibility model where the investment will pay off with the occurrence of any prolonged drought given the associated financial and economic losses, which will accompany any water supply scheduling or interrupting, aside the social and health related challenges. Depending on the available wastewater volume also ecosystem services can be supported with reclaimed water, beyond what Figure 250 indicates, although under severe drought highest priority is usually given to immediate socio-economic needs and benefits.

References and further readings

Martin-Ortega J., González-Eguino, M. and Markandya, A. 2012. The costs of drought: The 2007/2008 case of Barcelona. Water Policy 14: 539–560.

Molle, F. and Berkoff, J. 2006. Cities versus agriculture: Revisiting intersectoral water transfers, potential gains and conflicts. (Comprehensive Assessment Research 8). Colombo, Sri Lanka: IWMI (International Water Management Institute).

WHO. 2006. Guidelines for the safe use of wastewater, excreta and greywater, volume 2: Wastewater use in agriculture. Geneva: World Health Organization.

WHO. 2015. Sanitation safety planning manual (Manual for the safe use and disposal of wastewater, greywater and excreta). Geneva: World Health Organization.

CASE
Growing opportunities for Mexico City to tap into the Tula aquifer (Mexico)

Pay Drechsel, George K. Danso and Manzoor Qadir

Supporting case for Business Model 21	
Location:	Mezquital Valley, Mexico; Mexico City
Waste input type:	Urban wastewater
Value offer:	Agricultural and potable wastewater use
Organization type:	Public and private partners
Status of organization:	Irrigation since 1912; new treatment plant since 2016; potable reuse expansion to Mexico City under review
Scale of businesses:	Large
Major partners:	National Water Commission (CONAGUA), local, state and federal Government; Mezquital Valley Farmers and Water User Associations

Executive summary

This business case describes the double value proposition of (i) producing annually crops worth USD 400m through wastewater irrigation; and (ii) generating nearly potable water through the combination of conventional and natural wastewater treatment (aquifer recharge).

The Mezquital (or Tula[1]) Valley of Mexico is well-known for its large-scale wastewater irrigation on about 90,000 ha and its time (over 100 years) of operation which make the case in many textbooks a unique example of wastewater use in the global context. Until recently, the water was to 90% untreated depending on natural treatment processes which could not eliminate risks for the environment and human health. This situation has now been improved through the construction of new wastewater treatment plants, including the 800 million gallon-per-day (35m^3/s) Atotonilco mega plant which is one of the largest in the world, cleaning about 60% of the urban wastewater released from a population equivalent of 10.5 million people of the Greater Mexico City.

Although the value of irrigated food production received so far most attention, the significant rise of the groundwater level in the valley is shifting the attention to the use of its aquifer for supplying aside local communities in the valley also Mexico City with water. The city faces a long severe water crisis, and is running out of cost-effective options for its freshwater supply. The government's allocation of USD 255 million for tapping into the Tula aquifer to reduce the water deficit of Mexico City will make the city its own downstream water user to the direct and indirect benefit of several million urban dwellers.

CASE: OPPORTUNITIES FOR MEXICO CITY

KEY PERFORMANCE INDICATORS (ONLY ATOTONILCO WWTP, STATUS 2016)					
Land use:	159 ha (plant area)				
Water use:	35m^3/s wastewater treated				
Capital investment:	USD 786 million (numbers vary with source and reference year)				
O&M:	USD 81m per year				
Output:	Up to 90,000 ha of irrigated fodder and food crops, aside large-scale wastewater driven aquifer recharge of about 25–39m^3/s, which is retrieved in the valley for different purposes including domestic water supply (6.2m^3/s envisioned for Mexico City)				
Potential social and/or environmental impact:	Job creation along the value chain, savings for advanced treatment, increased soil fertility and water supply for addressing urban food needs; nutrient recycling (reducing additional N and P fertilizer needs), aquifer recharge and the provision of drinking water within the valley and for Mexico City				
Financial viability indicators:	Payback period:	N.A.	Post-tax IRR (Atotonilco):	14.2	Gross margin:	N.A.

Context and background

About 70% of the 21 million urban dwellers of Mexico City depend on groundwater as a source of drinking water. Overexploitation of groundwater by at least 117% resulted within the city in soil subsidence at the rate of 5–40cm annually, increasing the cost of water supply and urban drainage, affecting transport (metro) and built infrastructure. Alternative options to improve urban water supply are long-distance water import and a large-scale leakage control program. Both options face their own challenges, making wastewater reuse, either directly after treatment, or after use in irrigation from the recharged aquifer, cost-effective complementary measures (Jimenez, 2014). Already today, the Tula aquifer, which derived to 90–100% from former wastewater, supplies the local population with drinking water (17%), while supporting agriculture (38%), industry (33%) and other uses (12%).

Irrigation, especially with water rich in nutrients and organic matter, is in high demand as the climate is semi-arid and soils are poor. On the request of local farmers around 1920, the government supported a complex irrigation system in the valley, which constituted recognition, although informal, of the use of non-treated wastewater to irrigate crops. Later, the farmers requested the concession of 26m^3/s of Mexico City's wastewater – the entire quantity available at that time. Consent was granted by the President in 1955 (Jiménez, 2009). The use of wastewater quickly became a source of livelihoods as it enabled crops to be grown all year round. Land with access to wastewater costs more than twice the rent (USD 1,000/ha) than land with access to rain water (USD 400/ha) only.

Irrigation water quality in the valley varies regionally, with about 10,000 ha using raw wastewater, 35,000 ha diluted wastewater, 25,000 ha partially treated wastewater, and other areas benefitting from aquifer recharge (Navarro et al., 2015). These shares will change towards increased safety with the newly installed treatment capacity which can absorb 60% of Mexico City's wastewater and will release 23m^3/s directly for irrigation, while 12m^3/s will support indirect reuse, local reservoirs and the environmental flow of the Tula river.

Due to the high irrigation rate as well as storage and transport of wastewater in unlined dams and channels, the aquifer is unintentionally being recharged on a vast area at a rate between 25 and 39m^3/s which is exceeding natural recharge 13 times, and led to an increase of the groundwater level between 1938 and 1990 by 15–30m with new springs appearing and a higher water volume in the Tula river through groundwater inflow (Jimenez, 2014).

Market environment

There are two complementary water markets, Mexico City and the Tula Valley. While the valley needs the urban wastewater for its economy, the city needs the valley to absorb with limited costs its effluent.

a) According to Jiménez (2014), Mexico City uses about 86m^3/s of water derived from local wells (57m^3/s), long distance transfer (20m^3/s), surface water (1m^3/s) and is using all its reclaimed water (7.7m^3/s). Water consumption is mostly for domestic use (74%), local irrigation (Mexico Valley, 16%) and industrial and other uses (3%). For 2010, a water deficit of 15–38m^3/s was estimated to supply the increasing population and control soil subsidence within the city. Among the measures to close the gap are a long-term leakage control program and the careful protection of the inner-urban aquifer. Additional long distance supply will remain a critical component but is increasingly opposed by local population at the source, or faces very high pumping costs, not because of the distance, but 1,000–1,500m differences in altitude to reach Mexico City. Extending wastewater reuse from the Tula aquifer would offer at much lower vertical difference, and is increasingly considered a feasible and cost competitive option, although post-treatment is required to eliminate remaining water quality concerns (Jiménez, 2014; Navarro et al., 2015).

b) The Tula Valley receives on average about 60m^3/s of Mexico City's wastewater. Irrigation to supply Mexico City with food is the economic backbone of the area, as the additional water allows to grow two to three crops instead of one, and achieves 67–150% higher yields compared to freshwater irrigation. Direct and (via aquifer recharge) indirect wastewater use in the valley supports also other economic activities. Although water quality from the Tula aquifer appears in large better than of conventional wastewater treatment, the newly commissioned WWTPs are expected to further reduce gastrointestinal diseases (Contreras et al., 2017), and support market demand.

Both (rural and urban) markets are not mutually exclusive if the extraction points are well distributed, given that groundwater recharge is exceeding local water needs. The transfer of about 5m^3/s consisting of groundwater from the Mezquital (Tula) Valley to Mexico City has been initiated under Mexico's National Infrastructure Program and was in February 2017 under review (CONAGUA, 2017). If successful, higher water volumes are available.

Macro-economic environment

One of the main aims of Mexico's current National Water Program is to treat and reuse wastewater. In recent years, the percentage of collected wastewater that is treated has risen from 23% to 36%, and the goal is to reach 100% of municipal wastewater by 2020 and industrial wastewater by 2025. The gap is not caused by missing water demand, but treatment capacity. The use of **un**treated wastewater for irrigation is already supporting the livelihoods of several hundred thousand people. Agricultural production for 2011–2012 in the two main irrigation districts of the Tula Valley generated about USD 400 million in crop outputs (CONAGUA, 2013). To replace untreated with treated wastewater, the government catalyzed a multi-billion US Dollar investment program to improve urban water supply, drainage and wastewater treatment. Currently, Mexico City is using 100% of its reclaimed wastewater, making the city in relative terms one of the global reuse leaders. The new investments are paving the way to become also in absolute numbers a global leader given that the new treatment plants will multiply the amount of reclaimed water of immediate use in the Tula Valley. The additional allocation of USD 255 million for tapping into the Tula aquifer to reduce the urban water deficit makes Mexico City its own downstream water user. Aside water imports from Mezquital, also transfer from other basin remains important, but is increasingly objected due to negative local impacts like reduced irrigation areas.

Business model

The main 'value proposition' was and is the use of wastewater for crop production, turning an unwanted discharge into a resource which is mobilizing annually a value of several hundred million US Dollar (see above). A small part of these revenues is spent on O&M of the irrigation infrastructure, complemented by CONAGUA subsidies.

With 90–100% of the valley's aquifer being formed by Mexico City's wastewater, groundwater use for various economic activities offers a second waste-based value proposition. The treatment provided by the Atotonilco plant will support surface and groundwater quality, although for potable reuse further membrane filtration before reuse has been suggested.

Supplying Mexico City with groundwater from Tula could generate about USD 150m/year based on the upper water tariff, which is however unlikely to cover the operational costs, while the expected economic benefits will be far beyond this value. Taking as example the Gutzamala long-distance water transfer, which however requires more energy for a much higher difference in elevation, its annual operational cost is covered to 48% through user fees and 52% by federal funds. Without changes in water tariffs, the business model (Figure 251) will remain foremost a social one, subsidized by the municipal and federal governments, which is however well justified by the magnitude of reduced externalities, like damages to buildings, streets, sidewalks, sewers, storm water drains and other infrastructure due to land subsidence, as well as the magnitude of community benefits due to appropriate water supply.

Value chain and position

The traditional value chain of transforming urban wastewater into an agricultural asset, involving local farmers, water user associations and traders, is since decades common reality in the Tula Valley (Figure 252). To transport water from the replenished aquifer back to Mexico City appears cost effective and is under review (CONAGUA, 2017). It could potentially face institutional obstacles in view of water entitlements (FAO, 2010) although in general all goods found beneath the surface in Mexico belong to the country according to the Mexican Constitution, with CONAGUA in charge of groundwater management. While in other remote areas where CONAGUA is sourcing water for Mexico City, water competition is increasing and so local resistance, there should be less reason for competition in view of the boosted Tula aquifer.

Institutional environment

Mexico's National Water Law, passed in 1992, provides the legal framework for water management in Mexico. It states that the use of the nation's water or the right to discharge wastewater will be carried out by concessions from the Federal Executive Branch, through the National Water Commission (CONAGUA). CONAGUA also allocates the water-related budget for the 32 states in Mexico. The budget for water is approximately 60% of the total environmental budget in Mexico. One of the states is the State of Mexico, which includes the large majority of Greater Mexico City (Mexico City Metropolitan Area) with its 21 million people that is composed of 16 Municipalities, as well as a larger number of adjacent municipalities. Governmental responsibilities are complex given the stakes of the Federal Government, the government of Mexico City, and the government of the State of Mexico, resulting in fragmented responsibilities[2]:

- The Federal government is in charge of regulating the use of water resources, contributing to the financing of investments and supplying bulk water from other basins through CONAGUA.
- CONAGUA which is operating under the Ministry of the Environment and Natural Resources is also responsible for upstream parts of the wastewater irrigation infrastructure in the Tula Valley and its operation, while local water user associations (WUA) are in charge of downstream irrigation management and user tariffs.

FIGURE 251. BUSINESS MODEL CANVAS – WASTEWATER FOR IRRIGATION AND GROUNDWATER RECHARGE

KEY PARTNERS	KEY ACTIVITIES	VALUE PROPOSITIONS	CUSTOMER RELATIONSHIPS	CUSTOMER SEGMENTS
- National and State Water Commissions - Municipalities - Farmers and their WUAs - SACMEX - ATVM - Banks	- Wastewater transport and treatment - Reclaimed water distribution, post-treatment and sale	- Turning wastewater into water for irrigation and potable reuse at scale	- Formal agreements between all partners - Automatic billing of domestic water users	- Farmers located in the Mezquital Valley - Urban dwellers in the Valley and of Mexico City
	KEY RESOURCES - Urban wastewater - Finance - High level engineering skills - Positive public perception		**CHANNELS** - Water delivery through pipes and irrigation canals - Water bills can be paid at most banks, major grocery stores, some convenience stores, or the water company	

COST STRUCTURE	REVENUE STREAMS
- Capital investment and O&M of water treatment, long distance transport and irrigation schemes - Health risk monitoring and reduction costs	- Bulk water sales to WUAs, and fees charged to individual farmers - Bulk water sales to municipal water suppliers, and household charges via water bill

SOCIAL & ENVIRONMENTAL COSTS	SOCIAL & ENVIRONMENTAL BENEFITS
- Likely health cost for farmers and potentially also for consumers through diseases derived from unremoved contaminants transferred via the food chain or reclaimed water - Risk of long term soil contamination (e.g. by heavy metals)	- Benefits for rural and urban food and water security with value generation across sectors and value chains - Reduction in public health expenditure through increased conventional and natural wastewater treatment and water availability - Improvement in livelihoods and ecosystem services

- In Mexico State, the State Water Commission (CAEM) buys bulk water from CONAGUA, transmits it through its own bulk water infrastructure and sells it on to its municipalities. CAEM also monitors water quality, operates wastewater pumping stations and several wastewater treatment plants.
- The municipal governments in the State of Mexico and Hidalgo are in charge of water distribution and sanitation for their constituents. In Mexico City, for example, the water operator (Sistema de Aguas de la Ciudad de México or SACMEX) provides potable water, drainage, sewerage, wastewater treatment and water reuse services.

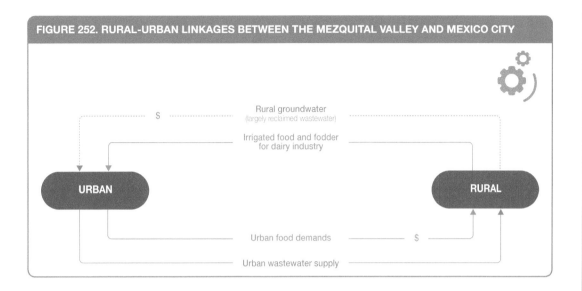

FIGURE 252. RURAL-URBAN LINKAGES BETWEEN THE MEZQUITAL VALLEY AND MEXICO CITY

The major program governing recent water developments is the Water Sustainability Program of the Valley of Mexico, which envisages a series of infrastructure investments supported by the drainage, water supply and wastewater treatment to serve the Mexican capital. The program is supported the National Infrastructure Program but relies heavily on private sector funding. One target is to increase the city's water supply by 14m^3/s with about 5m^3/s consisting of groundwater from the Tula Valley, at an estimated cost of USD 255 million[3]. The second-largest source of additional water will be mobilized through an exchange of treated wastewater for clean water at present used for green area irrigation (4m^3/s), at a cost of 140 million. Another 3m^3/s is envisioned to be gained through rehabilitation measures (Cutzamala system) and 2m^3/s would be made available from the Guadelupe dam in Mexico state[4]. The Sustainability Program governs also the construction of the Emisor Oriental (Western Sewer) and the Atotonilco wastewater treatment plant. The plant has been constructed and will now be operated by the Aguas Tratadas del Valle de Mexico (ATVM) private sector consortium.

Technology and processes

Discharge of wastewater from the Greater Mexico City into its sewer network is estimated around 41 to 44 m^3/s. Considering rainfall, the total average flow managed by the sewer system in the Metropolitan Area is around 60 +/– 15m^3/s. This wastewater is sent by gravity and pumping via five artificial exits to the Tula Valley. The latest tunnel, the East Emitter (Emisor Oriente) which was end of 2016 still in construction has a capacity of 150m^3/s and is 62 km long. Discharge from the tunnels will be primarily treated at the Atotonilco wastewater treatment plant which has a total treatment capacity of 35 m^3/s (Figure 253), with an additional hydraulic capacity of 20% to manage storm water that mixes with the wastewater, giving a maximum capacity of 42 m^3/s in rainy periods. Till the East Emitter is operational, the plant receives the flows from the older Central Emitter.

The treated water will support direct and indirect wastewater irrigation, and based on farmers' request try to limit the removal of crop nutrients during wastewater treatment. The sludge produced by the Atotonilco plant will be stabilized by anaerobic digestion and the gas produced will be used for power cogeneration, providing according to different sources 60–80% of the plant's own electricity requirements. The plant has an estimated lifespan of 50 years. There are also several smaller wastewater treatment plants in construction which together will add another 10m^3/s treatment capacity.

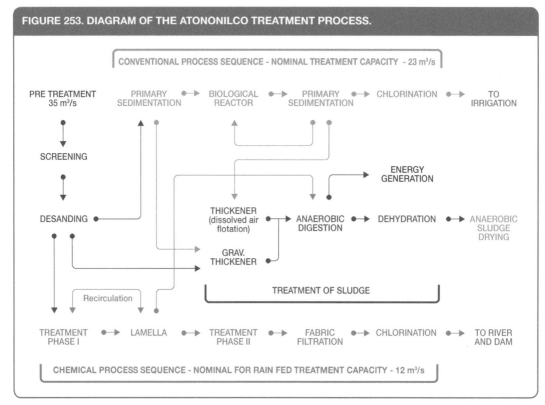

FIGURE 253. DIAGRAM OF THE ATONONILCO TREATMENT PROCESS.

Source: CONAGUA, 2017.

Due to unlined water reservoirs and irrigation channels, the Tula aquifer is unintentionally being recharged at a rate of (more than) 25m³/s, exceeding natural aquifer recharge multiple times. Aside local groundwater use, a part of the excess groundwater has been proposed to be pumped from twelve batteries of extraction wells in the Mezquital Valley over 80 km and an altitude difference of about 500 m back to Mexico City. Flow rate of extraction will be about 6.4m³/s, and at the destination at least 4.2m³/s (CONAGUA, 2017). Treatment before reuse to address potential health risks is highly recommended, especially if the water is used like in this case for potable purposes.

Funding and financial outlook

Local financing for water infrastructure comes from federal, state and municipal resources. CONAGUA which channels federal (governmental) funding to municipal and rural projects, and the National Bank of Public Works (BANOBRAS) which provides financing, subordinated debt and capital. States, municipalities and local authorities have very limited financing capacity for new infrastructure. CONAGUA is also a fiscal authority, charging duties for the use of Mexico's water resources which includes water supply as well as (the use of water receiving) wastewater discharges.

Irrigation: CONAGUA manages irrigation water supply across Mexico through local WUAs or smaller operators which are charging their farmers for O&M of the irrigation infrastructure. The tariff is to be calculated every year to cover O&M costs of the irrigation system. Fees are assessed by total area, by irrigated area, by type of crop, and by cultivated area, and only in a few cases by water volume. A part of the fee supports CONAGUA's maintenance of upstream infrastructure, which remains otherwise subsidized.

Rural-urban water supply: After construction of the planned pipeline, its operation might be – like in similar cases – with the Mexico Valley basin agency (OCAVM) for CONAGUA, supplying CAEM and SACMEX with water for the supply of communities and households. Water tariffs are set locally by the authorities of each municipality depending on the provisions of each state's legislation, and include fixed costs, proportional costs according to the water used, with or without costs for sewerage and wastewater treatment and taxes.

Wastewater treatment: The Atotonilco project was assigned by CONAGUA to a private sector consortium for a design-build-operate-transfer (DBOT) contract, with a 25-year construction and operating period. The ATVM consortium partners financed 20% with equity, and 31% with credit from BANOBRAS, while the National Development Fund of Mexico (FONADIN) contributed a subsidy of 49%. The winning bid was chosen on the basis of the lowest consumer tariff requested. The concessionaire is repaid, however, from CONAGUA's budget. CONAGUA is charging water use and discharge duties and is paid for the provision of bulk water, which the municipalities supply to households. The household water bill usually includes a share for sanitation/wastewater management (Figure 254). These tariffs are generally not sufficient to cover the costs of providing the services.

More information on financing water services (capital and operational costs) can be found in CONAGUA (2010).

Socio-economic, health and environmental impact

Mexico City was for over a century taking advantage of natural wastewater treatment in the Tula Valley, saving costs otherwise required for treatment infrastructure. This system appeared in large as a win-win situation as the city got rid of the water while the local economy in the Tula valley transformed the wastewater in an economic asset via additional crop harvests, higher yields per hectare, etc. To control possible health risks, legislations requesting crop restrictions are in place, though with limited enforcement, resulting in a long history of increased diarrheal diseases linked to water exposure (Contreras et al., 2017). Risks will also remain after the Atotonilco wastewater treatment plant is fully operating as it will only treat 60% of the wastewater released in the valley. However, it is a giant step forward given that before only 6–11% were treated. The treatment plant is supposed to benefit 700,000 people in the Mezquital valley, of which 300,000 live in irrigation.

Especially for aquifer recharge, natural land treatment will remain important. So far, the water passing the soil and unsaturated zone above the Tula aquifer is resulting in groundwater of a quality exceeding the one of conventionally treated wastewater. The higher groundwater table and the appearance of new springs are supporting different economic sectors including potentially several million households back in Mexico City once the long-distance transfer is in place. Water extraction from the Tula aquifer can also positively influence groundwater induced soil salinity in the valley. However, soil characteristics and hydro-geology vary regionally and so their filter characteristics. In fact, it is not known when the natural filter system might reach saturation. There is also the risk that the new treatment plants will remove organic matter from the wastewater which is needed to absorb contaminants when passing the soil. There could also be safety concerns due to the use of agro-chemicals by farmers. Thus, for potable use, additional membrane filtration has been recommended, especially in view of 'emerging contaminants', such as pharmaceutical or pesticide residues with so far unknown threshold values (Navarro et al., 2015).

Finally, in line with the recommendations of the National Water Plan (2012–2018), the Atotonilco wastewater treatment plant is covering to a large percentage its own water (92%) and energy (60–80%) needs and reducing greenhouse gas (GHG) emissions by an average of 400,000 tons of CO_2e per year.

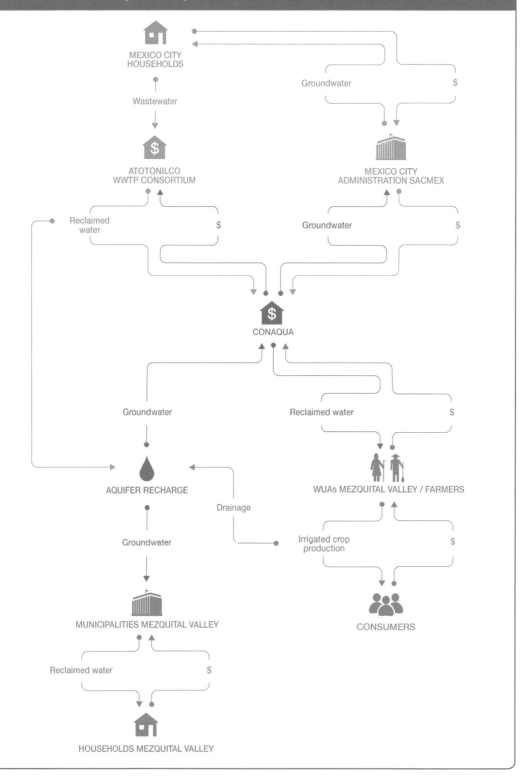

FIGURE 254. RURAL–URBAN WATER AND WASTEWATER TRANSACTIONS BETWEEN MEXICO CITY AND THE MEZQUITAL VALLEY (SIMPLIFIED)

The project is investing in reforestation using native plant species, with the aim of recovering and improving the quality of environmental services on the site.

Scalability and replicability considerations

This business case describes a rural-urban win-win situation with a double value proposition of (i) producing annually crops worth USD 400 million through the use of unwanted wastewater; and (ii) generating nearly potable water through the combination of conventional and natural wastewater treatment (aquifer recharge), resulting potentially in USD 150 million revenues through the water tariff.

The key drivers for the business which are also common in other regions are:
- Rapid urbanization resulting in large volumes of unwanted wastewater discharge and groundwater recharge.
- Water scarcity resulting in high demand for surface and groundwater for multiple financial and economic benefits.

Other drivers which are not always common:
- Governmental capital investments and subsidies based on expected large economic benefits.
- Government consent in providing farmers with (untreated or partially treated) wastewater and irrigation infrastructure.
- Vast aquifer with very high natural recharge rate.
- Scale of reuse making it a powerful economy.
- Alternative options for upgrading urban water supply face increasingly challenges.
- Significant research on health risks and options for risk reduction.
- World class engineering (wastewater treatment and long-distance/high elevation water transfer).

A major issue associated with this model is the continuous use of in part untreated wastewater for irrigation and groundwater recharge. However, there are various options to limit related risks for human health, which can be tailored to the actual water use and its quality requirements.

Summary assessment – SWOT analysis

The described model is very promising because water is in high demand in both the Mezquital Valley and Mexico City and both locations are short in alternative options to direct or indirect wastewater use. Minimizing possible health risks will be the key to a sustainable rural–urban partnership where the economic benefits of water for domestic use, agriculture, industry and the environment will easily justify the capital investment as well as O&M costs. Figure 255 shows a condensed SWOT analysis for this business case in Mexico.

FIGURE 255. SWOT ANALYSIS FOR WASTEWATER REUSE INCLUDING RURAL–URBAN WATER TRANSFER FROM THE MEZQUITAL VALLEY, MEXICO

	HELPFUL TO ACHIEVING THE OBJECTIVES	**HARMFUL** TO ACHIEVING THE OBJECTIVES
INTERNAL ORIGIN ATTRIBUTES OF THE ENTERPRISE	**STRENGTHS** - Supply of large amounts of wastewater secured - Farmers, authorities and legislations value wastewater - Strong water demand from rural and urban users - Financial support from the government available - High engineering capacity	**WEAKNESSES** - Wastewater treatment not covering all generated wastewater and types of contaminants - Farmers' concerns that further treatment will remove crop nutrients - High absolute cost of water transport and treatment
EXTERNAL ORIGIN ATTRIBUTES OF THE ENVIRONMENT	**OPPORTUNITIES** - Lower costs to supply Mexico City than via other long distance/high altitude transfers - Strong and growing urban demand for expanding business - Likely health risks for water consumers can be controlled - Reduced soil salinization through groundwater use	**THREATS** - Remaining health risks for farmers, consumers and potable water users have to be controlled to protect market demand - Potential conflict due to unclear wastewater use rights - Future competition for water reuse if acceptance grows

Contributors
Dr. Christina Siebe
Ing. Carlos A. Paillés
Dr. Munir A. Hanjra

References and further readings

CONAGUA. 2010. Financing water resources management in Mexico. https://www.gob.mx/cms/uploads/attachment/file/110801/Financing_Water_Resources_Management_in_Mexico.pdf (accessed 6 Nov. 2017).

CONAGUA. 2013. Estadísticas agrícolas de los Distritos de Riego. Año agrícola 2011–2012. www.conagua.gob.mx/CONAGUA07/Publicaciones/Publicaciones/SGIH-4-13.pdf (accessed 6 Nov. 2017).

CONAGUA. 2017. Strategic projects: Drinking water, sewerage, sanitation. National Infrastructure Program 2014–2018. (October 2017 update). 96pp. https://www.gob.mx/conagua/documentos/proyectos-estrategicos-28811 (accessed 6 Nov. 2017).

Contreras, J. D., Meza, R., Siebe, C., Rodríguez-Dozal, S., Silva-Magaña, M. A., Vázquez-Salvador, N., López-Vidal, Y. A., Mazari-Hiriart, M., Pérez, I. R., Riojas-Rodríguez, H. and Eisenberg, J. N. S. 2017. Health risks from exposure to untreated wastewater used for irrigation in the Mezquital Valley, Mexico: A 25-year update. Water Research 123: 834–850.

Jiménez, B. 2008a. Unplanned reuse of wastewater for human consumption: The Tula Valley, Mexico. Chapter 23 in: Jiménez, B. and Asano, T. (eds) Water reuse: An international survey of current practice issues and needs. London: IWA Publishing, Inc. pp. 414–433.

Jiménez, B. 2008b. Wastewater risks in the urban water cycle. In: Jiménez, B. and Rose, J. (eds) Urban water security: Managing risks. 324 pp. Geneva: Taylor and Francis Group.

Jiménez, B. 2009. Water and wastewater management in Mexico City. In: Mays, L. (ed) Integrated urban water management: Arid and semi-arid regions. UNESCO-IHP, CRC Press, pp. 81–112.

Jiménez, B. 2014. The unintentional and intentional recharge of aquifers in the Tula and the Mexico Valleys: The megalopolis needs mega solutions. In: Garrido, A. and Shechter, M. (eds) Water for the Americas: Challenges and opportunities. London and New York: Routledge, pp. 156–174.

Jiménez, B. and Chávez, A. 2004. Quality assessment of an aquifer recharged with wastewater for its potential use as drinking source: "El Mezquital Valley" case. Water Sci Technol. 50 (2): 269–276.

Navarro, I., Chavez, A., Barrios, J.A., Maya, C., Becerril, E., Lucario, S. and Jiménez, B. 2015. Wastewater reuse for irrigation – Practices, safe reuse and perspectives. In: Javaid, M.S. (ed) Irrigation and drainage – Sustainable strategies and systems. Rijeka, Croatia: InTech Publishing, pp. 33–54. www.intechopen.com/books/irrigation-and-drainage-sustainable-strategies-and-systems/wastewater-reuse-for-irrigation-practices-safe-reuse-and-perspectives (accessed 6 Nov. 2017).

Tortajada, C. 2006. Water management in Mexico City metropolitan area. Water Resources Development 22 (2): 353–376.

Water 21. 2012. Reuse solution for the valley of Mexico's water challenges. Magazine of the International Water Association.

Wikipedia. Water management in Greater Mexico City. https://en.wikipedia.org/wiki/Water_management_in_Greater_Mexico_City (accessed 6 Nov. 2017).

Case descriptions are based on primary and secondary data provided by case operators, literature, insiders or other stakeholders, and reflect our best knowledge at the time of the assessments 2016/17. As business operations are dynamic data can be subject to change.

Note

1 The Mezquital valley is located in the Tula Valley in the State of Hidalgo, 100 km north of Mexico City. In this case study the name Tula valley is mostly used except where a distinction is required.
2 https://en.wikipedia.org/wiki/Water_management_in_Greater_Mexico_City (accessed 12 Sept. 2016).
3 ibid.
4 ibid.

CASE

Revival of Amani Doddakere tank (Bangalore, India)

George K. Danso, Doraiswamy R. Naidu and Pay Drechsel

Supporting case for Business Model 21	
Location:	Hoskote[1], Bangalore, India
Waste input type:	Urban sewage (diluted with storm water)
Value offer:	Treated wastewater for irrigation, domestic use and restoration of ecosystem services
Organization type:	Public
Status of organization:	Fully operational: 2011
Scale of businesses:	Medium
Major partners:	Karnataka Department of Water Resources (Minor Irrigation); farmers at the Amani Doddakere tank. Indirectly: farmers along the lift irrigation transfer and the Hoskote Municipality

Executive summary

This business case describes the transformation of urban wastewater into an asset for peri-urban farmers and households through inter-sectorial water transfer for groundwater recharge. Excess water from Bangalore's highly polluted Yelemallappa Shetty tank[2] (YMST) is redirected over about 6.2 km to the Amani Doddakere tank (ADT) at Hoskote, reducing pressure on the sewage-fed YMST while partially restoring the ADT, a tank that was for over 18 years dried up.

The lift irrigation system was planned in the late nineties but only realized a decade later. The original idea was to directly feed the water in the irrigation channels at the ADT. Due to illegal tapping into the transfer canal and pipe, the water arriving at the ADT is however insufficient for this objective and most farmers benefitting from the transfer can be found between the YMST and ADT. However, through aquifer recharge, groundwater tables which had dropped below 1,000 feet (ca. 305m) in ADT vicinity, can now be accessed again, providing farmers and households quality water, either directly from wells or through water vendors with well access. The Hoskote Municipality started almost a 24/7 water supply after mandatory water treatment (chlorination). Before this, piped water was only available for short periods all few days. Capital and operational costs, the latter mostly for pumping (lifting) the water out of the YMST are moderate given the achieved benefits. Although the project might present primarily a social business model with still unvalued, social and environmental costs and benefits, operational cost recovery of up to 25% from farmers appears possible, while options on how to charge private water tankers remain to be explored. Although in this case, the recharged groundwater appeared to be of excellent quality and public perception very positive, for any replication of the model care has to be taken that the characteristics of the receiving

CASE: REVIVAL OF AMANI DODDAKERE TANK

aquifer are known, and a well-defined institutional and legal framework provides capacity for dedicated environmental impact assessments (EIA) and water quality monitoring in view of long-term impacts.

KEY PERFORMANCE INDICATORS (2014/15)	
Land use:	20 km of wastewater pipeline / open canal
Water requirements:	Lifting capacity of 0.26m^3 per second
Capital investment:	USD 674,000
Labor requirements:	Low in public sector, but high among benefiting farmers and private sector
O&M:	USD 3000–3500 per month (mostly for pumping)
Output:	5-6 MCM per year for up to 171 ha under irrigation
Potential social and/or environmental impact:	200–500 farmers between the YMST and ADT. Direct and indirect supply also for several thousand households without well via piped and tanker water supply. Improved ecosystem services through biodiversity increase after lake restoration
Financial viability indicators:	Payback period: Not available (N.A.) Post-tax IRR: N.A. Gross margin: N.A.

Context and background

Bangalore (Bengaluru), the capital city of India's Karnataka state, is with a total population of over 11.5 million people, the third most populous city of India. Bangalore's water demand-supply gap was estimated to be 750 million litres a day (MLD) in 2013, and is expected to increase to 1,300 million litres a day by 2026 (McKinsey and CII, 2014). The escalating water demands resulted in unsustainable groundwater extraction and correspondingly high wastewater generation. Although Bangalore is one of the most advanced cities in India with 3610 km of sewage lines and 14 sewage treatment plants, the sewer network is outdated, and less than half of the generated wastewater is captured and/or gets treated. The mix of untreated and treated wastewater pollutes local streams and [cascades of] freshwater reservoirs in and around the city. One of the largest tanks, the Yelemallappa Shetty tank (YMST) in north-eastern Bangalore, is such an example of an ecologically dying lake, increasingly filled up with city run-off, garbage and construction debris. Like 17 other (originally irrigation) tanks on the city outskirts, the YMST is under the management of the Minor Irrigation Department.

Further away from Bangalore, dried-up lakes are common. Despite an average of 800–900mm rain, many irrigation tanks have disappeared and their land was transformed for other use. In the case of Hoskote, a large county with 333 villages in Bangalore's vicinity, the local Amani Doddakere tank (ADT) dried up about 20 years ago, with groundwater levels dropping[3] over the same period by several hundred feet to a depth of more than 1,000 feet (Scharnowski, 2013). In the Hoskote municipality, the extracted 3.36 MLD of drinking water were by far not adequate to meet the 9.37 MLD water demand by Hoskote's ca. 60,000 inhabitants. To support all citizens, the government introduced a scheduled water supply, and made the process of getting permission to sink new borewells (bore holes) difficult.[4] However, under increasing water demand, owners of existing borewells started selling water to tanker companies.

With their livelihoods threatened, farmers from Hoskote requested in the late nineties from the Government of Karnataka to lift water from the YMST to the ADT in support of irrigation, a plan which was drafted in 1999, but only realized a decade later. By that time, the YMST had become a highly polluted water body. End of 2011, the scheme started transferring about 5–6 million cubic meter (MCM) of water from the YMST towards Hoskote. The original estimated cost was USD 579,000 which rose to USD 674,000 due to delays in completion. The ADT had an original capacity of 22.6MCM, with a command area of 940 ha and max. water surface area of 1,100–1,300 ha. The aim of the YMST lift irrigation scheme was to revive the ADT, support its irrigation channels and to recharge groundwater and wells in the area. The wastewater which flows from the YMST to the ADT in part through a pipeline, in part through an open channel, attracted farmers to illegally tap at four to five locations into the resource to fill their tanks and enable ground water recharge for drinking and irrigation, fish rearing and cattle feeding. This resulted in significantly less water eventually arriving in the Doddakere tank, in particular not enough to supply the irrigation channels.[5] Still a part of the ADT got filled with about six feet of water, improving noticeably the groundwater table in lake vicinity from recently 1,000–1,200 feet to 800 feet or much higher. Based on the expected inflow of polluted water, authorities banned direct water use from the ADT, while indirect use via the aquifer provided water fit for irrigation.

Market environment
Under the common water scarcity and dependency on dwindling groundwater, demand for water, water transfers and groundwater replenishment are very high in Karnataka and beyond, and more lift irrigation projects of similar nature are under discussion (see below). Aside supporting agriculture, the 'new' water is also replenishing groundwater for domestic use, making the local water supplying agency as well as private water vendors key customers of any water transfer. All actual as well as potential beneficiaries expressed a high willingness to pay for water (Scharnowski, 2013) as all alternatives are more expensive, from buying water or motor pumps to borewell construction. Well construction is in fact farmers' main cost item as farmers enjoy a broad spectrum of subsidies such as free electricity (pumping), subsidized fertilizer and seeds (The World Bank, 2012). Farmers who lost access to water either had to buy it from other farmers, change their cropping to only rainfed systems or abandon agriculture.

Macro-economic environment
Although some governmental statistics might indicate a large number of households connected to piped water supply, water pressure, for example in Bangalore, is usually very low and access sporadic. A similar mismatch of statistics and reality is found in the sanitation sector where installed treatment capacities are not supported by the sewer network which is outdated and large amounts of wastewater end in streams and lakes. Thus, water supply remains a key challenge, and water transfer and reuse remain high on the policy agenda, also as lake restoration is strongly promoted in Karnataka.

However, implementation of water transfers is not straight forward. Although governmental programs and policies call for wastewater reuse, treatment at the right (reuse) location is seldom, and (untreated) informal wastewater irrigation remains most common (Amerasinghe et al., 2013; Gupta et al., 2016). Also aquifer recharge with wastewater falls in a grey area. Karnataka's first groundwater law, which came into effect in 2011, introduced regulations to monitor the number of bore wells and groundwater use, and that commercial bore wells could be subject to tariffs and caps on water withdrawal. However, law implementation and registrations remained limited (Borthakur, 2015), partly due to missing incentives to register as well as lack of clarity over the exact mandates of different authorities (Bangalore Water Supply and Sewerage Board, Department of Mines and Geology, Department of Water Resources), not to mention water quality issues where freshwater lakes turned into sewage ponds, or options for charging for water abstraction. Moreover, recent suggestions for lift irrigation schemes in Karnataka (e.g. the replenishment of 29 minor tanks around Hoskote and Chikkaballapur) got stalled due to objections raised by the neighboring state of Tamil Nadu fearing that these projects will affect Tamil Nadu's access to water in the shared Dakshina Pinakini River basin. Competition for water, independently of its quality, is high in the region.[6]

Business model
This is primarily a social business model with a potentially high pay off. The city is trying to reduce pressure on lakes with high sewage and storm water inflows in support of groundwater recharge in the water-scarce hinterland, allowing indirect (waste)water reuse for irrigation, household and environmental needs with ecological, economic and social benefits.

Revenues are theoretically collected by the Department of Irrigation, charging farmers per hectare, while households connected to meters pay for drinking water supply. Field surveys showed that farmers between the YMST and ADT would be willing to pay significantly above the current water rates if they could rely on the wastewater flow. The amounts would allow to cover about 25% of the operational and maintenance cost of the lift scheme (Scharnowski, 2013).[7]

The originally unintended primary beneficiaries of the water transfer are those institutions whose obligatory functions as per the Constitution of India is to provide drinking water to the people. However, there are no systems (yet) in place to fund the lift irrigation from revenues accruing in other sectors, like charging water vendors (or farmers) for abstracting replenished groundwater for sale.[8] Changes in tariffs for water use or electricity (pumping) are being discussed, also in light of regulating water abstraction than only revenue generation. Given the low water tariffs, the project is unlikely to financially break even, while the expected economic returns in terms of environmental and livelihood benefits are probably surpassing both, the investment and running costs of the lift irrigation scheme which easily justifies the social character of the business model (Qadir et al., 2014).

Due to immense water demand around cities, and the success of the Hoskote case, the social business concept has a strong replication potential, especially if water access between source and target can be considered in the project design. For the business model to be sustainable, it has to be based on principles of integrated water resources management (IWRM) with full stakeholder participation beyond the irrigation sector, and geo-hydrological assessments including continuous groundwater quality monitoring. Figure 256 shows the business model canvas.

Value chain and position
The value chain (Figure 257) shows current services as well as actually possible and potential (dotted line) revenue streams.

Institutional environment
What was originally planned as a simple transfer of normal irrigation water (and correspondingly did not involve other stakeholders) became much more complex when the system eventually started, and the lift irrigation scheme evolved into a complex system of wastewater use, lake rehabilitation, groundwater recharge and drinking water extraction, elements which concerns a range of departments, authorities, and initiatives in the state of Karnataka. Overlap in responsibilities as well as reassignment of responsibilities are common features. The construction, maintenance and monitoring of minor irrigation projects, i.e. those with a 'culturable command area' of 2,000 ha or less are under the purview of the Minor Irrigation Department. Most of the Department's projects focus on surface water schemes while ground water schemes are dealt with in collaboration with the Department Mines & Geology (Groundwater Wing). The Minister for Minor Irrigation is also the chairperson of the governing council of the new (2015) Lake Conservation and Development Authority, which has members from several departments.

The Bangalore Water Supply and Sewerage Board (BWSSB) is responsible for providing drinking water supply to Bangalore City. The Karnataka Urban Water Supply and Sewerage Board is responsible for providing drinking water supply to urban areas throughout the state of Karnataka.

The legal framework influencing the extraction of groundwater are the Karnataka Groundwater (Regulation and Control of Development and Management) Bill (2009) and Act (2011) which basically lay down the application procedure for new borewells, process of registering and costs involved. Groundwater is considered the property of the government, and the drilling of borewells requires in the Hoskote area, like in several others harshly affected by groundwater depletion, official approval from the district committee. This resulted in a ban on new drilling of deep borewells in the area.[9]

FIGURE 256. BUSINESS MODEL CANVAS FROM THE PERSPECTIVE OF THE WATER PROVIDING DEPARTMENT OF WATER RESOURCES

KEY PARTNERS	KEY ACTIVITIES	VALUE PROPOSITIONS	CUSTOMER RELATIONSHIPS	CUSTOMER SEGMENTS
• Hoskote town municipal council • Urban Water Supply and Sewerage Board • National Bank for Agriculture and Rural Development (NABARD) • BMS College of Engineering, Bangalore	• To transfer water from YMST to Doddakere tank • Operation and maintenance **KEY RESOURCES** • YMST surplus water • Lift irrigation pump, pipeline and canal • Financing • Receiving tank	Providing reliable access to water for irrigation and other needs through tank revival and groundwater recharge with significant livelihood benefits	Formal relationship between the farmers and the Department of Water Resources (Minor Irrigation) **CHANNELS** • Direct contact with the farmers	Directly • Farmers Indirectly • Municipal water supply • Water traders • Households • Fishermen

COST STRUCTURE	REVENUE STREAMS
• Capital investment by Government • Operation and maintenance by Government	• Limited or no revenue from farmers but willingness to pay by farmers and other beneficiaries is high • Revenue systems for other water users under discussion

SOCIAL & ENVIRONMENTAL COSTS	SOCIAL & ENVIRONMENTAL BENEFITS
• Health risks likely for farmers accessing (illegally) untreated wastewater on the way to Hoskote or from the tank • Groundwater quality development over time unclear, including possible increase in health related costs • More mosquito related diseases in Hoskote	• Increase in irrigated farming, crop yield and food security • Increase in tank biodiversity (flora and fauna) and related activities (e.g. bird watching) • Private sector support (water vendors) • Recharge of municipal wells for drinking water supply • Livestock support through irrigated fodder production

Although water reuse is encouraged, questions around the ownership of the wastewater vis-à-vis the recharged groundwater, and the modalities for institutions to charge for groundwater abstraction remain subjects of discussion. The situation is complex as small farmers who are charged per irrigated area take advantage of their aquifers for selling water to tanker operators. Also in Bangalore and its vicinity private water supply is rampant filling gaps in the public supply system, while legislations to limit groundwater abstraction are hard to implement, especially where farmers can make easier money from selling water than via irrigation.

A 'larger' institutional challenge of the water transfer is that the river basin is shared by the states of Karnataka and Tamil Nadu. There are strong objections by the state of Tamil Nadu over Karnataka building permanent structure to divert water for its own needs while Tamil Nadu continues suffering from water scarcity. Thus, initiating any project even to utilize the wastewater for any existing tank command area needs clearance from the Central Water Commission.

CASE: REVIVAL OF AMANI DODDAKERE TANK

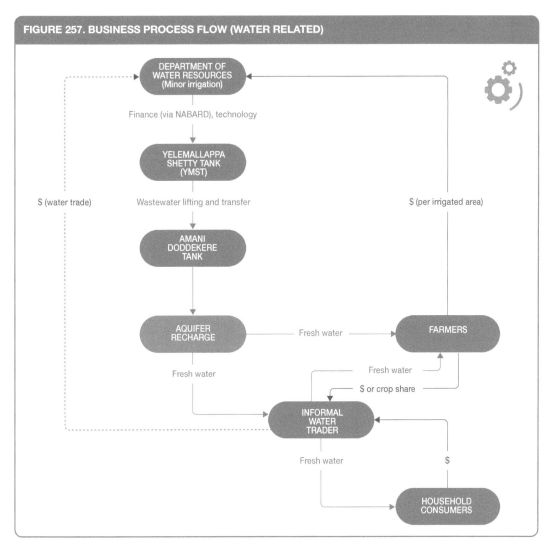

FIGURE 257. BUSINESS PROCESS FLOW (WATER RELATED)

The YMST lift irrigation scheme is one of several planned lake-to-lake inter-sectorial water transfers around Bangalore, for which models for institutional collaboration and ownership have been described (ICRA, 2012).

Technology and processes

Aside the initial lift pumping, the technology is based on physical, chemical and biological processes of natural water treatment (sedimentation, filtration, sun exposure, etc.) above and below ground along the 20 km water passage into the YMST, and between YMST and Hoskote. The potential of natural water treatment should in this context neither be over- nor underestimated. In the current case, the 6.2 km long wastewater overland transfer after leaving the YMST occurs partly piped, partly open, before the water enters the Amani Doddekere tank and gets filtered while percolating through 200 meters of rock to reach the groundwater table. About half of the passage requires pumping, half follows gravity flow. The water is not running continuously as pumping is sometimes stopped over hours or days. While the recharged groundwater at Hoskote appears to be of excellent (potable) quality, as tested by the BMS College of Engineering, Bangalore, any change in transport distance, groundwater table, type of rock, etc. can influence the final water quality. Therefore, water quality monitoring is important, also as there are no data how the natural treatment will continue over the years. To minimize health risks, other planned water transfer schemes around Bangalore recommend wastewater treatment before the final reservoir (ICRA, 2012).

Funding and financial outlook

The financial cost estimate for the YMST lift irrigation system was USD 579,000 with financing from National Bank for Agriculture and Rural Development (NABARD). The actual cost incurred, including additional works was USD 674,000. Charges for irrigation are marginal, about USD 2.6/season/ha for horticulture and floriculture, with free electricity for pumping groundwater.

The Department bears the operation and maintenance cost of at least USD 3,000 per month. Current irrigation water charges for horticultural crops (ca. USD 5.4/ha/yr) generate maximal USD 930 per year, or 2–3% of the annual O&M costs if all transferred water will end on farms which are charged and not be lost/redirected on the way to Hoskote. These charges are much lower than what farmers are willing to pay, which could cover up to 25% of the ongoing operation and maintenance costs as shown by IWMI in the Hoskote area (Scharnowski, 2013). Applying water charges to other users, especially water vendors, will be difficult as the market is informal and hard to monitor. However, for the success of the project, the present policy framework (the 2003 guidelines for lift irrigation) estimates the project benefits through the achieved agricultural yield increase, not through financial cost recovery.

Socio-economic, health and environmental impact

Due to surface water scarcity, groundwater access is most crucial around Bangalore. Nearly 99% of all farmers in rural Bangalore depend on tube wells. The water transfer allows farmers now to cultivate more land or more than one crop per year, or crops with a higher return on the urban market. According to local media, the water table in about 30 villages surrounding Hoskote has increased to the benefit of up to 500 farmers.[10]

The situation also improved water supply to households in Hoskote Municipality which had before the scheme only water for once a day to once in ten days for few hours. Now, up to 60,000 inhabitants are reported as beneficiaries, either directly via own borewells or indirectly via local water vendors (tankers). Improved water access is in particular helping women, given the gendered nature of water collection (Borthakur, 2015).

Also dairy development is among the benefits of the project due to the increased availability of fodder from the wastewater reuse. The 'new' water in the tank revived local fish farming and lured various species of birds to the revived wetland, creating a regional hot spot for birding.[11]

The positive impacts could also extend to the YMST if the lifting of larger water volumes for Hoskote and other lakes could be realized. However, aside some initial groundwater testing, neither, soil, water, crop or fish quality is being monitored, and health risks are high, especially as farmers (without well) might use the wastewater directly, and not via groundwater as seen at other polluted lakes. Safeguards are also needed to ensure that possible negative long-term impacts are under control.

Scalability and replicability considerations

The key drivers for the business model are:
- Water scarcity and high water demand catalyzing public investments.
- Strong policy support for lake conservation and development.

As both drivers are omnipresent in the region, already other lift irrigation schemes for water transfer are under discussion such as for replenishing 29 minor tanks around Chikkaballapur and Hoskote towns, using in this case treated wastewater. As part of an IWRM strategy for Bangalore, McKinzie and CII (2014) proposed a programme of lake regeneration to improve urban groundwater supply. Starting with 38 lakes, each one should be linked to a sewage treatment plant to clean lake inflow. These 38 lakes could increase Bangalore's water availability by an estimated 180 MLD. A comprehensive tank rejuvenation project was undertaken for example for the Jakkur Lake in the northern part of Bangalore at the cost of Indian Rupees 215 million (USD 3.37 million). The lake was dewatered, de-silted and all sewage inflows were diverted to a 10 MLD sewage treatment plant. The treated wastewater flows then through a constructed wetland before entering the lake itself. The result has been an increase in biodiversity, fishing and groundwater recharge (Evans, 2016).

While the use of wastewater for lake regeneration and aquifer recharge has been accepted in the case of Hoskote and shows favorable environmental and economic benefits, this does not have to be the case in other locations as water quality varies significantly and so the risks and public acceptance of indirect wastewater use is also not universal. Therefore, full stakeholder participation and information appear as important as water quality monitoring. Stakeholder inclusion is also needed for the discussion of options for cost recovery from the various beneficiaries, and modalities on how to address illegal water abstraction from the transfer canals.

For any replication of the reuse model, in particular in Karnataka, a legal and institutional framework with clear responsibilities would be beneficial. The same will be an institutional challenge in many other regions, given that such a water transfer links multiple sectors, i.e. urban and rural authorities in charge of surface and groundwater, sanitation, health, drinking water and agriculture. Regulations are also required to prevent that lift irrigation schemes eventually harm agriculture because of farmers becoming water vendors. In recent years, there has been a surge in the conversion of agricultural wells on the outskirts of Bangalore to supply urban consumers because agriculture is less profitable than selling water (and businesses can profit from the subsidized electricity afforded to rural landowners).

The resulting water loop appears to reflect an increasing reality of the circular economy between urban and rural areas in India, where the urban hinterland functions as a 'kidney' for urban water reuse.

Summary assessment – SWOT analysis

The business case focuses on mitigating the economic impact of water scarcity by providing water to farmers for irrigation through the use of (waste)water for groundwater recharge. Additional benefits were observed for household water supply and ecosystem services. Taking advantage of natural water treatment processes, the city saves on treatment and disposal cost for wastewater while farmers and others benefit from 'new' water for their economic activities. There are also substantial benefits for the informal water market through the sale of groundwater to farmers, industry and households. The observed and largely praised success of the project could have even been larger as a significant water volume got lost due to illegal wastewater extraction before the water reached the targeted ADT. There are various options for revenue generation from different beneficiaries who would pay for a reliable water supply. However, already the significant welfare benefits and their downstream impacts on regional economic performance make this social business highly worthwhile.

A long term impact on groundwater quality can be expected, and close water quality monitoring is highly recommended. A better alternative would be to treat all water entering the YMST. Another challenge will be to steer the right hydrological balance between formal aquifer recharge and formal and informal water extraction. Figure 258 shows the SWOT analysis for this business case example.

FIGURE 258. SWOT ANALYSIS OF BANGALORE GROUNDWATER RECHARGE, INDIA

	HELPFUL TO ACHIEVING THE OBJECTIVES	**HARMFUL** TO ACHIEVING THE OBJECTIVES
INTERNAL ORIGIN ATTRIBUTES OF THE ENTERPRISE	**STRENGTHS** • Strong policy support for tank rehabilitation and protection • Governmental support of lift irrigation for groundwater recharge • High demand for (waste)water • Recharged ground water suitable for drinking purpose • High economic and ecosystem service benefits, making the system worthwhile without cost recovery	**WEAKNESSES** • User participation was limited to farmers • Unexpected wastewater diversion along the water passage to Hoskote not covered in any EIA • Insufficient water quality monitoring • As the system is not designed for cost recovery, rigorous options to monitor and regulate water abstraction are missing
EXTERNAL ORIGIN ATTRIBUTES OF THE ENVIRONMENT	**OPPORTUNITIES** • Potential to establish clientele/revenue relationship with various water users, beyond farmers • Value of wastewater (farmers, water suppliers through tankers and local bodies) 3 to 12 times higher than the current water charges	**THREATS** • Conflicts with Tamil Nadu over water infrastructure could stall replication of the model • Aquifer pollution risks resulting in health issues for farmers and other water users, unless YMST water gets treated • Lift irrigation might serve more sale of drinking water than agriculture and eventually result in farmers stop farming

Contributors
Mr. Philipp Scharnowski, University of Bonn, Germany
Dr. Priyanie Amerasinghe, IWMI, India
Dr. Munir A. Hanjra, IWMI, South Africa

References and further readings

Amerasinghe, P., Bhardwaj, R.M., Scott, C., Jella, K. and Marshall, F. 2013. Urban wastewater and agricultural reuse challenges in India. Colombo, Sri Lanka: International Water Management Institute (IWMI). 36p. (IWMI Research Report 147).

Borthakur, N. 2015. Urban water access: Formal and informal markets: A case study of Bengaluru, Karnataka, India. School of Public Policy and Governance. Hyderabad: Tata Institute of Social Science. Working paper series no. 1. www.academia.edu/19848935/Urban_Water_Access_Formal_and_Informal_Markets_A_Case_Study_of_Bengaluru_Karnataka_India.

Doraiswamy, R. 2012. A Business Model To Revitalize Irrigation Tanks As Well As Recharge Ground Water Aquifers, Using Wastewater From Cities: The Case of Bangalore City, Karnataka State, India. Report prepared for the International Water Management Institute.

Evans, A.E.V. 2016. Institutional arrangements for resource recovery and reuse in the wastewater sector. PhD thesis, Loughborough University, UK.

Gupta, M., Ravindra, V. and Palrecha, A. 2016. Wastewater irrigation in Karnataka, an exploration. Water Policy Research Highlight 04. IWMI TATA Water Policy Program. http://iwmi-tata.blogspot.in.

ICRA. 2012. Institutional strengthening & sector Inventory for PPP mainstreaming in Directorate of Municipal Administration (DMA). Preliminary Feasibility Report – Hosakote Town Municipal Council. Infrastructure Development Department (IDD), Government of Karnataka.

IDECK. 2009. Development of lake conservation projects, Karnataka. Final pre-feasibility report. Infrastructure Development Corporation (Karnataka). Limited. www.iddkarnataka.gov.in/docs/23.Prefea_lake_cons.pdf.

McKinsey [McKinsey & Company] and CII [Confederation of Indian Industry]. 2014. Integrated water management strategy for Bengaluru. New Delhi, India: CII.

Nyamathi, S.J., Puttaswamy, Nataraju, C. and Ranganna, G. 1999. Studies on groundwater recharge potential of north and south taluks of Bangalore, Bangalore Urban Districts, Karnataka. National Conference on Water Resources and Management, S.V. University, Tirupathi.

Qadir, M., Boelee, E., Amerasinghe, P. and Danso, G. 2014. Costs and benefits of using wastewater for aquifer recharge. In: Drechsel, P., Qadir, M. and Wichelns, D. (eds) Wastewater: Economic asset in an urbanizing world. Dordrecht, Netherlands: Springer, pp. 153–167.

Rajashekar, A. 2015. Do private water tankers in bangalore exhibit "mafia-like" behavior? MSc thesis, Department of Urban Studies and Planning, Massachusetts Institute of Technology.

Scharnowski, P. 2013. Farmers' willingness to pay for groundwater recharge with urban wastewater: A contingent valuation study in rural Bangalore, India. MSc Thesis, University of Bonn, Germany.

World Bank. 2012. Deep wells and prudence: Towards pragmatic action for addressing groundwater overexploitation in India. Washington D.C., USA: The International Bank for Reconstruction and Development/The World Bank.

See also (accessed 21 Feb 2017):

www.deccanherald.com/content/227529/hoskote-reuses-bangalores-refuse-ends.html.
www.deccanherald.com/content/382200/hoskote-still-uses-city039s-sewage.html.
http://bangalore.citizenmatters.in/articles/bangalore-suburbs-sewage-flow-mechanism.

Case descriptions are based on primary and secondary data provided by case operators, local insiders or other stakeholders, and reflect our best knowledge at the time of the assessments 2013–2015. As business operations are dynamic data are likely subject to change.

Notes

1. Also spelled Hosakote.
2. In South Asia, the term 'tank' is used for man-made water reservoirs (lakes) which are often centuries-old, constructed for rain/surface water storage, mostly for irrigation but also other community needs. Several tanks can be interconected.
3. Groundwater overexploitation at Hoskote is reported as 144%. http://timesofindia.indiatimes.com/city/bangalore/Water-table-in-Bangalore-South-drying-up/articleshow/7838020.cms?referral=PM (accessed 4 Nov. 2017).
4. http://reliefweb.int/report/india/drought-hit-karnataka-regulates-borewells (accessed 4 Nov. 2017).
5. Observation during field work in 2012.
6. www.deccanherald.com/content/244394/tn-now-lays-claim-city.html (accessed 4 Nov. 2017).
7. This would require that those farmers who are illegally tapping into the water transfer will be charged. In fact, the Department of Minor Irrigation and Revenue Department are not charging farmers of the ADT, firstly as the tank was for nearly two decades dry and farmers invested big money on tube wells, and even the 'new' water pumped from YMST has not risen above the sluice level to carry water in the irrigation channels.
8. Tube well owners expressed their willingness to support the water transfer with a monthly rate, as they see a clear relation between tank water level and tube wells, usually with four to five days of delay. A revenue system for tanker operators could be based on number of tankers and their volumes (usually 4,000–6,000 liters), while neither actual pumping (tanker filling) nor water delivery are easy to monitor.
9. http://timesofindia.indiatimes.com/city/bengaluru/Depleting-water-table-could-hit-city-outskirts-hard/articleshow/50665373.cms (accessed 4 Nov. 2017).
10. www.deccanherald.com/content/227529/hoskote-reuses-bangalores-refuse-ends.html (accessed 4 Nov. 2017).
11. As the lake is, with its about 940 ha, rather large, the water inflow creates a patchwork of grassland and water bodies ideal for many kinds of birds.

BUSINESS MODEL 21
Cities as their own downstream user
(Towards managed aquifer recharge)

Munir A. Hanjra and Pay Drechsel

Key characteristics

Model name	Cities as their own downstream user
Waste stream	Treated and partially treated wastewater recharging local aquifers
Value-added waste product	Reclaimed groundwater for domestic and agricultural use
Geography	Water stressed urban areas with suitable peri-urban conditions for aquifer recharge
Scale of production	Medium to very large (depending on aquifer characteristics and urban demand)
Supporting cases in the book	Mexico City, Mexico; Bangalore, India
Objective of entity	Cost-recovery [ ]; For profit [ ]; Social enterprise [X]
Investment cost range	Depending on wastewater volume and scale from USD 500,000 to 700 million for wastewater treatment and conveyance (the water recovery from the aquifer will only be a fraction of this)
Organization type	Public, public-private, or mixed formal/informal sector arrangements
Socio-economic impact	Increased water security, reduced treatment costs, supported ecosystem services, but also health risks for farmers and urban consumers depending on final water quality
Gender equity	Beneficial to women and children due to increased water security and time savings for accessing water

Business value chain

The model builds on the common trajectory of cities that are addressing growing water demand by first exploiting urban ground- and surface-water resources, and then start tapping into peri-urban and rural water resources while releasing all the time their wastewater into the urban periphery. Over time, surface and groundwater reservoirs in urban proximity become increasingly dependent on the urban return flow, making cities eventually their own downstream user. As there are multiple sources for water supply and possibilities for wastewater release, the city can turn this usually unplanned development into a development effort to a) target particularly suffering peri-urban areas for groundwater replenishment; and/or b) particular aquifers for underground storage serving the city itself.

The aquifer replenishing capacity can be remarkable as the two cases from India and Mexico showed. The two cases are, however, only examples of a diverse set of surface-groundwater interactions taking place in an increasing number of rural-urban corridors in low-income countries. Common characteristics are missing institutional responsibilities, limited water quality monitoring and wastewater treatment, and an increasing dependency of urban water supply on informal water markets. As mentioned in the model introduction, the presented cases are thus not success stories according to best practices and standards, as presented by Lazarova et al. (2013) but reflect situations and challenges on the trajectory to a more planned and managed aquifer recharge (Jiménez, 2014). To build on the positive potential of the cases, Model 21 has to emphasize risk management. The same applies to informal wastewater irrigation in the following Models 22 to 24.

The business concept is to turn the need for waste disposal, with its related costs and potential environmental hazards, into an opportunity to generate 'new' water in water scarce environments, which allows generating revenue from the value that water has to farmers and other users (Figure 259). The concept builds on the potential of managed aquifer recharge to maximize social and economic benefits including the protection of public health. The business concept can be implemented by public water/wastewater utilities, or in public-private partnership. The model acknowledges that conventional wastewater treatment in most low-income countries will in the short and medium term not be able to treat all urban wastewater to the standards needed for irrigation and/or domestic use. It builds therefore on the cost-saving additional treatment capacity of natural processes taking place during wastewater conveyance in open channels and infiltration in the soil for (deep) aquifer recharge. Part of the wastewater stored in aquifers can be retained for longer periods during wet years and used in drier years, to supply – depending on the achieved water quality – water to those users (domestic, farmers) who would otherwise face water scarcity challenges.

As the case studies showed, the urban return flow can with time constitute a major share in the local aquifer. Among several monitoring needs, a particular challenge for controlling water withdrawal and quality, concerns the informal water sector, which is usually accessing the aquifer for community water supply in a non-transparent manner. A related challenge is how far these commercial water traders could be charged for the volume of water they are abstracting as the water volumes are difficult to monitor. Moreover, the private sector would probably try to recover any abstraction fees from the served households, which would further disadvantage the poor depending on the informal sector.

Business model
The business model tries to support in dry climates peri-urban and rural areas with depleted aquifers by channelling the urban return flow to unlined reservoirs, forest plantations, etc. for targeted aquifer recharge. Depending on wastewater quality, the model provides a set of benefits:
- Costs for additional treatment or water disposal can be avoided.
- With restrictions, also farmland or forests can be used for aquifer recharge while providing water and nutrients, e.g. to fodder crops and ornamentals.
- Replenished aquifers can support local and urban water needs and economies, including small industries and irrigated crop production.
- Support of ecosystem services and biodiversity in dried-up reservoirs turned wetlands.

In case that also farms are used for aquifer recharge, farming and irrigation practices must follow appropriate safety guidelines and be such that they facilitate water infiltration (e.g. high irrigation rate in the area, storage and transport of wastewater in unlined dams and channels), and the underground suitable for water treatment (supportive soil texture and geo-hydrological conditions).

With most direct revenues deriving from the sale of 'reclaimed' waste/groundwater to urban and industrial users (Figure 260), a positive benefit-cost ratio could be expected if avoided costs are considered, like for wastewater treatment, storage or obtaining water from other sources (Perrone and Merri Rohde, 2016; Vanderzalm et al., 2015). As shown from those cases of managed aquifer recharge, any additional environmental and social benefits will help outweighing costs. A main incentive for this

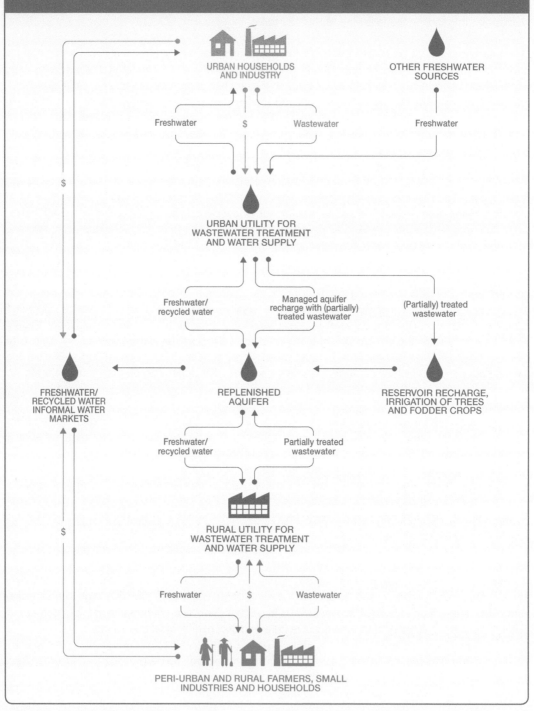

FIGURE 259. SIMPLIFIED VALUE CHAIN WITH ONLY ONE URBAN AND RURAL UTILITY REPRESENTING POTENTIALLY MORE PUBLIC ENTITIES ENGAGED IN THE MANAGEMENT OF WATER AND (TREATED) WASTEWATER FOR GROUNDWATER RECHARGE FOR REUSE

BUSINESS MODEL 21: CITIES AS THEIR OWN DOWNSTREAM USER

FIGURE 260. BUSINESS MODEL CANVAS FROM THE PERSPECTIVE OF THE MAIN PUBLIC UTILITY RESPONSIBLE FOR WASTEWATER MANAGEMENT IN THE CASE OF AQUIFER RECHARGE AND REUSE

KEY PARTNERS	KEY ACTIVITIES	VALUE PROPOSITIONS	CUSTOMER RELATIONSHIPS	CUSTOMER SEGMENTS
• Community representatives from recharge area • Other urban and rural water utilities and operators (health, agriculture, geology, . . .) • Private (informal) water market	• Managing wastewater (treatment, conveyance) • Selling treated wastewater and freshwater/groundwater • Aquifer management and water quality control	• Improving the cross-seasonal access to water for irrigation and domestic needs through underground storage and reuse of treated wastewater	• Formal water allocation contracts	• Households and farmers in rural, peri-urban and urban areas in need of water
	KEY RESOURCES • Land with hydro-geological conditions suitable for water treatment and aquifer recharge • Treatment plants and water conveyance infrastructure/capital • Aquifer management expertise		**CHANNELS** • Automatic billing • Water transfer through pipes and canals • Stakeholder platform	

COST STRUCTURE
- Wastewater treatment
- Water pumping for transfer and withdrawal
- Water quality monitoring

REVENUE STREAMS
- Sale of freshwater including reclaimed water from the aquifer
- Savings in potable water access from alternative sources
- Savings in wastewater treatment

SOCIAL & ENVIRONMENTAL COSTS
- Possible health risks/costs from insufficiently treated wastewater
- Possible soil contamination through long term wastewater infiltration

SOCIAL & ENVIRONMENTAL BENEFITS
- Increased water and food security
- Economic stability also in dry seasons and under climate change
- Reduced impact of water stress levels on livelihoods
- Ecosystem service support

social business model, like in the water swap model (Chapter 17, Business Model 20), should however be the potential costs for the society at large under extended periods of droughts as well as the costs of possible epidemics in the business-as-usual situation vis-à-vis investments in quality monitoring.

Thus, to propose a sustainable model, possible health hazards have to be controlled. This requires clear institutional responsibilities and regulations across the rural-urban boundary, the common administrative freshwater – sanitation divide, and acknowledgement of informal water markets to start a dialog on 'best practices'. All of this constitutes in many low-income countries significant challenges in need of multi-stakeholder platforms (Londhe et al., 2004; Foster et al., 2010; Foster and Vairavamoorthy, 2013; Yuan et al., 2016).

Potential risks and mitigation

In designing any optimized business model based on case studies, it is assumed that generic business risks are known and will be taken care of. However, some risks might be more model specific and will be acknowledged in the following:

Market risks: The market risk is small as long as no other water sources are available cheaply for farmers or households and groundwater ownership is clearly defined, allowing the utility to charge for water use or sell water entitlements. Market risk may however arise due to risks associated with the use of unsafe water, and customers losing trust in the replenished groundwater. The informal water market is likely to take advantage of the replenished aquifer (even with farmers entering the water trade) and its water withdrawal requires regulations and innovative ways of monitoring.

Competition risks: A risk could arise if the water receiving households or farmers get in seasons of high water demands access to a cheaper alternative water source. This is however unlikely as any additional water would also be sold by the same utility and the informal market sells at higher prices than the utility. There could be competition between sectors within the same community if for whatever reason the wastewater transfer is interrupted and so the aquifer supply. In this case, municipalities might compete against farmers to acquire their groundwater abstraction rights to harness the economic benefits and revenue gains that the business model offers to domestic and industrial users.

Technological risk: There seems to be limited risk due to the low-technology status as long as land is available, and the recharge is based on hydro-geological feasibility and environmental impact assessments.

Political and regulatory risks: The business requires a) well defined groundwater and wastewater related water rights or entitlements; b) reuse guidelines based on water quality; and c) monitoring mechanisms related to both requirements. Building on the currently available global water reuse regulations and guidelines for MAR with reclaimed water, a standardized approach should be developed and can be used by regulatory agencies, municipalities, and other water providers if their own regulatory framework is inadequate, as suggested e.g. by Yuan et al. (2016).

Social equity related risks: Like the other rural–urban water exchange model, the model links different interest groups in need of water, and this across administrative boundaries and sectors and thus needs an inclusive process of planning and implementation where all parties can express their interests during fair contract negotiations.

The additional availability of water is considered a particular advantage for women in charge of water acquisition with multiple social and health benefits.

Safety, environmental and health risks: The health risks connected to this business model depend strongly on the treatment in place, before and after aquifer recharge. Given the interaction of wastewater and drinking water, the WHO *Water Safety Planning* and *Sanitation Safety Planning* manuals could be applied, but also guidelines particularly developed for aquifer recharge (Yuan et al., 2016). The latter also applies to environmental protection and long-term accumulation of contaminants in the soil. When sound regulations are in place, managed aquifer recharge can be safe and offer simple, low-tech and cost-effective treatment systems for developing countries.

The address possible safety and health risks, standard safety precautions should be applied to water withdrawn from the recharged aquifer (Table 60).

TABLE 60. POTENTIAL HEALTH AND ENVIRONMENTAL RISK AND SUGGESTED MITIGATION MEASURES FOR BUSINESS MODEL 21

RISK GROUP	EXPOSURE					REMARKS
	DIRECT CONTACT	AIR/ ODOR	INSECTS	AQUIFER/ SOIL	RECLAIMED WATER	
Community				High risk		Multi-barrier approach recommended and application of WHO's *Water and Sanitation Safety Planning* manuals (WHO 2009, 2015)
Farmer			Medium risk	High risk		
Consumer					Medium risk	
Mitigation measures	🧤	😷 🌳	🧴	Pb Hg Cd	🚿 🍲 Pb Hg Cd	

Key: ☐ NOT APPLICABLE   ☐ LOW RISK   ▨ MEDIUM RISK   ■ HIGH RISK

Business performance

This model ranks highest on **social impacts** as the reuse of wastewater for domestic supply and crop production offers significant benefits to urban consumers and agricultural communities as long as safety requirements are met. While in water scarce regions the model will probably be profitable compared to alternative options, its larger benefits are the prevention of drought related costs for the society at large, which can exceed the investment costs multiple times. While there are several plans for replicating the Indian **case** at other locations around Bangalore, it will require clear institutional mandates and regulations to implement the required safety monitoring system for human and environmental health. Assuming these are in place, also possible ecosystem benefits can become substantial (Figure 261).

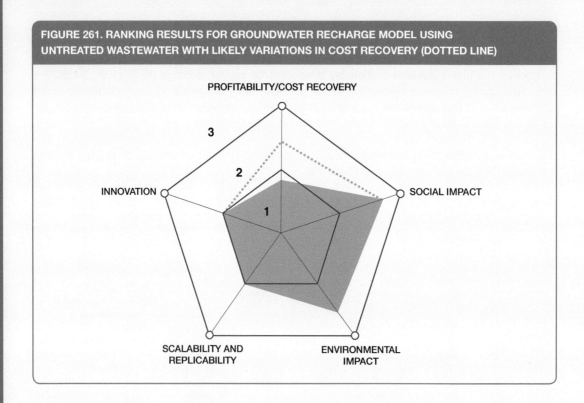

FIGURE 261. RANKING RESULTS FOR GROUNDWATER RECHARGE MODEL USING UNTREATED WASTEWATER WITH LIKELY VARIATIONS IN COST RECOVERY (DOTTED LINE)

References and further readings

Foster, S., Hirata, R., Misra, S. and Garduno, H. 2010. Urban groundwater use policy – Balancing the benefits and risks in developing nations. GW•MATE Strategic Overview Series 3. Washington DC, USA: World Bank.

Foster, S. and Vairavamoorthy, K. 2013. Urban groundwater. Policies and institutions for integrated management. Global Water Partnership Perspectives Paper.

Jiménez, B. 2014. The unintentional and intentional recharge of aquifers in the Tula and the Mexico Valleys: The megalopolis needs mega solutions. In: Garrido, A. and Shechter, M. (eds) Water for the Americas: Challenges and opportunities. London and New York: Routledge, pp. 156–174.

Lazarova, V., Asano, T., Bahri, A. and Anderson, J. 2013. Milestones in water reuse: The best success stories. IWA. 408pp.

Londhe, A., Talati, J., Singh, L.K., Vilayasseril, M., Dhaunta, S., Rawlley, B., Ganapathy, K.K. and Mathew, R.P. 2004. Urban-hinterland water transactions: A scoping study of six class I Indian cities. Draft paper of the IWMI-Tata Water Policy Program Annual Partner's Meeting 2004. 21p.

Perrone, D. and Merri Rohde, M. 2016. Benefits and economic costs of managed aquifer recharge in California. San Francisco Estuary and Watershed Science 14(2): Article 4.

Vanderzalm, J.L., Dillon, P.J., Tapsuwan, S., Pickering, P., Arold, N., Bekele, E.B., Barry, K.E., Donn, M.J. and McFarlane, D. 2015. Economics and experiences of managed aquifer recharge (MAR) with recycled water in Australia. Australian Water Recycling Centre of Excellence.

WHO. 2009. Water safety plan manual. Step-by-step risk management for drinking-water suppliers. Geneva, Switzerland: IWA and World Health Organization (WHO).

WHO. 2015. Sanitation safety planning: Manual for safe use and disposal of wastewater, greywater and excreta. Geneva, Switzerland: IWA and World Health Organization (WHO).

Yuan, J., Van Dyke, M.I. and Huck, P.M. 2016. Water reuse through managed aquifer recharge (MAR): Assessment of regulations/guidelines and case studies. Water Quality Research Journal of Canada 51 (4): 357–376.

18. BUSINESS MODELS FOR INCREASING SAFETY IN INFORMAL WASTEWATER IRRIGATION

Introduction

The challenge in view of wastewater reuse is not only to increase the reuse of **treated** wastewater as targeted for example in SDG 6.3, but to make the already ongoing informal irrigation which is on millions of hectares directly or indirectly using **untreated** wastewater safer. Untreated wastewater is released in large volumes across the developing world into rivers, used for irrigation purposes. The indirect reuse of this water for crop production, like any direct wastewater use, allows water-borne diseases which affect farmers in the field to turn into food-borne diseases affecting consumers, with a potentially significant economic impact. This informal wastewater reuse sector which support millions of livelihoods in and around four of five cities in the developing world occupies about 30 times the area than the one in our records where treated wastewater is used (Scott et al., 2010; Thebo et al., 2017). There is a significant need for business models to move from informal to formal reuse, despite inability of most developing countries to progress as fast as needed with wastewater collection, treatment capacity, or ability to enforce regulations.

Informal reuse of wastewater is a booming economic activity that benefits farmers and irrigators privately and also the local economies and food supply, but also entails significant health costs, mostly borne by the public. The **social** nature of these costs justifies public investments in incentives to promote safe reuse of wastewater and minimize risk along the entire value chain to turn this **unsafe** informal activity into a **safe** and formal one with shared rewards for all the stakeholders. But how to finance such investments where public budgets cannot keep pace with population growth and wastewater generation?

Examples of answers are provided in a set of different business models which are (like all models in this catalogue) based on empirical cases. The variety represents regionally different drivers and pathways to catalyze individual or institutional behavior change from informal to formal reuse (Saldias, 2016). The change can be based on direct or indirect incentives, increased risk awareness or on a dialog between key stakeholders on their needs, the analysis of costs and benefits and business contract. Another pathway is to seek technical synergies between private and public interests, and where the public sector has limited capacity or resources to engage, also the private sector can offer the incentives needed for behavior change. In all these cases, the value proposition is increased food and occupational safety. Investments in these business models can save USD 5 in consumer health care for every dollar spent on risk reduction from 'farm to fork' (Drechsel and Seidu, 2011; Keraita et al., 2015).

While the benefits outweigh the costs, risk awareness and the incidence of costs and benefits do not fall evenly across stakeholders along the value chain and this creates difficulty for incentive design and financing the investment. In other words, not all who could change the game might directly benefit from this, or understand the need for change. In this context it is for example important to distinguish between the direct reuse of raw, undiluted wastewater and the much more common indirect reuse. Direct reuse is usually a planned activity where farmers lack alternative water sources and/or seek the nutrient value of the water. In contrast, the indirect use of wastewater after it got mixed with other water sources is usually not driven by farmers' interest in wastewater, but simply in water. How far farmers experience and realize water pollution depends on the degree of wastewater dilution. As a result, many farmers would not consider themselves as wastewater users and also do not see anything wrong in their professional activities, in contrast to scientific risk assessments (Keraita et al., 2010).

The limited risk awareness is an important factor for the implementation of risk mitigation strategies, and calls for a mix of approaches with financial, regulatory as well as social incentives and awareness creation to support behavior change (Drechsel and Karg, 2013). From a technical perspective, there are many low-cost options available which can on their own or better in combination significantly reduce the risks on- and off-farm especially from pathogens. Such a multi-barrier approach is fully supported by the WHO (Figure 262). However, given the missing direct incentives and risk awareness of those who should implement these safety measures, they have to be easily adoptable, low-cost but highly cost-effective (Drechsel and Seidu, 2011).

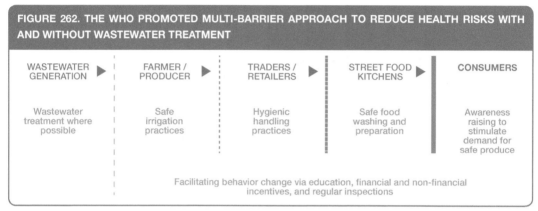

FIGURE 262. THE WHO PROMOTED MULTI-BARRIER APPROACH TO REDUCE HEALTH RISKS WITH AND WITHOUT WASTEWATER TREATMENT

Source: Amoah et al., 2011. The approach is based on the Hazard Analysis and Critical Control Point (HACCP) concept.

The informal irrigation sector where the use of wastewater or polluted water is common in and around four of five cities in the global South, shows a mosaic of situation and business model and multiple pathways towards formally recognized and supported wastewater use (Raschid-Sally and Jayakody, 2008).

Variations exist in terms of water quality, i.e. level of treatment or dilution, scale of use, water access, related costs/fees, market penetration, risk awareness along the value chain, enforcement of safety measures, etc. In most cases of indirect use, i.e. where wastewater and freshwater are mixed, the water is perceived as a natural and allocated to framers according to freshwater rules. Where farmers use raw wastewater they might pay a fee for the water, which is usually lower than the one they would pay for freshwater, or their rent for land with access to wastewater is higher than of land without water access, or the wastewater user rights might be auctioned to farmer groups.

Among the multitude of cases and situations in the informal reuse sector, three types were selected where different drivers support change towards a higher degree of safety. The three cases/models are each presented as **hybrids** (case-cum-model) as there would be too much overlap if presented separately. The three show options how the informal use of wastewater (be it polluted fresh water or diluted or raw wastewater) could become safer even under the common circumstances of missing treatment capacities and unenforced or absent water reuse regulations and standards:

a) **Business Model 22** is based mostly on examples from Ghana and shows how private sector driven corporate social responsibility (CSR) initiatives can be a driver of change, in particular in the informal food supply sector.
b) **Business Model 23** is based on examples from Pakistan and India. It shows how contractual agreements allow turning informal reuse into formal reuse, with the potential to introduce safety measures. In this example, wastewater is auctioned to farmer associations.

c) **Business Model 24** is based on a case from Southern Ghana. It shows options on how farmers' investments in low-cost infrastructure to access and store water can be combined with the WHO promoted multi-barrier approach, i.e. using farmers' innovation capacity to support reuse safety. Farmers' innovation capacity is well known (Reij and Waters-Bayer, 2001) and has been reported also from other countries where wastewater irrigation is common (Buechler and Mekala, 2005).

A model with focus on improving the safety of informal (sludge) reuse which combines elements from Models 23 and 24 was presented in the Nutrient Recovery section (Business Model 15).

Many variations of these models are possible and can be supported through various incentives such as land security, training, certification schemes for safe farming, access to loans or subsidies etc. Assistance is also needed in view of compliance monitoring as farmers will not be able to finance water or produce analysis for comparison with safety standards. The WHO supported *Sanitation Safety Planning* manual (WHO, 2015) will be a useful guidance document in this situation.

Our three examples are not exhaustive. There are other options, especially where with increased wastewater treatment and enforced regulations wastewater use became a business sector on its own. For an analysis of possible trajectories from informal to formal water reuse, see Saldias (2016) and for success stories from the formal reuse sector, for example, Lazarova et al. (2013).

References and further readings

Buechler, S. and Mekala, G.D. 2005. Local responses to water resource degradation in India: Groundwater farmer innovations and the reversal of knowledge flows. Journal of Environment and Development 14: 410–438.

Drechsel, P. and Seidu, R. 2011. Cost-effectiveness of options for reducing health risks in areas where food crops are irrigated with wastewater. Water International 36 (4): 535–548.

Drechsel, P. and Karg, H. 2013. Motivating behaviour change for safe wastewater irrigation in urban and peri-urban Ghana. Sustainable Sanitation Practice 16: 10–20.

Keraita, B., Drechsel, P., Seidu, R., Amerasinghe, P., Cofie, O. and Konradsen, F. 2010. Harnessing farmers' knowledge and perceptions for health risk reduction in wastewater irrigated agriculture. In: Drechsel, P., Scott, C.A., Raschid-Sally, L., Redwood, M. and Bahri, A. (eds) Wastewater irrigation and health: Assessing and mitigation risks in low-income countries. UK: Earthscan-IDRC-IWMI, pp. 337–354.

Keraita, B., Medlicott, K., Drechsel, P. and Mateo-Sagasta, J. 2015. Health risks and cost-effective health risk management in wastewater use systems. In: Drechsel, P., Qadir, M. and Wichelns, D. (eds) Wastewater: Economic asset in an urbanizing world. Dordrecht, the Netherlands: Springer, pp. 39–54.

Lazarova, V., Asano, T., Bahri, A. and Anderson, J. 2013. Milestones in water reuse: The best success stories. IWA. 408pp.

Raschid-Sally, L. and Jayakody, P. 2008. Drivers and characteristics of wastewater agriculture in developing countries: Results from a global assessment. Colombo, Sri Lanka: International Water Management Institute (IWMI). 29pp. (IWMI Research Report 127).

Reij, C. and Waters-Bayer, A. (eds) 2001. Farmer innovation in Africa: A source of inspiration for agricultural development. London: Earthscan.

Saldias, C.Z. 2016. Analyzing the institutional challenges for the agricultural (re)use of wastewater in developing countries. PhD Dissertation, University of Ghent. 263pp.

Scott, C., Drechsel, P., Raschid-Sally, L., Bahri, A., Mara, D., Redwood, M. and Jiménez, B. 2010. Wastewater irrigation and health: Challenges and outlook for mitigating risks in low-income countries. In: Drechsel, P., Scott, C.A., Raschid-Sally, L., Redwood, M. and Bahri, A. (eds) Wastewater irrigation and health: Assessing and mitigation risks in low-income countries. UK: Earthscan-IDRC-IWMI, pp. 381–394.

Thebo A.L., Drechsel, P., Lambin E.F. and Nelson, K.L. 2017. A global, spatially-explicit assessment of irrigated croplands influenced by urban wastewater flows. Environmental Research Letters 12: 074008.

WHO. 2015. Sanitation safety planning manual (Manual for the safe use and disposal of wastewater, greywater and excreta). Geneva: World Health Organization.

BUSINESS MODEL 22
Corporate Social Responsibility (CSR) as driver of change

Pay Drechsel

Key Characteristics

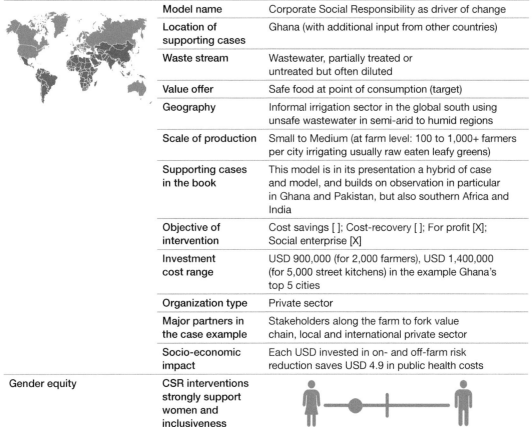

	Model name	Corporate Social Responsibility as driver of change
	Location of supporting cases	Ghana (with additional input from other countries)
	Waste stream	Wastewater, partially treated or untreated but often diluted
	Value offer	Safe food at point of consumption (target)
	Geography	Informal irrigation sector in the global south using unsafe wastewater in semi-arid to humid regions
	Scale of production	Small to Medium (at farm level: 100 to 1,000+ farmers per city irrigating usually raw eaten leafy greens)
	Supporting cases in the book	This model is in its presentation a hybrid of case and model, and builds on observation in particular in Ghana and Pakistan, but also southern Africa and India
	Objective of intervention	Cost savings [ ]; Cost-recovery [ ]; For profit [X]; Social enterprise [X]
	Investment cost range	USD 900,000 (for 2,000 farmers), USD 1,400,000 (for 5,000 street kitchens) in the example Ghana's top 5 cities
	Organization type	Private sector
	Major partners in the case example	Stakeholders along the farm to fork value chain, local and international private sector
	Socio-economic impact	Each USD invested in on- and off-farm risk reduction saves USD 4.9 in public health costs
Gender equity	CSR interventions strongly support women and inclusiveness	

Business value chain

A major social challenge are public health risks from the very common use of untreated wastewater in the informal irrigation sector of most low- and middle-income countries (Box 12). Many success stories on the trajectory from informal to formal reuse come from countries which succeeded in enhancing their treatment capacities and enforcing crop restriction, either as a result of epidemics or supported by high public risk awareness, such as in Israel, Chile, Jordan, Tunisia or Mauritius (Saldias, 2016). However, where capacities for wastewater treatment or the enforcements of crop restrictions are missing or only emerging, and also public risk perceptions are low, alternative strategies are needed for successful interventions in the usually highly profitable (wastewater) irrigation business.

In this situation, where the public sector is facing its limits, private sector driven corporate responsibility models can play an important role, and support occupational and consumer safety.

As discussed in Chapter 1, corporate responsibility can have different levels of buy-in, and even where environment values have been adopted, CSR drivers can range between 'selfish' investments in resource and cost efficiency to investments in longevity of the business in its protected environment:

> Corporate social responsibility (CSR) refers to business practices involving initiatives that benefit society. A business's CSR can encompass a wide variety of tactics, from giving away a portion of a company's proceeds to charity, to implementing "greener" business operations.[1]

The here presented model for improved water quality and food safety remains in large still hypothetical but is based on promising examples found in Western and Southern Africa as well as Pakistan. The model is highly compatible with the multi-barrier approach promoted by WHO in its 2006 wastewater reuse guidelines and further developed in the *Sanitation Safety Planning Manual* (WHO, 2006, 2015). However, the model is not an end in itself as it largely depends on behavior change and has to be supported by educational and regulatory measures to achieve its potential.

Box 12. The challenge of informal wastewater use

Reuse of raw or diluted wastewater for irrigation of field crops is practiced around most cities of the global South on a total area of up to 35 million hectares (Raschid-Sally and Jayakody, 2008; Thebo et al., 2017). Most of this irrigation is using untreated or at best partially treated wastewater, at a thirty times larger scale than the known areas using treated wastewater (Scott et al., 2010; Thebo et al., 2017.

This informal wastewater use is probably the most common 'business model' of resource recovery and reuse where waste is turned directly into an asset, however, without the required treatment to assure occupational safety or protect consumers of irrigated produce. The practice spreads without facilitation, driven by a reliable water supply and high demand for irrigated cash crops from growing urban markets, a demand which can lift farmers out of poverty (Drechsel et al., 2006). Where the wastewater is raw, farmers might also appreciate its nutrient content; while in those locations where it is diluted, farmers might not know about the invisible risks of their water source for human health and the environment (WHO, 2006).

As informal wastewater reuse flourishes especially in low-income countries where not only wastewater treatment capacities are limited, but also regulations weak and banning of wastewater irrigation neither practical nor feasible, the challenge is how to implement safety measures in this situation. From a technical perspective, there are many low-cost options available for on-farm and post-harvest risk reduction which can on their own or ideally in combination (multi-barrier approach) significantly reduce health risks for farmers and consumers, especially from pathogens. However, due to low risk awareness in the population there is limited market demand and financial incentive for safety measures. A high adoption rate is however required for any larger impact and cost-effectiveness (Drechsel and Seidu, 2011).

BUSINESS MODEL 22: CSR AS DRIVER OF CHANGE

Business model

Among different drivers of CSR, the here presented canvas has the focus on increasing product safety as value proposition (Figure 263). Protecting public health within and beyond the food chain can take place at different risk barriers, like (a) wastewater treatment, (b) on-farm, and (c) in the post-harvest and food processing sector as supported internationally by the Hazard Analysis and Critical Control Points (HACCP) concept. Some options related to the interface of water quality and reuse are illustrated below.

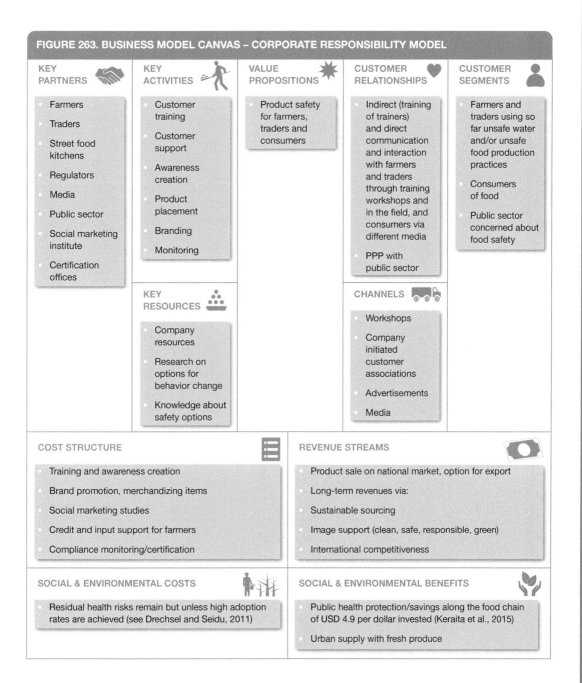

FIGURE 263. BUSINESS MODEL CANVAS – CORPORATE RESPONSIBILITY MODEL

Support of wastewater treatment

Companies, e.g. Nestle, are using wherever possible municipal wastewater treatment facilities, but where these are non-existent or not efficient enough, the company invests in own facilities, returning treated water to the environment according to local legislation or their internal standards, *whichever is more stringent*.[2] The corporate responsibly model has thus the potential that it can catalyze treatment development also where public sector capacities or existing legislations are still in development. Moreover, many companies invest in the reuse of their own wastewater as part of their corporate social responsibility program. Box 13 shows the strong motivation of the textile sector in Pakistan to comply with international safety standards, independently of national demands and regulations.

A similar benchmarking peer-pressure can also be applied to public utility providers or their operational partners including those responsible for wastewater treatment (Danilenko et al., 2014) and their international suppliers, which gives the WHO supported *Sanitation Safety Plan* entry points for its institutionalization if it can become an internationally accepted tool for compliance monitoring.

Box 13. Corporate responsibility as driver for change

There is a common and natural overlap between "corporate responsibility" and "business interests" and while for some companies CSR might be more a marketing factor, it becomes essential for company growth or even survival for others, especially in the highly competitive supply sector. In Pakistan, for example, the textile industry tries to double its USD 13 billion export volume through different initiatives of which a key one is to provide *environment friendly clothing* to the world, in particular the European market. This target requires that Pakistan's cotton factories are fully compliant with international standards, including sound chemical management and wastewater treatment. Until now, many textile manufacturers use substandard or banned chemicals and dyes. However, international conventions signed by Pakistan strictly restrict the use of unapproved raw materials, including their disposal to environment without proper treatment. As European buyers increasingly demand compliance, such as the Sweden Textile Water Initiative[3] or the Partnership for Sustainable Textiles[4] which has brought together almost half of the German textile industry with policy-makers and civil society, this provides a strong incentive for the textile industry in Pakistan, Bangladesh, India, etc. to invest in responsible sourcing and water quality. To this effect, entrepreneurs associated with different sections of the textile chain offered for example financial assistance to the Pakistani government for establishment of combined industrial **wastewater treatment plants**. To reduce the use of harmful raw materials, training in resource use efficiency and alternative materials is being provided. Eventually, both, the compliance with safety standards and a more efficient resource use will be crucial components for company acceptance on international markets, or to meet the benchmarking targets for corporate environmental compliance performance. A first result of increased compliance among 44,000 licensed cotton farmers in Pakistan in 2011–2012 was a significantly reduced environmental footprint, like a 20% lower water use in irrigation, 38% less pesticide use and 33% less commercial fertilizer use while farmers' profitability increased by 35% (Shaikh, 2013). Such as strong incentive as provided in this case by the European customer is needed as in general companies remain cautious, especially in view of in-house water reuse which is a common part of corporates' 'good water stewardship' but has often trade-offs between water and energy savings (Newborne and Dalton, 2016).

Support of farm-based interventions

Supermarket chains are subscribing increasingly to international codes of conduct, like the Global Social Compliance Programme (GSCP) supported by the Foreign Trade Association (FTA) and its Business Environmental Performance Initiative (BEPI), the latter serving retailers, importers and brands committed to improving environmental performance in supplying factories and farms worldwide. Supermarkets or wholesale companies engaging out-grower schemes can opt for compliances with a 'responsible sourcing policy' or other best practices or codes of conduct to meet international quality and sustainability standards, and to remain internationally competitive. For instance, in Botswana and South Africa industries, bulk buyers and supermarket chains (Figure 264) are directly sourcing their crops from urban and peri-urban vegetable, grapevine or olive farmers to secure a continuous year-round supply, guaranteed by the use of (partially) treated wastewater for irrigation (Hanjra et al., 2017).

For risk reduction, farmers use drip irrigation and the companies put post-harvest measures in place to clean the crops from possible pathogenic contamination. This is in line with WHO's emphasis on health-based targets, where the irrigation water quality is less critical as long as measures to minimize exposure of crops and consumers are put in place (WHO, 2006). Thinking beyond the farm is also important, as even where irrigation water is safe, post-harvest contamination can be severe. Food safety interventions in markets, street restaurants and households are therefore of equal, if not higher importance, to safeguard consumers. This is even truer from an impact perspective, as the relationship between the supermarket and its farmers might only benefit the (middle and upper class) consumers of the supermarket and not the general public buying crops via traditional market chains.

Support of post-harvest interventions

Social responsibility programs can be very powerful change agents in the post-harvest sector. In Ghana, for example, about 90% of the wastewater irrigated vegetables are sold raw as supplement to popular fast food dishes in the urban street food sector. For authorities and NGOs it is a challenge to enter or control this informal sector. However, the situation can be different for the private sector. Nestle, for example, supplies the street restaurant sector across West Africa with ingredients, like Maggi™ bouillon cubes, and uses its branding power to (i) maintain close links within the sector; and (ii) use it to advertise its brand. As part of Nestle's consumer service program, the company initiated in Ghana the formation of trader associations, the Maggi™ Fast Food (Seller) Association (MAFFAG) which is today the strongest association in Ghana's street food sector. MAFFAG regularly provides training in food preparation, cooking, environmental hygiene and food safety throughout the country, which combining elements of corporate responsibility with branding, free merchandise and product promotion. Compared with governmental workshops, the MAFFAG events attract large crowds, and their training programs are very well positioned for addressing food-safety concerns across the sector (Figure 265).

As the Maggi™ colors are today prominent in West Africa's street food sector, the high degree of brand recognition also implies responsibility to maintain the company's quality image. This motivation facilitated in Ghana the strong interest of MAFFAG in training in safe vegetable washing to minimize any food-related risk including those from vegetable irrigation (Amoah et al., 2009).

Based on the degree of educational efforts and risk awareness creation through public and private sector activities, the hope is that market demand for safe produce will slowly increase and catalyze further demand driven change. Wholesaler, trader or supermarkets can support this process through contracts with farmer cooperatives which allow them to secure a reliable crop supply while offering

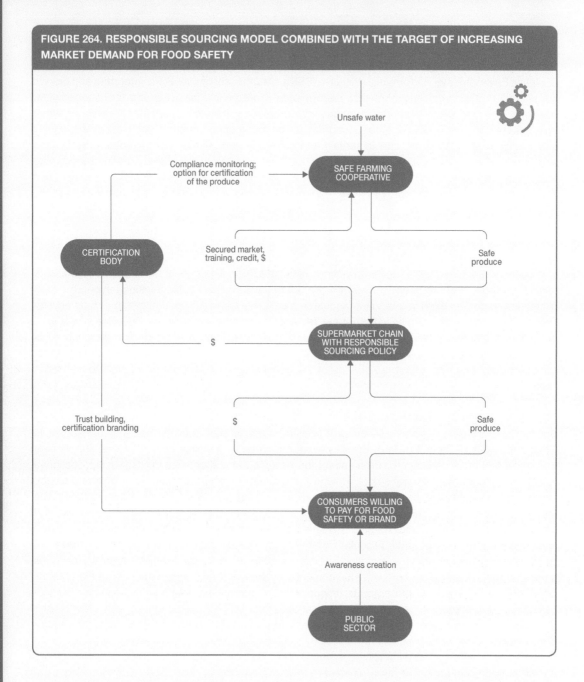

FIGURE 264. RESPONSIBLE SOURCING MODEL COMBINED WITH THE TARGET OF INCREASING MARKET DEMAND FOR FOOD SAFETY

inputs, training or credit. To qualify for such contracts best practices like safety measures could be made mandatory.

'Safe produce' branding could be an additional incentive mechanism for farmers and traders and support premium pricing. This could offer opportunities for third parties with capacities to perform compliance and quality monitoring to issue **quality certificates** as it is well established in the 'organic food' sector (Keraita and Drechsel, 2015).

BUSINESS MODEL 22: CSR AS DRIVER OF CHANGE

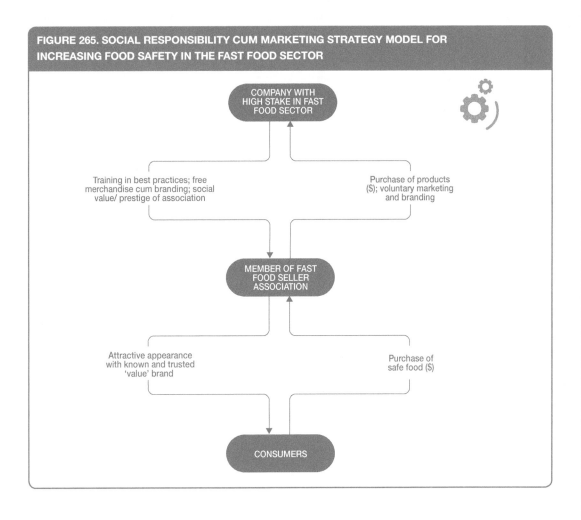

FIGURE 265. SOCIAL RESPONSIBILITY CUM MARKETING STRATEGY MODEL FOR INCREASING FOOD SAFETY IN THE FAST FOOD SECTOR

Private sector support is not only important where the public sector struggles but also where initially only a minority of consumers with better education will support a safe food niche market. Although it can be anticipated that consumer demand will continuously increase through awareness creation, market based incentives might not be sufficient for success at scale and have to be complemented with other triggers for the adoption of safety practices to achieve compliance, e.g. with WHO Sanitation Safety Plans (Box 14).

Box 14. Triggering behavior change

Where health risk awareness is low and stakeholders along the food chain do not see a reason for engaging in safety practices, they might however change their behavior for other values or benefits which can contractually be agreed on. Examples are:

Tenure security: Many users of wastewater farm along streams on public land with limited tenure security if any, and constant fear of eviction. Land release, zoning and tenure security are thus powerful incentives when demanding the implementation, e.g. of best practices, especially those which require farm-based infrastructure (Keraita et al., 2014).

Credit on condition: As similar incentive is the provision of low-interest credit to farmers who are applying safe irrigation methods.

In both cases, it remains the duty of the authorities to monitor farmers' compliance with their contractual obligations.

Fear of exposure: Where safety regulations cannot be monitored by authorities, media exposure (naming and shaming) can be a powerful alternative to steer compliance. Urban farmers and food restaurants in Ghana feared media exposure as it can trigger ad hoc policy response like eviction from the land or business closure.

Social values: Households might embark on safety measures if the right triggers and drivers can be identified and promoted through social marketing as it was successfully demonstrated in handwash and end-open-defecation campaigns. This might not be 'health' per se, but feeling of 'comfort', 'status', 'disgust', etc. Like in handwash campaigns, women in charge of food preparation should be a key target group.

Social marketing offers particular opportunities as it is only a relatively small step from the promotion of hand washing to salad washing (Drechsel and Karg, 2013). Also here private sector participation can be powerful as for example the Public-Private Partnership to Promote Handwashing between UNICEF and UNILEVER in West Africa has shown (see www.unicef.org/wcaro/overview_2765.html).

Potential risks and mitigation

In designing any business model, it is assumed that generic business risks are known and will be taken care of. However, some risks might be more model specific and will be acknowledged in the following:

Market risks: Household demand for the safer food is theoretically high, but does so far not translate in a different purchasing behavior (Keraita and Drechsel, 2015) although it can be influenced as the handwash campaign example (Box 14) shows. A larger risk is that the CSR company might not engage in the support of the farming communities using wastewater as long as they can source safer supply chains. Such (freshwater using) alternatives are however increasingly seldom in urban proximity.

Competition risks: Unsafe produce can have a price advantage. Awareness creation and social marketing flagging the difference between safe and unsafe produce can decrease the market demand/share of unsafe produce. Care has to be taken that safe and potentially still unsafe marketing channels are kept separate.

Technological risk: The involved technology for farmers, traders or restaurants is basic and in general affordable (Amoah et al., 2011).

Social equity related risks: Supporting women is a core element of many CSR programs. Social marketing campaigns, training, the formation of 'brand' association, etc. have a high potential to support women and gender inclusiveness. As urban vegetable farming on open spaces offers employment opportunities for rural migrants, any support through the private sector would be an important step towards social integration and poverty alleviation.

Political and regulatory risks: Corporate responsibility models by definition comply with local regulations. As the public sector is partner in the model, compliance will be monitored depending on local capacity. However, a challenge can come from a regulatory framework which is not supporting, as suggested by WHO (2006), a step-wise and multi-barrier HACCP approach to move towards safer wastewater irrigation or food safety in general.

Safety, environmental and health risks: The model helps to reduce risks where treatment systems are lacking and farmers use directly or indirectly untreated, partially treated or diluted wastewater. It builds on safety measures as recommended by WHO (2006) for this situation. Although these best practices target first of all pathogenic risk, the model can also address chemical risks if the sources can be controlled by the participating private sector entities through source pre-treatment and a 'zero waste' policy. The model follows the WHO recommendation of a step-wise and stakeholder inclusive approach to risk mitigation which is an intermediate step until (a) more comprehensive wastewater collection and treatment systems are in place, and (b) stricter safety guidelines can be implemented and enforced.

As the model is based on incentivizing human behavior change and a high degree of compliance with risk mitigation measures, **risks will remain** and have to be addressed through conventional mitigation measures (Table 61) supported by further awareness creation, capacity development and incentive systems (Drechsel and Karg, 2013).

TABLE 61. POTENTIAL HEALTH AND ENVIRONMENTAL RISK AND SUGGESTED MITIGATION MEASURES FOR BUSINESS MODEL 22

RISK GROUP	EXPOSURE					REMARKS
	DIRECT CONTACT	AIR/ODOR	INSECTS	WATER/SOIL	FOOD	
Farmer	HIGH	LOW	LOW	MEDIUM	LOW	WHO Sanitation Safety Plans with multi-barrier approach recommended along food chain, complemented with risk mitigation measures by the corporate sector.
Community	LOW	LOW	LOW	LOW	LOW	
Consumer	LOW	N/A	LOW	LOW	MEDIUM	
Mitigation measures	🧤	😷	🧴	Pb Hg Cd	🚿 / Pb Hg Cd	

Key: ☐ NOT APPLICABLE ░ LOW RISK ▒ MEDIUM RISK ■ HIGH RISK

SWOT analysis and business performance

The model is suggested for the common situation in sub-Saharan Africa, Asia and parts of Latin America where informal wastewater use is potentially threatening public health while local authorities have limited capacity to enforce restriction or change the situation, e.g. wastewater treatment is not available. The model builds on the rapidly growing opportunity of corporate social and environmental responsibility principles of the private sector (Figure 266) with related investments in the value chain. It argues for additional support to address key weaknesses of the model in particular in view of public awareness creation and the exploration of social marketing to catalyze behavior change and market demand. The model can support best practices on farm, but might have a wider outreach where it can target the post-harvest sector including consumers. The model offers possible incentives for making the crop value chain safer in the common situation where general risk awareness is still too low to rely on self-protection.

In comparison to other performance indicators the business model scores particularly high on social factors, via reduced expenditure on public health while supporting the informal irrigation sector which is often dominated by rural migrants or other social minorities looking for quick cash (Figure 267). Given the novelty of using CSR models to increase food safety, the model has certainly innovation potential. On the other hand, it requires more experience and practical examples before the scalability and replicability can be assessed. Given that social marketing requires context specific research, it is certainly not easily transferable. Also its environmental impact is limited as long as the focus is on human exposure and behavior change, and does not catalyze more wastewater treatment systems.

FIGURE 266. SWOT ANALYSIS OF THE CORPORATE RESPONSIBILITY MODEL TO IMPROVE FOOD SAFETY IN THE INFORMAL IRRIGATION SECTOR AND ITS VALUE CHAIN

	HELPFUL TO ACHIEVING THE OBJECTIVES	**HARMFUL** TO ACHIEVING THE OBJECTIVES
INTERNAL ORIGIN ATTRIBUTES OF THE ENTERPRISE	**STRENGTHS** • CSR targets can catalyze change also where public sector policies and regulations are only emerging • Irrigated farming in city vicinity is usually for the market, and not subsistence, thus private sector interest is likely • Risk reducing options have been researched and translated into training materials, videos, etc. • High social impact potential	**WEAKNESSES** • Risk reduction depends on degree of best practice adoption; missing adoption targets can result in backlash • Awareness creation among consumer and thus overall demand will take time • Local trading companies without international visibility might feel no need for CSR • Capacities in social marketing research only emerging • Safe produce might not reach the poor
EXTERNAL ORIGIN ATTRIBUTES OF THE ENVIRONMENT	**OPPORTUNITIES** • Social and environmental responsibility commitments sharply increasing • Existing PPPs (like global handwash campaign) could be expanded or at least provide lessons • Piggy-backing on existing trader or farmer association with high market penetration	**THREATS** • Unsafe produce can have a price advantage until awareness creation and social marketing leverage this in a positive sense • Company might source alternative farming communities using clean water sources • Some authorities may see private sector engagement as an intrusion into their domain

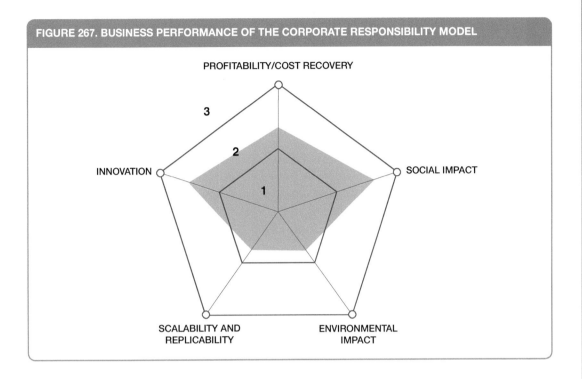

FIGURE 267. BUSINESS PERFORMANCE OF THE CORPORATE RESPONSIBILITY MODEL

Contributors
Jasper Buijs, Sustainnovate (http://sustainnovate.nl)
Munir Hanjra, IWMI
Juergen Hannak, GIZ (on the Germany-supported Partnership for Sustainable Textiles)

References and further readings

Amoah, P., Drechsel, P., Schuetz, T. Kranjac-Berisavjevic, G. and Manning-Thomas, N. 2009. From world cafés to road shows: Using a mix of knowledge sharing approaches to improve wastewater use in urban agriculture. Knowledge Management for Development Journal 5 (3): 246–262.

Amoah, P., Keraita, B., Akple, M., Drechsel, P., Abaidoo, R.C. and Konradsen, F. 2011. Low cost options for health risk reduction where crops are irrigated with polluted water in West Africa. IWMI Research Report 141. Colombo: IWMI.

Danilenko, A., Van den Berg, C., Macheve, B. and Joe Moffitt, L. 2014. The IBNET water supply and sanitation blue book 2014. Washington, DC: World Bank.

Drechsel, P. and Seidu, R. 2011. Cost-effectiveness of options for consumer health risk reduction from wastewater irrigated crops. Water International 36 (4): 535–548.

Drechsel, P., Graefe, S., Sonou, M. and Cofie O.O. 2006. Informal irrigation in urban West Africa: An Overview. IWMI Research Report Series 102. Colombo: IWMI.

Drechsel, P. and Karg, H. 2013. Motivating behaviour change for safe wastewater irrigation in urban and peri-urban Ghana. Sustainable Sanitation Practice 16: 10–20.

Hanjra, M. A., Drechsel, P. and Masundire, H. M. 2017. Urbanization, water quality and water reuse in the Zambezi river basin. In: Lautze, J., Phiri, Z. and Smakhtin, V. (eds.) Zambezi River Basin: Water and Sustainable Development. Routledge – Earthscan. pp.158–174.

Keraita, B. and Drechsel, P. 2015. Consumer perceptions of fruit and vegetable quality: Certification and other options for safeguarding public health in West Africa. IWMI Working Paper 164. Colombo, Sri Lanka: International Water Management Institute (IWMI). 32p.

Keraita, B., Drechsel, P., Klutse, A. and Cofie, O. 2014. On-farm treatment options for wastewater, greywater and fecal sludge with special reference to West Africa. Colombo, Sri Lanka: International Water Management Institute (IWMI). CGIAR Research Program on Water, Land and Ecosystems (WLE). 32p. (Resource Recovery and Reuse Series 1).

Keraita, B., Medlicott, K., Drechsel, P. and Mateo-Sagasta Dávila, J. 2015. Health risks and cost-effective health risk management in wastewater use systems. In: Drechsel et al. (eds). Wastewater: Economic asset in an urbanizing world. Springer, pp. 39–54.

Newborne, P. and Dalton, J. 2016. Water management and stewardship: Taking stock of corporate water behaviour. Gland, Switzerland: IUCN and London, UK: ODI. 132pp.

Raschid-Sally, L. and Jayakody, P. 2008. Drivers and characteristics of wastewater agriculture in developing countries: Results from a global assessment. IWMI Research Report 127. Colombo, Sri Lanka: International Water Management Institute (IWMI). 29pp.

Saldias, C.Z. 2016. Analyzing the institutional challenges for the agricultural (re)use of wastewater in developing countries. PhD Dissertation, University of Ghent. 263pp.

Scott, C., Drechsel, P., Raschid-Sally, L., Bahri, A., Mara, D., Redwood, M. and Jiménez, B. 2010. Wastewater irrigation and health: Challenges and outlook for mitigating risks in low-income countries. In: Drechsel, P., Scott, C.A., Raschid-Sally, L., Redwood, M. and Bahri, A. (eds) Wastewater irrigation and health: Assessing and mitigation risks in low-income countries. UK: Earthscan-IDRC-IWMI, p. 381–394.

Shaikh, A.F. 2013. Keynote address. Pakistan Textile Industry. First International Textile & Clothing (ITC) Conference, Lahore, Monday, 9 December, 2013.

Thebo, A.L., Drechsel, P., Lambin E.F. and Nelson, K.L. 2017. A global, spatially-explicit assessment of irrigated croplands influenced by urban wastewater flows. Environmental Research Letters 12: 074008.

WHO. 2006. Guidelines for the safe use of wastewater, excreta and greywater, volume 2: Wastewater use in agriculture. Geneva: World Health Organization.

WHO. 2015. Sanitation safety planning manual (Manual for the safe use and disposal of wastewater, greywater and excreta). Geneva: World Health Organization.

Notes

1 www.businessnewsdaily.com/4679-corporate-social-responsibility.html (accessed 4 Nov. 2017).
2 www.nestle.com/csv (accessed 4 Nov. 2017).
3 http://stwi.se/ (accessed 4 Nov. 2017).
4 www.textilbuendnis.com/en/ (accessed 4 Nov. 2017).

BUSINESS MODEL 23
Wastewater as a commodity driving change

Munir A. Hanjra, Krishna C. Rao, George K. Danso, Priyanie Amerasinghe and Pay Drechsel

In memory of Jeroen Ensink

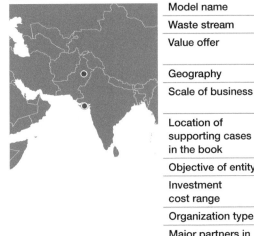

Model name	Wastewater as a commodity driving change
Waste stream	Domestic wastewater
Value offer	Untreated, partially treated and treated wastewater for auctioning to farmers
Geography	Arid and semi-arid regions
Scale of business	Medium (function of irrigation demand and wastewater supply)
Location of supporting cases in the book	This model is in its presentation a hybrid of case & model and builds on observation in particular in Faisalabad, Pakistan and Gujarat, India
Objective of entity	Cost recovery [X]; Safety [X]
Investment cost range	Varies largely with type and size of treatment plant (USD 2–4 m for 100,000 inhabitants)
Organization type	Public (utility) and private (farmers)
Major partners in the case example	Water and Sanitation Agency (WASA), Chakera Farmers
Socio-economic impact	The auctioning process is socially inclusive. Wastewater main source of income and food security; health risks can be controlled and are accepted as offset by revenues
Gender equity	In many countries, auctioning will favour men as women have less access to land and capital 

Executive summary

This business model has been informed by two cases where wastewater auctioning is common, in Pakistan and India. With sufficient information being only available from the situation in Pakistan, the presentation of its model here follows a **hybrid of the business model and case templates** focussing on mostly Faisalabad/Pakistan.

With insufficient supply of freshwater of low salinity to support irrigated crop production, farmers in the dry climate of Faisalabad in Pakistan overcame organizational, infrastructure and legal obstacles to secure access to urban wastewater. Like for freshwater (canal water), also wastewater became a marketable commodity farmers pay for[1]. The wastewater provider, the Water and Sanitation Agency (WASA) in Faisalabad, uses public auctions for bulk sale of its wastewater to farmers, a system which keeps WASA's transaction costs low, and is also reported from Gujarat, India. The farmers organize themselves into groups and the highest bidder gets the annual rights to reuse the wastewater and resell surplus water to other farmers. As the annual auction attracts several interested bidders, a floor price guarantee is not required and wastewater auction price is determined through a near competitive market. Despite the common experience that wastewater can only be sold cheaper than freshwater, it is not uncommon that farmers pay for the wastewater on top of their fees for canal water and up to 50% more for untreated wastewater than treated wastewater, given the lower nitrogen and higher salinity levels of the latter. The auction process allows WASA to cover its pumping costs and to maintain administrative control over the wastewater. Most of all, the process turns informal wastewater use into formal wastewater use and gives farmers and authorities a platform for dialog. This is missing in many countries where informal wastewater use is a grey sector. The dialog offers opportunities for negotiating health risk mitigation via alternative treatment options, which match farmers' needs while enhancing safety (WHO, 2006). The high market value of wastewater offers opportunities to introduce incentives (like extra water allocations) in support of the compliance with safety measures.

KEY PERFORMANCE INDICATORS OF THE PAKISTAN CASE (2012–2014)						
Land use:	Around Chakera: About 456 ha (71% of farm area) under untreated wastewater irrigation, ca. 25–35 ha under treated wastewater, 12–15 ha with freshwater supply, ca. 85–90 ha with partially treated or mixed sources, and 44 ha without irrigation					
Water treated:	Ca. 37,000m^3 per day (the design capacity is about 90,000m^3 per day. About 79,000m^3 of wastewater enter the plant premises, of which about 53% are redirected for irrigation before reaching the first treatment ponds					
Capital investment:	N.A.					
Labor employment:	N.A.					
O&M costs:	About USD 350,000 sewage system O&M costs (Faisalabad West) in 2012–2014, of which the O&M budget of the treatment plant is about USD 30,000 (WASA, 2014)					
Annual revenue:	Sewerage charges as part of water bill, property tax and non-tariff income from leasing of land to farmers and wastewater auctioning of around USD 6,000–7,000					
Output:	Secondary treated wastewater					
Potential social and/or enviornmental impact:	Job creation/income along the sanitation value chain (i.e., farmers, market sellers, and input suppliers); possible health risks for farmers and urban consumers					
Financial viability indicators:	Payback period:	Data not available (N.A.)	Post-tax IRR:	N.A.	Gross margin:	N.A.

Context and background

This business case pertains to wastewater reuse as seen on the example of Chakera, located on the western outskirts of Faisalabad City in Pakistan. Faisalabad, the third largest metropolis in Pakistan, has an estimated population of 3.6 million (WASA, 2014). Like in other larger cities in Pakistan, the semi-autonomous Water and Sanitation Agency (WASA) is responsibility for water supply and sanitation. It provides about 65–75% of the city area with a sewerage network which is linked to larger channels for final disposal of the wastewater in the Chenab and Ravi rivers. Faisalabad's wastewater treatment plant at Chakera can treat about 25–30% of the collected wastewater. The system consists of a series of waste stabilization ponds (WSP). Given the city's largely flat topography, 31–40% of WASA's O&M costs per year are for electricity (pumping).

Under the arid climate with annually about 350 mm rainfall, irrigation is most common and traditionally supported by canals fed by River Chenab. Due to increasing demands, water supply is declining since a few decades and inadequate for many regions and/or year round production. Common crops are wheat, sugarcane, cotton, vegetables, fruits and fodder crops (clover, Lucerne, barley, etc.). Chakera is a typical suburban village where irrigation canals stopped providing the needed water years ago (ca. 2002). As groundwater is saline, the only remaining option is wastewater, which is abundant in Chakera given its proximity to the city of Faisalabad. However, initially WASA did not support farmers' request for waste and it took some court cases before WASA agreed. The current system is that WASA auctions the wastewater, and this can be untreated, partially treated or treated wastewater. WASA also rents out land with wastewater access (Weckenbrock et al., 2011).

It is noteworthy that more than 90% of the farm households choose untreated wastewater although treated wastewater is in same proximity and cheaper. Freshwater alternatives like canal water and groundwater are too scarce and can only serve less than 5% of the water market. The reasons for choosing untreated wastewater are of agronomic nature, directly affecting farmers' livelihoods: the treatment process is increasing water salinity (see below) beyond what some of their crops can tolerate while significantly reducing the nitrogen and phosphorus content of the water. In fact, 70% of farmers using untreated wastewater reported that they stopped applying fertilizers, and 24% only used very targeted fertilizer applications (Ensink, 2006; Clemett and Ensink, 2006).

Market environment

Demand for irrigation water in Punjab province of Pakistan has increased steadily over the past decades, far beyond what can be supplied. Chakera is one of the villages with irrigation canals but hardly any water. However, farmers in the eastern boundary of the village began to use wastewater coming from Faisalabad. With visible gains in production and income, this informal practice spread among farmers in other parts of the village and beyond. With the increase in demand for wastewater, farmers constructed new irrigation canals to make wastewater available in more parts of the village and the existing canal-water infrastructure was modified to facilitate wastewater irrigation. Over the years, wastewater has become the most important source of water for irrigation with benefits clearly exceeding risks, while the majority of farmers having in fact no alternative (Clemett and Ensink, 2006; Weckenbrock, 2011). There are about 2,200 ha irrigated with wastewater around Faisalabad, and 32,500 across Pakistan (Weckenbrock et al., 2011).

Wastewater farmers around the WSP have, since the construction of the treatment plant, organized themselves and gone to court to establish their rights to use wastewater. For the water use, they pay WASA which allows WASA to maintain control over the resource as a service provider. The water fee ranges from USD 10 to USD 62 per hectare per year depending on the quantity and the quality of wastewater. The payment is on top of what farmers pay for freshwater[2]. The highest fees were paid for untreated wastewater with lower fees paid for wastewater from anaerobic ponds (Clemett and Ensink, 2006). A unique feature of this business case is that the wastewater is sold annually in bulk by WASA through an open auction to the highest bidder from the village, and the winning farmer resells the surplus wastewater to fellow farmers (Box 15). Through the bulk auction of wastewater, WASA outsources likely transaction costs of dealing with individual farmers such as water allocation, monitoring, compliance and collecting the wastewater fees, which is reducing the service provision largely to the energy costs for water pumping, the cost item WASA tries to recover from the business.

Macro-economic environment

The agriculture sector continues to play a central role in Pakistan's economy. It is the second largest sector, accounting for over 21% of gross domestic product (GDP), and remains by far the largest employer, absorbing 45% of the country's total labor force. Given the low precipitation, major investments in water resources management are required to prepare Pakistan for its growing population

> **Box 15. Wastewater auctioning process**
>
> The auction process for treated/untreated wastewater starts with the announcement of the bidding date to farmers. The farmers organize themselves into small bidding consortiums/ groups and each group nominates a bid leader, after background negotiations on the maximum bid amount and the exit strategy should the bid amount go higher than their expectations and upper ceiling. On the bidding day, farmers congregate at the venue and group leaders contest the bid in the open auction. Opening bid price is generally the last year's auction price, and then the bid amounts are raised gradually upwards through calling the amounts publically. Only one bid is left with the auctioneer at a time, and a punt by another group leader raises the bid with the hope to snap up the wastewater. Group rivalry and market competition are all at play along with some pride and political capital in winning the bid and this often leads to intense 'bidding wars' across the group leaders. Bids are conveyed to the auctioneer through various gestures, including waving of hand or cloth. The highest bidder wins the auction and WASA auctioneer announces the name of the winning bidder and the completion of the wastewater sale. The winning bidder/ purchaser of the wastewater is generally a wealthier farmer within the village – not an investor or company, and not a speculator. Further, the bid amounts are never undisclosed such that water prices/charges become known to all farmers spot on. Strong cooperation and greater understanding on water allocation rules among farmers ensures that all farmers get water and no one is excluded from using the wastewater, at payment to the winning bidder, like USD 6 per year for a one-hour allocation every ten days. These can be paid in different instalments, which benefits less wealthy farmers unable to pay the fee in a single instalment. There are no reported cases of abuse of power (Weckenbrock et al., 2011; Clemett and Ensink, 2006).

and risks through floods and droughts. One part of this is the promotion of water recovery and reuse. Water quality monitoring in the irrigation sector is generally lax. A main challenge will be to develop local guidelines for cost effective risk mitigation measures, which consider actual exposure and help to optimize the gains to farmers from reuse while reducing risks to actors along the wastewater value chain.

Business model

As a service provider, the enterprise (WASA) charges for the wastewater it is collecting, managing and treating as far as its facilities allow. The wastewater is not given away for free and WASA remains in control of its allocation. Although the revenue stream from agriculture is a minor one given WASA's overall budget, the WSP has limited maintenance costs, and the revenue allows to cover a good part of WASA's pumping costs in the Chakera area.

The auction model has direct and indirect advantages: WASA transfers the water rights (and related pro-poor obligations) as well as the transaction costs of reaching out to individual farmers to the winning bidder who is in charge of supporting all farmers who agree to a transparent pro rata price. There is no penalty to any farmer for breaking the informal contract with the winning bidder, due to high water demand, and collections from the farmers far exceed the bid amount. This allows the business to remain viable at bidder's end, and even provides for the maintenance of the water courses and seasonal canal cleaning. Also, a maintenance charge is factored into the price of wastewater farmers pay. The winning bidder pays WASA on a quarterly basis and collects water charges from the individual farmers in convenient rates. Further, WASA uses a price discrimination model to encourage the reuse of treated wastewater, i.e. the untreated wastewater is auctioned at highest prices followed by partially treated and treated at the lowest price. WASA also earns revenue from leasing of its land (wastewater access priced in). The business has a long history of cooperation and **turns informal wastewater use into formal wastewater use**, which opens space for dialog to address, e.g. farmers' problems with the current wastewater treatment system. A summary of the key elements of the Business model canvas is outlined in Figure 268 below.

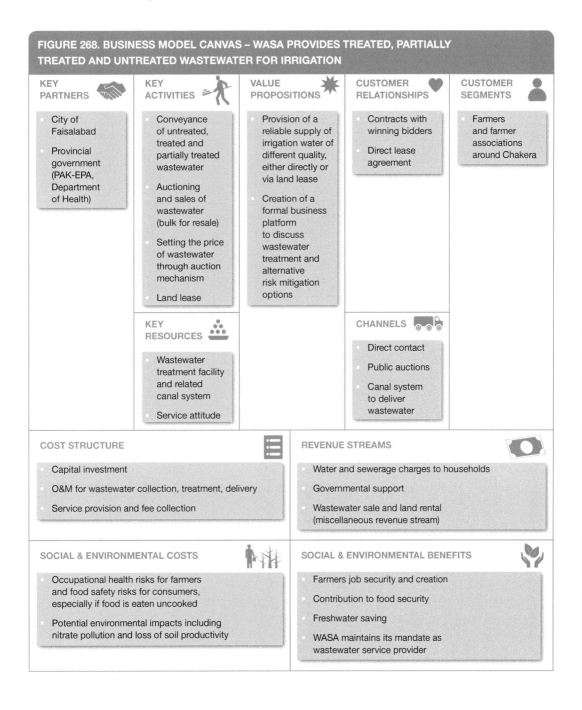

FIGURE 268. BUSINESS MODEL CANVAS – WASA PROVIDES TREATED, PARTIALLY TREATED AND UNTREATED WASTEWATER FOR IRRIGATION

KEY PARTNERS	KEY ACTIVITIES	VALUE PROPOSITIONS	CUSTOMER RELATIONSHIPS	CUSTOMER SEGMENTS
• City of Faisalabad • Provincial government (PAK-EPA, Department of Health)	• Conveyance of untreated, treated and partially treated wastewater • Auctioning and sales of wastewater (bulk for resale) • Setting the price of wastewater through auction mechanism • Land lease	• Provision of a reliable supply of irrigation water of different quality, either directly or via land lease • Creation of a formal business platform to discuss wastewater treatment and alternative risk mitigation options	• Contracts with winning bidders • Direct lease agreement	• Farmers and farmer associations around Chakera
	KEY RESOURCES • Wastewater treatment facility and related canal system • Service attitude		CHANNELS • Direct contact • Public auctions • Canal system to deliver wastewater	

COST STRUCTURE	REVENUE STREAMS
• Capital investment • O&M for wastewater collection, treatment, delivery • Service provision and fee collection	• Water and sewerage charges to households • Governmental support • Wastewater sale and land rental (miscellaneous revenue stream)

SOCIAL & ENVIRONMENTAL COSTS	SOCIAL & ENVIRONMENTAL BENEFITS
• Occupational health risks for farmers and food safety risks for consumers, especially if food is eaten uncooked • Potential environmental impacts including nitrate pollution and loss of soil productivity	• Farmers job security and creation • Contribution to food security • Freshwater saving • WASA maintains its mandate as wastewater service provider

Value chain and position

WASA is responsible for water supply and sanitation services in the city and has a natural monopoly on supplying any wastewater to the farmers. With the increasing water scarcity and the need to produce more vegetables to supplement local production of food crops, WASA has come under increasing pressure to provide wastewater for irrigation across locations. This can be treated, partially treated and untreated wastewater. However, only farmers with access to untreated wastewater were able to save on fertilizers and achieve higher cropping intensities as well as year-round cultivation which allowed them to earn on average USD 600 per hectare per year more than farmers who used regular irrigation water, easily absorbing the higher water fees set by the WASA (Clemett and Ensink, 2006). By charging farmers for the use of wastewater, WASA is able to recover some of its O&M costs for the wastewater treatment plant, while its main revenue stream (ca. 60%) are water supply and sewerage charges (2:1) billed to the residential, commercial and industrial sectors (Figure 269).

Institutional environment

In Pakistan's Punjab province, water services in the largest cities are provided by publicly-owned Water and Sanitation Agencies (WASAs). WASAs are accountable to both local- and provincial-level authorities, i.e. the respective City Development Authorities, and the Housing Urban Development and Public Health Engineering Department of the Government of Punjab.

The WASA in Faisalabad was created in 1978. In the 1980s WASA bought land at the outskirt of the city (Chakera village) for the purpose of building a WSP and its operation started in the late 1990s. During this period, there was a lengthy court case between the farmers and WASA over the right to use wastewater. The first court case accepted the lack of alternative water sources, and gave farmers the right to use wastewater for irrigation and to generate income to support their families. The court provided this ruling because there was no alternative canal water for the farmers to use for irrigation. The WASA appealed this decision and the court granted them the water rights, banning wastewater irrigation. As this decision was hard to implement, both parties had to reach an agreement whereby farmers agreed to pay WASA for the use of wastewater. This provides farmers with some legal standing of the practice and institutional support for wastewater reuse while WASA remains in charge as service provider (Clemett and Ensink, 2006). That wastewater can be sold for agricultural purposes is supported by the National Sanitation Policy (Ministry of Environment, 2006). However, there is no universal public acceptance of wastewater use and it is not supported, e.g. by the Ministry of Health.

Technology and processes

The wastewater treatment plant in Faisalabad was built in 1998 and designed for an inflow of nearly 90,000m^3 per day of (mostly) domestic wastewater at a site where untreated wastewater had been used for the past 50 years for the cultivation of vegetables, fodder, wheat and sugarcane. The plant is a basic waste stabilization pond system (WSP), consisting of six anaerobic, two facultative and four maturation ponds (Figure 270). Its operational costs are low while performance in terms of the removal of pathogens, BOD and TSS, as well as ammonia and phosphorus is good. It was expected that WASA sells the reclaimed quality water to farmers to recover some of its O&M costs. In practice however, the majority of farmers continued using the untreated wastewater, which they take from wastewater channels not passing the plant but also the main channel just before it reaches the first WSP ponds (Clemett and Ensink, 2006). Thus, from the expected average daily inflow into the treatment plant of 79,300m^3 per day, about half is diverted before it enters the plant. As a result of the lower flow, retention periods increased and so evaporation and water salinity, which stopped farmers being willing to pay for treated wastewater.

A much smaller quantity of water is also diverted from the anaerobic and facultative ponds. Because of the limited demand for final effluent by farmers[3], much of the treated wastewater has to be disposed

BUSINESS MODEL 23: WASTEWATER AS A COMMODITY

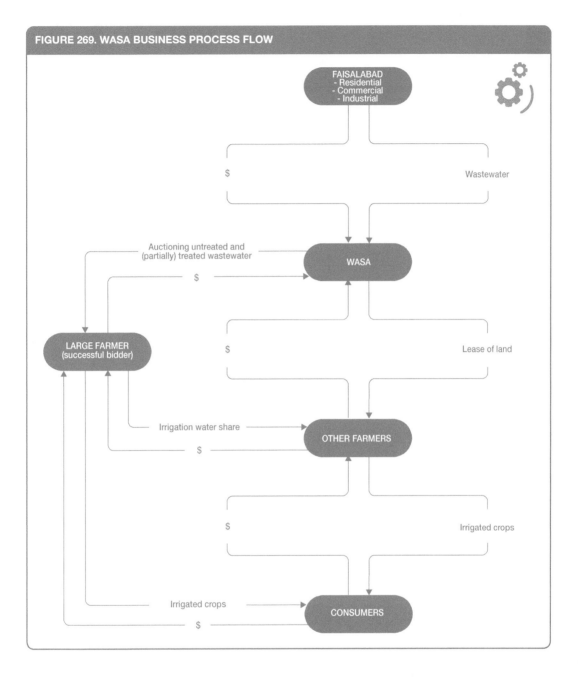

FIGURE 269. WASA BUSINESS PROCESS FLOW

of into the next drain. Other farmers access wastewater which is not flowing to the treatment plant. In total about 60,000m³ of wastewater is used per day by about 200–300 farmers in Chakera.

Given farmers' dissatisfaction with the treated wastewater generated by pond-based treatment systems, alternative treatment options as well as other risk mitigation measures could be introduced to farmers (WHO, 2006), and linked to the sale of wastewater (or as incentive to a free water allocation).

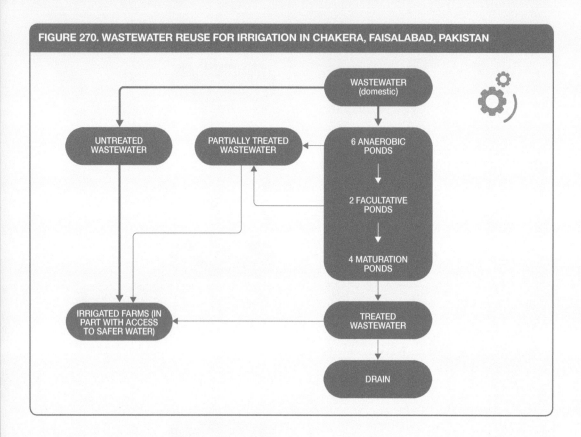

FIGURE 270. WASTEWATER REUSE FOR IRRIGATION IN CHAKERA, FAISALABAD, PAKISTAN

Funding and financial outlook

In general, the Water and Sanitation Agencies across Pakistan struggle to collect sufficient operating revenue to pay for their operating costs. Also the contribution of wastewater auctioning (Table 62) is financially negligible but allows WASA to maintain control over the wastewater use and achieve some O&M cost recovery with limited transaction costs. The annual income from leasing of land and auction of wastewater were in 2012–2013 about USD 45,000 which is more than the O&M budget of the pond-based sewage treatment system in Faisalabad and about 15% of the total O&M costs of the related sewer system in Faisalabad West. Compared to the revenues from sewerage charges which were in 2012–2013 about USD 2m, the amount is however very modest (WASA, 2014).

Because of the application of nutrient-rich wastewater, wastewater farmers in Chakera save on fertilizer; and although more pesticides are needed, the net gains are so substantial, that WASA could increase its water fees.

Socio-economic, health and environmental impact

The auctioning process as described in Box 15 has been considered as socially fair as it helps also small holders to access a share of the available water. As long as the water is not sufficiently treated, health risks remain. This is in part an accepted professional risk, as the benefit-cost ratio was significantly higher for wastewater than freshwater farmers, as well as farmers using untreated compared to treated wastewater (IWMI, 2009; Baig et al., 2011; Clemett and Ensink, 2006). Similar situations were also reported from other cities (Van der Hoek et al., 2002; Hussain et al., 2002).

TABLE 62. SUMMARY OF IRRIGATION WATER SOURCES AND INSTRUMENTS USED TO ALLOCATE WATER

SOURCE OF WATER	INSTRUMENT USED	AREAS IRRIGATED (HA)	NUMBER OF FARMERS
Canal (paid to Irrigation Department)	Flat rate per acre (varies by season)	12–15	5
Treated wastewater	Priced via land (unit) lease	25–35	Ca. 20
Partially treated (incl. mixed wastewater / fresh water)	Auction (bulk)	85–90	40–50
Untreated	Auction (bulk)	457	150–175

Sources: Weckenbrock, 2011; IWMI, 2014, unpublished.

Overall production costs were highest for freshwater farmers, especially if ground water pumping was required, and lowest were where the wastewater replace fertilizer needs. In addition, farmers using wastewater were able to produce more crops per year, including vegetables which require daily watering and care, and created more jobs. Where vegetables are grown, they are usually cooked which is reducing possible health risk for consumers.

Negative externalities relate mostly to pathogens (especially hookworm infections) and too high nitrogen levels for certain crops, like root crops. Risk reduction measures against hookworm infections, like protective footgear or de-worming campaigns are so far insufficiently used to reduce risks for farmers.

Scalability and replicability considerations

With increasing population and food demands, it appears inevitable that demand for water and its reuse will expand in Pakistan. Given the slow growth of the wastewater treatment sector, untreated or partially treated wastewater will continue to be the leading source of water (Ensink et al., 2004ab). The lessons from the well-researched Faisalabad case offer authorities an opportunity for engagement with farmers to provide regulatory oversight and bring options for health risk reduction into the business discussion. Especially the auction model has a high potential for this given its low transaction costs. Thus, WASA could be flexible in view of financial cost recovery, while targeting farmers' buy-in in risk reduction options which will probably easily pay off for the city in terms of reduced public health expenditures. Replication of wastewater auctions have been seen in other villages across WASA's jurisdiction as well as in India (Box 16). The wastewater auction model could thus be scaled across the region in all those locations where authorities see the livelihood and food security opportunities that wastewater with/out treatment offers as long as farmers and authorities can jointly work on modalities like risk mitigation. Dialog between authorities and farmers can also address other issues: The increased water and nutrient availability calls for changes in cropping patterns and fertilizer rates where farmers might need assistance. On the other hand, many treatment plants are poorly sited when it comes to optimizing their water reuse potential.

Support by the private sector will be needed where industrial effluent is mixed with domestic wastewater, which is according to Weckenbrock et al. (2011) so far not a problem. While pathogenic risks from domestic wastewater can be addressed also on farm via so called 'non-treatment' options (Amoah et al., 2011; Mara et al., 2010), chemical contaminants from industrial origin require conventional treatment, ideally at source. The willingness of the Pakistani private sector to accept this responsibility is high given its wish to comply with European import requirements (see **Corporate Responsibility Model 22** in this publication).

Box 16. Direct and indirect wastewater auctioning

Also in **India's Gujarat**, many cities reportedly sell access to treated and untreated wastewater for use in agriculture, and also auctioning is common (Palrecha et al., 2012). The use of wastewater is recognized by the Government of Gujarat and water charges are being collected at the same rates as applicable for lifting water from notified rivers. Competition is high for its assured availability and nutrient value. Wastewater auctions are held annually, for example in the Kutch district. In the villages of Anadpur (Yaksh), Mota Dhavda and Sanyara, wastewater is auctioned annually at USD 100–200 for irrigating 2 to 6 ha in each village. With increasing demand for freshwater in cities, there have been trade-offs between farmers and the cities for availing freshwater in exchange for wastewater. There exists an MOU, which was signed between the farmers of a wastewater cooperative in Rajkot and the Rajkot Municipal Corporation since around 1970 according to which farmers are not allowed to lift water from Lalpari Lake for irrigation to allow supply to Rajkot city. In exchange, wastewater is supplied to the farmers by the Corporation, similar to the Iran case under Model 20 in the catalogue. This MOU is still operational.

Like in Pakistan, many farmers in Gujarat irrigate with wastewater despite having the option of groundwater irrigation because they see wastewater as (a) more reliable and accessible throughout the year; (b) cheaper to lift; and/or (c) more profitable because of its nutrient value leading to higher yields and savings in fertilizer input costs (Palrecha et al., 2012; www.peopleincentre.org/PiC/?p=748). Also the nutrient value of fecal sludge has been recognized, with reported sludge auctioning in Karnataka, India (see Business Model 17).

Auctioning replenished groundwater has been described from other countries as a new perspective for financing water reuse with the issuance of groundwater credits where treated municipal wastewater effluent is used to recharge groundwater. In Prescott Valley, in the state of Arizona in the **USA**, groundwater credits created in such a manner was auctioned in November 2007 with the help of WestWater Research. The winning bidder, Water Property Investors LLC (a New York-based water resource investment firm) agreed to pay USD 67 million in total annually (USD 20.16/m^3) for the right to withdraw 3.3 million m^3/yr from the ground. Prior to finalization of bids, Aqua Capital Management LP agreed to pay USD 53 million for the equivalent rights, which guaranteed the floor price of the auction (GWI, 2010).

Potential risks and mitigation

In designing any business model, it is assumed that generic business risks are known and will be taken care of. However, some risks might be more model specific and will be acknowledged in the following:

Market risks: Most farming locations where wastewater is formally or informally used are in close proximity to major urban markets and well positioned to respond quickly to market needs, save on transport costs and deliver high-value crops also in the lean season when revenues peak. As crops produced with wastewater or freshwater are mixed in markets and risk awareness along the food chain is commonly low, market related risks are limited.

Competition risks: Only with increasing risk awareness, the potential of competition from freshwater farmers could be growing. So far this awareness is in most low-income countries limited.

Technology and performance risks: WHO approved low cost wastewater treatment and non-treatment options are available which either treat the water before reuse, or on-farm accompanied by safe irrigation practices and post-harvest safety measures (Amoah et al., 2011). Care has to be taken that employed technologies like in the presented case study, do not have side effects (nutrient loss, salinity increase) which are not accepted by farmers.

Political and regulatory risks: The regulatory framework has to acknowledge wastewater as a commodity with value, which water authorities can market. This can give authorities bargaining power to lobby for safety practices.

Social equity related risks: The share of men and women in the informal irrigation sector differs between countries and cultures. In Pakistan, women have significant difficulties accessing irrigation water, and less responsibilities in irrigation than men. In other countries, it can be the opposite, with changing roles along the value chain. However, under the current auctioning conditions, no (additional) gender related discrimination has been reported, although the process is male dominated while there are value chain advantages for women.

Safety, environmental and health risks: The model applies the WHO (2006) recommendation of a step-wise and stakeholder inclusive approach to risk mitigation which is an intermediate step until (a) more comprehensive wastewater collection and treatment systems are in place, which farmers also can accept; and (b) stricter safety guidelines can be implemented and enforced. Within this trajectory, risks, risk monitoring and risk mitigation measures remain important (Table 63) even if the dialog between authority and farmers will lead to the adoption of farm-based risk reduction measures (safer irrigation practices, on farm water treatment, crop restrictions).

TABLE 63. POTENTIAL HEALTH AND ENVIRONMENTAL RISK AND SUGGESTED MITIGATION MEASURES FOR BUSINESS MODEL 23

RISK GROUP	EXPOSURE					REMARKS
	DIRECT CONTACT	AIR/ ODOR	INSECTS	WATER/ SOIL	FOOD	
Farmers	HIGH RISK	MEDIUM RISK	MEDIUM RISK			After introduction of farm based risk reduction measures, their adoption has to be monitored.
Community		MEDIUM RISK	MEDIUM RISK			
Consumer					HIGH RISK	
Mitigation measures	🧤	😷 🌳	🦟	Pb Hg Cd	🚿 🍲 Pb Hg Cd	

Key: ☐ NOT APPLICABLE ☐ LOW RISK ■ MEDIUM RISK ■ HIGH RISK

SWOT analysis and business performance

The strength of this business case is its ability to develop cooperative working relationship with farmers based on principles of mutual interest (Figure 271). By this approach, the authorities from the

WASA are able to negotiate land rent with farmers and implicitly determine the price of wastewater with the farmers leasing land from WASA. Another significant strength is the application of auction mechanism in setting the price of wastewater and thus following market principles in determining the price of wastewater.

While joint business between wastewater suppliers and users might be common where treated wastewater is offered to farmers, the case of WASA auctioning untreated wastewater should allow negotiating crop restrictions and other on-farm risk mitigation options. Another entry point for safety regulations is that farmers actively engaged in the discussion about water, organized themselves and did not shy away from legal battles. Also this offers opportunities to formalize the otherwise informal wastewater use.

The institutional linkages between both parties go far beyond other situations and countries where the use of untreated wastewater is considered illegal or authorities do not engage at all, or only with disciplinary action. WASA's engagement, however, and the acceptance of wastewater as a marketable commodity provides authorities with an instrument for the **introduction of a variety of possible risk mitigation options**, which are recommended by the WHO (2006) for the safe use of wastewater

FIGURE 271. SWOT ANALYSIS OF CHAKERA VILLAGE BUSINESS CASE DESCRIPTION

	HELPFUL TO ACHIEVING THE OBJECTIVES	**HARMFUL** TO ACHIEVING THE OBJECTIVES
INTERNAL ORIGIN ATTRIBUTES OF THE ENTERPRISE	**STRENGTHS** • Wastewater accepted as commodity • Low O&M of the treatment plant • Cooperative working relationship (dialog between partners) • Service attitude • Auction pricing mechanism • Continuous supply of wastewater	**WEAKNESSES** • Inter-authority dialog on a joint reuse strategy for minimizing risks while maximizing benefits missing • Treatment process (e.g. retention time) needs to be adjusted to avoid unnecessary changes in salinity and nitrogen levels which do not serve reuse objectives (farmers prefer untreated wastewater over treated wastewater)
EXTERNAL ORIGIN ATTRIBUTES OF THE ENVIRONMENT	**OPPORTUNITIES** • High agricultural wastewater demand and legal support of water reuse • Wastewater sharing arrangements • Freshwater savings • Auctioning gives an instrument for dialog on risk reduction • Urban population growth resulting in increasing food (and agricultural water) demand • Occupational and consumer health risks from pathogens can be minimized even without treatment plants if WHO (2006) guidelines are followed	**THREATS** • Farmers exposed, e.g. to hookworms • Possible human health risk may decrease demand for crops • Potential conflicts due to worsening water scarcity and higher agric. demand than domestic wastewater supply • Changes in policy on reuse • Untreated industrial effluents may increasingly comingled with domestic wastewater • Urbanization and loss of farmland

in agriculture. The market value of wastewater allows in addition the introduction of incentives (like extra water allocations) in exchange of compliance with safety measures.

WHO had realized with their 2006 guidelines that their 1989 water quality thresholds for wastewater irrigation as main risk barrier was in many countries unachievable due to the overwhelming challenge of wastewater collection and treatment. Therefore, WHO (2006) started emphasizing alternative risk barriers to protect farmers and consumers. So even where untreated wastewater is used, there are still multiple options for risk minimization, which include safe irrigation practices (Amoah et al., 2011; Mara et al., 2010).

A particular challenge in Faisalabad is that farmers prefer untreated wastewater compared to treated effluent. While untreated wastewater is considered high in nutrients and low in salinity, treated wastewater is considered low in nitrogen and – due to evaporation in the treatment ponds – high in salinity. In fact, the few recorded negative perceptions related to wastewater usually concern more plant growth than human health (Weckenbrock, 2011). A dialog between authorities and farmers should address these perceptions targeting a redesign of the local treatment plant's retention time. The hope is that participatory planning will lead to mutually acceptable standards for water quality and solutions for wastewater risk reduction which could become part of the business deal (Clemett and Ensink, 2006).

The innovative capacity of the model lies in the opportunity of a dialog on the trajectory from unsafe to safe wastewater use, with a relatively high scalability and replication potential due to its low costs. With neighbouring villages investing in pipes to access the wastewater, the models appear scalable as far as water is available, and replicable where policies support a dialog that helps negotiating risk mitigating measures (Figure 272). While the auctioning is not influencing water quality and the environment, it could help the stakeholder dialog which is central to any Sanitation Safety Plan (WHO 2015).

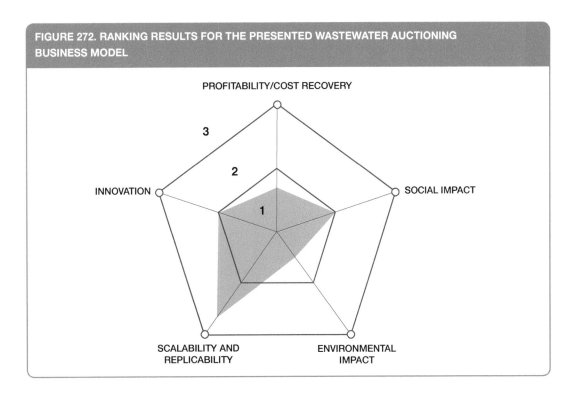

FIGURE 272. RANKING RESULTS FOR THE PRESENTED WASTEWATER AUCTIONING BUSINESS MODEL

References and further readings

Amoah, P., Keraita, B., Akple, M., Drechsel, P., Abaidoo, R.C. and Konradsen, F. 2011. Low cost options for health risk reduction where crops are irrigated with polluted water in West Africa. IWMI Research Report Series 141. Colombo: IWMI.

Baig, I.A., Ashfaq, M., Hassan, I., Javed, M.I., Khurshid, W. and Asghar, A. 2011. Economic impacts of wastewater irrigation in Punjab, Pakistan. Journal of Agricultural Research 49 (2): 261–270.

Clemett, A.E.V. and Ensink, J.H.J. 2006. Farmer driven wastewater treatment: A case study from Faisalabad, Pakistan. 32nd WEDC International Conference, Sustainable Development of Water Resources, Water Supply and Environmental Sanitation, Colombo, Sri Lanka, proceedings, p. 99–103.

Ensink, J.H.J.; Mahmood, T.; van der Hoek, W.; Raschid-Sally, L.; Amerasinghe, F. 2004a. A nationwide assessment of wastewater use in Pakistan: an obscure activity or a vitally important one? Water Policy, 6(3):197–206.

Ensink, J.H.J., Simmons, R. and van der Hoek, W. 2004b. Wastewater use in Pakistan: The cases of Haroonabad and Faisalabad. In Scott, C. A.; Faruqui, N. I.; Raschid-Sally, L. (Eds.), Wastewater use in irrigated agriculture: Confronting the livelihood and environmental realities. CABI Publishing. pp.91–99.

Ensink, J.H.J. 2006. Wastewater quality and the risk of hookworm infection in Pakistani and Indian sewage farmers. PhD thesis, London University, London, UK.

EPMC. 2002. Master plan for urban wastewater (municipal and industrial) treatment facilities in Pakistan. Final Report. Lahore: Engineering, Planning and Management Consultants.

Global Water Intelligence (GWI). 2010. Municipal water reuse markets 2010. Oxford: Media Analytics, Ltd.

Hussain, I., Raschid, L., Hanjra, M.A., Marikar, F. and van der Hoek, W. 2002. Wastewater use in agriculture: review of impacts and methodological issues in valuing impacts. Colombo, Sri Lanka: International Water Management Institute (IWMI). v, 55p. (IWMI Working Paper 037).

International Water Management Institute (IWMI). 2009. Ensuring health and food safety from rapidly expanding wastewater irrigation in South Asia. Final Project Report. Colombo, Sri Lanka: IWMI. http://publications.iwmi.org/pdf/H042649.pdf.

Mara, D., Hamilton, A., Sleigh, A. and Karavarsamis, N. 2010. Updating the 2006 WHO guidelines. More appropriate tolerable additional burden of disease. Improved determination of annual risks. Norovirus and Ascaris infection risks. Extended health-protection control measures: Treatment and non-treatment options. Information kit for the third edition of the WHO Guidelines for the safe use of wastewater, excreta and greywater in agriculture and aquaculture. Vol. 2. Geneva: WHO-FAO-IDRC-IWMI. www.who.int/water_sanitation_health/publications/human_waste/en/ (accessed 6 Nov. 2017).

Ministry of Environment. 2006. National sanitation policy. Islamabad, Pakistan: Government of the Islamic Republic of Pakistan.

Murtaza, G. and Zia, M.H. 2012. Wastewater production, treatment and use in Pakistan. Background paper prepared for UN-Water SUWA (Safe Use of Wastewater in Agriculture). www.ais.unwater.org/ais/pluginfile.php/232/mod_page/content/134/pakistan_murtaza_finalcountryreport2012.pdf (accessed 6 Nov. 2017).

Palrecha, A., Kapoor, D. and Malladi, T. 2012. Wastewater irrigation in Gujarat. An exploratory study. IWMI-TATA Water Policy Research Highlight Vol. 30. https://iwmi-tata.blogspot.in/2012/11/2012-highlight-30.html and https://youtu.be/qZUxptX1018 (accessed 6 Nov. 2017).

Syed, A.R., Yasir, A. and Farhan, M. 2012. Role of agriculture in economic growth of Pakistan. International Research Journal of Finance and Economics 82. https://mpra.ub.uni-muenchen.de/32273 (accessed 6 Nov. 2017).

Van der Hoek, W., Hassan, M.U., Ensink, J.H.J., Feenstra, S., Raschid-Sally, L., Munir, S., Aslam, R., Ali, N., Hussain, R. and Matsuno, Y. 2002. Urban wastewater in Pakistan: A valuable resource for agriculture. Research Report 63. Colombo, Sri Lanka: IWMI.

WASA. 2014. Budget 2013–14, and revised 2012–13. Faisalabad, Pakistan: Water and Sanitation Agency (WASA). http://wasafaisalabad.gop.pk/assets/uploads/Documents/6/wasa_budget_2013-14_final.xls.pdf (accessed 6 Nov. 2017).

Weckenbrock, P., Evans, A., Majeed, M., Ahmad, W., Bashir, N. and Drescher, A. 2011. Fighting for the right to use wastewater: What drives the use of untreated wastewater in a peri-urban village of Faisalabad, Pakistan? Water International 36 (4): 522–534.

Weckenbrock, P. 2011. Making a virtue of necessity – Wastewater irrigation in a periurban area near Faisalabad, Pakistan. A GIS based analysis of long-term effects on agriculture. Freiburger Geographische Hefte, Heft 68: 198pp.

WHO. 2006. Guidelines for the safe use of wastewater, excreta and greywater, volume 2: Wastewater use in agriculture. Geneva: World Health Organization.

WHO. 2015. Sanitation safety planning manual (Manual for the safe use and disposal of wastewater, greywater and excreta). Geneva: World Health Organization.

Case descriptions are based on primary and secondary data provided by case operators, insiders or other stakeholders, and reflect our best knowledge at the time of the assessments 2014/15. As business operations are dynamic data are likely subject to change.

Notes

1 Groundwater is of poor quality in the area of Chakera. In general, accessing groundwater is costlier than wastewater, either because of expensive tube well installation, maintenance and pumping fuel prices, or because of paying tubewell owners (Weckenbrock, 2011).
2 When asked about the reason why farmers still pay for freshwater in spite of not having received any for decades, several farmers indicated that the amounts are low and they simply preferred not to instigate trouble (Weckenbrock, 2011).
3 Some farners can only access treated wastewater due to the local topography.

The case has been dedicated to Dr Jeroen Ensink (1974–2015) who worked with IWMI in Faisalabad on safe wastewater irrigation for many years.

www.justgiving.com/fundraising/The-Jeroen-Ensink-Memorial-Fund

BUSINESS MODEL 24
Farmers' innovation capacity as driver of change

Sena Amewu, Solomie Gebrezgabher and Pay Drechsel

	Model name	Farmers' innovation capacity as driver of change
	Waste stream	Domestic grey water, wastewater-polluted stream water
	Value offer	Partially treated wastewater for crop irrigation
	Geography	Suburban low/wetlands used by farmers
	Scale of business	Small scale (community)
	Location of supporting cases	This model is in its presentation a hybrid of case and model and builds on observation in particular in Southern Ghana
	Objective of entity	Social/Environmental enterprise [X]
	Investment cost range	USD 15,000–25,000
	Organization type	Community based organization
	Major partners in the case example	Farmer association, Friends of Ramsar Site (NGO), Environmental Protection Agency, Wildlife Division of Forestry Commission, local assemblies, UNEP, Ministry of Food and Agriculture (MoFA)
	Socio-economic impact	Fresh food for urban households. Every USD invested in on-farm treatment and post-harvest safety returns up to USD 4.9 in public health cost savings
Gender equity	Generally balanced, but gender roles vary along value chain	

Executive summary

This business model has been informed by observations from wastewater using farming communities in India and West Africa where farmers show a significant innovative spirit to adapt either to declining water quality (Buechler and Mekala, 2005) or challenges in accessing water (IWMI, 2008a). The here presented model is based on a distinct example from Ghana and follows in its presentation a **hybrid of the business model and case templates**.

This example derives from Ghana's coastal region where farmers struggle with poor water quality and their irrigation infrastructure supports natural water remediation processes. Although risk reduction is not the main driver, the system supports the public interest in water (food) safety and forms a first step transition from informal to formal wastewater use. The farming site is located between several smaller, essentially temporary streams, which feed into the Sakumo lagoon in the densely populated

Accra-Tema mega-polis of Southern Ghana. There is very limited sewerage and wastewater treatment in this suburb and the streams carry highly polluted water from a wider urban catchment area, generated by households, trade and small industry. Since 1992, the 1360 ha wetland area around the lagoon is protected under the Ramsar convention[1]. About 414 ha of the land are used for irrigating traditional vegetables, with increasing shares of rainfed maize in the rainy season. The informal irrigation system as designed by the farmers combines gravity flow (also by blocking streams), canals or PVC pipes, and smaller storage ponds (dugouts), as well as portable water pumps. The system is designed to reduce the burden of carrying water over longer distances. Based on farmers' original efforts of creating storage facilities, the local community based NGO Friends of Ramsar Site (FORS) suggested in 2011 to upgrade the created canals and ponds into a designed natural treatment system. Farmers invested labor and cash to the tune of USD 3,600 while FORS secured from UNEP an additional amount of about USD 13,200 to upgrade the system with four smaller constructed wetland lagoons. Currently, more than 200 farmers are settled around the site, supported by a much larger number of seasonal labor. It is estimated that farmers generate a gross revenue of about USD 200,000 annually from the production of crops on the overall site with a high benefit-cost ratio[2]. As only a section (max. 30%) of the farmers was able to connect to the treatment system, FORS plans its extension. This has, however, to be accompanied with awareness creation on health risks, for farmers and consumers, to create more market demand for safer produce as further incentive for the farmers to engage in the innovation. In 2014, due to severe flooding and damage of infrastructure, the system stopped functioning and was still not operational again early 2017.

KEY PERFORMANCE INDICATORS OF THE CASE IN GHANA (2014)					
Land use:	414 hectares (1022 acres) of irrigated land of which about 30% were connected to the treatment system in 2012/13				
Water treated:	0.6–1.2 million cubicmeter (MCM) per year assuming 1–2 60-day cropping cycles				
Capital investment:	Ca. USD 16,800				
Labor requirements:	12–20 people needed for dredging (dredging done 2–3 times a month)				
O&M:	Up to USD 1,200 per season distributed over the local farmer association				
Output:	Partially treated wastewater for irrigation and in part livestock watering				
Potential social and/or environmental impact:	With the planned extension up to 200 crop farmers (80–90% men) and an estimated 400 seasonal laborers (60% women) could benefit from access to partially treated water. The production of safer food benefits consumers in Tema and Accra, especially for food items eaten uncooked, and the overall site which as a traditional as well as tourist value				
Viability indicators:	Payback period:	N.A.	Post-tax IRR:	N.A.	Gross margin:	N.A.

Context and background

Due to limited wastewater collection and treatment, urban streams are across sub-Saharan Africa heavily polluted and mostly conveying domestic greywater, solid waste but also overflow from septic tanks, pushing especially pathogenic water quality indicators far above acceptable levels. The poor water quality is an increasing burden for farmers who depend on irrigation, as well as the environment as also shown on the example of the Sakumo Lagoon (Asmah et al., 2008; Agbemehia, 2014) near Accra. This wetland of international importance, which was declared in 1992 as a Ramsar site[3], covers an area of about 1,360 hectares and is situated between Ghana's capital city and Tema in the Greater Accra Region of Ghana. The size of the open lagoon varies between 100 and 350 hectares depending on the season. Four sub-basins are supporting the freshwater supply of the site: the major ones, named after their streams, are the Mamahuma and Onukpa Wahe (at the western side) and the

Dzorwulu and Gbagbla-(An)konu (situated at the northern end). The Eastern and Southern subbasins constitute minor inflows. The main feeder streams, the Dzorwulu and Mamahuma have been dammed upstream of the Ramsar site and re-channeled for local irrigation, such as the Dzorwulu stream which supports the well-known Ashaiman reservoir and irrigation scheme. The damming has resulted in very little influx of freshwater, that especially during the dry season wastewater dominates the flow. The streams are draining a wide urban catchment area capturing mostly domestic wastewater and storm water, but also effluents from lighter industry.

Ramsar administrative authority in Ghana is the Wildlife Division of the Forestry Commission. Farming and fishing are permitted and date back as long as farmers can recall. In 2010, the local farmer association *'Resource Users Association'* invested major efforts in improving water access, especially in the dry season, including a larger storage pond which can be connected to several farms. Farmers contributed labor and USD 3,600 in cash. In a subsequent development, the Friends of Ramsar Site (FORS), a non-profit organization, mobilized about USD 13,200 from UNEP to upgrade the treatment potential of the canals and pond system the farmers put in place via constructed wetlands (lagoons). The potential for high synergies between infrastructure in farmers' interests and natural pathogen elimination have been described for other sites in Accra by IWMI (2008a,b) and by Keraita et al. (2014), which offers a possible pathway in support of a gradual transition towards safer wastewater irrigation as supported by WHO (2006).

There are about 600 ha under farming of which around 414 ha are irrigated by at least 200 farmers supported by about 400 seasonal laborers. The major crops grown include fresh vegetables such as cucumber and green pepper, local vegetables (like okra, pepper, onion, tomatoes, ayoyo) and maize that are all in high demand in Accra. About 30% of the farmers were so far connected to the natural treatment system while the majority continues using untreated wastewater, but there are plans by FORS to increase the number of users by expanding the treatment system. The type of water used by farmers still depends mostly on convenience and pumping costs, not on risk awareness. Urban farmers are generally more concerned with visible trash (e.g. plastic) in the water while missing knowledge of invisible contaminants (Keraita et al., 2008). However, farmers at Sakumo indicated that the appearance and bad smell sometimes emanating from the wastewater is a challenge to them that they stopped irrigating a few days before harvest[4]. Sensory attributes such as the crop appearance, neatness and size rather than possible invisible health risks are also common among traders and consumers and reflect the common educational status (Keraita and Drechsel, 2015).

The Sakumo area received annually about 800 mm rain and has high educational (e.g. bird watching) and recreational value, being one of the few 'green' areas left in the rapidly expanding Accra-Tema metropolitan area. The lagoon is moreover regarded as a fetish by the local people and the local Black Heron bird is considered sacred.

Macro-economic environment

With an upsurge of both wastewater generation and irrigated urban farming, options which can increase produce and farmers' safety are needed across sub-Saharan Africa.

Urbanization and the growing urban demand for food are driving year-round food production which requires irrigation in the dry season(s). While some crops can be produced in irrigation schemes in rural areas and with safe freshwater, other crops are easily perishable and urban proximity is favored due to the lack of cold transport and storage but also as shorter food chains give financial advantages. However, urban proximity has also disadvantages. As at least 80% of the wastewater generated in Ghana's urban centers is released into the environment in its untreated form, making it nearly impossible for farmers to find any unpolluted water source (Drechsel and Keraita, 2014). Groundwater access could be one option but seldom in ocean vicinity and also not at Sakumo (Agyepong, 1999). In Ghana, there are no data to tell where along natural streams contamination levels exceed irrigation thresholds. Without ability to monitor water quality or offer farmers a viable alternative, irrigated urban farming with its obvious benefits but also health risks remains in a state of "laissez-faire" without enforced restrictions or serious assistance (Drechsel et al., 2006; Drechsel and Keraita, 2014). The national irrigation policy (MoFA, 2011) permits safe wastewater use in line with the 2006 edition of the WHO-FAO-UNEP water reuse guidelines which demand for situations without treatment plant alternative risk barriers from 'farm to fork' (Amoah et al., 2011). The importance of urban farming in this context should not be underestimated: Lydecker and Drechsel (2010) estimated that in Accra more wastewater is 'treated' on-farm than in designated treatment plants.

Business model

The business is run by the Resource Users Association, a commercial farmers group producing crops for the local market. The value proposition of their and FORS co-investment is improved water access combined with reduced health risk despite the use of polluted irrigation water (Figure 273). Although the initial main driver of this business model was to access water for irrigation all year around, the private sector-NGO partnership added the safety objective.

The drive to get access to water has catalyzed farmers to invest jointly in the pond and canal system, a system which supports natural water remediation processes and can easily be combined with further safety enhancing features (cf. IWMI, 2008a,b; Keraita et al., 2014). The partnership with FORS created a win-win situation whereby the irrigation water receives a pre-treatment, farmers who like to join the association get access to water also in the dry season, and consumers are a step closer to safer food. The farmer association can be considered as owners of the wastewater treatment system as they invested both cash and labor for the construction of the system and are paying for its O&M costs. The farmers' association is now registered in the registrar general and has a constitution which explains the responsibilities of each member with regards to the wastewater treatment system[5]. The cost to maintain the system are borne by farmers as the situation arises, i.e. they don't pay regular fees for using the water but when there is a need, farmers are required to contribute. This is the case after seasonal flooding when the self-made dams blocking the river are commonly destroyed. If the farmer fails to contribute, the association will give a warning to the farmer to make the payment.

Normally, farmers understand that if the system does not work, they will not be able to get water. But in instances where a farmer fails to contribute to the maintenance of the system, the association can seize the farmer's water pump.

Value chain and position

The Ramsar wetland is used for different productive uses such as for crop farming, livestock rearing and fishing (Figure 274). Initially farmers had no alternative to using highly polluted stream water. An alternative option was created by the Resource Users Association and FORS which enabled farmers to use partially treated wastewater, and the lagoon to receive less floating debris. Although so far not all farmers at the Ramsar site can connect to the treated wastewater and traders still receive crops produced with untreated water, there are plans by FORS to increase the number of users by expanding the treatment system.

FIGURE 273. BUSINESS MODEL CANVAS – FARMERS' INNOVATION AS DRIVER OF CHANGE

KEY PARTNERS
- Friends of Ramsar Site (FORS)
- Environmental Protection Agency (EPA)
- Wildlife division
- UNEP
- Ministry of Food and Agriculture (MoFA)
- Local community of Sakumono
- Research partner or technical advisor

KEY ACTIVITIES
- Installing and maintaining a water storage cum treatment facility
- Advocacy by FORS
- Ensuring that farmers adhere to rules stipulated in the farmer's association's constitution

KEY RESOURCES
- Diluted wastewater
- Technical expertise
- Constructed wetland
- Macrophytes for natural water treatment
- Financing, labor

VALUE PROPOSITIONS
- Year-round access to safer water for irrigation than so far available

CUSTOMER RELATIONSHIPS
- Formal relationship between farmers and the farmer association
- Personal relationships with crop buyers (traders), indirect with consumers

CHANNELS
- Direct use of wastewater by farmers
- Depending on crop direct marketing or on-farm sale of produce

CUSTOMER SEGMENTS
- Farmers cultivating irrigated crops
- Public/authorities calling for safe produce
- Crop traders

COST STRUCTURE
- Capital investment by the Resource Users Association
- Capital investment by FORS
- Operation & Maintenance cost by farmers

REVENUE STREAMS
- Cash and in kind contribution by farmers for system set up and O&M; no payment for water from streams which is free in Ghana

SOCIAL & ENVIRONMENTAL COSTS
- Possible increase in mosquito bites due to constructed wetlands (but as the whole area is a wetland, the added risk is marginal)

SOCIAL & ENVIRONMENTAL BENEFITS
- Compared with 'no intervention' possible risks to farmers, soils, crops and consumers will be reduced, but not eliminated
- Enhanced food security, possibility of connecting more farmers
- Partial removal of plastic waste which will benefit tourists and the local community around the main Lagoon

BUSINESS MODEL 24: FARMERS' INNOVATION AS DRIVER OF CHANGE

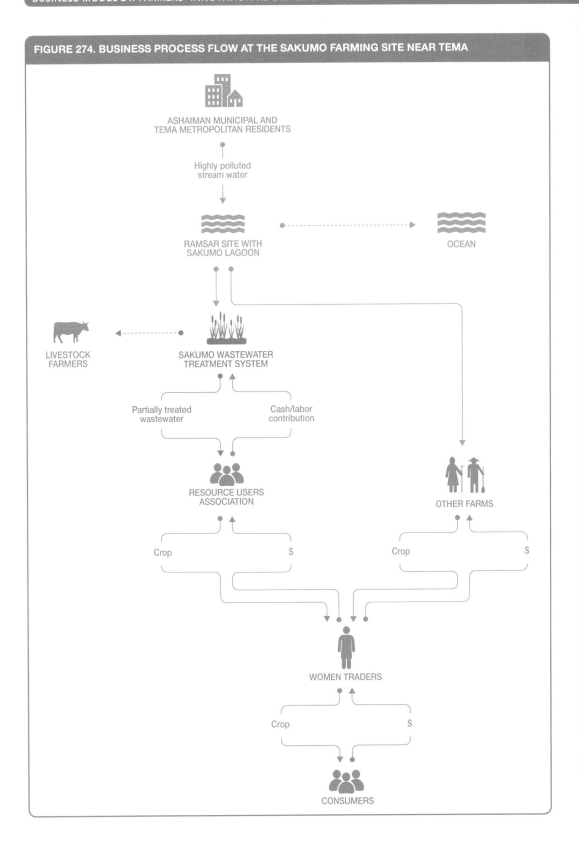

FIGURE 274. BUSINESS PROCESS FLOW AT THE SAKUMO FARMING SITE NEAR TEMA

Institutional environment

A set of policies and development plans provides the legal context for the institutional arrangements at the Sakumo Ramsar site near the community of Sakumono. The Ramsar site was created in 1992 by the legislative instrument (LI) 1659 and classified as an environmentally sensitive area under the Ghana Environmental Assessment (EA) regulation, legislative instrument (LI) 1652 of 1999. The National Land Policy of 1999 allows for the agricultural cultivation of wetlands provided its productivity is sustained. The Ministry of Local Government and Rural Development under the Ghana National Urban Policy Action Plan of 2012 recommends the development and use of open spaces, green belts and ecologically sensitive areas for urban farming. The common use of 'wastewater' in this context has been acknowledged in Ghana's National Irrigation Policy, Strategies and Regulatory Measures which recognized the relevance of the informal irrigation sector, and recommends compliance with the WHO (2006) wastewater use guidelines. Guidelines for the protection of the wetland are given in Ghana's National Wetlands Conservation Strategy and Action Plan (2007–2016).

The various institutions involved at the site and their roles include:
- *The Wildlife Division of the Forestry Commission under the Ministry of Land Forestry and Natural Resources* – responsible for the management of the Ramsar site, and helps to resolve conflicts between resident and seasonal farmers.
- *The Environmental Protection Agency* – responsible for monitoring and preventing of the pollution from the surrounding areas also as the Ramsar site is officially an environmentally sensitive area[6].
- *Tema Metropolitan Assembly* – is the city authority responsible for enforcing laws/bylaws and legislations concerning the site.
- *The Ministry of Food and Agriculture* – provides extension services to the farmers to guide and provide advice on agricultural input use and farming practices.
- *Resource Users Association* – a farmer association which had in 2014 about 75 members (13% women) use partially treated wastewater for irrigation at the site and which contributed in the construction and maintenance of the treatment system[7].
- *Friends of Ramsar Site* (FORS) – a non-governmental organization and advocacy group that helped to construct the wastewater treatment system, is responsible for its management and actively lobbies for the protection of the Sakumo site.
- *UNEP* – co-funded the construction of the wastewater treatment system and local tree planting[8].
- The surrounding communities such as Klagon, Sakumono, Community 3 and 19, and Nungua; their assemblies and traditional chiefs.

The local NGO FORS plays in this case a prominent role as broker between the different parties. However, for any replication of the case, FORS represents only one of many opportunities of local communities to engage and support their wetland and open farming areas in an urbanizing environment based on their various direct and indirect benefits (see also Lydecker and Drechsel, 2010).

Technology and processes

The water treatment at the Ramsar site (Figure 275) is based on natural processes (pathogen die-off, sedimentation, nutrient uptake, physical barriers, ...) where stream water is temporarily blocked and redirected through channels to four treatment ponds (100m^2 lagoons). The macrophytes *Pistia* (water lettuce), *Ipomoea* (water spinach) and *Ludwigia* (water primrose) are growing in the first three lagoons respectively while the fourth lagoon exposes the polluted water to sunlight. Eventually the water flows into a reservoir from where it is pumped onto the farms while excess water flows through a canal into the Sakumo lagoon and then into the sea. From time to time, the macrophytes are harvested and composted to fertilize the soil.

BUSINESS MODEL 24: FARMERS' INNOVATION AS DRIVER OF CHANGE

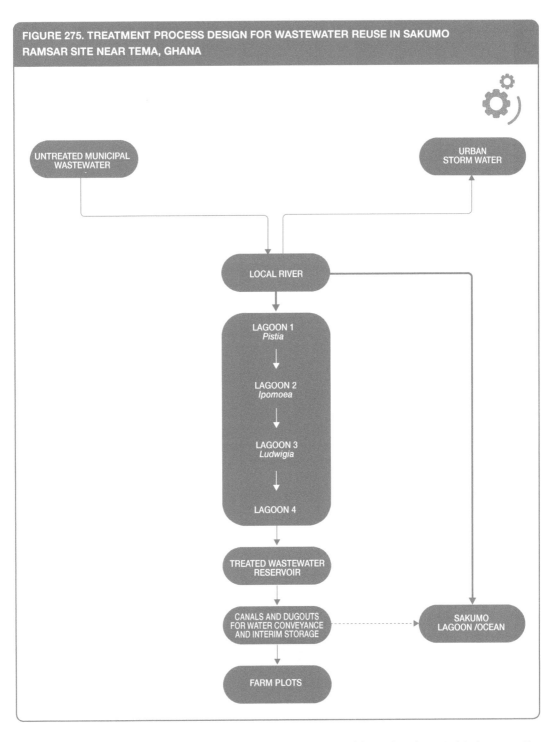

FIGURE 275. TREATMENT PROCESS DESIGN FOR WASTEWATER REUSE IN SAKUMO RAMSAR SITE NEAR TEMA, GHANA

First laboratory data showed that the system could be improved (retention time etc.) to increase the treatment quality. On another site in Accra at La, a farmer based cascade of small reservoirs showed a positive impact on pathogen levels (IWMI, 2008a). FORS is actively seeking collaboration with research institutions to optimize the system.

There are other examples, e.g. from India, showing how typical irrigation infrastructure can support water treatment processes, in particular the removal of pathogens (Ensink et al., 2010).

Funding and financial outlook

The generated capital investment for the wastewater treatment system was about USD 16,800, contributed by farmers and FORS. The investment took place in three phases:
- In 2010 a total of about USD 700 and labor for dredging was contributed by farmers.
- In 2011 about USD 2,900 was contributed by the farmers.
- In 2011 UNEP provided funding of USD 13,200 via EPA to FORS to work on the treatment ponds.

Maintenance of the system is done by the farmers. Dredging and removal of floating waste takes place two to three times a month depending on how chocked the system is, which varies between seasons. To dredge, 12–20 farmers work together. In addition to dredging, sacks are filled with sand to divert wastewater from the main river course into the constructed lagoons. Farmers estimated that about 150 sacks priced at USD 0.50 per sack are needed (i.e. total USD 75). Following heavy rains, the man-made dams usually need repair or reconstruction, and this is done three to four times a month. Over four months of rain, maintenance costs can exceed USD 1000. The contribution to maintain the wastewater is done by farmers as the situation arises, i.e. regular fees are low but when there is a need to work on the system, farmers are required to add additional money, with differences on where one's farm is located, i.e. how much farmers benefit. Farmers who were interviewed confirmed that despite these investments, their returns are multiple times higher than their costs[9].

In June and July 2014, severe flooding and sedimentation damaged the system, and its operation was paused[10]. A revised treatment system has been proposed by FORS to expand the present capacity of treatment and also improve the efficiency of the system. The new design will expand the size of the planted lagoons and intends to increase the share of water flowing by gravity to individual farms instead of being pumped. Buying or renting portable pumps also increases the initial investment of farmers especially those whose farms are located farther away from the treatment lagoons on top of investing in PVC pipes which can reach USD 500.

In an attempt to protect the site, improve the revenue streams and also maintain the ecology of the site, FORS in collaboration with UNEP and EPA has planted about 1,500 coconut seedlings at the site.

Socio-economic, health and environmental impact

Most of the farmers operating on the open wetland area practice commercial agriculture and produce fresh vegetables and cereals for sale in the city. The availability of water throughout the year gives them a competitive advantage. Although 90% of the about 200 farmers are men, more than the same number of women find employment as field workers for planting, weeding and harvesting; and women dominate trade and retail of most perishable vegetables.

The use of highly polluted water poses risks to farmers and consumers, and the initial mitigation measures by farmers are only one step on a longer journey. A microbial risk assessment estimated a possible loss of about 12,000 disability-adjusted life years (DALYs) annually in Ghana's major cities through the consumption of salad prepared from wastewater-irrigated lettuce (Drechsel and Seidu, 2011). This figure represents nearly 10% of the World Health Organization (WHO)-reported DALY loss occurring in urban Ghana due to various types of water, sanitation and hygiene-related diarrhea. Thus, the shift to partially treated irrigation water has been appreciated although more awareness creation on health benefits is needed to establish a related "safer food" value chain where premium prices make investment and behavior change of traders worthwhile (Keraita and Drechsel, 2015). So far, farmers appreciate the increased water proximity, storage and separation of solid waste more than possible health benefits. However, farmers also indicated their support for treatment measures improving the smell of the water. Farmers' willingness to invest in better water was also confirmed by Amponsah et al. (2016) in Kumasi (Ghana) showing that 60% of surveyed open space commercial vegetable farmers were willing to pay for reclaimed water for irrigation.

Women traders who were interviewed appreciated farmers' efforts at Sakumo as it has created a good image that the vegetables are cleaner. However, this does not prevent traders from mixing vegetables produced under safer and unsafer conditions. More consumer awareness is needed as well as public controls to keep the two value chains separate. The investment would pay off as every USD spent in on-farm treatment and post-harvest safety returns up to USD 4.9 in public health cost savings (Keraita et al., 2015).

Farmer support of waste management in this area will have benefits beyond the farms. The wetland provides valuable products and services, which include the provision of important spawning and nursery grounds for many fish species. It is absorbing floodwaters and protecting biodiversity. The wetland also serves as roosting, nesting and feeding sites for many species of birds (Entsua-Mensah et al., 2000). The site is rated the third most important for seashore birds along Ghana's coast. More than sixty bird species have been identified including six internationally important species.

Scalability and replicability considerations

Farmers' innovation capacity is well known (Reij and Waters-Bayer, 2001) and has been reported also from other countries where wastewater irrigation is common (Buechler and Mekala, 2005). The innovation requires relatively low investment costs and can easily be replicated on similar (peri)urban farming sites. Depending on the scale of local risk awareness, capacity development and further incentives would be supportive. The key drivers for the Sakumo case are:

- A business advantage for farmers to engage (as an organized group) in on-farm intervention, driven in this case by their desire to channel the water closer to their plots, create storage facility for periods of low flow, filter floating (plastic) debris, and remove bad water smell. A very similar situation exists, e.g. on the La farming site in Accra.
- An advantage for the local community interested in the protection and image of their wetland which has both a traditional role as well as a potential value for recreation and tourism (bird watching), and the formation of a related interest group (FORS) supporting the farmers.
- An enabling environment where policies, authorities and international agencies are supportive of the community efforts.
- A favourable cost-benefit ratio based on the additional cultivation area (and less production risks).
- Knowledge on technical options able to link farmers' interest with water quality treatment.
- Sense of ownership of the infrastructure by farmers and willingness to contribute to its O&M.

This business case presents a low-cost effort where simple technology provided a first step towards safer water reuse and there are more irrigation infrastructure options, in particular weirs (Ensink et al., 2010), which support natural remediation processes, independently if implemented with or without risk awareness.

However, to maintain and extend the treatment process, risk awareness supported by demand for safer food would be helpful. Value chains linking to dedicated outlets, like particular 'food quality' markets could be a start. The model would also gain in sustainability if EPA or MoFA could regularly monitor water quality and support farmers and traders complying with on- and off-farm safety protocols. The WHO (2015) *Sanitation Safety Planning Manual* provides a framework for such a process, which will facilitate further up- and out-scaling.

Potential risks and mitigation

In designing any business model, it is assumed that generic business risks are known and will be taken care of. However, some risks might be more model specific and will be acknowledged in the following:

Market risks: Like in the here presented case of Accra, most farming locations where wastewater is informally used are in close proximity to major urban markets and well positioned to respond quickly to market needs, save on transport costs, and deliver high-value crops also in the lean season when revenues peak. As crops produced with wastewater or freshwater are with few exceptions mixed in markets and risk awareness along the food chain is commonly low, market incentives for safe production remain limited, while urban demand for vegetables is high.

Competition risks: This is only possible where with increasing risk awareness along the food chain, the potential of competition from freshwater farmers is growing. So far this awareness is in most low-income countries limited and competition is stronger from the other end, i.e. farmers using raw wastewater without any investments (extra costs) in safety.

Technology and performance risks: The employed technologies are low-cost and mostly based on manual work, where one-time or seasonal investments in irrigation infrastructure pay off through reduced operational (labor) costs. As wetlands in coastal areas also function as buffer for flooding, the system has to withstand flash floods.

Political and regulatory risks: A significant challenge can come from the regulatory framework if this is not supporting. While in Accra, the use of wastewater for crop production is forbidden by local byelaws, Ghana's national irrigation policy is supporting the WHO (2006) guidelines which recommend a step-wise approach to move towards safer wastewater irrigation (Drechsel and Keraita, 2014).

Social equity related risks: The share of men and women in the informal irrigation sector differs between countries and cultures from mostly female, e.g. in Sierra Leone, to mostly male, e.g. in

TABLE 64. POTENTIAL HEALTH AND ENVIRONMENTAL RISK AND SUGGESTED MITIGATION MEASURES FOR BUSINESS MODEL 24

RISK GROUP	EXPOSURE					REMARKS
	DIRECT CONTACT	AIR/ ODOR	INSECTS	WATER/ SOIL	FOOD	
Farmers	HIGH	MEDIUM	MEDIUM	HIGH		After introduction of farm based risk reduction measures, their adoption has to be monitored
Community		MEDIUM	MEDIUM	LOW		
Consumer					HIGH	
Mitigation measures	🧤	😷 / 🌫️	🦟	Pb Hg Cd	🥬 / 🍲 Pb Hg Cd	

Key: NOT APPLICABLE | LOW RISK | MEDIUM RISK | HIGH RISK

Senegal (Drechsel et al., 2006). There is no difference in innovation capacity although some of the innovations are very labor intensive. In the presented case study, both gender are equally presented within the overall value chain from farm to market.

Safety, environmental and health risks: The model follows the WHO (2006) recommendation of a step-wise and stakeholder inclusive approach to risk mitigation which is an intermediate step until (a) more comprehensive wastewater collection and treatment systems are in place; and (b) stricter safety guidelines can be implemented and enforced. In this sense, there are significant risks remaining – although less than without farmers' innovative efforts – which need to be controlled (Table 64). While pathogen loads can be reduced through on-farm treatment, other health risks will not be eliminated and additional preventive measures are required.

SWOT analysis and business performance

While this business case focused originally on supporting urban agriculture with better access to irrigation water, the installed pond system has the potential to improve also water quality and food safety. If combined with awareness creation and monitoring, incentives can be created to expand the system to progress gradually from informal to formal wastewater use. Similar synergies between

FIGURE 275. SWOT ANALYSIS OF SAKUMO WASTEWATER TREATMENT CASE, GHANA

	HELPFUL TO ACHIEVING THE OBJECTIVES	**HARMFUL** TO ACHIEVING THE OBJECTIVES
INTERNAL ORIGIN ATTRIBUTES OF THE ENTERPRISE	**STRENGTHS** • Strong farmer's association with formal rules • Willingness of farmers to invest in the set-up and maintenance of the treatment system • Partnership with an NGO (FORS) able to advocate farmers interest and leverage funds • Involvement of different institutions such as EPA, Forestry commission, UNEP, local chiefs • Low O&M cost and the system can easily be upgraded • Higher safety than in a business-as-usual scenario	**WEAKNESSES** • The achieved treatment level is only an initial step in the right direction • Farmers are more concerned with visible trash (e.g. plastic blocking pumps) than pathogens in the water, and might bypass designed treatment process • Difficulty in expanding the scheme in the region due to only emerging 'safe food' awareness and marketing • Inadequate monitoring of water quality to verify/improve treatment quality • System reconstruction requires resources if severely damaged through flooding
EXTERNAL ORIGIN ATTRIBUTES OF THE ENVIRONMENT	**OPPORTUNITIES** • Treatment system addresses demands related to water quantity and quality • Opportunity for higher yields/extra harvest • Environmental benefits from reduced trash and wastewater in the Sakumo lagoon • New revenue option for (male) farmers and (female) traders based on increasing awareness for food safety • Farmers' occupational health risks are controllable	**THREATS** • Despite multiple strong stakeholders, and public interest in food safety, no institution accepts so far responsibility to assist farmers regularly • Remaining crop contamination risks • Remaining farmer exposure • Urban encroachment on the site • Septage operators dumping raw sludge into the wetland and wastewater inflow continues to increase • Flooding destroying nature based treatment ponds

private and public interests are possible in view of the timing of irrigation (see above) and other farming practices (IWMI 2008a, b). This creates potentially a win-win situation whereby the city's wastewater undergoes a first treatment and farmers get access to more and safer irrigation water than without the intervention, resulting in higher returns and relatively safer food for consumers than in a business as usual scenario. However, the Sakumo water treatment system will not eliminate health risks and other risk mitigation measures have to be added between 'farm and fork' (Amoah et al., 2011). Figure 275 shows the SWOT analysis for the business case, while Figure 276 shows the impact potential of a farmer innovation model for increasing food safety.

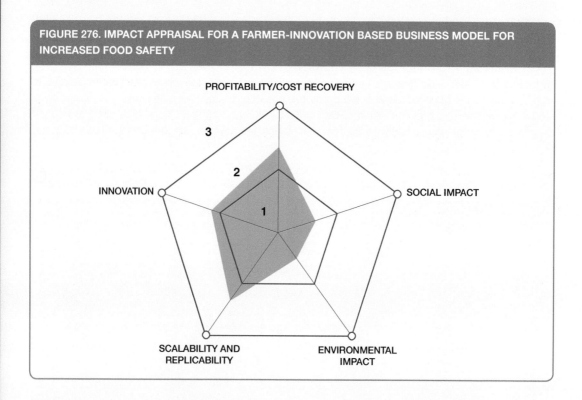

FIGURE 276. IMPACT APPRAISAL FOR A FARMER-INNOVATION BASED BUSINESS MODEL FOR INCREASED FOOD SAFETY

As the model is only a building block on the trajectory from unsafe to safe wastewater use, its impact remains modest. Although the technical innovation is down to earth, the effort to create a win-win situation between farmers' initial interests and safeguarding public health is very innovative. Where this engagement can be supported, the model will rank well in view of scalability and replicability without undermining the profitability of the business for farmers (Figure 276). The cost-benefit balance might shift through the introduction of more advanced and capital or maintenance intensive on-farm technologies. Thus, any replication or expansion should be aligned with the support of a value chain which targets (the increase of) market segments cherishing food safety.

Contributors

Paul Achulivor, Wildlife Division of the Forestry Commission
Members of the Resource Users Association at the site
Richard Agopa, Friends of Ramsar Site, Tema, Ghana

References and further readings

Agbemehia, K. 2014. Effects of industrial waste effluents discharged into Sakumo II lagoon in Accra, Ghana. MSc thesis. KNUST, College of Science, Kumasi.

Amponsah, O., Håkan, V., Braimah, I., Schou, T.W. and Abaidoo, R.C. 2016. The policy implications of urban open space commercial vegetable farmers' willingness and ability to pay for reclaimed water for irrigation in Kumasi, Ghana. Heliyon 2. www.heliyon.com/article/e00078/pdf (accessed 6 Nov. 2017).

Amoah, P., Keraita, B., Akple, M., Drechsel, P., Abaidoo, R.C. and Konradsen, F. 2011. Low cost options for health risk reduction where crops are irrigated with polluted water in West Africa. IWMI Research Report Series 141. Colombo: IWMI.

Asmah, R., Dankwa, H., Biney, C.A. and Amankwah, C.C. 2008. Trends analysis relating to pollution in Sakumo Lagoon, Ghana. African Journal of Aquatic Science 33 (1): 87–93.

Agyepong, G.T. 1999. Coastal wetlands management project. Management plan for Ramsar site. Prepared for Game and Wildlife, Government of Ghana. 45pp.

Buechler, S. and Mekala, G.D. 2005. Local responses to water resource degradation in India: Groundwater farmer innovations and the reversal of knowledge flows. Journal of Environment and Development 14: 410–438.

Drechsel, P., Graefe, S., Sonou, M. and Cofie, O.O. 2006. Informal irrigation in urban West Africa: An Overview. IWMI Research Report Series 102. Colombo: IWMI. www.iwmi.cgiar.org/Publications/IWMI_Research_Reports/PDF/pub102/RR102.pdf (accessed 6 Nov. 2017).

Drechsel, P. and Keraita, B. (eds). 2014. Irrigated urban vegetable production in Ghana: Characteristics, benefits and risk mitigation. 2nd ed. Colombo, Sri Lanka: International Water Management Institute (IWMI). 247pp.

Drechsel, P. and Seidu, R. 2011. Cost-effectiveness of options for reducing health risks in areas where food crops are irrigated with wastewater. Water International 36 (4): 535–548.

Ensink, J.H.J., Scott, C.A., Brooker, S. and Caincross, S. 2010. Sewage disposal in the Musi River, India: Water quality remediation through irrigation infrastructure. Irrigation and Drainage Systems 24: 65–77.

Entsua-Mensah, M., Ofori-Danson, P.K. and Koranteng, K.A. 2000. Management issues for the sustainable use of lagoon fish resources. In: Abban, E.K., Casal, C.M.V., Falk, T.M. and Pullin, R.S.V. (eds) Biodiversity and sustainable use of fish in the coastal zone. ICLARM Conf. Proc. 63. pp. 24–27.

IWMI. 2008a. Health risk reduction in a wastewater irrigation system in urban Accra, Ghana. DVD. http://youtu.be/f_EnUGa_Gdm (accessed 6 Nov. 2017).

IWMI. 2008b. Good farming practices to reduce vegetable contamination. Awareness and training video for wastewater farmers. CPWF, DVD. www.youtube.com/watch?v=Aa4u1_RblfM (accessed 6 Nov. 2017).

Keraita, B. and Drechsel, P. 2015. Consumer perceptions of fruit and vegetable quality: Certification and other options for safeguarding public health in West Africa. (IWMI Working Paper 164). Colombo, Sri Lanka: International Water Management Institute (IWMI). 32pp.

Keraita, B., Drechsel, P., Klutse, A. and Cofie, O. 2014. On-farm treatment options for wastewater, greywater and fecal sludge with special reference to West Africa. Colombo, Sri Lanka: International Water Management Institute (IWMI). CGIAR Research Program on Water, Land and Ecosystems (WLE). 32pp. (Resource Recovery and Reuse Series 1).

Keraita, B., Drechsel, P. and Konradsen, F. 2008. Perceptions of farmers on health risks and risk reduction measures in wastewater-irrigated urban vegetable farming in Ghana. Journal of Risk Research 11(8): 1047–1061.

Keraita, B., Konradsen, F., Drechsel, P. and Abaidoo, R.C. 2007. Reducing microbial contamination on lettuce by cessation of irrigation before harvesting. Tropical Medicine & International Health 12 suppl. 2: 8–14.

Keraita, B., Medlicott, K., Drechsel, P. and Mateo-Sagasta Dávila, J. 2015. Health risks and cost-effective health risk management in wastewater use systems. In: Drechsel et al. (eds) Wastewater: Economic asset in an urbanizing world. Springer, pp. 39–54.

Lydecker, M. and Drechsel, P. 2010. Urban agriculture and sanitation services in Accra, Ghana: The overlooked contribution. International Journal of Agricultural Sustainability 8 (1): 94–103.

MoFA (Ministry of Food and Agriculture). 2011. National irrigation policy, strategies and regulatory measures. MoFA, Government of Ghana.

Reij, C. and Waters-Bayer, A. (eds) 2001. Farmer innovation in Africa: A source of inspiration for agricultural development. London: Earthscan.

WHO. 2006. Guidelines for the safe use of wastewater, excreta and greywater, volume 2: Wastewater use in agriculture. Geneva: World Health Organization.

WHO. 2015. Sanitation safety planning manual (Manual for the safe use and disposal of wastewater, greywater and excreta). Geneva: World Health Organization.

Case and model descriptions are based on primary and secondary data provided by case operators, insiders or other stakeholders, and reflect our best knowledge at the time of the assessments 2015. As business operations are dynamic data are likely subject to change.

Notes

1 www.ramsar.org/about-the-ramsar-convention (assessed 4 Nov. 2017).
2 GTV news video (www.youtube.com/watch?v=CGZVW4nb7cc; assessed 4 Nov. 2017).
3 https://rsis.ramsar.org/ris/565 (assessed 4 Nov. 2017).
4 This is an interesting example where farmers changed behavior, probably to avoid traders to reject their 'smelly' produce, which in fact supports the natural die-off of pathogens as recommended by WHO (Keraita et al., 2007).
5 As an association, farmers have an increased ability to offer traders a higher and more reliable supply at lower contracting costs (one-stop-shop). Moreover, a registered association can easier access agricultural loans and possibly use its cooperative capital as collateral for fund raising.
6 While the protection of the wetland has to start upstream where pollution is generated, EPA struggles with the lack of sewage collection and treatment.
7 In 2014, the Resource Users Association and local fishermen registered as an official association under the companies act and the name of "Sakumo Ramsar Conservation and Resource Users Association". www.ghananewsagency.org/social/users-of-sakumo-wetland-form-association--76109 (assessed 4 Nov. 2017).
8 Ghana's Environmental Protection Agency (EPA) and the United Nation Environment Programme (UNEP) initiated in 2013–2014 an afforestation project of planting mangoes and coconuts in the wetland area. The trees should provide income and prevent further encroachment and land degradation.
9 See also www.youtube.com/watch?v=CGZVW4nb7cc (assessed 4 Nov. 2017).
10 Famers continued using the treatment infrastructure for their own advantage, including abstracting water also from the treatment lagoons nearest to their farm. At the time of writing in late 2016, FORS was still seeking support for system repair and extension.

SECTION V – ENABLING ENVIRONMENT AND FINANCING

Photo: istockphoto.com/andykazie

19. THE ENABLING ENVIRONMENT AND FINANCE OF RESOURCE RECOVERY AND REUSE[1]

Luca di Mario,[2] Krishna C. Rao, and Pay Drechsel

THE ENABLING ENVIRONMENT AND FINANCE OF RESOURCE RECOVERY AND REUSE

Businesses are influenced by policies, plans and regulations, trade agreements, institutional setups and strength, access to finance and subsidies, technology, matching partners and availability of land, and local infrastructure, which all may facilitate or hinder business sustainability and scalability, also in view of resource recovery and reuse (RRR) (IFC, 2013; Otoo et al., 2016). In addition to these more formal factors, social norms, business culture, as well as local preferences, expectations, environmental awareness, knowledge and perceptions can be powerful aspects of the business enabling environment. The creation of an enabling environment by central, provincial, state and local governments, the private sector, and civil society organizations thus provides the necessary basis for a business to grow. As this goes beyond what the business model canvas does address, it deserves its own chapter.

Common drivers of success for investments in RRR are (i) market demand, driven by resource scarcity, like declining water reserves or soil fertility, and (ii) environmental legislations demanding safer and more environmentally sound waste management. Examples are waste disposal limitations, recycling obligations, and carbon emission reduction targets. Both drivers are important but seldom sufficient conditions for RRR success. The first usually creates a market space for the sale of RRR products, while the second pushes for waste prevention, less dumping on landfills, and investments in alternative practices to safeguard environmental and human health. These elements can be seen as 'push-and-pull' forces for RRR and their relative importance may vary according to the context or the business model. However, compared to resources like glass, metal or plastic, the organic waste recycling sector is less driven by (in/formal) market mechanisms. Competition from (subsidized) chemical fertilizers and fossil fuels is fierce and hinders the development of market opportunities for compost and renewable energy (Matter et al., 2015). In the case of water reuse, the commonly subsidized freshwater tariffs strongly limit revenue expectations from wastewater sales. All this puts additional weight on the role of policy incentives to support the valorization of nutrients, water and biomass for a circular economy.

Table 65 shows how selected enabling conditions for municipal solid waste (MSW) composting can differ among countries in South Asia. While data will have changed, the comparison shows the importance of country specific information.

Another important factor is that the RRR value chain cuts across various sectors, which include sanitation/solid waste management, environmental protection, health, renewable energy, food security, and private sector development involving different ministries and levels of governance. Table 66 shows examples of governmental responsibilities in the domain of organic waste management, which however, are seldom organized under a dedicated framework.

In the following sections, four groups of key factors of the **enabling environment** for RRR will be addressed in more detail:
- Policies, regulations, and guidelines.
- Finance and financial incentives.
- Technologies matching resource constraints.
- Local capacities and stakeholder acceptance.

19.1 Policies, regulations and guidelines

Over the last few years, Europe is spearheading policies and regulations in direct support for the circular economy (Box 17). Although in most low- and middle-income countries the value of water reuse is equally recognized in national water policies, for example, they often fail to define related standards, guidelines, or national targets. Also, international support is limited, with the exception of the World Health Organization's guidelines (WHO, 2006a, b) and the related Sanitation Safety Planning

> **Box 17. Netherlands approves Circular Economy 2050 strategy**
>
> On 5th October 2016, the **Netherlands** national Circular Economy program till 2050 was presented to Parliament. The program fixes an interim objective of 50% reduction in raw materials use (minerals, metals, fossil fuels) by 2030, and an objective of 100% sustainable, non-polluting use of raw materials by 2050. 'Biomass and food' is one of the five priority areas identified in the program. Actions specified to address this priority area include reducing food waste, sustainable agri-food- and biomass value chains, development of alternative protein sources, recycling of food industry residues, soil quality and increasing soil carbon, precision farming, and closing the loop for nutrients.
>
> But also, other European countries are getting active: **Switzerland** was in 2016 the first European country to make phosphorus recovery and recycling from sewage sludge and slaughterhouse waste obligatory. **Germany** is expected from 2018 on to make phosphorus recovery obligatory for larger sewage works, and **Austria** drafted in 2017 a new legislation on P recovery from municipal wastewater (see Chapter 13).
>
> Source: https://www.government.nl/documents/policy-notes/2016/09/14/a-circular-economy-in-the-netherlands-by-2050 (accessed 4 Nov. 2017)

Manual (WHO, 2015) for the safe reuse of wastewater, greywater and excreta in agriculture and aquaculture. This situation is likely to change under the peer pressure of the Sustainable Development Goals; in particular targets 6.3 and 12.5 demand more attention to waste reduction, recycling and reuse, in support of a circular economy.

As the WHO example shows, policies and regulations are not only important to support the business side of RRR but also to maintain operational safety for workers, customers and the environment wherever resources are extracted from potentially harmful waste. As many waste-related policies were originally designed to protect the public and the environment, they can be very cumbersome in their support of RRR. A 2016 report for the European Commission[3], identified such regulatory barriers within EU directives, legislations, and regulations, such as the lack of (clear) definitions and an overemphasis on safety than resource recovery.

When biosolids are, for example, defined as solid waste rather than as a renewable fuel, it will be difficult for biosolids-to-energy projects to benefit from renewable energy incentives. In a similar way, the European Waste Framework Directive does not recognize the potential fertilizer value of the digestate of anaerobic biogas production, while the EU Fertilizer Regulations are so far missing to recognize organic fertilizers (a gap which is under revision).

Another example is the encouragement of phosphorus (P) recovery but the slow pace of acceptance of recovered P (in the form of struvite) as fertilizer. In fact, what is needed in all RRR cases are clear criteria to determine the "**End of Waste status**" of the recovered resource (see Chapter 13 and Model 16) that for example, struvite can also be traded across borders as a (new) raw material without the need to comply every time again with sanitary regulations.

The International Solid Waste Association (ISWA, 2015) call in this context for a new 'regulatory construct,' moving from waste as a harmful substance for disposal to the management of 'materials

TABLE 65. STATUS OF ENABLING ACTIVITIES IN BANGLADESH, INDIA, NEPAL, AND SRI LANKA

COUNTRY	COORDINATION COMMITTEE	STRATEGY, POLICY, RULES, AND STANDARDS	PROMOTION OF SOURCE SEGREGATION	FEED-IN TARIFF FOR WASTE TO BIOGAS OR ELECTRICITY
Bangladesh	Proposed	Rules under preparation	Pilot Project started	No
India	Proposed	Yes	Partial	Yes
Nepal	No	No	No	No
Sri Lanka	Yes	Yes	No	No

IPNS = Integrated Plant Nutrient System, PPP = public-private partnership, RDF = refuse-derived fuel.

Source: ADB, 2011.

TABLE 66. COMMON ROLES AND RESPONSIBILITIES OF SELECTED RELEVANT MINISTRIES

MINISTRY OR ORGANIZATION	COORDINATION COMMITTEE	STRATEGY, POLICY, RULES, AND STANDARDS	PROMOTION OF SOURCE SEGREGATION OF WASTE	FEED-IN TARIFF FOR WASTE TO BIOGAS OR ELECTRICITY
MOA	X	X		
MOE	X	X		X
MOEF	X	X	X	
MOF	X			
MOI	X		X	
MOUD/MOLG	X	X	X	
ULBs	X	X	X	

IPNS = Integrated Plant Nutrient System, MOA = Ministry of Agriculture, MOE = Ministry of Energy, MOEF = Ministry of Environment and Forests, MOF = Ministry of Finance, MOI = Ministry of Information, MOLG = Ministry of Local Government, MOUD = Ministry of Urban Development, PPP = public-private partnership, RDF = refuse-derived fuel, ULB = urban local body.

Source: ADB, 2011; referencing Waste Concern.

TIPPING FEE PAID TO OPERATOR FOR ORGANIC WASTE RECYCLING	PROVISION OF FREE GOVERNMENT LAND FOR ESTABLISHING ORGANIC WASTE-RECYCLING PLANTS	STANDARD FOR COMPOST AND SLURRY, PROMOTION OF IPNS, AND CO-MARKETING OF COMPOST WITH CHEMICAL FERTILIZERS	PROMOTING RDF	PPP GUIDELINES	CAPACITY BUILDING INSTITUTES ON ORGANIC WASTE MANAGEMENT
No	In some cases	Enforced	No	Yes	In some cases
No	Yes	Yes, not enforced	Yes	Yes	Yes
No	No	No	No	No	No
No	Yes	Yes, not enforced	No	No	Yes

IN ORGANIC WASTE MANAGEMENT FOR RRR THROUGHOUT SOUTH ASIA

TIPPING FEE FOR ORGANIC WASTE-RECYCLING PLANT OPERATORS	PROVISION OF LAND FOR ESTABLISHMENT OF ORGANIC WASTE-RECYCLING PLANTS	STANDARD FOR COMPOST AND SLURRY, PROMOTION OF IPNS, AND CO-MARKETING OF COMPOST WITH CHEMICAL FERTILIZERS	PROMOTION OF USE OF RDF	GUIDELINE ON PPP	INCENTIVES ON COMPOST, BIOGAS, AND RDF	CAPACITY BUILDING OF STAKE-HOLDERS
		X				X
	X			X		X
						X
				X	X	
						X
X	X		X	X		
X	X				X	X

in transition,' including 'End-of-Waste' criteria for recovered raw materials. For these secondary products, quality standards and specifications are needed to ensure market confidence, and not only standards assessing their potential harm.

While there are also an increasing number of examples of supporting regulations and mechanisms, experiences from the RRR private sector on their actual accessibility and performance in low- and middle-income countries are largely missing. One of the indicator in the 2017 report of the World Bank on 'Enabling the Business of Agriculture' is the time, cost, and regulation for fertilizer registration. This concerns the import of new fertilizers but is also a key indicator for any enterprise engaged in the creation of new fertilizers or composts from resources embedded in waste. As stated by Muspratt (2016b) there is a significant gap between the speed successful startups bring to markets and which can give them a competitive edge, and the way administrations in many low- and middle-income countries work. While, for example, according to the World Bank (2017), the global average to register a new fertilizer is below one year, significant variation was found across countries with respect to the efficiency and complexity in registering fertilizer products. The time and cost to register a new fertilizer product are lowest on average in OECD high-income and upper-middle-income countries, and highest in low-income countries (Figure 277[4]). This harsh difference in time is driven principally by lengthy field testing.

Many countries in particular in Sub-Saharan Africa, only have rudimentary regulatory frameworks for registering fertilizer. And even where legal frameworks are strong and elaborate, the registration process can be very time-consuming and discouraging, or only allows the public sector to register new fertilizer products (Figure 278). In the case of Ghana, where the International Water Management Institute (IWMI) registered its fortified waste compost with the Ministry of Food and Agriculture, the process, which is similar to the one, described for other countries (Box 18) took 36 months (Nikiema, personal communication), i.e., two years longer than one would expect in Ghana (World Bank, 2017). Where the public sector is partner of the compost production, ADB (2011) recommends that the public

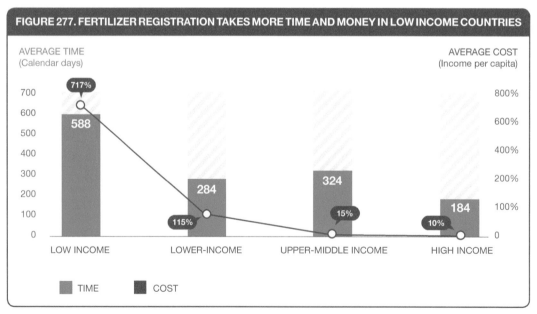

FIGURE 277. FERTILIZER REGISTRATION TAKES MORE TIME AND MONEY IN LOW INCOME COUNTRIES

Source: EBA Database; World Bank 2017.

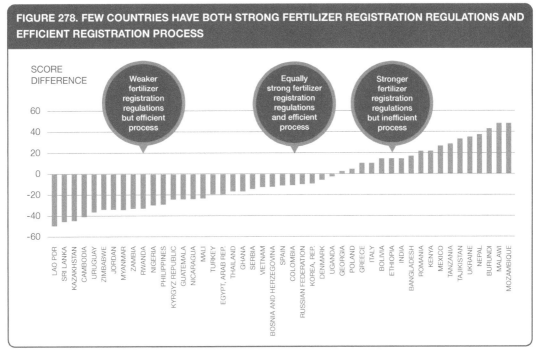

Source: EBA Database; World Bank 2017.

Box 18. Waste compost registration in Bangladesh

To market a new compost commercially in Bangladesh, the compost manufacturer must obtain licenses for the product and its brand name. Approval for licensing is a two-stage process. First, the compost produced by the manufacturer is tested in government laboratories. Subsequent to compliance with the national standards, the compost is sent for a field trial on crops for two agricultural seasons. If field trial results demonstrate that use of the compost lowers the need for chemical fertilizer and increases yields, the compost and its brand name will be approved. After this stage, the Department of Agricultural Extension issues a license to the compost manufacturer. In Bangladesh, the process for product and manufacturer registration takes approximately 1–2 years. Although the entire process is lengthy, the final government approval of the product indirectly assists in marketing the compost. The Government of Bangladesh is promoting the use of compost as part of its Integrated Plant Nutrient System program through field-level agricultural extension officers who are encouraging farmers to use registered, government-approved compost. As a regulatory requirement, the compost producer has to send monthly production data and compost quality data to the Department of Agricultural Extension. Moreover, the department also randomly undertakes quality control tests on the compost for laboratory analysis to ensure compliance. If major deviations from the approved standard are detected, then the government may cancel the license to market compost. The Ministry of Agriculture instructs fertilizer shops all over the county to market only government-approved compost.

Source: Waste Concern (ADB, 2011)

sector starts buying the compost for city greening programs in parks, landscaping, and roads, where quality requirements are low, while the registration process continues. Without income from the sale of compost, the plant's cash flow will be affected, and the private partner/investor might not even engage or jump off.

Particularly helpful can be regulations that combine disincentives, e.g. for environmental pollution with policies that encourage RRR. In Rwanda, for example, the government has regulations to reduce deforestation while at the same time it provides policies for promoting renewable energy (like waste to briquetting enterprises). An often-cited example from Europe is described in Box 19.

Box 19. Disincentives as driver

One of the most influential legal drivers of RRR in Europe was the establishment of the **Landfill Directive** (1999/31/EC), which defines landfills as the last option for waste treatment and disposal. The directive imposes, for example, staged landfill reduction targets for the biodegradable fraction of municipal solid waste. Member states are obliged to devise national strategies to meet the targets. Examples of a national strategy are Sweden's 2002 ban on landfilling of separated combustible waste and the 2005 ban on landfilling of organic waste. Because the Landfill Directive limits the landfill capacity, it has pushed the market to find alternative waste/material management options. To comply with the Landfill Directive, countries have introduced various measures to disincentivize landfilling, e.g. by increasing the gate fee, making landfilling not the cheaper but the more expensive option. Other regulatory factors that can steer the financial viability of RRR are: emissions caps, carbon taxes or carbon trading schemes; incentives related to the share of alternative (green) energy (see also Box 22).

California's policy prohibiting landfilling of untreated sewage sludge drove its beneficial reuse as 'Class A biosolids.' In Quebec, Canada, plans are under way to stop landfilling of all organic wastes, including sewage sludges, sorted municipal food waste, green wastes, paper industry sludges, etc. (Hasanbeigi et al., 2012; MDDEP, 2012).

Without regulatory support, enterprises might find it easier to grow in the informal sector, which can become a significant challenge for society. In urban Ghana, for example, up to 800,000 city dwellers eat every day exotic vegetables produced with raw or diluted wastewater (Drechsel and Keraita, 2014). Although the informal sector had been acknowledged in Ghana's latest national irrigation policy, capacity development, e.g. in safer irrigation methods, remains an exception. Such support, however, would eventually help in reducing the size of the informal sector while promoting growth. A strong lead Ministry with a clearly defined role for promoting private sector development would be helpful in this regard.

19.2 Finance and financial incentives

Resource Recovery and Reuse infrastructure financing varies to some extent between the water, energy, and nutrient/biomass sectors. Many waste to energy projects are commercially driven, with venture financing and bank loans, supported by governmental programs (green growth, renewable energy, rural electrification, etc.) and the carbon market. Financing wastewater reuse, on the other hand, refers mostly to the provision of treatment infrastructure, which is commonly relying on public finance, supported in low-income countries by foreign aid, and with increasing size of the project by multiple partners for risk sharing, including the private sector. Nutrient and biomass recovery from municipal solid waste, like in composting projects, shows components of both often relying on 2–4 finance sources (Kaza et al., 2016).

The following text will first address common challenges and options for RRR infrastructure financing, looking separately at nutrient, water, and energy recovery, followed by a cross-cutting section on financing operations and maintenance.

Infrastructure investment

Organic waste composting: Almost all municipally-run compost systems have benefitted from international, national, state, or local funding to some degree, from the most developed models in high-income countries to burgeoning and innovative models in low-income countries (Table 67). The most common role the private sector plays in composting projects is in operations providing technical, managerial, and process know-how. However, private organizations may also provide loans, grants, equity, and venture capital, when counterparties are creditworthy and there is a clear mechanism to recoup costs and earn a return within an acceptable risk. In countries without developed credit rating systems, a mechanism to garner lenders' trust is through providing key documents such as a comprehensive business plan with detailed assumptions, market and feasibility studies, land concession/provision, environmental permit to operate, feedstock supply agreements, product offtake agreements, and financial forecasts (World Bank, 2013; Oliver, 2016; Kaza et al., 2016).

Water and water reuse: The traditional governance and finance models, especially of public sector water utilities, like in Africa, showed a decreasing trend of bankable projects for funding by both the private financial institutions and the international finance institutions (Kruger, 2017). While governments

TABLE 67. SUMMARY OF COMMON FINANCE INSTRUMENTS FOR RRR INFRASTRUCTURE, IN PARTICULAR COMPOST PROJECTS

FINANCING STRUCTURE	LENDING INSTITUTION	LENDER REQUIREMENTS	ADDITIONAL DETAILS
Equity	Banks, private individual investors, venture capital, NGOs, for-profit companies, and business partners	• Share of profits proportional to ownership in entity (performance-based), though principle does not need to be repaid • Clear revenue potential (market analysis, carbon credit value, feedstock supply, offtake agreement, diversified revenues), such as through a business plan	• Venture capital funds may require majority ownership and major involvement in operations • Decision-making authority decreases as external ownership of the company increases
Debt Financing	Banks, credit unions, savings institutions	• Typically require 20–30% owner equity (cash, stocks, bonds, inventory, land/ equipment, angel investor or venture capital fund (*if project is high capital*)) • Personal guarantees of debt repayment by business officers and owners (e.g., pledged assets) • References, credit rating, detailed pro-forma and business plan, financial statements • Signed feedstock and offtake agreements (letters of intent or contracts) diversified across multiple customers	• Lender may require borrower to demonstrate cash flow to debt ratio of 1.5 times the value of the loan

TABLE 67 CONTINUED

FINANCING STRUCTURE	LENDING INSTITUTION	LENDER REQUIREMENTS	ADDITIONAL DETAILS
International Aid	International and multilateral development banks, national development agencies, NGOs, other humanitarian organizations	• Promotion of economic development and welfare • May require co-investment by recipient government • Alignment of objectives with recipient policy environment • Clear metrics for success • Political stability	• International grants should only be used to support capital costs; operational and maintenance costs require a clear path to financial sustainability
Government Financing	Tax credits (equipment tax credit or property tax credit), grants, direct loans through a third party intermediary, repayment guarantees on bank loads, issuing bonds (for public sector projects)	• Mission alignment to national policy objectives • Proof of long-term financial sustainability through business plan and contracts • Social and environmental benefits, including jobs, carbon emission reductions, and cost mitigation	• Grants typically cover capital costs and are not used to cover annual O&M • Government guaranteed loans can be more expensive than traditional financing and be more onerous to obtain and manage

Source: Kaza et al. (2016), modified

often do not have the financial means required for large scale investments, they also struggle to provide the guarantees to mitigate investment risk. Based on low water pricing and inadequate fiscal transfers from the central government, their creditworthiness is low, making **reforms** and **credit enhancements** a high priority. This is a common water sector challenge and not RRR specific. Finance mechanisms used today by cities, states, provinces, and countries can range from commercial or non-commercial bank lending, to green bonds, taxes, or pooled financing arrangements. Some countries, such as Mexico, Brazil, South Africa, India and the Philippines, have used municipal bonds (Platz, 2009; ADB, 2014). In the example of the Philippines, the Local Government Unit Guarantee Corporation provides credit guarantees for municipalities that seek to finance infrastructure projects through debt issuances. In Dakar, Senegal, on the other hand, the failed setup of a city-level bond without a central government guarantee resulted in important lessons for other African cities (Gorelick, 2017). Innovative models of capital cost co-financing through pension funds (see below) or by the benefiting households (Box 20) have been reported from South Africa, Latin America and India.

One stimulus to encourage private investments in water infrastructure has come from low-carbon and climate-resilient (green) bonds where verifiable standards can guide investors. The process can be supported through the setup of Water Financing Facilities or Urban/Municipal National or State Financial Intermediaries, which are assisting utilities in preparing bankable project plans (Oliver et al., 2016). The concept has been proven to be successful also in countries like the Philippines, India, and Colombia. The contribution of RRR in view of energy recovery, water reuse, and carbon cycling offers options to ensure that, in particular, wastewater treatment plants can show green bond features in support of climate change adaptation and mitigation.

The trend towards public-private partnerships (PPPs) in the wastewater sector goes beyond operational support but targets private financing initiatives (Mandri-Perrot and Stiggers, 2013). Many middle-income countries are committed to the promotion of private sector participation and particularly interested in the financial schemes that use public funding to leverage private investment. Typically, the private companies involved in delivering the project provide the initial equity, although

> **Box 20. Public deposits levied on households**
>
> Using a mix of grants and loans from central, state and local governments, most wastewater projects in the Indian state of Tamil Nadu between 2006 and 2014 were implemented through either Design Build Operate and Transfer (DBOT) or Design Build Finance Operate and Transfer (DBFOT) models as opposed to BOT models. A DBOT model encouraged technology firms to participate in project execution, and improve the overall design to minimize the cost of the projects. A unique feature was that a portion (14–32%) of the capital expenditure was funded through collection of public deposits levied on households, which is the 'one-time non-refundable deposit' obtained from the users. The advantages of this deposit contribution from the public have been: (i) accountability on the part of the local urban body to provide quality services; (ii) ensuring that households connect to the network upon completion of the construction; and (ii) as the deposit formed the public equity in the project, it reduced debt servicing costs and therefore the monthly user charge (WSP, 2016).

they may invite financial investors to participate either in the initial fund raising or subsequently when the construction risk has passed and it is possible to sell on the project equity at a higher price (GWI, 2010).

Official development assistance (ODA) from international development agencies remains particularly strong for solid waste and wastewater management in low-income countries, although for more costly wastewater infrastructure, finance can be significantly more complex with multi-partner cost sharing, investment guarantees, and related risk management arrangements. For wastewater treatment, public financed and owned facilities established by the private partner under a design-build-operate (DBO) or build-operate-transfer (BOT) model are often preferred. The operator is taking no or minimal financing risk on the capital but remains responsible for smooth operations. Also, other types of arrangements are common, including build-own-operate-transfer (BOOT) or build-own-operate (BOO) models where the private project development company has or even continues ownership of the facility after the contractual period (GWI 2010). In this catalogue, the BOT case of As Samra (Business Model 19) has been elaborated in more detail given its importance for many RRR projects which share high socioeconomic and environmental benefits, but carry significant political/macro-economic, sector specific, or project related risks, in particular uncertain cost recovery. There are many examples, especially in the purview of financing, which can add risks and encourage or discourage both the public and private sector to invest in RRR (ADB, 2011; Bjornali and Ellingsen, 2014; World Bank, 2016). The same applies to the operational site and the only slowly emerging support of cost recovery through different forms of direct or indirect (green growth) subsidies in low- and middle-income countries.

International project finance for the water sector is generally available subject to its (minimum) size. According to Winpenny (2003) a typical minimum project size is USD 50–100 million. Below that level, returns to scale generally tend to make project financing uneconomic, and projects will have to be addressed by the corporate or municipal sectors. For project finance to be a viable option, project revenues and returns to equity must be acceptable, though this does not preclude the use of aid to reduce the debt or equity burden of the project. However, there is a project size (USD 10,000 to USD 100,000) which is often too small for the corporate sector and too large for micro loans (Winpenny, 2003).

From the RRR perspective, a common bottleneck in many developing countries is the lack of local capital markets that provide long-term financing for small- and middle-scale infrastructure projects

(Muspratt, 2016ab), although there are encouraging example of financial instruments in support of local enterprises and business start-ups. In Singapore, for example, the government has initiated 35+ funding and incentive schemes related to clean energy, green buildings and construction, water and environmental technologies, waste minimization and recycling, environmental initiatives, and so forth. This also includes the Clean Development Mechanism (CDM), funding for water recycling, and use of alternative sources of water.[5] Other finance examples of case studies across the waste and sanitation sector are described, e.g. by Ali (2004), ADB (2011, 2014, 2016), Beltramello et al. (2013) and the World Bank (2016).

Energy recovery: Aside from wind and solar energy, the recovery of energy from biomass and waste constitutes important components of total renewable energy investments, which have been shifting towards developing countries for several years (IRENA, 2012). The sector struggled with the failure of energy pricing to account for externalities or the environmental and social costs of production, which has made renewable energy technologies to look more expensive than they really are, compared to fossil fuels. To assess project eligibility with an eye on externalities, the European Investment Bank (EIB), like also other larger banks, provides an interesting example of accounting via the net carbon footprint. The absolute carbon footprint of a project is compared with the carbon emissions in absence of the project. Then a net carbon impact of projects is calculated, using advanced models including industry-specific ones (Griffith-Jones et al., 2012). There are many other examples of trust funds, etc. which can support projects on renewable energy addressing climate change (see e.g. www.adb.org/site/funds/funds).

In general, there is no "one size fits all" finance formula for renewable energy (IRENA, 2012). Every national market is unique, and effective finance strategy requires a holistic approach that is tailored to the local context. That said, governments should generally seek to perform two broad functions: first, create overarching regulatory frameworks that shift incentives onto a macro level; and second, use targeted public financing to fill or overcome niche gaps and barriers. Regulatory frameworks can employ both energy policy (e.g. feed-in tariffs, energy auctions, and self-supply regulation) and finance policy (e.g. banking regulation and other measures that incorporate sustainability into financial decision-making). The most effective public finance programs will employ a flexible package of financing mechanisms rather than relying on any single mechanism or fixed set of mechanisms. These packages may employ credit lines to local finance institutions; project debt financing; loan softening programs; guarantees to mitigate lending risk; grants and contingent grants for project development costs; equity, quasi-equity, and venture capital; or carbon finance facilities (IRENA, 2012). Even more than in the water sector, **green bonds** designed for climate resilient cities are in high demand for the production of environmentally friendly energy as well as reduced methane emissions from better waste management. Green bonds which had in 2015 a total value of around USD 44 billion are commonly issued by larger development banks, but increasingly also by real estate companies, municipalities (Johannesburg), international corporations, and commercial banks (Oliver, 2016). Institutional investors also include pension funds. While smaller pension funds require pooled investment vehicles, larger pension funds have the capacity to invest directly in infrastructure projects (Box 21).

As investment in renewable energy (but also wastewater reuse and composting) can take years, or even decades, to yield good returns, 'patient capital' is needed. Unfortunately, however, this is the type of investment that is most difficult to attract in most developing countries, given typical short-term horizons of private capital markets. Exceptions can be funds that have long-term liabilities such as sovereign wealth funds (SWFs) and pension funds (Griffith-Jones et al., 2012) or e.g., pooled local currency bonds with a 15 to 23 years tenor as reported from the water sector (Oliver et al., 2016), aside from support from multilateral development banks.

Box 21. Pension funds for green infrastructure investments

The South African Government Employees Pension Fund (GEPF) is Africa's largest pension fund, with over USD 138 billion in assets under management. The GEPF is also the single largest investor in the Johannesburg Stock Exchange-listed companies. The fund has set aside 5 percent of its portfolio for investing in developmental projects – mostly infrastructure projects supporting positive economic, social, and environmental outcomes for South Africa over the long term. During 2010/11, the fund accelerated investments in developmental projects in different areas including water infrastructure, alternative energy, and environmental projects. In each area, the aim is to maintain a balance between social impact and financial returns. Interesting vehicles to assist smaller pension funds to invest in the infrastructure sector have been developed in some Latin American countries, such as in Chile via infrastructure bonds with insurance guarantees, in Mexico and Peru via investment trusts, and in Brazil via a joint-owed infrastructure company. Common investments are in housing, roads, or renewable energy (OECD, 2012).

Facilitating operational cost recovery

Among the different RRR sub-sectors, it is mostly the compost sector's image that suffered from a large array of failed or underperforming projects, which began with high amounts of grant funding but ultimately collapsed due to inability to support their operational costs (Kaza et al., 2016). Common reasons are the selection of a too complex technology for which repairs and maintenance costs become unmanageable, and limited understanding of the compost market. Public-private partnerships can offer in this regard not only private capital, but also market know-how and the private sector's technical and managerial expertise. This is crucial as experience shows that compost as well as wastewater treatment plants owned or operated by private sector companies are usually functioning better than municipally-operated ones (e.g. ADB, 2011; Murray and Drechsel, 2011). A key criterion are incentives, and operational cost recovery is a strong one, which can be supported by different forms of direct or indirect subsidies. There are many options for RRR related financial instruments to keep the private sector engaged (ADB 2011; ADB, 2016; Eyraud et al., 2011):

Tax holidays. Entrepreneurs setting up a compost plant as part of a joint venture or within the private sector could be considered for a tax holiday for a number of years, like an exemption on customs duty, excise duty, value-added tax, sales tax, or other local taxes on equipment, machinery, etc. Tax exemptions could also include the waste derived products, to support sales, while, e.g. tax penalties could be used to prevent resource wastage or tax credits to support, e.g. renewable energy (Box 22).

Capital subsidies. Entrepreneurs could be considered for example, for a capital subsidy (or Viability Gap Funding) provided to support infrastructure projects that are economically justified but fall short of financial viability. In the same way, different types of PPPs could be supported with different shares of grant subsidy, equity, and debt from the government or a financial intermediary. Moreover, for any project financed by banks, lower interest rates could be supported by the government, along with a long loan term.

Tipping fees. A private sector entity operating organic waste-recycling facilities such as compost or waste-to-energy plants should not be asked to pay royalties to the city. Alternatively, tipping fees should be paid by the city for each ton of waste processed by the entrepreneur because waste recycling reduces the landfilling cost. To promote RRR, legislation should however enable the private sector to be paid for every ton of quality waste recycled and sold, not only for every ton collected.

> **Box 22. Green taxations for a Circular Economy**
>
> Green taxation is being increasingly used to push the circular economy. KPMG identified in a 21-country survey 200 green tax incentives and penalties of which 30 appeared just in the last two years before the survey in 2013. These include landfill taxes, incineration gate fees, accelerated asset depreciation, tax credits, VAT refunds linked to secondary materials purchase, reduced VAT, or VAT refunds on recycled goods (e.g. in China and South Korea).
>
> Taxation can be applied at different levels: resource recovery, first industrial use (e.g. fertilizer production) or final consumption (e.g. fertilizer use). A possible taxation package in support of phosphorus recovery and reuse could have different entry points for tax support or penalties:
> - To secure long-term availability and reduce import dependency.
> - To reduce phosphorus losses/disposal into surface waters and ultimately the oceans.
> - To close the phosphorus cycle as far as possible, reducing inputs and outputs and developing recycling.
>
> An example of a penalty taxation is the one on nutrient surpluses (over-fertilization) in the Netherlands. Tax penalties to support renewable energy concern, e.g. the use of conventional fossil fuels. Such taxes only exist so far in developed countries. Developing or emerging economies appear to avoid taxing conventional fuel, presumably on the basis that such penalties could damage development and growth prospects. Other options are disposal taxes which are typically imposed per ton of waste resource landfilled or incinerated to catalyze firms' investments in waste reduction and recycling.
>
> Sources: KPMG, 2013, Dubois et al., 2015

Concessionary rates for utilities. RRR companies should be considered like other utilities, and be able to access similar concessionary/commercial rates for electricity, fuel, and water supply, if available.

Creating parity with chemical fertilizers. Although governments might promote compost use, they are usually providing direct or indirect subsidies to chemical fertilizer companies to the detriment of organic fertilizer and compost manufacturers. Given the environmental benefits of compost, including greenhouse gas emission (GHG) reduction, it is either recommended to include waste compost at par into the governmental fertilizer subsidy program, as it was introduced in 2016 for example in Ghana, or reduce subsidies to the chemical fertilizer companies, or of fossil fuels which in turn can support green investments, also in renewable energy.

Co-marketing compost with chemical fertilizers. Fertilizer companies can be asked to adopt a 'basket approach,' entailing the co-marketing of compost with chemical fertilizers as is now the law in India, supported by a governmental subsidy. The regulation helps, in particular, compost stations that fail to penetrate existing marketing chains. For larger-scale compost plants, the use of fertilizer marketing companies for distribution and sale of compost provides a great advantage. A possible marketing ratio of chemical fertilizer bags versus bags of certified registered compost was also discussed in India (Box 23) and has also been implemented in parts of Sri Lanka.

Power purchasing agreements. In most developing countries electricity is regulated by the government and utilities are typically owned by the government especially infrastructure for transmission and

> **Box 23. Co-marketing directive**
>
> To promote the acceptance of city compost, in early 2016 the **Indian** Cabinet approved a policy on Promotion of City Compost. The Ministry of Urban Development in consultation with the Ministry of Chemicals and Fertilizers agreed to subsidize compost sale at Indian Rupee (INR) 1,500 (USD 22.5) per ton of city compost. This market development assistance will be paid to fertilizer companies with the expectation of co-marketing city compost with chemical fertilizers. The co-marketing details will be decided by the Department of Fertilizers depending on supply and demand (Government of India, 2016). Earlier suggestions for co-marketing were, for example, to sell one bag of municipal compost with every two bags of chemical fertilizer, or that only a co-marketing arrangement gives access to the subsidy on chemical fertilizer. Such a directive would urge fertilizer companies to seek compost from compost stations; turning the common situation around where compost plants have to seek customers.

distribution of electricity. However, in many countries, such as Vietnam, Sri Lanka, and Uganda, the private sector is encouraged to generate (non-conventional) renewable electricity, including municipal waste energy by providing an attractive feed-in-tariff (FIT) and a long-term power purchasing agreement. Supportive agreements accept alternative energy throughout the year and not only in times of peak demand. According to KPMG[6], the total number of countries with feed-in tariffs globally is over 50. While the goals of FITs are the same in developed and developing countries, there are particular features of the latter that require consideration. For example, FITs in developed countries are generally funded by a premium placed on all energy bills, while in low-income countries external finance might be needed (Griffith-Jones et al., 2012).

Carbon market. Energy recovery from MSW or wastewater opens options for earning carbon credits through the CDM, while nutrient-rich wastewater can help to sequester carbon in fast-growing trees. However, the process of CDM registration and certification can take time and have significant transaction costs (Michaelowa and Jotzo, 2005). Examples of additional revenues from carbon credits are the Kinoya Sewerage Treatment Plant in Fiji and the National Biodigester Program in Cambodia (ADB, 2016). Due to the weakening of the carbon market, new initiatives for carbon finance were established (Box 24). Other alternatives could be watershed protection schemes or payments for ecosystem services.

Government promotion and awareness raising. Within the market purview, a campaign is needed to generate awareness of the new products and encourage the use of compost and organic fertilizer on its own and as a supplement to chemical fertilizers. This can also be incorporated in the extension activity of the ministry of agriculture and agricultural departments. Where applicable, a media campaign has to be undertaken to encourage source segregation of waste, as a key activity for successful organic waste management. Moreover, the ministry of energy can encourage the use of briquettes, for example, for certain types of industries, promote generation of electricity from biogas, and provide special rates for such electricity.

Results-based financing (RBF). Performance-based subsidies are disbursed based on the delivery of pre-agreed outputs and after independent verification (Trémolet, 2011). This ensures that facilities are constructed according to specifications and based on the desired quality. An application for carbon finance is presented in Box 24. Output-based Aid (OBA) is a form of RBF designed to enhance access to and delivery of infrastructure and social services for the poor through the use of performance-based incentives, rewards or subsidies. OBA was applied for sanitation investments in Nepal and in increasing household access to domestic sanitation in Sri Lanka (ADB, 2016).

Relevant information and statistics provide the foundation for monitoring returns on investments. Measuring progress towards green growth in low- and middle-income countries requires some special considerations as these countries face different challenges than other countries, such as a much lower statistical capacity. The OECD has therefore developed a measurement framework for green growth that provides countries around the globe with a robust tool that can be adapted to different national circumstances and priorities. The measurement framework combines the main features of green growth with the basic principles of accounting and the pressure-state-response model. It gives countries the flexibility to focus on the indicators that reflect their own green growth objectives, such as building economic and environmental resilience and ensuring that growth is inclusive (OECD, 2011).

Box 24. New carbon financing initiatives (Kaza et al., 2016)

Since the 1990s and early 2000s, carbon markets have been a supplementary source of income for those RRR projects that reduce greenhouse gas emissions. Carbon markets generate funding through sales of carbon offsets or credits (e.g. tons of CO_2 reductions) in open markets. However, obtaining carbon credits is time and resource intensive, with a registration process taking between 200 and 800 days. Not only is it costly to register within the carbon market, but the process of calculating and validating greenhouse gas emissions reductions requires consultation and validation with a third party. Therefore, using the carbon markets to fund, for example composting projects, may only be feasible when done on a large scale and may generally be more appropriate for middle- than low-income countries.

As of 2017, more than 700 CDM projects converting biomass to energy were listed by UNEP, compared to 281 engaged in wastewater treatment and 46 in composting, both reducing methane production (UNEP, 2017). A challenge, aside from the registration process, is that carbon markets fluctuate over time, with prices peaking at €30 per ton in 2006 and 2008, but dramatically lowering to about €5 per ton between 2012–2016, although there can be significant differences between countries and carbon pricing initiatives (World Bank et al., 2016). Composting projects that have received funding through the CDM include Waste Concern in Bangladesh, earning USD 1.5 million in carbon credits, and the Temesi integrated resource recovery center in Bali. For the latter, while a USD 1.5 million revenue in credit sales was expected, USD 70,000 was required upfront for financing quantification, certification, and registration in the CDM program which makes the CDM not well suited for small scale, community-based composting (Mitchell and Kusumowati, 2013).

In response to the weakening carbon credit market, a range of new results-based climate finance (RBCF) initiatives emerged including the World Bank's Pilot Auction Facility (World Bank et al., 2016). RBCF is particularly adept at helping to build an international carbon market. It is an approach where funding is conditional upon the verified achievement of, e.g. predefined emission reductions. This provides assurance to the funder and a continued financing flow for the recipient. The auction facility is a payment mechanism that sets a floor price on the future price of carbon through a public auction. The agreement is facilitated through a tradeable put option, which provides buyers the right, but not obligation, to sell carbon at the agreed-upon price at a future date. The auction encourages private sector investment in particular in methane reduction projects while efficiently disbursing limited public funds. Three successful auctions took place between 2015 and early 2017 (www.pilotauctionfacility.org; World Bank, 2015).

19.3 Technologies matching resource constraints

Infrastructure and technologies which acknowledge local opportunities and constraints are very important for any investments in low and middle-income countries if an innovation is to survive and succeed. One of the major drawbacks of composting but also wastewater treatment, as mentioned above, was reliance on technology not matching local capacities. Another typical barrier for the implementation of, e.g. organic waste processing is the lack of available land in urban vicinity, ideally free of cost given the social and environmental services RRR provides. Peri-urban land is a precious asset, also for the public sector, and not easily available for various reasons including negative public perceptions of waste management facilities. A secured long-term lease is however important as most RRR investments have a long payback period of, for example, at least seven years for organic waste recycling (ADB, 2011), and for wastewater treatment even longer. Only investments in energy efficiency, like in wastewater treatment facilities, can have much shorter payback periods of less than one to a maximal three years (Barry, 2007). The required land area for a RRR facility will depend on the size of the urban community to be served (i.e. the quantity of possibly available waste), the type of the waste volume matching technology (i.e. gravity thickening or drying beds for liquid fecal sludge), as well as the required peri-urban farm area to 'absorb' the recovered resources, based on a carefully stratified demand analysis (Otoo et al., 2016).

Different recovered resources have in this regard very different requirements, even if derived from the same number of households. Figure 279 shows, as an example, a two order of magnitude difference in the basic land requirement for the reuse of wastewater, urine, and fecal sludge (FS) for food production. Urine application requires the most land area given its high nitrogen concentration and alkalinity, which can easily be harmful to plants and limits the application rates. To support for example 100 ha of urban agriculture city-wide, the amounts generated daily in some public toilets might already be sufficient. On the other hand, wastewater-fed aquaculture and wastewater irrigation require the least land for making use of the volume of waste generated by 100,000 person equivalents (pe) of waste producers (Murray et al., 2011).

As compost is first of all a soil ameliorant with an impact on the soil lasting longer than any inorganic fertilizer, its application is often limited to once per year. In a comparative analysis, the low frequency makes compost reuse land-intensive with 400–500 ha required, assuming one application of 14 tonnes ha^{-1} yr^{-1}. Biogas production, for comparison, might require for the same waste input 200–300 ha, with less than 1 ha for the gas generation and storage, and the majority for farm land if the digestate is returned as soil conditioner. Similar to biogas itself, the reuse 'area' of briquettes is decentralized over many households and comparatively insignificant (Murray et al., 2011).

Land and energy demands can be negatively correlated as known from wastewater treatment where, e.g. Waste Stabilization Ponds (WSP) take in India about ten times the area than other treatment systems, while only using about 1–2 percent of their energy needs for the same amount of wastewater. Apart from the land-demanding WSP systems, Upflow Anaerobic Sludge Blanket (UASB) treatment requires relatively more land but the least amount of energy compared with other treatment systems

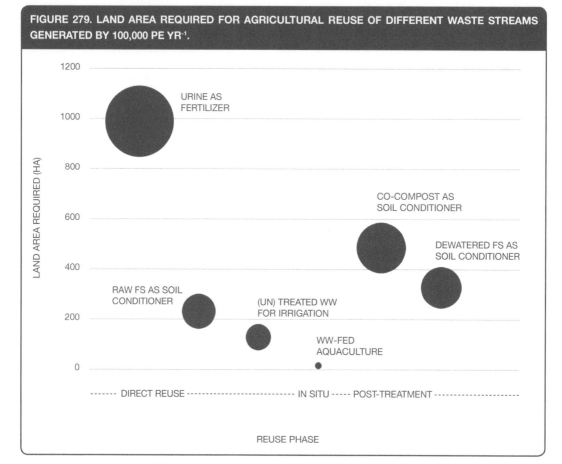

FIGURE 279. LAND AREA REQUIRED FOR AGRICULTURAL REUSE OF DIFFERENT WASTE STREAMS GENERATED BY 100,000 PE YR^{-1}.

Note: Land area for wastewater (WW) assumes three crop cycles ha^{-1} yr^{-1}; urine assumes two crop cycles ha^{-1} yr^{-1}; raw FS, dewatered FS and co-compost assume one application ha^{-1} yr^{-1}. Circle sizes reemphasize land area (Murray et al., 2011, modified).

common in India, opposite, e.g. to membrane based systems. Closely related are the O&M costs which are to 40–45% reflecting energy needs (CPCB, 2013), and can be strongly steered by choice of technology (Libhaber and Orozco-Jaramillo, 2013). A key criterion for O&M of wastewater treatment is uninterrupted energy supply, which the infrastructure especially in many low-income countries does not provide, resulting in a trajectory to failure for many treatment systems (Murray and Drechsel, 2011).

Thus, the availability of land and technologies with low energy demands or ideally in-house energy recovery constitute important components in support of an enabling environment for technology choice. An example is the different energy requirement for phosphorus (P) recovery, which drive the costs of P removal/recovery. However, as Figure 280 shows, there are also high recovery rates possible at lower costs.

In-house energy recovery is most appropriate for countries which can not provide uninterrupted power supply. Moreover, it has the highest potential for cost savings as verified in an increasing number of wastewater treatment plants worldwide (Lazarova et al., 2012).

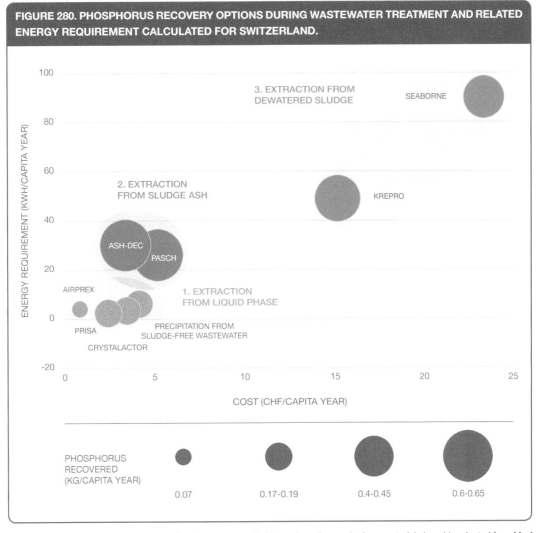

FIGURE 280. PHOSPHORUS RECOVERY OPTIONS DURING WASTEWATER TREATMENT AND RELATED ENERGY REQUIREMENT CALCULATED FOR SWITZERLAND.

Note: The energy requirements consider also the energy needed to produce the required raw materials (graphic adapted from Morf and Koch (2009); USD 1 = CHF 1.1).

19.4 Local capacities and stakeholder acceptance

Depending on the nature and size of the RRR project, private partners can range from local enterprises to international companies. Smaller enterprises are more common in the bio-energy and waste-to-compost sectors than in wastewater treatment and often struggle with low credit history, limited capacity to present a bankable project and to realize the right mix of debt and equity investment. In fact, in many low- and middle-income countries, private partners with innovation experience in green technology and knowledge about business development, finance access, and reuse markets are still an exception, in particular at the local government level. The reason for the generally underdeveloped business capacity is that, especially in the municipal waste and sanitation sectors, only one business model determines so far, the game: the public partner pays. As a result, innovation capacity remains low and many tenders only attract the usual sanitation- or waste management-based enterprises with very limited experience, e.g. with the agricultural reuse market, carbon financing, etc. PPP matchmaking remains a significant challenge and private sector participation is not a guarantee for a viable business, in particular in Africa but also large parts of Asia.

Capacity related limitations do not only concern possible business partners, or their capacity to maintain more advanced technology, but also other stakeholders along the value chain, especially where the business requires environmental awareness as the adoption of eco-innovations is heavily dependent on consumers' education and attitudes (Beltramello et al., 2013). Source separation, for example, remains a significant challenge for supplying compost station with quality waste, and where recycled waste products enter the market, social perceptions can function both as a promoter and as a barrier for resource recovery and market access, for example where customers have a non-waste derived alternative. Results-based-financing can support in this respect RRR by catalyzing the design of projects that go beyond resource recovery but provide incentives to households for behavior change, for example, towards waste separation and recycling (Banna et al., 2014). Comparing common experience with the acceptance of waste-derived resources across the presented case studies in sections II to IV, biogas and electricity have the best reception, followed by dry fuel, waste compost, and finally treated wastewater which is often facing the largest acceptance challenges. These challenges might not only relate to possible health risks but also the basic reuse suitability, for example in farming. In the case of Pakistan, treated wastewater got rejected due to its higher salinity than raw wastewater. In the case of compost, farmers might expect more a fertilizer than a soil ameliorant, and struvite crystals are (only) a slow-release fertilizer. Therefore, it is important to understand market expectations, adjust as far as possible the recovered resource to users' preferences, and accompany the introduction of recovered resources with information and training in their use.

In view of reclaimed water, public acceptance depends on the kind of reuse, with more distant uses, such as landscape irrigation, being the most preferred option, while potable reuse receives most hesitation (Drechsel et al., 2015). Different factors come into play when promoting social acceptance and behavior change, such as household and gender specific knowledge, trust, attitudes toward the environment, as well as the availability of (perceived safer) alternatives. These social aspects can include religious, cultural, and aesthetic values. As any waste-related facility can generate public questions and opposition, a high level of transparency and public disclosure of information is required. Stakeholder participation in re-use planning, awareness and capacity development programs including market surveys are therefore crucial to address any possible adoption challenges (ADB, 2011; Dolnicar and Saunders, 2006; Holmgren et al., 2015).

References and further readings

Ali, M. (ed.). 2004. Sustainable composting: Case studies and guidelines for developing countries. Leicestershire, UK: WEDC and SANDEC/EAWAG.

Asian Development Bank (ADB). 2014. From toilets to rivers: Experiences, new opportunities, and innovative solutions. Manila, Philippines: Asian Development Bank.

Asian Development Bank (ADB). 2016. Financing mechanisms for wastewater and sanitation. Manila, Philippines: Asian Development Bank.

Asian Development Bank (ADB). 2011. Toward sustainable municipal organic waste management in South Asia: A guidebook for policy makers and practitioners. Manila, Philippines: Asian Development Bank.

Banna, F.M., Bhada-Tata, P., Ho, R. Y-Y., Kaza, S. and Lee, M. 2014. Results-based financing for municipal solid waste (Vol. 2): Main report (Urban development series knowledge papers vol. 20). Washington, DC: World Bank Group. http://documents.worldbank.org/curated/en/237191468330923040/Main-report (accessed 5 Nov. 2017).

Barry, J.A. 2007. WATERGY: Energy and water efficiency in municipal water supply and wastewater treatment. Washington, D.C.: The Alliance to Save Energy.

Beltramello, A., Haie-Fayle, L. and Pilat, D. 2013. Why new business models matter for green growth (OECD Green Growth Papers, 2013-01). Paris: OECD Publishing.

Bjornali, E.S. and Ellingsen, A. 2014. Factors affecting the development of clean-tech start-ups: A literature review. Energy Procedia 58: 43–50.

Central Pollution Control Board (CPCB). 2013. Performance evaluation of sewage treatment plans in India funded under NRCD. Delhi: Central Pollution Control Board.

Dolnicar, S. and Saunders, C. 2006. Recycled water for consumer markets: A marketing research review and agenda. Desalination 187(1–3): 203–214.

Drechsel, P. and Keraita, B. (eds.). 2014. Irrigated urban vegetable production in Ghana: Characteristics, benefits and risk mitigation (2nd ed.). Colombo, Sri Lanka: International Water Management Institute (IWMI).

Drechsel, P., Mahjoub, O. and Keraita, B. 2015. Social and cultural dimensions in wastewater use. In Drechsel, P., Qadir, M., Wichelns, D. (eds.). Wastewater: Economic asset in an urbanizing world. Dordrecht, Netherlands: Springer. pp. 75–92.

Dubois, M., Hoogmartens, R., Van Passel, S., Van Acker, K. and Vanderreydt, I. 2015. Innovative market-based policy instruments for waste management: A case study on shredder residues in Belgium. Waste Management & Research 33(10): 886–893.

Eyraud, L., Zhang, C., Wane, A.A. and Clements, B.J. 2011. Who's going green and why? Trends and determinants of green investment (IMF Working Paper WP/11/296). Washington D.C.: International Monetary Fund. www.imf.org/en/Publications/WP/Issues/2016/12/31/Who-s-Going-Green-and-Why-Trends-and-Determinants-of-Green-Investment-25440 (or https://goo.gl/d4ApCK; accessed 6 Nov. 2017).

Global Water Intelligence (GWI). 2010. Municipal water reuse markets 2010. Media Analytics, Ltd., Oxford.

Gorelick, J. 2017. Financing Africa's cities. Bonds and Loans. 9 March 2017. www.bondsloans.com/news/article/1282/financing-africas-cities (accessed 6 Nov. 2017).

Government of India. 2016. Office memorandum: Policy on promotion of city compost (F. No. 11026/14/2015-M&E). 10 February 2016. New Delhi: Ministry of Chemicals and Fertilizer.

Griffith-Jones, S., Antonio Ocampo, J. and Spratt, S. 2012. Financing renewable energy in developing countries: Mechanisms and responsibilities. Report by the Overseas Development Institute (ODI) in partnership with the Deutsches Institut für Entwicklungspolitik (DIE) and the European Centre for Development Policy Management (ECDPM). www.stephanygj.net/papers/Financing_Renewable_Energy_in_Developing_Countries.pdf (accessed 6 Nov. 2017).

Hasanbeigi, A., Lu, H., Williams, C. and Price, L. 2012. International best practices for pre-processing and co-processing municipal solid waste and sewage sludge in the cement industry. Ernest Orlando Lawrence Berkeley National Laboratory. http://eetd.lbl.gov/sites/all/files/publications/co-processing.pdf (accessed 6 Nov. 2017).

Holmgren, K.E., Li, H., Verstraete, W. and Cornel, P. 2015. Sate of the art compendium report on resource recovery from water. London, UK: IWA resource recovery cluster.

International Finance Corporation (IFC). 2013. Mobilizing public and private funds for inclusive green growth investment in developing countries. Washington D.C.: International Finance Corporation.

International Renewable Energy Agency (IRENA). 2012. Financial mechanisms and investment frameworks for renewables in developing countries. Abu Dhabi, UAE.

International Solid Waste Association (ISWA). 2015. Six reports on the circular economy and waste management. www.iswa.org/resourcemanagement (accessed 6 Nov. 2017).

Kaza, S., Yao, L. and Stowell, A. 2016. Sustainable financing and policy: Models for municipal composting (Urban development series, Knowledge Papers 24). Washington, D.C.: World Bank Group.

KPMG. 2013. The KPMG Green Tax Index 2013: An exploration of green tax incentives and penalties. https://assets.kpmg.com/content/dam/kpmg/pdf/2013/08/kpmg-green-tax-index-2013.pdf (accessed 6 Nov. 2017).

Kruger, A. 2017. How to make a water project bankable: Strengthen the financial capital planning. Waterfront 1 (May): 14.

Lazarova, V., Choo, K.H. and Cornel, P. (eds.). 2012. Water-Energy interactions in water reuse. London: IWA Publishing.

Libhaber, M. and Orozco-Jaramillo, A. 2013. Sustainable treatment of municipal wastewater. IWA's Water 21 (October 2013): 25–28.

Mandri-Perrot, C. and Stiggers, D. 2013. Public private partnerships in the water sector. London: IWA Publishing.

Matter, A., Ahsan, M., Marbach, M. and Zurbrügg, C. 2015. Impacts of policy and market incentives for solid waste recycling in Dhaka, Bangladesh. Waste Management 39: 321–328.

Michaelowa, A. and Jotzo, F. 2005. Transaction costs, institutional rigidities and the size of the clean development mechanism. Energy Policy 33(4): 511–523.

Ministère du Développement Durable, de l'Environnement et des Parcs (MDDEP). 2012. Bannissement des matières organiques de l'élimination au Québec: état des lieux et prospectives. Direction des matières résiduelles et des lieux contaminés, Service des matières résiduelles www.mddelcc.gouv.qc.ca/matieres/organique/bannissement-mat-organ-etatdeslieux.pdf (accessed 6 Nov. 2017).

Mitchell, C. and Kusumowati, J. 2013. Is carbon financing trashing integrated waste management? Experience from Indonesia. Climate and Development 5: 268–276.

Morf, L. and Koch, M. 2009. Synthesebericht für interessierte Fachpersonen, Zürcher Klärschlammentsorgung unter besonderer Berücksichtigung der Ressourcenaspekte. Baudirection Kanton Zuerich.

Murray, A. and Drechsel, P. 2011. Why do some wastewater treatment facilities work when the majority fail? Case study from the sanitation sector in Ghana. Waterlines 30(2), 135–149.

Murray, A., Cofie, O. and Drechsel, P. 2011. Efficiency indicators for waste-based business models: Fostering private sector participation in wastewater and faecal-sludge management. Water International 36(4): 505–521.

Muspratt, A. 2016a. Make room for the disruptors: While desperate for innovation, the sanitation sector poses unique structural challenges to startup companies. www.linkedin.com/pulse/make-room-disruptors-while-desperate-innovation-sector-muspratt (accessed 6 Nov. 2017).

Muspratt, A. 2016b. How do we leverage the speed and innovation of small companies in the inherently slow and bureaucratic sanitation sector? www.linkedin.com/pulse/how-do-we-leverage-speed-innovation-small-companies-slow-muspratt? (accessed 6 Nov. 2017).

Organisation for Economic Co-operation and Development (OECD). 2011. Towards green growth: Monitoring progress. Paris: OECD. www.oecd.org/greengrowth/48224574.pdf (accessed 6 Nov. 2017).

Organisation for Economic Co-operation and Development (OECD). 2012. G20/OECD policy note on pension fund financing for green infrastructure and initiatives. Paris: OECD. www.oecd.org/g20/

topics/energy-environment-green-growth/S3%20G20%20OECD%20Pension%20funds%20 for%20green%20infrastructure%20-%20June%202012.pdf (accessed 6 Nov. 2017).

Oliver, P. 2016. Green bonds for cities: A strategic guide for city-level policymakers in developing countries. Climate Policy Initiative. https://climatepolicyinitiative.org/wp-content/uploads/2016/12/Green-Bonds-for-Cities-A-Strategic-Guide-for-City-level-Policymakers-in-Developing-Countries.pdf (or https://goo.gl/WwsdYL; accessed 6 Nov. 2017).

Oliver, P., Mazza, F. and Wang, D. 2016. Water financing facility. Global Innovation Lab for Climate Finance. http://climatefinancelab.org/wp-content/uploads/2016/01/160623-Lab-WFF-Report.pdf (accessed 6 Nov. 2017).

Otoo, M., Drechsel, P., Danso, G., Gebrezgabher, S., Rao, K. and Madurangi, G. 2016. Testing the implementation potential of resource recovery and reuse business models: From baseline surveys to feasibility studies and business plans (Resource Recovery and Reuse Series 10). Colombo, Sri Lanka: International Water Management Institute (IWMI); CGIAR Research Program on Water, Land and Ecosystems (WLE).

Platz, D. (2009), Infrastructure finance in developing countries: The potential of sub-sovereign bonds (DESA Working Paper 76). New York: United Nations Organization. www.un.org/esa/desa/papers/2009/wp76_2009.pdf (accessed 6 Nov. 2017).

Trémolet, S. 2011. Identifying the potential for results-based financing for sanitation (WSP Working Paper). World Bank, Water and Sanitation Program (WSP) Scaling Up Rural Sanitation Initiative. Washington, DC: The World Bank.

United Nations Environment Programme (UNEP). 2017. CDM projects by type. www.cdmpipeline.org/cdm-projects-type.htm (accessed 6 Nov. 2017).

Water and Sanitation Program (WSP). 2016. Approaches to capital financing and cost recovery in sewerage schemes implemented in India: Lessons learned and approaches for future schemes. New Delhi: Water and Sanitation Program Guidance Note.

Winpenny, J. 2003. Financing water for all: Report of the World Panel on Financing Water Infrastructure. World Water Council; Global Water Partnership. www.oecd.org/greengrowth/21556665.pdf (accessed 6 Nov. 2017).

World Bank, Ecofys, and Vivid Economics. 2016. State and trends of carbon pricing 2016 (October). Washington, DC: World Bank.

World Bank. 2013. Financing sustainable cities: How we're helping Africa's cities raise their credit ratings. Washington, D.C.: World Bank Group. www.worldbank.org/en/news/feature/2013/10/24/financing-sustainable-cities-africa-creditworthy (accessed 6 Nov. 2017).

World Bank. 2015. First pilot auction to capture methane a success. http://www.worldbank.org/en/news/press-release/2015/07/17/first-pilot-auction-to-capture-methane-a-success (accessed 6 Nov. 2017).

World Bank. 2016. Sustainable financing and policy models for municipal composting (Urban Development Series Knowledge Papers 24). Washington, DC: World Bank.

World Bank. 2017. Enabling the business of agriculture 2017. Washington, DC: World Bank. doi:10.1596/978-1-4648-1021-3. License: Creative Commons Attribution CC BY 3.0 IGO.

World Health Organization (WHO). 2006a. WHO guidelines for the safe use of wastewater, excreta and greywater (Vol II: Wastewater use in agriculture). Geneva: WHO.

World Health Organization (WHO). 2006b. WHO guidelines for the safe use of wastewater, excreta and greywater (Vol III: Wastewater and excreta use in aquaculture). Geneva: WHO.

World Health Organization (WHO). 2015. Sanitation safety planning manual for safe use and disposal of wastewater, greywater and excreta. Geneva: WHO.

Notes

1. The authors acknowledge the input of Chris Zurbrügg (SANDEC/EAWAG) and Ashley (Murray) Muspratt (Pivot Ltd.) in an earlier version of the chapter.
2. The views expressed in this document are those of the author, and do not necessarily reflect the views and policies of the Asian Development Bank (ADB), its Board of Directors, or the Governments they represent.
3. http://ec.europa.eu/DocsRoom/documents/19742 (accessed Nov. 4, 2017).
4. Figure 277 and 278 are adaptations of an original work by the World Bank. Views and opinions expressed in the adaptation are the sole responsibility of the authors of the adaptation and are not endorsed by the World Bank.
5. www.greenfuture.sg/2015/02/16/2015-guide-to-singapore-government-funding-and-incentives-for-the-environment (accessed Nov. 4, 2017).
6. https://assets.kpmg.com/content/dam/kpmg/pdf/2013/08/kpmg-green-tax-index-2013.pdf (accessed Nov. 4, 2017).

FRUGAL INNOVATIONS FOR THE CIRCULAR ECONOMY: AN EPILOGUE

Jaideep Prabhu

Jawaharlal Nehru Professor of Indian Business & Enterprise, and Director of the Centre for India & Global Business (CIGB) at the Judge Business School, University of Cambridge, England

FRUGAL INNOVATIONS FOR THE CIRCULAR ECONOMY: AN EPILOGUE

Planet Earth is at a crossroads. On the one hand, over 4 billion people, many of whom live in developing countries, face unmet needs in core areas such as food, energy, housing and transport. On the other hand, meeting the needs of these large populations (while continuing to satisfy the needs of the other 3 billion who live in developed economies) poses a threat to the finite resources available on the planet.

Can we sustain the growing economic, social and environmental pressures caused by increases in global population, urbanization, food consumption and waste generation? If so, how?

One thing is clear: the 20th century linear model of urban metabolism is no longer sustainable. That model has already created many environmental and health problems in hungry and thirsty cities. So, policies and investments will be needed to transform this linear model into a circular one. Indeed, the UN's Sustainable Development Goals (SDGs) recognise this and focus on ensuring sustainable consumption and production patterns, and promoting greater recycling, recovery and reuse of resources. Moving to a circular economy model will not only help mend broken geochemical and hydrologic cycles, but it will also determine how society and economies cope with increasingly important rural-urban interdependencies.

The transition to a more balanced interplay of environmental and economic systems can be achieved through closing-the-loop of production patterns within an economic system. This will also help increase the efficiency of limited resources. But such a transition will require innovation, and moreover an approach to innovation that is itself frugal, flexible and inclusive. Such frugal and inclusive technical, social or economic innovations will be more responsive to limits on resources and sensitive to their management, whether financial, material or institutional. They will also require the use of a range of methods designed to turn constraints into opportunities. For instance, by minimising the use of resources in development, production and delivery, and by supporting resource recovery and reuse, such a frugal approach to innovation can result in significantly lower cost products and services. Successful frugal innovations will be able to outperform their alternatives, and can be made available at large scale, as the many cases and models presented in this catalogue show. Often, but not always, frugal innovations also have an explicitly social mission – increasing the number of customers/members and expanding the service for greater social impact. This is of particular relevance to the sanitation and waste sectors.

Thus, there are obvious and strong links between frugal innovation and the circular economy. This begins with learning how to create value out of what others consider "waste". Thus, entrepreneurs as well as urban utilities can lead the transition towards a circular economy and become resource stewards by employing frugal innovation techniques and mindsets. The rise of the circular economy can, for instance, help with the adoption of nutrient, water and energy recovery models in support of agriculture and other sectors that the urban metabolism depends on. Cost competitive and successful business entities are very likely to lead the adoption of integrative approaches for multi resource recovery, while capitalizing on pathway drivers, enablers and boosters. Yet, they typically continue to face roadblocks in the shape of draconian regulatory environments and opaque market conditions. Even in highly developed regulatory environments such as in Europe, it is only now that concerted efforts are being made to create regulatory frameworks which see waste as a resource and not only a hazard.

Meanwhile, in particular in middle-income countries, resource recovery and reuse initiatives abound. However, sustainable programs at scale are rare. In particular, financial and institutional sustainability is rare and remains a critical challenge. In these countries, an enabling environment is only slowly emerging, especially with respect to financial incentives for green businesses; however, these initiatives are likely to gain pace in the future.

This catalogue is intended to serve as a vital input for decision-making in resource recovery and reuse businesses in the urban sanitation-agriculture interface. It presents business models and cases developed from new thinking in different geographical settings in Asia, Africa and Latin America. In these settings, an enabling environment for frugal innovation in the circular economy is only now emerging. Typically, the trajectory from informal to formal, unplanned to planned and unsafe to safe use remains unclear with an often under-tapped potential for frugal innovation. The challenge in many of these cases is to balance success with safeguarding public health, and how best to support these emerging leaders of a circular economy in their capacity-building needs. To that end, this catalogue is designed to be adopted as a handbook at universities for graduate training in applied economics, business schools, resource economics, marketing, environmental studies, civil engineering, international development and public policy. The catalogue is also a compendium of business cases to support departments at universities in Europe and elsewhere that have recently launched Masters programs on the SDGs, to train the next generation of experts for the implementation of SDGs where most efforts are needed, i.e., in low- and middle-income countries, for greater success.

It is my strong belief that this handbook is a vital resource for all those seeking to help the world grow sustainably and equitably through the 21st century and beyond. I am confident that it will soon become the standard reference for all those who study and practice these important issues, in developed and developing countries alike.

Index

Page numbers in italics refer to figures.
Page numbers in bold refer to tables.
Page numbers with "n" refer to notes.

Agriquatics (Bangladesh) 608–616
   PRISM Bangladesh 607,
      608–610, 612, 616n1
   see also duckweed-based
      wastewater treatment
agro-waste, on-site combined heat
   and power 278–283
agro-waste, power 215–221
   see also Ravikiran Power project
Amani Doddakere tank (ADT), revival 710–719
   environmental impact assessment
      (EIA) 710–711
   Hoskote Municipality 710, 711,
      712, 713, 715–717, 719n3
   lift irrigation system 710, 716
   Yelemallappa Shetty tank (YMST)
      710–712, 713, 715, 716, 717, 719n7
Amman, Jordan see As Samra wastewater
   treatment plant (WWTP)
animal waste 163, 166, 183, 448
   swine manure, power from see
      3S Program (Brazil)
   see also ProBio Humibac (ProBio); SuKarne
      methane recovery project; Santa
      Rosillo, Peru, rural electrification
Animal Waste Management System
   (AWMS) 174–175, 448, 475
aquaculture (water reuse) 12, 607, 793, *794*
   Agriquatics (Bangladesh) 606–616
   fishfeed production 607–608, 610, 614,
      615
   value chain **20**, *548*, 549–550, 605,
      614, 622, 631–638, *633*

wastewater 12, 109, 548–551, 605,
   607–608, 610, 614, 617–630,
   631–632, 635, 636, 779, 793, **794**
wastewater treatment (Ghana)
   228, 617–630, *628*
aquifer recharge see managed aquifer recharge
Asian Development Bank (ADB) 151,
   259, 782, 788, 800n2
Athi Water Service Board (AWSB) 115, 118–119
A2Z Infrastructure Private Limited
   (A2Z-PL) 381–390
   Fertilizer Control Order (FCO) 383–384, 386
   municipal solid waste (MSW)
      381–382, 383, 385, 386, 388
   open-windrow composting system 386
   process *387*
   public-private partnership (PPP)
      381, 382, 383, 390
As Samra wastewater treatment plant
   (WWTP) 549–550, 554, 641,
   642–653, 656–657, 660
   Jordan Valley Authority (JVA) 647
   King Talal Reservoir (KTR) 643, 652

bagasse, combined heat and power from
   see Mumias Sugar Company Ltd
   (MSC); Shri Someshwar Sahakari
   Sakhar Karkhana (SSSSK)
Balangoda, Sri Lanka see Balangoda
   Compost Plant (BCP)
Balangoda Compost Plant (BCP) 341–350
   chemical and organic fertilizers
      342–343, 345, 346
   fecal sludge (FS) 341–350
   MSW-based compost 346–348
   non-degradable waste 341, 342, 343
   Sri Lanka Standards Institution (SLS) 346

INDEX

Bangalore Honey Suckers 508–515
Bangalore Water Supply and Sewerage Board (BWSSB) 510
Bangalore, India 667, 725
   see also Amani Doddakere tank (ADT), revival; Bangalore Honey Suckers; Karnataka Compost Development Corporation Limited (KCDC); Terra Firma Biotechnologies Limited (Terra Firma); Wipro Employees Canteen
Bangkok, Thailand see Thai Biogas Energy Company (TBEC)
Bangladesh 111, 550, 631, 635, 636, 736, **780**, 783, 792
   see also Dhaka, Bangladesh; Mirzapur, Bangladesh
Barcelona, Spain see Llobregat delta, Spain
Below Poverty Line (BPL) 205
beyond cost recovery, wastewater 605, 637, 695
bidding, wastewater 578, 705, 748, 754
Bihar, India 382
   see also Husk Power Systems Inc. (HPS)
Bihar Renewable Energy Development Agency (BREDA) 208–209
Bill and Melinda Gates Foundation (BMGF) 119
biochar 73, 77, 205, 210, 218, 220–221
Biochemical Oxygen Demand (BOD) 109
'biocycle economy' 5, 6
bio-ethanol
from cassava waste see ETAVEN C.A. (ETAVEN)
bio-ethanol and chemical products, from agro and agro-industrial waste 307–313
biogas 34–35, 36, 150–151, 793
   from agro-industries wastewater see Thai Biogas Energy Company (TBEC)
   from agro-waste 215, 278–279, 308
   from alcohol production 297
   from animal and slaughterhouse waste see SuKarne methane recovery project; Nyongara Slaughter House
   from cassava waste 289
   from fecal sludge 93, 94–95, 98, 100, 103–113, 114–123; see also biogas from fecal sludge (community level); International Committee of Red Cross (ICRC); Sulabh International Social Service Organization; TOSHA 1
   internal consumption 133–141
   from kitchen waste 93, 95, 97, 98, 100, 143–148; see also biogas from kitchen waste; Wipro Employees Canteen
   from municipal solid waste (MSW) 411–412, 413, 417, 418
   policies, regulations, and guidelines 778–784, **785–786**
   in prisons 93–102
   and wastewater 550, 591, 647, 652–653
   see also International Committee of Red Cross (ICRC); TOSHA 1
biogas from fecal sludge (community level) 124–132
   toilet complex model 124–126, *125*
   see also waste management and sanitation
biogas from kitchen waste 143–148
   BOOT arrangements 143, 144, 145
Biogas Sector Partnership Nepal (BSP-N) 94, 97
Biogas Support Programme (BSP) 97
bio-methanation 133, 137, 174, 175, 199, 215, 411, 417
biosolids, and wastewater for fruit trees see National Sanitation Utility (ONAS)
blended wastewater 643
bonds *646*, 786, 788, 789
Brazil 266, 288, 294, 786, 789
   see also Concordia, Brazil
Brazilian Development Bank (BNDES) 166, 168
briquettes from municipal solid waste (MSW) 35, 61–71, 82–90,
   compost production 63, *64*, **68**, 77, 78, 86, *86*, 87, 88, 89
   manufacturing of improved cook stoves 87
   see also Coopérative Pour La Conservation De L'Environement (COOCEN); Eco-Fuel Africa (EFA)
briquettes from agro-waste 52–60
   compost production 56
   manufacturing of improved cook stoves and ovens 56
   see also Eco-Fuel Africa (EFA); Kampala Jellitone Suppliers (KJS); municipal solid waste (MSW)
Bruhat Bengaluru Mahanagara Palike (BBMP) 401, 402, 412, 413, 415
Build, operate and transfer (BOT) model 36, 103, 106, 186, 275, 554, 558, 572, 640, 656, 787

viability gap funding 642, 644,
  645, **646**, 647, 651
Burkina Faso 111
  see also ECOSAN-EU project
business cases and models, defining
    and analyzing 17–30
  assessment of business cases 17–22, *19*, **20**
  business risks and risk mitigation
    24–27, **25–26**
  categorization 20
  definitions 17
  development of business models 22–23
  limitations 30
  performance potential 27–29, **27–29**, *29*
Business Model Canvas (BMC) 22–23

Cairo, Egypt 556–568
capital purchase business model 540
capital subsidies 227, 789
Carabobo, Venezuela see ETAVEN
    C.A. (ETAVEN)
Carbon Disclosure Project (CDP) 10
carbon financing 48, 150, 792
carbon market 55, 218, 429, 600, 791, 792–793
Central Electricity Authority (CEA) 205
Centre for Innovations and Technology
    Transfer (CITT) 98
certification 87, 337, 376, 443, 468,
    476, 502, 600, 731, 791
  branding 356
  Fair Trade 451, 452, 456, 457
  guidelines for 444, 473, 482
  KEBS 364, 366, 370, 455
  process of 169
  of product, third-party 322, 328, 373,
    375, 444, 448, 470, 473, 480, 482
  standards 349
Certified Emission Reductions (CERs)
    172–173, 175, 193, 194, 195,
    200, 242, 393, 398, 433n1
charcoal 40, 41–50, 62–70, 72–80
Chemical Oxygen Demand (COD) 109
circular economy 4–12, 802–803
  biocycle economy 5, 6
  business cases, opportunities, and
    business models 10–12
  corporate social and environmental
    responsibility 8–10
  cost savings 5, 7, 8, 11

  externalities 11–12
  food chain 10
  food sector 9
  organic waste, potential 6, 7
  probability of replication 10–11
  World Health Organization
    (WHO) 12, 778–779
  see also waste management and sanitation
civil society organization (CSO) 114
Class A biosolids 784
Clean Development Mechanism (CDM) 56,
    85, 127, 145, 162, 172–173, 182, 235
co-composting 347–348, 506, 525, 584, 589
co-marketing **781**, 790, 791
combined heat and power (CHP) units 177–178
Comercial Industrial Delta SA (CIDELSA) 153
Commissariat Régional de Développement
    Agricole (CRDA) 569, 571,
    572–575, 576, 577–578
commodity driving change,
    wastewater **20**, 745–759
Communal Energy Services Unit
    (USEC) 155–156
community-based organizations
    (CBOs) 362–370
  sanitation services 124, 127
  TOSHA 1 114–115, 116, **120**, 121, *122*
  see also under specific case studies
composting
  large-scale for revenue generation
    see composting, large-scale
    for revenue generation
  fecal sludge from public toilets see
    Rwanda Environment Care (REC)
  municipal solid waste (MSW) see under
    municipal solid waste (MSW)
composting, partially subsidized
    (district level) 351–358
  carbon credits 356
  large-scale operation as public-
    private partnership 355
composting, large-scale for revenue
    generation 379–380, 434–446
  consultancy services and franchising
    438–439, *440–441*
  energy generation and carbon credit
    sales 439–443, *442–443*
  see also A2Z Infrastructure
    Private Limited (A2Z-PL)

INDEX

composting, subsidy-free community-based *see* subsidy-free community-based composting
compost production for sustainable sanitation service delivery 496–503
   franchise model 498–500, *500–501*
   see also ProBio Humibac (ProBio)
Concordia, Brazil 162
   see also 3S Program (Brazil)
Confined Animal Feeding Operations (CAFOs) 174–175
Coopérative Pour La Conservation De L'Environement (COOCEN) 61–71
   Kigali 62, 63, 65, 66, 68, 69
   public-private partnership (PPP) 61, 62, 63
   Rwanda development strategy 62–63
Corporate Social Responsibility (CSR) 8–9, 10, 134–135, 136–137, 141
Corporate Social Responsibility (CSR), business model for 733–744
   behavior change, triggering 739–740
   Business Environmental Performance Initiative (BEPI) 737
   farm-based interventions 737
   Foreign Trade Association (FTA) 737
   Global Social Compliance Programme (GSCP) 737
   informal wastewater use, challenge of 734
   post-harvest interventions 737–740
   responsibility and business interests 736
   risk barriers 735
   triggers 739–740
   wastewater treatment, support 736
cost recovery (wastewater) 7, 11, **20**, 319, 341, 605, 637, 640, 641, 651, 677, 686, 710, 716, 717, 752, 753, 787, 789
   agriculture, forestry and aquaculture 549, 550, 551, 554–555
   and biosolids, for fruit trees 569, 572, 574, 578, 580–581
   design for reuse and replication 586, 589, 592
   for fruit and wood production 556–557, 558, 559, 561, 563, 565, 566
   fruit and wood production (Egypt) 558, 559, 561, 565, 566 for greening the desert 596, 599, 602
   inter-sectoral water exchange 691, 695, 697
   private sector investment at scale 659, 661
   for production of fish feed 612, 614
   suburban wastewater treatment for reuse 584, 586, 590, 592, 593
cost sharing and risk minimization 640–641
   private sector investment 656–663
   viability gap funding 642–653
Covered Lagoon Bio-Reactor (CLBR) technology 274
Culiacan, Mexico *see* ProBio Humibac (ProBio); SuKarne methane recovery project

Dagorretti, Kenya *see* Nyongara Slaughter House
Design Build Finance Operate and Transfer (DBFOT) 787
Design Build Operate and Transfer (DBOT) 787
Deutsche Gesellschaft für Internationale Zusammenarbeit
   see Gesellschaft für Internationale Zusammenarbeit (GiZ)
Dhaka, Bangladesh 607, 611
   see also Waste Concern Group (Bangladesh)
Dhaka City Corporation (DCC) 423
Drarga, Morocco *see* Drarga wastewater treatment plant
Drarga wastewater treatment plant 584–594, 599
   facilitators/dealmakers 589
   National Electricity and Water Company (ONEE) 584, 589
   Souss-Massa River Basin (SRB) 585
   Water Resources Sustainability (WRS) project 585
driver of change, wastewater **20**, 733–744, 745–759, 760–774
dry fuel 69, 796
duckweed 605, 606–614, 631, 634–637
   see also Agriquatics (Bangladesh)
duckweed-based wastewater treatment 607, 608, 610–611, 612, 613, 614
DuduTech 450–458
   Integrated Pest Management (IPM) 450
   Kenya Bureau of Standards (KEBS) 455

Eco Biosis S.A (Eco Biosis) 296–306
   BioDisperSis VC® 296, 298, 299, 302–303, 304
   Tembec 298
   WestRock 298

Eco-Fuel Africa (EFA) 35, **43**, 72–81
   eco-fuel press machine 72, 74
   prices of briquettes **73**
ecological sanitation 7
ecological sanitation (EcoSan) system 528
Economic Development and Poverty Reduction Strategy (EDPRS) 95
ECOSAN-EU project 527–537
   community-based organization (CBO) 528, 529, 530–531, 532, 533–534, 535–536
   German Corporation for International Cooperation (GIZ) 527, 528
   National Water and Sanitation Authority (ONEA) 527, 528
   urine diversion dehydrating toilet (UDDT) 528, 529, 531, 532–534, 535
   Water and Sanitation for Africa (WSA) 528
Egypt 554, **570**, 580, 600
   see also Cairo, Egypt
El Berka wastewater and agroforestry system (Egypt) 556–568
   Egyptian Environmental Affairs Agency (EEAA) 563, 565
   Greater Cairo Sewage Water Company (GCSWC) 557, 561
   Holding Company for Water and Wastewater (HCWW) 557, 558–559, 561, 562, 566
   Ministry of Health and Population (MHP) 563
   Ministry of State for Environmental Affairs (MSEA) 563
   sewage sludge 565, 597
   wastewater, supply of 564–565
enabling environment 778–796, 802–803
end-of-waste 543, 781–782
energy recovery 92 (subject of whole Section II)
   overview of business models 34-37
   see also briquettes from agro-waste; waste, recovering energy
Energy Regulatory Commission (ERC) 252
Energy Service Companies (ESCOs) 208
enriched compost production, from sugar industry waste see Pondicherry Agro Service and Industries Corporation Limited (PASIC)
environmental and health risk assessment 8–9, 24–26
   see within every Business Model under Safety, Environmental and Health Risks

environmental impact assessment (EIA) 119, 710–711
Environmental Protection Agency (EPA)
   see under Ghana Environmental Protection Agency and United States Environmental Protection Agency US EPA
ETAVEN C.A. (ETAVEN) 286–295
   Andean Community of Nations (CAN) 289–291
   Petroleos de Venezuela S.A. (PDVSA) 288
   Yaretanol 286, 287, 288, 289, 294
excreta 93, 104–110, 319, 502–503, 524–525, 528–535

Faisalabad, Pakistan 745–757, **746**, 752
   see also driver of change, wastewater
farmers as drivers of change 760–774
fecal sludge (FS) 343
   see also biogas from; biogas from fecal sludge (community level); TOSHA 1
   to compost, from public toilets see Rwanda Environment Care (REC)
   see also Bangalore Honey Suckers; composting, partially subsidized (district level); Rwanda Environment Care (REC)
   see septage
fecal sludge treatment, outsourcing 505, 506, 516–522
   see also fecal sludge (FS)
feed-in-tariff 265, 791
financial instruments 24, 788, 789
firewood 40, 41–49, 52, 56, 73–77, 88, 94–100
Flush Compost toilet (FCT) 104
forestry, wastewater 548–551
Friends of Ramsar Site (FORS) 761–762, 763, 766–768, 769, 774n10
Frugal innovation 802–803
Fuel from Wastes Research Centre (FWRC) 41
fuelwood see firewood

gasifiers 54, 83, 199, 203–211
gender 24, 27, 28, 796
   see within every Business Model under Key Characteristics and Social Equity Related Risks
Gesellschaft für Internationale Zusammenarbeit (GIZ) 41, 98, 119, 527-528, 533, 563

Ghana 111, 505, 506, 737, 740, 782, 784, 790
  farmers as drivers of change 760–774
  see also Kumasi, Ghana
Global Environmental Facility's Small Grants Programme (GEF-SGP) 62
Ghana Environmental Protection Agency 622, 626, 764, 768, 774n
Global Water Intelligence (GWI) 549, 605
Government Employees Pension Fund (GEPF) 789
Greater Cairo Sewage Water Company (GCSWC) 557, 561
Greenfield Crops (GC) 333–340
  "Pillisaru" project 334, 338
  public-private partnership (PPP) 333, 334, 335, 339
  Sri Lanka Standards Institution (SLS) 337
greenhouse gas (GHG) 4, 173–174, 324
green taxation 790
Groupement de Développement Agricole (GDA) 572–575
Grupo Viz (GV) 172–173
Gulburga Electricity Supply Company (GESCOM) 194

Health risks
  see within every Business Model under Safety, Environmental and Health Risks
high-performance temperature-controlled (HPTC) 253
Holding Company for Water and Wastewater (HCWW) 557, 558–559, 561, 562, 566
Husk Power Systems Inc. (HPS) 203–214
  biomass gasification 205–206, 208–209
  Rajiv Gandhi Grameen Vidyutikaran Yojana (RGGVY) Programme 205, 208–209
hydraulic retention time (HRT) 99, 119

Improved Cook Stove (ICS) 66
independent power producers (IPPs) 151, 205, 208, 250
India 505, 541, 550, 610, **780**, 786–787
  bagasse-based cogeneration 151
  co-marketing compost 790–791
  Companies Act 2013 8–9
  growth of urban population 667
  Sulabh International Social Service Organisation 119
  wastewater 550–551, 610, 667, 710–719, 736, 745–746, 753, 754, 786, 787, 793–794
  see also Bangalore, India; Bihar, India; Infrastructure Leasing and Financial Service Okhla composting plant (IL&FS Okhla); Koppal, India; Ludhiana, India; Pondicherry Agro Service and Industries Corporation Limited (PASIC); Pune Municipal Corporation (PMC); Shri Someshwar Sahakari Sakhar Karkhana (SSSSK)
Indian Renewable Energy Development Agency (IREDA) 151
indoor air pollution 34, 56, 79, 87, 93, 211
informal sector **20**, 78, 402, 437, 485, 508, 515, 558, 721, 737, 784
informal wastewater irrigation 551, 712, 721
  Sakumo wastewater treatment 760–774
  Water and Sanitation Agency (WASA) 745–759
  see also Corporate Social Responsibility (CSR)
infrastructure investment 785–789
Infrastructure Leasing and Financial Service Okhla composting plant (IL&FS Okhla) 435, 391–399
  Fertilizer Control Order (FCO) 391, 392, 393
  GHG emissions 391, 397–398
  Municipal Corporation of Delhi (MCD) 391, 392, 392, 396, 398
in-house biogas production 92
innovation capacity **20**, 731, 760–774, 795
integrated resource recovery facilities (IRRF) 381–382
integrated waste management facility (ISWM) 412
Intended Nationally Determined Contribution (INDC) 242–243
internal combustion (IC) 105
International Committee of Red Cross (ICRC) 93–102
  fixed dorm digester *99*
  Kigali Institute of Science Technology and Management (KIST) 94, 96, 98, 101
  SNV Netherlands Development Organisation (SNV) 97–98
International Solid Waste Association (ISWA) 779–782
inter-sectoral water exchange 691–697

Iran 666, 667, 693, 695, 754
   see also Mashhad Plain, Iran
irrigation 22, 100, 118, 166, 203, 254, 257, 265, 282, 292, 452–454, 456, 473, 549, 551, 554, 605, 613, 631, 665, 667–668, 721, 729, 730, 731, 782, 793, 796
   Amani Doddakere Tank 710–719
   As Samra wastewater treatment plant 642, 643, 644, 647–650
   CSR as driver of change 733–737, 740, 741, 742
   farmers' innovation as driver of change 760–763, 766, 768–772
   fixed wastewater-freshwater swap 672, 674, 675, 677
   flexible wastewater-freshwater swap 680–688
   forest carbon offset 599
   inter-sectoral water exchange 693, 695,
   suburban wastewater treatment for reuse 584, 586–587, 589, 591, 592
   Tula aquifer 698–707
   wastewater and biosolids for fruit trees 569–583
   wastewater as a commodity **746**, 746–750, *752*, **753**, 754–755, 757
   wastewater for fruit and wood production 556–557, 558, 563, 564–565, 566
   wastewater for greening the desert 599, 601

Japan Carbon Finance Ltd (JCF) 239
Jordan 550, 580, 659, 733
   see also As Samra wastewater treatment plant (WWTP)

Kampala, Uganda 82, 328
   see also Kampala Jellitone Suppliers (KJS); Uganda
Kampala Jellitone Suppliers (KJS) 41–51
Karnataka Compost Development Corporation Limited (KCDC) 400–410
   compost products 405–406
   Terra-Firma Biotechnologies Limited 403
Karnataka Renewable Energy Development Ltd (KREDL) 197, 199
Karnataka State Cooperative Marketing Federation (KSCMF) 401
Karnataka State Pollution Control Board (KSPCB) 197, 199

Kenya 111, 150–151
   see also Mumias Sugar Company Ltd (MSC); Nairobi, Kenya; Naivasha, Kenya; Kakuru, Kenya; Nyongara Slaughter House
Kenya Bureau Standard Board (KEBS)
   certification 364, 366, 367
Kenya Electricity Generating Company (KENGEN) 239
Kigali, Rwanda 82, 94, 98, 496
   see also Coopérative Pour La Conservation De L'Environement (COOCEN); Rwanda Environment Care (REC)
Kigali Institute of Science Technology and Management (KIST) 94, 96, 98, 101
Koppal, India 193–199
Kumasi, Ghana 769
   see also Waste Enterprisers Ltd. (WE)

landfill directive 784
large scale wastewater treatment **20**, 550, 656–663,
leapfrogging, wastewater **20**, 550, 555, 605, 631–638
Liquefied Petroleum Gas (LPG) 42, 88, 130, 133, 146
livestock waste, for compost production
   see ProBio Humibac (ProBio)
Llobregat delta, Spain 666, 676, 679–690, 693
Ludhiana, India 435
   see also A2Z Infrastructure Private Limited (A2Z-PL)
Ludhiana Municipal Corporation (LMC) see A2Z Infrastructure Private Limited (A2Z-PL)
Lugazi Town, Uganda 72, 77
   see also Eco-Fuel Africa (EFA)

Maharashtra Electricity Regulatory Commission (MERC) 227
Maharashtra Energy Development Agency (MEDA) 259
Maharashtra, India see Pune Municipal Corporation (PMC); Shri Someshwar Sahakari Sakhar Karkhana (SSSSK)
Mailhem Wipro see Wipro Employees Canteen
managed aquifer recharge 720–727
manure, power from 182–192
   BOT model 186
   carbon credit 183, *183*, 184–185, *185–186*
   centralized biogas systems 186, *187*, 188

INDEX

rural electrification 183–184, *184*, 185–186, *187*
Mashhad Plain, Iran 666, 670–678
Matara, Sri Lanka *see* Greenfield Crops (GC)
Mbale, Uganda 351
  see also Mbale Municipal Composting Plant (MCP)
Mbale Municipal Composting Plant (MCP) 324–332
  cost-recovery 324–331
methane (CH_4) 110–111, 139, 174–175, **174**, 448
  see also municipal solid waste (MSW), power from; Nyongara Slaughter House; Pune Municipal Corporation (PMC); SuKarne methane recovery project
Mexico 150, 667, 786, 789
  see also Eco Biosis S.A (Eco Biosis); Mexico City, Mexico; Mezquital Valley, Mexico; ProBio Humibac (ProBio); SuKarne methane recovery project; Tula aquifer (Mexico)
Mexico City, Mexico 667, 698–709
  see also Tula aquifer (Mexico)
Mezquital Valley, Mexico 667, 698, *703*, 704, 705, 707, *708*, 709n1
Middle East and Northern Africa (MENA) 21, 554, 558, 559
Millennium Development Goals (MDGs) 116, 119
Ministry of New and Renewable Energy (MNRE), Indian 137, 208
Ministry of Non-Conventional Energy Sources (MNES) 151, 259, 263
Ministry of Water and Irrigation (MWI), Kenyan 118
Mirzapur, Bangladesh 605, 382
  see also Agriquatics (Bangladesh)
Morocco 270, 554, 599
  see also Drarga wastewater treatment plant
Multilateral Investment Guarantee Agency (MIGA) 640
Mumias District, Kenya *see* Mumias Sugar Company Ltd (MSC)
Mumias Sugar Company Ltd (MSC) 151, 238–247
municipal solid waste (MSW) 36, 150, 778
  briquettes from *see* briquettes from municipal solid waste (MSW)
  compost and carbon credits for profit *see* Infrastructure Leasing and Financial Service Okhla composting plant (IL&FS Okhla)
  composting, cooperative model *see* Nakuru Waste Collectors and Recyclers Management Cooperative Society (NAWACOM)
  composting, and fecal sludge for cost recovery *see* Balangoda Compost Plant (BCP)
  composting, for cost recovery *see* Mbale Municipal Composting Plant (MCP)
  composting for profit, franchising approach to *see* Terra Firma Biotechnologies Limited (Terra Firma)
  composting for profit, inclusive, public-private partnership-based *see* A2Z Infrastructure Private Limited (A2Z-PL)
  composting for profit, socially-driven model *see* Waste Concern Group (Bangladesh)
  composting, partnership-driven *see* Karnataka Compost Development Corporation Limited (KCDC)
  composting, public-private partnership-based *see* Greenfield Crops (GC)
  power from *see* Pune Municipal Corporation (PMC)
  see also composting, partially subsidized (district level); municipal solid waste (MSW), power from
municipal solid waste (MSW), power from 232–237

Nairobi, Kenya 105, 111, 114–123, 248–249, 254, 451
Naivasha, Kenya 451
  see also DuduTech
Nakuru, Kenya 451
  see also Nakuru Waste Collectors and Recyclers Management Cooperative Society (NAWACOM)
Nakuru Waste Collectors and Recyclers Management Cooperative Society (NAWACOM) 362–370
  community-based organizations (CBOs) 362–364, 366, 367, 369–370
National Centre of Organic Farming (NCOF) 383

National Domestic Biogas Program (NDBP) 98
National Environment Management
    Authority (NEMA), Kenyan 118, 119
National Sanitation Utility (ONAS) 569–583
    Commissariat Régional de Développement
        Agricole (CRDA) 569, 571, 572, 574–578
    government bodies and institutions 576
    Groupement de Développement
        Agricole (GDA) 569, 572–575
nutrient and organic matter recovery 322–323,
    379–380 (subject of whole section III)
    overview of business models 316–320
nutrient recovery, from own agro-
    industrial waste 448, 478–483
Nyongara Slaughter House 248–256
    Dagorretti 248, 249, 254

Ouardanine, Tunisia 569–582
Okhla, India see Infrastructure Leasing
    and Financial Service Okhla
    composting plant (IL&FS Okhla)
organic matter recovery see nutrient
    and organic matter recovery
Ostara 319, 525, 538–546, 549
Ouagadougou, Burkina Faso see
    ECOSAN-EU project
Ouardanine tree crop system (Tunisia) see
    National Sanitation Utility (ONAS)
outsourcing, of fecal sludge treatment see
    fecal sludge treatment, outsourcing

partial guarantees 640
palm oil mill effluent (POME) 270, 272
participatory urban appraisal (PUA) 117
pay-and-use toilets 103, 104
pension funds 640, 786, 788–789
Peru 289, 789
    see also Santa Rosillo, Peru,
        rural electrification
Petroleos de Venezuela S.A. (PDVSA) 288
phosphorous 316, 347, 524, 549,
    661, 747, 779, 790
    see also phosphorous recovery
        from wastewater
phosphorous recovery from
    wastewater 538–546
regulations and obstacles in Europe 543
    see also wastewater treatment
        plants (WWTP)

Pondicherry, India see Pondicherry
    Agro Service and Industries
    Corporation Limited (PASIC)
Pondicherry Agro Service and Industries
    Corporation Limited (PASIC) 459–467
Pondicherry Cooperative Sugar Mill (PCSM)
    459, 460, 461, 463, 465, 466
power from manure see manure, power from
power from livestock and agro-waste see
    Santa Rosillo, Peru, rural electrification
power purchase agreements (PPAs) 238, 252
private sector investment, in large
    scale wastewater treatment see
    wastewater treatment, enabling
    private sector investment
ProBio Humibac (ProBio) 468–477
    see also SuKarne methane recovery project
Programmable Logical Controller (PLC) 167
Provincial Electricity Authority
    (PEA) 270, 272, 274
public-private partnerships (PPPs) 17, 35,
    61, 82, 116, 232–223, 355, 659
    see also A2Z Infrastructure Private
        Limited (A2Z-PL); Greenfield
        Crops (GC); Infrastructure Leasing
        and Financial Service Okhla
        composting plant (IL&FS Okhla)
public toilets, composting fecal sludge from
    see Rwanda Environment Care (REC)
Pune Municipal Corporation (PMC) 222–231
    Solid Waste Collection and Handling
        (SWaCH) 222, 223, 225, 226, 228–229
pyrolysis 57, 78, 88

Rajiv Gandhi Grameen Vidyutikaran
    Yojana (RGGVY) 205, 208–209
Ravikiran Power project 193–202
    Certified Emission Reductions (CERs) 195
    Greenko Group 193, 194, 197, 199
    Gulburga Electricity Supply
        Company (GESCOM) 194
reclaimed water 549, 558, 565, 566,
    569, 600–601, 605, 634, 635, 636,
    644, 646, 647, 649–650, 661,
    665, 666, 700, 724, 769, 796
    fixed wastewater-freshwater swap 670–677
    flexible wastewater-freshwater
        swap **681**, 681–689
    inter-sectoral water exchange 691–697

wastewater and biosolids for
fruit trees 570–582
recovery
of nutrients and organic matter *see*
nutrient and organic matter
recovery (section III)
of energy from waste *see* waste,
recovering energy (section II)
of water from wastewater see (section IV)
refuse-derived fuel (RDF) 382, 388,
393, 398, 437, **780–781**
resource recovery and reuse (RRR) 6,
7–8, 17, 29, 36–37, 549, 554
see also business cases and models,
defining and analyzing
resource recovery and reuse (RRR),
environment and finance 778–796
finance and financial incentives 784–793
local capacities and stakeholder
acceptance 795–796
operational cost recovery,
facilitation 789–792
policies, regulations, and guidelines
778–784
technologies matching resource
constraints 793–794
results-based financing (RBF) 791–792
risk mitigation and business risks, RRR
see under business cases and
models, defining and analyzing
rural electrification *see* Santa Rosillo,
Peru, rural electrification
rural–urban linkages, wastewater **703**
rural-urban water trading 665–669
see also Amani Doddakere tank
(ADT), revival; inter-sectoral water
exchange; managed aquifer
recharge; Tula aquifer (Mexico);
wastewater-freshwater swap, fixed;
wastewater-freshwater swap, flexible
Rwanda 35–36, 92, 93–101, 111, 784
see also Kigali, Rwanda
Rwanda Correctional Services (RCS) 96
Rwanda Development Board (RDB) 66
Rwanda Environment Care (REC) 487–495
Rwanda Environment Management
Authority (REMA) 66
Rwanda Utilities Regulatory Agency (RURA)
66

Safety
see within every Business Model under
Safety, Environmental and Health Risks
Sanitation Development Fund (SANDEF) 122
Sanitation Safety Planning (SSP) 12
see also under World Health Organization
(WHO); Sanitation Safety Planning (SSP)
Santa Rosillo, Peru, rural electrification 152–161
Communal Energy Services
Unit (USEC) 155–156
septage 318, 319, 485, 497, 505, 506,
516, 517, 524, 525, 771
see also fecal sludge
sewage treatment plant (STP) 111, 717
see also Wastewater; Wastewater Treatment
Plant, Waste Stabilization Ponds
Shri Someshwar Sahakari
Karkhana (SSSSK) 257–267
bagasse co-generation 259–260
ethanol from molasses 260
slaughterhouse waste *see* Nyongara
Slaughter House
small and medium enterprises (SMEs) 140
social business model 7, 676,
710, 712, 723–724
social capital 317, 361
Solid Waste Collection and Handling (SWaCH)
222, 223, 225, 226, 228–229
Spain 691, 693
see also Llobregat delta, Spain
Sri Lanka 356, 451, 506, **780**, 790–792
see also Greenfield Crops (GC);
Balangoda Compost Plant (BCP)
Sri Lanka Standards Institution
(SLS) 337–338, 346
see also Balangoda Compost Plant
(BCP); Greenfield Crops (GC)
subsidy-free community-based
composting 360–361, 371–377
see also Nakuru Waste Collectors
and Recyclers Management
Cooperative Society (NAWACOM)
sugar industry waste
combined heat, power and ethanol
see Shri Someshwar Sahakari
Sakhar Karkhana (SSSSK)
enriched compost production *see*
Pondicherry Agro Service and Industries
Corporation Limited (PASIC)

SuKarne methane recovery project 172–181
    Grupo Viz (GV) 172–173
Sulabh Effluent Treatment (SET) 109
Sulabh, India 103–112, 119
Sulabh International Social Service
    Organization 103–113
    public toilets 103–106, **108**, 109, 110–111
sustainable and renewable power
    generation 150
    see also Santa Rosillo, Peru,
        rural electrification
Sustainable Development Goals
    (SDG) 5, 34, 640
Sustainable Environment Management
    Programme (SEMP) 429
sustainable sanitation service delivery 485
    see also compost production for
        sustainable sanitation service delivery
sustainable sourcing 9

tax holidays 151, 195, 252, 259, 355, 789
technologies matching resource
    constraints 793–794
tender 559, 659, 661, 795
Terra Firma Biotechnologies Limited
    (Terra Firma) 403, 411–421
textile industry, wastewater 736
Thai Biogas Energy Company (TBEC) 268–277
    palm oil mill effluent (POME) 270, 272
    Private Energy Market Fund (PEMF) 270
    Provincial Electricity Authority
        (PEA) 270, 272, 274
    Very Small Power Producer
        Program (VSPP) 272
Thailand 295
    see also Thai Biogas Energy
        Company (TBEC)
3S Program (Brazil) 162–171
    Sadia Institute (SI) 36, 164–169
tipping fees 335, 339, 355, 398, 407, 418, 789
TOSHA 1 114–123
    community-based organization (CBO) 114
    Umande Trust 114, 115, 116–117,
        119–120, 121–122
treatment fee model 540
Tula aquifer (Mexico) 698–709
    Atotonilco wastewater treatment
        plant 703, 705
    CONAGUA 701–702, 704–705

Tunisia 549, 554–555, 596
    see also Ouardanine, Tunisia

Uganda 111, 239, 246, 451, 791
    see also Kampala, Uganda; Lugazi
        Town, Uganda; Mbale, Uganda
Uganda National Bureau of
    Standards (UNBS) 78
UNFCCC Annex I countries 379, 441
United Nations Development
    Programme (UNDP) 62, 488
United States Agency for International
    Development (USAID) 259, 335, 561,
    563, 585, 593, 642, 645, 652
United States Environmental Protection
    Agency (US EPA) 173, 593
upflow anaerobic sludge blanket (UASB)
    137, 227, 652, 793–794
urine diverting dry toilets (UDDTs) 485,
    497, 498, 502, 503, 524–525,
    528–529, 531–535, 542
urine and fecal matter collection
    see ECOSAN-EU project

value-driven business model 44
Venezuela see ETAVEN C.A. (ETAVEN)
Veracruz, Mexico see Eco Biosis
    S.A (Eco Biosis)
Viability Gap Funding (VGF) see As Samra
    wastewater treatment plant (WWTP)
Voluntary Emission Reductions (VERs)
    56, 85, 127, 145, 193, 195

Waste Concern Group (Bangladesh) 422–432
    CDM/carbon trading model 428–429
    partnership model 426–428
Waste Enterprisers Ltd. (WE) 617–630
    Kumasi Metropolitan Assembly (KMA)
        617, 618, 620–621, 622, 630n1
waste management and sanitation
    4–8, 10–12, 322
    see also Corporate Social
        Responsibility (CSR)
waste, recovering energy 34–37
Waste Stabilization Ponds (WSP) 750, 793
wastewater
    agro-industrial, combined heat and
        power from see Thai Biogas
        Energy Company (TBEC)

biosolids for fruit trees *see* National Sanitation Utility (ONAS)
fish feed production *see* Agriquatics (Bangladesh)
fruit and wood production *see* El Berka wastewater and agroforestry system (Egypt)
phosphorous recovery *see* phosphorous recovery from wastewater
treatment, designed for reuse and replication *see* Drarga wastewater treatment plant
wastewater as change-driving commodity 745–759
wastewater auctioning 748, 754
wastewater for greening desert 595–603
    cost recovery through accounting 599
    forest carbon offset 599–600
wastewater-freshwater swap, fixed 670–678
wastewater-freshwater swap, flexible 679–690
    Catalonian Water Agency (ACA) 679, 683
    integrated water resources management (IWRM) 679, 681, 682, 687
    treatment for nature 685
wastewater treatment, enabling private sector investment 656–663
    public-private partnership (PPP) agreement 657, 659, 661
    Viability Gap Funding (VGF) 658
wastewater treatment plant (WWTP) 549, 550, 644
    see also sewage treatment plant, waste stabilization ponds, cost recovery (wastewater); phosphorous recovery from wastewater
water exchange, wastewater **20**

fixed wastewater-freshwater swap 671, **673**, 674, 675, 676, 677
flexible wastewater-freshwater swap 679, 681–682, 687, 689
inter-sectoral 665–667, **666**, 687, 691–697
rural–urban 667, 724
water recovery 548, 652, 748 (subject of whole Section IV)
    overview of business models 548-552
Water Services Trust Fund (WSTF) 119
welfare/profit maximization **20**
    profit maximization 17, 218, 352, 379, 435
win-win situation 6, 605, 614, 632, 674, 705, 707
Wipro Employees Canteen 133–141
    Corporate Social Responsibility (CSR) 134–135, 136–137, 141
wood production, wastewater 556–568, 597, 602
word of mouth 204, 362, 364, 368, 377
World Health Organization (WHO) 26, 40, 66, 104, 579, 581, 584, 613, 650, 762, 768
    guidelines 558, 566, 627, 631, 636, 644, 695, 734, 757, 763, 766, 770
    multiple barrier approach 601, 730, **730**, 731, 741
    recommendations 551, 576, 626, 741, 755, 756–757, 771, 774n4
    Sanitation Safety Planning (SSP) 12, 566, 601, 676, 695, 725, 731, 736, 739, 769, 778–779
    standards 589, 613, 644

Yelemallappa Shetty tank (YMST) *see* Amani Doddakere tank (ADT), revival